Role of Microbes in Sus
Development

R. C. Sobti • Ramesh Chander Kuhad •
Rup Lal • Praveen Rishi
Editors

Role of Microbes in Sustainable Development

Human Health and Diseases

Springer

Editors
R. C. Sobti
INSA Senior Scientist and Professor Emeritus
Panjab University
Chandigarh, Chandigarh, India

Ramesh Chander Kuhad
Former Vice Chancellor
Central University
Haryana, India

Rup Lal
Professor Emeritus
University of Delhi
New Delhi, Delhi, India

Praveen Rishi
Department of Microbiology
Panjab University
Chandigarh, Chandigarh, India

ISBN 978-981-99-3125-5 ISBN 978-981-99-3126-2 (eBook)
https://doi.org/10.1007/978-981-99-3126-2

This Springer imprint is published by the registered company Springer Nature Singapore Pte Ltd.
The registered company address is: 152 Beach Road, #21-01/04 Gateway East, Singapore 189721, Singapore

Paper in this product is recyclable.

Preface

The incredible power of the invisible kingdom of microbes is continuously gaining appreciation for their vital role in health and disease. Recent advances in research revealed that microbes constitute nearly 20% of the total living biomass and hence can benefit mankind in many ways. In other words 'Microorganisms will give you anything you want if you know how to ask them'. This statement by Kinichiro Sakaguchi rightfully exemplifies the unequivocal contribution of microorganisms in the functioning of virtually every aspect of life on earth. The ubiquitous nature of microbes, their co-evolution, co-diversification, and the function they perform together with humans have laid the foundation for the concept of the human microbiome. Microbiome has become a critical component of the human existence, with a diverse array of roles in human health as well as the development of sustainable ecosystem. While studies on the diversity of human microbes date back to the 1860s when Antonie van Leeuwenhoek studied and compared his oral and faecal microflora, we now know that microbiome is a complex ecosystem consisting of archaea, bacteria, protozoa, viruses and fungi. The composition of microbial communities or microbiome varies not only in different individuals but in different organs of the same individual. Thus, microbiome-based treatment specific to each population may take the conventional line of treatment towards personalized medical intervention. In the last decade, we have witnessed rapid progress in research on various aspects of the human microbiome, particularly the gut microbiome and its influences on and vital role in modulating key markers of health and disease.

Under normal conditions, microflora particularly of the gut are involved in a symbiotic association with the host wherein the latter provides the gut microbiome with a nutrient-rich environment for its survival as well as proliferation, and gut microbes, in turn, perform important functions within the host that human genes cannot perform. The production of essential vitamins, absorption of iron, induction of immunity to confer protection from various pathogenic organisms along with directly competing with these harmful bacteria by production of various antibacterial substances are some of the examples of functions that can only be carried out by microbial genes in the human body. This association is quintessential in determining the host health status, and any deviation from this leads to a state of dysbiosis. There is growing evidence that dysbiosis can lead to several manifestations ranging from irritable bowel syndrome and obesity to diabetes, cardiovascular diseases and even the development of cancers.

Besides the role of the microbiome or commensals in maintaining homeostasis or symbiosis, the contribution of microorganisms to human health can also be traced back to the discovery of microbially derived antibiotics, i.e. penicillin, which marked the beginning of the antibiotic era. Since then, a wide range of life-saving antibiotics have been produced using microbes. Antibiotics prove to be the miraculous drugs for treating several infections; however clinical medicine started facing challenges in view of the emergence of antimicrobial resistance, for which the potential use of beneficial microbes is being looked into.

The potential of microbes is being exploited for developing vaccines against several infections. The current era of the Covid-19 pandemic has underpinned the fundamental role played by microorganisms in shaping human life. While the disease itself is caused by a microbial agent (SARS CoV-2), the most effective weapons developed by humans against this virus, i.e. the vaccines, are also microbially derived. Thus, there have been a wide range of microbes-driven applications in disease prophylaxis and treatment, be it in the form of probiotics, vaccines, and antibiotics, or protein therapeutics.

Microorganisms are believed to be smarter than humans, but, if mankind exploits their potential more intelligently, problems being faced by the world today including food, health and well-being can be adequately taken care of. Keeping this essence, the present book aims to present the reader with recent advances being made in the field.

New insights concerning the role of microbiome in human health as well as in diseases have been discussed in some chapters. The section discusses how a balanced microbiota composition confers health benefits to the human host and the role of probiotics in maintaining the gut homeostasis. Further, the role of microbiome in various manifestations such as metabolic disorders, reproductive diseases, cardiovascular diseases and cancer has been elaborated.

Chapters on the emerging microbial technologies in microbiome research, such as Crisper-Cas, genome engineering, and development of nanoparticles, have also been included. The possible exploration of such technologies to interrogate the potential of gut microbiota in human health has been discussed.

The book includes chapters on the application of microbial products that impact human health. These provide relevant information about the exploitation of microbes in the production of vaccines and microbes induced calcite precipitation approach. Since good health starts with good food, any human deprived of daily nutritional requirements cannot remain healthy and is more prone to diseases. From this perspective, the contribution of microbial functional foods and nutraceuticals to human health has also been described.

In addition to chapters on the applications of beneficial bacteria, the concept of fungi and bacteriophages is envisaged for the treatment of various infectious diseases.

The volume has been prepared with the wholehearted support of all the contributors.

R.C. Sobti acknowledges the support of the Indian National Science Academy (INSA), New Delhi, for providing a platform as a Senior Scientist and Panjab University, Chandigarh, as Emeritus Professor to continue his academic pursuits. He is obliged to his family members Dr. Vipin Sobti (Wife), Er. Aditi and Dr. Aastha (Daughters). Er. Vineet and Er. Ankit (Sons-in Law) and granddaughter Irene for their continuous encouragement for completing the book.

Chandigarh, India R. C. Sobti
Mahendragarh, India Ramesh Chander Kuhad
New Delhi, India Rup Lal
Chandigarh, India Praveen Rishi

Contents

Editors and Contributors

About the Editors

Ranbir Chander Sobti Former Education Consultant, Governor of Bihar, Senior Scientist (Indian National Science Academy), Former Vice Chancellor, Babasaheb Bhimrao Ambedkar University, Lucknow (UP), and Panjab University, Chandigarh, is, indeed, a polymath—a renowned academician, distinguished scientist, dynamic administrator and a visionary gifted with an immensely optimistic disposition and integrity of character, words and action. He has published more than 350 high-impact research publications, about 50 plus chapters in books, 45 books and 23 sponsored research projects.

Professor Sobti is a Fellow of the Third World Academy of Sciences (TWAS), National Academy of Sciences, Indian National Science Academy, National Academies of Medical Sciences and Agricultural Sciences and of the Canadian Academy of Cardiovascular Diseases and is associated with many other academic associations and institutions in the domain of higher education and research. The litany of honours showered upon him includes, among others, the INSA Young Scientist Medal (1978), UGC Career Research Award, Punjab Rattan Award, JC Bose Oration Award and the Life Time Achievement Award of the Punjab Academy of Sciences, Zoological Society of India and the Environment Academy of India, besides many other medals and awards from various reputable national and international organizations. As a protean public intellectual, Professor Sobti is a remarkable leader in the domain of science/scientific knowledge—both in theory and practice. In recognition of his enormous contribution to higher education, he has been bestowed with a large number of prestigious awards and medals that adorn his academic and professional career. The Government of India bestowed upon him the honour of 'PADMASHRI' in 2009 by way of rightfully acknowledging his great contribution to higher education in India. On 2 July 2016, he was also honoured with 'Bharat Gaurav', Life Time Achievement Award at the House of Commons, British Parliament, London, UK. He had been General President of Indian Science Congress in the session 101. He is a widely travelled scientist.

Ramesh Chander Kuhad is Chairman/Member of Executive Council/Academic Council/Court/Governing Board of various universities, UGC and NAAC. In the past, he has also served as a member of the University Grants Commission (UGC) and a member of Executive Council and Governing Body of UGC NAAC. He is a Fellow of the Academy of Microbiological Sciences (FAMSc), Fellow of the National Academy of Sciences India (FNASc), Fellow of the National Academy of Agricultural Sciences (FNAAS) and Fellow of Biotech Research Society (FBRS). He has also served as a Professor at the University of Delhi and as Vice Chancellor at Central University of Haryana, Mahendergarh. He is the Founder Chairman of the Indian Academy of Microbiological Sciences and has served as the Joint Director of the Institute of Life Long Learning (ILLL), University of Delhi. He has also been the General Secretary (2004–2010) and President (2011–2012) of the Association of Microbiologists of India. He is an editor/associate editor of various journals. He is the recipient of prestigious awards and recognitions, a few of which are lifetime achievement awards, AMI (2019), the prestigious Prof. G. S. Rangaswamy Memorial Award (Biennial) (2017), Dr. G. B. Manjrekar Award (2014), AMI-Platinum Jubilee Life Time Achievement Award (2013), AMI-Titan Biotech Award (2011), Short Term Biotechnology Overseas Research Associateship Award (2002–03), AMI-Alembic Award (1999), Biotechnology Long Term Overseas Research Associateship Award (1995–96), UNIDO/ICGEB Fellowship (1994) and Commonwealth Scholarship for Post-Doctoral Research.

Rup Lal is presently the INSA Senior Scientist at Acharya Narendra Dev College, Delhi University. He is the Fellow of Indian National Science Academy, National Academy of Sciences, National Academy of Agricultural Sciences and Academy of Microbiological Sciences. Presently, he is also the Senior Ambassador of Indian region for ISME (International Society for Microbial Ecology), the Netherlands, and Federation of European Microbiological Societies (FEMS). He is also the Advisor to IMiLI-SAC (International Microbial Literacy Initiative-South Asia Center) to spearhead the microbial literacy campaign in India as well as abroad. He has over 40 years of teaching, administrative and research experience in the area of microbial diversity, genomics and metagenomics. He has supervised 75 PhD thesis and trained more than 150 UG/PG students. He has over 250 publications in peer-reviewed journals with 9147 ISI citations and h-index 47. He has delivered more than 300 lectures in the area of biotechnology, microbiology, genomics and metagenomics at various national and international platforms. Professor Lal is the recipient of several prestigious fellowships including the Alexander von Humboldt Fellowship (Germany), DBT Overseas Fellowship and Indo-US-ASM Professorship in Microbiology, Endeavour executive fellowship (2018-2019), Moselio Schaechter Distinguished Service Award from American Society for Microbiology and many more.

Praveen Rishi is a Professor in the Department of Microbiology, Panjab University, Chandigarh. She has also been the Dean, Faculty of Science, Panjab University. She has been also the President of the Association of Microbiologists of India (AMI). Dr. Rishi has to her credit 150 research publications in international and national journals of high repute. She is a Fellow of the Academy of Microbiological

Sciences (FAMSc) as well as of the Indian Association of Biomedical Sciences (FABMS). She has also been the recipient of DBT overseas Associateship, INSA Visiting Fellowship and Dr. Y.S. Narayana Oration Award by the Indian Council of Medical Research (ICMR), New Delhi. Recently she has been awarded the Prof. S.R. Vyas Memorial Award as well by the Association of Microbiologists of India (AMI).

Contributors

Amrik Singh Ahluwalia Department of Botany, Akal College of Basic Sciences, Eternal University, Baru Sahib, Himachal Pradesh, India

Kanwaljit Kaur Ahluwalia Department of Zoology, Eternal University, Baru Sahib, Himachal Pradesh, India

Younis Ahmed Department of Life Sciences and Allied Health Sciences, Sant Baba Bhag Singh University, Jalandhar, Punjab, India

Mohammad Khusni Bin Ahmat Amin School of Science, Monash University, Selangor, Malaysia

Achhada Ujalkaur Avatsingh Department of Microbiology, Akal College of Basic Sciences, Eternal University, Baru Sahib, Himachal Pradesh, India

Arush Behl Laboratory of Enzymology and Gut Microbiology, Department of Microbiology, Maharshi Dayanand University, Rohtak, Haryana, India

Neha Rani Bhagat DRDO-Defence Institute of High Altitude Research (DIHAR), Ladakh, India

Meghali Bharti Fish Molecular Biology Laboratory, Department of Zoology, University of Delhi, Delhi, India

Kawalpreet Kaur Bhatia Department of Botany, SGGSC, Sector 26, Chandigarh, India

Musadiq Husain Bhat Central National Herbarium, Botanical Survey of India, Howrah, Calcutta, India

Richa Bhatnagar Royal Berkshire NHS Foundation Trust, Bracknell Lighthouse Laboratory, Bracknell Healthspace, Bracknell, UK

Avesh Choubey Faculty of Applied Sciences and Biotechnology, Shoolini University of Biotechnology and Management Sciences, Solan, Himachal Pradesh, India

Vincent T. K. Chow Department of Microbiology and Immunology, Yong Loo Lin School of Medicine, National University of Singapore, Singapore, Singapore

Priyanka Dahiya Department of Microbiology, Maharshi Dayanand University, Rohtak, Haryana, India

Amita Suneja Dang Centre for Medical Biotechnology, Maharshi Dayanand University, Rohtak, Haryana, India

Inderpal Devgon School of Bioengineering and Biosciences, Lovely Professional University, Phagwara, Punjab, India

Rubee Devi Department of Biotechnology, Dr. Khem Singh Gill Akal College of Agriculture, Eternal University, Baru Sahib, India

Prabu Dhandapani Department of Microbiology, Dr. ALM PG IBMS, University of Madras, Chennai, India

Rakhi Dhankhar Department of Microbiology, Maharshi Dayanand University, Rohtak, India

Tejpal Dhewa Department of Nutritional Biology, Central University of Haryana, Mahendergarh, Haryana, India

Gauri Garg Dhingra Department of Zoology, Kirori Mal College, University of Delhi, New Delhi, India

Diksha Division Zoology, Department of Biosciences, Career Point University, Hamirpur, Himachal Pradesh, India

Padma Dolkar Fish Molecular Biology Laboratory, Department of Zoology, University of Delhi, Delhi, India

Isra Ahmad Farouk School of Science, Monash University, Selangor, Malaysia

Junaid Furhan Department of Microbiology, SKIMS Medical College & Hospital, Srinagar, India

Yogalakshmi Ganapathy Department of Microbiology, Dr. ALM PG IBMS, University of Madras, Chennai, India

Arup Giri Department of Zoology, Baba Mastnath University, Rohtak, Haryana, India

Gunjan Goel Department of Microbiology, Central University of Haryana, Mahendergarh, India

Kajol Goria Department of Environmental Sciences, Central University of Jammu, Samba, Jammu and Kashmir, India

Pooja Gulati Department of Microbiology, Maharshi Dayanand University, Rohtak, India

Parakriti Gupta Department of Microbiology, Government Medical College Hospital, Chandigarh, India

Ravi Kr. Gupta Department of Environmental Microbiology, Babasaheb Bhimrao Ambedkar University, Lucknow, India

Varsha Gupta Department of Microbiology, Government Medical College Hospital, Chandigarh, India

Younis Ahmad Hajam Department of Life Sciences and Allied Health Sciences, Sant Baba Bhag Singh University, Jalandhar, Punjab, India

Akinobu Hayashi Department of Oncologic Pathology, Mie University Graduate School of Medicine, Tsu, Mie, Japan
Department of Diagnostic Pathology, Mie University Hospital, Tsu, Japan

Princy Hira Department of Zoology, Maitreyi College, University of Delhi, New Delhi, India

Yoshifumi Hirokawa Department of Oncologic Pathology, Mie University Graduate School of Medicine, Tsu, Mie, Japan

Nikita Jangra Department of Microbiology, Maharshi Dayanand University, Rohtak, India

Ishwerpreet Kaur Jawanda Department of Microbiology, Panjab University, Chandigarh, India

Amey Jedhe Faculty of Applied Sciences and Biotechnology, Shoolini University of Biotechnology and Management Sciences, Solan, Himachal Pradesh, India

Deshraj Deepak Kapoor Department of Environmental Microbiology, Babasaheb Bhimrao Ambedkar University, Lucknow, India

Arun Karnwal School of Bioengineering and Biosciences, Lovely Professional University, Phagwara, Punjab, India

Jasvinder Kaur Department of Zoology, Gargi College, University of Delhi, New Delhi, India

Rajinder Kaur Centre of Experimental Medicine and Surgery, Banaras Hindu University, Varanasi, India

Tanvir Kaur Department of Biotechnology, Dr. Khem Singh Gill Akal College of Agriculture, Eternal University, Baru Sahib, India

Tejinder Kaur Department of Zoology, DAV University, Jalandhar, India

Anubhuti Kawatra Department of Microbiology, Maharshi Dayanand University, Rohtak, India

Azhar Khan Faculty of Applied Sciences and Biotechnology, Shoolini University of Biotechnology and Management Sciences, Solan, Himachal Pradesh, India

Sofia Shareif Khan Department of Biotechnology, Shri Mata Vaishno Devi University, Katra, Jammu and Kashmir, India

Khushboo School of Bioengineering and Biosciences, Lovely Professional University, Phagwara, Punjab, India

Richa Kothari Department of Environmental Sciences, Central University of Jammu, Samba, Jammu and Kashmir, India

Divjot Kour Department of Microbiology, Akal College of Basic Sciences, Eternal University, Baru Sahib, Himachal Pradesh, India

Gagandeep Kour Department of Environmental Sciences, Central University of Jammu, Samba, Jammu and Kashmir, India

Ivneet Kour Department of Microbiology, Government Medical College Hospital, Chandigarh, India

Anil Kumar Centre for Medical Biotechnology, Maharshi Dayanand University, Rohtak, India

Ashwani Kumar Department of Nutrition Biology, Central University of Haryana, Mahendergarh, India

Pradeep Kumar Department of Microbiology, Maharshi Dayanand University, Rohtak, Haryana, India

Rajesh Kumar Department of Biosciences, Himachal Pradesh University, Shimla, Himachal Pradesh, India

Roshan Kumar Post-Graduate Department of Zoology, Magadh University, Bodh Gaya, Bihar, India

Sachin Kumar Faculty of Applied Sciences and Biotechnology, Shoolini University of Biotechnology and Management Sciences, Solan, Himachal Pradesh, India

Vikram Kumar Department of Basic and Applied Sciences, National Institute of Food Technology and Entrepreneurship Management, Sonipat, Haryana, India

Vinay Kumar Department of Environmental Microbiology, Babasaheb Bhimrao Ambedkar University, Lucknow, Uttar Pradesh, India

Seema Kumari Department of Microbiology, Panjab University, Chandigarh, India

Rup Lal Professor Emeritus, University of Delhi, New Delhi, Delhi, India

Sunil Kumar Lal School of Science, Monash University, Selangor, Malaysia

Nishu Lohan Department of Biochemistry, Panjab University, Chandigarh, India

Shanmugaraja Meenakshi Faculty of Allied Health Sciences, Chettinad Hospital and Research Institute, Chettinad Academy of Research and Education, Chennai, Tamil Nadu, India

Sonakshi Modeel Fish Molecular Biology Laboratory, Department of Zoology, University of Delhi, Delhi, India

Shreya Vishwas Mohite Laboratory of Enzymology and Gut Microbiology, Department of Microbiology, Maharshi Dayanand University, Rohtak, Haryana, India

Nandhini Muthusamy Sridhar Department of Microbiology, Dr. ALM PG IBMS, University of Madras, Chennai, India

Shekhar Nagar Fish Molecular Biology Laboratory, Department of Zoology, University of Delhi, Delhi, India
Deshbandhu College, University of Delhi, Delhi, India

Neelam Department of Life Sciences and Allied Health Sciences, Sant Baba Bhag Singh University, Jalandhar, Punjab, India

Neeraj Department of Life Sciences, Arni School of Basic Sciences, Arni University, Kathgarh, Himachal Pradesh, India

Raheshwari Negi Department of Biotechnology, Dr. Khem Singh Gill Akal College of Agriculture, Eternal University, Baru Sahib, India

Ram Krishan Negi Fish Molecular Biology Laboratory, Department of Zoology, University of Delhi, Delhi, India

Tarana Negi Department of Zoology, Government College, Jhajjar, India

Nisha School of Bioengineering and Biosciences, Lovely Professional University, Phagwara, Punjab, India

Kalash Pandey Faculty of Applied Sciences and Biotechnology, Shoolini University of Biotechnology and Management Sciences, Solan, Himachal Pradesh, India

Souparno Paul Department of Microbiology, Central University of Haryana, Mahendergarh, India

Vijay Prabha Department of Microbiology, Panjab University, Chandigarh, India

Aman Prakash Department of Molecular and Cellular Engineering, Jacob Institute of Biotechnology and Bioengineering, Sam Higginbottom University of Agriculture, Technology and Sciences, Naini, India

Priya Department of Microbiology, Maharshi Dayanand University, Rohtak, Haryana, India

Samridhi Pushkarna Centre for Medical Biotechnology, Maharshi Dayanand University, Rohtak, India

Khem Raj Department of Microbiology, Panjab University, Chandigarh, India

S. Ramadevi Faculty of Allied Health Sciences, Chettinad Hospital and Research Institute, Chettinad Academy of Research and Education, Chennai, Tamil Nadu, India

Abhishek Rana Jindal Global Law School, OP Jindal Global University, Sonipat, Haryana, India

Ananya Rana Department of Basic and Applied Sciences, National Institute of Food Technology and Entrepreneurship Management, Sonipat, Haryana, India

Pushpa Rani Department of Genetics, Maharshi Dayanand University, Rohtak, Haryana, India

Simran Rani Department of Microbiology, Maharshi Dayanand University, Rohtak, Haryana, India

Nayan Rishi Department of Microbiology, Central University of Haryana, Mahendergarh, India

Mohammad Riyaz Department of Microbiology, Panjab University, Chandigarh, India

Rohan Samir Kumar Sachan School of Bioengineering and Biosciences, Lovely Professional University, Phagwara, Punjab, India

Neha Salaria Department of Life Sciences, Arni School of Basic Sciences, Arni University, Kathgarh, Himachal Pradesh, India

Kaveesha Senasinghe School of Science, Monash University, Selangor, Malaysia

Anshika Sharma School of Science, Monash University, Selangor, Malaysia

Anuj Sharma Faculty of Applied Sciences and Biotechnology, Shoolini University of Biotechnology and Management Sciences, Solan, Himachal Pradesh, India

Krishna Kant Sharma Laboratory of Enzymology and Gut Microbiology, Department of Microbiology, Maharshi Dayanand University, Rohtak, Haryana, India

Sukesh Chander Sharma Department of Biochemistry, Panjab University, Chandigarh, India

Pratik Balwant Shinde Laboratory of Enzymology and Gut Microbiology, Department of Microbiology, Maharshi Dayanand University, Rohtak, Haryana, India

Gupta Shruti Department of Biotechnology, Himachal Pradesh University, Shimla, Himachal Pradesh, India

M. C. Sidhu Department of Botany, Panjab University, Chandigarh, India

Ashish Kumar Singh Centre of Experimental Medicine and Surgery, Banaras Hindu University, Varanasi, India

Dinesh Kumar Singh District Hospital, Chandauli, India

Kanwar Shamsher Singh Department of Biotechnology, Himachal Pradesh University, Shimla, Himachal Pradesh, India

Nasib Singh Department of Microbiology, Akal College of Basic Sciences, Eternal University, Baru Sahib, Himachal Pradesh, India

Samer Singh Centre of Experimental Medicine and Surgery, Banaras Hindu University, Varanasi, India

Lipika Singhal Department of Microbiology, Government Medical College Hospital, Chandigarh, India

Sneha Siwach Fish Molecular Biology Laboratory, Department of Zoology, University of Delhi, Delhi, India

R. C. Sobti Panjab University, Chandigarh, Chandigarh, India

Thomson Soni Department of Microbiology, Panjab University, Chandigarh, India

Utkarsh Sood Department of Zoology, Kirori Mal College, University of Delhi, New Delhi, India

Ankita Srivastava Department of Environmental Microbiology, Babasaheb Bhimrao Ambedkar University, Lucknow, Uttar Pradesh, India

Pooja Suneja Department of Microbiology, Maharshi Dayanand University, Rohtak, Haryana, India

Neetu Kumra Taneja Department of Basic and Applied Sciences, National Institute of Food Technology and Entrepreneurship Management, Sonipat, Haryana, India

Manish Thakur Department of Microbiology, DAV University, Jalandhar, India

Santosh Kumar Tiwari Department of Genetics, Maharshi Dayanand University, Rohtak, Haryana, India

Eri Usugi Bio-Bank Center, Mie University Hospital, Tsu, Japan

Miki Usui Department of Diagnostic Pathology, Mie University Hospital, Tsu, Japan

Digvijay Verma Department of Environmental Microbiology, Babasaheb Bhimrao Ambedkar University, Lucknow, Uttar Pradesh, India

Akanksha Vishwakarma Department of Environmental Microbiology, Babasaheb Bhimrao Ambedkar University, Lucknow, Uttar Pradesh, India

Masatoshi Watanabe Department of Oncologic Pathology, Mie University Graduate School of Medicine, Tsu, Mie, Japan
Bio-Bank Center, Mie University Hospital, Tsu, Japan
Department of Diagnostic Pathology, Mie University Hospital, Tsu, Japan

Ajar Nath Yadav Department of Biotechnology, Dr. Khem Singh Gill Akal College of Agriculture, Eternal University, Baru Sahib, Himachal Pradesh, India

Asha Yadav Laboratory of Enzymology and Gut Microbiology, Department of Microbiology, Maharshi Dayanand University, Rohtak, Haryana, India

Pankaj Yadav Fish Molecular Biology Laboratory, Department of Zoology, University of Delhi, Delhi, India

Sheetal Yadav Fish Molecular Biology Laboratory, Department of Zoology, University of Delhi, Delhi, India

Shilpi Yadav Department of Biotechnology, Babasaheb Bhimrao Ambedkar University, Lucknow, India

Microbial Diversity and Their Role in Human Health and Diseases

1

Musadiq Husain Bhat, Younis Ahmad Hajam, Neelam,
Rajesh Kumar, and Diksha

Abstract

Micro-organisms are as old as human civilization and have co-evolved with human beings by inhabiting either inside or outside their body for deriving food and nutrition. Since time immemorial, these tiny animals/microbes have played a vital role in human health both in positive and negative manners. There is a close association between human health and microbial disease; however, the survival of these microbes/pathogens in the bodies of human beings gets affected in several ways, one of which is an environmental factor that immensely interferes with human–microbial interaction. Modern studies/research on micro-organisms affecting human health have become much more refined and inclusive. Therefore, in depth research should be conducted on the human–microbe interactions that may lead to understanding the role of microbiota inhuman health and disease and provide new therapeutic goals and treatment approaches in clinical practices. The present chapter has been designed to summarize the role of human health in the development of major human diseases, such as liver

M. H. Bhat
Central National Herbarium, Botanical Survey of India, Howrah, Calcutta, India

Y. A. Hajam (✉) · Neelam
Department of Life Sciences and Allied Health Sciences, Sant Baba Bhag Singh University, Jalandhar, Punjab, India

R. Kumar
Department of Biosciences, Himachal Pradesh University, Shimla, Himachal Pradesh, India
e-mail: rkbios@hpuniv.ac.in

Diksha
Department of Biosciences, Division Zoology, Career Point University, Hamirpur, Himachal Pradesh, India

R. Sobti et al. (eds.), *Role of Microbes in Sustainable Development*,
https://doi.org/10.1007/978-981-99-3126-2_1

diseases, gastrointestinal, metabolic, respiratory, mental or psychological, and autoimmune disorders.

Keywords

Gut microbiota · Human · Health · Diseases · Applications

1.1 Introduction

Microbes, due to their distinctive capability to adapt to extreme situations, have so far been found in almost every imaginable environment. Advanced techniques have revealed enormous diversity within microorganisms. This group now represents three distinct life domains, viz. Bacteria, Archaea, and Eucarya, and perhaps still hiding many undiscovered types (Woyke and Rubin 2014; Fuerst 2014). Therefore, methods usually used for the characterization of bacteria and archaea in natural communities are nowadays also applied to uncultured eukaryotes and viruses. Microbial diversity is actually staggering; however, all these micro-organisms can be classified into five groups, viz. Bacteria, Archaea, Fungi, Protozoa, and Viruses (Cuvelier et al. 2010; Breitbart et al. 2002). Indeed, microorganisms are pioneer colonizers and have had a profound effect on climate and environment, over geologic time. The limitations of microbial life are still unknown.

1.2 Main Types of Microorganisms

Microorganisms may be grouped into five major types that include Bacteria, Archaea, Fungi, Protozoa, and Viruses.

Bacteria: Bacteria are found in almost every habitat on earth, including the human body. Besides showing the least complex structures, they have the highest metabolic flexibility and are highly diverse. It is estimated that there are more than 50 phyla of this group, which have so far been estimated on the basis of the conserved 16S rRNA sequence analysis. Bacteria are prokaryotic with cell walls containing mainly peptidoglycan. Bacteria are usually described according to their general shapes that "include rod-shaped (bacillus), spherical (coccus), or curved (spirochete, spirillum, or vibrio)". The characteristics of these bacteria are high replication rate, high surface area, and genetic flexibility. The basic simple features of the bacterial cell enable it to respond quickly to adaptations to extreme environmental circumstances (Schloss and Handelsman 2004; Roane et al. 2009). Mostly, bacteria are harmless or beneficial, but some are pathogenic and cause various diseases in animals including humans.

Archaea: Archaea are microbes that resemble bacteria in shape and size when observed under the light microscope but they are distinct at the genetic and bio-chemical levels. They may be the oldest life form on Earth as they appear to be the simplest form of life. On the basis of their habitat, Archaeans can be divided into halophiles (salty environment), psychrophiles (cold-temperature habitat), and thermophiles (extremely hot temperatures). It is believed that *Archaea* reside in extreme environments and, hence, were called extremophiles, but recent observations suggest that they inhabit a variety of normal environments as well including soil, seawater, or even sewage (Giovannoni and Stingl 2005). So far, no pathogenic archaea have been isolated (Cavicchioli et al. 2003).

Fungi: The cell wall of fungi is composed of chitin rather than cellulose. Fungi comprise a large of eukaryotic micro-organisms, having the maximum biomass. About 1.5 million fungal species have been reported in the world; however, only 7% of them explored so far. Based on morphology, fungi are further grouped as moulds, mushrooms, and yeasts.

Protozoa: Protozoa are unicellular eukaryotic micro-organisms having relatively complex structures. All protozoa live in water bodies and mostly inhabit freshwater and marine habitats, although some are terrestrial in moist soils and gastrointestinal tracts of animals. Nearly 50,000 species of protozoa have been explored so far, which are mostly free-living organisms. It has been revealed that the single-celled protozoa are polyphyletic eucaryotic microorganisms. Parasitic protozoa in humans are usually less than 50 μm in size. The smallest is 1–10 μm long. *Balantidium coli* may measure 150 μm.

Viruses: Viruses are acellular microorganisms (not composed of cells) consisting of proteins and genetic material (DNA or RNA), which are inert outside of their host. These are biological entities having nucleic acid encapsulated within a protein coat called capsid in different sizes and shapes. Their growth cycle can be defined in five steps, viz. adsorption, penetration, replication, maturation, and release.

Trillions of symbiotic microorganisms live on and within the body of human beings and contribute to human health and disease. The microbiota of the human body, particularly the gut microbiota, carries about 150 times more genes than that of the entire human genome (Ursell et al. 2014). Advanced studies have shown that the microbiome is involved in regulating epithelial development, amending the metabolic phenotype, and inducing innate immunity (Savage 1977; Whitman et al. 1998; Ley et al. 2006, b; Wang and Li 2015). Chronic diseases, such as diabetes mellitus, obesity, metabolic syndrome, inflammatory bowel disease, atherosclerosis, alcoholic liver disease, cirrhosis, non-alcoholic fatty liver disease, and hepatocellular carcinoma, have been associated with the human microbiota (Ley et al. 2006, b; Wang et al. 2016) (Figure 1.1). In recent times, evidence has strongly suggested the vital role of the microbiota in human health and disease (Ley et al. 2006, b; Gill et al. 2006; Roberfroid et al. 1995; Cash et al. 2006; Hooper et al. 2003; Schauber et al. 2003; Bouskra et al. 2008; Rakoff-Nahoum and Medzhitov 2008; Macpherson and Harris 2004; Sekirov et al. 2010; Sartor 2008; Liu et al. 2004; Scanlan et al. 2008; Verhulst et al. 2008; Finegold et al. 2002; Wen et al. 2008) through a number of mechanisms. The microbiota contains more versatile genes as compared to those found in the human genome, which provides the human body with exceptional and specific enzymes and pathways (Gill et al. 2006). Furthermore, a large number of metabolic microbiotic processes that are beneficial to the host include the metabolism of undigested sugars and vitamin synthesis (Roberfroid et al. 1995). The human

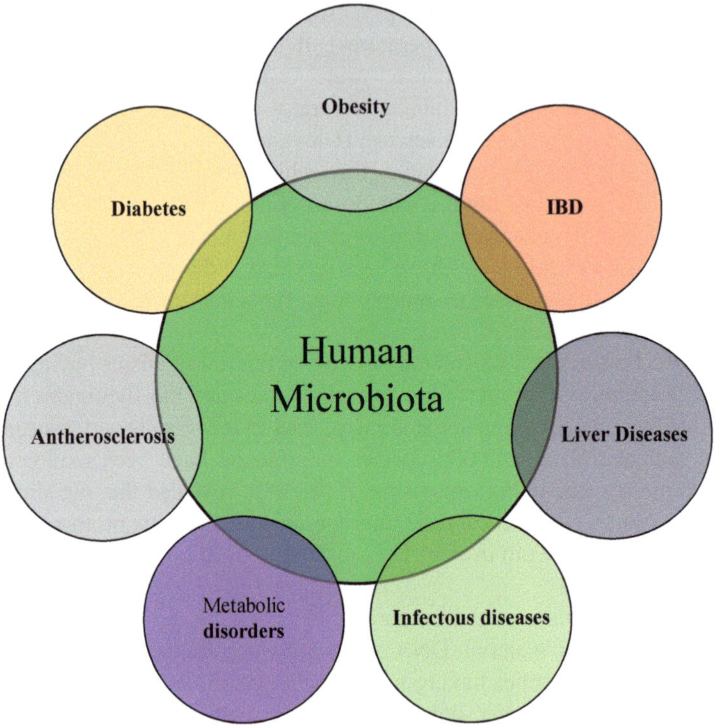

Fig. 1.1 Human microbial interaction causing various diseases

microbiota also provides a barrier, protecting the host body against pathogens by producing antimicrobial substances and competitive exclusion (Cash et al. 2006; Hooper et al. 2003; Schauber et al. 2003). Moreover, studies have shown the crucial role of microbial symbiosis in the development of many diseases, such as infections, liver diseases, gastrointestinal malignancy, metabolic disorders, respiratory diseases, mental disorders, and autoimmune diseases (Ley et al. 2006, b; Sartor 2008; Liu et al. 2004; Scanlan et al. 2008; Verhulst et al. 2008; Finegold et al. 2002; Wen et al. 2008). In this chapter, we provide an overview of the role of micro-organisms in human health and diseases, and potential advancement in the improvement of clinical applications to prevent and treat human diseases.

1.3 Microorganisms in Human Health

The microbiome in humans affects the host body structure to an exquisite extent. Trillions of microbes in the form of colonies are present in the human body, which includes bacteria, archaea, viruses, and eukaryotic ones. Microbial composition and function vary with host location, age, race, sex, and diet (Hollister et al. 2014). Symbiotic bacteria colonize the host immediately after birth. This community slowly

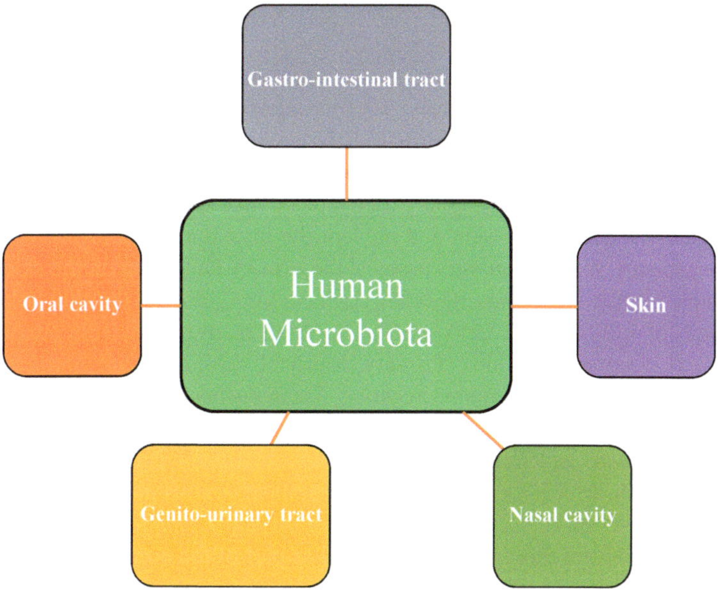

Fig. 1.2 Interaction of microbiota with the human body

evolves into a highly diverse ecosystem along with the development of hosts. With time, the host–bacteria relationship has evolved into a favourable one. Commensal bacteria digest indigestible compounds, supply essential nutrients, and prevent colonization by foreign pathogens (Round and Mazmanian 2009). For instance, the human gut microbiota is involved in the digestion of certain indigestible in the stomach and small intestine and plays a vital role in energy homeostasis maintenance. These foods include fibres, e.g. xyloglucans, usually found in vegetables and digested by certain species of Bacteroidetes (Larsbrink et al. 2014). Other indigestible fibres like fructo-oligosaccharides and oligosaccharides are consumed by beneficial microbes like Lactobacillus and Bifidobacterium (Goh and Klaenhammer 2015). Several studies have revealed the role of gut microbes in protein and lipid homeostasis and the biosynthesis of some essential vitamins (Morowitz et al. 2011). However, all microbes are not beneficial for health, and some microbes induce inflammation under certain conditions.

The human body is inhabited by a large number of microorganisms distributed in various body parts. It is estimated that the abundance of microbiota is higher than the cell number in the body and directly affects the immune system and better utilization of food (Fig. 1.2).

1.4 Microbes on Skin

Skin is the largest human organ that is in contact with the world and is the home of diverse microbe populations. Nearly, 1000 species of bacteria, fungi, viruses, and other microorganisms inhabit the skin, mostly they are beneficial for human beings. Formation of colonies on the skin shows variations depending on endogenous host factors, topographical location, and exogenous environmental variables. Symbiotic microorganisms occupy the skin and prevent the invasion of pathogens or harmful micro-organisms. *Bacillus subtilis* protects the skin by bacitracin on the skin, which fights to protect the skin from different microbes. Microbiota of the skin helps in the production of billions of T cells and keeps them ready against the attack of the pathogen (Cogen et al. 2008). Primary bacterial colonizers are *Staphylococcus epidermidis* and other coagulase-negative *Staphylococci.* Other microorganisms that are generally regarded as skin colonizers are species of *Corynebacterium, Propionibacterium,* and *Brevibacterium.* Most commonly isolated fungal species is *Malassezia* sp., which is especially prevalent in sebaceous gland areas of the skin The Demodex mites, viz. *Demodex folliculorum* and *Demodex brevis* are microscopic arthropods and these are also regarded as part of the normal skin flora (Grice and Segre 2011).

1.5 Microbes in the Nasal Cavity

Evidence suggests that microbes of the nasal cavity contribute to the importing of the reactions of the mucosal and immune systems. Studies reported that Gram-negative bacteria are absent in the nasal cavity but are present in the pharynx. However, *viridans* type Streptococci is sporadically present in the nasal cavity. Conversely, species of *Corynebacterium, Aureobacterium, Staphylococcus,* and *Rhodococcus* have been found predominately present. This suggests that microbiota present in the nasal cavity of adult humans is extremely different from the pharynx (Rasmussen et al. 2000). A study by Savolainen and his co-workers (Savolainen et al. 1986) showed that out of examined nasal cavities of 97 young healthy persons all had aerobic bacteria and 76.5% cavities had anaerobic bacteria. They concluded that the most common aerobic bacteria found was *Staphylococcus epidermidis* in 79% of cases, whereas diphtheroids were found in 41% of cases and *Staphylococcus aureus* in 34% of examined cases. They also found *Haemophilus influenzae* and *Streptococcus pneumoniae* in 5% and 0.5% of cases, respectively. The anaerobic bacteria found were *Propionibacterium acnes* and *Peptococcus magnus* in 74.5% and 3.5% of cases.

1.6 Microbes in the Oral Cavity

The oral microbiota comprises over 600 species. Most organisms that colonize the oral cavity are beneficial to human health but some transit from a commensal relationship to pathogenicity, which may be due to environmental changes or personal hygiene (Dewhirst et al. 2010). The oral microbiota includes the species of Actinobacteria, Fusobacteria, Bacteroidetes, Chlamydiae, Proteobacteria, Firmicutes, Synergistetes, Chloroflexi, Spirochaetes, Euryarchaeota, Streptococcus, and Tenericutes.

Both agonist and antagonist interactions have been found among these microbes. For instance, the relationships between *Actinomyces naeslundii* and *Streptococcus gordonii* are both agonists and antagonists in nature as both are involved in biofilm production. The *A. naeslundii* allows *S. gordonii* to grow in the absence of arginine and removes hydrogen peroxide from cultures, which leads to decreased protein oxidation in *S. gordonii*. On the other hand, hydrogen peroxide produced by *S. gordonii* obstructs the growth of *A. naeslundii* (Jakubovics et al. 2008).

1.7 Microbes in Human Gut

In the gut of humans, various microbes are present which assist in digestion and prevent pathogenic microbes. In newborns, the diversity of microbiota of the intestines is very less diverse and dominance of the phyla Proteobacteria and Actinobacteria. Initially, microbial colonization of the gut in infants seems to be dependent on delivery mode. Vaginally delivered babies obtain microbiota similar to those of the mother's vagina, which is mostly dominated by *Lactobacillus* and *Prevotella* (Rutayisire et al. 2016). The type of delivery influences the diversity and pattern of colonization in the gut during the infants' early years. Feeding habits also affect the diversity and forms in the gut microbiota. It is found that at the age of 2.5 years, the microbiota composition in the gut resembles fully that of an adult. The gut microbiota becomes more diverse with the advent of the dominance of *Clostridia* and *Bacilli*, *Bacteroides fragilis* and *B. thetaiotaomicron*, which resemble the adult microbiota composition (Hsiao et al. 2008). The microbiota composition is influenced by age, diet, and socio-economic situations. Host–microbe interactions have been well studied. For instance, the ability of bacterial disaccharidases to recover unabsorbed dietary sugars and convert them into short chain fatty acids, which are later utilized as an energy source by the colonic mucosa (Wong et al. 2006; den Besten et al. 2013). The gut microbiota also plays a vital role in guarding the host against pathogenic colonization. Intestinal bacteria produce a number of substances, like fatty acids, peroxides, and bacteriocins, which can inhibit potentially pathogenic bacteria. Some strains also produce proteases having the potential of denaturing bacterial toxins (Quigley 2013).

1.8 Microbes in the Reproductive Tract

Successful reproduction in humans is possible only due to a healthy microbial community in the reproductive tract. Microbes exist throughout the entire female reproductive tract in variable compositions and densities. These play an important role in gametogenesis, pregnancy, reproductive cyclicity, and successful delivery of babies. Vaginal microbiota plays an important role in the prevention of various diseases like bacterial vaginosis, sexually transmitted diseases, yeast infections, urinary tract infections, and human immunodeficiency virus (Haldar et al. 2016). *Lactobacillus* sps. in the vagina is believed to play the protective role as they lower the pH through lactic acid production and by producing some bacteriocidal and bacteriostatic substances or through exclusion. The levels of *Lactobacilli* fluctuate with menstrual periods with a prominent increase in *Gardnerella vaginalis*, which may be attributed to increased iron because of the presence of vaginal blood. A healthy vagina contains a low number of *Lactobacillus* species with *L. crispatus, L. jensenii, L. iners,* and *L. gasseri* being predominant. The microbiota of the male genital tract has an important role to play in the reproductive life of the person. Studies have revealed the presence of microbiota in the urethra and coronal sulcus of the male genital tract. Mostly, the upper genital tract has been reported to be free from these microorganisms. Allhough, the presence of coryneform bacteria has been reported, which plays a key role in the male urogenital tract. The interactions between the microbiota of male and female genital tracts have not been observed (Mändar 2013).

1.9 The Microorganisms in the Disease

Microorganisms and Infectious Diseases

1.9.1 Infection with Clostridium difficile

The bacterium originally known as *Bacillus difficilis* was later renamed *Clostridium difficile* due to the difficulty related with its isolation in the laboratory. *C. difficile* results in human-associated diarrhoea, which is now called *C.difficile* infection (CDI), the main causes of nosocomial infection, mainly occur in developed countries. Alterations in antibiotics strategies result in the development of toxins that are responsible for the higher mortality rates associated with CDI; hence, the epidemiology and seriousness of CDI have been changed. In addition to CDI, *C. difficile* is also responsible for conditional infections in the gut of humans generally when the microbiota of gut is interrupted by a wide spectrum of antibiotics. The disruption of normal microbiota occurs as spores of *C. diffcile* germinate and spread in the gut mucosa by binding with the mucosal layer of the epithelium through flagella where various proteins are associated with the binding of *C. diffcile*.

Clostridum difficile is a Gram-positive, anaerobic, spore-bearing, toxin-forming bacillus, which was formally renamed in 2016 to *Clostridioides difficile*. Spores of *C. dfficile* spread through the faecal–oral route, and the pathogen is broadly present in the environment. The potential source of *C. difficile* includes asymptomatic carriers, infected patients, the polluted environment, and intestinal tract of animals (canine, feline, porcine, avian). About 5% of adults and 15–79% of infants are colonized by *C. difficile*, and the colonizing ability is many times higher in patients admitted in hospitals or those admitted nursing homes (Czepiel et al. 2019). "Prior to the insertion of antibiotics, the prevalence of *C. difficile* in the pathogenesis of large intestine diseases rises".

CDI is one of the main causes of nosocomial infectious disease which mainly occurs in developed countries (Khanna et al. 2012; Hung et al. 2013; Loo et al. 2011). It has been reported that 75% of antibiotic-related cases are caused by *C. difficle*, and 90–100% are caused by pseudomembranous colitis (Eyre et al. 2013). Initially, for certain years, *C. difficle* has remained vague; however, after a decade, this causative agent shown a striking increase in CDI cases, and then after, the cases of CDI have become a very common cause of nosocomial diarrhoea, especially in developed countries. However, a higher prevalence of CDI was recorded in North America, Europe, Australia, and various parts of Asia (Oren and Garrity 2019; Khanna et al. 2012; Clabots et al. 1992). The Centres for Disease Control and Prevention (CDC; Atlanta, GA) estimated that 2,50,000 individuals were found infected per year, and 14000 died due to CDI in the United States alone (Clabots et al. 1992).

A sudden rise in the prevalence and pathogenicity has been found to increase in other parts of the world. This consistent increase in CDI is related to hospital emergence in Europe. *Clostridiodes difficile* infection in children (CDI) is like a consistent or severe disease. Most children suffering from *Clostridiodes difficile* infection (CDI) have been observed with mild to moderate diarrhoea and with the administration of antibiotics having a continuous clinical cure. Whereas in some patients who suffer from other complications, 20–30% suffer continuous and serious, or fulminant CDI has led to poor outcomes in children.

1.9.2 Infection with Helicobacter pylori

In our gastrointestinal tract, a diverse type of micro-biome is present, which constitutes 80% of the overall microbial biomass. Previously, it was believed that the stomach is sterile. In the 1980s, Marshall and Warren isolated *Helicobacter pylori* (*H. pylori*) from gastric biopsies collected from patients suffering from chronic gastritis and peptic ulceration and were awarded Nobel Prize in 2005 in Physiology or Medicine for their discovery (Pattison et al. 1997). Moreover, other than *H. pylori,* other micro-organisms are also present in the stomach, which mutually constitutes a microbial environment in the stomach and might be associated with the presence of various gastrointestinal diseases. It has been examined that approximately 1014 bacteria are pre-set in the intestines of a normal healthy person, which creates a microbiome and maintains an dynamic equilibrium ecosystem (Honda and Littman 2012). Commonly, the number of bacteria varies from stomach to intestines and is present in large numbers; however, the concentration of bacteria increases successively. However, in the buccal cavity, a number of bacteria are higher, whereas in the stomach, the bacterial count is very less. In the small intestine, a large number of bacterial colonies are present, whereas in the colon part of the intestine constitutes, a higher number, viz. 1010 times higher number than the stomach (Hattori and Taylor 2009; Arweiler and Netuschil 2016). Few studies have reported that microbes present in the stomach and duodenum are mostly because of specific physiological processes like gastric acid secretion, secretion of bile, and motility of gastrointestinal tract.

1.10 Interactions Between Microbes in the Stomach

The gastrointestinal micro-ecosystem constitutes a unity developed by the interaction and effect of the gastrointestinal flora, its host, and also external environment. The function of gastrointestinal flora maintains the stability and equilibrium of the gastrointestinal micro-ecosystem via several regulatory systems and signalling pathways. If this balance is disturbed, this results in an imbalance in microecology and causes various diseases. Recent studies reported that non-*H. pylori* gastric micro-organisms are associated with gastric diseases, and *H. pylori* gastric micro-organisms also interact with other gastric micro-organisms. When *H. pylori* forms colonies within the gastric mucosa, it alters the environment in the gastric system through the decomposition of the mucosal layer and alkalizing gastric juice (Kelly et al. 1993). Other studies explored that non-*H. pylori* of gastric micro-organisms are associated with the development of gastric cancer. When Ins-GAS mice were administered antibiotics, they developed gastric cancer (Lofgren et al. 2011). Alterations in the gastric eco-microbiological environment following *H. pylori* elimination revealed that *H. pylori* influences the association of other micro-organisms in the stomach, probably promoting the development of inflammation and cancer in patients (Sung et al. 2020). By the elimination of *H. pylori* it is able to prevent the development of gastric mucosal lesions (Sung et al. 2000; Doorakkers et al. 2018); however, in some patients, continuous development of precancerous lesions, such as gastric atrophy (GA) and intestinal metaplasia (IM), following the radical treatment with *H. pylori* was observed (Soeta et al. 2009). It has been reported that only 3% or less than that who were found infected with *H . pylori* develop gastric cancer (Hooi et al. 2017) and approximately 20% of patients suffering from chronic gastritis were found negative for *H .pylori*, which indicates that other micro-organisms could induce gastritis and sometimes gastric cancer (Zavros et al. 2002).

Even though *H. pylori* initiates the inflammation in the gastric tract, and other micro-organisms having the ability to secrete pro-inflammatory cytokines might be playing an essential role in the maintenance of the development of inflammation and unusual hyperplasia, which, in turn, results in the progression of gastric cancer.

Recent studies reported that *H. pylori* and EBV combined develop gastritis, peptic ulcer, dyspepsia, and gastroesophageal reflux disease (GERD). During the coinfection process caused by *H. pylori* and EBV, the deployment of immune cells at the site of infection considerably enhances, hence exacerbating gastric inflammation and damaging tissues also (Dávila-Collado et al. 2020). Monochoramine is a good antioxidant produced in the stomach when *H. pylori* causes infection, which is able to induce the transition of EBV from the latent phase to the cleavage phase (Minoura-Etoh et al. 2006), and *H. pylori* results in the secretion of interferon γ (INF-γ), which promotes the inflammatory process and aggravates the seriousness of the disease (Dávila-Collado et al. 2020). The microbiota in the intestines can control infection caused by *H .pylori*, and vice versa, *H. pylori* modulates the composition of the microbiota of the stomach (Espinoza et al. 2018; Coker et al. 2018; Alarcón et al. 2017). However, the microbiota present in various human niches directly or indirectly affects the infection caused by viruses including EBV and human papillomavirus (Wakabayashi et al. 2019).

Recently, changes in lifestyle such as dieting habits are associated with increased blood pressure, and the prevalence of *H. pylori* infection and gastric ulcer has become very high. It has been assumed that *H. pylori* infection is a major risk factor for the occurrence of gastric cancer and gastric ulcer, and it may also change into a malignant state. Hence, *H. pylori* infection and gastric ulcer have become very common and life-threatening health burden that needs to be treated.

1.10.1 Bacterial Vaginosis

The microbiome of humans is essential for health and diseases (Bernabe et al. 2018). The micro-environment of the vagina is a complex one in females, as it is playing a crucial contribution to reproduction and to protect against diseases (Guven-Maiorov et al. 2017). The disruption of vaginal microbiota affects the reproductive health of a woman and sometimes may cause infertility (Ma et al. 2019). The microbiota of a normal vagina includes aerobic bacteria, particularly *Lactobacillus* is predominantly present in the vagina, it suppresses the growth and colony formation of pathogenic micro-organisms by the production of lactic acid, H_2O_2, bacteriocins, probiotics, biofilms and bio-surfactants, and so on. The unusual interruption of the vaginal microbiota induces alterations in hormonal profile, smoking, use of antibiotics, and recurrent sexual intercourse could cause a considerable decrease in the population of *Lactobacillus* and elevate the chances of vaginal infection. These infections might be caused due to conditional pathogens, such as *Bacterial Vaginosis* (BV) (Hawes et al. 1996), *aerobic vaginitis* (Donders et al. 2002), and sexually transmitted infections. BV is one of the most widespread reproductive infections in females having a broad prevalence of about 21.2 million (29.2%) (Javed et al. 2019). Moreover, BV is most widespread in black females in comparison to their white counterparts, whereas its geographical distribution is barely affected (Javed et al. 2019). The features of BV include environmental shifts from a *Lactobacillus*-dominating mixture into the facultative and stringent anaerobic polymicrobial mixture, such as *Gardnerella,*

Atopobium, Mobiluncus, Prevotella, Mycoplasma, Urea plasma, and *Sneathia.* Moreover, the prevalence of BV is comparatively higher in infertile females (Verstraelen et al. 2016, b). Females having infertility are more prone to asymptomatic vaginosis and a higher degree of BV-associated bacteria than their healthy ones (Babu et al. 2017).

It has been reported that BV is associated with other reproductive problems, such as miscarriage, premature birth, preterm premature rupture of membrane and other issues. Other studies also reported that bacterial infections in the vagina spread into the uterus, and the microbiota of the vagina is essential for the formation of bacterial colonies within the endometrium in non-pregnant women; however, the endometrium contains some peculiar types of bacterial species (Verstraelen et al. 2016; Chen et al. 2017) (Table 1.1).

1.11 Infection with HIV

The human gastrointestinal (GI) tract contributes a particular role in structural and immunological protection against exposure from outside the environment. In the GI tract, the mucosal surface comes in contact with food, environmental antigens, as well as with micro-organisms like bacteria, fungi, and viruses. Due to this reason, the GI tract is an essential immunological site for the maintenance of the equilibrium between tolerance and reactivity. The connection between innate and adaptive immunity in gut-related lymphoid tissues (GALT) is crucial for quick protection from pathogens for a prolonged time. Studies reported that HIV-1 (HIV) and its non-human primate corresponding partners SIV target GALT as principal sites for the transmission of the virus, its replication and seeding, and depletion of $CD4^{+}T$ cells During HIV and SIV infection, $CD4^{+}T$ cells become depleted at a very higher rate than their depletion in the peripheral blood During HIV and SIV infection, several factors are responsible for the rapid depletion of $CD4^{+}T$ cells, such as (IL-1β induced pyroptosis, Fas-ligand, and TNF-α) and viral factor (HIV-1 Tat and Nef).

HIV infection alters the constitution of the microbiota of the gut rapidly after the infection, which damages the intestinal barriers and also changes mucosal immune responses. HIV infection results in the shift of gut microbiota or dysbiosis until the treatment with anti-retroviral therapy (ART). ART actively inhibits the rate of replication in HIV; however, it does not eliminate the HIV pool or wholly reverse the CD4+T cells in the gut as well as in the periphery in HIV-infected persons. In HIV-infected individuals, an unconventional microbiota has the ability to systemic inflammation or may also maintain it. The alterations in microbiota take place in two ways in HIV-infected patients. The first trend is related to sexual activity, independent of HIV infection. HIV-induced dysbiosis was found related to the decreased alpha diversity, abundance of Gamma proteobacteria, such as Enterpbacteriaceae, and reduction in Lachinospiraceae and Ruminococcaceae (Vujkovic-Cvijin et al. 2020). HIV-induced decrease in alpha diversity is predominantly found in patients having low CD4 count (Monaco et al. 2016; Parbie et al. 2021; Ishizaka et al. 2021). Enterobacteriaceae and Desulfovibrionaceae are related to different chronic

Table 1.1 Showing the antimicrobial activity of different antibiotics against different types of microbes

S. no.	Alternative treatments for BV	Example	Effect
1.	Antimicrobials	Rifaximin	Suppresses the growth of different bacterial species (*Atopobiumvaginae, Prevotella, Megasphaera, Mobiluncus,* and *Sneathia*)
		Secnidazole	Shows antimicrobial activity against the BV-associated bacterial species
2.	Antimicrobial peptides	Retrocyclin	Suppresses the function of vaginolysin and reduces the formation of biofilm
		Subtilisin	Suppresses the growths of *G. vaginalis* by the formation of pores in the plasma membrane
3.	Antiseptics	Dequalinium chloride	Shows antibacterial activity due to their non-specific binding on the plasma membrane and its disruption
		Povidone iodide	
		Hydrogen peroxide	
		Chlorhexidine	
		Octenidine	
4.	Surfactants	Coamphopropionate	Shows a broad spectrum of antibacterial activities
5.	Natural compounds	"*Prangosferulacea, Myrtus communis, Berberis vulgaris,* "*Zataria multiflora Thymus vulgaris, Eugenia caryophyllata, Calendula officinalis*", *Tribulus terrestres,* "*Myrtus communis, Foeniculum vulgare, Tamarindus indica, Thymbra capitata, Artemisia princeps* Pamp"	suppression of growth in "*G. vaginalis*, antibiofilm activity or improves the symptoms like itching of vagina, burning sensation, odour, dysuria and dyspareunia)"
6.	Acidifying agents	Vitamin C, lactic acid, polycarbophil and boric acid	These are having potential to decrease the pH below 4.5, normalizes the pH of vagina
7.	Probiotics	*L. reuteri,* "*L. rhamnosus, L. plantarum, L. crispatus, L. fermentum, L. gasseri, L. brevis and L. acidophilus*"	Re-creates saprophytic vaginal flora
8.	Prebiotics:	Fructo oligosaccharides, sugar alcohols, galacto-oligosacchrides, lactulose, raffinose	Supports the selective growth of beneficial micro-organisms

(continued)

Table 1.1 (continued)

S. no.	Alternative treatments for BV	Example	Effect
9.	Other substances	Nifuratel	Shows antimicrobial activity against *G. vaginalis* and *A. vaginae*
10.		Benzoyl peroxide	Shows effectiveness against *G. vaginalis*
		DNase	Suppresses the growth of *G. vaginalis*

inflammatory conditions like inflammatory bowel disease (Figliuolo et al. 2017), suggesting that there exists a potential relationship between HIV-associated dysbiosis and chronic inflammation.

1.12 Microorganisms and Liver Diseases

In human beings, the microbiota of the gut comprises trillions of microbes and plays an important role in health and diseases via different mechanisms. It has been reported that numerous novel activities of gut microbiota associate the gut with the liver (Tripathi et al. 2018; Sanduzzi Zamparelli et al. 2017). Amendments in microbiota cause various diseases like inflammatory bowel disease (IBD), colon cancer, diabetes, and hepatic diseases (Marchesi et al. 2016). Hepatic diseases cause about two million deaths every year in the world (Eshraghian et al. 2020), which includes cirrhosis and hepatocellular carcinoma and are the main cause of mortality and morbidity. Chronic liver diseases result due to different aetiologies, such as viral infections, alcoholic liver diseases (ALD), and non-alcoholic fatty liver diseases (NAFLD). Chronic liver disease causes fibrosis and cirrhosis in the liver having a higher threat of hepatocellular carcinoma. About 75% of blood for the liver is supplied by the portal vein, which receives blood from the intestines and spleen. This unique physiological constitution of the liver reflects its association with the gut-resident microorganisms and their metabolites.

1.13 Non-Alcoholic Fatty Liver Disease (NAFLD)

Non-alcoholic fatty liver disease (NAFLD) comprises a broad spectrum of hepatic diseases, ranging from steatosis to non-alcoholic steatohepatitis (NASH), liver fibrosis, and cirrhosis. The pathological process of NASH was first hypothesized due to the result of two hits, which are specified by hepatic steatosis and lipid peroxidation (Day and James 1998). However, this theory was later modified into multiple hits consisting of signals received from the gut or the adipose tissues like endotoxins, adiponectin, IL-6, and TNF-α (Tilg and Moschen 2010). In addition to dietary and genetic factors, the microbiota of gut plays an important role in the pathogenesis of liver diseases involving all multiple hits (Tilg and Moschen 2010; Wieland et al. 2015).

1.14 Alcoholic Liver Diseases

Alcohol has been considered as a potent hepatotoxin, and alcoholic liver disease is one of the major causes of chronic liver disease in the world (Bataller et al. 2019). The ALD pathogenesis involves the toxicity of acetaldehyde, hepatic steatosis, and inflammation. Other contributory factors involved in liver diseases are genetic and non-genetic, contributing to inter-individual changes in the development of diseases. Gut microbiota contributes significantly to the pathogenesis of alcohol-induced liver injury. Regular consumption of alcohol accelerates the excess growth of bacteria, dysbiosis, and interrupts the function of the gut barrier. The consequent transport of bacterial products via the portal vein initiates hepatic inflammation and metabolic disorders (Vassallo et al. 2015).

1.14.1 Cirrhosis

Cirrhosis is the last phase of chronic hepatic disease, the features of these hepatic diseases include the formation of nodules, regeneration of parenchyma, and sometimes goes with portosystemic shunts. Other factors associated with the pathogenesis of liver dysfunction included gender and socio-economic status (Eshraghian et al. 2020).

1.14.2 Liver Cancer

The progress of hepatic cancer is a multiple stage process that involves various factors genetics, environmental factors, metabolism, and the immune system. Changes in the microbiota contribute to the progression of cancer and modulate the potency of cancer therapy (Fessler et al. 2019). In patients suffering from cirrhosis along with HCC, faecal microbiota dysbiosis is signified by the excessive growth of *E. coli* (Grąt et al. 2016). In human beings, *Helicobacter* has been detected in hepatic carcinoma; however, *Helicobacter* was not found in patients devoid of malignancy (Huang et al. 2004). In addition to the faecal microbiota, microbiota from other origins are also associated with hepatic diseases, including saliva (Bajaj et al. 2015, b), colonic mucosa (Bajaj et al. 2012), sigmoid mucosa (Bajaj et al. 2015, b), and the small intestine (Chen et al. 2016). In addition to this, studies also reported that other constituents of the gut microbiota, such as fungi, archaea, and viruses, contribute to liver diseases (Lloyd-Price et al. 2016). Fungal dysbiosis is associated with liver cirrhosis (Bajaj et al. 2018). However, with the administration of antibiotics, the population of fungi has been found to reduce associated with the incidence of Candida (Bajaj et al. 2018). Patients suffering from alcoholic liver diseases also show a reduction in fungal diversity. An excessive growth of Candida was discovered in various stages of alcoholic liver diseases, such as alcoholic hepatitis and cirrhosis patients (Yang et al. 2017). Viruses comprise an essential group of microorganisms including microbes of the gut. Histologically, the seriousness of NAFLD is related to the reduction in gut virome diversity (Lang et al. 2020).

In addition to live microorganisms, portions and extracts from bacteria, recognized as 'postbiotics', have been predominantly influencing the health of the host (Aguilar-Toalá et al. 2018). Postbiotics are bioactive compounds produced by food-grade micro-organisms. The postbiotics activities include anti-inflammatory, anti-hypertensive, and antioxidant (Haileselassie et al. 2016). The association between dysbiosis of the gut microbiome and several hepatic diseases had been explained at clinical and preclinical levels. Various microbes play an important role in different stages of liver diseases. Nutrients and metabolites produced by commensal gut bacteria reach into the hepatocytes via the portal vein and affect the metabolism of hepatocytes and ultimately cause liver injuries (Tripathi et al. 2018; Ridlon and Bajaj 2015). Bacterial and bacterial-derived products like lipo-polysaccharides are associated with the progression of local and systemic immunity and contribute to the development of hepatic diseases by disrupting the gut barrier and activating the inflammatory processes (Lambertucci et al. 2018).

1.15 Microbiota in Therapeutics of Liver Diseases

Current researchers are engaged in the development of therapeutic molecules for the prevention of hepatic diseases. Treatment with antibiotics (Bergheim et al. 2008), prebiotics (Li et al. 2003), and probiotics (Cani et al. 2007; Wagnerberger et al. 2013) plays an important role in the inhibition of NAFLD development. The mechanism associated with these molecules includes hindrance in the translocation of bacterial endotoxins and the simultaneous stimulation of Kupffer cells and induction of TNF-α.

1.16 Potential Impact of Microbiota on Liver Injury in COVID-19 Patients

Various studies reported that microbiota and hepatic diseases have some kind of relationship between them. Particular and specific microbes of the gut contribute to the pathogenesis and therapeutic of hepatic diseases (Oh et al. 2020). Screening through computational techniques and experimental studies has evidenced that a combined analysis between microbiota along with assessment at a clinical level will serve as a standard to confirm the liver disease, its stage, and the sensitivity of the disease. Moreover, an intrahepatic bacterial metataxonomic signature has been found in NAFLD patients who provided further evidence for the association of microbiota in the pathogenesis of diseases and associated mechanisms.

1.16.1 Gastric Cancer

Gastric cancer (GC) is generally diagnosed as cancer and is the fifth leading cause of death in the entire world (Majewski et al. 2022, b). It has been reported that GC caused 780,000 deaths (8.8% among all cancer deaths) in 2018. World Health Organization (WHO) reported that in 2020, 1,089,103 cases of GC were registered in the world, which constitutes about 5.6% among all cancer cases in the world. GC is a heterogenic type of cancer and is diagnosed at a late stage, and the preventive measures for this type of cancer are very complicated (Patel et al. 2017; Lee and Derakhshan 2013). There are various risk factors and causes of GC (Yang et al. 2020; Sitarz et al. 2018) changes in the constitution of gastric and intestinal microflora (micro-biological dysbiosis and *Helicobacter pylori* also participate), inappropriate diet, environmental and genetic factors, and weakening of immunological homeostasis.

H. pylori is most commonly associated with the development of cancer (Primo et al. 2022), which comprises 89% of non-cardia GC (Flood et al. 2022). Some studies reported that GC includes intermediate stages, such as premalignant lesions (atrophy, metaplasia. and dysplasia), considering the intestinal-type of GC these are identified as stages of carcinogenic development as stated by Correa's hypothesis (Taja-Chayeb et al. 2022). *H. pylorus* is one of the predominant bacteria found in normal gastric mucosa as well as in non-atrophic gastritis (Luo et al. 2018). However, following the infection by *H. pylori*, the population of microbiome decreases parallelly with the reduction in *H. pylori*, like premalignant lesions (Kwak et al. 2020; Becerril-Rico et al. 2021). Lower pH (acidic) in the stomach creates a probable barrier for various microorganisms (Chen et al. 2019). In the stomach, the result of the common gastric microbiome is designated by an acidic micro-environment.

Elimination of *H. pylori* decreased the rate of metachronous GC after the endoscopic resection during the initial phase of GC or higher grade of adenoma (Taylor et al. 2015) (Yiannakou et al. 2022). Various studies reported that the prevalence of *H. pylori* infection has declined in different countries of the world, partially constantly with the decrease in GC (Zhu et al. 2013). Chronic infection of *H. pylori* causes gastritis, and in some patients, peptic ulcer disease in approximately 1% causes gastric cancer (Kumar et al. 2021; Zhang et al. 2016).

1.16.2 Colorectal Cancer

According to the World health organization (WHO), colorectal cancer has become a big and life-threatening health issue in the entire world. In the year 2011, cancer became a leading cause of death followed by stroke and coronary diseases. About half of these deaths were reported due to lung, stomach, and colon, breast cancer. Colorectal cancer (CRC) has been ranked the third disease having a higher prevalence in males followed by females (1.4 million in the year 2004). It has been ranked third as it has caused 6.9 lakh deaths annually (Jacques Ferlay et al. 2014).

Moreover, this disease is age dependent about 75% of the CRC deaths have occurred in people with age more than 65 years, affecting mostly males. In Europe, the prevalence of CRC showed a considerable decrease in countries such as Austria, France, Ireland, Sweden, and Norway, unlike Eastern and Mediterranean countries including Spain, Italy, and France (Bosetti et al. 2011). There are significant differences in the prevalence of CRC across the countries, which is mainly due to diet.

There are various environmental and person-specific factors associated with the development of CRC. In 1981, it has been reported that 90% of gastrointestinal cancers are due to variations in diet (Ames and Gold 1998). In the African race, the prevalence of CRC is rare, which is associated with the avoidance of consumption of animal flesh and fat within this ethnic group (O'Keefe et al. 1999). It has been evidenced that people migrate from low-prevalence countries to higher-prevalence areas showing higher rates of CRC (McMichael and Giles 1988). The essential contribution of environmental factors was also found in the age structure of CRC in developing countries. These countries are having a prevalence of CRC during their young age and with lower levels in the old group. There are various factors associated with the increased prevalence of CRC, such as a sedentary lifestyle, and a diet rich in meat, fat, and low in fibre. Least physical activity, higher consumption of caloric diet, and alteration in eating habits were found in more Western people (Eastern Asia, Europe, and North Africa) and were found to increase the rate of obesity and CRC in these emerging countries (Center et al. 2009; Minami et al. 2006). Previous studies have reported that patients suffering from obesity and metabolic disorder are having a higher risk of adenomas and CRC, which is related to economic growth and globalization (Kim et al. 2012; Yang et al. 2013). It revealed that insulin-like growth factors and IGF-1 are found more in obese people with insulin resistance, promoting cellular growth and suppressing signalling pathways of apoptosis leading to carcinoma (Becker et al. 2009).

1.16.3 Oesophageal Cancer

Oesophageal cancer (EC) is a common type of upper gastrointestinal cancer, producing a global threat of cancer burden, particularly in developing countries. EC has been ranked as the 9th most common type of cancer and the 6th leading cause of death in the entire world, about 572,034 cases and 508,585 deaths have been recorded in 2018 (Bray et al. 2018); in this whole scenario, China has contributed 50% of EC cases in the world (Abnet et al. 2018). The pathogenesis and progression of EC is a very complex process; however, in the majority of cases, EC remains does not show any symptoms till cancer reached an advanced stage (Torre et al. 2015). Hence, an initial diagnosis of EC shall immensely improve cancer treatment and increase the rate of survival. It is essential to design non-invasive biomarkers having higher sensitivity and that are precisely specific. The human intestinal tract has the capacity to host around 36,000 bacterial species constituting about 100 trillion micro-organisms, collectively known as the microbiota of intestines (Frank et al.

2007; Neish 2009; Qin et al. 2010). Gut microbiota plays a significant role in the maintenance of host health (Nardone and Compare 2015), which includes absorption of nutrients, maintenance of mucosal integrity, and the regulation of homeostasis in intestines (Nelson et al. 2015). Some studies reported that disorders of intestinal microbiota are associated with various diseases including diabetes, non-alcoholic fatty liver diseases, inflammatory bowel diseases, and some other cancers (Cani et al. 2008; Henao-Mejia et al. 2012; Parkes et al. 2012; Gao et al. 2015; Pushalkar et al. 2018; Goedert et al. 2015). Definitely, some researchers have also examined the microbial features of oesophageal diseases, including reflux-associated oesophagitis, Barrett's oesophagus, and oesophageal adenocarcinoma. In the whole intestinal network, the colon contains the highest microbiota in the whole body (Neto et al. 2016), there are numerous proofs that reveal that the microbes in the gut play an essential role in the development of cancer and patho-physiology of extra-intestinal cancers. It has been assumed that the bacteria present in the gut affect the progression of breast cancer by modulating the metabolism of oestrogen (Kwa et al. 2016; Flores et al. 2012). It has been reported that microbial diversity decreases in patients suffering from leukaemia (Rajagopala et al. 2016). In animal models, it has been studied that the microbiota of intestines promotes the development of hepatocellular carcinoma via the gut microbiota-liver axis (Dapito et al. 2012; Schnabl and Brenner 2014). Moreover, in the mouse model, it has been studied that there exists a relationship between microbiota-immune crosstalk and the progression of lung cancer (Jin et al. 2019). Some studies reported that a particular gut and tumour microbiome in murine models of pancreatic ductal adenocarcinoma suggests a strong bacterial translocation from the intestinal tract into the peri-tumoural milieu (Pushalkar et al. 2018).

Oesophageal cancer has been ranked as 7th most common cause of cancer associated with death in the entire world, constituting more than 400,000 deaths per year (Fitzmaurice et al. 2018). It is divided into two types on the basis of histology, viz. oesophageal squamous carcinoma (OSCC) and oesophageal adeno-carcinoma (OAC), having distinct prevalence patterns and risk factor profiles (Lagergren et al. 2017; Arnold et al. 2015). Both of these histological types share a weak prognosis and an estimated 5 year survival below 20%, and this has been low in patients diagnosed at advanced phases of tumour (Lagergren et al. 2017). Prevention and initial identification might be important in decreasing the load of this cancer. The prevalence of oesophageal squamous cell carcinoma (OSCC) constitutes more than 80% of all oesophageal cancer in the world and has attained a dominant position in less developed countries (Arnold et al. 2015). The prevalence of OSCC has been found to be highest in Central and South-East Asia, where there was about 80% of all OSCC cases; in the year 2012, 80% of the cases of OSCC were recorded (Arnold et al. 2015). However, 50% of these cases of OSCC in the world have been reported in China per year (Arnold et al. 2015; Abnet et al. 2018). The prevalence of OSCC in males showed reduction during the past certain decades in the world. In females, the prevalence of OSCC has been found considerably reduced or stabilized in most regions; however, in some countries, cases of OSCC showed an increase, such as Japan, the Netherlands, New Zealand, Norway, and Switzerland (Wang et al. 2018).

The prevalence of OSCC increased considerably in rural areas in comparison to urban areas in China and in some other Asian populations (Pan et al. 2017; Xie and Lagergren 2018).

1.17 Incidence of Oesophageal Adenocarcinoma

Over the past four decades, a significantly increased prevalence of OAC has been observed in western regions such as Europe, Northern America, and Australia and the prevalence of OAC has increased than those of OSCC in various Western countries (Edgren et al. 2013; Coleman et al. 2018; Xie et al. 2017, b). The prevalence of OAC has considerably more in whites than in other races in the United States (Xie et al. 2017, b). However, OAC affects predominantly males, the ratio of OAC in males to females is 6:1, and in some populations, the ratio is 8:10 (United States) (Coleman et al. 2018; Xie and Lagergren 2016).

1.18 Risk Factors for Oesophageal Squamous Cell Carcinoma

1.18.1 Socio-Economic Status

Lower socio-economic status revealed by low income or education has been continuously been found related to the increased threat of oesophageal squamous cell carcinoma (OSCC) as per research conducted in developing and developed countries (Abnet et al. 2018; Xie and Lagergren 2018). A control study was conducted and reported that 4-fold increased risk of OSCC in skilled manual workers in comparison to professionals, after change in tobacco smoking and consumption of alcohol (Jansson et al. 2005). These differentiations in socioeconomic position might play an essential role in the rural–urban dis-similarity in the prevalence of OSCC.

1.18.2 Tobacco Smoking and Smoking Cessation

Tobacco smoking is another main risk factor for OSCC in developed as well as in developing countries, even though the relation looks to be very powerful in people living in developed countries. Meta-analysis of 52 studies showed that around five-fold higher risk of OSCC in current smokers than non-smokers in Western countries, and an equivalent Three-fold higher threat in Asia and South America (Qiao-Li et al. 2017). About 50% of the OSCC cases are those who are tobacco smokers in the United States (Freedman et al. 2007; Engel et al. 2003). The consumption of tobacco products other than cigarettes might increase the threat of OSCC with apparently the same intensity of relation as cigarette use but it is more precisely population-specific (Abnet et al. 2018, b). About 41 research reports revealed a 66% reduction in risk of OSCC in patients who stopped smoking from last 20 years ago (odds ratio 0.34, 95%

confidence interval [CI] 0.25e0.47) compared with current smokers (Qiao-Li et al. 2017).

1.18.3 Alcohol Overconsumption

Addictions of alcoholic beverages are associated with a higher threat of OSCC. The relation looks very strong in lower incidence Western countries in comparison to higher prevalence Asian countries (Abnet et al. 2018). It has been reported that in the United States, higher consumption of alcohol is related to a Nine-fold increased threat of OSCC and might describe the higher incidence of OSCC cases (Engel et al. 2003; Morris Brown et al. 2001). In Asian countries, the risk associated with the OSCC ranges from 1.6 to 5.3 in people consuming alcohol (Abnet et al. 2018). This variation in the threat of OSCC might be due to the differences in exposure strategies, susceptibility of the considered population, study design, and capability to adjust for cofounders (Pan et al. 2017; Xie and Lagergren 2018).

1.18.4 Dietary Factors

Numerous studies carried out around the entire world have examined the relationship between different dietary factors and the threat of OSCC, whereas, the present literature is restricted to relating OSCC threat with any particular nutrient of food stuff. This might be due to the complex nature of dietary assessment, specifically inter-correlations among individual items and confusing from other threatening agents, e.g. tobacco smoking, alcohol consumption, and socioeconomic status. Various studies reported that a higher threat of OSCC is related to a higher consumption of pickled vegetables and low consumption of fresh fruits and vegetables (Qiao-Li et al. 2017; Islami et al. 2009; Liu et al. 2013), but substantial evidence from high-quality potential studies remained highly lacking. International Agency for Research on Cancer (IARC) has reported that research reports conducted on humans have been restricted about the carcinogenicity of pickled vegetables and the evidence from experimental animal models are not sufficient (Loprieno 1975).

People consuming hot food and beverages have been found associated with OSCC threat (Kamangar and Freedman 2018). Some metaanalysis studies revealed that a Two-fold higher threat of OSCC associated with the consumption of hot food or drinks, and pooled odds ratios were found in developing countries (2.80, 95% CI 2.05e4.02) in comparison to developed countries (1.65, 95% CI 1.17e2.33) (Andrici and Eslick 2015).

1.19 Microorganisms and Metabolic Disorders

1.19.1 Type-2 Diabetes

The characteristics of diabetes (Type 1 and Type 2) are the same, i.e. hyperglycaemic state of the body due to the irregular synthesis of insulin, insulin resistance, or both. In the case of type 2 diabetes, insulin resistance is a major cause of hyperglycaemia. Various evidence have associated type 2 diabetes with the modulation in the diversity of gut microbiota. Hence, gut microbiota has become an emerging target for the management of type 2 diabetes. This gut microbiota included micro-organisms (bacteria, fungi, archaea, and viruses) present in the gastro-intestinal tract (Thursby and Juge 2017). Among the entire microbiota, bacteria are gaining more interest, it has been reported that around a hundred trillion bacterial species inhabit the gastrointestinal tract in humans (Cani 2018). These bacteria are classified into more than 1000 species; however, 99% of them belong to 30–40 species (Almeida et al. 2019). However, the distributions of these bacteria are not uniform. The most predominant bacterial species present in the gut of human beings are Firmicutes, Bacteroidetes, Actinobacteria, and Proteobacteria. Mostly these bacteria belong to the genera *Bacteroides, Clostridum, Faecali bacterium, Eubacterium, Ruminococcus, Peptococcus, Peptostreptococcus, and Bifidobacterium* (Arumugam et al. 2011). Some researchers consider microbiota as a special organ that contributes a significant role in the maintenance of human health. Similar to the other organs, gut microbial diversity may cause loss of homeostasis that in turn leads to various diseases like allergies, Celiac's disease, gastric cancer, and obesity (Carding et al. 2015). During the last two decades, research at a large scale has been conducted to associate gut micro-biota with T2D (Cardoso 2021). It has been reported that intestinal flora affects T2D by regulating bacterial metabolites. Bacterial metabolites play an essential role in flora-mediated sugar metabolism. It has been reported that imidazole propionate, a histidine-derived metabolite generated by micro-organisms, might be acting as a bridge connecting intestinal flora with T2D. Increased imidazole propionate level in the circulatory blood has been observed in T2D patients and might be a causative agent for the destruction of glucose tolerance and insulin signallingpathways (Koh et al. 2018).

Another key mechanism through which intestinal flora influences T2D is the metabolism of BCAAs, which includes three essential amino acids, viz Leucine, isoleucine, and valine. An essential signalling molecule for nutritional metabolism for the human body and BCAA synthesis is directly associated with the position of the intestinal flora (He and Li 2020). BCAAs and aromatic amino acids (AAAs) are related to insulin resistance and the progression of diabetes (Holeček 2018). Recent studies evidenced revealed that the alteration in the composition of the gut microbiota can amend BCAA metabolism and hence contributes to the progression of diabetes. Ultimately, the intestinal flora is having the ability to influence T2D through the regulation of LPS-induced inflammation. Different pro-inflammatory cytokines are responsible for the induction of T2D pathogenesis (Tsalamandris et al. 2019). Dysfunction in the intestinal flora generates a huge quantity of LPS (Fuke et al. 2019), which consequently results in various changing activities in the biosystem. Li and co-authors observed two kinds of lactobacilli, G15 and Q14, which can considerably decrease the number of Gram-negative bacteria, decrease the level of LPS and inflammatory cytokines, and improve T2D (Li et al. 2017).

1.20 Future Perspectives

The human microbiota plays a pivotal role in the well-being of humans and actively participates in the development of numerous diseases. Due to the extensive influence of micro-organisms on the human body, it is proposed that extensive research on host–microbiota interactions should be carried out beyond the characterization of the community composition. Besides the structure and function of the microbiota, future research should focus to explain their causalities. With the advent of new techniques for the prediction of the microbiota function, novel analytical and simulation methods and advanced microbiota interaction models should be used that may further help to clarify the human–microbiota interactions and their possible role in the development of various diseases. The central role of the human microbiota should be investigated critically; proper diagnoses and advanced strategies for treatment should be used in the future.

References

Abnet CC, Arnold M, Wei WQ (2018) Epidemiology of esophageal squamous cell carcinoma. Gastroenterology 154(2):360–373

Aguilar-Toalá JE, Garcia-Varela R, Garcia HS, Mata-Haro V, González-Córdova AF, Vallejo-Cordoba B, Hernández-Mendoza A (2018) Postbiotics: an evolving term within the functional foods field. Trends Food Sci Technol 75:105–114

Alarcón T, Llorca L, Perez-Perez G (2017) Impact of the microbiota and gastric disease development by Helicobacter pylori. Curr Top Microbiol Immunol 400:253–275

Almeida A, Mitchell AL, Boland M, Forster SC, Gloor GB, Tarkowska A, Lawley TD, Finn RD (2019) A new genomic blueprint of the human gut microbiota. Nature 568(7753):499–504

Ames BN, Gold LS (1998) The causes and prevention of cancer: the role of environment. Biotherapy 11(2):205–220

Andrici J, Eslick GD (2015) Hot food and beverage consumption and the risk of esophageal cancer: a meta-analysis. Am J Prev Med 49(6):952–960

Arnold M, Soerjomataram I, Ferlay J, Forman D (2015) Global incidence of oesophageal cancer by histological subtype in 2012. Gut 64(3):381–387

Arumugam M, Raes J, Pelletier E, Le Paslier D, Yamada T, Mende DR, Fernandes GR, Tap J, Bruls T, Batto JM, Bertalan M (2011) Erratum: enterotypes of the human gut microbiome (Nature (2011) 473 (174-180)). Nature 474(7353):666

Arweiler NB, Netuschil L (2016) The oral microbiota. Adv Exp Med Biol 902:45–60

Babu G, Singaravelu BG, Srikumar R, Reddy SV (2017) Comparative study on the vaginal flora and incidence of asymptomatic vaginosis among healthy women and in women with infertility problems of reproductive age. J Clin Diagn Res 11(8):DC18

Bajaj JS, Betrapally NS, Hylemon PB, Heuman DM, Daita K, White MB, Unser A, Thacker LR, Sanyal AJ, Kang DJ, Sikaroodi M (2015) Salivary microbiota reflects changes in gut microbiota in cirrhosis with hepatic encephalopathy. Hepatology 62(4):1260–1271

Bajaj JS, Betrapally NS, Hylemon PB, Thacker LR, Daita K, Kang DJ, White MB, Unser AB, Fagan A, Gavis EA, Sikaroodi M (2015) Gut microbiota alterations can predict hospitalizations in cirrhosis independent of diabetes mellitus. Sci Rep 5(1):1–9

Bajaj JS, Hylemon PB, Ridlon JM, Heuman DM, Daita K, White MB, Monteith P, Noble NA, Sikaroodi M, Gillevet PM (2012) Colonic mucosal microbiome differs from stool microbiome in cirrhosis and hepatic encephalopathy and is linked to cognition and inflammation. Am J Physiol Gastrointest Liver Physiol 303(6):G675–G685

Bajaj JS, Liu EJ, Kheradman R, Fagan A, Heuman DM, White M, Gavis EA, Hylemon P, Sikaroodi M, Gillevet PM (2018) Fungal dysbiosis in cirrhosis. Gut 67(6):1146–1154

Bataller R, Arteel GE, Moreno C, Shah V (2019) Alcohol-related liver disease: time for action. J Hepatol 70(2):221–222

Becerril-Rico J, Alvarado-Ortiz E, Toledo-Guzmán ME, Pelayo R, Ortiz-Sánchez E (2021) The cross talk between gastric cancer stem cells and the immune microenvironment: a tumor-promoting factor. Stem Cell Res Ther 12(1):1–14

Becker S, Dossus L, Kaaks R (2009) Obesity related hyperinsulinaemia and hyperglycaemia and cancer development. Arch Physiol Biochem 115(2):86–96

Bergheim I, Weber S, Vos M, Krämer S, Volynets V, Kaserouni S, McClain CJ, Bischoff SC (2008) Antibiotics protect against fructose-induced hepatic lipid accumulation in mice: role of endotoxin. J Hepatol 48(6):983–992

Bernabe BP, Cralle L, Gilbert JA (2018) Systems biology of the human microbiome. Curr Opin Biotechnol 51:146–153

Bosetti C, Levi F, Rosato V, Bertuccio P, Lucchini F, Negri E, La Vecchia C (2011) Recent trends in colorectal cancer mortality in Europe. Int J Cancer 129(1):180–191

Bouskra D, Brézillon C, Bérard M, Werts C, Varona R, Boneca IG et al (2008) Lymphoid tissue genesis induced by commensals through NOD1 regulates intestinal homeostasis. Nature 456(7221):507–510

Bray F, Ferlay J, Soerjomataram I, Siegel RL, Torre LA, Jemal A (2018) Global cancer statistics 2018: GLOBOCAN estimates of incidence and mortality worldwide for 36 cancers in 185 countries. CA Cancer J Clin 68(6):394–424

Breitbart M, Salamon P, Andresen B, Mahaffy JM, Segall AM, Mead D, Azam F, Rohwer F (2002) Genomic analysis of uncultured marine viral communities. Proc Natl Acad Sci U S A 99:14250–14255

Cani PD (2018) Human gut microbiome: hopes, threats and promises. Gut 67(9):1716–1725

Cani PD, Delzenne NM, Amar J, Burcelin R (2008) Role of gut microflora in the development of obesity and insulin resistance following high-fat diet feeding. Pathol Biol 56(5):305–309

Cani PD, Neyrinck AM, Fava F, Knauf C, Burcelin RG, Tuohy KM, Gibson GR, Delzenne NM (2007) Selective increases of bifidobacteria in gut microflora improve high-fat-diet-induced diabetes in mice through a mechanism associated with endotoxaemia. Diabetologia 50(11): 2374–2383

Carding S, Verbeke K, Vipond DT, Corfe BM, Owen LJ (2015) Dysbiosis of the gut microbiota in disease. Microb Ecol Health Dis 26(1):26191

Cardoso AM (2021) Can changes in gut Microbiota predict progression toward diabetes? J Explor Res Pharmacol 6(4):168–169

Cash HL, Whitham CV, Behrendt CL, Hooper LV (2006) Symbiotic bacteria direct expression of an intestinal bactericidal lectin. Science 313(5790):1126–1130

Cavicchioli R, Curmi P, Saunders N, Thomas T (2003) Pathogenic archaea: do they exist? Bioessays 25:1119–1128

Center MM, Jemal A, Ward E (2009) International trends in colorectal cancer incidence rates. Cancer Epidemiol Biomarkers Prev 18(6):1688–1694
Chen Y, Ji F, Guo J, Shi D, Fang D, Li L (2016) Dysbiosis of small intestinal microbiota in liver cirrhosis and its association with etiology. Sci Rep 6(1):1–9
Chen L, Shi Y, Zhu X, Guo W, Zhang M, Che Y, Tang L, Yang X, You Q, Liu Z (2019) IL-10 secreted by cancer-associated macrophages regulates proliferation and invasion in gastric cancer cells via c-Met/STAT3 signaling. Oncol Rep 42(2):595–604
Chen C, Song X, Wei W, Zhong H, Dai J, Lan Z, Li F, Yu X, Feng Q, Wang Z, Xie H (2017) The microbiota continuum along the female reproductive tract and its relation to uterine-related diseases. Nat Commun 8(1):1–11
Clabots CR, Johnson S, Olson MM, Peterson LR, Gerding DN (1992) Acquisition of Clostridium difficile by hospitalized patients: evidence for colonized new admissions as a source of infection. J Infect Dis 166(3):561–567
Cogen AL, Nizet V, Gallo RL (2008) Skin microbiota: a source of disease or defence? Br J Dermatol 158:442–455
Coker OO, Dai Z, Nie Y, Zhao G, Cao L, Nakatsu G, Wu WK, Wong SH, Chen Z, Sung JJ, Yu J (2018) Mucosal microbiome dysbiosis in gastric carcinogenesis. Gut 67(6):1024–1032
Coleman HG, Xie SH, Lagergren J (2018) The epidemiology of esophageal adenocarcinoma. Gastroenterology 154(2):390–405
Cuvelier ML, Allen AE, Monier A, McCrow JP, Messié M, Tringe SG, Woyke T, Welsh RM, Ishoey T, Lee JH (2010) Targeted metagenomics and ecology of globally important uncultured eukaryotic phytoplankton. Proc Natl Acad Sci 107:14679–14684
Czepiel J, Dróżdż M, Pituch H, Kuijper EJ, Perucki W, Mielimonka A, Goldman S, Wultańska D, Garlicki A, Biesiada G (2019) Clostridium difficile infection. Eur J Clin Microbiol Infect Dis 38(7):1211–1221
Dapito DH, Mencin A, Gwak GY, Pradere JP, Jang MK, Mederacke I, Caviglia JM, Khiabanian H, Adeyemi A, Bataller R, Lefkowitch JH (2012) Promotion of hepatocellular carcinoma by the intestinal microbiota and TLR4. Cancer Cell 21(4):504–516
Dávila-Collado R, Jarquín-Durán O, Dong LT, Espinoza JL (2020) Epstein–Barr virus and Helicobacter pylori co-infection in non-malignant gastroduodenal disorders. Pathogens 9(2): 104
Day CP, James OF (1998) Steatohepatitis: a tale of two "hits"? Gastroenterology 114(4):842–845
den Besten G, van Eunen K, Groen AK, Venema K, Reijngoud DJ et al (2013) The role of short-chain fatty acids in the interplay between diet, gut microbiota, and host energy metabolism. J Lipid Res 54:2325–2340
Dewhirst FE, Chen T, Izard J, Paster BJ, Tanner AC et al (2010) The human oral microbiome. J Bacteriol 192:5002–5017
Donders GG, Vereecken A, Bosmans E, Dekeersmaecker A, Salembier G, Spitz B (2002) Definition of a type of abnormal vaginal flora that is distinct from bacterial vaginosis: aerobic vaginitis. BJOG 109(1):34–43
Doorakkers E, Lagergren J, Engstrand L, Brusselaers N (2018) Helicobacter pylori eradication treatment and the risk of gastric adenocarcinoma in a Western population. Gut 67(12): 2092–2096
Edgren G, Adami HO, Weiderpass E, Nyrén O (2013) A global assessment of the oesophageal adenocarcinoma epidemic. Gut 62(10):1406–1414
Engel LS, Chow WH, Vaughan TL, Gammon MD, Risch HA, Stanford JL, Schoenberg JB, Mayne ST, Dubrow R, Rotterdam H, West AB (2003) Population attributable risks of esophageal and gastric cancers. J Natl Cancer Inst 95(18):1404–1413
Eshraghian A, Taghavi A, Nikeghbalian S, Malek-Hosseini SA (2020) Reduced rate of hospital admissions for liver-related morbidities during the initial COVID-19 outbreak. Lancet Gastroenterol Hepatol 5(9):803–804

Espinoza JL, Matsumoto A, Tanaka H, Matsumura I (2018) Gastric microbiota: An emerging player in Helicobacter pylori-induced gastric malignancies. Cancer Lett 414:147–152

Eyre DW, Griffiths D, Vaughan A, Golubchik T, Acharya M, O'Connor L, Crook DW, Walker AS, Peto TE (2013) Asymptomatic Clostridium difficile colonisation and onward transmission. PloS One 8(11):e78445

Fessler J, Matson V, Gajewski TF (2019) Exploring the emerging role of the microbiome in cancer immunotherapy. J Immunother Cancer 7(1):1–15

Figliuolo VR, Dos Santos LM, Abalo A, Nanini H, Santos A, Brittes NM, Bernardazzi C, de Souza HSP, Vieira LQ, Coutinho-Silva R, Coutinho CMLM (2017) Sulfate-reducing bacteria stimulate gut immune responses and contribute to inflammation in experimental colitis. Life Sci 189: 29–38

Finegold SM, Molitoris D, Song Y, Liu C, Vaisanen ML, Bolte E et al (2002) Gastrointestinal microflora studies in late-onset autism. Clin Infect Dis 35(Suppl 1):S6–S16

Fitzmaurice C, Akinyemiju TF, Al Lami FH, Alam T, Alizadeh-Navaei R, Allen C, Alsharif U, Alvis-Guzman N, Amini E, Anderson BO, Aremu O (2018) Global, regional, and national cancer incidence, mortality, years of life lost, years lived with disability, and disability-adjusted life-years for 29 cancer groups, 1990 to 2016: a systematic analysis for the global burden of disease study. JAMA Oncol 4(11):1553–1568

Flood MP, Lawrentschuk N, Heriot AG (2022) Malignant gastrointestinal conditions. In: Men's health and wellbeing. Springer, Cham, pp 331–341

Flores R, Shi J, Fuhrman B, Xu X, Veenstra TD, Gail MH, Gajer P, Ravel J, Goedert JJ (2012) Fecal microbial determinants of fecal and systemic estrogens and estrogen metabolites: a cross-sectional study. J Transl Med 10(1):1–11

Frank DN, St. Amand AL, Feldman RA, Boedeker EC, Harpaz N, Pace NR (2007) Molecular-phylogenetic characterization of microbial community imbalances in human inflammatory bowel diseases. Proc Natl Acad Sci 104(34):13780–13785

Freedman ND, Abnet CC, Leitzmann MF, Mouw T, Subar AF, Hollenbeck AR, Schatzkin A (2007) A prospective study of tobacco, alcohol, and the risk of esophageal and gastric cancer subtypes. Am J Epidemiol 165(12):1424–1433

Fuerst JA (2014) Microorganisms—A Journal and a Unifying Concept for the Science of Microbiology. Microorganisms 2:140–146

Fuke N, Nagata N, Suganuma H, Ota T (2019) Regulation of gut microbiota and metabolic endotoxemia with dietary factors. Nutrients 11(10):2277

Gao Z, Guo B, Gao R, Zhu Q, Qin H (2015) Microbiota disbiosis is associated with colorectal cancer. Front Microbiol 6:20

Gill SR, Pop M, Deboy RT, Eckburg PB, Turnbaugh PJ, Samuel BS et al (2006) Metagenomic analysis of the human distal gut microbiome. Science 312(5778):1355–1359

Giovannoni SJ, Stingl U (2005) Molecular diversity and ecology of microbial plankton. Nature 427: 343–348

Goedert JJ, Jones G, Hua X, Xu X, Yu G, Flores R, Falk RT, Gail MH, Shi J, Ravel J, Feigelson HS (2015) Investigation of the association between the fecal microbiota and breast cancer in postmenopausal women: a population-based case-control pilot study. J Natl Cancer Inst 107(8):djv147

Goh YJ, Klaenhammer TR (2015) Genetic mechanisms of prebiotic oligosaccharide metabolism in probiotic microbes. Annu Rev Food Sci Technol 6:137–156

Grąt M, Wronka KM, Krasnodębski M, Lewandowski Z, Kosińska I, Grąt K, Stypułkowski J, Rejowski S, Wasilewicz M, Gałęcka M, Szachta P (2016) Profile of gut microbiota associated with the presence of hepatocellular cancer in patients with liver cirrhosis. Transplant Proc 48(5): 1687–1691

Grice EA, Segre JA (2011) The skin microbiome. Nat Rev Microbiol 9:244–253

Guven-Maiorov E, Tsai CJ, Nussinov R (2017) Structural host-microbiota interaction networks. PLoS Comput Biol 13(10):e1005579

Haileselassie Y, Navis M, Vu N, Qazi KR, Rethi B, Sverremark-Ekström E (2016) Postbiotic modulation of retinoic acid imprinted mucosal-like dendritic cells by probiotic Lactobacillus reuteri 17938 in vitro. Front Immunol 7:96

Haldar S, Kapil A, Sood S, Sengupta S (2016) Female reproductive tract microbiome in gynecological health and problems. J Reprod Health Med 2:S48–S54

Hattori M, Taylor TD (2009) The human intestinal microbiome: a new frontier of human biology. DNA Res 16(1):1–12

Hawes SE, Hillier SL, Benedetti J, Stevens CE, Koutsky LA, Wølner-Hanssen P, Holmes KK (1996) Hydrogen peroxide—producing lactobacilli and acquisition of vaginal infections. J Infect Dis 174(5):1058–1063

He FF, Li YM (2020) Role of gut microbiota in the development of insulin resistance and the mechanism underlying polycystic ovary syndrome: a review. J Ovarian Res 13(1):1–13

Henao-Mejia J, Elinav E, Jin C, Hao L, Mehal WZ, Strowig T, Thaiss CA, Kau AL, Eisenbarth SC, Jurczak MJ, Camporez JP (2012) Inflammasome-mediated dysbiosis regulates progression of NAFLD and obesity. Nature 482(7384):179–185

Holeček M (2018) Branched-chain amino acids in health and disease: metabolism, alterations in blood plasma, and as supplements. Nutr Metab 15(1):1–12

Hollister EB, Gao C, Versalovic J (2014) Compositional and functional features of the gastrointestinal microbiome and their effects on human health. Gastroenterology 146(6):1449–1458

Honda K, Littman DR (2012) The microbiome in infectious disease and inflammation. Annu Rev Immunol 30:759

Hooi JK, Lai WY, Ng WK, Suen MM, Underwood FE, Tanyingoh D, Malfertheiner P, Graham DY, Wong VW, Wu JC, Chan FK (2017) Global prevalence of Helicobacter pylori infection: systematic review and meta-analysis. Gastroenterology 153(2):420–429

Hooper LV, Stappenbeck TS, Hong CV, Gordon JI (2003) Angiogenins: a new class of microbicidal proteins involved in innate immunity. Nat Immunol 4(3):269–273

Hsiao WW, Metz C, Singh DP, Roth J (2008) The microbes of the intestine: an introduction to their metabolic and signaling capabilities. Endocrinol Metab Clin North Am 37:857–871

Huang Y, Fan XG, Wang ZM, Zhou JH, Tian XF, Li N (2004) Identification of helicobacter species in human liver samples from patients with primary hepatocellular carcinoma. J Clin Pathol 57(12):1273–1277

Hung YP, Lin HJ, Wu TC, Liu HC, Lee JC, Lee CI, Wu YH, Wan L, Tsai PJ, Ko WC (2013) Risk factors of fecal toxigenic or non-toxigenic Clostridium difficile colonization: impact of Toll-like receptor polymorphisms and prior antibiotic exposure. PloS One 8(7):e69577

Ishizaka A, Koga M, Mizutani T, Parbie PK, Prawisuda D, Yusa N, Sedohara A, Kikuchi T, Ikeuchi K, Adachi E, Koibuchi T (2021) Unique gut microbiome in HIV patients on antiretroviral therapy (ART) suggests association with chronic inflammation. Microbiol Spectr 9(1): e00708–e00721

Islami F, Ren JS, Taylor PR, Kamangar F (2009) Pickled vegetables and the risk of oesophageal cancer: a meta-analysis. Br J Cancer 101(9):1641–1647

Jacques Ferlay IS, Dikshit R, Eser S, Mathers C, Rebelo M, Parkin DM, Forman D, Bray F (2014) Cancer incidence and mortality worldwide: sources, methods and major patterns in GLOBOCAN 2012. Int J Cancer 136:29

Jakubovics NS, Gill SR, Vickerman MM, Kolenbrander PE (2008) Role of hydrogen peroxide in competition and cooperation between Streptococcus gordonii and Actinomyces naeslundii. FEMS MicrobiolEcol 66:637–644

Jansson C, Johansson AL, Nyrén O, Lagergren J (2005) Socioeconomic factors and risk of esophageal adenocarcinoma: a nationwide Swedish case-control study. Cancer Epidemiol Biomarkers Prev 14(7):1754–1761

Javed A, Parvaiz F, Manzoor S (2019) Bacterial vaginosis: An insight into the prevalence, alternative treatments regimen and it's associated resistance patterns. Microb Pathog 127:21–30
Jin C, Lagoudas GK, Zhao C, Bullman S, Bhutkar A, Hu B, Ameh S, Sandel D, Liang XS, Mazzilli S, Whary MT (2019) Commensal microbiota promote lung cancer development via γδ T cells. Cell 176(5):998–1013
Kamangar F, Freedman ND (2018) Hot tea and esophageal cancer. Ann Intern Med 168(7): 519–520
Kelly SM, Crampton JR, Hunter JO (1993) Helicobacter pylori increases gastric antral juxtamucosal pH. Dig Dis Sci 38(1):129–131
Khanna S, Pardi DS, Aronson SL, Kammer PP, Orenstein R, Sauver JLS, Harmsen WS, Zinsmeister AR (2012) The epidemiology of community-acquired Clostridium difficile infection: a population-based study. Am J Gastroenterol 107(1):89
Kim BC, Shin A, Hong CW, Sohn DK, Han KS, Ryu KH, Park BJ, Nam JH, Park JW, Chang HJ, Choi HS (2012) Association of colorectal adenoma with components of metabolic syndrome. Cancer Causes Control 23(5):727–735
Koh A, Molinaro A, Ståhlman M, Khan MT, Schmidt C, Mannerås-Holm L, Wu H, Carreras A, Jeong H, Olofsson LE, Bergh PO (2018) Microbially produced imidazole propionate impairs insulin signaling through mTORC1. Cell 175(4):947–961
Kumar S, Patel GK, Ghoshal UC (2021) Helicobacter pylori-induced inflammation: possible factors modulating the risk of gastric cancer. Pathogens 10(9):1099
Kwa M, Plottel CS, Blaser MJ, Adams S (2016) The intestinal microbiome and estrogen receptor–positive female breast cancer. J Natl Cancer Inst 108(8):djw029
Kwak Y, Seo AN, Lee HE, Lee HS (2020) Tumor immune response and immunotherapy in gastric cancer. J Pathol Trans Med 54(1):20–33
Lagergren J, Smyth E, Cunningham D, Lagergren P (2017) Oesophageal cancer. Lancet 390(10110):2383–2396
Lambertucci F, Arboatti A, Sedlmeier MG, Motiño O, de Luján Alvarez M, Ceballos MP, Villar SR, Roggero E, Monti JA, Pisani G, Quiroga AD (2018) Disruption of tumor necrosis factor alpha receptor 1 signaling accelerates NAFLD progression in mice upon a high-fat diet. J Nutr Biochem 58:17–27
Lang S, Demir M, Martin A, Jiang L, Zhang X, Duan Y, Gao B, Wisplinghoff H, Kasper P, Roderburg C, Tacke F (2020) Intestinal virome signature associated with severity of nonalcoholic fatty liver disease. Gastroenterology 159(5):1839–1852
Larsbrink J, Rogers TE, Hemsworth GR, McKee LS, Tauzin AS, Spadiut O et al (2014) A discrete genetic locus confers xyloglucan metabolism in select human gut Bacteroidetes. Nature 506(7489):498–502
Lee YY, Derakhshan MH (2013) Environmental and lifestyle risk factors of gastric cancer. Arch Iran Med 16(6):358–365
Ley RE, Peterson DA, Gordon JI (2006) Ecological and evolutionary forces shaping microbial diversity in the human intestine. Cell 124(4):837–848
Ley RE, Turnbaugh PJ, Klein S, Gordon JI (2006) Microbial ecology: human gut microbes associated with obesity. Nature 444(7122):1022–1023
Li KK, Tian PJ, Wang SD, Lei P, Qu L, Huang JP, Shan YJ, Li BL (2017) Targeting gut microbiota: Lactobacillus alleviated type 2 diabetes via inhibiting LPS secretion and activating GPR43 pathway. J Funct Foods 38:561–570
Li Z, Yang S, Lin H, Huang J, Watkins PA, Moser AB, DeSimone C, Song XY, Diehl AM (2003) Probiotics and antibodies to TNF inhibit inflammatory activity and improve nonalcoholic fatty liver disease. Hepatology 37(2):343–350
Liu Q, Duan Z, Ha D, Bengmark S, Kurtovic J, Riordan SM (2004) Synbiotic modulation of gut flora: effect on minimal hepatic encephalopathy in patients with cirrhosis. Hepatology 39(5): 1441–1449
Liu J, Wang J, Leng Y, Lv C (2013) Intake of fruit and vegetables and risk of esophageal squamous cell carcinoma: a meta-analysis of observational studies. Int J Cancer 133(2):473–485

Lloyd-Price J, Abu-Ali G, Huttenhower C (2016) The healthy human microbiome. Genome Med 8(1):1–11

Lofgren JL, Whary MT, Ge Z et al (2011) Lack of commensal flora in Helicobacter pylori-infected INS-GAS mice reduces gastritis and delays intraepithelial neoplasia. Gastroenterology 140(1): 210–220

Loo VG, Bourgault AM, Poirier L, Lamothe F, Michaud S, Turgeon N, Toye B, Beaudoin A, Frost EH, Gilca R, Brassard P (2011) Host and pathogen factors for Clostridium difficile infection and colonization. N Engl J Med 365(18):1693–1703

Loprieno N (1975) International Agency for Research on Cancer (IARC) monographs on the evaluation of carcinogenic risk of chemicals to man:" relevance of data on mutagenicity". Mutat Res 31(3):210

Luo W, Fedda F, Lynch P, Tan D (2018) CDH1 gene and hereditary diffuse gastric cancer syndrome: molecular and histological alterations and implications for diagnosis and treatment. Front Pharmacol 9:1421

Ma D, Chen Y, Chen T (2019) Vaginal microbiota transplantation for the treatment of bacterial vaginosis: a conceptual analysis. FEMS Microbiol Lett 366(4):fnz025

Macpherson AJ, Harris NL (2004) Interactions between commensal intestinal bacteria and the immune system. Nat Rev Immunol 4(6):478–485

Majewski M, Mertowska P, Mertowski S, Smolak K, Grywalska E, Torres K (2022) Microbiota and the immune system—actors in the gastric cancer story. Cancer 14(15):3832

Mändar R (2013) Microbiota of male genital tract: impact on the health of man and his partner. Pharmacol Res 69:32–41

Marchesi JR, Adams DH, Fava F, Hermes GD, Hirschfield GM, Hold G, Quraishi MN, Kinross J, Smidt H, Tuohy KM, Thomas LV (2016) The gut microbiota and host health: a new clinical frontier. Gut 65(2):330–339

McMichael AJ, Giles GG (1988) Cancer in migrants to Australia: extending the descriptive epidemiological data. Cancer Res 48(3):751–756

Minami Y, Nishino Y, Tsubono Y, Tsuji I, Hisamichi S (2006) Increase of colon and rectal cancer incidence rates in Japan: trends in incidence rates in Miyagi Prefecture, 1959-1997. J Epidemiol 16(6):240–248

Minoura-Etoh J, Gotoh K, Sato R, Ogata M, Kaku N, Fujioka T, Nishizono A (2006) Helicobacter pylori-associated oxidant monochloramine induces reactivation of Epstein–Barr virus (EBV) in gastric epithelial cells latently infected with EBV. J Med Microbiol 55(7):905–911

Monaco CL, Gootenberg DB, Zhao G, Handley SA, Ghebremichael MS, Lim ES, Lankowski A, Baldridge MT, Wilen CB, Flagg M, Norman JM (2016) Altered virome and bacterial microbiome in human immunodeficiency virus-associated acquired immunodeficiency syndrome. Cell Host Microbe 19(3):311–322

Morowitz MJ, Carlisle EM, AlverdyJC. (2011) Contributions of intestinal bacteria to nutrition and metabolism in the critically ill. Surg Clin North Am 91(4):771–785

Morris Brown L, Hoover R, Silverman D, Baris D, Hayes R, Swanson GM, Schoenberg J, Greenberg R, Liff J, Schwartz A, Dosemeci M (2001) Excess incidence of squamous cell esophageal cancer among US Black men: role of social class and other risk factors. Am J Epidemiol 153(2):114–122

Nardone G, Compare D (2015) The human gastric microbiota: Is it time to rethink the pathogenesis of stomach diseases? United European Gastroenterol J 3(3):255–260

Neish AS (2009) Microbes in gastrointestinal health and disease. Gastroenterology 136(1):65–80

Nelson MH, Diven MA, Huff LW, Paulos CM (2015) Harnessing the microbiome to enhance cancer immunotherapy. J Immunol Res 2015:368736

Neto AG, Whitaker A, Pei Z (2016) February. Microbiome and potential targets for chemoprevention of esophageal adenocarcinoma. Semin Oncol 43(1):86–96

O'Keefe SJ, Kidd M, Espitalier-Noel G, Owira P (1999) Rarity of colon cancer in Africans is associated with low animal product consumption, not fiber. Am J Gastroenterol 94(5): 1373–1380
Oh TG, Kim SM, Caussy C, Fu T, Guo J, Bassirian S, Singh S, Madamba EV, Bettencourt R, Richards L, Ruth TY (2020) A universal gut-microbiome-derived signature predicts cirrhosis. Cell Metab 32(5):878–888
Oren A, Garrity GM (2019) List of new names and new combinations previously effectively, but not validly, published. Int J Syst Evol Microbiol 69(1):5–9
Pan R, Zhu M, Yu C, Lv J, Guo Y, Bian Z, Yang L, Chen Y, Hu Z, Chen Z, Li L (2017) Cancer incidence and mortality: a cohort study in China, 2008–2013. Int J Cancer 141(7):1315–1323
Parbie PK, Mizutani T, Ishizaka A, Kawana-Tachikawa A, Runtuwene LR, Seki S, Abana CZY, Kushitor D, Bonney EY, Ofori SB, Uematsu S (2021) Dysbioticfecal microbiome in HIV-1 infected individuals in Ghana. Front Cell Infect Microbiol 11:646467
Parkes GC, Rayment NB, Hudspith BN, Petrovska L, Lomer MC, Brostoff J, Whelan K, Sanderson JD (2012) Distinct microbial populations exist in the mucosa-associated microbiota of sub-groups of irritable bowel syndrome. Neurogastroenterol Motil 24(1):31–39
Patel TN, Roy S, Ravi R (2017) Gastric cancer and related epigenetic alterations. Ecancermedicalscience 11:714
Pattison CP, Combs MJ, Marshall BJ (1997) Helicobacter pylori and peptic ulcer disease: evolution to revolution to resolution. Am J Roentgenol 168(6):1415–1420
Primo BDAA, Dantas CKD, Ferreira CWS (2022) Impacts of COVID-19 on physiotherapy care for women with breast cancer. Fisioter Mov 35:e35605
Pushalkar S, Hundeyin M, Daley D, Zambirinis CP, Kurz E, Mishra A, Mohan N, Aykut B, Usyk M, Torres LE, Werba G (2018) The pancreatic cancer microbiome promotes oncogenesis by induction of innate and adaptive immune suppression microbiome influences pancreatic oncogenesis. Cancer Discov 8(4):403–416
Qiao-Li W, Shao-Hua X, Wen-Tao L, Lagergren J (2017) Smoking cessation and risk of esophageal cancer by histological type: systematic review and meta-analysis. J Natl Cancer Inst 109(12):1
Qin J, Li R, Raes J, Arumugam M, Burgdorf KS, Manichanh C, Nielsen T, Pons N, Levenez F, Yamada T, Mende DR (2010) A human gut microbial gene catalogue established by metagenomic sequencing. Nature 464(7285):59–65
Quigley EM (2013) Gut bacteria in health and disease. Gastroenterol Hepatol (NY) 9:560–569
Rajagopala SV, Yooseph S, Harkins DM, Moncera KJ, Zabokrtsky KB, Torralba MG, Tovchigrechko A, Highlander SK, Pieper R, Sender L, Nelson KE (2016) Gastrointestinal microbial populations can distinguish pediatric and adolescent Acute Lymphoblastic Leukemia (ALL) at the time of disease diagnosis. BMC Genomics 17(1):1–10
Rakoff-Nahoum S, Medzhitov R (2008) Innate immune recognition of the indigenous microbial flora. Mucosal Immunol 1(Suppl 1):S10–S14
Rasmussen T, Kirkeby LP, Poulsen K, Reinholdt J, Kilian M (2000) Resident aerobic microbiota of the adult human nasal cavity. Apmis 108:663–675
Ridlon JM, Bajaj JS (2015) The human gut sterolbiome: bile acid-microbiome endocrine aspects and therapeutics. Acta Pharmaceutica Sinica B 5(2):99–105
Roane TM, Reynolds KA, Maier RM, Pepper IL (2009) Microorganisms. Environ Microbiol:9–39
Roberfroid MB, Bornet F, Bouley C, Cummings JH (1995) Colonic microflora: nutrition and health. Summary and conclusions of an International Life Sciences Institute (ILSI) [Europe] workshop held in Barcelona, Spain. Nutr Rev 53(5):127–130
Round JL, Mazmanian SK (2009) The gut microbiota shapes intestinal immune responses during health and disease. Nat Rev Immunol 9(5):313–323
Rutayisire E, Huang K, Liu Y, Tao F (2016) The mode of delivery affects the diversity and colonization pattern of the gut microbiota during the first year of infants' life: a systematic review. BMC Gastroenterol 16:86

Sanduzzi Zamparelli M, Rocco A, Compare D, Nardone G (2017) The gut microbiota: A new potential driving force in liver cirrhosis and hepatocellular carcinoma. United European Gastroenterol J 5(7):944–953
Sartor RB (2008) Microbial influences in inflammatory bowel diseases. Gastroenterology 134(2): 577–594
Savage DC (1977) Microbial ecology of the gastrointestinal tract. Annu Rev Microbiol 31(1): 107–133
Savolainen S, Ylikoski J, Jousimies-Somer H (1986) The bacterial flora of the nasal cavity in healthy young men. Rhinology 24:249–255
Scanlan PD, Shanahan F, Clune Y, Collins JK, O'Sullivan GC, O'Riordan M et al (2008) Culture-independent analysis of the gut microbiota in colorectal cancer and polyposis. Environ Microbiol 10(3):789–798
Schauber J, Svanholm C, Termén S, Iffland K, Menzel T, Scheppach W et al (2003) Expression of the cathelicidin LL-37 is modulated by short chain fatty acids in colonocytes: relevance of signalling pathways. Gut 52(5):735–741
Schloss PD, Handelsman J (2004) Status of the microbial census. Microbiol Mol Biol Rev 68:686–691
Schnabl B, Brenner DA (2014) Interactions between the intestinal microbiome and liver diseases. Gastroenterology 146(6):1513–1524
Sekirov I, Russell SL, Antunes LC, Finlay BB (2010) Gut microbiota in health and disease. Physiol Rev 90(3):859–904
Sitarz R, Skierucha M, Mielko J, Offerhaus GJA, Maciejewski R, Polkowski WP (2018) Gastric cancer: epidemiology, prevention, classification, and treatment. Cancer Manag Res 10:239
Soeta N, Terashima M, Gotoh M, Mori S, Nishiyama K, Ishioka K, Kaneko H, Suzutani T (2009) An improved rapid quantitative detection and identification method for a wide range of fungi. J Med Microbiol 58(8):1037–1044
Sung JJ, Coker OO, Chu E, Szeto CH, Luk STY, Lau HCH, Yu J (2020) Gastric microbes associated with gastric inflammation, atrophy and intestinal metaplasia 1 year after Helicobacter pylori eradication. Gut 69(9):1572–1581
Sung JJ, Lin SR, Ching JY, Zhou LY, To KF, Wang RT, Leung WK, Enders KW, Lau JY, Lee YT, Yeung CK (2000) Atrophy and intestinal metaplasia one year after cure of H. pylori infection: a prospective, randomized study. Gastroenterology 119(1):7–14
Taja-Chayeb L, Vidal-Millán S, Trejo-Becerril C, Pérez-Cárdenas E, Chávez-Blanco A, Domínguez-Gómez G, González-Fierro A, Romo-Pérez A, Dueñas-González A (2022) Hereditary diffuse gastric cancer (HDGC). An overview. Clin Res Hepatol Gastroenterol 46(4):101820
Taylor VM, Ko LK, Hwang JH, Sin MK, Inadomi JM (2015) Gastric cancer in asianamerican populations: a neglected health disparity. Asian Pac J Cancer Prev 15(24):10565–10571
Thursby E, Juge N (2017) Introduction to the human gut flora. Biochem J 474(11):1823–1836
Tilg H, Moschen AR (2010) Evolution of inflammation in nonalcoholic fatty liver disease: the multiple parallel hits hypothesis. Hepatology 52(5):1836–1846
Torre LA, Bray F, Siegel RL, Ferlay J, Lortet-Tieulent J, Jemal A (2015) Global cancer statistics, 2012. CA Cancer J Clin 65(2):87–108
Tripathi A, Debelius J, Brenner DA, Karin M, Loomba R, Schnabl B, Knight R (2018) The gut–liver axis and the intersection with the microbiome. Nat Rev Gastroenterol Hepatol 15(7): 397–411
Tsalamandris S, Antonopoulos AS, Oikonomou E, Papamikroulis GA, Vogiatzi G, Papaioannou S, Deftereos S, Tousoulis D (2019) The role of inflammation in diabetes: current concepts and future perspectives. Eur Cardiol Rev 14(1):50
Ursell LK, Haiser HJ, Van Treuren W, Garg N, Reddivari L, Vanamala J et al (2014) The intestinal metabolome: an intersection between microbiota and host. Gastroenterology 146(6):1470–1476

Vassallo G, Mirijello A, Ferrulli A, Antonelli M, Landolfi R, Gasbarrini A, Addolorato G (2015) alcohol and gut microbiota-the possible role of gut microbiota modulation in the treatment of alcoholic liver disease. Aliment Pharmacol Ther 41(10):917–927

Verhulst SL, Vael C, Beunckens C, Nelen V, Goossens H, Desager K (2008) A longitudinal analysis on the association between antibiotic use, intestinal microflora, and wheezing during the first year of life. J Asthma 45(9):828–832

Verstraelen H, Vilchez-Vargas R, Desimpel F, Jauregui R, Vankeirsbilck N, Weyers S, Verhelst R, De Sutter P, Pieper DH, Van De Wiele T (2016) Characterisation of the human uterine microbiome in non-pregnant women through deep sequencing of the V1-2 region of the 16S rRNA gene. PeerJ 4:e1602

Vujkovic-Cvijin I, Sortino O, Verheij E, Sklar J, Wit FW, Kootstra NA, Sellers B, Brenchley JM, Ananworanich J, Loeff MSVD, Belkaid Y (2020) HIV-associated gut dysbiosis is independent of sexual practice and correlates with noncommunicable diseases. Nat Commun 11(1):1–15

Wagnerberger S, Spruss A, Kanuri G, Stahl C, Schröder M, Vetter W, Bischoff SC, Bergheim I (2013) Lactobacillus casei Shirota protects from fructose-induced liver steatosis: a mouse model. J Nutr Biochem 24(3):531–538

Wakabayashi R, Nakahama Y, Nguyen V, Espinoza JL (2019) The host-microbe interplay in human papillomavirus-induced carcinogenesis. Microorganisms 7(7):199

Wang B, Jiang X, Cao M, Ge J, Bao Q, Tang L et al (2016) Altered fecal microbiota correlates with liver biochemistry in nonobese patients with non-alcoholic fatty liver disease. Sci Rep 6:32002

Wang B, Li L (2015) Who determines the outcomes of HBV exposure? Trends Microbiol 23(6): 328–329

Wang QL, Xie SH, Wahlin K, Lagergren J (2018) Global time trends in the incidence of esophageal squamous cell carcinoma. Clin Epidemiol 10:717

Wen L, Ley RE, Volchkov PY, Stranges PB, Avanesyan L, Stonebraker AC et al (2008) Innate immunity and intestinal microbiota in the development of Type 1 diabetes. Nature 455(7216): 1109–1113

Whitman WB, Coleman DC, Wiebe WJ (1998) Prokaryotes: the unseen majority. Proc Natl Acad Sci U S A 95(12):6578–6583

Wieland A, Frank DN, Harnke B, Bambha K (2015) Systematic review: microbial dysbiosis and nonalcoholic fatty liver disease. Aliment Pharmacol Ther 42(9):1051–1063

Wong JM, De Souza R, Kendall CW, Emam A, Jenkins DJ (2006) Colonic health: fermentation and short chain fatty acids. J Clin Gastroenterol 40:235–243

Woyke T, Rubin EM (2014) Evolution searching for new branches on the tree of life. Science 346: 698–699

Xie SH, Lagergren J (2016) The male predominance in esophageal adenocarcinoma. Clin Gastroenterol Hepatol 14(3):338–347

Xie SH, Lagergren J (2018) Social group disparities in the incidence and prognosis of oesophageal cancer. United European Gastroenterol J 6(3):343–348

Xie SH, Mattsson F, Lagergren J (2017) Incidence trends in oesophageal cancer by histological type: an updated analysis in Sweden. Cancer Epidemiol 47:114–117

Xie SH, Rabbani S, Petrick JL, Cook MB, Lagergren J (2017) Racial and ethnic disparities in the incidence of esophageal cancer in the United States, 1992–2013. Am J Epidemiol 186(12): 1341–1351

Yang AM, Inamine T, Hochrath K, Chen P, Wang L, Llorente C, Bluemel S, Hartmann P, Xu J, Koyama Y, Kisseleva T (2017) Intestinal fungi contribute to development of alcoholic liver disease. J Clin Invest 127(7):2829–2841

Yang MH, Rampal S, Sung J, Choi YH, Son HJ, Lee JH, Kim YH, Chang DK, Rhee PL, Kim JJ, Rhee JC (2013) The association of serum lipids with colorectal adenomas. J Am Coll Gastroenterol 108(5):833–841

Yang L, Ying X, Liu S, Lyu G, Xu Z, Zhang X, Li H, Li Q, Wang N, Ji J (2020) Gastric cancer: Epidemiology, risk factors and prevention strategies. Chin J Cancer Res 32(6):695

Yiannakou I, Barber LE, Li S, Adams-Campbell LL, Palmer JR, Rosenberg L, Petrick JL (2022) A prospective analysis of red and processed meat intake in relation to colorectal cancer in the Black Women's Health Study. J Nutr 152(5):1254–1262
Zavros Y, Rieder G, Ferguson A, Merchant JL (2002) Gastritis and hypergastrinemia due to Acinetobacter lwoffii in mice. Infect Immun 70(5):2630–2639
Zhang RG, Duan GC, Fan QT, Chen SY (2016) Role of Helicobacter pylori infection in pathogenesis of gastric carcinoma. World J Gastrointest Pathophysiol 7(1):97
Zhu H, Yang X, Zhang C, Zhu C, Tao G, Zhao L, Tang S, Shu Z, Cai J, Dai S, Qin Q (2013) Red and processed meat intake is associated with higher gastric cancer risk: a meta-analysis of epidemiological observational studies. PloS One 8(8):e70955

Part I

Gut Microbes and Perspectives

Emerging Microbial Identification Technologies in the Era of OMICS and Genome Editing

2

Mohammad Riyaz and Khem Raj

Abstract

The field of microbial identification is witnessing significant technological advancements in biomedical sciences. The current state of microbial identification research and future directions are very much dependent on the integration of OMICS technologies, such as genomics, proteomics, and metabolomics. These technologies have revolutionized microbial identification by providing a comprehensive understanding of microorganisms' genetic composition, protein expression, and metabolic profiles. Combining contemporary techniques, including matrix-assisted laser desorption/ionization time-of-flight, polymerase chain reaction, and spectroscopic methods, for the identification of microbial targets and biomarkers enhances the sensitivity and specificity.

There is enormous potential for clustered regularly interspaced short palindromic repeats (CRISPR)-based assays, which utilize the CRISPR system, for microbial identification. CRISPR enables precise targeting and modification of nucleotide sequences and has shown promise in molecular diagnostics and therapeutics. Various signal readout techniques, such as fluorescence, colorimetric, lateral flow, electrochemical biosensors, and nanopore sensors, are explored to enhance the efficiency of CRISPR-based assays.

Moreover, the integration of DNA-based identification techniques, including 16S rRNA gene sequencing, whole genome sequencing, metagenomic sequencing, and comparative genomics, is a robust identification intervention. These methods provide accurate and rapid identification of microorganisms in medical, environmental, and industrial contexts.

M. Riyaz (✉) · K. Raj
Department of Microbiology, Panjab University, Chandigarh, India

R. Sobti et al. (eds.), *Role of Microbes in Sustainable Development*,
https://doi.org/10.1007/978-981-99-3126-2_2

While these advancements have significant implications for medical diagnostics, food safety, and environmental monitoring, further research is necessary to fully comprehend the capabilities and limitations of these techniques. Responsible utilization of these tools can help in harnessing their potential in microbial identification and disease management.

Keywords

CRISPR · OMICS · Microbial identification · Health and Environment

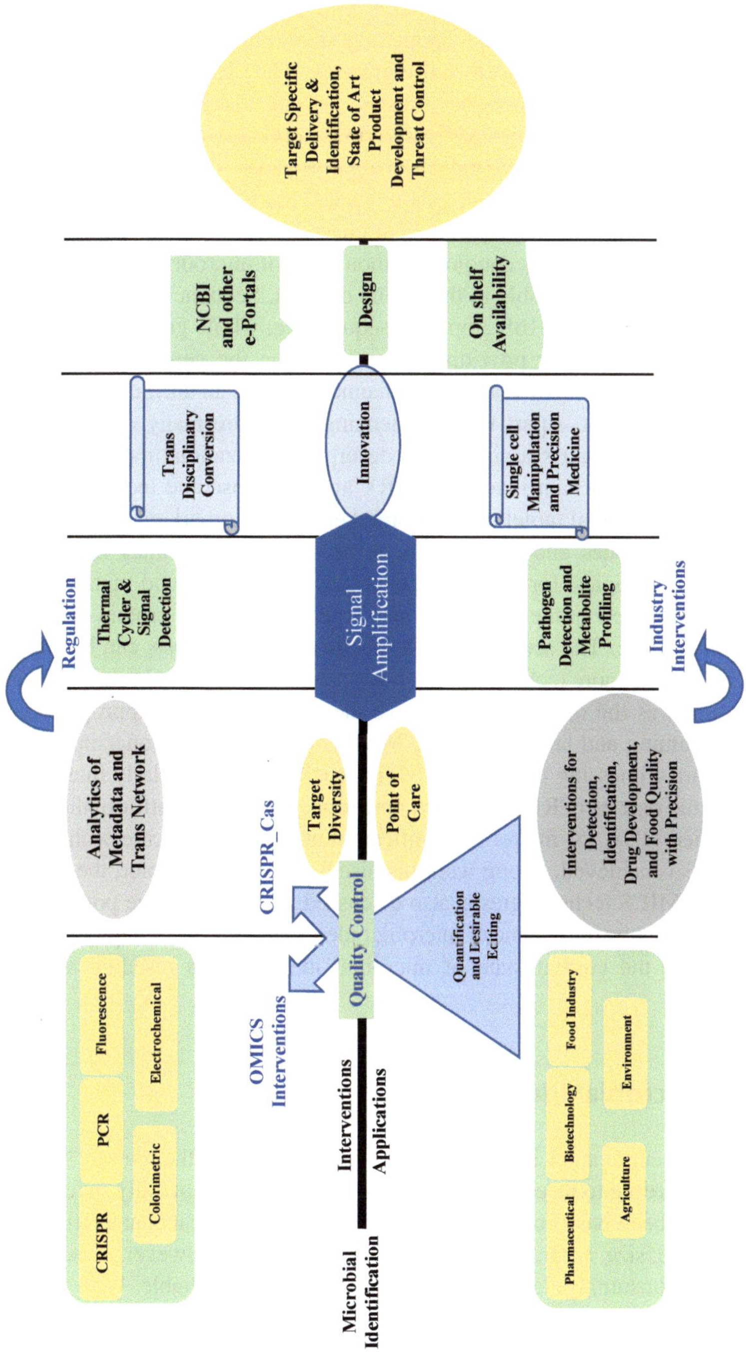
Microbial Identification
Interventions
Applications
CRISPR
PCR
Fluorescence
Colorimetric
Electrochemical
Pharmaceutical
Biotechnology
Food Industry
Agriculture
Environment
OMICS Interventions
CRISPR_Cas
Quality Control
Quantification and Desirable Eciting
Analytics of Metadata and Trans Network
Target Diversity
Point of Care
Interventions for Detection, Identification, Drug Development, and Food Quality with Precision
Regulation
Industry Interventions
Thermal Cycler & Signal Detection
Signal Amplification
Pathogen Detection and Metabolite Profiling
Trans Disciplinary Conversion
Innovation
Single cell Manipulation and Precision Medicine
NCBI and other e-Portals
Design
On shelf Availability
Target Specific Delivery & Identification, State of Art Product Development and Threat Control

2.1 Introduction

As technological advances are happening in biomedical sciences, parallelly the global microbes disease burden is also growing. The identification and characterization of microorganisms is a crucial aspect of any research of biotechnology and microbiology. These domains address various applications in various fields, such as medical, environmental, and industrial microbiology. The recent development of new OMICS technologies has greatly advanced the field of microbial identification, allowing for more accurate and efficient characterization of microorganisms. After the advent of sequencing technologies, thousands of microorganisms are sequenced, and the database is available in the public domain. Sequencing involves hypervariable region sequencing, 16S amplicon sequencing, whole genome sequencing, and metagenome sequencing providing information on the genome. This information can be harnessed to achieve a rational diagnosis and is an attribute to global disease pattern management. Apart from sequencing, there are many other contemporary technologies, like matrix-assisted laser desorption/ionization time-of-flight (MALDI TOF), polymerase chain reaction (PCR), and spectroscopic techniques, which do identify the microbial target or the biomarkers. The biomarkers can be identified as proteins or peptides and also as unique secondary metabolites of the organism. OMICS technologies, for example, genomics, proteomics, and metabolomics, provide a holistic view of the microorganism's genetic makeup, protein expression, and metabolic profile. The sensitivity and specificity of these tests can be enhanced by coupling two or more techniques. In conventional methods, culturing and biochemical testing was the only way to achieve the identification in appropriation. It was time-consuming and labor-intensive. Now, with the rapid advancement of technology, we are witnessing the genome editing tools; cluster regularly interspaced short palindromic repeats (CRISPR); when performed on a nano platform, the detection of unamplified DNA for microbe identification at the point of care with efficiency at pico levels is achieved. Along with this discussion in this chapter, the integration of multiple OMICS technologies is also discussed, highlighting the potential of multi-omics approaches for enhanced microbial identification. This chapter aims to give a summary of the current state of microbial identification research and its future directions.

2.2 Microbial Identification

Microbial identification is the procedure for figuring out the identity of a microorganism according to its characteristics and properties. Historically, microbial identification has been based on morphological and physiological characteristics, such as colony shape, size, color, and growth on specific media. However, these methods are often time-consuming, labor-intensive, and not always reliable. With the advent of molecular biology techniques, DNA-based microbial identification methods have become increasingly popular, providing a more accurate and efficient way to identify microorganisms (Ghazi et al. 2022; Branysova et al. 2022; van Dongen et al. 2020; Rittmann et al. 2008; Nguyen et al. 2021).

DNA-based methods rely on the analysis of genetic markers, such as all bacteria and archaea containing the 16S rRNA gene. The identification of microorganisms at the genus or species level is possible by the sequencing of the 16S rRNA gene. The development of whole genome sequencing has also allowed for even more detailed identification, as the entire genome of a microorganism can be sequenced and compared to known genomes in databases (1,2,6). Protein-based identification methods, such as mass spectroscopy with MALDI-TOF, rely on the analysis of the microorganism's protein expression pattern, while metabolite-based methods, such as metabolomics, focus on the analysis of the microorganism's metabolic profile (Ashfaq et al. 2022; Tsuchida and Nakayama 2022). A more complete picture of the features of the microbe can be provided by combining many OMICS technologies, such as multi-omics techniques, which can also increase the precision of microbial identification (Jiang et al. 2022; Rajczewski et al. 2022).

2.3 OMICS Interventions

The term "OMICS technologies" refers to a group of methods that enable the simultaneous study of numerous biological molecules, including proteins, metabolites, DNA, and RNA. These technologies have greatly advanced the field of microbial identification, providing a more comprehensive view of the microorganism's characteristics (Jiang et al. 2022; Rajczewski et al. 2022).

Studying an organism's entire DNA, including its genes, genome structure, and function, is known as genomics. Whole genome sequencing, which involves sequencing the entire genome of a microorganism, has developed into a potent technique for identifying microorganisms, as it allows for the comparison of a microorganism's genome to known genomes in databases (Rajczewski et al. 2022; Chiu and Miller 2019). Comparative genomics is a technique that compares the genomes of different microorganisms to identify similarities and differences, which can be used for identification. The study of an organism's entire protein repertoire, including its structure, function, and connections, is known as proteomics (Rajczewski et al. 2022; Chiu and Miller 2019). Protein-based identification methods, such as MALDI-TOF mass spectrometry (MS), rely on the analysis of the microorganism's protein expression pattern and can provide a rapid and accurate way to identify microorganisms (Ashfaq et al. 2022; Tsuchida and Nakayama 2022).

The comprehensive study of an organism's tiny molecules involved in metabolism, such as lipids and carbohydrates, is called metabolomics. Metabolite-based identification methods, such as metabolic fingerprinting, rely on the analysis of the microorganism's metabolic profile and can provide a unique way to identify microorganisms. Integrative analysis of multiple OMICS technologies, such as multi-omics approaches, can provide a more comprehensive view of the microorganism's characteristics and improve the accuracy of microbial identification (Kumar et al. 2022).

2.4 CRISPR

Precisely targeting nucleotide, identifying, and modification are enabled because of the cluster regularly interspaced short palindromic repeats (CRISPR) and CRISPR associated (Cas) proteins or effectors system. These are programmable nucleases guided by RNA, identifying the nucleotide and then modifying it. This ability can be harnessed in the field of molecular diagnostics and therapeutics. To date, two CRISPR Cas systems are divided into two classes, i.e., Class 1 and Class 2, which are further divided into I, III, and IV, and II, V, and VI, respectively. Class 1 degrades DNA by guided multiple RNA and class 2 cleaves DNA using a single large RNA-guided Cas effector. These VI types are subdivided into 19 subtypes. CRISPR sensitively and specifically differentiates between the single nucleotide variants (SNVs) and can detect SNVs instead of doing complete sequencing (Lino et al. 2018; Bock et al. 2022). Technological interventions are enhancing the specificity of detection at every level from DNA recognition to readout signal detector sensitivity. CRISPR-based assays are being clubbed with different signal readout techniques, such as colorimetric, electrochemical, electronic, fluorescence, lateral flow, and microfluidics (Katti et al. 2022; Phan et al. 2022).

CRISPR can be used for microbial identification by targeting and cutting specific sequences in the DNA of a microbe that are unique to that organism. The CRISPR-Cas system is a prokaryotic adaptive immune system that provides resistance to foreign genetic elements, such as those present within plasmids and phages (Barrangou and Marraffini 2014).

In the context of microbial identification, the CRISPR array identify specific sequences in the DNA of a microbe that are unique to that organism. The process begins by extracting DNA from a sample, amplifying specific regions of the genome using PCR, and then using the CRISPR-Cas system to target and cleave specific sequences within the amplified DNA (Kumar et al. 2022).

CRISPR-based microbial identification has several advantages over traditional methods. For example, it is highly specific, as it targets unique sequences within the genome of a microbe, and it is also very sensitive, as it can detect even small amounts of DNA. Additionally, it is relatively simple and inexpensive to perform, making it a valuable tool for both research and clinical applications (van Dongen et al. 2020).

However, there are also limitations to CRISPR-based microbial identification. For example, it can only detect microbes that have been previously characterized and whose DNA sequences are available in databases, and it is also not able to distinguish between live and dead microbes. Furthermore, the specificity of the CRISPR-Cas system can be affected by variations in the spacer sequences of the CRISPR array, which can lead to false positives or negatives (Pursey et al. 2018; Li et al. 2022).

CRISPR-based microbial identification is a powerful tool for identifying and tracking microorganisms. It has the potential for applications in medical, food safety, and environmental monitoring. However, more research is needed to fully understand its potential and limitations and to ensure that it is used responsibly.

2.4.1 PCR Clubbed Detection

Direct probe-based detection frequently produces misleading results and non-specific binding of the probes. Therefore, polymerase chain reaction-based diagnostics have been extensively used for nucleic acid detection because of the target amplification. Adding to this, if PCR is clubbed with CRISPR, it can enhance the specificity by amplifying the probes. Cas9 effector-based diagnostic platform NASBACC (nucleic acid sequence-based amplification CRISPR cleavage) was used for the ZIKA virus detection, which then further explored the use of Cas 12, Cas 13, and Cas14 (Wang et al. 2021a). By trans cleaving fluorescently labeled nucleic acid probes, Cas effectors recognize the complementary activator sequence after being activated by gRNA. The cas13-based ssRNA sequence-based platform is named SHERLOCK (Specific High-Sensitivity Enzymatic Reporter UnLOCKing), a rapid and sensitive test that can differentiate the ZIKA and Dengue virus at the strain level (Kellner et al. 2019; Li et al. 2019).

2.4.2 Fluorescence Clubbed Detection

Ease of fluorescence intensities detection makes it a commonly clubbed technique to the CRISPER. These fluorescence intensities are either fluorophore-quencher labeled or fluorophore-biotin labeled ssDNA/ssRNA. This sensitive output suits well to the Cas effector, as a strong fluorescent signal of the readout is found following the reporter probe's trans-cleaving action. Fluorescence is preferred but in resource-limited conditions and at point-of-care not feasible. Biochemical readout and UV/Vis use along with the different Cas CRISPR systems can curb these limitations. Plate readers that use fluorescence detection may adjust to higher throughput (Ang et al. 2022; Choi et al. 2021).

2.4.3 Colorimetric Assays

Colorimetric CRISPR-based assay uses a color change to indicate the presence or absence of a specific target sequence. These assays are often used in field settings because they do not require specialized equipment or a UV/blue light source for fluorescence detection. They also often use structured single-stranded DNA reporters, such as hairpin-DNA or G-triplex DNA, which can make the reaction more sensitive. Overall, colorimetric assays are a simple, cost-effective, and easy-to-use method for detecting specific sequences in a sample (Cheng et al. 2021; Zhang et al. 2021).

2.4.4 Lateral Flow Assays

Lateral flow CRISPR-based assay uses a visual readout to indicate the presence or absence of a specific target sequence. These assays are often used as a point-of-care option, as they are simple to use and can provide quick results. Gold nanoparticles (AuNPs) are used to capture fluorescently labeled ribonucleoprotein (RNP) biotinylated substrate molecules that are bound or cleaved, often on a strip of paper. As a result, a clear test line appears on the paper strip, indicating the target sequence is present. LFAs are a practical and simple procedure, although they can be time consuming when working with a lot of samples (Kumar et al. 2022).

2.4.5 Electrochemical Biosensors

Electrochemical biosensor CRISPR-based assay uses electronic or electrochemical signals to indicate the presence or absence of a specific target sequence. These assays are particularly useful for high-throughput applications because they have high sensitivity and can detect very low levels of target sequences. Studies have reported the use of various electrochemical biosensors, such as graphene-based field-effect transistors (gFET), toehold switch sensors, hairpin-DNA electrochemical reporters, and electrochemical microfluidic biosensors. These sensors can detect target sequences in the range of femtomolar to attomolar concentrations. The use of electrochemical biosensors can increase the throughput of the assay, and reusing or regenerating the biosensor platforms can also contribute to waste reduction (Chen et al. 2021; Xu et al. 2022).

There are several ways to increase the sensitivity of electrochemical biosensors without the need for amplification steps. One strategy is to decrease the reaction volume and raise the target concentration, which can be achieved by using droplet microfluidics technology. This technology uses picoliter-sized systems or microchamber-array technologies to more than a thousand times raise the local target concentration and boost the detection sensitivity, enabling the identification of even a single molecule. The assay's sensitivity can also be increased by utilizing more sensitive electrical or electrochemical biosensors and by amplifying the output signal through cascade or circuitry phases. These highly sensitive systems can provide absolute target nucleic acid quantification, which is useful for applications, such as disease diagnosis, pathogen detection, and genetic analysis (Fang et al. 2022; Pan et al. 2022). Such techniques offer absolute target nucleic acid quantification in addition to being extremely sensitive.

Electrochemical biosensors are a promising option for high-throughput CRISPR-based detection assays. They have the ability to detect substrates at very low concentrations and can be easily integrated into microfluidic systems to increase sensitivity even further. Another example of an electrochemical biosensor is the E-CRISPR, which monitors Cas12a trans-cleavage activity using a three-electrode sensor and a ssDNA reporter connected to methylene blue (MB). These techniques provide absolute target nucleic acid quantification and have a high sensitivity (Kato et al. 2022).

2.4.6 Nanopore Sensor

Nanopore sensors are a direct detection method that can detect single molecules without the need for nucleic acid amplification. They can be combined with microfluidics to enable the simultaneous quantification of several target RNAs without the need for amplification (Zhang et al. 2022). Target DNA activator molecules are used in the CRISPR-based improved electrochemical DNA system as a double-stranded DNA reporter for Cas9 or Cas12a-mediated cis-cleavage, providing a signal by the dissociation of the electrochemical tag from the probe. Another example is the use of the CRISPRChip, a graphene-based field-effect transistor that can quickly and accurately detect unamplified genomic DNA and catalytically inactive, dead Cas9 (dCas9) (Balderston et al. 2021).

2.4.7 Identification and Applications

With just one base change, the Cas systems can distinguish between different sequences, which has opened new possibilities for the detection of single nucleotide variations (SNVs) and subsequent analysis of genetic variations. These systems have been used to develop point-of-care diagnostics for SNV detection across DNA/RNA substrates. The first method for SNV detection using Cas systems depended on mutant or mismatched nucleotides being present at the cleavage site or inside the PAM sequence. However, this method has limited applicability as the PAM sequence is occasionally missing from the target DNA/RNA sequences. Cas13a has been used to find a single-nucleotide difference between the American and African Zika virus strains. Different CRISPR-based SNV detection platforms have been created by researchers using various Cas effectors and various detection readouts. However, based on the base-pairing off-target sequences and modified gRNA, the created gRNAs with mismatched nucleotides may develop a new set of potential off-targets. Therefore, when employing changed gRNA for in vitro or in vivo SNP detection, it is crucial to look for off-targets. Human HERC2 gene DNA substrates with an eye color-related SNP can be distinguished using Cas14a. For CRISPR-based single nucleobase discrimination, a very sensitive graphene-based field-effect transistor (gFET) has also been created. The gFET-based readout can distinguish even minute differences between the electrical responses that Cas9 binds to wild-type and mutant sequences. This detection technology, known as the SNP-Chip, is quick, inexpensive, and small, but it needs specialized instrumentation, which can restrict its use in environments with limited resources.

2.4.8 Addons and Interventions to Enhance the Readability

Various methods have been developed to increase the sensitivity of CRISPR-based detection, including cascade signal amplification, droplet microfluidics, electrochemical biosensors, nanopore sensors, and the use of Cas effectors to create positive feedback circuits. These methods have been used to achieve highly sensitive detection of target nucleic acids, even at the single-molecule level, without the need for amplification steps. To increase the sensitivity of CRISPR-based detection, a variety of methods have been investigated, including the use of Csm6 RNA endonucleases complex with Cas13a, chemically stabilized activators, ST-HP activator, casCRISPR system, the use of ssDNA collateral cleavage with more sensitive readouts like SERS, and dis-aggregation of gold nanoparticles (Kellner et al. 2019).

There are various strategies that have been developed to enhance the sensitivity of CRISPR-based biosensors, including signal amplification, increasing the target concentration, and using more sensitive detection methods. To achieve high sensitivity and substrate detection in the range of femtomolar to attomolar concentrations. Another strategy that has been explored is the use of nanopore sensors, which have the ability to directly detect single molecules without the need for nucleic acid amplification (Bock et al. 2022). Furthermore, multiple target RNAs can be quantified simultaneously without the need for amplification by integrating an electrochemical biosensor with microfluidics.

2.4.9 Cas Integration and Application

Different Cas effector activities are coupled in Cas-based cascade signal amplification to produce a louder signal. This is accomplished either by employing a specifically created ST-HP activator for the casCRISPR system, which can boost the detection limit by 1000 times when compared to using Cas13a alone or by combining the Csm6 RNA endonucleases complex with Cas13a. Signal amplification is also possible when a Cas effector is used to build a positive feedback circuit, as in the CONAN technique. In order to enable single molecule direct DNA detection, new amplification-free Cas12a nucleic acid assays have been created by combining ssDNA collateral cleavage with more sensitive readouts including SERS (surface-enhanced Raman scattering) and disaggregation of gold nanoparticles (Azhar et al. 2021; Wang et al. 2021b).

Nucleic acid detection assays based on CRISPR have lately gained popularity because of their quick, one-pot reaction enhanced sensitivity and throughput. For a variety of uses, including genotyping of viral and bacterial strains, agriculture, pathogens and infection diagnostics, food safety, and industrial biotechnology, they have been altered, improved, and customized. The sensitivity of these assays can be increased by using different techniques.

In Cas systems for detecting nucleic acids using the RNA-targeting and cleavage ability of Cas13a a diagnostic platform called SHERLOCK has been developed (Specific High-Sensitivity Enzymatic Reporter UnLOCKing) (Kellner et al. 2019).

This platform can distinguish between several strains and detect the presence of viruses like Zika and Dengue through a three-step process: direct amplification or reverse transcription–amplification, treatment with a Cas effector, and signal readout from the assay.

The sensitivity of the various Cas effectors to reaction temperatures varies, which can affect the selection of amplification methods. These methods can be classified as isothermal or temperature-dependent amplification techniques.

Temperature-dependent amplification methods are those in which the reaction temperature is crucial for the amplification process to work. These methods typically require a thermal cycler to change the temperature of the reaction at specific intervals. The Cas/crRNA effector and the amplification step are done as a single reaction, so the active temperature range of the Cas protein must match the temperature range required for the amplification method (Zhang et al. 2021).

Isothermal amplification methods, on the other hand, do not require a thermal cycler and can be done at a single constant temperature. This can be beneficial for point-of-care testing as it eliminates the need for specialized equipment (Ang et al. 2022). However, the sensitivity of the Cas protein to temperature must still be considered when selecting an isothermal amplification method as it can affect the performance of the assay.

In addition to the amplification step, the signal readout from the assay can also affect the sensitivity and scalability of the CRISPR-based molecular diagnostic assays. Different methods, such as fluorescence, colorimetric, or electrochemical detection, can be used depending on the specific application and the sensitivity required.

Overall, CRISPR-based molecular diagnostic assays for nucleic acid detection may become more sensitive and scalable by combining various amplification techniques with post-detection signal readouts.

2.5 DNA-Based Identification Techniques

DNA-based microbial identification techniques are methods used to identify and differentiate between different microorganisms using their DNA sequences. These techniques include PCR-based methods, such as 16S rRNA gene sequencing and whole genome sequencing, as well as DNA–DNA hybridization methods, such as DNA–DNA reassociation and DNA–DNA hybridization. These techniques allow for rapid and accurate identification of microorganisms, providing important information for applications, such as medical diagnosis, food safety, and environmental monitoring.

2.5.1 16S rRNA Gene Sequencing

This involves the sequencing of the 16S rRNA gene, which is present in all bacteria and archaea. The sequencing of the 16S rRNA gene allows for the identification of microorganisms at the genus or species level. The 16S rRNA gene is highly conserved in microorganisms, and small variations in the gene sequence can be used to distinguish various microorganisms (Fraser et al. 2000). This technique is widely used in medical and environmental microbiology, as it can identify microorganisms in a sample without the need for culturing them.

2.5.2 Whole Genome Sequencing

Whole genome sequencing is sequencing the entire genome of a microorganism and comparing it to known genomes in databases. Whole genome sequencing provides a more detailed identification, as it allows for the comparison of the microorganism's entire genetic makeup (Rittmann et al. 2008). Whole genome sequencing can provide information on a microorganism's virulence factors, antibiotic resistance genes, and other genetic elements that can be used for identification.

2.5.3 Metagenomic Sequencing

It involves sequencing the DNA of a mixed microbial population, such as a sample from the soil, water, and fecal. Metagenomic sequencing allows for the identification of multiple microorganisms present in a sample as well as providing information on the microbial community structure and function (Chiu and Miller 2019). Metagenome sequencing is widely used in environmental microbiology to study the diversity of microorganisms in different ecosystems. Along with this to study the gut microbiome, metagenome sequencing is used very frequently.

2.5.4 Comparative Genomics

Comparative genomics involves comparing the genomes of different microorganisms to identify similarities and differences, which can be used for identification. Comparative genomics can be used to identify new microorganisms as well as to classify microorganisms that are difficult to identify using traditional methods. This technique is widely used in medical and industrial microbiology, as it can identify microorganisms that are responsible for infections and product spoilage (Chen et al. 2019).

Overall, DNA-based identification techniques provide a more accurate and efficient way to identify microorganisms, as they rely on the analysis of genetic markers that are unique to each microorganism. These techniques can be used to identify microorganisms in a wide range of samples, including clinical, environmental, and industrial samples.

2.6 Protein-Based Identification Techniques

Protein-based identification techniques rely on the analysis of the microorganism's protein expression pattern, providing a rapid and accurate way to identify microorganisms. These techniques include the following:

2.6.1 Matrix-Assisted Laser Desorption/Ionization Time-of-Flight Mass Spectrometry (MALDI-TOF MS)

It involves the analysis of the microorganism's protein expression pattern through the use of a laser to ionize the proteins, and a mass spectrometer to measure the mass-to-charge ratio of the ions. The resulting protein expression pattern, also known as a "fingerprint," can be compared to a database of known fingerprints to identify the microorganism. MALDI-TOF MS is a rapid and accurate method for identifying bacteria and yeasts and is widely used in medical and industrial microbiology (Ashfaq et al. 2022; Tsuchida and Nakayama 2022).

2.6.2 SDS-PAGE Gel Electrophoresis

SDS-PAGE involves the separation of proteins based on their size and charge through the use of an electric field and a gel matrix. The separated proteins can then be visualized using a variety of staining methods, such as Coomassie Brilliant Blue or silver staining. SDS-PAGE gel electrophoresis can provide information on the microorganism's protein expression pattern and can be used for identification in conjunction with other techniques (Su et al. 2021).

2.6.3 2D Gel Electrophoresis

2D gel electrophoresis is an extension of SDS-PAGE gel electrophoresis and involves the separation of proteins based on both their size and charge in two dimensions. The separated proteins can then be visualized using a variety of staining methods, such as Coomassie Brilliant Blue or silver staining. 2D gel electrophoresis can provide a more detailed view of the microorganism's protein expression pattern and can be used for identification in conjunction with other techniques.

2.6.4 Proteomic Analysis

It involves the analysis of the microorganism's complete set of proteins, including their structure, function, and interactions. Proteomic analysis can provide a comprehensive view of the microorganism's protein expression pattern and can be used for identification in conjunction with other techniques (Dayon et al. 2022).

Protein-based identification techniques provide a rapid and accurate way to identify microorganisms, as they rely on the analysis of the microorganism's protein expression pattern, which is unique to each microorganism. These techniques can be used to identify microorganisms in a wide range of samples, including clinical, environmental, and industrial samples. It is important to note that while these techniques are very useful in identifying microorganisms, they are not 100% accurate and are best used in conjunction with other techniques, such as DNA-based identification to confirm and validate the result (Rajczewski et al. 2022).

2.7 Metabolite-Based Identification Techniques

Metabolite-based microbial identification techniques are methods for identifying microorganisms based on the unique metabolic profiles or fingerprints that they produce. These techniques can include mass spectrometry-based methods, such as Fourier transform infrared spectroscopy (FTIR) and matrix-assisted laser desorption/ionization-time of flight mass spectrometry (MALDI-TOF MS), as well as chromatography-based methods, such as gas chromatography (GC) and liquid chromatography (LC) (Zhang et al. 2012). Metabolite-based identification methods are useful for identifying microorganisms in complex samples, such as in clinical or environmental samples, and can provide rapid, accurate identification of microorganisms without the need for culturing.

2.7.1 Metabolomics

Metabolomics is the study of an organism's complete set of small molecules, such as lipids and sugars, which are involved in metabolism. Metabolomics allows for the identification of the microorganism's metabolic pathways and can provide a unique metabolic fingerprint that can be used for identification (Tsuchida and Nakayama 2022). Metabolomics can be used to identify microorganisms in a wide range of samples, including clinical, environmental, and industrial samples.

2.7.2 Lipidomics

Lipidomics is a subfield of metabolomics that focuses on the analysis of lipids, which are a diverse group of molecules that are important for energy storage, cell membrane structure, and signaling. Lipidomics allows for the identification of the microorganism's lipid metabolism and can provide a unique lipid fingerprint that can be used for identification (Han and Gross 2022). Lipidomics can be used to identify microorganisms in a wide range of samples, including clinical, environmental, and industrial samples.

2.7.3 Metabolic Fingerprinting

Metabolic fingerprinting involves the analysis of a microorganism's metabolic profile and can provide a unique metabolic fingerprint that can be used for identification. Metabolic fingerprinting can be used to identify microorganisms in a wide range of samples, including clinical, environmental, and industrial samples.

One of the advantages of metabolite-based identification techniques is that they can be used to identify microorganisms that cannot be cultured or that have not been previously described. For example, metabolomics can be used to identify microorganisms in environmental samples, such as soil or water, where traditional culture-based methods may not be effective. Additionally, metabolomics can be used to identify microorganisms in clinical samples, such as blood or urine, where traditional culture-based methods may not be sensitive enough to detect the microorganism (Ashfaq et al. 2022).

Another advantage of metabolite-based identification techniques is that they can provide information on the microorganism's metabolic pathways, which can be used to infer its ecological role or potential virulence. For example, the identification of specific metabolic pathways in a microorganism can indicate its ability to degrade certain compounds and can be used to infer its role in a particular ecosystem. Additionally, the identification of specific metabolic pathways in a microorganism can indicate its ability to produce certain virulence factors and can be used to infer its potential virulence (Rajczewski et al. 2022).

However, metabolite-based identification techniques also have some limitations. One limitation is that the analysis of the microorganism's metabolic profile can be complex and time consuming. Additionally, the interpretation of the data can be challenging, as the metabolic profile of a microorganism can be affected by a variety of factors, such as the growth conditions or the presence of other microorganisms.

Metabolomics has been used to identify microorganisms responsible for infections and to study the metabolic changes that occur during infection. This can aid in the diagnosis of infections and the development of new treatments. Additionally, metabolomics has been used to study the metabolic changes that occur during the development of antibiotic resistance and to identify potential targets for new antibiotics (Kok et al. 2022).

Environmental metabolomics defines the diversity and function of microorganisms in various ecosystems. This can provide insight into the role of microorganisms in nutrient cycling and the degradation of pollutants. Additionally, metabolomics has been used to study the impact of environmental changes, such as climate change, on the diversity and function of microorganisms.

The metabolic profile of a microorganism can be affected by a variety of factors, such as the growth conditions or the presence of other microorganisms, which can make the identification process more difficult. To overcome these limitations, it is essential to combine metabolite-based identification techniques with other OMICS technologies, such as genomics and proteomics, to provide a more comprehensive view of the microorganism's characteristics and improve the accuracy of microbial identification. The integration of metabolite-based identification techniques with other OMICS technologies can provide a more comprehensive view of the microorganism's characteristics and improve the accuracy of microbial identification.

2.8 Multi-Omics Identification Techniques

Multi-omics identification techniques involve the integration of multiple OMICS technologies, such as genomics, proteomics, and metabolomics, to provide a more comprehensive view of the microorganism's characteristics and improve the accuracy of microbial identification. These techniques include the following.

2.8.1 Genome–Metabolome Integration

It involves the integration of whole genome sequencing with metabolomics to provide a more comprehensive view of the microorganism's characteristics. The genome provides information on the microorganism's genetic makeup, while the metabolome provides information on the microorganism's metabolic profile (Gonzalez-Covarrubias et al. 2022). The integration of these two data sets can provide a more detailed view of the microorganism's characteristics and improve the accuracy of microbial identification.

2.8.2 Proteome–Metabolome Integration

It gives the integration of proteomics with metabolomics to provide a more comprehensive view of the microorganism's characteristics. The proteome provides information on the microorganism's protein expression pattern, while the metabolome provides information on the microorganism's metabolic profile (Man et al. 2021). The integration of these two data sets can provide a more detailed view of the microorganism's characteristics and improve the accuracy of microbial identification.

2.8.3 Genome–Proteome–Metabolome Integration

It involves the integration of whole genome sequencing, proteomics, and metabolomics to provide a comprehensive view of the microorganism's characteristics. The genome provides information on the microorganism's genetic makeup, the proteome provides information on the microorganism's protein expression pattern, and the metabolome provides information on the microorganism's metabolic profile. The integration of these three data sets can provide a most detailed view of the microorganism's characteristics and improve the accuracy of microbial identification (Jiang et al. 2022; Man et al. 2021).

Multi-omics identification techniques provide a more comprehensive view of the microorganism's characteristics, which can improve the accuracy of microbial identification. For example, the integration of whole genome sequencing with metabolomics can provide information on the microorganism's genetic makeup, its metabolic pathways, and potential virulence factors. Additionally, the integration of proteomics with metabolomics can provide information on the microorganism's protein expression pattern, its metabolic pathways, and potential virulence factors. In this way, multi-omics identification techniques can provide a more complete picture of the microorganism's characteristics, leading to a more accurate identification (Ares-Arroyo et al. 2022).

Furthermore, multi-omics identification techniques have the potential to provide new insights into the ecology and physiology of microorganisms. For example, the integration of whole genome sequencing with metabolomics can provide information on the microorganism's genetic makeup, its metabolic pathways, and potential virulence factors, which can be used to infer its ecological role or potential virulence (Tsuchida and Nakayama 2022). Additionally, the integration of proteomics with metabolomics can provide information on the microorganism's protein expression pattern, its metabolic pathways, and potential virulence factors, which can be used to infer its ecological role or potential virulence.

Despite the advantages, multi-omics identification techniques still have some limitations. One limitation is that the analysis of multiple omics data sets can be complex and time consuming. Additionally, the interpretation of the data can be challenging, as the integration of multiple omics data sets requires advanced bioinformatics and computational skills. Additionally, the cost of multi-omics identification techniques can be high, making it difficult for some laboratories to implement them.

Multi-omics identification techniques, such as genome–metabolome integration, proteome–metabolome integration, and genome–proteome-metabolome integration, provide a more comprehensive view of the microorganism's characteristics and improve the accuracy of microbial identification.

2.9 Machine Learning-Based Identification Techniques

Machine learning-based identification techniques involve the use of artificial intelligence algorithms to analyze and interpret large sets of data, such as genomics, proteomics, and metabolomics data, to identify microorganisms.

2.9.1 Random Forest

Random forest is a type of machine learning algorithm that can be used to classify a dataset of microorganisms based on their genomic, proteomic, or metabolomic information. Random forest is a supervised learning algorithm that can learn from a training dataset and classify new microorganisms based on their characteristics (Weis et al. 2020).

2.9.2 Deep Learning

Deep learning is a type of machine learning algorithm that can be used to classify microorganisms based on their genomic, proteomic, or metabolomic data. Deep learning algorithms learn from a training dataset and classify new microorganisms based on their characteristics (genomic, proteomic, or metabolomic). Deep learning can provide high accuracy in microbial identification and can also identify microorganisms that cannot be identified by traditional methods (Yu et al. 2022).

2.9.3 Support Vector Machine

Support vector machine (SVM) is a type of machine learning algorithm similar to random forest and deep learning that can be used to classify microorganisms based on their genomic, proteomic, or metabolomic data. It also learns from a training dataset and classifies new microorganisms based on their characteristics.

Machine learning-based identification techniques provide a powerful tool for the identification and characterization of microorganisms. These techniques can analyze and interpret large sets of data, such as genomics, proteomics, and metabolomics data, to identify microorganisms (Weis et al. 2020).

One of the advantages of machine learning-based identification techniques is that they can handle large sets of data and can identify patterns that are not visible to the human eye. Additionally, machine learning-based identification techniques can be trained to identify microorganisms based on different characteristics, such as genetic makeup, protein expression pattern, or metabolic profile. This allows for the identification of microorganisms based on multiple criteria, which can improve the accuracy of microbial identification (Man et al. 2021).

Another advantage of machine learning-based identification techniques is that they can be updated and re-trained with new data. This allows for the identification of new microorganisms that may not have been included in the original training dataset. Additionally, machine learning-based identification techniques can be used to identify microorganisms in real-time, which can be useful in medical, environmental, and industrial microbiology (Ashfaq et al. 2022).

However, machine learning-based identification techniques also have some limitations. One limitation is that the accuracy of the algorithm depends on the quality and quantity of the training dataset. Additionally, machine learning-based identification techniques require advanced computational skills and specialized software, which can be difficult for some laboratories to implement.

In conclusion, machine learning-based identification techniques, such as random forest, deep learning, and support vector machine, provide a powerful tool for the identification and characterization of microorganisms. These techniques can analyze and interpret large sets of data, such as genomics, proteomics, and metabolomics data, to identify microorganisms.

2.9.4 Single-Cell Identification Techniques

Single-cell identification techniques involve the analysis of individual microorganisms at the single-cell level rather than analyzing populations of microorganisms.

2.9.5 Single-Cell Genomics

Single-cell genomics involves the isolation and sequencing of individual microorganisms at the single-cell level. This technique allows for the identification of microorganisms that are present in low numbers or are difficult to culture and can also provide information on the microorganism's genetic makeup, including the presence of virulence genes or antibiotic resistance genes.

2.9.6 Single-Cell Proteomics

Single-cell proteomics involves the isolation and analysis of individual microorganisms at the single-cell level, with a focus on the microorganism's protein expression pattern. This technique can provide information on the microorganism's protein expression pattern, which can be used to infer its ecological role or potential virulence (Kashima et al. 2020).

2.9.7 Single-Cell Metabolomics

Single-cell metabolomics involves the isolation and analysis of individual microorganisms at the single-cell level, with a focus on the microorganism's metabolic profile. This technique can provide information on the microorganism's metabolic pathways, which can be used to infer its ecological role or potential virulence.

One of the advantages of single-cell identification techniques is that they can provide information on the genetic diversity within a population of microorganisms. This can be particularly useful in the study of microbial populations in complex environments, such as the human gut microbiome, where traditional methods may not be able to accurately identify the diversity of microorganisms present (Abril et al. 2023).

Another advantage of single-cell identification techniques is that they can provide information on the ecological role and potential virulence of individual microorganisms rather than just the population as a whole. This can be particularly useful in the study of infectious diseases, where the identification of individual virulent microorganisms can aid in the development of new treatments.

2.10 DNA Microarray

DNA microarray is a powerful tool used in molecular biology and genomics, which allows for the simultaneous analysis of the expression levels of thousands of genes in a single experiment. This is achieved by immobilizing a large number of DNA probes, each corresponding to a specific gene, on a solid surface, such as a glass slide or a silicon chip. These probes are then hybridized with labeled target DNA, which is typically obtained from a biological sample of interest.

The process of hybridization occurs when the target DNA binds to the probes that have a complementary sequence. The degree of binding is measured by the amount of label that is detected on the microarray. This measurement provides a quantitative measure of the expression level of the corresponding gene in the sample (Ang et al. 2022).

There are two types of microarray, cDNA and oligonucleotide arrays. cDNA arrays are made by reverse transcribing mRNA into cDNA and then printing the cDNA on the array surface. Oligonucleotide arrays are made by printing the short pieces of DNA on the array surface.

Microarrays also have applications in the study of epigenetics by using ChIP-on-chip to identify specific regions of the genome that are bound by particular proteins.

2.11 Nanopore-Based Technologies

Nanopore-based technologies are a third-generation, ultrasensitive, label-free sequencing technique used to analyze nucleic acids, proteins, metabolites, and protein–DNA complexes. This technology works by analyzing characterized molecules under ionic current and frequency modulation in a geometrically confined system. The detectable results are achieved when the nanopore dimensions and analyte are in range and beyond the noise level.

The biological nanopore is created by self-assembling the analyte or analyte complex in an atomically precise dimension of a lipid bilayer or biopolymer. This allows the detection of smaller molecules on the 1–10 nm scale. Larger molecules of a hundred or thousand nanometer scale can also be detected using thin plastic or inorganic membranes. This is achieved through the dielectric breakdown of micro or nano membranes, laser-based optical stacking, and ion/electron milling. While the detection size can be manipulated, it cannot be reduced to the size of the natural nanopore (Zhang et al. 2022).

Nanopore technology can also be used beyond sequencing, as it can detect molecular heterogeneity by capturing numerous single molecules at high rates and analyzing the ionic current signals, even detecting enantiomeric level changes. It can discriminate more than one analyte, is label-free, and hence provides ease in detecting organisms that have less signal amplification labels. Nanopore systems also provide scaffolds for biomimetic systems and can track analyte interactions and clarify ongoing biological processes. Additionally, it can align chemical entities spatially in the protein environment to achieve nanoreactors of proteins.

2.12 Microbial Identification in Clinical Settings

Microbial identification is an essential aspect of clinical microbiology, as it allows for the accurate diagnosis of infectious diseases and the selection of appropriate treatments. In recent years, there have been significant advances in microbial identification techniques, which have led to the development of faster and more accurate methods for identifying microorganisms.

2.12.1 MALDI-TOF Mass Spectrometry

Matrix-assisted laser desorption/ionization time-of-flight mass spectrometry (MALDI-TOF MS) is a rapid and accurate method for identifying microorganisms. MALDI-TOF MS uses the protein patterns of microorganisms to identify them and can provide results in as little as 15 min. This technology is widely used in clinical microbiology laboratories and has been shown to have high accuracy and specificity for the identification of microorganisms.

2.12.2 Next-Generation Sequencing

Next-generation sequencing (NGS) is a powerful tool for identifying microorganisms in clinical settings. NGS can provide high-resolution data on the microorganism's genetic makeup and can be used to identify microorganisms that cannot be identified by traditional methods. NGS can also be used to identify antibiotic resistance genes and virulence factors, which can be useful for selecting appropriate treatments.

2.12.3 Microbial Fingerprinting

Microbial fingerprinting is a technique that involves the analysis of the microorganism's metabolic profile, which can be used to identify microorganisms. This technique can provide information on the microorganism's metabolic pathways, which can be used to infer its ecological role or potential virulence. Microbial fingerprinting has been used to identify microorganisms that are difficult to culture and can be useful for the identification of microorganisms in clinical settings.

2.13 Microbial Identification in Environmental Settings (Agriculture)

Microbial identification is an essential aspect of environmental settings, as it allows for the understanding of the diversity and complexity of microorganisms in various environments. Recent advances in microbial identification techniques have led to the development of more efficient and accurate methods for identifying microorganisms in environmental settings.

2.13.1 Metagenomics

Metagenomics is a powerful tool for identifying microorganisms in environmental settings. Metagenomics involves the sequencing of the entire microbial genome present in a sample and can provide information on the genetic makeup of microorganisms as well as their functional roles and interactions. This technique can be used to identify microorganisms that are difficult to culture and can also be used to study the diversity of microorganisms in different environments, such as soil, water, and air.

2.13.2 Metaproteomics

Metaproteomics is a technique that involves the analysis of the protein expression patterns of microorganisms in a sample. This technique can provide information on the functional roles of microorganisms in an environment and can also be used to study the interactions between microorganisms. Metaproteomics can be used to identify microorganisms that are difficult to culture and can also be used to study the diversity of microorganisms in different environments, such as soil, water, and air.

2.13.3 Metabolomics

Metabolomics is a technique that involves the analysis of the metabolic profile of microorganisms in a sample. This technique can provide information on the metabolic pathways of microorganisms as well as their functional roles and interactions. Metabolomics can be used to identify microorganisms that are difficult to culture and can also be used to study the diversity of microorganisms in different environments, such as soil, water, and air.

2.14 Microbial Identification in Industrial Settings (Food and Beverage Industry)

Microbial identification is an essential aspect of industrial microbiology, as it allows for the control of microorganisms that can cause contamination or degradation of products, as well as the optimization of microorganisms used in industrial processes. In recent years, there have been significant advances in microbial identification techniques, which have led to the development of more efficient and accurate methods for identifying microorganisms in industrial settings.

2.14.1 PCR-Based Methods

Polymerase chain reaction (PCR) is a powerful tool for identifying microorganisms in industrial settings. PCR-based methods, such as PCR-DGGE and PCR-TTGE, can be used to identify microorganisms based on their genetic makeup and can be used to detect specific microorganisms that may cause contamination or degradation of products.

2.14.2 Biochemical Tests

Biochemical tests are a traditional method for identifying microorganisms in industrial settings. These tests involve the analysis of the microorganism's metabolic profile and can be used to identify microorganisms based on their ability to ferment or hydrolyze specific substrates.

2.14.3 Flow Cytometry

Flow cytometry is a technique that can be used to identify microorganisms in industrial settings. This technique involves the analysis of the microorganism's physical and metabolic characteristics and can be used to identify microorganisms based on their size, shape, and fluorescence. Flow cytometry can be used to identify microorganisms that are difficult to culture and can also be used to analyze large numbers of microorganisms in a short period of time.

2.15 Conclusion

The emerging microbial identification technologies in the era of omics have revolutionized the way we study and understand microbial organisms. These technologies involve the use of various high-throughput and sensitive methods to analyze different aspects of microbial biology, such as genetics, transcriptomics, proteomics, and metabolomics.

Genomics, for example, allows for the sequencing and annotation of entire microbial genomes. This information can be used to identify the presence of specific genes or genetic markers associated with certain microbial species or functional traits. Techniques, such as PCR and qPCR, can also be used to specifically target and amplify specific genetic sequences, making it possible to detect and quantify specific microbial targets.

Transcriptomics, on the other hand, allows for the study of the entire set of RNA molecules expressed by a microorganism, providing insights into the functional activities and metabolic processes of a microbial population. Proteomics, similarly, allows for the identification and characterization of the entire set of proteins expressed by a microorganism.

Next-generation sequencing technologies, such as Illumina and PacBio, have made it possible to sequence entire microbial genomes at high-throughput and high-resolution. This has allowed for the identification of new microbial species, the detection of rare microbial populations, and the study of microbial diversity at a global scale.

Nanopore-based technologies is a third generation ultrasensitive label-free sequencing technique that can analyze nucleic acid, protein, metabolites, and protein DNA complexes. It works stochastically by analyzing the characterized molecules under ionic current and frequency modulation in a geometrically confined system. This technology is highly sensitive and can detect molecules in the range of 1–10 nm scale.

In conclusion, the advancements in microbial identification technologies have greatly expanded our understanding of the microbial world and have provided powerful tools for identifying and characterizing microbial populations. These technologies are increasingly being used in various ecological and biomedical situations to study the diversity, function, and interactions of microbial communities.

References

Abril AG, Carrera M, Sánchez-Pérez Á, Villa TG (2023) Gut microbiome proteomics in food allergies. Int J Mol Sci 24(3):3. https://doi.org/10.3390/ijms24032234

Ang YS, Qiu X, Yam HM, Wu N, Lanry Yung L-Y (2022) Enzyme-free and isothermal discrimination of microRNA point mutations using a DNA split proximity circuit with turn-on fluorescence readout. Biosens Bioelectron 217:114727. https://doi.org/10.1016/j.bios.2022.114727

Ares-Arroyo M et al (2022) Genomics, transcriptomics, and metabolomics reveal that minimal modifications in the host are crucial for the compensatory evolution of ColE1-like plasmids. mSphere 7(6):e00184–e00122. https://doi.org/10.1128/msphere.00184-22

Ashfaq MY, Da'na DA, Al-Ghouti MA (2022) Application of MALDI-TOF MS for identification of environmental bacteria: a review. J Environ Manag 305:114359. https://doi.org/10.1016/j.jenvman.2021.114359

Azhar M et al (2021) Rapid and accurate nucleobase detection using FnCas9 and its application in COVID-19 diagnosis. Biosens Bioelectron 183:113207. https://doi.org/10.1016/j.bios.2021.113207

Balderston S et al (2021) Discrimination of single-point mutations in unamplified genomic DNA via Cas9 immobilized on a graphene field-effect transistor. Nat Biomed Eng 5(7):713–725. https://doi.org/10.1038/s41551-021-00706-z

Barrangou R, Marraffini LA (2014) CRISPR-Cas systems: prokaryotes upgrade to adaptive immunity. Mol Cell 54(2):234–244. https://doi.org/10.1016/j.molcel.2014.03.011

Bock C et al (2022) High-content CRISPR screening. Nat Rev Methods Primer 2:8. https://doi.org/10.1038/s43586-021-00093-4

Branysova T, Demnerova K, Durovic M, Stiborova H (2022) Microbial biodeterioration of cultural heritage and identification of the active agents over the last two decades. J Cult Herit 55:245–260. https://doi.org/10.1016/j.culher.2022.03.013

Chen IMA et al (2019) IMG/M v.5.0: an integrated data management and comparative analysis system for microbial genomes and microbiomes. Nucleic Acids Res 47(D1):D666–D677. https://doi.org/10.1093/nar/gky901

Chen Z et al (2021) CRISPR/Cas12a and immuno-RCA based electrochemical biosensor for detecting pathogenic bacteria. J Electroanal Chem 901:115755. https://doi.org/10.1016/j.jelechem.2021.115755

Cheng M, Xiong E, Tian T, Zhu D, Ju H, Zhou X (2021) A CRISPR-driven colorimetric code platform for highly accurate telomerase activity assay. Biosens Bioelectron 172:112749. https://doi.org/10.1016/j.bios.2020.112749

Chiu CY, Miller SA (2019) Clinical metagenomics. Nat Rev Genet 20:341. https://doi.org/10.1038/s41576-019-0113-7

Choi J-H, Lim J, Shin M, Paek S-H, Choi J-W (2021) CRISPR-Cas12a-based nucleic acid amplification-free DNA biosensor via au nanoparticle-assisted metal-enhanced fluorescence and colorimetric analysis. Nano Lett 21(1):693–699. https://doi.org/10.1021/acs.nanolett.0c04303

Dayon L, Cominetti O, Affolter M (2022) Proteomics of human biological fluids for biomarker discoveries: technical advances and recent applications. Expert Rev Proteomics 19(2):131–151. https://doi.org/10.1080/14789450.2022.2070477

Fang D et al (2022) Electrochemical sensor based on micromotor technology for detection of ox-LDL in whole blood. Biosens Bioelectron 217:114682. https://doi.org/10.1016/j.bios.2022.114682

Fraser CM, Eisen JA, Salzberg SL (2000) Microbial genome sequencing. Nature 406(6797):799–803. https://doi.org/10.1038/35021244

Ghazi AR, Münch PC, Chen D, Jensen J, Huttenhower C (2022) Strain identification and quantitative analysis in microbial communities. J Mol Biol 434(15):167582. https://doi.org/10.1016/j.jmb.2022.167582

Gonzalez-Covarrubias V, Martínez-Martínez E, del Bosque-Plata L (2022) The potential of metabolomics in biomedical applications. Meta 12(2):2. https://doi.org/10.3390/metabo12020194

Han X, Gross RW (2022) The foundations and development of lipidomics. J Lipid Res 63(2):100164. https://doi.org/10.1016/j.jlr.2021.100164

Jiang N, Wu R, Wu C, Wang R, Wu J, Shi H (2022) Multi-omics approaches to elucidate the role of interactions between microbial communities in cheese flavor and quality. Taylor & Francis, New York, pp 1–13. https://doi.org/10.1080/87559129.2022.2070199

Kashima Y, Sakamoto Y, Kaneko K, Seki M, Suzuki Y, Suzuki A (2020) Single-cell sequencing techniques from individual to multiomics analyses. Exp Mol Med 52(9):1419. https://doi.org/10.1038/s12276-020-00499-2

Kato K et al (2022) Structure and engineering of the type III-E CRISPR-Cas7-11 effector complex. Cell 185(13):2324–2337.e16. https://doi.org/10.1016/j.cell.2022.05.003

Katti A, Diaz BJ, Caragine CM, Sanjana NE, Dow LE (2022) CRISPR in cancer biology and therapy. Nat Rev Cancer 22:259. https://doi.org/10.1038/s41568-022-00441-w

Kellner MJ, Koob JG, Gootenberg JS, Abudayyeh OO, Zhang F (2019) SHERLOCK: nucleic acid detection with CRISPR nucleases. Nat Protoc 14:2986. https://doi.org/10.1038/s41596-019-0210-2

Kok M, Maton L, van der Peet M, Hankemeier T, van Hasselt JGC (2022) Unraveling antimicrobial resistance using metabolomics. Drug Discov Today 27(6):1774–1783. https://doi.org/10.1016/j.drudis.2022.03.015

Kumar M, Maiti S, Chakraborty D (2022) Capturing nucleic acid variants with precision using CRISPR diagnostics. Biosens Bioelectron 217:114712. https://doi.org/10.1016/j.bios.2022.114712

Li Y, Li S, Wang J, Liu G (2019) CRISPR/Cas systems towards next-generation biosensing. Trends Biotechnol 37(7):730–743. https://doi.org/10.1016/j.tibtech.2018.12.005

Li Y, Man S, Ye S, Liu G, Ma L (2022) CRISPR-Cas-based detection for food safety problems: current status, challenges, and opportunities. Compr Rev Food Sci Food Saf 21(4):3770–3798. https://doi.org/10.1111/1541-4337.13000

Lino CA, Harper JC, Carney JP, Timlin JA (2018) Delivering CRISPR: a review of the challenges and approaches. Drug Deliv 25(1):1234–1257. https://doi.org/10.1080/10717544.2018.1474964

Man L, Klare WP, Dale AL, Cain JA, Cordwell SJ (2021) Integrated mass spectrometry-based multi-omics for elucidating mechanisms of bacterial virulence. Biochem Soc Trans 49(5):1905–1926. https://doi.org/10.1042/BST20191088

Nguyen PQ et al (2021) Wearable materials with embedded synthetic biology sensors for biomolecule detection. Nat Biotechnol 39(11):1366–1374. https://doi.org/10.1038/s41587-021-00950-3

Pan D, Lin Y, Liu X, Xin Y, Tian Q, Zhang J (2022) Ultrasensitive and preprocessing-free electrochemical biosensing platform for the detection of cancer-derived exosomes based on spiky-shaped aptamer-magnetic beads. Biosens Bioelectron 217:114705. https://doi.org/10.1016/j.bios.2022.114705

Phan QA, Truong LB, Medina-Cruz D, Dincer C, Mostafavi E (2022) CRISPR/Cas-powered nanobiosensors for diagnostics. Biosens Bioelectron 197:113732. https://doi.org/10.1016/j.bios.2021.113732

Pursey E, Sünderhauf D, Gaze WH, Westra ER, van Houte S (2018) CRISPR-Cas antimicrobials: challenges and future prospects. PLoS Pathog 14(6):e1006990. https://doi.org/10.1371/journal.ppat.1006990

Rajczewski AT, Jagtap PD, Griffin TJ (2022) An overview of technologies for MS-based proteomics-centric multi-omics. Expert Rev Proteomics 19(3):165. https://doi.org/10.1080/14789450.2022.2070476

Rittmann BE, Krajmalnik-Brown R, Halden RU (2008) Pre-genomic, genomic and post-genomic study of microbial communities involved in bioenergy. Nat Rev Microbiol 6(8):604–612. https://doi.org/10.1038/nrmicro1939

Su M, Zhang Z, Zhou L, Han C, Huang C, Nice EC (2021) Proteomics, personalized medicine and cancer. Cancers 13(11):11. https://doi.org/10.3390/cancers13112512

Tsuchida S, Nakayama T (2022) MALDI-based mass spectrometry in clinical testing: focus on bacterial identification. Appl Sci 12(6):770. https://doi.org/10.3390/app12062814

van Dongen JE, Berendsen JTW, Steenbergen RDM, Wolthuis RMF, Eijkel JCT, Segerink LI (2020) Point-of-care CRISPR/Cas nucleic acid detection: recent advances, challenges and opportunities. Biosens Bioelectron 166:112445. https://doi.org/10.1016/j.bios.2020.112445

Wang S et al (2021a) Highly sensitive and specific detection of hepatitis B virus DNA and drug resistance mutations utilizing the PCR-based CRISPR-Cas13a system. Clin Microbiol Infect 27(3):443–450. https://doi.org/10.1016/j.cmi.2020.04.018

Wang R et al (2021b) opvCRISPR: one-pot visual RT-LAMP-CRISPR platform for SARS-cov-2 detection. Biosens Bioelectron 172:112766. https://doi.org/10.1016/j.bios.2020.112766

Weis CV, Jutzeler CR, Borgwardt K (2020) Machine learning for microbial identification and antimicrobial susceptibility testing on MALDI-TOF mass spectra: a systematic review. Clin Microbiol Infect 26(10):1310–1317. https://doi.org/10.1016/j.cmi.2020.03.014

Xu Y et al (2022) Tetrahedral DNA framework based CRISPR electrochemical biosensor for amplification-free miRNA detection. Biosens Bioelectron 217:114671. https://doi.org/10.1016/j.bios.2022.114671

Yu T, Shixuan S, Jing H, Zhang J, Xianyu Y (2022) A new strategy for microbial taxonomic identification through micro-biosynthetic gold nanoparticles and machine learning—Yu—2022. Adv Mat 34(11):2109365. https://doi.org/10.1093/nar/gky901

Zhang A, Sun H, Wang P, Han Y, Wang X (2012) Modern analytical techniques in metabolomics analysis. Analyst 137(2):293–300. https://doi.org/10.1039/C1AN15605E

Zhang WS et al (2021) Reverse transcription recombinase polymerase amplification coupled with CRISPR-Cas12a for facile and highly sensitive colorimetric SARS-CoV-2 detection. Anal Chem 93(8):4126–4133. https://doi.org/10.1021/acs.analchem.1c00013

Zhang S et al (2022) Detection of small-sized DNA fragments in a glassy nanopore by utilization of CRISPR-Cas12a as a converter system. Analyst 147(5):905–914. https://doi.org/10.1039/D1AN02313F

Gut Microbiome: Perspectives and Challenges in Human Health

3

Neha Salaria, Neeraj, Junaid Furhan, and Rajesh Kumar

Abstract

The phrase "gut microbiome" refers to the huge array of symbiotic bacteria in the gastrointestinal tract of humans as well as their genomes that interact collectively. The latest research suggests that the gut microbes mount multiple crucial biochemical roles for the host as well as those microbiome abnormalities are linked to a wide range of human disease processes. Trillions of microorganisms (altogether referred as the gut microbiota) live in the gastrointestinal system and perform critical roles which are related to host physiology and health. Pathogenesis research has found certain species, bacterial genes, as well as metabolites that have roles in different illnesses and medicinal targets. The gut microbiota has a functional part in macronutrient metabolism, immune system development, and the synthesis of pro or anti-inflammatory signalling molecules and peptides. It has been demonstrated that the gut microbiome has a role in the development of a number of systemic disease states, including obesity and cardiovascular disease, as well as intestinal disorders like inflammatory bowel disease. Active investigation on the roles of the microbes along with the processes underpinning host–microbe interactions will result in a higher understanding function of the microbiota in health moreover illness. Thus, knowing microbiome action is critical for the creation of future customised healthcare methods, and possibly giving novel potential pharmacological targets and research pathways in this

N. Salaria (✉) · Neeraj
Department of Life Sciences, Arni School of Basic Sciences, Arni University, Kathgarh, Himachal Pradesh, India

J. Furhan
Department of Microbiology, SKIMS Medical College & Hospital, Srinagar, India

R. Kumar
Department of Biosciences, Himachal Pradesh University, Shimla, Himachal Pradesh, India

R. Sobti et al. (eds.), *Role of Microbes in Sustainable Development*,
https://doi.org/10.1007/978-981-99-3126-2_3

rapidly increasing sector in terms of future personalised healthcare strategies. The present chapter focuses on composition, manipulation, and biological factors related to gut microbiota. It also discusses technical challenges related to the gut microbiome and host interactions. Furthermore, future perspectives and utilisations of the gut microbiome are also highlighted.

Keywords

Gut microbiome · Microorganisms · Bacteria · Metagenomic · Probiotics · Human health

3.1 Introduction

The mammalian gastrointestinal system is a suitable habitat to variety of microbes identified as the gut microbiota. The makeup of this microbial population is host-specific, varying during a person's lifespan in addition to sensitive to both exogenous as well as endogenous alterations (Grond et al. 2018). The recent resurgence of study in the anatomy as well as the role of this "organ" has shed light on its essential role in health along with sickness. Numerous aspects of typical host physiology are impacted by the microbiota, including behaviour, nutritional status, as well as stress response. Furthermore, they could be a primary or secondary reason for many illnesses, affecting both nearby as well as distant organ systems (Moya and Ferrer 2016).

The balance of the gut's microbial population generally, and the presence and absence of certain strains that might trigger specific responses, is critical in guaranteeing or preventing homeostasis at the gut mucosa (Chow et al. 2010). The methods via which microbes exerts helpful or adverse outcomes are largely unknown, although they include the synthesis of signalling molecules along with the ability of intestinal epithelial and mucosal immune cells to recognise bacterial epitopes (Sekirov et al. 2010). Advancements in gut microflora modelling and examination will expand our understanding of their function in health and illness, enabling for the customisation of present and future therapeutic and preventive methods (Kostic et al. 2014). A population of microorganisms called a microbiota, which includes archaea, viruses, bacteria along with certain unicellular eukaryotes, resides in a particular habitat. Metagenomics is the discipline of molecular study that explores the intricacies of microbiome, whereas a microbiome is the whole of all the genetic components within a given microbiome (Sood et al. 2022).

The curiosity of researchers in microbiota has been increased over the last 15 years. Despite the finding that intestinal bacteria have researched for several years, research into the function of microbes that live within human gut has attained considerable interest outside of typical infectious disorders. Numerous studies have found changes in the gut microbiome throughout not just liver disease, diabetes as well as obesity, but also neurological illnesses and cancer (Cani 2018). The microbiome of the human intestine possibly a reservoir of novel medications.

Intestinal bacteria in humans are believed to be potential sources of new treatments and have a profound effect in the metabolism of host (Cani and de Vos 2017).

In this context, the human microbiota describes the whole community of microbes that inhabit outside as well as within bodies (Barr 2017). The microbial load in the stomach is substantially smaller, and it rises exponentially along the digestive tract out of the stomach through the duodenum, jejunum, along with ileum, and at last to the colon, which comprises around 10^9 and 10^{13} microorganisms (Friedman et al. 2018). These communities play critical roles in digestion, immune system development, human physiology as well as detoxification processes (Tanaka and Nakayama 2017). Furthermore, a few of these bacteria in the gut produce proteins and enzymes important for the digestion of some undigested dietary components, among other essential functions for the host health (Flint et al. 2012).

As a result, we have two genomes in microbiome; one is inherited from parents and another is acquired (Rosenberg and Zilber-Rosenberg 2011). This idea underpins the classification of living being as "superorganisms." First most crucial distinction among both of these genomes, is that, the inherited genome is still mostly steady throughout life and the microbiome is very active and may be changed with a variety of circumstances, including age, nutrition, hormone cycles, travel, medicines, and sickness (D'Argenio and Salvatore 2015).

3.2 Biological Factors to Consider While Determining a Healthy Gut Microbiome

The gut of an organism has a particular environment that supports specialised microbes (Kwong and Moran 2016). The gut microbiome, which is frequently regarded as a complex property, is made up of several aspects, and its traits are affected by a mixture of both internal and external forces. Some of them are covered below.

3.2.1 Infections

The microbiota in the stomach influences viral and bacterial illnesses, these microbial illnesses continue to cause significant morbidity and mortality worldwide (Dudek-Wicher et al. 2018). Several microbes can cause an immune response to be triggered in order to eradicate the microbes. Nonetheless, there is emerging proof that aberrant actions are to blame for catastrophic consequences as well as other inflammatory disorders (Bartold and Van Dyke 2013; Bergström and Lindholm 2000). Patients with severe illness have elevated concentrations of inflammatory cytokines as well as inflammation-related indicators in their blood plasma, for instance, IL 6, 8, and 10, and C reactive protein (CRP) as well as numerous enzymes that represent immunological response and infection-related tissue damage (Kasperska-Zajac et al. 2011). Recent increases in these disorders are almost certainly the consequence of complex and multidimensional external reasons such as changes in climate, greater commodities and people mobility, as well as a quick demographic shift. Parallel to these extrinsic influences, a greater knowledge of the inside components related with immunity against viruses has been acquired. The gut microbiome tract is acknowledged as an imperative component of the host system of defence, operating like a critical regulator of host protection mechanisms along with immune systems (Patel et al. 2022).

3.2.2 Genetics

The genetic makeup of the host, which impacts host metabolism, as well as health, influences the population of specific bacteria found in microbiome of gut (Boulund et al. 2022; Maglione et al. 2021). Although the variety of gut microbiome varies greatly among humans, members of same family are frequently found to have comparable microbiota than members belong to different family. Familial resemblances are typically linked to common environmental effects, such as food choice, which is a significant shaper of microbiome makeup (Rodríguez-Frías et al. 2021). However, related people have a greater degree of genetic identity, suggesting the likelihood that familial microbiome commonalities are due to shared genetic makeup. Living beings and the associated microorganisms that live in them have developed considerably, resulting in a diverse assemblage of thousands of microbial species coexisting in their digestive tracts in a mutualistic manner (Hernandez and Moeller 2021). Therefore, because gut microbiome is acquired in the environment from birth, it may operate as both an environmental element which interact with the DNA of the host to create phenotypic and a genotypic resolute feature which is influenced via communicates with the host (Maglione et al. 2021; Sedghi et al. 2021). Because the microbes may be manipulated for medicinal purposes, it is an appealing target for modification. In fact, a carefully balanced interaction among microbes as well as human physiologies can have a variety of effects on health and growth, making dysbiosis frequently linked to disease (Rogers et al. 2016; Elias-Oliveira et al. 2020). Due to this, growing data suggests that human genetic variability affects the composition and regulation of their gut flora.

3.2.3 Drugs

Pharmacological reactions and effectiveness in living being had now been linked to their gut microbiota, as well as the chemicals in these medications could also impact the gut microbes (Weersma et al. 2020). Understanding pharmacological processes and how specific drug side effects occur requires a considerate of the relationship among pharmaceuticals as well as the makeup of gut microbiome. Antibiotics have long been recognised to modify the gut microbiome's constitution, but research in population-based cohorts has revealed connections between certain drug classes and specific gut microbe's patterns (Francino 2016; Schwartz et al. 2020; Patangia et al. 2022). Antibiotics are routinely administered treatments that have prevented millions of living beings from illnesses; nonetheless, medications significantly affect the normal gut flora. The impact is immediate and sometimes long-lasting.

Specifically, an elevation in bacteria that make fatty acids of short chain that has been linked to shifts in the intestinal microbial makeup including both vivo and in animals after taking the drug Metformin, a type II diabetes medication that is often used (Foretz et al. 2019). A modified gut microbiome composition has been associated with number of medicines, according to a latest report in a general population dataset (Wang et al. 2020). In a similar vein, an in vitro examination of more than thousand commercially available medications revealed that non-antibiotic medicines can indeed hinder the development of gut bacterial strains (Maier et al. 2018). The relationship within intestinal microbiota as well as routinely prescribed non-antibiotic medications is complicated as well as multidirectional: the gut flora can affect a person's response toward a medication via enzymatically changing the drug's architecture thus affecting its bioactivity, bioavailability, even toxicity (Le Bastard et al. 2021; Lindell et al. 2022). Indirectly, the gut flora can also affect a person's response towards immunotherapy for the cancer therapy (Xavier et al. 2020).

3.3 Delivery Methods at Birth

Right from birth to early life, the gut flora is crucial. Health of neonatal may be impacted by gut microbiome populations because they guard against harmful pathogens (Martin and Sela 2013), assisting in the metabolisation as well as digestion of breast milk but also formula milk (Koropatkin et al. 2012), promoting immune system development (Ge et al. 2021), sustaining intestinal homeostasis (Chen 2014), as well as having an impact on neurodevelopment (Collins et al. 2012). Early in childhood, there is a progressive and dynamic process of microbiome succession. Neonatals tend to have fewer microbial communities than an adolescent because the makeup and complexity of the gut flora settle throughout time (Dicks et al. 2018; Sumich et al. 2022). Microbes in the gut of newborns are mainly made up of the phylum *Actinobacteria, Proteobacteria,* as well as *Bacteroidetes*, which comprise the genera *Escherichia, Clostridium, Bifidobacterium, Lactobacillus, Prevotella* along with *Bacteroides* (Pushpanathan et al. 2019). Numerous elements, such as delivery method, newborn food (breast milk vs. formula milk), usage of antibiotics, as well as geographical area, may affect the variety of the newborn microbiota (Marques et al. 2010). Sixteen percent of caesarean sections (CS) are performed globally. Microbes from the mother's vagina as well as gut are initially present in the newborns that are born vaginally (VD). In fact, children delivered by caesarean sections have a higher percentage of genes associated with antibiotic resistance and are invaded by bacteria discovered on the mother's body (Rutayisire et al. 2016). Birth through caesarean sections may influence the gut microbes of the infant as well as encourage the invasion of harmful microorganisms. In later life, it could increase the chance of immunological and metabolic illnesses such obesity, type II diabetes, and allergies (Takiishi et al. 2017).

3.4 Nourishment of Infant

Breastfeeding can also have a substantial impact on the bacterial makeup of an infant's gut. Breast milk is mostly constituted of carbohydrates, particularly HMOs that are human milk oligosaccharides, which encourage Bifidobacterium development in the infant's stomach (Lawson et al. 2020). Numbers of studies have found that newborn gut flora contains more Bifidobacterium and Lactobacillus during nursing, resulting in a consistent colonisation model when contrasted with the gut microbiome of infants who were not nursed by breastfed (Selvamani et al. 2021; Taylor et al. 2021). As breast feed is not sterile, newborn nutrition is another crucial aspect in creating the bacterial population in the gut (Khor et al. 2020). Breast feeding has been found to be a provider of commensal and possibly probiotics bacterial agents that hinder growth of the newborn gut flora. Human breast milk includes about 700 different types of bacteria. Whereas the bacterial populations in breast milk are frequently diverse as well as differ from person to person, the average microbial load across time is 10^6 bacterial cells/mL (Wen and Duffy 2017). Thus, nursing newborn ingesting 800 mL breast milk per day might consume up to 8×10^8 bacterial population per day, which is hundred times greater than earlier predictions, as well as an alteration in composition throughout lactation (Hale and Rowe 2016). Even when not exposed to antibiotics, infants have higher amounts of antibiotic resistance genes (ARG) in their gut microbiome than adults (Gibson et al. 2015). Consequently, the shift of antibiotic-resistant bacteria, as well as other assortment mechanisms other than antibiotic therapy, may be driving ARG enrichment in neonates and infants (Ahmad et al. 2021). However, the influence of selection pressure-causing substances other than antibiotics on resistance loading in neonates and infants is little understood. Mobile genetic element (MGEs) transmits ARGs across bacteria, potentially spreading ARGs in the newborn gut microbiota, and ARGs can be passed vertically from the mother or acquired in the hospital environment (Hildebrand et al. 2021). Feed type can have a substantial consequence on microbiota; earlier investigation suggests that food can change the profusion of certain ARGs in the gut of a newborn.

3.5 Composition of Gut Microbiota

The gut microbiome only contains a small number of phyla. The majority of the gut microbiota is made up of *Firmicutes* (60–80%), Gram-positive bacteria with more than 200 genera, the most significant of which are *Mycoplasma, Ruminococcus, Eubacterium, Lactobacillus, Clostridium, Faecalibacterium* along with *Roseburia*), while *Bacteroidetes* (20–30%), Gram-negative bacteria with genera *Bacteroides, Xylanibacter* as well as *Prevotella*, are also significant (Szablewski 2018). *Proteobacteria* (1% Gram-negative bacteria with the genera *Enterobacteriace, Escherichia* along with *Desulfovibrio*), *Actinobacteria* (10% Gram-negative bacteria with species *Bifidobacterium*), as well as *Verrucomicrobia* are detected in negligible levels (with genus *Fusobacteria, Akkermansia*, along with *Cyanobacteria*) (Dinan and Cryan 2015).

In dissimilar parts of the gastrointestinal tract, the luminal contents and microbial composition vary. The main phyla of bacteria that can be found in the gut are *Firmicutes, Actinobacteria, Bacteroidetes, Verrucomicrobia* along with *Proteobacteria* (Bruning et al. 2020). The quantity of microbes within body can exceed 1.5 kg or 2% of a typical 75-kg person's weight, and the bacterial genes number in the stomach is 150 times bigger than those in the genetic material (Farooqui 2021). In animals that are sterile, it has been proved that alterations in the microbiome of gut perform a significant position in the progression of diseases, including diabetes and obesity. There is proof that the gut microbiota's composition effects on gut permeability, inflammation, along with energy balance, all of which are connected to obesity and other associated illnesses such as T2D (Geurts et al. 2014). Gastric bypass surgery, which causes 70% weight loss and an improvement in glucose metabolism, has been authorised by the American Diabetes Association as an efficient treatment for obesity as well as type 2 diabetes. The original intent of this surgical procedure was to limit meal intake and prevent calorie absorption. The gut flora is significantly altered by gastric bypass, according to studies, which may aid in weight loss (Murphy et al. 2017). Duodenum samples from bypass patients show a substantial difference in the gut flora between those with and those without diabetes, with obese people with T2D having smaller strains of bacteria (Tai et al. 2015). Although the gut microbiota's makeup may alter as a result of bariatric surgery, weight loss after bypass surgery may not be primarily attributed to the microbiota.

3.6 Manipulation of Gut Microbiota

Although gut microbes can have a significant impact on its hosts, efforts have been made to comprehend that how diversity in the initial colonisation's source and timing affects an animal's function and health throughout ontogeny (Warne et al. 2019). Animals' immune systems as well as metabolic processes may be affected for the rest of their lives by early colonisation during developmental windows such as birth (Houghteling and Walker 2015). For instance, caesarean delivery in humans and antibiotic-induced microbial community disruption in young mice are linked to higher consequences of metabolic disorders as well as obesity, respectively (Zhou et al. 2019). Although microorganisms often appear to recover from interruptions by antibiotics or other elements, altered host metabolic phenotypes may persist if the interruption occurred early in life, supporting the possibility that important developmental windows may influence both microbial and host interactions. Non-model wildlife offers special opportunity to improve understanding of microbial as well as host interactions in contexts that are appropriate to ecology and evolution, even if the large part of this study regarding mammals and how they relate to agronomic practices as well as people's health (Le Roux et al. 2016). Additionally, a variety of hosts' fitness-related features, such as growth and development, behaviour, and vulnerability to infectious diseases, could be profoundly impacted by the gut microbiome diversity throughout ontogeny, which has not yet been thoroughly investigated (Bosch 2013).

3.6.1 Probiotics

Several bacteria and yeasts have long been revered for their health-promoting characteristics (Lordan et al. 2020). The term probiotics was originally described as live microorganisms, when consumed in adequate quantities, impart advantages for the host's health (Sánchez et al. 2017). The concept distinguishes between live microorganisms employed as processing aids or sources of valuable chemicals and those supplied solely for health advantages (Gurung et al. 2013).

Lactobacilli, streptococci, and *bifidobacteria* are now the most common probiotics bacteria utilised in human diets, supplements, and/or animal feed (Gibson and Roberfroid 1995). Additionally, *E. coli* and *Saccharomyces boulardii* are routinely used (Martins et al. 2009). The European Food Safety Authority (EFSA) acknowledged the advantages of using various probiotics as parts of animal feed for a variety of bacteria, whereas the sole recognised claim for humans at the moment is the advantage on lactose digesting (De Simone 2019). However, scientific evidence is mounting that certain probiotics strains or combinations of strains could be useful in a variety of conditions. Effects might be direct or indirect, depending on whether the probiotics bacteria interact with the commensal microbiota (Timmerman et al. 2007).

3.6.2 Prebiotics

Prebiotics were initially characterised in 1995 and the recent, improved meaning specifies that a prebiotic is "a preferentially fermented element which leads in particular alteration in the makeup as well as the gut microbiota's functions, imparting advantages on host health" (Cani and Delzenne 2009). Rather than focusing exclusively on the "bifidogenic impact," this enlarged definition aims to include changes in additional advantageous gut microbiota members (Gibson et al. 2017). Prebiotics that are now in use are primarily poorly digested carbohydrates with a relatively short chain length that are categorised based on their molecular weight. Prebiotics are classed as mono, oligo, or polysaccharides based on their degree of polymerisation. Carbohydrates, on the other hand, can be classed as digestible or non-digestible depending on their physiological and biochemical features (Yoo et al. 2012). Non-digestible carbohydrates are more commonly used to modify the gut microbiota than digestible carbohydrates because they can have several effects mediated by diverse metabolic pathways, such as glucose and lipid metabolism, inflammatory responses, and even changes in appetite control (Louis et al. 2014). Furthermore, the prebiotic index, which measures the absolute increase in *bifidobacteria* without mentioning the impact on other microbial inhabitants of the gut, is still often used to describe prebiotic effectiveness (Ouwehand et al. 2005).

Prebiotics work as cultivation medium to specifically raise the figures along with performance of specific microbes present in the colon, enhancing their development or participation. Information from Metagenomic research contrasting the gut flora of healthy and ill persons allows for the identification of bacterial species that are suppressed beneath certain illness states (Grice and Segre 2012). Specific growth tests can then be conducted under increasingly sophisticated settings to uncover substrates that might specifically encourage the functioning or development of specific bacteria, when prebiotics are provided; this will eliminate the dysbiosis related to the condition (Roberts and Darveau 2015).

3.6.3 Antibiotics

Since the discovery of antibiotics in the 1940s and the start of their mass production, the consequences of these medications on the native gut microorganisms have been extensively studied (Modi et al. 2014). Antibiotics have been used for decades to inhibit the spread of bacterial pathogens and, as a result, to treat bacterial illnesses (Unemo and Shafer 2011). They are used to boost the effectiveness of animal feeding. However, the rapid development of virulence genes in bacteria allows them to withstand these antibiotics. Recent research has proven that antibiotic usage has an influence on gut health (Tuchscherr et al. 2020). These have a lot of negative health implications. Even though antibiotics save millions of lives, they also reduce residential bacteria, which are essential for gut health (Fijan 2014).

Nonetheless, almost the entirety of this research has been on how certain antibiotics influence bacteria that have been grown in a laboratory or on particular kinds of bacteria grown on hosts that have received antibiotic treatment (Valli et al. 2020). Furthermore, the majority of these investigations used rather high doses of these medications compared to their normal proportions in microbial populations that are present in nature, with a particular emphasis on disease causing bacterial species (Engel and Moran 2013). Because of this, the majority of our knowledge of antibiotics consequences is focused pertaining to killing processes as well as particular resistance genotypes as well as phenotypes in the setting of a restricted subgroup of the gut flora separated from the general population.

In the past decade or so, investigation on gut microbiota and antibiotics has shifted to a more ecological as well as systemic viewpoint (Sanders et al. 2019). Current research facilities and healthcare institutions frequently use ecological concepts as well as molecular methods. More investigation regarding the effect of antibiotics on the human gut bacterial population, stressing the implications for lateral transmission of resistance genes is to be focused on.

3.6.4 Faecal Microbial Transplant (FMT)

The importance of gut flora, also known as microbiota, in preserving human health is becoming better recognised. Beginning with birth, our microbiota develops a lifelong close relationship with our nutrition and environment, as well as determining the post-natal anatomical along with purposeful development of the gut (Van Belkum et al. 2020). Furthermore, essential connections between the microbiota and our metabolic processes, as well as the immune machinery that serves as our primary defence against foreign antigens, persist throughout life (Bronzo et al. 2020). The trouble in the gut flora has been associated to a rising diversity of gastrointestinal as well as non-gastrointestinal illness. It has been established that faecal microbiota transplantation (FMT) may repair the dysbiosis that defines various persistent illnesses, resulting in an apparently safe, very affordable, and quickly effective treatment in the great majority of patients treated (Person and Keefer 2021).

Various additional gastrointestinal as well as non-gastrointestinal issues have also been resolved using FMT, however expertise with these diseases is limited. More work is needed with faecal microbial transplant to guarantee its security and appropriate administration route (Sbahi and Di Palma 2016). There is a conceptual shift occurring in regard to bacteria, from being infections to being essential for maintaining health in a dynamic society. Potential research will very certainly reduce the range of organisms that can be administered to patients to treat a variety of disorders. FMT is only the initial stage in this process.

3.7 Technical Challenges in Studying Gut Microbiome and Host Interactions

We must emphasise that, in addition to bacteria, the stomach includes archeae, viruses, phages, yeast, and fungus (Gurung et al. 2019). These microbes, which are assumed to regulate the host's activities, more significantly, the activity of the microbes, have been thoroughly studied and could be just as essential as bacteria (Engel and Moran 2013). As a result, our knowledge of host–microorganism interactions is expanded by the addition of the phageome, archaea, mycobiome as well as virome. For example, phages not only exceed bacteria (e.g., phages outnumber bacteria tenfold), however, they are also fresh participants in these intricate connections. Consistent methodologies for analysing faecal phageomes using Metagenomics, on the other hand, have just lately been discovered (Sutton and Hill 2019). As a result, it will take longer time for major basic advancements in this field to be translated into widespread applications for the general public. The following are some of the technological challenges in studying gut microbiota.

3.7.1 Metagenomics

The term "Metagenome" was coined in 1998, primarily concentrating on the soil bacterial genomes (Yan and Yu 2011). Metagenomic frequently work with complicated microbiome mixtures in comparison to genomics, and as a result, relies primarily on the depth of the sequencing. The fast improvement of sequencing technologies, notably the announcing the use of next-generation sequencing, brought about a decrease in the amount of money and time needed for each sequencing despite increasing the sequencing depth (Bharti and Grimm 2021). As a result, Metagenomic shotgun sequencing has been widely employed in investigations of host–microbes association as well as gut microflora (Karaduta et al. 2021). Metagenomics, as opposed to 16S rRNA sequencing, provides more taxonomic details as well as resolution, particularly at the level of species. Numerous research studies have yet used Metagenomic to analyse microbiota dispersed at the species level (Hillmann et al. 2018). Additionally, Metagenomic can give a straight evaluation of the purposeful potential of the microbiota (Walker et al. 2014). Through Metagenomic, Ye et al. (2018) discovered an altered gut microbe makeup in patients of Behcet's disease (BD), as well as that processes Patients who suffer from BD have elevated levels of these systems, including the transport mechanism for capsular polysaccharides and the oxidation-reduction process. Nonetheless, Metagenomic has several limits. To begin with, Metagenomic reads the genomes of all bacteria among the sample; therefore, it is impossible to determine whether the microorganisms are alive, dormant, or dead (Quince et al. 2017). Second, during the study, DNA loss, deterioration, as well as contamination may occur, resulting in significant variance. Then, the commonly used new generation sequencing technologies, such Illumina (Solexa) sequencing as well as ABI SOLiD sequencing, cannot prevent space construction in cases where the read length is smaller than repeated sequence (Bao et al. 2011). Furthermore, Metagenomic cannot determine whether or not a gene is expressed, nor does it provide the location and breadth of expression.

3.7.2 Metatranscriptomics

Metatranscriptomics involves the global expression of RNA in the microbiome (Yost et al. 2015; Duran-Pinedo 2021). Metatranscriptomics is likely utilised to analyse gene expression regulation at transcription level, since DNA is transcribed into RNA sequences during transcription, which allows the researchers better understand the role as well as metabolic pathway of the gut microbes (Bashiardes et al. 2016). When compared to Metagenomic, Metatranscriptomics likewise depends on NGS but has higher sensitivity and reproducibility (Shi et al. 2010). In contradictory to the higher variances found in Metagenomic data, Gosalbes et al. (2012) conducted a Metatranscriptomics examination on ten healthy young adults and found that the genetic makeup of functional gut microbiota and prospective functional level was rather similar within people. They proposed a paradigm for investigating the association between the health status and functional microbes, as well as contrasting the makeup of gut microbes under dissimilar physiologic circumstances. Nevertheless, because of mRNA's brief half-life, enzymatic destruction of mRNA, and difficulties in identifying responses to external stimuli, Metatranscriptomics has certain challenges (Walworth et al. 2021). Additionally, host RNA often pollutes the rRNA sequences of the gastrointestinal microbiota, raising the expense of sequencing and complicating data processing (Morgan and Huttenhower 2014). Finally, whereas Metatranscriptomics analyses the intestinal microbes, at the mRNA level, it does not accurately predict the protein-level, and while mRNA often indicates the presence of protein, this is not always the case.

3.7.3 Metabolomics

A complete picture of microbial metabolism as well as host–microbiota interactions is provided by Metabolomics. Metabolomics can be used to identify the alive, inactive, or dead states of microbes (Bashiardes et al. 2016). The overall metabolite composition can be evaluated using Metabolomics and to analyse individual metabolite (Withers et al. 2020). Mass spectrometry (MS) as well as nuclear magnetic resonance spectroscopy (NMR) are the two techniques used in Metabolomics the most frequently (Marshall and Powers 2017). Small compounds generated by the gut microbiota are identified and quantified using these approaches. A number of researchers have used ^{1}H NMR to analyse the metabolic composition of faecal water from babies born with hypoxic-ischemic encephalopathy (Watkins et al. 2017). Identification of distinct Metabolomics patterns in intestinal lumen, serum, and hippocampus that correspond with seizure defence, counting decreases in systemic gamma-glutamylated amino acids along with higher hippocampal GABA/glutamate levels have also been done (Ghaffari et al. 2022).

3.7.4 Metaproteomics

The term "Metaproteomics" was first used to describe the collection of proteins in an environment in 2004 (Schiebenhoefer et al. 2019). Expression of protein variations in gut microbiota could be monitored using Metaproteomics. With hundreds of distinct microbiotas available in a particular sample, the Metaproteomics can be exceedingly complicated (Gonzalez 2020). As a result, the development of Metaproteomics has been more difficult and slower than that of standard proteomics (Wilmes et al. 2015). Weak peptide identifications, limited protein yields, as well as database difficulties are among the key hurdles in Metaproteomics. Metaproteomics has only recently become increasingly pertinent to the investigation of the gut microbes due to advancements in specimen preparation, the advent of elevated Mass spectrometry and the application of novel bioinformatics techniques (Isaac et al. 2019). It can, in particular, indicate alteration in the taxonomy, function, along with metabolic pathway of gut microbes. Recent investigated stated that

Metaproteomics was extra useful as compared to Metagenomic in studying the significance of the intestinal microbiota in well-being as well as illness (Lee et al. 2017). Furthermore, not only the gut microbiota's makeup in addition to metabolic alterations in disease states is studied, other than the dynamic alterations of biochemical pathways linked with illnesses.

3.8 DNA Sequencing-Based Methodologies

One of the first critical stages in ensuring the integrity and consistency of the obtained material is sample collecting methods (Dakappagari et al. 2017). However, a previous study indicated that there was no variation in community structure following field data collecting; numerous studies demonstrated the significance of the collection procedure, with reliability being boosted by rapid downstream processing (Panek et al. 2018). Faeces is among the most complicated biological resources for the isolation bacterial DNA because it carries residues of human being DNA, dietary DNA, as well as several constraints that impede future PCR amplification as well as New Generation Sequences techniques (Eisenhofer et al. 2019). Extracting DNA from faeces is a critical action in acquiring high-quality DNA along with accurately identifying microbial composition and relative abundance (Wagner Mackenzie et al. 2015). As a result, necessary to analyse the various approaches for enhance procedures and techniques for extraction of bacterial DNA from faeces that offer an adequate volume, clarity, and the DNA's integrity, resulting in a good quality sample for additional research (Maghini et al. 2021).

The past 10 years have outlined a gradual transition in sequencing techniques from traditional Sanger sequencing to New Generation Sequence and the technologies available using sequencers from Roche, Illumina, Pacific Bioscience, and Thermo Fischer Scientific They have all been utilised with effectiveness for the examination of a variety of biological materials (Morganti et al. 2019).

Therefore, each stage in the experimental pipeline contributes heterogeneity that influences the final result, there is an unmet demand for technique standardisation that would allow for accurate and repeatable investigation of important samples of human biological material for studying intestinal bacteria.

3.9 Imaging Strategies

The microbial makeup varies from niche to niche even inside the digestive system (Min et al. 2020). Faecal samples cannot disclose such variances. Recent research reveals that to analyse the distribution and colonisation of bacteria in the intestines, a limited range of bacterial species may be fluorescently labelled by genetic and chemical engineering and introduced into uninfected mice through stomach or rectal injection (Zhang et al. 2021). This method, however, necessitates the sacrifice of experimental animals, making it impossible to analyse the regular spatial as well as temporal structure of gut microbe in real time inside the identical mammal prior to as well as after an investigational modification (Grond et al. 2018). Furthermore, as only a small number of gut bacterial species may be genetically modified for fluorescent labelling and the intestinal consists of relatively a significant amount of non-culturable microbes, existing fluorescent labelling approaches are only suited for a subset of gut bacteria (Daliri et al. 2017). Furthermore, it is uncertain how long it takes for colonisation to achieve the usual steady-state position, along with sampled surgically may damage the normal, bacterial species' three-dimensional dispersion in the gut microbiota (Cryan et al. 2019). Even when fluorescent dyes are used to designate indigenous animals in their native habitat, the fluorescent light absorption is frequently too insufficient to expand beyond tiny-animal investigations into diagnostic and clinical research on humans (Yao et al. 2014).

Thus, new approaches are therefore needed for in vivo imaging of the gut microbes in human beings, with minimum disruption to the natural microbiota.

3.10 Future Perspectives and Utilisations of Gut Microbiome

Numerous human metabolic processes and our well-being are impacted by the relationship between the human microbiome as well as system of defence (Althani et al. 2016; Lasselin et al. 2016). Furthermore, the connections among people in addition to microbes can be crucial in deciding whether or not a person is healthy or ill (Dorrestein et al. 2014). Several disorders, including cutaneous, inflammatory, metabolic, and neurological problems, are associated with dysbiosis (Scotti et al. 2017). Accurate diagnosis and therapy of many disorders depend on a deeper comprehension of the host–microbe relationship. Because of the consequence of the microbes to host health, novel therapeutic approaches have emerged that concentrate on the recommended alteration of the host microbiome, either through the removal of negative taxa or the restoration of positive taxa and the functional roles they play (Gilbert and Lynch 2019). Large numbers of microbial taxa are difficult to cultivate in the laboratory, if not impossible (Stewart 2012). As a result, it is very challenging to list each individual microbiome member and to comprehend how microbial population work and affect host–pathogen interactions (Dillard et al. 2021). Numerous Metagenomic research studies are now being conducted thanks to recent improvements in sequencing technology and computational tools. These investigations have offered crucial information about the human microbiome as well as numerous other microbial communities in different environments (Langille et al. 2013).

A healthy microbiome is crucial for host organisms, because it facilitates the efficient execution of crucial physiological functions (Foster et al. 2017). Indeed, host organisms and their microbiota have coevolved, with some commensals becoming pathobionts and others becoming symbionts (Ruff et al. 2020). In the human gut, some commensals release signals that encourage appropriate immune system development. Within distinct hosts, as well as various physical surroundings, microbial communities develop a distinct structure (Zheng et al. 2020). Due to this, research on the host–microbiome is primarily concerned with identifying and characterising the bacteria that live there, their distinctive host phenotypes, as well as the biochemical mechanisms with which these microorganisms affect their hosts (Henry et al. 2021).

Investigations into host–microbe interactions can disclose the relationship's essential properties, such as identification, categorisation, profile prediction, and interaction processes (Starr et al. 2018). Although these microbes' structure, function, dynamics, and interactions are critical in human metabolism, their identification, measurement, as well as characterisation can be difficult (Rowland et al. 2018). The majority of microbial communities are incredibly varied, and the vast the predominance of each species has yet to be grown. Second, their contact with one another, as well as their proclivity to construct sophisticated networks, makes it impossible to forecast their behaviour (Fierer 2017). Establishing molecular links between gut microbiota and its function provides a new dimension to insight the biology of complex microbial consortia. To research host–microbe interactions, culture-dependent techniques have been the mainstay of traditional approaches to microbial ecology (Gilbert et al. 2016). Even though these are culture-specific methodologies produced intriguing data sets, they also produced a distorted image of microbiota. However, for identifying microorganisms in both qualitative and quantitative ways, a variety of culture-independent approaches, mostly the methods based on PCR, have recently emerged (Hameed et al. 2018). These approaches have completely altered people's perceptions of the living being microbes and opened the ground for the development of Metagenomic (Lepage et al. 2013). Metagenomic research is expanding our understanding of interactions between host and pathogenic by identifying genes that may permit microorganisms to impact their hosts in unforeseen ways. Pathogen surveillance, biotechnology, host–microbe interactions, functional dysbiosis, and evolutionary biology can all benefit from Metagenomic research of host–microbe interactions (Cruz et al. 2022). Recent Metagenomic investigations of host microbiome has provided important insights into host–microbe interactions.

3.11 Conclusion

For all forms of life on Earth, microorganisms (host and parasite) are necessary. They both are defined by their surroundings. However, our knowledge of host–pathogen systems is still quite limited. Technologies for bioinformatics as well as sequencing have evolved significantly over the past two decades, creating it possible to investigate microbial populations living inside of various hosts. There is widespread agreement that the diversification of microbes found in harsh environments has mostly gone untouched. Novel approaches are necessary to acquire new information about this "latent" microflora. NGS technology has enabled the quick and low-cost generation of sequencing data, as well as the development of sequencing systems that may be employed in both big genome-sequencing centres and individual laboratories. Updated versions of their specific DNA sequencing technologies have been revealed by Illumina, PacBio, and Applied Biosystems. These enhancements will increase read length as well as high-throughput capability yet dramatically reducing the amount of sequencing each base. Such innovations will considerably benefit research scientists moreover offer them intriguing novel prospects. To address concerns regarding the ecology as well as the complexity of the microbial flora, multiple techniques for biological investigations must be integrated. The enormous amounts of genomic information that are going to be created will provide novel obstacles intended for data analysis, storage, and transport. Sequencing of genome facilities as well as laboratories will grow more reliant on information technology and bioinformatics. To analyse vast volumes of data as well as extract the data for relevant knowledge about bacterial populations, bioinformatics knowledge will become more important. Metagenomic will become increasingly important in health, biological as well as environmental research. A detailed representation and knowledge of the functional microbiome are required for future rational therapies that use the gut microbiome to change host phenotype. Experiments with representative models will be crucial for assessing the influence of various microbial activities on human physiology and explaining their mechanism of action. The authors believe that this chapter provides a thorough summary of gut microbiome of humans, as well as presents sequencing technologies and their prospects, along with their high value and limitations.

References

Ahmad I, Malak HA, Abulreesh HH (2021) Environmental antimicrobial resistance and its drivers: a potential threat to public health. J Glob Antimicrob Resist 27:101–111

Althani AA, Marei HE, Hamdi WS, Nasrallah GK, El Zowalaty ME, Al Khodor S, Al-Asmakh M, Abdel-Aziz H, Cenciarelli C (2016) Human microbiome and its association with health and diseases. J Cell Physiol 231(8):1688–1694

Bao S, Jiang R, Kwan W, Wang B, Ma X, Song YQ (2011) Evaluation of next-generation sequencing software in mapping and assembly. J Hum Genet 56(6):406–414

Barr JJ (2017) A bacteriophage journey through the human body. Immunol Rev 279(1):106–122

Bartold PM, Van Dyke TE (2013) Periodontitis: a host-mediated disruption of microbial homeostasis. Unlearning learned concepts. Periodontol 2000 62(1):203–217

Bashiardes S, Zilberman-Schapira G, Elinav E (2016) Use of metatranscriptomics in microbiome research. Bioinform Biol Insights 10:BBI-S34610

Bergström J, Lindholm B (2000) What are the causes and consequences of the chronic inflammatory state in chronic dialysis patients? In: Seminars in dialysis, vol 13. Blackwell Science Inc., Boston, pp 163–164

Bharti R, Grimm DG (2021) Current challenges and best-practice protocols for microbiome analysis. Brief Bioinform 22(1):178–193

Bosch TC (2013) Cnidarian-microbe interactions and the origin of innate immunity in metazoans. Annu Rev Microbiol 67(1):499–518

Boulund U, Bastos DM, Ferwerda B, van den Born BJ, Pinto-Sietsma SJ, Galenkamp H, Levin E, Groen AK, Zwinderman AH, Nieuwdorp M (2022) Gut microbiome associations with host genotype vary across ethnicities and potentially influence cardiometabolic traits. Cell Host Microbe 30(10):1464–1480

Bronzo V, Lopreiato V, Riva F, Amadori M, Curone G, Addis MF, Cremonesi P, Moroni P, Trevisi E, Castiglioni B (2020) The role of innate immune response and microbiome in resilience of dairy cattle to disease: the mastitis model. Animals 10(8):1397

Bruning J, Chapp A, Kaurala GA, Wang R, Techtmann S, Chen QH (2020) Gut microbiota and short chain fatty acids: influence on the autonomic nervous system. Neurosci Bull 36(1):91–95

Cani PD (2018) Human gut microbiome: hopes, threats and promises. Gut 67(9):1716–1725

Cani PD, de Vos WM (2017) Next-generation beneficial microbes: the case of *Akkermansia muciniphila*. Front Microbiol 8:1765

Cani PD, Delzenne NM (2009) The role of the gut microbiota in energy metabolism and metabolic disease. Curr Pharm Des 15(13):1546–1558

Chen GY (2014) Role of Nlrp6 and Nlrp12 in the maintenance of intestinal homeostasis. Eur J Immunol 44(2):321–327

Chow J, Lee SM, Shen Y, Khosravi A, Mazmanian SK (2010) Host–bacterial symbiosis in health and disease. Adv Immunol 107:243–274

Collins SM, Surette M, Bercik P (2012) The interplay between the intestinal microbiota and the brain. Nat Rev Microbiol 10(11):735–742

Cruz N, Abernathy GA, Dichosa AE, Kumar A (2022) The age of next-generation therapeutic-microbe discovery: exploiting microbe-microbe and host-microbe interactions for disease prevention. Infect Immun 90:e0058921

Cryan JF, O'Riordan KJ, Cowan CS, Sandhu KV, Bastiaanssen TF, Boehme M, Codagnone MG, Cussotto S, Fulling C, Golubeva AV, Guzzetta KE (2019) The microbiota-gut-brain axis. Physiol Rev 99:1877

D'Argenio V, Salvatore F (2015) The role of the gut microbiome in the healthy adult status. Clinicachim Acta 451:97–102

Dakappagari N, Zhang H, Stephen L, Amaravadi L, Khan MU (2017) Recommendations for clinical biomarker specimen preservation and stability assessments. Bioanalysis 9(8):643–653

Daliri EBM, Wei S, Oh DH, Lee BH (2017) The human microbiome and metabolomics: current concepts and applications. Crit Rev Food Sci Nutr 57(16):3565–3576

De Simone C (2019) The unregulated probiotic market. Clin Gastroenterol Hepatol 17(5):809–817

Dicks LMT, Geldenhuys J, Mikkelsen LS, Brandsborg E, Marcotte H (2018) Our gut microbiota: a long walk to homeostasis. Benefic Microbes 9(1):3–20

Dillard LR, Payne DD, Papin JA (2021) Mechanistic models of microbial community metabolism. Mol Omics 17(3):365–375

Dinan TG, Cryan JF (2015) The impact of gut microbiota on brain and behaviour: implications for psychiatry. Curr Opin Clin Nutr Metab Care 18(6):552–558

Dorrestein PC, Mazmanian SK, Knight R (2014) Finding the missing links among metabolites, microbes, and the host. Immunity 40(6):824–832

Dudek-Wicher RK, Junka A, Bartoszewicz M (2018) The influence of antibiotics and dietary components on gut microbiota. Prz Gastroenterol 13(2):85–92

Duran-Pinedo AE (2021) Metatranscriptomics analyses of the oral microbiome. Periodontol 2000 85(1):28–45

Eisenhofer R, Minich JJ, Marotz C, Cooper A, Knight R, Weyrich LS (2019) Contamination in low microbial biomass microbiome studies: issues and recommendations. Trends Microbiol 27(2): 105–117

Elias-Oliveira J, Leite JA, Pereira ÍS, Guimarães JB, Manso GMDC, Silva JS, Tostes RC, Carlos D (2020) NLR and intestinal dysbiosis-associated inflammatory illness: drivers or dampers? Front Immunol 11:1810

Engel P, Moran NA (2013) The gut microbiota of insects–diversity in structure and function. FEMS Microbiol Rev 37(5):699–735

Farooqui AA (2021) The contribution of microbiota, cerebral blood flow, and sleep deprivation in the pathogenesis of Alzheimer's disease. In: Gut microbiota in neurologic and visceral diseases. Academic Press, New York, pp 143–158

Fierer N (2017) Embracing the unknown: disentangling the complexities of the soil microbiome. Nat Rev Microbiol 15(10):579–590

Fijan S (2014) Microorganisms with claimed probiotic properties: an overview of recent literature. Int J Environ Res Public Health 11(5):4745–4767

Flint HJ, Scott KP, Duncan SH, Louis P, Forano E (2012) Microbial degradation of complex carbohydrates in the gut. Gut Microbes 3(4):289–306
Foretz M, Guigas B, Viollet B (2019) Understanding the glucoregulatory mechanisms of metformin in type 2 diabetes mellitus. Nat Rev Endocrinol 15(10):569–589
Foster JA, Rinaman L, Cryan JF (2017) Stress & the gut-brain axis: regulation by the microbiome. Neurobiol Stress 7:124–136
Francino MP (2016) Antibiotics and the human gut microbiome: dysbioses and accumulation of resistances. Front Microbiol 6:1543
Friedman ES, Bittinger K, Esipova TV, Hou L, Chau L, Jiang J, Mesaros C, Lund PJ, Liang X, FitzGerald GA, Goulian M (2018) Microbes vs. chemistry in the origin of the anaerobic gut lumen. Proc Natl Acad Sci 115(16):4170–4175
Ge Y, Wang X, Guo Y, Yan J, Abuduwaili A, Aximujiang K, Yan J, Wu M (2021) Gut microbiota influence tumor development and alter interactions with the human immune system. J Exp Clin Cancer Res 40(1):1–9
Geurts L, Neyrinck AM, Delzenne NM, Knauf C, Cani PD (2014) Gut microbiota controls adipose tissue expansion, gut barrier and glucose metabolism: novel insights into molecular targets and interventions using prebiotics. Benefic Microbes 5(1):3–17
Ghaffari S, Abbasi A, Somi MH, Moaddab SY, Nikniaz L, Kafil HS, Ebrahimzadeh Leylabadlo H (2022) Akkermansia muciniphila: from its critical role in human health to strategies for promoting its abundance in human gut microbiome. Crit Rev Food Sci Nutr:1–21
Gibson GR, Roberfroid MB (1995) Dietary modulation of the human colonic microbiota: introducing the concept of prebiotics. J Nutr 125(6):1401–1412
Gibson MK, Crofts TS, Dantas G (2015) Antibiotics and the developing infant gut microbiota and resistome. Curr Opin Microbiol 27:51–56
Gibson GR, Hutkins RW, Sanders ME, Prescott SL, Reimer RA, Salminen SJ, Scott K, Stanton C, Swanson KS, Cani PD, Verbeke K (2017) The International Scientific Association for Probiotics and Prebiotics (ISAPP) consensus statement on the definition and scope of prebiotics. Nat Rev Gastroenterol Hepatol 14:491–502
Gilbert JA, Lynch SV (2019) Community ecology as a framework for human microbiome research. Nat Med 25(6):884–889
Gilbert JA, Quinn RA, Debelius J, Xu ZZ, Morton J, Garg N, Jansson JK, Dorrestein PC, Knight R (2016) Microbiome-wide association studies link dynamic microbial consortia to disease. Nature 535(7610):94–103
Gonzalez CG (2020) Revealing host-gut microbiome interactions: a metaproteomic perspective. Stanford University, Stanford
Gosalbes MJ, Abellan JJ, Durban A, Pérez-Cobas AE, Latorre A, Moya A (2012) Metagenomics of human microbiome: beyond 16S rDNA. Clin Microbiol Infect 18:47–49
Grice EA, Segre JA (2012) The human microbiome: our second genome. Annu Rev Genomics Hum Genet 13:151–170
Grond K, Sandercock BK, Jumpponen A, Zeglin LH (2018) The avian gut microbiota: community, physiology and function in wild birds. J Avian Biol 49(11):e01788
Gurung N, Ray S, Bose S, Rai V (2013) A broader view: microbial enzymes and their relevance in industries, medicine, and beyond. Biomed Res Int 2013:329121
Gurung K, Wertheim B, Falcao Salles J (2019) The microbiome of pest insects: it is not just bacteria. Entomol Exp Appl 167(3):156–170
Hale TW, Rowe HE (2016) Medications and mothers' milk 2017. Springer Publishing Company, Berlin
Hameed S, Xie L, Ying Y (2018) Conventional and emerging detection techniques for pathogenic bacteria in food science: a review. Trends Food Sci Technol 81:61–73
Henry LP, Bruijning M, Forsberg SK, Ayroles JF (2021) The microbiome extends host evolutionary potential. Nat Commun 12(1):1–13

Hernandez CJ, Moeller AH (2021) The microbiome: a heritable contributor to bone morphology? In: Seminars in cell & developmental biology, vol 123. Academic Press, New York, p 82

Hildebrand F, Gossmann TI, Frioux C, Özkurt E, Myers PN, Ferretti P, Kuhn M, Bahram M, Nielsen HB, Bork P (2021) Dispersal strategies shape persistence and evolution of human gut bacteria. Cell Host Microbe 29(7):1167–1176

Hillmann B, Al-Ghalith GA, Shields-Cutler RR, Zhu Q, Gohl DM, Beckman KB, Knight R, Knights D (2018) Evaluating the information content of shallow shotgun metagenomics. mSystems 3(6):e00069-18

Houghteling PD, Walker WA (2015) Why is initial bacterial colonization of the intestine important to infants' and children's health? J Pediatr Gastroenterol Nutr 60(3):294–307

Isaac NI, Philippe D, Nicholas A, Raoult D, Eric C (2019) Metaproteomics of the human gut microbiota: challenges and contributions to other OMICS. Clin Mass Spectrom 14:18–30

Karaduta O, Dvanajscak Z, Zybailov B (2021) Metaproteomics—an advantageous option in studies of host-microbiota interaction. Microorganisms 9(5):980

Kasperska-Zajac A, Sztylc J, Machura E, Jop G (2011) Plasma IL-6 concentration correlates with clinical disease activity and serum C-reactive protein concentration in chronic urticaria patients. Clin Exp Allergy 41(10):1386–1391

Khor GL, Tan SS, Stoutjesdijk E, Ng KWT, Khouw I, Bragt M, Schaafsma A, Dijck-Brouwer DJ, Muskiet FA (2020) Temporal changes in breast milk fatty acids contents: a case study of Malay breastfeeding women. Nutrients 13(1):101

Koropatkin NM, Cameron EA, Martens EC (2012) How glycan metabolism shapes the human gut microbiota. Nat Rev Microbiol 10(5):323–335

Kostic AD, Xavier RJ, Gevers D (2014) The microbiome in inflammatory bowel disease: current status and the future ahead. Gastroenterology 146(6):1489–1499

Kwong WK, Moran NA (2016) Gut microbial communities of social bees. Nat Rev Microbiol 14(6):374–384

Langille MG, Zaneveld J, Caporaso JG, McDonald D, Knights D, Reyes JA, Clemente JC, Burkepile DE, Vega Thurber RL, Knight R, Beiko RG (2013) Predictive functional profiling of microbial communities using 16S rRNA marker gene sequences. Nat Biotechnol 31(9): 814–821

Lasselin J, Alvarez-Salas E, Grigoleit JS (2016) Well-being and immune response: a multi-system perspective. Curr Opin Pharmacol 29:34–41

Lawson MA, O'Neill IJ, Kujawska M, Gowrinadh Javvadi S, Wijeyesekera A, Flegg Z, Chalklen L, Hall LJ (2020) Breast milk-derived human milk oligosaccharides promote Bifidobacterium interactions within a single ecosystem. ISME J 14(2):635–648

Le Bastard Q, Berthelot L, Soulillou JP, Montassier E (2021) Impact of non-antibiotic drugs on the human intestinal microbiome. Expert Rev Mol Diagn 21(9):911–924

Le Roux F, Wegner KM, Polz MF (2016) Oysters and vibrios as a model for disease dynamics in wild animals. Trends Microbiol 24(7):568–580

Lee PY, Chin SF, Neoh HM, Jamal R (2017) Metaproteomic analysis of human gut microbiota: where are we heading? J Biomed Sci 24(1):1–8

Lepage P, Leclerc MC, Joossens M, Mondot S, Blottière HM, Raes J, Ehrlich D, Doré J (2013) A metagenomic insight into our gut's microbiome. Gut 62(1):146–158

Lindell AE, Zimmermann-Kogadeeva M, Patil KR (2022) Multimodal interactions of drugs, natural compounds and pollutants with the gut microbiota. Nat Rev Microbiol 20:1–13

Lordan C, Thapa D, Ross RP, Cotter PD (2020) Potential for enriching next-generation health-promoting gut bacteria through prebiotics and other dietary components. Gut Microbes 11(1): 1–20

Louis P, Hold GL, Flint HJ (2014) The gut microbiota, bacterial metabolites and colorectal cancer. Nat Rev Microbiol 12(10):661–672

Maghini DG, Moss EL, Vance SE, Bhatt AS (2021) Improved high-molecular-weight DNA extraction, nanopore sequencing and metagenomic assembly from the human gut microbiome. Nat Protoc 16(1):458–471
Maglione A, Zuccalà M, Tosi M, Clerico M, Rolla S (2021) Host genetics and gut microbiome: perspectives for multiple sclerosis. Genes 12(8):1181
Maier L, Pruteanu M, Kuhn M, Zeller G, Telzerow A, Anderson EE, Brochado AR, Fernandez KC, Dose H, Mori H, Patil KR (2018) Extensive impact of non-antibiotic drugs on human gut bacteria. Nature 555(7698):623–628
Marques TM, Wall R, Ross RP, Fitzgerald GF, Ryan CA, Stanton C (2010) Programming infant gut microbiota: influence of dietary and environmental factors. Curr Opin Biotechnol 21(2): 149–156
Marshall DD, Powers R (2017) Beyond the paradigm: combining mass spectrometry and nuclear magnetic resonance for metabolomics. Prog Nucl Magn Reson Spectrosc 100:1–16
Martin MA, Sela DA (2013) Infant gut microbiota: developmental influences and health outcomes. In: Building babies. Springer, New York, pp 233–256
Martins FS, Silva AA, Vieira AT, Barbosa FH, Arantes RM, Teixeira MM, Nicoli JR (2009) Comparative study of Bifidobacterium animalis, Escherichia coli, Lactobacillus casei and Saccharomyces boulardii probiotic properties. Arch Microbiol 191(8):623–630
Min S, Kim S, Cho SW (2020) Gastrointestinal tract modeling using organoids engineered with cellular and microbiota niches. Exp Mol Med 52(2):227–237
Modi SR, Collins JJ, Relman DA (2014) Antibiotics and the gut microbiota. J Clin Invest 124(10): 4212–4218
Morgan XC, Huttenhower C (2014) Meta'omic analytic techniques for studying the intestinal microbiome. Gastroenterology 146(6):1437–1448
Morganti S, Tarantino P, Ferraro E, D'Amico P, Duso BA, Curigliano G (2019) Next generation sequencing (NGS): a revolutionary technology in pharmacogenomics and personalized medicine in cancer. In: Translational research and onco-omics applications in the era of cancer personal genomics. Springer, Berlin, pp 9–30
Moya A, Ferrer M (2016) Functional redundancy-induced stability of gut microbiota subjected to disturbance. Trends Microbiol 24(5):402–413
Murphy R, Tsai P, Jüllig M, Liu A, Plank L, Booth M (2017) Differential changes in gut microbiota after gastric bypass and sleeve gastrectomy bariatric surgery vary according to diabetes remission. Obes Surg 27(4):917–925
Ouwehand AC, Derrien M, de Vos W, Tiihonen K, Rautonen N (2005) Prebiotics and other microbial substrates for gut functionality. Curr Opin Biotechnol 16(2):212–217
Panek M, ČipčićPaljetak H, Barešić A, Perić M, Matijašić M, Lojkić I, Vranešić Bender D, Krznarić Ž, Verbanac D (2018) Methodology challenges in studying human gut microbiota—effects of collection, storage, DNA extraction and next generation sequencing technologies. Sci Rep 8(1): 1–13
Patangia DV, Anthony Ryan C, Dempsey E, Paul Ross R, Stanton C (2022) Impact of antibiotics on the human microbiome and consequences for host health. Microbiol Open 11(1):e1260
Patel M, McAllister M, Nagaraju R, Al Badran SSF, Edwards J, McBain AJ, Barriuso J, Aziz O (2022) The intestinal microbiota in colorectal cancer metastasis–passive observer or key player? Crit Rev Oncol Hematol 180:103856
Person H, Keefer L (2021) Psychological comorbidity in gastrointestinal diseases: update on the brain-gut-microbiome axis. Prog Neuro-Psychopharmacol Biol Psychiatry 107:110209
Pushpanathan P, Mathew GS, Selvarajan S, Seshadri KG, Srikanth P (2019) Gut microbiota and its mysteries. Indian J Med Microbiol 37(2):268–277
Quince C, Walker AW, Simpson JT, Loman NJ, Segata N (2017) Shotgun metagenomics, from sampling to analysis. Nat Biotechnol 35(9):833–844
Roberts FA, Darveau RP (2015) Microbial protection and virulence in periodontal tissue as a function of polymicrobial communities: symbiosis and dysbiosis. Periodontol 2000 69(1): 18–27

Rodríguez-Frías F, Quer J, Tabernero D, Cortese MF, Garcia-Garcia S, Rando-Segura A, Pumarola T (2021) Microorganisms as shapers of human civilization, from pandemics to even our genomes: villains or friends? A historical approach. Microorganisms 9(12):2518
Rogers GB, Keating DJ, Young RL, Wong ML, Licinio J, Wesselingh S (2016) From gut dysbiosis to altered brain function and mental illness: mechanisms and pathways. Mol Psychiatry 21(6): 738–748
Rosenberg E, Zilber-Rosenberg I (2011) Symbiosis and development: the hologenome concept. Birth Defects Res C Embryo Today 93(1):56–66
Rowland I, Gibson G, Heinken A, Scott K, Swann J, Thiele I, Tuohy K (2018) Gut microbiota functions: metabolism of nutrients and other food components. Eur J Nutr 57(1):1–24
Ruff WE, Greiling TM, Kriegel MA (2020) Host–microbiota interactions in immune-mediated diseases. Nat Rev Microbiol 18(9):521–538
Rutayisire E, Huang K, Liu Y, Tao F (2016) The mode of delivery affects the diversity and colonization pattern of the gut microbiota during the first year of infants' life: a systematic review. BMC Gastroenterol 16(1):1–12
Sánchez B, Delgado S, Blanco-Míguez A, Lourenço A, Gueimonde M, Margolles A (2017) Probiotics, gut microbiota, and their influence on host health and disease. Mol Nutr Food Res 61(1):1600240
Sanders ME, Merenstein DJ, Reid G, Gibson GR, Rastall RA (2019) Probiotics and prebiotics in intestinal health and disease: from biology to the clinic. Nat Rev Gastroenterol Hepatol 16(10): 605–616
Sbahi H, Di Palma JA (2016) Faecal microbiota transplantation: applications and limitations in treating gastrointestinal disorders. BMJ Open Gastroenterol 3(1):e000087
Schiebenhoefer H, Van Den Bossche T, Fuchs S, Renard BY, Muth T, Martens L (2019) Challenges and promise at the interface of metaproteomics and genomics: an overview of recent progress in metaproteogenomic data analysis. Expert Rev Proteomics 16(5):375–390
Schwartz DJ, Langdon AE, Dantas G (2020) Understanding the impact of antibiotic perturbation on the human microbiome. Genome Med 12(1):1–12
Scotti E, Boué S, Sasso GL, Zanetti F, Belcastro V, Poussin C, Sierro N, Battey J, Gimalac A, Ivanov NV, Hoeng J (2017) Exploring the microbiome in health and disease: implications for toxicology. Toxicol Res Appl 1:2397847317741884
Sedghi L, DiMassa V, Harrington A, Lynch SV, Kapila YL (2021) The oral microbiome: role of key organisms and complex networks in oral health and disease. Periodontol 2000 87(1): 107–131
Sekirov I, Russell SL, Antunes LCM, Finlay BB (2010) Gut microbiota in health and disease. Physiol Rev 90:859
Selvamani S, Mehta D, Deb C, El Enshasy H, Abomoelak B (2021) Role of the gut microbiota in the digestive tract diseases. In: Probiotics, the natural microbiota in living organisms. CRC Press, Boca Raton, pp 30–57
Shi W, Syrenne R, Sun JZ, Yuan JS (2010) Molecular approaches to study the insect gut symbiotic microbiota at the 'omics' age. Insect Sci 17(3):199–219
Sood U, Dhingra GG, Anand S, Hira P, Kumar R, Kaur J, Verma M, Singhvi N, Lal S, Rawat CD, Singh VK (2022) Microbial journey: Mount Everest to Mars. Indian J Microbiol 62:1–15
Starr AE, Deeke SA, Li L, Zhang X, Daoud R, Ryan J, Ning Z, Cheng K, Nguyen LV, Abou-Samra E, Lavallée-Adam M (2018) Proteomic and metaproteomic approaches to understand host–microbe interactions. Anal Chem 90(1):86–109
Stewart EJ (2012) Growing unculturable bacteria. J Bacteriol 194(16):4151–4160
Sumich A, Heym N, Lenzoni S, Hunter K (2022) Gut microbiome-brain axis and inflammation in temperament, personality and psychopathology. Curr Opin Behav Sci 44:101101
Sutton TD, Hill C (2019) Gut bacteriophage: current understanding and challenges. Front Endocrinol 10:784
Szablewski L (2018) Human gut microbiota in health and Alzheimer's disease. J Alzheimers Dis 62(2):549–560

Tai N, Wong FS, Wen L (2015) The role of gut microbiota in the development of type 1, type 2 diabetes mellitus and obesity. Rev Endocr Metab Disord 16(1):55–65
Takiishi T, Fenero CIM, Câmara NOS (2017) Intestinal barrier and gut microbiota: shaping our immune responses throughout life. Tissue Barriers 5(4):e1373208
Tanaka M, Nakayama J (2017) Development of the gut microbiota in infancy and its impact on health in later life. Allergol Int 66(4):515–522
Taylor SL, Simpson JL, Rogers GB (2021) The influence of early-life microbial exposures on long-term respiratory health. Paediatr Respir Rev 40:15–23
Timmerman HM, Niers LE, Ridwan BU, Koning CJ, Mulder L, Akkermans LM, Rombouts FM, Rijkers GT (2007) Design of a multispecies probiotic mixture to prevent infectious complications in critically ill patients. Clin Nutr 26(4):450–459
Tuchscherr L, Löffler B, Proctor RA (2020) Persistence of Staphylococcus aureus: multiple metabolic pathways impact the expression of virulence factors in small-colony variants (SCVs). Front Microbiol 11:1028
Unemo M, Shafer WM (2011) Antibiotic resistance in Neisseria gonorrhoeae: origin, evolution, and lessons learned for the future. Ann N Y Acad Sci 1230(1):E19–E28
Valli RX, Lyng M, Kirkpatrick CL (2020) There is no hiding if you Seq: recent breakthroughs in Pseudomonas aeruginosa research revealed by genomic and transcriptomic next-generation sequencing. J Med Microbiol 69(2):162–175
Van Belkum M, Mendoza Alvarez L, Neu J (2020) Preterm neonatal immunology at the intestinal interface. Cell Mol Life Sci 77(7):1209–1227
Wagner Mackenzie B, Waite DW, Taylor MW (2015) Evaluating variation in human gut microbiota profiles due to DNA extraction method and inter-subject differences. Front Microbiol 6:130
Walker AW, Duncan SH, Louis P, Flint HJ (2014) Phylogeny, culturing, and metagenomics of the human gut microbiota. Trends Microbiol 22(5):267–274
Walworth NG, Saito MA, Lee MD, McIlvin MR, Moran DM, Kellogg RM, Fu FX, Hutchins DA, Webb EA (2021) Why environmental biomarkers work: transcriptome–proteome correlations and modeling of multistressor experiments in the marine Bacterium Trichodesmium. J Proteome Res 21(1):77–89
Wang J, Xiong K, Zhao S, Zhang C, Zhang J, Xu L, Ma A (2020) Long-term effects of multi-drug-resistant tuberculosis treatment on gut microbiota and its health consequences. Front Microbiol 11:53
Warne RW, Kirschman L, Zeglin L (2019) Manipulation of gut microbiota during critical developmental windows affects host physiological performance and disease susceptibility across ontogeny. J Anim Ecol 88(6):845–856
Watkins C, Murphy K, Yen S, Carafa I, Dempsey EM, O'Shea CA, Vercoe EA, Ross RP, Stanton C, Ryan CA (2017) Effects of therapeutic hypothermia on the gut microbiota and metabolome of infants suffering hypoxic-ischemic encephalopathy at birth. Int J Biochem Cell Biol 93:110–118
Weersma RK, Zhernakova A, Fu J (2020) Interaction between drugs and the gut microbiome. Gut 69(8):1510–1519
Wen L, Duffy A (2017) Factors influencing the gut microbiota, inflammation, and type 2 diabetes. J Nutr 147(7):1468S–1475S
Wilmes P, Heintz-Buschart A, Bond PL (2015) A decade of metaproteomics: where we stand and what the future holds. Proteomics 15(20):3409–3417
Withers E, Hill PW, Chadwick DR, Jones DL (2020) Use of untargeted metabolomics for assessing soil quality and microbial function. Soil Biol Biochem 143:107758
Xavier JB, Young VB, Skufca J, Ginty F, Testerman T, Pearson AT, Macklin P, Mitchell A, Shmulevich I, Xie L, Caporaso JG (2020) The cancer microbiome: distinguishing direct and indirect effects requires a systemic view. Trends Cancer 6(3):192–204
Yan Q, Yu Y (2011) Metagenome-based analysis: a promising direction for plankton ecological studies. Sci China Life Sci 54(1):75–81

Yao J, Yang M, Duan Y (2014) Chemistry, biology, and medicine of fluorescent nanomaterials and related systems: new insights into biosensing, bioimaging, genomics, diagnostics, and therapy. Chem Rev 114(12):6130–6178

Ye Z, Zhang N, Wu C, Zhang X, Wang Q, Huang X, Du L, Cao Q, Tang J, Zhou C, Hou S (2018) A metagenomic study of the gut microbiome in Behcet's disease. Microbiome 6(1):1–13

Yoo HD, Kim D, Paek SH (2012) Plant cell wall polysaccharides as potential resources for the development of novel prebiotics. Biomol Ther 20(4):371

Yost S, Duran-Pinedo AE, Teles R, Krishnan K, Frias-Lopez J (2015) Functional signatures of oral dysbiosis during periodontitis progression revealed by microbial metatranscriptome analysis. Genome Med 7(1):1–19

Zhang Z, Xu D, Fang J, Wang D, Zeng J, Liu X, Hong S, Xue Y, Zhang X, Zhao X (2021) In situ live imaging of gut microbiota. mSphere 6(5):e0054521

Zheng D, Liwinski T, Elinav E (2020) Interaction between microbiota and immunity in health and disease. Cell Res 30(6):492–506

Zhou X, Du L, Shi R, Chen Z, Zhou Y, Li Z (2019) Early-life food nutrition, microbiota maturation and immune development shape life-long health. Crit Rev Food Sci Nutr 59(Suppl 1):S30–S38

Probiotics: A Healthy Treasure

4

Yogalakshmi Ganapathy, Nandhini Muthusamy Sridhar, and Prabu Dhandapani

Abstract

Probiotics are live microorganisms that enhance the health benefits of the host. They are capable of colonizing the gastrointestinal tract by surviving gastric and bile acids. It strengthens the host's immunity and lowers the risk of urogenital infections, allergies, cancer, and other intestinal issues that are common among people. The mechanism of their action is through the production of antimicrobial substances, and increased adhesion to the intestinal mucosa. It has several immunomodulatory properties which enhance humoral and cell-mediated immunity. In this chapter, we will learn about the health benefits of probiotics and their potential therapeutic application.

Keywords

Probiotic · Immunity · *Lactobacillus* · Microbiota

4.1 Introduction

Probiotics are live bacteria that provide health benefits to the host when administered in appropriate doses (FAO/WHO 2002). Probiotic bacteria are safe for human consumption and provide health benefits to the host, such as improving gut health, boosting immunity, reducing urogenital infections, allergies, cancer and other intestinal problems that affect a majority of the world's population. Although there is limited data to support their health benefits in humans, probiotics are an essential component of functional foods. To maintain a healthy gut, specific beneficial

Y. Ganapathy · N. Muthusamy Sridhar · P. Dhandapani (✉)
Department of Microbiology, Dr. ALM PG IBMS, University of Madras, Chennai, India
e-mail: dprabu@unom.ac.in

R. Sobti et al. (eds.), *Role of Microbes in Sustainable Development*,
https://doi.org/10.1007/978-981-99-3126-2_4

bacterial strains must be introduced into the gastrointestinal tract (GIT) to prevent disease. In the early 1900s, Elle Metchnikoff hypothesized that beneficial bacteria may be administered to replace harmful microorganisms with healthy ones. In the 1960s, Lilly and Stillwell coined the phrase probiotic, which means 'for life'. Probiotics, on the other hand, contain live microbial supplements that improve the microbial balance in the intestine of the host (Fuller 1989). *Lactobacillus, Bifidobacterium, Escherichia coli, Enterococcus, Bacillus*, *Streptococcus* and *Saccharomyces* are some of the most often used probiotic microbes.

Microorganisms present on the skin, in the mouth and in the gastrointestinal tract coexist with humans. The GIT, covering more than 400 square metres of surface area, has the highest concentration of commensal microorganisms. Probiotic bacteria present in the GIT and establish themselves rapidly after birth, are relatively stable throughout life and are essential for human homeostasis. Interactions between the intestinal microflora and the host during the establishment of the microbiota result in the evolution of a unique and distinct intestinal immune system. Probiotic foods account for 65% of the global functional food market, resulting in a wide range of probiotic food products available in supermarkets and health food stores. Probiotic supplements have been introduced to the market since they protect against infections, reduce lactose intolerance, lower blood cholesterol levels and activate the immune system. This has prompted more research into the identification of foods and food components that provide unique consumer benefits. Prebiotic, for example, is a non-digestible food component that benefits the body by activating one or more probiotic bacteria in the colon. When a product has both probiotics and prebiotics, it is referred to as a synbiotic.

The first probiotic to gain widespread clinical attention was *Lactobacillus rhamnosus GG* strain (LGG). It was discovered in 1985, and used in the dairy sector in the fermentation process. LGG boosts intestinal immunity by increasing the number of cells that secrete IgA and other immunoglobulins in the intestinal mucosa. It stimulates the local interferon release and facilitates antigen transport to underlying lymphoid cells, resulting in increased antigen uptake in Peyer's patches. Bile, hydrochloric acid and pancreatic juice have no effect on it and it possesses anti-carcinogenic properties (Gupta and Garg 2009).

4.2 Probiotic Microorganisms and Their Characteristics

Lactobacillus and *Bifidobacterium*, two major gram-positive bacteria species, are widely utilized as probiotics. Other genera, such as *Escherichia, Enterococcus, and Saccharomyces*, have been promoted as probiotics, while safety concerns persist (Holzapfel et al. 2001). Probiotics are strain-specific, therapeutic benefits attributed to one strain cannot be expected to be delivered by another, even if they are from the same species.

According to the Food and Agriculture Organization (FAO) and the World Health Organization (WHO), probiotics have the following characteristics (FAO/WHO, 2001):

1. It must be able to survive passage through the gut by withstanding gastric acids and bile exposure.
2. It should be able to survive both conjugated and deconjugated bile acids, as well as low pH and high bile acid concentrations.
3. It should be able to colonize and grow in the gastrointestinal tract.
4. The immune system should be able to withstand it.
5. It must be free of any infectious, allergic or mutagenic/carcinogenic properties.
6. It must have the administration sites that are desired.
7. They must be safe as well as effective.
8. It should remain effective and potent for the duration of the product's shelf life.

4.3 Mechanism of Action

Probiotics compete with pathogens for nutrition by inhibiting bacterial growth, as a result, infections in the gastrointestinal tract become less common. Probiotic protective mechanisms include epithelial barrier improvement, immunomodulatory properties, antibacterial activity and competitive adhesion to the intestinal mucosa (Bermudez-Brito et al. 2012). A recent study finds that introducing probiotics to the intestine may prevent the spread of potential infections by lowering luminal pH, producing bacteriocins, releasing antimicrobial defensins in the intestinal crypts and preventing bacterial adhesion to epithelial cells (Plaza-Diaz et al. 2019).

4.3.1 Production of Antimicrobial Substances

A good example of probiotic antibacterial action is the effect of *Lactobacillus* species on *Helicobacter pylori* infection of the gastric mucosa (Homan and Orel 2015). Acid-resistant *Lactobacillus* species and commensal organisms persist longer in the stomach than other bacteria. *Lactobacillus acidophilus* produces two bacteriocin compounds: lactacin B and acidolin, which may interfere with *H. pylori* infection or inflammation. An in vitro study found that lactacin and acidolin inhibit the growth of *H. pylori* and enteropathogenic bacteria, respectively. De-conjugated bile acids, which are bile salt derivatives produced by probiotic bacteria, have better antibacterial properties than the bile salts produced by the human host (Ridlon et al. 2016). Gut bacteria produce a wide range of health-promoting fatty acids (Conlon and Bird 2014). *Bifidobacterium* and *Lactobacillus* in the intestine have been found to produce conjugated linoleic acid (CLA), a potent anti-carcinogenic agent (Den Hartigh 2019). CLA-producing *L. plantarum* was reported to have an anti-obesity effect in rats having diet-induced obesity (Lee et al. 2007). In mouse models, CLA-producing *Bifidobacterium* and *Lactobacilli* were shown to regulate the fatty acid content of the host's liver and adipose tissue (Rosberg-Cody et al. 2011). *Lactobacillus* produces fungicidal compounds like benzoic acid, methyl hydantoin, mevalolactone and short-chain fatty acids. *Lactobacillus coryniformis*, for example, produces antifungal proteinaceous compounds (Magnusson and Schnürer 2001).

4.3.2 Increased Adhesion to the Intestinal Mucosa

Inflammatory bowel disease is caused whenever a pathogen or foodborne antigen breaches the intestinal barrier and penetrates the intestinal sub-mucosa. Hence, the effectiveness of the probiotics relies on their ability to adhere and colonize the human stomach, which in turn strengthens the host immune system (Chichlowski and Hale 2008). Many *Lactobacillus* bacteria maintain intestinal barrier integrity by regulating and increasing the expression of genes, such as E-cadherin/β-catenin, involved in tight junction signalling (Hummel et al. 2012). Several lactobacillus proteins have been found to improve surface and mucosal adhesion by integrating with glycoproteins secreted by intestinal epithelial cells, such as mucin, resulting in the competitive exclusion of mucosal pathogens (Wang et al. 2016). Attachment to the intestinal mucosa is required for colonization in host–probiotic interaction. *Lactobacillus* possesses a mannose-specific adhesion mechanism that allows it to adhere to human intestinal epithelial cells. When the probiotic adheres to the cell, it triggers a cascade of biological events, including the production of cytokines and chemokines. As a result, secondary functions in the host, such as activating mucosal and systemic immune responses, were activated (Plaza-Díaz et al. 2020). Probiotics increase gut mucin levels, which inhibit pathogen binding and induce epithelial cells to release defensins. These small proteins help to maintain the integrity of the intestinal barrier by exhibiting antibacterial, antifungal and antiviral properties (Gong et al. 2021).

4.3.3 Probiotics and the Immune System

Probiotic bacteria have long been known to have immunomodulating properties. These bacteria can communicate with epithelial cells, dendritic cells (DCS), monocytes/macrophages and lymphocytes. It can regulate the immune system by increasing endogenous host defense mechanisms, including humoral, cellular and non-specific immunity (Belkaid and Hand 2014). According to a recent study, probiotics improve the natural killer cell activity and alter non-specific host defense mechanisms in the elderly (Aziz and Bonavida 2016). Probiotics play important roles in mucus production, macrophage activation, stimulation of secretory IgA and neutrophil, suppression of inflammatory cytokines and stimulation of peripheral immunoglobulins (Cristofori et al. 2021). Probiotics have different immune-regulatory effects in healthy and diseased persons (Yan and Polk 2011). Probiotics enhance phagocytosis in healthy persons, while they reduce it in allergy patients (Huang et al. 2022).

4.4 Health Benefits of Probiotics

Probiotics promote gastrointestinal health and immunity; prevent urogenital infections, allergies, cancer and certain bowel disorders (Binnendijk and Rijkers 2013). Several well-characterized *Lactobacillus* and *Bifidobacterium* strains are now available for human use to prevent and manage gastrointestinal (GI) infections. Probiotics provide a number of health benefits, including:

1. Improved gut health through microbiota control and immune system stimulation and development.
2. Increasing nutrient bioavailability by synthesizing and improving their absorption.
3. Alleviating lactose intolerance symptoms and lowering the risk of certain diseases.
4. It restores the normal gut flora, improving gut barrier function.

4.4.1 Disease Prevention and Treatment

Probiotic LGG, *Bifidobacterium lactis* BB-12 and *Lactobacillus reuteri* SD2222 were used to prevent and treat rotavirus diarrhoea in children (Shornikova et al. 1997). In children, the usage of LGG resulted in a significant reduction in diarrhoea symptoms as well as a shorter hospital stay (Li et al. 2019). For the treatment of rotavirus diarrhoea, LGG in milk or capsule form was used as an adjuvant to oral rehydration therapy (Guandalini et al. 2000).

Necrotizing enterocolitis is a leading cause of morbidity and mortality in premature infants. In premature babies, a combination of two probiotic strains, *Lactobacillus acidophilus* and *Bifidobacterium infantis*, was reported to reduce the risk of necrotizing enterocolitis and death (Patel and Underwood 2018).

Probiotics play an effective role in the prevention and treatment of *Helicobacter pylori* infection and other inflammatory bowel diseases, such as Crohn's disease and irritable bowel syndrome (Verna and Lucak 2010). Intestinal microbiota plays a critical role in the pathogenesis of bowel inflammatory disease, and the use of a combination probiotic strain proves effective in alleviating the symptoms (Hold et al. 2014)

Probiotics reduce cholesterol levels by deconjugating the bile salt. The bile salt is deconjugated, which makes it less soluble and absorbable in the intestines and excreted in the stool. Cholesterol is then used to make new bile acids, reducing blood cholesterol levels in a homeostatic reaction. In vitro studies proved that *Lactobacillus gasseri*c reduce cholesterol in the laboratory media by adhering to its cellular surfaces (Kumar et al. 2012). Probiotics, both living and dead, work in a similar way to reduce cholesterol levels. Growing cells, on the other hand, eliminated more cholesterol than dead cells. During the developmental phase, some probiotics *Lactococcus* strain lower cholesterol levels in the blood through the absorption and integration of cholesterol into their cellular membranes. Cholesterol incorporation into the cellular membrane benefits the bacterial strain by improving the membrane strength, which increases cellular resistance to lysis (Ooi and Liong 2010).

The presence of *Neisseria gonorrhoeae* and *Chlamydia trachomatis* is indicated by the absence of *Lactobacilli* in the vagina (Wiesenfeld et al. 2003). Direct instillation of *Lactobacilli* resulted in a significant reduction of *Escherichia coli* growth, as well as a reduction in the severity of inflammation and risk of recurrent urinary tract infection (Asahara et al. 2001). Probiotic *Lactobacilli* taken orally have been shown to reduce the incidence of urinary tract infection, bacterial vaginosis and candidiasis (Reid et al. 2003).

4.5 Role of Probiotics in Healthy Individuals

Recent studies have shown that probiotics can modestly decrease the incidence and duration of common upper respiratory tract infections in children and adults (King et al. 2014). People who have a low amount of the intestinal enzyme i8-galactosidase, the disaccharide lactose can cause severe intestinal irritation, including bloating, gas and stomach pain. It restricts the use of calcium-rich dairy products at a time when the elderly are in desperate need of them owing to bone loss. Lactose digestion in lactose-intolerant people can be improved by probiotics, with the majority of data suggesting that lactose taken in yoghurt containing alive cultures is more easily digested than lactose consumed without live cultures (Oak and Jha 2019). *Lactobacilli* produce lactase, which hydrolyses lactose in dairy products to glucose and galactose during fermentation, potentially benefiting lactose-intolerant persons (Saqib et al. 2017).

The animals were given probiotics to help them gain weight on a liquid diet of milk, yoghurt or milk fermented with *S. thermophilus*. The animal gut microbiota was altered, resulting in a significant increase in weight gain (Goldin 1998).

The composition of the microbiota can influence the development of mucosal and systemic immunity. Probiotics colonize the oral cavity naturally, reducing the negative effects of bacterial infections, while also improving the inflammatory cytokine system. Plaque growth can be inhibited by neutralizing free electrons, affecting the systemic immune system and controlling mucosal permeability. *Lactobacillus bulgaricus, L. acidophilus, L. casei, L. Helveticus* and *L. lactis* are the most commonly used oral probiotics (Parul et al. 2020). Probiotics that stabilize the oral flora are used to treat gingivitis (Gupta 2011). Acidic probiotic bacteria like *Lactobacilli*, *Streptococcus* and *Bifidobacterium* produce antimicrobial substances that suppress pathogens by aggregation, toxic by-product release and competition for substrates (Markowiak and Śliżewska 2017). Probiotics contribute to the treatment of periodontitis by maintaining the oral microbiota. The majority of *Lactobacillus* strains inhibit the spread of dermabrasion periodontal diseases (Haukioja 2010). Recent studies have shown that the use of beneficial bacteria in addition to root resurfacing inhibits the re-colonization of periodontal pathogens, which reduces overall pocket depth and improve clinical adhesion (Retamal-Valdes et al. 2021).

Microbiota in the intestine produce enzymes such as glycosidase, nitroreductase, azoreductase and β-glucuronidase, which control the onset of carcinogenesis and protect against carcinogenic activity (Molska and Reguła 2019). *L. casei* strain shirota was found to decrease the recurrence rate of superficial bladder cancer, indicating that probiotic consumption had a positive influence on cancer risk (Mutoh et al. 2020).

The composition of a mother's vaginal microbiota affects a baby's asthmatic condition. The administration of LGG to pregnant women and newborns resulted in a significant reduction in the prevalence of early atopic illness in neonates (Sestito et al. 2020).

The administration of LGG increased the levels of rotavirus-specific IgM-secreting cells than the control and placebo groups (Maidens et al. 2013). Similarly, the LGG group had significantly increased the IgA and IgM levels in the control group. As a result, LGG is used as an adjuvant in a rotavirus vaccination for children.

Acute liver diseases sometimes lead to liver failure. Probiotics significantly reduce acute liver injury by up-regulating the tight junction protein gene, increasing microbial flora, lowering bacterial translocation, minimizing endotoxin infiltration and reducing pro-inflammatory cytokines (Xu et al. 2021).

4.6 Drawbacks

Probiotics are living microorganisms that can infect the host. The risk of probiotic-induced sepsis should be evaluated against the risk of pathogenic-bacterial-induced sepsis. Future studies should validate the benefits of probiotics such as the selection of a suitable probiotic agent, dosage standardization, a complete understanding of their therapeutic effects and long-term storage. Although more promising potential health effects of probiotics are being observed in ongoing research, probiotic strains should be examined for antibiotic susceptibility patterns, toxin production, metabolic and haemolytic activities, infectivity in immunocompromised animal models and adverse events in humans.

References

Asahara T, Nomoto K, Watanuki M, Yokokura T (2001) Antimicrobial activity of intraurethrally administered probiotic *Lactobacillus casei* in a murine model of Escherichia coli urinary tract infection. Antimicrob Agents Chemother 45(6):1751–1760

Aziz N, Bonavida B (2016) Activation of natural killer cells by probiotics. For Immunopathol Dis Therap 7(1–2):41–55

Belkaid Y, Hand TW (2014) Role of the microbiota in immunity and inflammation. Cell 157(1): 121–141

Bermudez-Brito M, Plaza-Díaz J, Muñoz-Quezada S, Gómez-Llorente C, Gil A (2012) Probiotic mechanisms of action. Ann Nutr Metab 61(2):160–174

Binnendijk KH, Rijkers GT (2013) What is a health benefit? An evaluation of EFSA opinions on health benefits with reference to probiotics. Benef Microbes 4(3):223–230

Chichlowski M, Hale LP (2008) Bacterial-mucosal interactions in inflammatory bowel disease: an alliance gone bad. Am J Physiol Gastrointest Liver Physiol 295(6):G1139–G1149

Conlon MA, Bird AR (2014) The impact of diet and lifestyle on gut microbiota and human health. Nutrients 7(1):17–44

Cristofori F, Dargenio VN, Dargenio C, Miniello VL, Barone M, Francavilla R (2021) Anti-inflammatory and immunomodulatory effects of probiotics in gut inflammation: a door to the body. Front Immunol 12:578386

den Hartigh LJ (2019) Conjugated linoleic acid effects on cancer, obesity, and atherosclerosis: a review of pre-clinical and human trials with current perspectives. Nutrients 11(2):370

FAO/WHO (2001) Evaluation of health and nutritional properties of powder milk and live lactic acid bacteria. Report from FAO/WHO Expert Consultation, pp 1–4

Fuller R (1989) Probiotics in man and animals. J Appl Bacteriol 66(5):365–378

Goldin BR (1998) Health benefits of probiotics. Br J Nutr 80(4):S203–S207

Gong T, Fu J, Shi L, Chen X, Zong X (2021) Antimicrobial peptides in gut health: a review. Front Nutr 8:751010

Guandalini S, Pensabene L, Zikri MA, Dias JA, Casali LG, Hoekstra H, Kolacek S, Massar K, Micetic-Turk D, Papadopoulou A, de Sousa JS, Sandhu B, Szajewska H, Weizman Z (2000) Lactobacillus GG administered in oral rehydration solution to children with acute diarrhea: a multicenter European trial. J Pediatr Gastroenterol Nutr 30(1):54–60

Gupta G (2011) Probiotics and periodontal health. J Med Life 4(4):387–394

Gupta V, Garg R (2009) Probiotics. Indian J Med Microbiol 27(3):202–209

Haukioja A (2010) Probiotics and oral health. Eur J Dent 4(3):348–355

Hold GL, Smith M, Grange C, Watt ER, El-Omar EM, Mukhopadhya I (2014) Role of the gut microbiota in inflammatory bowel disease pathogenesis: what have we learnt in the past 10 years? World J Gastroenterol 20(5):1192–1210

Holzapfel WH, Haberer P, Geisen R, Björkroth J, Schillinger U (2001) Taxonomy and important features of probiotic microorganisms in food and nutrition. Am J Clin Nutr 73(2 Suppl):365S–373S. https://doi.org/10.1093/ajcn/73.2.365s. PMID: 11157343

Homan M, Orel R (2015) Are probiotics useful in *Helicobacter pylori* eradication? World J Gastroenterol 21(37):10644–10653

Huang J, Zhang J, Wang X, Jin Z, Zhang P, Su H, Sun X (2022) Effect of probiotics on respiratory tract allergic disease and gut microbiota. Front Nutr 9:821900

Hummel S, Veltman K, Cichon C, Sonnenborn U, Schmidt MA (2012) Differential targeting of the E-cadherin/β-catenin complex by gram-positive probiotic lactobacilli improves epithelial barrier function. Appl Environ Microbiol 78(4):1140–1147

Joint FAO/WHO Working Group and Joint FAO/WHO Working Group (2002) Guidelines for the evaluation of probiotics in food. World Health Organization/Food and Agriculture Organization, London/Ottawa

King S, Glanville J, Sanders ME, Fitzgerald A, Varley D (2014) Effectiveness of probiotics on the duration of illness in healthy children and adults who develop common acute respiratory infectious conditions: a systematic review and meta-analysis. Br J Nutr 112(1):41–54

Kumar M, Nagpal R, Kumar R, Hemalatha R, Verma V, Kumar A, Chakraborty C, Singh B, Marotta F, Jain S, Yadav H (2012) Cholesterol-lowering probiotics as potential biotherapeutics for metabolic diseases. Exp Diabetes Res 2012:902917

Lee K, Paek K, Lee HY, Park JH, Lee Y (2007) Antiobesity effect of trans-10,cis-12-conjugated linoleic acid-producing Lactobacillus plantarum PL62 on diet-induced obese mice. J Appl Microbiol 103(4):1140–1146

Li YT, Xu H, Ye JZ, Wu WR, Shi D, Fang DQ, Liu Y, Li LJ (2019) Efficacy of *Lactobacillus rhamnosus* GG in treatment of acute pediatric diarrhea: a systematic review with meta-analysis. World J Gastroenterol 25(33):4999–5016

Magnusson J, Schnürer J (2001) *Lactobacillus coryniformis* subsp. *coryniformis* strain Si3 produces a broad-spectrum proteinaceous antifungal compound. Appl Environ Microbiol 67(1):1–5

Maidens C, Childs C, Przemska A, Dayel IB, Yaqoob P (2013) Modulation of vaccine response by concomitant probiotic administration. Br J Clin Pharmacol 75(3):663–670

Markowiak P, Śliżewska K (2017) Effects of probiotics, prebiotics, and synbiotics on human health. Nutrients 9(9):1021

Molska M, Reguła J (2019) Potential mechanisms of probiotics action in the prevention and treatment of colorectal cancer. Nutrients 11(10):2453

Mutoh M, Yoshimura K, Fujii G, Nakamura T, Takeshita T, Wakabayashi K, Sakai T, Ishikawa H (2020) Very long-term treatment with a *Lactobacillus* probiotic preparation, *Lactobacillus casei* strain shirota, suppresses weight loss in the elderly. Nutrients 12(6):1599

Oak SJ, Jha R (2019) The effects of probiotics in lactose intolerance: a systematic review. Crit Rev Food Sci Nutr 59(11):1675–1683

Ooi LG, Liong MT (2010) Cholesterol-lowering effects of probiotics and prebiotics: a review of in vivo and in vitro findings. Int J Mol Sci 11(6):2499–2522

Parul C, Renuka D, Anuradha S, Neeru B, Mahesh SD (2020) Critical appraisal of the effects of probiotics on oral health. J Funct Foods 70:103985

Patel RM, Underwood MA (2018) Probiotics and necrotizing enterocolitis. Semin Pediatr Surg 27(1):39–46

Plaza-Diaz J, Ruiz-Ojeda FJ, Gil-Campos M, Gil A (2019) Mechanisms of action of probiotics. Adv Nutr 10(Suppl 1):S49–S66

Plaza-Díaz J, Solís-Urra P, Rodríguez-Rodríguez F, Olivares-Arancibia J, Navarro-Oliveros M, Abadía-Molina F, Álvarez-Mercado AI (2020) The gut barrier, intestinal microbiota, and liver disease: molecular mechanisms and strategies to manage. Int J Mol Sci 21(21):8351

Reid G, Jass J, Sebulsky MT, McCormick JK (2003) Potential uses of probiotics in clinical practice. Clin Microbiol Rev 16(4):658–672

Retamal-Valdes B, Teughels W, Oliveira LM, da Silva RN, Fritoli A, Gomes P, Soares GMS, Temporão N, Furquim CP, Duarte PM, Doyle H, Faveri M, Figueiredo LC, Feres M (2021) Clinical, microbiological, and immunological effects of systemic probiotics in periodontal treatment: study protocol for a randomized controlled trial. Trials 22(1):283

Ridlon JM, Harris SC, Bhowmik S, Kang DJ, Hylemon PB (2016) Consequences of bile salt biotransformations by intestinal bacteria. Gut Microbes 7(1):22–39

Rosberg-Cody E, Stanton C, O'Mahony L, Wall R, Shanahan F, Quigley EM, Fitzgerald GF, Ross RP (2011) Recombinant lactobacilli expressing linoleic acid isomerase can modulate the fatty acid composition of host adipose tissue in mice. Microbiology (Reading) 157(Pt 2):609–615

Saqib S, Akram A, Halim SA, Tassaduq R (2017) Sources of β-galactosidase and its applications in food industry. 3 Biotech 7(1):79. https://doi.org/10.1007/s13205-017-0645-5

Sestito S, D'Auria E, Baldassarre ME, Salvatore S, Tallarico V, Stefanelli E, Tarsitano F, Concolino D, Pensabene L (2020) The role of prebiotics and probiotics in prevention of allergic diseases in infants. Front Pediatr 8:583946

Shornikova AV, Casas IA, Mykkänen H, Salo E, Vesikari T (1997) Bacteriotherapy with *Lactobacillus reuteri* in rotavirus gastroenteritis. Pediatr Infect Dis J 16(12):1103–1107

Verna EC, Lucak S (2010) Use of probiotics in gastrointestinal disorders: what to recommend? Ther Adv Gastroenterol 3(5):307–319

Wang G, Xia Y, Song X, Ai L (2016) Common non-classically secreted bacterial proteins with experimental evidence. Curr Microbiol 72(1):102–111

Wiesenfeld HC, Hillier SL, Krohn MA, Landers DV, Sweet RL (2003) Bacterial vaginosis is a strong predictor of *Neisseria gonorrhoeae* and *Chlamydia trachomatis* infection. Clin Infect Dis 36(5):663–668

Xu S, Zhao M, Wang Q, Xu Z, Pan B, Xue Y, Dai Z, Wang S, Xue Z, Wang F, Xu C (2021) Effectiveness of probiotics and prebiotics against acute liver injury: a meta-analysis. Front Med (Lausanne) 8:739337

Yan F, Polk DB (2011) Probiotics and immune health. Curr Opin Gastroenterol 27(6):496–501

Different Generations of Probiotics: An Effective Way to Restore Gut Homeostasis

5

Nayan Rishi, Souparno Paul, Ashwani Kumar, and Gunjan Goel

Abstract

The gastrointestinal tract of humans harbours an ecological niche of different species of microbes known as gut microbiota. This balanced composition of microbiota confers health benefits to the human host. Various complex interactions among the different genera of gut microbiome play important roles such as the prevention of infections by intestinal pathogens, producing various nutrients for the host, enhancing immunological responses and other beneficial effects. However, the modern lifestyles, dependence on junk foods, stress conditions, lack of physical activities and other individual factors result in perturbations in this ecological system resulting in gut dysbiosis. As the probiotic candidate originates from the gut, therefore it is imperative to use probiotics to maintain this gut homeostasis. Various probiotic cultures reported so far are members of a group of lactic acid bacteria. These were mainly species of *Lactobacilli* and *Bifidobacteria*. Nevertheless, to achieve the maximum benefits of probiotics, the cultures should possess mandatory activities to remain viable until they reach the colon after ingestion. Therefore, different strategies have been employed to design probiotics in terms of increased survival and targeted benefits. The postbiotics (secretions from probiotics) and identification of next-generation probiotics are the novel strategies underway to develop probiotics as personalised medicine. This chapter details the basic concepts of probiotic science, their importance, mechanisms of action and health benefits.

N. Rishi · S. Paul · G. Goel (✉)
Department of Microbiology, Central University of Haryana, Mahendergarh, India
e-mail: ggoel@cuh.ac.in

A. Kumar
Department of Nutrition Biology, Central University of Haryana, Mahendergarh, India

R. Sobti et al. (eds.), *Role of Microbes in Sustainable Development*,
https://doi.org/10.1007/978-981-99-3126-2_5

5.1 Introduction

The human gastrointestinal tract (GI) is regarded as a superorganism as it is a niche to thousands of species of microorganisms which live in homeostasis and confer health benefits to the human host. This microbiota plays multifarious roles such as the regulation of host metabolism, immunomodulation, protection of humans against pathogens and many more. Various studies have reported that this gut microbiota comprises around 150–400 bacterial phyla, dominated by Bacteroidetes, Firmicutes, Proteobacteria and Actinobacteria. Apart from these bacterial species, the gut microbiota also contains fungi, viruses and archea. Among these groups of microbiota, a particular group comprising species of LAB (lactic acid bacteria) are regarded as probiotics which are defined conventionally as live microorganisms that perform a variety of vital tasks portraying a positive influence on human health. The FAO/ WHO defines Probiotics as "Live microorganisms which, when administered in adequate amounts, confers health benefits" (FAO 2002). These probiotics exert their beneficial effects through different mechanisms such as the modulation of gut microbiota, enhancing the immune status of the host, exclusion of pathogenic microorganisms through secretion of antimicrobial components, maintenance of intestinal membrane permeability, production of short chain fatty acids (SCFAs), vitamin synthesis and deconjugation of bile salts and others. The major genera reported for the probiotic attributes include different species of *Lactobacillus, Bifidobacterium, Saccharomyces, Streptococcus, Enterococcus, Escherichia and Bacillus.* Prebiotics are the substrate*s* in food, primarily the oligosaccharides which can promote the growth of probiotic bacteria. International Scientific Association for Probiotics and Prebiotics (ISAPP) defines a prebiotic as a substrate that is specifically and selectively used by host microorganisms that confer health benefits. The most common prebiotic substrates include fructo-oligosaccharides (FOS), galacto-oligosaccharides (GOS), inulin, glucans, pectins and lactulose. In relation to the term probiotic and prebiotics, synbiotics is also used for the combination of both (Markowiak and Śliżewska 2017). The synbiotic products are developed to improve the survival of probiotics during passage through the upper intestinal tract with the help of prebiotics. However, very few studies have been undertaken on synbiotics and their health benefits. Further, in continuation with probiotic research, the ISAPP stated that besides the probiotics, the metabolic by-products of probiotics, dead microorganisms or other microbe-based nonviable products may also exert health benefits. Currently, the term "postbiotic (metabiotics/parabiotics)" refers to such substances which are released by probiotics exhibiting biological activity (Fig. 5.1). These postbiotics are produced by the inactivation of probiotics which preserves the beneficial effects provided by probiotics in the living form. Various inactivation methods such as sonication, enzymatic processes, solvent extraction and chemicals (e.g. formalin) at both laboratory and industrial levels have been used whereby heat treatment is most commonly used. Since a plethora of research has been conducted in the area of probiotics, prebiotics and postbiotics, various databases have been developed to identify the genomic characteristics of the probiotic strains. Few of them are http://kwanlab.bio.cuhk.edu.hk/PBQ/, AEProbio

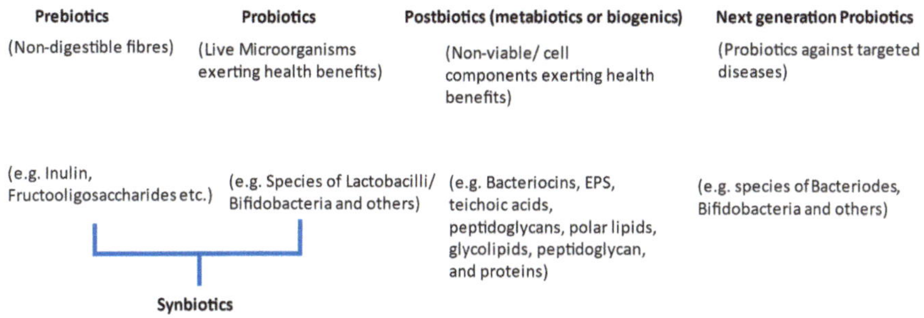

Fig. 5.1 Probiotics, prebiotics, synbiotics and postbiotics. (Adapted from Mohd Fuad et al. 2022)

(http://usprobioticguide.com/) and Optibac Probiotics (http://www.optibacprobiotics.com/uk/professionals/probiotics-database), https://probresist.com and http://probiogenomics.unipr.it/files/Probiotic_Bifidobacteria_DataBase.zip. In recent years, the researchers are emphasising on next-generation probiotics which are similar to postbiotics that offer targeted strategies for promoting health and managing a particular disease. These are also directly or indirectly involved in microbiota modulation, immune system activation, enzyme production or gut barrier interaction. This chapter provides a review of the recent updates on the probiotics, postbiotics and next-generation probiotics, guidelines to evaluate probiotic strains and their products, the mechanisms behind their health benefits and how far we have reached in probiotic research.

5.2 Selection Criteria for Probiotics

As per the guidelines provided by International Life Sciences Institute-India, any culture to be taken as probiotic it should fulfil the following criteria:

(a) Evaluation of exact taxonomy and probiotic activities of microorganisms.
(b) Evaluation of the safety of probiotics for humans.

As the probiotic effect of any microbial culture depends on the type of strain, therefore, the exact identification of the strain is the mandatory step before considering any strain as a probiotic. The strain should be of human origin which should be identified using phenotypic as well as newer genotypic tests. Recently, Mattarelli et al. (2014) described the "Recommended minimal standards for description of new taxa of the genera Bifidobacterium, Lactobacillus and related genera". However, whole genome sequencing is the "gold standard" method for correct strain identification. Once the strain has been identified, the strain should be assessed for the presence of virulence factors such as toxins, invasion and adhesion factors (EFSA 2020a). The strains should be tested for the presence of transferable antibiotic-resistant genes. The organism should fall under GRAS (Generally Recognised as

Safe) category. The uniqueness of the strain should be periodically verified before usage in foods to ensure that it has not mutated. For the probiotic attributes, the cultures should have the following properties tested through in vitro as well as in vivo tests. The following in vitro tests are suggested to ascertain their probiotic attributes:

- Resistance to gastric juice (low pH) or bile salts and pancreatic enzymes.
- Adherence to mucus and/or human epithelial cells and cell lines.
- Antimicrobial activity against potentially pathogenic bacteria.
- Assessment of certain metabolic activities (e.g. haemolytic activity, D-lactate, biogenic amines, bacteriocin production, hydrogen peroxide, etc.).
- Bile salt hydrolase activity.
  The above-said activities are only suggestive for a culture to be a probiotic, as these are not validated biomarkers for the functions of the probiotic strain. Therefore, the strain should be further tested using animal models (mice, rats and Guinea pigs).

Human clinical trials are carried out to confirm the specific health claims of a potential probiotic microbial strain. The safety of probiotic usage should be the top priority for general claims, especially if it is intended for extended usage. Usually, randomised control trials are conducted to support the health claims of any culture. As most of the LAB strains such as bifidobacteria and yeasts have a long history of use in food fermentation, these strains have been determined to be secure for usage in supplements and foods. Apart from these, the European Food Safety Authority (EFSA) has published a list of microorganisms presumed to be safe under the "Qualified Presumption of Safety" (QPS) concept since 2007 (EFSA 2007). The list has been prepared based on history and scientific literature reviews on the submissions of traditionally used microorganisms to EFSA for market approval (EFSA 2020b).

Apart from these safety attributes, if the potential probiotic culture is to be used in the food sector, then the culture should be highly productive, genetically stable, should survive the drying procedures used in food processing units, resistant to phages and should not change the organoleptic qualities of the finished product. Various bacterial cultures used as probiotics include species of Lactobacilli ("*L. acidophilus, L. rhamnosus L. gasseri, L. casei L. reuteri, L. plantarum, L. salivarius L. johnsonii, L. gallinarum, L. plantarum, L. fermentum L. helveticus, L. brevis L. murinus, L. crispatus, L. amylovorus*), Bifidobacteria (*B. infantis, B. longum B. lactis, B. adolascentis B. bifidum, B. animalis B. breve, B. thermophilum B. pseudolongum*), yeasts (*S. boulardii, S. lactis, S. carlsbergensis, Kluyveromyces marxianu, S. cerevisiae*)" and other bacterial species such as *Bacillus subtilis, B. lichemiformis*, few species of Enterococci, Leuconostocs, Pediococci and Streptococci.

5.3 Mechanism of Action of Probiotics and Health Benefits

The probiotics are well known for their activity as a modulator of intestinal microbial balance whereby these cultures maintain the number as well as the diversity of different gut microbes thereby maintaining gut homeostasis. Any disturbance in this microbial diversity and population results in gut dysbiosis. Under these conditions, probiotics are often recommended to maintain the ecological niche of the intestine. A tailored probiotic intervention may be utilised to treat or prevent intestinal infections with a full understanding of the regulation and mechanism of action of probiotics on the micro-ecology of intestinal infections. These cultures through this equilibrium are helpful in restoring the natural microbiota after antibiotic therapy. These probiotics are reported to exert a direct or indirect impact on the health of the human host. The direct effect includes antimicrobial activity through the competitive exclusion of pathogens, binding to the mucus membrane, production of antimicrobial substances, production of SCFAs and maintenance of membrane integrity (Li et al. 2021). The indirect effect includes immunomodulatory activities (Fig. 5.2).

5.3.1 Antimicrobial Activities

The antimicrobial activities of the probiotic are claimed due to different mechanisms such as release of antimicrobial compounds, competition with pathogens for adherence sites or nutrients, enhancement of immunological status of host and inhibition of production of toxins by pathogens, interfering with quorum sensing pathways of pathogens. The LABs are usually reported to produce organic acids, bacteriocins, hydrogen peroxides, biogenic amines, biosurfactants, extrapolymeric substances (EPS) and other metabolites exhibiting antimicrobial activities. Among organic acids, probiotics usually produce lactic acid as an antimicrobial agent. The organic

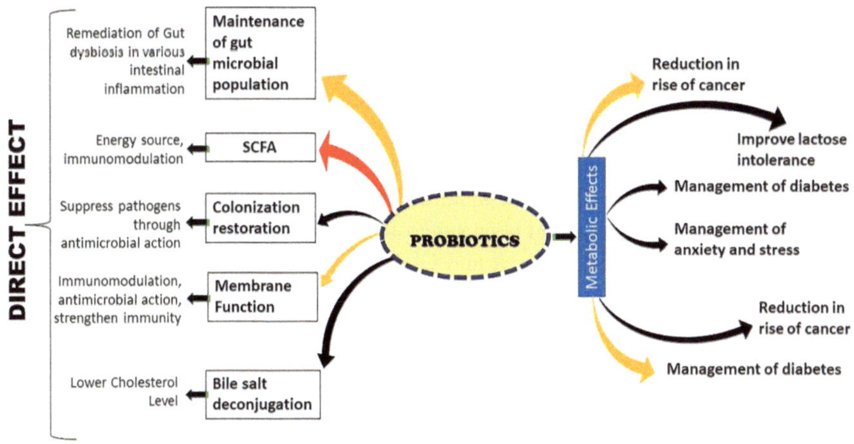

Fig. 5.2 Mechanisms of action of probiotics and their health benefits

acid, in an un-dissociated form, enters into the bacterial cells and gets dissociated in their cytoplasm. This results in lowering the intracellular pH and can lead to the death of pathogens. The metabolites such as diacetyl, fatty acids, reuterin (3-hydroxypropionaldehyde in various forms), methylhydantoin and mevalonolactone are often produced by probiotics in minor quantities (Mayorgas et al. 2021). The cultures also produce peptides such as defensins and bacteriocins which exert antimicrobial actions against pathogens. The SCFA produced as a result of fermentation can disrupt the outer membranes of Gram-negative pathogens. Bacteriocins are able to induce pores in the inner membrane of Gram-negative pathogens thereby inhibiting their survival in the host. Although there are several antimicrobial substances released by probiotics, however, majority of these are characterised partially and not brought to the commercial level which necessitates that there should be improvement in new tools and approaches to decipher their interactions with the host as well as the pathogen. Apart from the release of antimicrobial components, the probiotics also exclude the pathogens via competitive exclusion, whereby one species of bacteria, more strongly fights for a receptor site than another species in the intestinal tract. The mechanisms that are applied for exclusion or reduction of growth by bacteria can be the removal of available bacterial receptor sites, establishment of hostile micro-ecology, development and secretion of antimicrobials, and selective metabolites and competitive loss of vital nutrients. The competition for adhesion sites on epithelial cells, over mucus layer and nutrients, mediates the interaction between enteric bacteria to execute competitive exclusion. However, this competition is reported to be strain specific. Interfering with quorum sensing signalling pathways in pathogenic bacteria is another mechanism by which probiotics exert antimicrobial activities. The use of this cell-to-cell signalling mechanism by pathogens facilitates the regulation of important virulent traits and allows them to successfully colonise the host cells. The reports on the inhibition of quorum sensing pathways in pathogens by probiotics or their components are limited.

5.3.2 Maintaining Barrier Function

Any disease's progression is significantly influenced by the way the gut barrier works. Any disruptions to this epithelial barrier can have disastrous effects and may give rise to a wide range of disease syndromes. The up-regulation of "close junction signalling" gene expression had been suggested as a potential method for improving "intestinal barrier integrity" (Serek and Oleksy-Wawrzyniak 2021). The probiotics reduce these disturbances through the upregulation of TJ proteins and occludin proteins. The intestinal barrier's tiny scaffolding is made of TJ proteins. This activity of probiotics is reported through the ligand-mediated stimulation of several toll-like receptors such as TLR-2. Various studies have reported the role of surface molecules of probiotics including surface layer proteins, capsular polysaccharides, flagella and fimbriae in maintaining intestinal homeostasis and promoting gut health. Probiotics' proteins, metabolites including "indole, extracellular vesicles, short-chain fatty acids and bacteriocins" interact with certain epithelial receptors or directly encourage goblet cells to release mucus, or enhance the expression of tight junctions. In addition, the probiotics control the death of intestinal epithelial cells (IECs) and promote IEC proliferation to repair the mechanical integrity of the intestine.

5.3.3 Production of SCFA

The SCFAs are produced in millimolar quantities in the gut and are mainly the fermentation products of anaerobic microorganisms. These volatile SCFAs are 1–6 aliphatic carbon atoms which may exist as straight or branched. The probiotic cultures usually ferment the undigested part of the diets which includes fibres, resistant starch or non-starch polysaccharides (Cheng et al. 2022). These substrates are fermented to produce SCFA, viz. acetate, propionate and butyrate usually in the ratio of 3:1:1 constituting a total of 90–95% of SCFAs present in the human colon. The level and rate of production of these SCFA are strain specific and dependent on the type of substrate. These SCFAs are reported to activate G-coupled protein receptors (GPCRs) which are the binding partner of GPR41 and GPR43 including the inhibition of histone deacetylases with the blocking of free fatty-acid receptors 2 and 3 (FFAR2/3), and reportedly serve as energy substrates and can supply up to 10% of the host's daily caloric needs. All these confer beneficial effects on the physiological processes of the host. Apart from these, by reducing pH, SCFA also aids in the bioavailability of metals and acts as a barrier to prevent the colonisation of harmful microbes. Recently, SCFAs are also reported to affect neuronal excitability. The SCFAs also affect the integrity and function of the intestines as well as possess anti-inflammatory activity.

5.3.4 Modulation of the Immune System

The effect of probiotics on strengthening immunity has long been known. In both innate and acquired immunity, various types of cells such as epithelial and dendritic cells (DCs), monocytes, macrophages and lymphocytes help in the execution of immunomodulatory effects of the probiotics (Yeşilyurt et al. 2021). The presence of extracellular and intracellular receptors on the cells has been shown to transmit signals when bacteria are encountered. The IECs and DCs are reported to interact with probiotics through PPRs. Probiotics may lessen intestinal inflammation by reducing the expression of TLRs, blocking NF-$_{\text{K}}$B signalling and secreting substances that may prevent TNF from entering "peripheral blood cells" (Kaur and Ali 2022).

Apart from these major mechanisms, the probiotics also induce hydrolysis of toxins and receptors, exert antiapoptotic effects on enterocytes through the activation of brush border enzymes and the stimulation of glucose absorption. Additionally, certain intracellular pathways involved in virus replication are inhibited by the probiotics (such as mitogen-activated protein kinase (MAPK) and others).

5.4 Postbiotics

Although probiotics are well documented for their health benefits, however their use is limited as still we have a limited understanding of the molecular mechanisms of action of probiotics, their strain-specific effects, risks associated with gene transfer, irregularities in the maintenance of viability and stability in the probiotic product and some other issues. Moreover, the production of key metabolic components in lower concentration does not result in any benefits, therefore, postbiotic components represent a promising new class of dietary supplements for human health and well-being. The well-documented postbiotics include exopolysaccharide; EPS, teichoic acids on the bacterial membrane, peptidoglycans (cell wall component), polar lipids, glycolipids and various other proteins components (Moradi et al. 2021). Inactivating probiotic microorganisms results in postbiotic production, however one should ensure that the inactivation process used to generate postbiotics and parabiotics must be able to maintain the positive effects offered by the living form. As such, it has been suggested that it can be created or acquired via a variety of inactivation techniques, including sonication, enzymatic procedures, solvent extraction and chemicals. Paraprobiotics (ghost probiotics) is the related term to postbiotics which refer to the non-viable or inactivated probiotic strains, either in intact or ruptured forms. Various benefits of postbiotics over the probiotic are as follows:

- No risk associated with the transfer acquisition of antibiotic resistance genes.
- As all these derived components are non-viable components, therefore, these are less sensitive to environmental conditions.
- Longer shelf-life and enabling easy storage and transportation at ambient temperatures.
- No issues related to survival or maintenance of viability as with the live cultures and well-defined chemical composition.
- Better and direct interaction of the postbiotic components with the target cell/pathogen.

As far as the mechanisms of action of postbiotics are concerned, these components act mainly through four mechanisms enhancing intestinal tight connections, triggering the formation of mucus, enhancing barrier function or stimulating mucous production, maintaining gut microbial population and immunomodulatory activities. The recent studies clearly indicate the benefits of postbiotics in gastrointestinal diseases, metabolic and neurological disorders. However, the type of probiotic strain, culture conditions and method of inactivation have a major effect on the structure and function of postbiotics. Various commercial postbiotic preparations include Del-Immune V®, LLC Hylak® forte, CytoFlora®, Aktoflor C and Zakofalk (Moradi et al. 2021).

5.5 Next-Generation Probiotics

Thanks to advancements in culture methods, cheaper genome and metagenome sequencing, and powerful tools to alter and modify bacterial genomes, the next generation of probiotics has arisen. This new phase in probiotic research enables us to create customised probiotics that cater to particular consumer needs and problems. Although many of these are still in the very early stages of mechanistic investigation, the understanding of the variety of organisms with potential health advantages has greatly expanded thanks to the makeup and function of the human gut microbiome, which was also hastened by "massively parallel sequencing". The traditional or conventional probiotic cultures exert generalised health benefits due to their metabolic activity which is not specific against certain diseases. These microorganisms are also known as live biotherapeutic products (LBPs) and next-generation probiotics (NGPs) (Kumari et al. 2021). These NGPs are characterised as biological products that do not qualify as vaccines and contain living organisms like bacteria that can be used for the treatment or cure of human diseases. NGPs can be useful because the word emphasises how regulators are likely to perceive them differently from regular probiotics and acknowledges the potential that NGPs will also contain genetically engineered bacteria. A process for finding an NGP includes an evaluation of safety, growth dynamics, culture genetics and a thorough grasp of their targeted diseases in order to characterise an NGP. Modern platforms for bioinformatics and NGS (next generation sequencing) technology are mostly used to screen NGPs. Once a promising culture has been found, functional validation tests are carried out using in vitro, in vivo and human clinical trials. The likely path to market for LBPs and NGPs will therefore be marked by preclinical studies of mechanisms of "action, safety, pharmacokinetics, pharmacodynamics, and phase 1–3 trials", along with passing adequately timed regulatory licensing barriers. This is similar to the path taken by probiotics. The distinction and necessity of the NGP and LPB terminologies should also be taken into account. Operational differences, albeit not exclusively, make up the disparities: unlike NGPs, which are typically investigated by laboratories with experience in probiotic and gut bacteria investigation and often have a progression based on the probiotic observation in the laboratory, LBPs are generally evaluated by start-up biotech companies or the pharmaceutical industry with the expressed intention of asking permission for pharmaceutical marketing.

Taking GRAS organisms or commensals and using them as a delivery system for a bioactive chemical is an alternative method for creating some NGPs. The bacterial vehicle used in this strategy is understood not to create any virulence factors, and the host will tolerate it, and, if selected appropriately, may even fail to colonise the host. The amount of evidence for the majority of NGS is still in the early stages, and further research is required to determine the precise health advantages of these bacterial species. Table 5.1 represents some of the key NGPs against their targeted diseases. Among the tested strains of NGP, *Faecalibacterium prausnitzii*, *Akkermansia muciniphila*, *Bacteroides xylanisolvens* and *Eubacterium hallii* have been reported for their benefits in the amelioration of dysbiosis-associated diseases. However, these strains are more sensitive than conventional probiotics; therefore, to

Table 5.1 The next generation probiotics and their target diseases[a]

Organism	Source	Targeted disease
Bacteroides xylanisolvens DSM23694	Human	Cancer
Bifidobacterium longum	Human	Anti-tumour effect in mice model
Bacteroides ovatus D-6	Human	Cancer
Bacteroides *ovatus* V975	GMO (originally from human gut samples) expressing KGF-2	Intestinal inflammation
Bacteroides dorei D8	Human	Cardiovascular disease
Bacteroides fragilis ZY-312	Human	Antimicrobial activities against intestinal pathogens
Bacteroides acidifaciens JCM 10556(T)	Mouse	Antimicrobial activities against intestinal pathogens
Clostridium butyricum MIYAIRI 588	Human	Multiple targets including cancer, inflammation and infectious agents
Faecalibacterium prausnitzii	Human	Mainly inflammatory bowel disease
Lactococcuslactis elafin	GMO (host isolated from food)	Mainly inflammatory diseases such as IBD
Lactococcus lactis trefoil factor 1 or IL-10	GMO (host isolated from food)	Allergen sensitivity and autoimmune diseases—type 1 diabetes
Akkermansia muciniphila	Human	Anti-obesity
Bacteroides ovatus V975	GMO expressing TGF-β1	Intestinal inflammation

[a] The list is non-exhaustive

develop them for a personalised approach, these need to be stable and active in humans (Singh and Natraj 2021).

5.6 Conclusions and Future Prospects

It is clear from the cited literature that the gut microbiota contributes to many aspects of human health and maintenance of this gut homeostasis is essential for effective metabolic, physiological and each activity of the human host. Any disbalance in this homeostasis leads to several complications such as gastrointestinal inflammation, metabolic disorders such as diabetes, obesity and anxiety. Although probiotics have a long history of use in foods, most of these effects have been demonstrated in experimental studies and only a few in clinical studies. In all the studies, the probiotics or their metabolites have been reported to show positive/beneficiary effects towards a large number of gut-based diseases like inflammatory bowel disease, allergies, liver diseases, cardiovascular diseases, obesity and cholesterol and hence improve host's health. Further studies are warranted to increase their use in the field of food and pharmaceutical industries. The increasing trends in the application of probiotics for human application are linked to their natural or GRAS origin which fulfils the desire of customers to take probiotics as an alternate to medicines. Novel research in probiotics for identifying postbiotics or next-generation probiotics is emerging which emphasises the concrete mechanisms of probiotics against chronic diseases. This may also further be investigated to develop probiotics as personalised medicine using specific strains for specific conditions. The beneficial effects should be investigated with the use of a single bacterial strain or consortia. The studies should be focussed on identifying the benefits of using preclinical models and/or clinical trials. With an advancement in tools and software in Bioinformatics, using CRISPR/Cas9 there is a scope of editing the genome more precisely to binate gene expression or provide novel features to enhance host colonisation and promote human health.

References

Cheng Y, Liu J, Ling Z (2022) Short-chain fatty acids-producing probiotics: a novel source of psychobiotics. Crit Rev Food Sci Nutr 62(28):7929–7959

EFSA (2007) Introduction of a Qualified Presumption of Safety (QPS) approach for assessment of selected microorganisms referred to EFSA—Opinion of the Scientific Committee. EFSA J 5: 587

EFSA (2020a) EFSA statement on the requirements for whole genome sequence analysis of microorganisms intentionally used in the food chain. EFSA, Parma

EFSA (2020b) Scientific opinion on the update of the list of QPS recommended biological agents intentionally added to food or feed as notified to EFSA (2017–2019). EFSA J 18:5965

Food and Agricultural Organization of the United Nations and World Health Organization (2002) Joint FAO/WHO working group report on drafting guidelines for the evaluation of probiotics in food. Food and Agricultural Organization of the United Nations, Rome

Kaur H, Ali SA (2022) Probiotics and gut microbiota: mechanistic insights into gut immune homeostasis through TLR pathway regulation. Food Funct 13:7423

Kumari M, Singh P, Nataraj BH, Kokkiligadda A, Naithani H, Ali SA, Behare PV, Nagpal R (2021) Fostering next-generation probiotics in human gut by targeted dietary modulation: an emerging perspective. Food Res Int 150:110716

Li HY, Zhou DD, Gan RY, Huang SY, Zhao CN, Shang A, Xu XY, Li HB (2021) Effects and mechanisms of probiotics, prebiotics, synbiotics, and postbiotics on metabolic diseases targeting gut microbiota: a narrative review. Nutrients 13(9):3211

Markowiak P, Śliżewska K (2017) Effects of probiotics, prebiotics, and synbiotics on human health. Nutrients 9(9):1021

Mattarelli P, Holzapfel W, Franz CM, Endo A, Felis GE, Hammes W, Pot B, Dicks L, Dellaglio F (2014) Recommended minimal standards for description of new taxa of the genera Bifidobacterium, Lactobacillus and related genera. Int J Syst Evol Microbiol 64(Pt 4): 1434–1451

Mayorgas A, Dotti I, Salas A (2021) Microbial metabolites, postbiotics, and intestinal epithelial function. Mol Nutr Food Res 65(5):2000188

Mohd Fuad AS, Amran NA, Nasruddin NS, Burhanudin NA, Dashper S, Arzmi MH (2022) The mechanisms of probiotics, prebiotics, synbiotics, and postbiotics in oral cancer management. In: Probiotics and antimicrobial proteins. Springer, Berlin, pp 1–14

Moradi M, Molaei R, Guimarães JT (2021) A review on preparation and chemical analysis of postbiotics from lactic acid bacteria. Enzym Microb Technol 143:109722

Serek P, Oleksy-Wawrzyniak M (2021) The effect of bacterial infections, probiotics and zonulin on intestinal barrier integrity. Int J Mol Sci 22(21):11359

Singh TP, Natraj BH (2021) Next-generation probiotics: a promising approach towards designing personalized medicine. Crit Rev Microbiol 47(4):479–498

Yeşilyurt N, Yılmaz B, Ağagündüz D, Capasso R (2021) Involvement of probiotics and postbiotics in the immune system modulation. Biologics 1(2):89–110

Application of Potential Microbial Biotechnology for Sustainable Human Health

6

Neha Rani Bhagat, Younis Ahmed, Rajesh Kumar, and Arup Giri

Abstract

In the present era, the most important concern is food security for providing stability and sustainability to the growing population throughout the world. Therefore, there is an urgent need for the development of sustainable methods for the augmentation of agricultural products. This chapter discusses some microbiological aspects of sustainable agriculture that can help us to maintain human civilization in changing climatic conditions. In this aspect, an emphasis has been laid upon significance of microbiotechnology in agriculture and human sustainability along with microbial bio-fertilizers, growth-promoting rhizobacteria (PGPR) in plants, and microbial biopesticides for sustainable agriculture. It has been found that a lot of bacteria may act as PGPR for higher agricultural yields. Microplastics are a great hindrance to reducing agricultural products. Therefore, the role of microbiotechnology in the detection of pollution has been discussed. However, waste management strategies, especially for plastic waste, industrial wastewater treatment, attenuation of textile industry based dyes, and microbial strategies for the biotransformation of food waste into useful resources, have been elaborated. In the agro-animal sector, microbes play a great role in maintaining normal animal physiology for optimum production,

N. R. Bhagat
DRDO-Defence Institute of High Altitude Research (DIHAR), Ladakh, India

Y. Ahmed
Department of Life Sciences and Allied Health Sciences, Sant Baba Bhag Singh University, Jalandhar, Punjab, India

R. Kumar
Department of Biosciences, Himachal Pradesh University, Shimla, Himachal Pradesh, India

A. Giri (✉)
Department of Zoology, Baba Mastnath University, Rohtak, Haryana, India

R. Sobti et al. (eds.), *Role of Microbes in Sustainable Development*,
https://doi.org/10.1007/978-981-99-3126-2_6

like the flourishing of the aquaculture sector, the livestock industry, and especially the poultry industry. Microbiotechnology has a great impact on the maintenance of normal human physiology after improvement in biomedical sector activity. This technology has the potential for the generation of electricity and may act as the main source of omega-3 fatty acids as well as the required protein for future sustainability.

6.1 Introduction

Global agricultural production is growing day by day for the feeding of growing population (Köhler and Triebskorn 2013). There are a lot of methods to increase the plant and animal based product yield. The higher yield will provide the sustainability of human race. Meanwhile, among all those methods, microbial biotechnology is flourishing day by day for better agricultural management and yield.

Meanwhile, biotic and abiotic factors causes the stressful environment to plants (Tewari and Mishra 2018; Farooq et al. 2009; Gowtham et al. 2020; Tevini 2004; Sharma et al. 2017; Bharti and Barnawal 2019; Awasthi et al. 2015; Khan et al. 2017; Hameed et al. 2016; Ghori et al. 2019; Dubois et al. 2018). Therefore, agricultural yield reduced throughout the world (Vurukonda et al. 2016). It has been reported that due to abiotic stress, the yield has been reduced to 50–82% (Wang et al. 2003) (Fig. 6.1). It has been found that all the accumulated stress

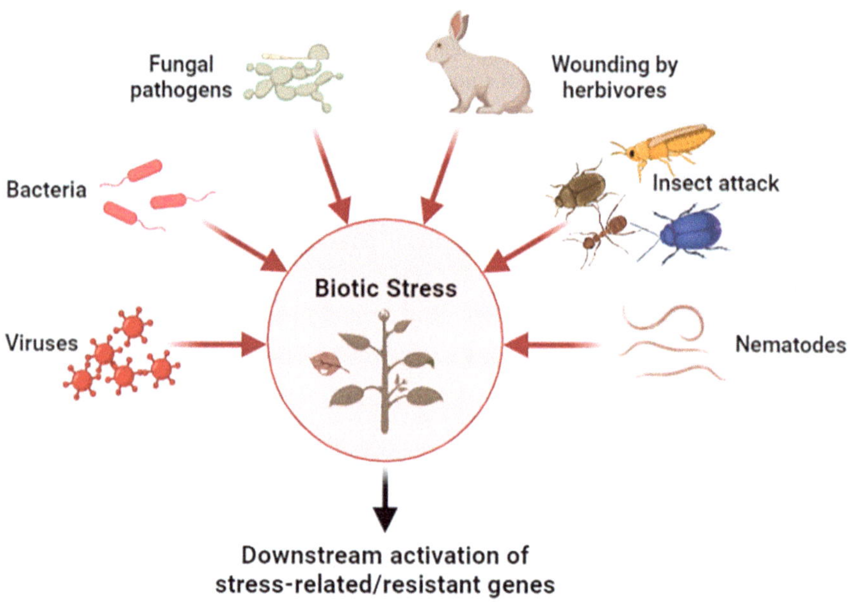

Fig. 6.1 Different types of stress that affects plant growth and development

caused reduction in plant growth and development after induction of reactive oxygen species formation in plants (Huang et al. 2019; Sharma et al. 2012).

Therefore, there is need of antidote development against different stress to plants. In these terms, growth-promoting rhizobacteria in plants has great role to improve the soil fertility and to maintain the soil environment so that optimum plant growth, development, and yield occur (Yang et al. 2009; Glick 2012; Vejan et al. 2016; Vurukonda et al. 2016; Gowtham et al. 2020).

This discussion covers various aspects of microbiotechnology in agriculture, including the use of microbial bio-fertilizers, plant growth-promoting rhizobacteria (PGPR), microbial biopesticides, and microbiotechnology's role in developing pollution detection and mitigation methods for agricultural and human sustainability.

6.2 Microbiotechnology in Agriculture and Human Sustainability

6.2.1 Plant Growth-Promoting Rhizobacteria for Sustainable Agriculture

6.2.1.1 Importance of PGPR

"Plant growth-promoting rhizobacteria (PGPR)" are population of rhizospheric bacteria which promote plant growth and development after rhizosphere engineering, nitrogen fixation, siderophore production, etc. (Bhattacharyya and Jha 2012) (Fig. 6.2).

Bacillus, Rhizobium, Azotobacter, Azospirillum, Frankia, Gluconacetobacter, Burkholderia, Azorhizobium, Beijerinckia, Cyanobacteria increased the nitrogen fixation and soil fertility (Zahran 2001; Govindasamy et al. 2010; Bhattacharyya and Jha 2012; Merzaeva and Shirokikh 2006; Jang et al. 2017; Islam et al. 2016; Ahmad et al. 2011). "*Arthrobacter, Burkholderia, Enterobacter, Microbacterium Pseudomonas, Bacillus, Erwinia, Rhizobium, Mesorhizobium, Flavobacterium, Rhodococcus, Serratia*" increase the phosphate solubilization in soil (Podile and Kishore 2007; Oteino et al. 2015). Siderophore production in soil is caused by *Pseudomonas, Bacillus, Rhizobium, Azotobacter, Enterobacter, Serratia* (Ansari et al. 2017). Different types of phytohormones are produced by *Bradyrhizobium*, *Burkholderia*, *Xanthomonas*, *Mesorhizobium*, *Bacillus*, *Rhizobium*, *Pantoea*, *Enterobacter*, *Agrobacterium*, *Azospirillum, and Arthrobacter Pseudomonas* (Egamberdieva et al. 2017; Tsukanova et al. 2017; Kalam et al. 2020; Cassán et al. 2001; Tahir et al. 2017; Barnawal et al. 2017). *Pseudomonas species, Bacillus species, Burkholderia, Brevibacterium, Streptomyces* help in antibiotic production (Jayaprakashvel and Mathivanan 2011; Zhou et al. 2019). Volatile metabolites which help in plant development may be produced by *Bacillus, Pseudomonas, Agrobacterium, Burkholderia, Xanthomonas,* and *Paenibacillus polymyxa* (Sharifi et al. 2018). However, *Bacillus*, and *Pseudomonas* species are responsible for lytic enzyme production (Mabood et al. 2014). Incorporation of *Pseudomonas*, *Bacillus*, *Pantoea*, *Burkholderia*, *Rhizobium* reduced stress tolerance (Jha and Subramanian

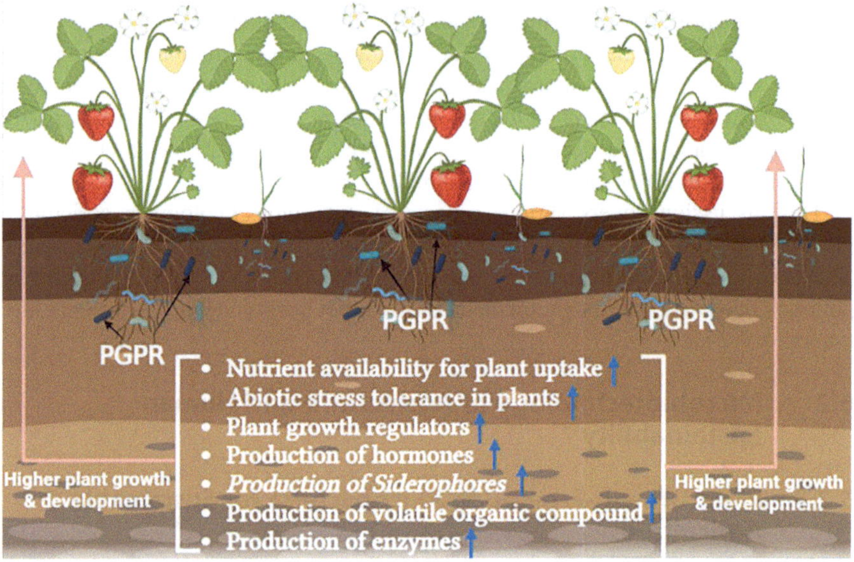

Fig. 6.2 Role of plant growth-promoting rhizobacteria (PGPR) for plant growth and development

2014; Egamberdieva et al. 2019). However, *Pseudomonas*, *Bacillus*, *Trichoderma* act as biocontrol agents (Saraf et al. 2014; Meena and Swapnil 2019). Drought stress in plants has been reduced by *Bacillus subtilis, Pseudomonas fluorescens DR11, Enterobacter hormaechei DR16, Rhizobium tropici, Phyllobacterium brassicacearum, Pseudomonas migulae DR35, Achromobacter piechaudii ARV8, Paenibacillus polymyxa,* and *Azospirillum brasilense* (Niu et al. 2018; de Lima et al. 2019; Bresson et al. 2013; Figueiredo et al. 2008; Yang et al. 2009; Ilyas et al. 2020). Salinity related stress to plants gets reduced by *Bacillus pumilus, Exiguobacterium oxidotolerans, Bacillus megaterium, Azospirillum* sp., *Achromobacter piechaudii, Enterobacter* sp. *PR14* (Mayak et al. 2004; Bharti et al. 2013; Marulanda et al. 2010; Fasciglione et al. 2015; Sagar et al. 2020). Different types of biotic stress to plants are reduced by *Paenibacillus xylanexedens, Bacillus amyloliquefaciens, Streptomyces* sp., *Ochrobactrum intermedium, Paenibacillus lentimorbus, Pseudomonas* spp. (Verma et al. 2016; Srivastava et al. 2016; de Vasconcellos and Cardoso 2009; Gowtham et al. 2016; Khan et al. 2012; Reshma et al. 2018). Meanwhile, nutrient absorption to plants increases by *Pantoea* sp. *S32, Paenibacillus polymyxa* (Chen and Liu 2019; Pii et al. 2015; Castillo-Aguilar et al. 2017). It has been found that seed germination also increases due to the presence of *Pseudomonas putida, Serratia marcescens, Bacillus subtilis, Pseudomonas fluorescens, Azospirillum lipoferum, Providencia* sp., *Brevundimonas diminuta* (Almaghrabi et al. 2014; Nezarat and Gholami 2009; Rana et al. 2011). Bioremediation in terms of lowering the heavy metals and other pollutants level is done by "*Ochrobactrum* sp., *Bacillus* spp., *Pseudomonas* spp., *Pseudomonas fluorescens, Bacillus cereus, Alcaligenes faecalis*

RZS2, Pseudomonas aeruginosa RZS3, Enterobacter sp. *RZS5"* (Pandey et al. 2013; Khan and Bano 2016; Das and Kumar 2016; Kalam et al. 2020; Patel et al. 2016; Sayyed et al. 2015).

6.2.1.2 Mechanism of Action

PGPR helps in different dimensions as stated above so that the agricultural environment is changed to most suitable condition. They are doing all these activities in two diverse ways: direct and indirect effects. In direct effects, plant growth augmentation happens due to higher nutrient acquisition to plants, which is helping plants to resist against different pathogens. This pathway mainly depends upon the bio-fertilization and phyto-stimulation. Conversely, in indirect effects, PGPR helps to reduce the completion effects from nearby plants and increase the antibiosis against adverse pathogens. Therefore, ultimately, PGPR mainly increased the pathogen resistance to plants (Mustafa et al. 2019).

6.2.1.3 Limitations of PGPR

Apart from all the usefulness of PGPR, some limitations also existed. Sometimes, PGRP is not suitable for a specific plant variety. In different climatic conditions and soil ecosystems, it may not act at optimum level. However, the algorithm on the plant–PGPR relation may reduce the limitation of PGPR use in different geographical conditions in different doses.

6.2.2 Microbial Biopesticides

Pesticides are used for higher production by reducing the pest attack—causes the reduction of 45% crop yield per annum (Mundt 2014; Yadav et al. 2015). Synthetic pesticides have great concern on soil health after long term use, higher ecotoxicity, reduced fungal activity in soil, reduction of rhizospheric bacteria, etc. (Bowles et al. 2014; Fenner et al. 2013; Kumar et al. 2017). Therefore, in the recent decades, "biopesticides are being used for pest management."

Bacillus thuringiensis (Bt) is used as biopesticide due to the presence of δ-endotoxins against insect. These δ-endotoxins secreted after the activation of cry genes (Kumar et al. 2021). Carbendazim is utilized to combat various plant diseases due to its properties as a benzimidazole fungicide (Yang et al. 2014; Zhang et al. 2009). There are a lot of bacterial strains which have significant impact as biopesticide (Table 6.1).

Some studies indicated that biopesticide may impact soil health. But it has been recorded that low dose of biopesticide reduces those bad impact on soil health (Czaja et al. 2015; Leahy et al. 2014).

Table 6.1 Different microbial biopesticides and their role (Kumar et al. 2019; Brownbridge and Buitenhuis 2019)

Types of microbes	Name of microbes	Activity against different pests
Bacteria	*Bacillus thuringiensis* (Bt) kurstaki and aizawai	Reduce caterpillar pests like tortricid leafrollers, *European corn borer*, etc.
Bacteria	*Bt galleriae* and *Bt tenebrionis*	Leaf beetle of potatoes, tomatoes
Bacteria	*Bt israelensis* (Bti) and *Lysinibacillus sphaericus*	Larvae of different insects like mosquito, blackfly
Bacteria	*Paenibacillus popilliae*	Caused milky disease in Japanese beetle
Bacteria	*Actinomycetales*	Thrips, caterpillars, and many other pests
Bacteria	*Burkholderia rinojensis* strain A396, *Chromobacterium subtsugae* strain PRAA4-1	Pest in aquatic environments
Bacteria	*Bacillus firmus* and *Pasteuria* spp.	Nematodes lives on plants
Fungi	*Beauveria bassiana*	Reduce the number of foliar pests
Fungi	*Metarhizium brunneum*	Weevils, mites, whiteflies, and thrips in different vegetables and ornamental plants
Fungi	*Isaria fumosorosea*	Aphids, mites, whiteflies, and thrips in different vegetables and ornamental plants
Fungi	*Purpureocillium lilacinum* and *Myrothecium verrucaria*	Nematodes living on plants
Fungi	*Paranosema locustae*	Mormon crickets and grasshopper in agricultural field as well as domestic premices
Fungi	Baculoviruses	Act against Lepidoptera larvae found in different vegetables in greenhouse gas condition
Fungi	Nucleopolyhedroviruses (NPV)	Beet armyworm, corn earworm, *Spodoptera exigua*, etc.
Fungi	Granuloviruses (GV)	Moth found in apple orchard
Virus	*Autographa californica* NPV	*Trichoplusia ni*
Nematode	*Steinernema feltiae*	Western flower thrips

6.3 Role of Microbiotechnology in Waste Management and Human Sustainability

The interaction between humans and the environment is a complex phenomenon. It's not solely dependent on meeting basic food needs but is also influenced by factors such as resource consumption levels, waste generation, and the use of various technologies across different applications (Selvam and Wong 2016). However, the continuous discharge of hazardous wastes such as plastic, industrial, food, household waste, into the environment poses the greatest threat to mankind in the long run, raising enormous concerns about environmental and human sustainability (Mondal and Palit 2019). Given the broad and intricate nature of sustainability goals and their various components, microbial technology can play a significant role in helping to achieve sustainability, as elaborated below (Timmis et al. 2017).

6.3.1 Microbiotechnology Against Plastic Waste

Plastics are employed in a wide variety of industries across the globe, including the medical, transportation, manufacturing, sanitation, food packaging, storage, and petroleum-based feedstock industries, worldwide. Plastics viz. polyhydroxybutyrate/polyhydroxybutyrate-*co*-valerate (PHB/PHBV), polyethylene (PE), poly(ε-caprolactone) "(PCL), poly(lactic acid) (PLA), polyester/poly(butylene succinate) (PES/PBS), Poly(ethylene terephthalate) (PET), Low-density polyethylene (LDPE), high-density polyethylene (HDPE), poly(vinyl alcohol)-linear low-density polyethylene)" (PVA-LLDPE), poly butyl succinate, etc. comprise 80% of plastic usage and has contributed significantly to the expansion of the global economy as well as employment for over 60 million people (Urbanek et al. 2018; Mazhandu et al. 2020).

Plastic offers numerous advantages, but its production, consumption, and disposal have led to the depletion of non-renewable resources. These patterns also contribute to environmental degradation, negatively impacting the sustainability of both humans and animals (Mazhandu et al. 2020). Besides these concerns, the rapidly growing demand and production of plastics in the past few decades give rise to huge generation of plastic waste in the environment that accounts for 54% of the total anthropogenic waste (Satti and Shah 2020). In terms of the numbers that are currently available, there is a persistent rise in plastics usage and has led to an increase in the generation of plastics waste. Furthermore, this waste is considered to be one of the most pervasive forms of pollution, and it has significant negative effects on the environment as a result of its composition and the presence of hazardous chemicals. These chemicals are known to cause cancer, chronic respiratory disorders, neurological disorders, and reproductive anomalies in humans and animals (Rajmohan et al. 2019; Satti and Shah 2020).

Efforts have been made to reduce plastic waste through methods like reuse, recycling, incineration, or landfilling. However, approximately 8300 million tonnes of plastic still end up in the environment, including landfills, oceans, and terrestrial areas (Mohanan et al. 2020). These plastics exhibit high resistance to natural biodegradation and can endure for many years. As a result, alternative methods such as photo-, bio-, or thermo-oxidative depolymerization and friction methods can be employed for plastic waste management (Mohanan et al. 2020). However, with growing concerns about environmental friendliness and economic effectiveness, microorganisms or their byproduct-based degradation has emerged as a potential option in recent years (Mohanan et al. 2020). Hence, several studies related to microorganisms capable of degrading synthetic plastics are discussed in Table 6.2.

Table 6.2 Various microbes and enzymes involved in biodegradation of different types of plastics

S. no.	Plastic	Microorganisms/enzymes	References
1.	Polyhydroxybutyrate/ polyhydroxybutyrate-*co*-valerate (PHB/PHBV)	**M/Os:** "*Bacillus megaterium N-18-25-9, Bacillus* sp. *AF3, Actinomadura* sp. *AF-555, Streptoverticillium kashmirense AF1, Streptomyces ascomycinicus, Streptomyces* sp. *AF-111," Catenibacterium thermophilum, Thermus thermophiles HB8 Paecilomyces* sp. *"1407; Penicillium* sp. *DS9701-D2; Aspergillus fumigatus NA 25; Emericellopsis minima W2"*	Satti and Shah (2020)
		Enzymes: PHA depolymerases	
2.	Polyethylene (PE)	*Zalerion maritimum, Bacillus amylolyticus, Bacillus subtilis, "Enterobacter* sp. *D1, Pseudomonas alcaligenes, Pseudomonas fluorescens, Pseudomonas putida, Pseudomonas putida MTCC 2475, Streptomyces SSP2, Streptomyces SSP4, Streptomyces SSP14, Acinetobacter ursingii"*	Urbanek et al. (2018), Asiandu et al. (2021)
3.	Poly(ε-caprolactone) (PCL)	**M/Os:** "*Lactobacillus plantarum, Brevundimonas* sp. *Strain MRL-AN1 Streptomyces thermoviolaceus* subsp. *Thermoviolaceus 76T-2, Ralstonia* sp.*"*	Satti and Shah (2020), Urbanek et al. (2018)
		Enzymes: Lipase, esterase, protease, cutinase, catalase, glucosidases	
4.	Poly(lactic acid) (PLA)	*"Sphingobacterium* sp. *S2, Chryseobacterium* sp. *S1, Pseudomonas aeruginosa strain S3 and S4, Amycolatopsis orientalis, Alcanivorax borkumensis ABO2449, Bordetella petrii PLA-3 Laceyella sacchari LP175"*	Satti and Shah (2020), Asiandu et al. (2021)
		Enzymes: Lipase, protease, cutinase	
5.	Polyester/poly(butylene succinate) (PES/PBS)	"Bacillus pumilus strain KT1012, *Pseudomonas* sp. AKS2" *Microbispora rosea* subsp.	Satti and Shah (2020)

(continued)

Table 6.2 (continued)

S. no.	Plastic	Microorganisms/enzymes	References
		Taiwanensis *Aspergillus clavatus* NKCM1003	
		Enzymes: Lipase, esterase, cutinase	
6.	Poly(ethylene terephthalate) (PET)	*"Pseudomonas* sp. *AKS2, Ideonella sakaiensis 201-F6 Saccharomonospora viridis AHK190, Thermobifida fusca KW3, Thermobifida cellulosilytica, Penicillium" funiculosum, Rhizopus delemar; Phormidium, Lewinella, Stanieria, Pseudophormidium; Bacillus subtilis; Staphylococcus pyogenes; Staphylococcus aureus*	Satti and Shah (2020), Asiandu et al. (2021)
		Enzymes: Lipase, esterase, cutinase	
7.	Low-density polyethylene (LDPE)	*Aspergillus versicolor, Aspergillus* sp., *Arcobacter Colwellia* sp., *Bacillus vallismortis bt-dsce01, "Aspergillus oryzae strain A5, Bacillus cereus strain A5, Trichoderma viride RH03, Aspergillus nomius RH06; Bacillus* sp., *Paenibacillus* sp."	Urbanek et al. (2018), Elahi et al. (2021), Asiandu et al. (2021)
8.	High-density polyethylene (HDPE)	*Ochrobactrum anthropic, Aspergillus flavus, Klebsiella pneumoniae CH001*	Urbanek et al. (2018), Elahi et al. (2021), Asiandu et al. (2021)
9.	"Poly (vinyl alcohol)-linear low-density polyethylene) (PVA-LLDPE)"	*Vibrio alginolyticus, Vibrio parahaemolyticus*	Urbanek et al. (2018)

6.3.2 Microbial Biotechnology for Industrial Wastewater Treatment

Industrial wastewater pollution has become more problematic over the world, when treatment and administration of industrial effluents are not being handled effectively. These industrial effluents comprise dangerous inorganic and organic contaminants that pollute water streams and the surrounding soil ecosystem, affecting entire living species (Maszenan et al. 2011; Rani et al. 2019). Effluent water and solid discharge constitute around one-third of total water pollution in India alone, and growing industrialization causes approximately "3.4 million people to suffer and die worldwide" (Rani et al. 2019). It is possibly harmful to the ecosystems and has garnered a lot of attention from research scientists for its management to maintain the environmental as well as human sustainability.

6.3.3 Microbial Biotechnology to Reduce the Textile Industry Based Dyes

The textile industry is accountable for the consumption of vast quantities of water, which is then released back into the atmosphere in the form of wastewater. This effluent consisting of textile-based colors (such as anthraquinone, azo dyes, and phthalocyanine), metals/metalloids, salts, and organic contaminants make it one of the most major contributors to the contamination of surface and groundwater bodies (Imran et al. 2015; Thanavel et al. 2019). In addition, in the years ahead, the annual growth rates of reactive dyes all over the world are going to expand as a direct result of the rising demand for items made of textiles. At the same time, the share of wastewater that is produced by the industry will also expand, "making it one of the principal sources for some of the most serious pollution problems in" recent times (Karim et al. 2018). Indeed, due to the carcinogenic and mutagenic properties of the dyeing agents, these dyes have become a severe hazard to all life (Thanavel et al. 2019). This has led to increased concerns over the removal of these dyes from the environment through the application of a variety of physicochemical and biological methods (Imran et al. 2015; Thanavel et al. 2019). Particularly, the biological strategies involving bacteria, yeasts, fungi, algae, and actinomycetes have drawn interest all over the world for their relative cost-effectiveness and environmentally friendly nature compared to the physicochemical methods (Thanavel et al. 2019). Subsequently, several microbial cultures and their microbial enzymes have been characterized and used for removal of dyes from simulated wastewater as discussed in Table 6.3.

6.3.4 Microbial Strategies for Bio-transforming Food Waste into Resources

As a result of shifting lifestyles and the rising urbanization round the globe, there is an increase in the amount of food that has been thrown away from a variety of industrial, agricultural, and domestic settings. Recently, "the Food and Agriculture Organization (FAO)" estimated that around one-third of all the food that is produced each year is lost or wasted. This raises significant concerns, not only because valuable resources are being wasted, but also because their disposal in the environment raises concerns. In fact, massive food waste disposal due to an increasing global population has prompted the hunt for sustainable food waste management systems to alleviate environmental issues. Traditional methods of food waste management, such as landfilling and incineration, pose substantial risks to both the environment and human health due to the generation of toxic fumes. Therefore, there is potential in utilizing microbial bioprocesses that offer economic and sustainable solutions for transforming food waste into high-value bioproducts. These strategies have appeared to be promising for a better human as well as environment sustainability (Ng et al. 2020). Thus, employing "biological methods for the treatment of" such waste offers a sustainable way for valorization (Sharma et al. 2020). In

Table 6.3 Microorganisms and their enzymes involved in treatment of textile industry based dyes

S. no.	Microorganisms	Most active enzyme	References
A. Bacteria			
1.	*Enterobacter* sp. GY-1	Azoreductase	Chen et al. (2011), Imran et al. (2015)
2.	*Pseudomonas putida* WLY	Azoreductase	Yang et al. (2011), Imran et al. (2015)
3.	*Bacillus* sp. VUS	"Lignin peroxidase, tyrosinase, NADH-DCIP reductase, azoreductase, and riboflavin reductase"	Dawkar et al. (2010), Imran et al. (2015)
4.	"*Lactobacillus casei* TISTR 1500"	Azoreductase	Seesuriyachan et al. (2007), Imran et al. (2015)
5.	*Galactomyces geotrichum* MTCC 1360	Lignin peroxidase and laccase activities	Jadhav et al. (2009), Imran et al. (2015)
6.	*Rhizobium radiobacter* MTCC 8161	"Azoreductase, lignin peroxidase, DCIP reductase, MG reductase, and aminopyrine *N*-demethylase"	Telke et al. (2008), Imran et al. (2015)
7.	*Bacillus cereus*	Azoreductase	Deng et al. (2008), Imran et al. (2015)
8.	*Basidiomycetous* isolate	Laccase	D'Souza et al. (2006), Imran et al. (2015)
9.	*Bacillus* sp.	Lignin peroxidase, laccase, and NADH–DCIP reductase	Dawkar et al. (2008), Imran et al. (2015)
10.	*Bacillus* sp.	–	Pourbabaee et al. (2006), Imran et al. (2015)
11.	*Bacillus* Mk-8	–	Cheunbarn et al. (2008), Imran et al. (2015)
B. Fungi			
1.	*Curvularia lunata* URM6179 and *Phanerochaete chrysosporium* URM 6181	"Laccase, lignin peroxidase, and Mn-peroxidase"	Imran et al. (2015)
2.	"Mixed culture consisting of *Pleurotus ostreatus* IBL-02 and *Coriolus versicolor* IBL-04"	"Laccase, lignin peroxidase, and Mn-peroxidase"	Asgher et al. (2010), Imran et al. (2015)
3.	*Bjerkandera adusta* (Willdenow) P. Karsten MUT 3060	–	Anastasi et al. (2011), Imran et al. (2015)
4.	*Ganoderma* sp. En3	Laccase	Zhuo et al. (2011), Imran et al. (2015)

(continued)

Table 6.3 (continued)

S. no.	Microorganisms	Most active enzyme	References
5.	*Penicillium ochrochloron* MTCC 517	Lignin peroxidase	Shedbalkar and Jadhav (2011), Imran et al. (2015), Abataneh et al. (2017)
6.	"*Myrothecium roridum IM 6482*"	Laccase, MG reductase	Jasińska et al. (2012, 2013), Abataneh et al. (2017)
7.	*Aspergillus lentulus*	–	Kaushik and Malik (2010), Imran et al. (2015)
8.	*Aspergillus niger*	–	Agarry and Ayobami (2011), Imran et al. (2015)
9.	*Ascomycetes*	–	Verma et al. (2010), Imran et al. (2015)
10.	*Basidiomycetous* fungus-PCK-3	–	Diwaniyan et al. (2010), Imran et al. (2015)
11.	*Ganoderma lucidum* IBL-05	–	Asgher et al. (2010), Imran et al. (2015)
12.	*Phanerochaete chrysosporium*	Lignolytic "enzymes such as lignin peroxidase and Mn-peroxidase"	Sedighi et al. (2009), Imran et al. (2015)
13.	"*Aspergillus fumigatus XC6*"	–	Jin et al. (2007), Imran et al. (2015)
14.	*Pleurotus florid*	–	Perumal et al. (2007), Imran et al. (2015)
15.	*Phanerochaete chrysosporium*	–	Cing et al. (2003), Imran et al. (2015)
16.	*Aspergillus niger*	–	Assadi and Jahangiri (2001), Imran et al. (2015)
C. Consortium			
1.	"*Providencia* sp. SDS and *Pseudomonas aeruginosa* strain BCH"	"Azoreductase, DCIP reductase, veratryl alcohol oxidase, and laccase enzymes"	Phugare et al. (2011), Imran et al. (2015)
2.	"*Sphingomonas paucimobilis*, *Bacillus* sp. and filamentous bacteria"	–	Ayed et al. (2011), Imran et al. (2015)
3.	*Pseudomonas* sp. SU-EBT	Intracellular laccase enzyme	Telke et al. (2010), Imran et al. (2015)
4.	*Pseudomonas* sp. SUK1, *Pseudomonas* sp. LBC2, and *Pseudomonas* sp. LBC3	Laccase and azoreductase	Jadhav et al. (2010), Imran et al. (2015)

(continued)

Table 6.3 (continued)

S. no.	Microorganisms	Most active enzyme	References
5.	*Sphingobacterium* sp. ATM, *Bacillus odysseyi* SUK3, and *Pseudomonas desmolyticum* NCIM 2112	"Laccase, veratryl alcohol oxidase, DCIP reductase, riboflavin reductase, and azoreductase"	Tamboli et al. (2010), Imran et al. (2015)
6.	*Proteus vulgaris* NCIM-2027 and *Micrococcus glutamicus* NCIM-2168	–	Saratale et al. (2010), Imran et al. (2015)
7.	*"B. subtilis strain NAP1, NAP2, NAP4"*	Peroxidases and dehydrogenases	Phulpoto et al. (2016), Abataneh et al. (2017)
8.	*Pycnoporus sanguineous, Phanerochaete chrysosporium, and Trametes trogii*	Laccase	Yan et al. (2014), Abataneh et al. (2017)
9.	*"Micrococcus luteus, Listeria denitrificans, and Nocardia atlantica"*	–	Hassan et al. (2013), Abataneh et al. (2017)
10.	*"Bacillus* spp. *ETL-2012, Pseudomonas aeruginosa, Bacillus pumilus HKG212"*	Azoreductase	Patel and Gupte (2016), Abataneh et al. (2017)
11.	*Exiguobacterium indicum, Exiguobacteriumaurantiacum, Bacillus cereus, and Acinetobacter baumannii*	–	Kumar et al. (2016), Abataneh et al. (2017)
12.	*"Bacillus firmus, Bacillus macerans, Staphylococcus aureus, and Klebsiella oxytoca"*	–	Adebajo et al. (2017)

this section, we provide a comprehensive discussion on the utilization of various food industry-generated wastes and the production of value-added bioactive compounds.

We also delve into the role of microorganisms in the eco-friendly biotransformation of food waste as a sustainable solution for its management (Table 6.4).

6.4 Application of Microbiotechnology to Control Pollution

6.4.1 Microbiotechnology Derived Biosensors for Pollution Monitoring

Currently, due to the tight association "between environmental pollution and human health/socioeconomic" development, "environmental monitoring has become one of the" top objectives on a global scale. In this arena, biosensor technologies have emerged as an important area for numerous applications and are recognized as an

Table 6.4 Various food industry generated wastes and their related biotransformed value-added products

S. no.	Food industry	Waste	Value-added bioactive compounds	References
1.	"Fruits processing industry"	Pomace, peel, and seeds, rag, rind, skin, stem	Ethanol, dietary fiber, grape seed oil, pomace oil, oleanolic acid, polyphenols (catechin, epicatechin, gallic acid, and resveratrol), anthocyanins (enocyanin), procyanidins, tartates, malates, citric acid, single cell protein, pectin, lactic acid, citric acid, aroma compounds, biogas, ethanol, butanol, pectinases, essential oil (limonene), antioxidants, flavonoids, starch, fiber, sterols, tocopherols, tannins, xanthones, alkylresorcinols, lignin, cellulose, hemicelluloses, procyanidins, and flavanols, carotenoids, bromelain, dietary fiber, phytochemicals, biofertilizer	Galanakis (2017), Schieber (2017), Kiran et al. (2014), Matharu et al. (2016), Rohm et al. (2015), Sharma et al. (2021)
2.	Vegetable processing industry	Shell, peel, core, skin	"Lysine, protein (patatin), steroidal alkaloids, cellulolytic enzymes, adsorption dyes, biopolymer films, carotenoids, dietary fiber, pectin, fructans, phenolic compounds"	Kiran et al. (2014), Matharu et al. (2016), Schieber (2017), Sharma et al. (2021)
3.	Cereal and pulses industry	Husk, bran, lignocellulosic biomass	"Carbohydrates, oligosaccharides, phenolic compounds, lipid soluble vitamins, folic acid, phytosterols, amino acids, and peptides"	Schieber (2017), Kiran et al. (2014), Parate and Talib (2015)
			"Proteins, lipids, dietary fiber, minerals, and antioxidants (vitamin E and oryzanol)"	
			"Activated carbon, proteins, lipids, fatty acids, vitamins, minerals, and phenolic compounds"	
4.	Beverage industry	Wastewater	"Antioxidants, vitamins, enzymes, cellulose, starch, lipids, proteins, pigments, citric acid, gibberellic acid, ethanol, biogas, dyes, and dietary fibers (Murthy and Naidu 2012; Kiran et al. 2014) 46 (cellulose, hemicelluloses, lignin, pectin, gums) Caffeine, polyphenols, triacontanol, and saponins"	Murthy and Naidu (2012), Kiran et al. (2014), Sui et al. (2019)
5.	Dairy industry	Dairy effluent (cheese whey)	"Biodiesel, ethanol, whey protein, lactose, baker's yeast, and minerals"	Kiran et al. (2014), Parashar et al. (2016)

6.	Meat and seafood processing industry	Feathers, blood, heads, bones, skin, viscera, and sometimes whole fish and parsley	"Fertilizer, animal feed, blood meal, meat and bone meal, feather meal, lactic acid, and probiotics, chitosan, and glycosaminoglycans, nutraceuticals (Astaxanthin), chitin, and chitinase"	Kiran et al. (2014), Ning et al. (2018), Yaakob et al. (2019), Prameela et al. (2017), Kumar et al. (2018), Yan and Chen (2015), Sharma et al. (2021)
			"Chitin, calcium carbonate, and protein, meat protein"	
7.	Edible oil industry	Wastewater, organic wastes	"Biosurfactants like rhamnolipids and glycolipids, biodiesel, tocopherols, sterols, squalene, and single cell protein; phenolic compounds, polyphenols, carotenoids, phytosterols, squalene, and dietary fiber"	Henkel et al. (2012), Kiran et al. (2014), Schieber (2017)

analytical and self-sufficient device, appropriate for environmental assessment-cum-monitoring due to its rapid detection property, high sensitivity, and cost-effectiveness in regard to their sensing and monitoring qualities (Oldach and Zhang 2014; Ali and Singh 2020). These biosensors are categorized by the type of biological receptors viz. genetic material, aptamers, enzymes, antibodies, proteins, microbes, etc. or by physicochemical transducers viz. optical, electrochemical, visual, thermal, piezo-electrical, etc. employed for the detection of toxic environmental contaminants (Verma and Bhardwaj 2015; Ali and Singh 2020). Most biosensors used for environmental monitoring are traditionally microbe-based, immunosensors, or enzymatic-based. However, there has been a recent increase in the development of aptamer-based biosensors, driven by their advantageous characteristics, including ease of modification, stability, in vitro synthesis, and target specificity. Moreover, due to the rise of awareness about the detrimental impact of environmental contaminants (organic pollutants, pathogens, pesticides, toxic elements, etc.) on human health, more studies are being encouraged on the biosensor's development for a more sustainable society establishment. In fact, the efficiency of existing biosensors is not so adequate and therefore, more studies are essential for the progression of a robust biosensing device that can successfully be used for the detection of pollutants directly from the complex environments. Therefore, some of the existing pollutants and their monitoring biosensor are discussed in Table 6.5.

6.4.2 Microbiotechnology for Remediation of Pollutants

Throughout the years, the growth of industrial activity and the spread of urbanization have had a significant impact on the environment. As a result, diverse pollutants from industrial, agricultural, and even domestic spheres have been discharged into the environment. This has led to an acceleration in the overall concentration of pollutants in the environment; consequently, the deterioration of environmental health and its adverse effects on living entities has become a major concern these years. In this context, microbial biotechnology has revolutionized the bioremediation field for environmental pollutants (xenobiotics, petroleum hydrocarbons, "polycyclic aromatic hydrocarbons (PAHs), organic pollutants," heavy metals, toxins, pesticides, etc.). Bioremediation involves removal/detoxification/transformation of the pollutants with the use of biological entity such as microorganisms, mainly the contaminants of soil, water, or sediment which may otherwise threaten public health (Wasi et al. 2013; Abataneh et al. 2017; Dangi et al. 2019).

Furthermore, systemic biology (omics biology) is also gaining attention as an attractive bioremediation method by determining the biological agents with respect to their intricate networks and their inter-relations in various biological processes at the cell/molecular, community, or ecological level, which will provide a clear and true picture of the bioremediation (Dangi et al. 2019). However, there is a need to expand the information of microbial genetics to upsurge the abilities of pollutants degradation at large scale and field experiments for advances in this field. Therefore,

Table 6.5 Microbiotechnology derived biosensing elements for environmental pollutants monitoring

S. no.	Analytes		Biosensing elements	References
1.	Heavy metals	Me, Cd, Ar	Urease enzyme	Pal et al. (2009), Negi and Choephel (2020)
		Me(II), Pb (II) ions	DNA	Knecht and Sethi (2009), Negi and Choephel (2020)
		Cd, Cu, Pb	Sol-gel-immobilized urease	Ilangovan et al. (2006), Negi and Choephel (2020)
		Cd, Pb	*Staphylococcus aureus*	Negi and Choephel (2020)
		Cu, Cr, Pb	*Escherichia coli*	Negi and Choephel (2020)
		Cd, Co, Pb	Monoclonal antibody	Negi and Choephel (2020)
		Ni, Co	*Ralstonia eutropha* with *lux*	Shin (2011)
		As	*E. coli* with *lux*, *lacZ*, *gfp*	Shin (2011)
		Hg, As, Cu, Zn, Pb, Cd	*E. coli* with *luc*, *lux*	Shin (2011)
		Hg	Peroxidase	Negi and Choephel (2020)
2.	Herbicides	"2,4-Dichloro-phenoxy acetic acid"	Acetylcholinesterase	Negi and Choephel (2020)
		Diuron, paraquat	Cyanobacteria	Negi and Choephel (2020)
3.	Phenolic compounds	Phenol	Mushroom tissue (tyrosinase)	Silva et al. (2010), Negi and Choephel (2020)
		Phenolics	*E. coli* with *lacZ*	Shin (2011)
		"Phenol, *p*-cresol, *m*-cresol, and catechol"	Polyphenol oxidase	Karim and Fakhruddin (2012), Negi and Choephel (2020)
		m-cresol or catechol	DNA	Claude et al. (2007), Negi and Choephel (2020)
4.	Pesticides	Parathion	Parathion hydrolase (biocatalytic)	Negi and Choephel (2020)
		Carbaryl	Acetylcholinesterase	Negi and Choephel (2020)
		Paraoxon	Alkaline, phosphatase	Negi and Choephel (2020)

(continued)

Table 6.5 (continued)

S. no.	Analytes		Biosensing elements	References
5.	Pathogens	*E. coli*	Electrochemical biosensors; mass-sensitive biosensors; optical biosensor-T4-bioprobe phage	Aliakbar Ahovan et al. (2020), Nnachi et al. (2022)
		P. aeruginosa	Mass-sensitive biosensors	Aliakbar Ahovan et al. (2020), Nnachi et al. (2022)
		Salmonella typhimurium	Immunoassay; aptamer-based assay	Aliakbar Ahovan et al. (2020), Nnachi et al. (2022)
		MRSA	Optical biosensor-BP-14-bioprobe phage	Aliakbar Ahovan et al. (2020)
		Salmonella enteritidis	Bioluminesence-SJ2 bioprobe phage	Aliakbar Ahovan et al. (2020)
		Yersinia pestis	Bioluminescence lux system-Phage A1122 with lux tag	Aliakbar Ahovan et al. (2020)

some of the common microorganisms used for bioremediation of different pollutants are discussed in Table 6.6.

6.5 Microbiotechnology in Livestock Management for Better Human Sustainability

6.5.1 Microbes for Improvement of Aquaculture

The population of human race is increasing in size with every passing year and will likely reach approximately 10^{10} individuals in the next 30 years (Bentzon-Tilia et al. 2016). This growing population needs a steady supply of high-quality protein, which to large extent can be supplied by non-vegetarian food/meat (Bentzon-Tilia et al. 2016). However, the growing demand for seafood may not be met by capture fisheries alone. Hence, production in cultured fisheries should be increased in coming time to fulfill the demand of seafood. Thus, aquaculture production practices have been intensified for fulfilling the global demand (Tuan et al. 2013; Zorriehzahra et al. 2016). Undoubtedly, aquaculture industry has significantly contributed in enhanced production of seafood; however, this industry is facing lots of many problems especially due to environmental and anthropogenic activities (Bentzon-Tilia et al. 2016). In addition, controlling the growth of pathogens using antimicrobials can pose a severe "risk to human health due to the spread of microbial antibiotic resistance (Cabello et al. 2013; Bentzon-Tilia et al. 2016)." These alarming hindrances provoked the aquaculture industry for exploring and developing approaches which may be correspondingly effective as antibiotics, sustainable and

Table 6.6 Microorganisms and different pollutants bioremediation

S. no.	Microorganisms	Degradable compound	References
1.	*"Penicillium chrysogenum"*	"Monocyclic aromatic hydrocarbons, benzene, toluene, ethyl benzene and xylene, phenol compounds"	Abataneh et al. (2017)
2.	*"P. alcaligenes, P. mendocina, P. putida, P. veronii, Achromobacter, Flavobacterium, Acinetobacter"*	"Petrol and diesel polycyclic aromatic hydrocarbons toluene"	Abataneh et al. (2017)
3.	*"Pseudomonas putida"*	"Monocyclic aromatic hydrocarbons, e.g. benzene and xylene"	Abataneh et al. (2017)
4.	*"Phanerochaete chrysosporium"*	"Biphenyl and triphenyl methane"	Abataneh et al. (2017)
5.	*"A. niger, A. fumigatus, F. solani, P. funiculosum"*	"Hydrocarbon"	Abataneh et al. (2017)
6.	*"Alcaligenes odorans, Bacillus subtilis, Corynebacterium propinquum, Pseudomonas aeruginosa"*	"Phenol"	Abataneh et al. (2017)
7.	*"Tyromyces palustris, Gloeophyllum trabeum, Trametes versicolor"*	Hydrocarbons	Abataneh et al. (2017)
8.	*"Candida viswanathii"*	"Phenanthrene, benzopyrene"	Abataneh et al. (2017)
9.	"Green algae and diatoms, *Cyanobacteria, Bacillus licheniformis"*	Naphthalene	Abataneh et al. (2017)
10.	"*Acinetobacter* sp., *Pseudomonas* sp., *Ralstonia* sp., *Microbacterium* sp."	Aromatic hydrocarbons	Abataneh et al. (2017)
11.	Gloeophyllum Striatum	"Pyrene, anthracene, dibenzothiophene, lignin peroxidase"	Abataneh et al. (2017)
12.	*"Naegleria, Vorticella, Arabidopsis, Asarum, and Populus* (metagenomics)"	Hydrocarbons	Dangi et al. (2019)
13.	*Bacillus subtilis* HUK15 (genomics)	Hexachlorocyclohexane (HCH)	Dangi et al. (2019)
14.	*Aspergillus nigera semo A, Talaromyces purpurogenus semo F, and Aspergillus flavus semo M* (transcriptomics)	Degradation of hydrocarbons	Dangi et al. (2019)

(continued)

Table 6.6 (continued)

S. no.	Microorganisms	Degradable compound	References
15.	*Rhizobiales, Burkholderiales, and Actinomycetales* (metaproteomics)	Polycyclic aromatic hydrocarbon	Dangi et al. (2019)
16.	*"Pseudomonas putida* KT2440 (fluxomics)"	Hydrocarbons	Dangi et al. (2019)
17.	*"Bacillus, Staphylococcus"*	Endosulfan	Abataneh et al. (2017)
18.	*"Enterobacter"*	Chlorpyrifos	Abataneh et al. (2017)
19.	*"Pseudomonas putida, Acinetobacter* sp., *Arthrobacter* sp."	FitorazWP76, Decis2.5EC, malathion	Abataneh et al. (2017)
20.	*Acinetobacter* sp., *Pseudomonas* sp., *Enterobacter* sp., *and Photobacterium* sp.	"Chlorpyrifos and methyl parathion"	Abataneh et al. (2017)
21.	*Koribacter, Acidimicrobium, Bradyrhizobium, Burkholderia, Solibacter, Singulisphaera, Desulfomonile* (metagenomics)	Organophosphorus-containing pesticides	Dangi et al. (2019)
22.	*Sphingomonas* sp. *GY2B* (proteomics)	Phenanthrene	Dangi et al. (2019)
23.	*Fusarium* sp.	Oil	Abataneh et al. (2017)
24.	*Pseudomonas aeruginosa, Corynebacterium propinquum, Bacillus subtilis, Alcaligenes odorans*	Oil	Abataneh et al. (2017)
25.	*Bacillus cereus* A	Diesel oil	Abataneh et al. (2017)
26.	"*Pseudomonas aeruginosa, P. putida, Arthobacter* sp., *and Bacillus* sp."	Diesel oil	Abataneh et al. (2017)
27.	*"Pseudomonas cepacia, Bacillus cereus, Bacillus coagulans, Serratia ficaria, Citrobacter koseri"*	Diesel oil, crude oil	Abataneh et al. (2017)
28.	*Saccharomyces cerevisiae*	Pb, Me, Ni	Abataneh et al. (2017)
29.	*"Cunninghamella Elegans"*	Heavy metals	Abataneh et al. (2017)

(continued)

Table 6.6 (continued)

S. no.	Microorganisms	Degradable compound	References
30.	"*Pseudomonas fluorescens* and *Pseudomonas aeruginosa*"	Fe, Zn, Pb, Mn, Cu	Abataneh et al. (2017)
31.	*Lysinibacillus sphaericusCBAM5*	Co, Cu, Cr, Pb	Abataneh et al. (2017)
32.	*Mycobacterium profundi strain S49T*	Fe	Abataneh et al. (2017)
33.	"*Aspergillus versicolor, A. fumigatus, Paecilomyces* sp., *Trichoderma* sp., *Microsporum* sp., *Cladosporium* sp."	Cd	Abataneh et al. (2017)
34.	*Geobacter* sp.	Fe(III), U(VI)	Abataneh et al. (2017)
35.	"*Bacillus safensis (JX126862) strain (PB-5 and RSA-4)*"	Cd	Abataneh et al. (2017)
36.	"*Pseudomonas aeruginosa, Aeromonas* sp."	U, Cu, Ni, Cr	Abataneh et al. (2017)
37.	"*Aerococcus* sp., *Rhodopseudomonas palustris*"	Pb, Cr, Cd	Abataneh et al. (2017)
38.	"*Brevibacterium epidermidis* EZ-K02 (metagenomics)"	As, Co, Cd	Dangi et al. (2019)
39.	*Pseudomonas aeruginosa* sanai (proteomics)	Cd	Dangi et al. (2019)

eco-consumer-friendly (Standen et al. 2013; Lazado et al. 2015; Zorriehzahra et al. 2016).

Studies have shown that managing microbial communities associated with fish and their habitat can provide an efficient solution to various issues. This management not only improves nutrient levels in water but also reduces fish pathogenic bacteria, enhances larval survival, eliminates the need for antimicrobials, and boosts nutrient utilization and immunity (Bentzon-Tilia et al. 2016; Zorriehzahra et al. 2016). Probiotics are among the strategies that can reduce the aquaculture industry's reliance on antibiotics and serve various functions, including nutrient recycling, organic matter degradation, and disease protection. They play a crucial role in not only reducing antibiotic use but also contributing to nutrient management and disease prevention (Bentzon-Tilia et al. 2016; Zorriehzahra et al. 2016). These beneficial activities of microbes on aquatic species have contributed immensely to the growth, development, and sustainability of aquaculture. Hence, the microbial communities in fish/aquaculture industry have immense potential for improving the aquaculture and most of these microbes with their beneficial effect on aquatic host are discussed in Table 6.7.

6.5.2 Probiotics for Livestock Animals: An Industry Level Exploration

With the fast-rising worldwide demand for animal food products, it has become crucial for livestock farmers to increase livestock productivity to fulfill the rising demands (Lambo et al. 2021). In this context, feed additives such as antibiotic growth promoters, enzymes, mineral supplements, probiotics, etc. provide a safe and healthy way to enhance the animal health, productivity, and their general well-being. However, due to the rise of antimicrobials resistance and their cost concerns, probiotics **gained popularity** in the livestock industry in both animals and consumers of animal products. Consequently, livestock industry has recognized them as non-toxic, non-pathogenic supplements/additives that could enhance disease resistance against the pathogens, improves the intestinal microbial balance, digestion, immunity, and performance in administered animals as discussed in the following table (Jiménez 2010; Lambo et al. 2021).

Therefore, probiotics stand a good chance of substituting the use of antibiotics in livestock industry and are believed to "have a significant impact on the animal nutrition, health, and productivity." However, the effect of probiotics as well as their responses of host animals varies among the published literature and this variation may be attributable to the kind/strains of microbes used single/multi H strain, their viability in gut; dosage; the physiological health of the host animal, the environment, diet, and the route of administration. In addition, cross-contamination of the human food chain by probiotics used in animal feed cannot be minimized, therefore for the food safety and security to be successful, the established safety measures in probiotic development must be adhered (Alayande et al. 2020). Also, additional research is required to comprehend the interaction processes between the single/combined bacteria and the host's gut microbiota, as well as their direct comparisons should be examined for optimal advantages (Table 6.8).

6.6 Microbiotechnology in Food Security

6.6.1 Production of Microbial Oils Rich in Omega-3

Lipids are natural molecules that can be grouped based on their ability to dissolve in certain solvents. They include substances like triacylglycerols (found in fats), phospholipids (containing phosphoric acid), glycolipids (with glucose), and sphingolipids. Lipids serve as a rich source of metabolic energy, providing 9 calories per gram. Lipids also are identified to be important additives of several herbal processes, which consist of power storage, cascade molecular signaling, plasma

Table 6.7 List of beneficial microorganisms and their beneficial effect on aquatic host (adapted from Zorriehzahra et al. 2016)

S. no.	Microorganism		For aquatic species	Beneficial impact	References
	Genus	Species			
1.	Aeromonas	*A. Hydrophila; A. sobria* GC2	*Oncorhynchus mykiss*	Reduce infections and protects against *Aeromonas salmonicida*; "*Lactococcus garvieae*; *Streptococcus iniae*; *Aeromonas bestiarum;* and *Ichthyophthirius multifiliis*"	Irianto and Austin (2002a, b), Pieters et al. (2008), Brunt and Austin (2005)
2.	"Agarivorans"	"*A. albus* F1-UMA"	"*Haliotis rufescens*"	Increases survivability	Silva-Aciares et al. (2011)
3.	Alteromonas	*A. Macleodii* 0444	*Perna canaliculus; Pecten maximus*	Protects against *Vibrio splendidus*, *V. coralliilyticus,* and *V. splendidus*	Kesarcodi-Watson et al. (2010, 2012)
4.	Burkholderia	*B. cepacia* Y021	*Crassostrea corteziensis; Nodipecten subnodosus*	Increases survivability	Granados-Amores et al. (2012)
5.	"Enterobacter"	"*E. amnigenus*"	*Oncorhynchus mykiss*	"Increases protection toward *Flavobacterium psychrophilum*"	Burbank et al. (2011)
6.	"Neptunomonas"	*Neptunomonas* 0536	"*Perna canaliculus*"	Protect against *V. splendidus*	Kesarcodi-Watson et al. (2010, 2012)
7.	Shewanella	*S. putrefaciens*	*Sparus aurata* L.	Improves growth	García de la Banda et al. (2012)
8.	Bacillus	*B. circulans PB7; B. subtilis; B. licheniformis; B. subtilis UTM 126; B. subtilis E20; B. megaterium*; *B. pumilus*; *Bacillus* P64; *Bacillus* 48	*Labeo rohita;* Trout; *Litopenaeus vannamei*; *Ictalurus punctatus; Pangasianodon hypophthalmus; P. japonicus; O. niloticus; Centropomus undecimalis*	Improves feed conversion ratio; growth; immunity status and protects against *A. hydrophila; Yersinia ruckeri; V. harveyi; Edwardsiella ictaluri*	Zorriehzahra et al. (2016)
9.	Brochothrix	*Thermosphacta* ba211	*Oncorhynchus mykiss*	Protects against *A. bestiarum*	Pieters et al. (2008)

(continued)

Table 6.7 (continued)

S. no.	Microorganism		For aquatic species	Beneficial impact	References
	Genus	Species			
10.	"Clostridium"	*Butyricum*	*Oncorhynchus mykiss; Miichthys miiuy*	Increases immunity and protects against vibriosis; *A. hydrophila;* and *V. anguillarum*	Pan et al. (2008)
11.	Enterococcus	*E. faecium* SF 68 *E. faecium* MC13	*"Anguilla anguilla Penaeus monodon"*	Protects against *Edwardsiellosis*; *V. harveyi;* and *V. parahaemolyticus*	Chang and Liu (2002), Swain et al. (2009)
12.	Kocuria	*Kocuria SM1*	*O. mykiss*	"Protects against *V. anguillarum and V. ordalii"*	Sharifuzzaman and Austin (2010)
13.	Lactobacillus	*L. acidophilus;* *L. rhamnosus;* *L. rhamnosus;* *L. fructivorans;* *L. plantarum; L. plantarum*	*Nile tilapia; Clarias gariepinus; O. mykiss; O. niloticus; S. aurata*	Increases growth performance, immunity, heat tolerance and protects against *P. fluorescens; S. iniae; A. salmonicida; E. tarda*; *L. garvieae*	Zorriehzahra et al. (2016)
14.	"Lactococcus"	*"lactis AR21"*	*Rotifers*	Improves growth, levels of biochemical constituents and protects against *V. anguillarum*	Seenivasan et al. (2012)
15.	"Leuconostoc"	*L. mesenteroides CLFP 196*	*O. mykiss*	Reduces mortality due to *L. garvieae*	Vendrell et al. (2008)
16.	Micrococcus	*M. luteus; Micrococcus MCCB 104*	*O. mykiss; M. rosenbergii*	Protects against *A. salmonicida*	Irianto and Austin (2002a), Jayaprakash et al. (2005)
17.	Pediococcus	*Pediococcus acidilactici*	*O. mykiss*	Reduces vertebral column compression syndrome	Aubin et al. (2005)
18.	Rhodococcus	*Rhodococcus SM2*	*O. mykiss*	Improves immunity and protects against *V. anguillarum*	

19.	"Streptococcus"	*"S. phocae P180; S. faecium"*	*P. monodon; Oreochromis niloticus (Nile tilapia)*	Increases growth and protects against *V. harveyi* infection; as growth promoters	Swain et al. (2009), Lara-Flores et al. (2003)
20.	Streptomyces	–	*"P. monodon"*	"Growth improved and water quality was also increased"	Das et al. (2006), Newaj-Fyzul et al. (2014)
21.	Vagococcus	*V. fluvialis*	*Sea bass*	"Protects against *V. anguillarum* infection"	Sorroza et al. (2012)
22.	"Dunaliella"	*Tertiolecta*	*Artemia*	Protects against *Vibrio campbellii* and *V. proteolyticus*	Marques et al. (2006)
23.	Yarrowia lipolytica	*Y. lipolytica*	*"Pinctada mazatlanica"*	Improves growth	Aguilar-Macías et al. (2010)

Table 6.8 Various probiotics and their beneficial effect on livestock animals (ruminants, poultry, and swine) (adapted from Mahesh et al. 2021; Lambo et al. 2021)

S. no.	Probiotics	Beneficial effect	References
A. Ruminants			
1.	*Bacillus foraminis, B. firmus, B. licheniformis, Staphylococcus saprophyticus bovis*	Reduces feed intake and acetate to propionate ratio	Yang et al. (2012)
2.	*"Lactobacillus plantarum, L. acidophilus, L. casei, L. salivarius, L. gallinarum, L. reuteri, Streptococcus bovis"*	Stimulates lactate utilizers, stabilizes rumen pH, reduces fecal shedding of *E. coli* O157, and improves the immune function	Mahesh et al. (2021)
3.	*"E. faecalis; L. rhamnosus"*	Improves weight gain, lowered gut pH and maintains the ecology ruminal microbiota	Lan et al. (2016)
4.	*"L. acidophilus; L. casei; B. thermophilum; E. faecium"*	Increases milk fat and organic acids viz. butyric, caproic acid	Hu et al. (2021)
5.	*"L. acidophilus, L. plantarum, B. bifidum, B. subtilis, A. oryzae"*	Increases dry matter intake, milk yield and composition, serum albumin, and reduced globulin during postpartum	Ishaq et al. (2015)
6.	*"L. farraginis, L. reuteri, L. rhamnosus"*	"Increases feed: milk ratio, dry matter intake, milk yield, % milk fat, and protein"	Salvedia et al. (2015)
7.	*"L. casei Zhang, L. plantarum P-8"*	"Improves milk production and milk IgG content, lactoferrin, lysozyme, and lactoperoxidase as well as the rumen fermentative and beneficial bacteria population"	Maake et al. (2021)
8.	*"L. acidophilus, S. cerevisiae, E. faecium, A. oryzae, B. subtilis"*	"Increases percentage of lymphocyte and genes associated with immunity and homeostasis"	Aalaei et al. (2018)
9.	*"L. fermentum, L. plantarum, M. elsdenii, S. cerevisiae"*	Improves "nutrient digestibility, rumen fermentation characteristics, and nitrogen retention"	Shreedhar et al. (2016)
10.	*Megasphaera elsdenii*	Enhances lactate utilization in rumen and synthesis of propionate from lactate	Mahesh et al. (2021)
11.	*Propionibacterium freudenreichii, P. jensenii, and P. acidipropionici*	Enhances lactate utilization in rumen, synthesis of propionate from lactate and regulates ruminal pH	Mahesh et al. (2021)
12.	*Prevotella bryantii*	Moderates' rumen fermentation	Mahesh et al. (2021)
13.	*Bacillus subtilis, B. licheniformis,* and *B. coagulans*	Supporting rumen microbes, moderates rumen fermentation, and improves feed degradation	Mahesh et al. (2021)

(continued)

Table 6.8 (continued)

S. no.	Probiotics	Beneficial effect	References
14.	*Ruminococcus flavefaciens*	Moderates' rumen fermentation	Mahesh et al. (2021)
15.	*Pediococcus acidilactici*	Faster development of rumen function	Mahesh et al. (2021)
B. Poultry			
1.	*"L. acidophilus, L. casei, E. faecium, B. thermophilum"*	Decreases "gizzard weight, abdominal fat, and increased antibody production"	Abou-Kassem et al. (2021)
2.	*"L. acidophilus, L. casei, E. faecium, B. bifidum"*	Improves body weight, and protection against "new castle disease and infectious bursal disease"	Goto et al. (2016)
3.	*"L. salivarius, L. reuteri. L. crispatus, L. johnsonii"*	Downregulates "the expression of cecal tonsils cytokine gene and enhanced antibody-mediated immune responses against a highly immunogenic T cell-dependent antigen"	Mostafa et al. (2014)
4..	*"L. acidophilus, B. subtilis DSM 17299, C. butyricum"*	Increases body weight, digestibility of ileal amino acid, and humoral immune response	Xu et al. (2017)
5.	*"L. acidophilus, L. casei, B. thermophilum, E. faecium"*	Reduces ileal *E. coli*	Adjei-Fremah et al. (2018)
6.	*"A. oryzae, B. subtilis, S. cerevisiae, L. plantarum, Rhodopseudomonas capsulate"*	Improves egg protein quality	Direkvandi et al. (2020)
7.	*"L. casei, L. acidophilus, Bifidobacterium"*	Improves "growth performance, carcass trait, antioxidant capacity gut microbiota, and immunity"	Talebi et al. (2008)
8.	*"L. casei, L. lactis, L. plantarum, Carnobacterium divergens, S. cerevisiae"*	Increases femur elongation and area, elastic strength of tibia and reduced bone strength	Tomaszewska et al. (2018)
C. Swine			
1.	*"L. acidophilus, B. subtilis, S. cerevisiae"*	Improves growth performance and gut microbiota	Liu et al. (2018)
2.	"*L. plantarum* L21, *L. plantarum* L80, *L. paraplantarum* L103, *B. subtilis, L. acidophilus, S. cerevisiae*"	Increases "growth performance, fecal lactobacillus population, and reduced fecal *E. coli*"	Ray et al. (2020)
3.	*"B. coagulans, B. licheniformis, B. subtilis, C. butyricum"*	Improves body weight, growth performance, nutrient digestibility, fecal lactobacilli, and meat quality	Ramlucken et al. (2020)
4.	*"L. amylovorus, L. reuteri LAB 26, L. reuteri LAB 49, L. johnsonii, L. salivarius, L. mucosae"*	Increases bacteria population in the jejunum and influenced the expression of specific intestinal mucosa cytokines	Biswas et al. (2022)

(continued)

Table 6.8 (continued)

S. no.	Probiotics	Beneficial effect	References
5.	*B. subtilis, B. licheniformis*	Increases weight gain, "improved nutrient digestibility in sows and reduced" *E. coli* population in sows	Chung et al. (2015)

membrane structure, and lots of others (Kannan et al. 2021). Glycerol, esters, and fatty acids are the primary constructing blocks of fat and oils.

Fatty acids can be categorized into saturated, monounsaturated, and polyunsaturated types. Saturated fatty acids have no double bonds and are in foods like meat and butter. Monounsaturated fatty acids have one double bond and are also produced by the body. On the other hand, polyunsaturated fatty acids (PUFAs), which have multiple double bonds, are considered essential and must be obtained from the diet. PUFAs are crucial for maintaining human health and play a significant role in preventing and treating various diseases (Masurkar et al. 2015).

Polyunsaturated fatty acids (PUFAs) are defined by their long carbon chains with a carboxyl group at one end and a methyl group at the other. They have multiple double bonds in their structure. Eating too many saturated fatty acids (SFAs) has been linked to health issues like arteriosclerosis, which involves the thickening of artery walls, chest pain called angina pectoris due to blocked heart arteries, and other circulation problems. This happens because SFAs are solid at room temperature and can become hard even when heated or pressurized. In contrast, unsaturated fatty acids remain liquid at room temperature and are less stable (Lee 2013).

Even there are various PUFAs in nature; just a few are critical from a physiological standpoint. Most of the pertinent ones fall beneath the omega series, such as omega-3, 6, 9 fatty acids. Linoleic acid (LA) is that this kind of essential PUFAs that would only be furnished through manner of diet, as modified into previously mentioned (Spector and Kim 2015). Another double bond will have to be added to the fatty acid chain for arachidonic acid (AA) (20:4), an EFA that is a member of the *n*-6 FAs, to be converted from LA (18:3, *n*-6) (Spector and Kim 2015). An analogue of LA, -linoleic acid (ALA) (18:3, *n*-3) is produced through manner of de novo synthesis and is a member of the omega-3 (*n*-3) family. Because of their exquisite useful functionality and numerous health advantages, "eicosapentaenoic acid (EPA) (20:5, *n*-3) and docosahexaenoic acid (DHA) (22:6, *n*-3) have turn out to be critical nutrients. Figure 6.3 illustrates many PUFA varieties."

6.6.1.1 Safety Profile of Omega-3 Fatty Acids

Eating more *n*-3 fatty acids may raise low-density lipids, which can lead to serious illness over time. So, despite their many benefits, it is important to know how much to eat. According to the National Institutes of Health, it is recommended to consume at least 650 mg of *n*-3 fatty acids per day (Oh 2005; Weylandt et al. 2015).

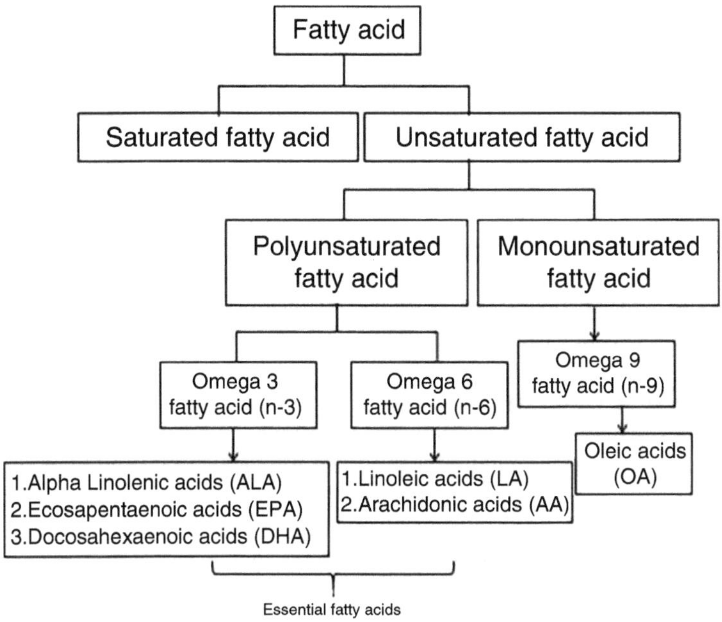

Fig. 6.3 Classification of fatty acids

6.6.2 Development of Microbial Protein for Future Sustainability

Microbes can help create clean energy and convert renewable resources into fuels. They also assist plants in absorbing more nutrients, which is known as "nutrient recycling." Microbes provide essential energy sources to plants, and in return, plants offer them their waste products as nourishment. Microbial proteins contain vitamins, carotene, and carbs in good amounts. These proteins can be produced under regular conditions, which means they aren't affected by land or environmental restrictions like drought or flood.

6.6.2.1 MP as Food

In the context of livestock and aquaculture nutrition, microbial protein offers a high-quality alternative to animal-based protein sources such as fish meal. When we consider human nutrition, microbial protein meets the essential amino acid requirements set by the FAO/WHO, making it a valuable direct dietary option for people as well. Historically, coastal populations have relied on algae as a consistent source of protein and nutrients for many centuries. Nowadays, algae and microalgae are also used in the food industry as ingredients in various foods and dietary supplements (Ravindra 2000; Becker 2007). Yeast has a long history of use in food production, dating back to the earliest bread baking and grape fermentation. It has been employed as a direct food source, such as during World War II when it was distributed to the military and the public. Today, yeast plays a significant role in the production of various microbial-based products for the food industry and beyond. Its primary applications include baking and alcohol fermentation, with an estimated global market value of up to 9.2 billion Euros in 2019 and an expected annual growth rate of 7.9%. Another notable food source is fungi, which have also been consumed by humans. One well-known type of fungal-based protein, known as mycoprotein, is marketed under the name Quorn™ and is available in approximately 15 different countries worldwide (Wiebe 2004). Mycoproteins excel in replicating the taste and texture of meat, making them a popular alternative to traditional animal-based products. Currently, the production of mycoprotein for Quorn™ products amounts to 25,000 tonnes per year, with a market value of approximately 214 million euros and an expected annual growth rate of 20% in the years ahead.

6.6.2.2 Added Value Applications

Microorganisms are having the potential to provide an extensive variety of added-cost merchandise which might be appropriate for each animal and human vitamin, similar to being an excellent supply of nutritious protein (Vandamme and Revuelta 2016). The average protein yield achievable from algae, fungi, and microorganisms, along with other potential value-added products, has been extensively researched and developed.

6.6.2.3 Production of Microbial Protein from Waste Streams

The call for animal-sourced food is expected to boom substantially over the following couple of decades, particularly in rising economies, because of populace increase and developing earnings levels (FAO 2018). In numerous low-income countries, animal-derived foods will continue to play a crucial role in addressing the micronutrient deficiency in the diets of young and vulnerable populations (Nelson et al. 2018). However, it is increasingly important to explore new, previously uncharted nonagricultural protein production methods to meet the rising demand for protein, both for direct human consumption and as inputs for animal production, while minimizing environmental impact. In this context, there is growing interest in the potential of microbial protein as an alternative protein production method (Jones et al. 2020; Matassa et al. 2020; Pikaar et al. 2018a). Microbial proteins encompass algae, yeast, bacteria, and fungi (Matassa et al. 2016). Organic waste streams, along with gaseous substrates such as methane, carbon dioxide, and hydrogen, can be harnessed to produce microbial proteins for use in human food, animal feed, and as slow-release natural fertilizers (Matassa et al. 2020; Pikaar et al. 2018a, b). Although, it is not however economically feasible to replace traditional soybeans with microbial proteins, mycoprotein is now a doable opportunity for fish meal in aquaculture and meat in human. Moreover, as energy charges fall, conventional feed charges rise, or environmental pollution is charged, extraordinary microbial protein production strategies also can fast start to look increasingly more appealing. Circular feeds are much more likely to be ordinary via way of means of the overall populace than different plant-primarily based totally proteins, which can be gaining a quick increasing marketplace among high-earnings consumers (Pikaar et al. 2018a, b). Several SDGs may be substantially impacted via way means of round feed technology, each favorably and unfavorably. For instance, microbial protein may also lower the call for soybean meal, that is presently basically used for animal feed, in addition to the profitability of the soybean industry, the growth of soybean cultivating regions (the latest motive force of land-use change), and greenhouse fuel line emissions. It can also have a fantastic effect on the SDG intention of biodiversity and the soybean sector's profitability (SDG). But soybeans do not simply yield protein. Soybeans do not just provide protein. Hence, a decrease in soybean oil supply might lead to increased palm oil production and consumption, which could have a ripple effect on deforestation (SDG) and potentially raise the prevalence of non-communicable diseases (NCDs; SDG) (Chen et al. 2011; Kadandale et al. 2019). Additionally, microbial protein may decrease the call for fish used as animal feed, which may gain fish populations (SDG). If broadly used, round feed ought to partly uncouple the manufacturing of protein-wealthy animal feed from land use, presenting a second course for decreasing greenhouse fuel line emissions, even as the results for the marketplace energy and pricing of feed deliver consolidation in the meals' gadget are unknown. Conversely, an ample and cheaper deliver of feed may decrease the value of elevating animals and lift the call for cattle products. This impact ought to result in better greenhouse fuel line emissions and probably greater obesity (You and Henneberg 2016; Wang and Beydoun 2009) and NCDs in regions in which meat consumption is already high (SDG). The dietary reputation and fitness of undernourished subpopulations, in particular youngsters and pregnant or lactating women, may be advanced via way of means of extra intake of livestock products (Shapiro et al. 2019). Small-scale cattle farmers' livelihoods can be impacted via way means of decrease feed (SDG). Furthermore, recycling food waste could also enhance the economic value of waste (SDG). Apart from potential trade-offs with existing livelihood options and their environmental impacts, such as the reduced availability of animal manure as a source of organic soil nutrients in mixed crop-livestock systems, this effect could create new sources of income from waste collection, distribution, and processing.

6.6.3 Microbial Fuel Cells for Generation of Electricity

Bioelectricity production is the process by which organisms generate energy through the production of electrons during their metabolic processes. These generated electrons can be collected to maintain a steady or continuous source of power generation. When provided with the right substances, bacterial cells can metabolize them and produce electrons that can be retrieved and used by connecting them through a circuit. These components can be combined to create an assembly called a 'microbial fuel cell' (MFC), which serves as an energy source. The microorganisms' anaerobic digestion of the substrate is crucial for the creation of the electrons resulting from their metabolism. The mentioned reactions illustrate the metabolic processes that were initially carried out by microorganisms without oxygen and later with oxygen (Moqsud et al. 2013).

6.6.3.1 Microbial Fuel Cell

Most MFCs consist of two separate chambers: the anodic chamber, housing the anode, and the cathodic chamber, housing the cathode. These chambers are separated by a proton exchange membrane (PEM). A suitable substrate is provided to the microbes in the anodic chamber, where it is aerobically broken down to release electrons. These electrons are then transported from the anode to the cathode through an external circuit. Meanwhile, the protons generated in this process selectively pass through the exchange membrane. Both of these byproducts created by the microorganisms in the anodic compartment migrate to the cathode and combine with oxygen to form water (Sharma and Li 2010).

MFCs, utilizing electrochemically active bacteria (EAB) to oxidize various carbon sources, including organic waste, are devices capable of converting chemical energy into electrical energy (Angenent et al. 2004, 2004; Lovley 2008; Logan 2009). The MFC chambers may be fabricated from Plexiglas, polycarbonate, or glass (Rhoads et al. 2005). Anode electrodes may be crafted from substances such as "carbon paper, carbon cloth, graphite, and graphite felt (Zhang et al. 2011; Zhuo et al. 2011; Wei et al. 2011; Sangeetha and Muthukumar 2013)." To keep the electrode's cardio nature, an air cathode is needed, and this may be built of platinum (Pt) or Pt-black catalyst substances. The natural substrates within side the anode chamber might be utilized by the microorganisms to supply electrons, in order to then tour thru the outside circuit to the cathode and be absorbed with the aid of using the answer within side the cathode chamber. The ion trade membrane lets in the produced "protons to head from the anode to the cathode (Wang et al. 2013)." Permanganate (MnO_4) and ferricyanide $[(Fe(CN_6))^3]$ answers can characteristic as green catholytes however aren't long-lasting (Jang et al. 2004; Wei et al. 2012).

6.6.3.1.1 Design

Depending on elements just like the quantity of chambers, the way of operation, etc., there are numerous designs for the constructing of an MFC. These classes specifically consist of:

6.6.3.1.1.1 Two-Chamber MFC

This format is an everyday MFC with chambers which might be separated through manner of approach of an ion extrude membrane. These are currently best applied in laboratories and normally characteristic in batch mode, though they can also "characteristic" in non-forestall mode (Du et al. 2007).

6.6.3.1.1.2 Single-Chamber MFC

An anodic chamber and an air cathode are the best additives of a single-chamber MFC, that is, wherein protons and electrons are exchanged. "For the constructing of a single-chamber MFC, diverse designs, including Table 6.9, were suggested."

6.6.3.1.1.3 Stacked MFC

A stack of MFCs related in collection or parallel collectively is known as a microbial fuel (Aelterman et al. 2006). By acquiring numerous configurations of the hydraulic waft and electrode, MFC may be layered. These are available in four special varieties; along with (i) "parallel electrode connections in parallel waft mode (ii), collection electrode connections in collection waft mode (iii), and (iv) collection electrode connections in parallel waft mode (Choi and Ahn 2013)." While treating wastewater (Choi and Ahn 2013), discovered that the use of parallel electrode connections (collection waft mode) elevated the general balance of the oxidation-discount potentials and led to more COD removal, Colombia efficiencies, and maximal strength densities.

6.6.3.1.2 Applications

6.6.3.1.2.1 Production of Bioelectricity

The use of herbal carbohydrate substrates that is derived from municipal, "industrial and agricultural wastes for the generation of bioelectricity are the precept feature of an MFC." Another advantage of MFCs is the direct conversion of fueling molecules into power without the arrival of heat. By averting the Carnot cycle, which lowers the overall performance of thermal energy conversion, a higher conversion overall performance (gt; 70%) is possible (Du et al. 2007). MFCs have made advances in power generation all through the previous couple of years; however, the reality is that they will be now not currently a cost-effective technology for producing power. Substrates which consist of domestic wastewater and glucose, respectively, had been used to create power outputs of 10–50 and 250–500 mW/m^2, respectively (Logan 2004). "A mixed consortium of microbial community" and a sincere substrate like glucose were utilized (Rabaey et al. 2003) to offer a power density of 3.6 W/m^2 (Rabaey et al. 2003). The use of the particular microorganism Rhodoferax ferrireducens, that could oxidize glucose to CO_2 without the requirement for electron mediators to move the electrons to the anode, end up suggested with the useful resource of the use of Chaudhuri and Lovely in 2003 (Rabaey et al. 2003). As a result, the need for a virtual mediator is removed, setting up the door for future changes to the MFC format that might boom its overall performance. Another approach that might be used to enhance power generation and examine simultaneously through several unique applications is stacking MFCs. Six separate MFC gadgets were joined in a stacked configuration with the useful resource of the use of Aelterman et al. (2006), yielding a maximum power output of 258 W/m^3 on an hourly average. Although the power output of MFC is not as right as that of various fueling cells, such as methanol-driven FCs, the fluctuation in substrate consumption gives the device some brought appeal (Rabaey and Verstraete 2005). Additionally, a self-retaining phototrophic MFC has been advanced that produces non-forestall power output beneath illumination without the need for substrate, such as organics or nutrients, and if improved, have to feature as a reliable possibility deliver of sustainable energy (He et al. 2008). The metabolic interest of *Rhodobacter sphaeroides* ended up being used to assemble an MFC for the in situ oxidation of photo-natural hydrogen for the producing of power. "Therefore, this MFC technology" has the functionality to be used as a sustainable energy deliver. The concept of a bio-battery that could recharge domestic system and exceptional devices requiring

Table 6.9 Reports on designs proposed for the construction of a single-chamber MFC

Design	Anode	Cathode
"Rectangular"	"Mn^{4+} graphite anode"	"Fe^{3+} graphite cathode"
"Cylindrical"	"Carbon paper without wet proofing"	"Carbon electrode/PEM assembly or rigid carbon paper without PEM"
"Tubular"	"Granular graphite matrix"	"Ferricyanide solution"
"Flat plate"	"Carbon paper"	"Carbon cloth"
"Concentric design"	"Graphite"	"Air porous made up of carbon/Pt catalyst"

low voltage, additionally may be used with MFC technology. The primary and crucial format of an MFC has long past through several revisions, developing a foundation for the development of new thoughts and applications.

6.6.3.1.2.2 Bio-hydrogen Production

Typically, in MFCs, microorganism performs as catalysts, oxidizing the substrates with inside the anodic chamber to provide protons and electrons which might be then added to the cathode through the wire (externally) and the PEM, respectively. They each integrate to make water, casting off the opportunity of hydrogen synthesis. The electrode that generates power or the microorganism that make decreased metabolites like methane or hydrogen fueling can function the remaining electron acceptor for the microorganism with inside the anodic compartment at some stage in substrate oxidation (Rabaey et al. 2004). In an MFC, outside capability software enabled the formation of hydrogen on the cathode with the aid of using overcoming the thermodynamic barrier via the interplay of electrons and protons. This gives a capability manner to alternate the MFCs operation in order that its miles centered on generating bio-hydrogen. In an MFC, superior microorganisms have been taken from the anodic chamber and blanketed microbial consortia that, upon identification, have been recognized to additionally consist of a few species able to developing hydrogen. In this investigation, while the electron switch fee increased, the hydrogen manufacturing rates, which have been to start with as much as 43 5%, reduced underneath the detection limit. As a result, this examines indicates that power manufacturing and hydrogen introduction are neither affected nor because of each other in MFCs and are extraordinarily implausible to take place on the identical time.

6.6.3.1.2.3 Wastewater Management

Industrial, municipal, and different wastewater effluent function a key useful resource for strength harvesting and on the equal time show to be a great substrate for bioremediation. The chronic trouble of wastewater control has been solved flawlessly via way of means of microbial gasoline era. Maximum strength density, Colombia efficiencies, and COD are the primary three elements that decide how green MFC era is. When the usage of natural substrates like acetate, glucose, and sucrose, MFC era continually achieves better most strength densities (494 mW/m^2) than while the usage of a complicated "substrate like wastewater (146 mW/m^2)

(Feng et al. 2008). Domestic wastewater (Liu et al. 2004), swine wastewater (Min et al. 2005), meat packing wastewater (Heilmann and Logan 2006)," meals processing wastewater, "hydrogen fermentation reactor effluent (Oh and Logan 2005)," and brewery wastewater have all been examined the usage of MFCs for strength generation (Kim et al. 2004). Complex natural substrates, inclusive of wastewater remnants from numerous sources, can also additionally pose capacity troubles that might intrude with the manufacturing of electricity, inclusive of toxicity because of an excessive ammonia attention or because of the manufacturing of risky acids at some stage in "hydrolysis and substrate fermentation (Min et al. 2005)".

6.6.3.1.2.4 Biosensors

In the sector of biosensors, MFCs are regularly used to evaluate the pollutant content material of numerous wastewater effluents. The organic oxygen calls for numerous remedy plant effluents may be calculated in the usage of MFC technology. Due to the proportionate dating among the Colombia yield and the organic electricity of wastewater, MFCs have this capability (Kim et al. 2003).

6.7 Conclusion

Human sustainability depends on a higher agricultural yield so that the population may be fed. A lot of hindrances exist in our nature due to anthropogenic activity. Some factors exist naturally; however, these factors are getting increased day by day due to the greedy nature of the human race. The entire web of the ecosystem on our planet is getting damaged due to the overexploitation of nature. Microbiotechnology, a flourishing arena, may help to create a better environment for higher plant production, reduce pollution, and develop new sustainable energy sources. Advanced technologies should be implicated at the field level and should be transferred at the farmer level so that all corners of our society may get benefitted and human civilization may sustain well for longer time period in sustainable manner.

References

Aalaei M, Khatibjoo A, Zaghari M, Taherpour K, Akbari Gharaei M, Soltani M (2018) Comparison of single- and multi-strain probiotics effects on broiler breeder performance, egg production, egg quality and hatchability. Br Poult Sci 59(5):531–538

Abataneh E, Gizaw B, Tsegaye Z, Wassie M (2017) The role of microorganisms in bioremediation—a review. Open J Environ Biol 2(1):38–46

Abou-Kassem DE, Elsadek MF, Abdel-Moneim AE, Mahgoub SA, Elaraby GM, Taha AE et al (2021) Growth, carcass characteristics, meat quality, and microbial aspects of growing quail fed diets enriched with two different types of probiotics (*Bacillus toyonensis* and *Bifidobacterium bifidum*). Poult Sci 100(1):84–93

Adebajo SO, Balogun SA, Akintokun AK (2017) Decolourization of vat dyes by bacterial isolates recovered from local textile mills in Southwest, Nigeria. Microbiol Res J Int 18.1–8

Adjei-Fremah S, Ekwemalor K, Asiamah EK, Ismail H, Ibrahim S, Worku M (2018) Effect of probiotic supplementation on growth and global gene expression in dairy cows. J Appl Anim Res 46(1):257–263

Aelterman P, Rabaey K, Pham HT, Boon N, Verstraete W (2006) Continuous electricity generation at high voltages and currents using stacked microbial fuel cells. Environ Sci Technol 40(10): 3388–3394

Agarry SE, Ayobami OA (2011) Evaluation of microbial systems for biotreatment of textile waste effluents in Nigeria: biodecolourization and biodegradation of textile dye. J Appl Sci Environ Manag 15(1):79–86

Aguilar-Macías OL, Ojeda-Ramírez JJ, Campa-Córdova AI, Saucedo PE (2010) Evaluation of natural and commercial probiotics for improving growth and survival of the pearl oyster, Pinctada mazatlanica, during late hatchery and early field culturing. J World Aquacult Soc 41(3):447–454

Ahmad M, Zahir ZA, Asghar HN, Asghar M (2011) Inducing salt tolerance in mung bean through coinoculation with rhizobia and plant-growth-promoting rhizobacteria containing 1-aminocyclopropane-1-carboxylate deaminase. Can J Microbiol 57(7):578–589

Alayande KA, Aiyegoro OA, Ateba CN (2020) Probiotics in animal husbandry: applicability and associated risk factors. Sustainability 12(3):1087

Ali R, Singh DV (2020) Biosensors: a biotechnological tool for monitoring environmental pollution. In: Bioremediation and biotechnology, vol 3. Springer, Cham, pp 331–348

Aliakbar Ahovan Z, Hashemi A, De Plano LM, Gholipourmalekabadi M, Seifalian A (2020) Bacteriophage based biosensors: trends, outcomes and challenges. Nanomaterials 10(3):501

Almaghrabi OA, Abdelmoneim TS, Albishri HM, Moussa TA (2014) Enhancement of maize growth using some plant growth promoting rhizobacteria (PGPR) under laboratory conditions. Life Sci J 11(11):764–772

Anastasi A, Parato B, Spina F, Tigini V, Prigione V, Varese GC (2011) Decolourisation and detoxification in the fungal treatment of textile wastewaters from dyeing processes. New Biotechnol 29(1):38–45

Angenent LT, Karim K, Al-Dahhan MH, Wrenn BA, Domíguez-Espinosa R (2004) Production of bioenergy and biochemicals from industrial and agricultural wastewater. Trends Biotechnol 22(9):477–485

Ansari RA, Mahmood I, Rizvi R, Sumbul A (2017) Siderophores: augmentation of soil health and crop productivity. In: Probiotics in agroecosystem. Springer, Singapore, pp 291–312

Asgher M, Noreen S, Bhatti HN (2010) Decolorization of dye-containing textile industry effluents using Ganoderma lucidum ibl-05 in still cultures. Water Environ Res 82(4):357–361

Asiandu AP, Wahyudi A, Sari SW (2021) A review: plastics waste biodegradation using plastics-degrading bacteria. J Environ Treat Tech 9(1):148–157

Assadi MM, Jahangiri MR (2001) Textile wastewater treatment by Aspergillus niger. Desalination 141(1):1–6

Aubin J, Gatesoupe FJ, Labbé L, Lebrun L (2005) Trial of probiotics to prevent the vertebral column compression syndrome in rainbow trout (*Oncorhynchus mykiss* Walbaum). Aquac Res 36(8):758–767

Awasthi R, Bhandari K, Nayyar H (2015) Temperature stress and redox homeostasis in agricultural crops. Front Environ Sci 3:11

Ayed L, Mahdhi A, Cheref A, Bakhrouf A (2011) Decolorization and degradation of azo dye Methyl Red by an isolated Sphingomonas paucimobilis: biotoxicity and metabolites characterization. Desalination 274(1–3):272–277

Barnawal D, Bharti N, Pandey SS, Pandey A, Chanotiya CS, Kalra A (2017) Plant growth-promoting rhizobacteria enhance wheat salt and drought stress tolerance by altering endogenous phytohormone levels and TaCTR1/TaDREB2 expression. Physiol Plant 161(4):502–514

Becker EW (2007) Micro-algae as a source of protein. Biotechnol Adv 25(2):207–210

Bentzon-Tilia M, Sonnenschein EC, Gram L (2016) Monitoring and managing microbes in aquaculture—towards a sustainable industry. Microb Biotechnol 9(5):576–584. https://doi.org/10.1111/1751-7915.12392

Bharti N, Barnawal D (2019) Amelioration of salinity stress by PGPR: ACC deaminase and ROS scavenging enzymes activity. In: PGPR amelioration in sustainable agriculture. Woodhead Publishing, Sawston, pp 85–106

Bharti N, Yadav D, Barnawal D, Maji D, Kalra A (2013) Exiguobacterium oxidotolerans, a halotolerant plant growth promoting rhizobacteria, improves yield and content of secondary metabolites in *Bacopa monnieri* (L.) Pennell under primary and secondary salt stress. World J Microbiol Biotechnol 29(2):379–387

Bhattacharyya PN, Jha DK (2012) Plant growth-promoting rhizobacteria (PGPR): emergence in agriculture. World J Microbiol Biotechnol 28(4):1327–1350

Biswas A, Dev K, Tyagi PK, Mandal A (2022) The effect of multi-strain probiotics as feed additives on performance, immunity, expression of nutrient transporter genes and gut morphometry in broiler chickens. Anim Biosci 35(1):64

Bowles TM, Acosta-Martínez V, Calderón F, Jackson LE (2014) Soil enzyme activities, microbial communities, and carbon and nitrogen availability in organic agroecosystems across an intensively-managed agricultural landscape. Soil Biol Biochem 68:252–262

Bresson J, Varoquaux F, Bontpart T, Touraine B, Vile D (2013) The PGPR strain Phyllobacterium brassicacearum STM 196 induces a reproductive delay and physiological changes that result in improved drought tolerance in Arabidopsis. New Phytol 200(2):558–569

Brownbridge M, Buitenhuis R (2019) Integration of microbial biopesticides in greenhouse floriculture: the Canadian experience. J Invertebr Pathol 165:4–12

Brunt J, Austin B (2005) Use of a probiotic to control lactococcosis and streptococcosis in rainbow trout, *Oncorhynchus mykiss* (Walbaum). J Fish Dis 28(12):693–701

Burbank DR, Shah DH, LaPatra SE, Fornshell G, Cain KD (2011) Enhanced resistance to coldwater disease following feeding of probiotic bacterial strains to rainbow trout (*Oncorhynchus mykiss*). Aquaculture 321(3–4):185–190

Cabello FC, Godfrey HP, Tomova A, Ivanova L, Dölz H, Millanao A, Buschmann AH (2013) Antimicrobial use in aquaculture re-examined: its relevance to antimicrobial resistance and to animal and human health. Environ Microbiol 15(7):1917–1942

Cassán FD, Lucangeli CD, Bottini R, Piccoli PN (2001) Azospirillum spp. metabolize [17,17-2H2] gibberellin A20 to [17,17-2H2] gibberellin A1 in vivo in dy rice mutant seedlings. Plant Cell Physiol 42(7):763–767

Castillo-Aguilar C, Zúñiga-Aguilar JJ, Guzmán-Antonio AA, Garruña R (2017) PGPR inoculation improves growth, nutrient uptake and physiological parameters of Capsicum chinense plants. Phyton 86:199

Chang CI, Liu WY (2002) An evaluation of two probiotic bacterial strains, Enterococcus faecium SF68 and Bacillus toyoi, for reducing Edwardsiellosis in cultured European eel, Anguilla anguilla L. J Fish Dis 25(5):311–315

Chen Q, Liu S (2019) Identification and characterization of the phosphate-solubilizing bacterium Pantoea sp. S32 in reclamation soil in Shanxi, China. Front Microbiol 10:2171

Chen BK, Seligman B, Farquhar JW, Goldhaber-Fiebert JD (2011) Multi-Country analysis of palm oil consumption and cardiovascular disease mortality for countries at different stages of economic development: 1980-1997. Glob Health 7(1):1–10

Cheunbarn TAPANA, Cheunbarn SIRAPORN, Khumjai THITINUT (2008) Prospects of bacterial granule for treatment of real textile industrial wastewater. Int J Agric Biol 10(6):89–92

Choi J, Ahn Y (2013) Continuous electricity generation in stacked air cathode microbial fuel cell treating domestic wastewater. J Environ Manag 130:146–152

Chung SH, Lee J, Kong C (2015) Effects of multi strain probiotics on egg production and quality in laying hens fed diets containing food waste product. Int J Poult Sci 14(1):19

Cing SDEO, Hamamci DA, Apohan ELİF, Yeşilada O (2003) Decolorization of textile dyeing wastewater by Phanerochaete chrysosporium. Folia Microbiol 48(5):639–642

Claude D, Houssemeddine G, Andriy B, Jean-Marc C (2007) Whole cell algal biosensors for urban waters monitoring. Ther Nova 7(3):1507–1514

Czaja K, Góralczyk K, Struciński P, Hernik A, Korcz W, Minorczyk M et al (2015) Biopesticides—towards increased consumer safety in the European Union. Pest Manag Sci 71(1):3–6

D'Souza DT, Tiwari R, Sah AK, Raghukumar C (2006) Enhanced production of laccase by a marine fungus during treatment of colored effluents and synthetic dyes. Enzym Microb Technol 38(3–4):504–511

Dangi AK, Sharma B, Hill RT, Shukla P (2019) Bioremediation through microbes: systems biology and metabolic engineering approach. Crit Rev Biotechnol 39(1):79–98

Das AJ, Kumar R (2016) Bioremediation of petroleum contaminated soil to combat toxicity on Withania somnifera through seed priming with biosurfactant producing plant growth promoting rhizobacteria. J Environ Manag 174:79–86

Das S, Lyla PS, Ajmal Khan S (2006) Application of Streptomyces as a probiotic in the laboratory culture of Penaeus monodon (Fabricius). Isr J Aquac Bamidgeh 58(3):198–204

Dawkar VV, Jadhav UU, Jadhav SU, Govindwar SP (2008) Biodegradation of disperse textile dye Brown 3REL by newly isolated Bacillus sp. VUS J Appl Microbiol 105(1):14–24

Dawkar VV, Jadhav UU, Tamboli DP, Govindwar SP (2010) Efficient industrial dye decolorization by Bacillus sp. VUS with its enzyme system. Ecotoxicol Environ Saf 73(7):1696–1703

de Lima BC, Moro AL, Santos ACP, Bonifacio A, Araujo ASF, de Araujo FF (2019) Bacillus subtilis ameliorates water stress tolerance in maize and common bean. J Plant Interact 14(1): 432–439

de Vasconcellos RLF, Cardoso EJBN (2009) Rhizospheric streptomycetes as potential biocontrol agents of Fusarium and Armillaria pine rot and as PGPR for Pinus taeda. BioControl 54(6): 807–816

Deng D, Guo J, Zeng G, Sun G (2008) Decolorization of anthraquinone, triphenylmethane and azo dyes by a new isolated Bacillus cereus strain DC11. Int Biodeterior Biodegradation 62(3): 263–269

Direkvandi E, Mohammadabadi T, Salem AZ (2020) Oral administration of lactate producing bacteria alone or combined with Saccharomyces cerevisiae and Megasphaera elsdenii on performance of fattening lambs. J Appl Anim Res 48(1):235–243

Diwaniyan S, Kharb D, Raghukumar C, Kuhad RC (2010) Decolorization of synthetic dyes and textile effluents by basidiomycetous fungi. Water Air Soil Pollut 210(1):409–419

Du Z, Li H, Gu T (2007) A state of the art review on microbial fuel cells: a promising technology for wastewater treatment and bioenergy. Biotechnol Adv 25(5):464–482

Dubois M, Van den Broeck L, Inzé D (2018) The pivotal role of ethylene in plant growth. Trends Plant Sci 23(4):311–323

Egamberdieva D, Wirth SJ, Alqarawi AA, Abd Allah EF, Hashem A (2017) Phytohormones and beneficial microbes: essential components for plants to balance stress and fitness. Front Microbiol 8:2104

Egamberdieva D, Wirth S, Bellingrath-Kimura SD, Mishra J, Arora NK (2019) Salt-tolerant plant growth promoting rhizobacteria for enhancing crop productivity of saline soils. Front Microbiol 10:2791

Elahi A, Bukhari DA, Shamim S, Rehman A (2021) Plastics degradation by microbes: a sustainable approach. J King Saud Univ Sci 33(6):101538

FAO (2018) The future of food and agriculture: alternative pathways to 2050. Food and Agriculture Organization of the United Nations, Rome

Farooq M, Wahid A, Kobayashi NSMA, Fujita DBSMA, Basra SMA (2009) Plant drought stress: effects, mechanisms and management. In: Sustainable agriculture. Springer, Dordrecht, pp 153–188

Fasciglione G, Casanovas EM, Quillehauquy V, Yommi AK, Goñi MG, Roura SI, Barassi CA (2015) Azospirillum inoculation effects on growth, product quality and storage life of lettuce plants grown under salt stress. Sci Hortic 195:154–162

Feng Y, Wang X, Logan BE, Lee H (2008) Brewery wastewater treatment using air-cathode microbial fuel cells. Appl Microbiol Biotechnol 78(5):873–880
Fenner K, Canonica S, Wackett LP, Elsner M (2013) Evaluating pesticide degradation in the environment: blind spots and emerging opportunities. Science 341(6147):752–758
Figueiredo MV, Burity HA, Martinez CR, Chanway CP (2008) Alleviation of drought stress in the common bean (Phaseolus vulgaris L.) by co-inoculation with Paenibacillus polymyxa and Rhizobium tropici. Appl Soil Ecol 40(1):182–188
Galanakis CM (ed) (2017) Handbook of grape processing by-products: sustainablc solutions. Academic Press, New York
García de la Banda I, Lobo C, Chabrillón M, León-Rubio JM, Arijo S, Pazos G et al (2012) Influence of dietary administration of a probiotic strain Shewanella putrefaciens on Senegalese sole (Solea senegalensis, Kaup 1858) growth, body composition and resistance to Photobacterium damselae subsp piscicida. Aquac Res 43(5):662–669
Ghori NH, Ghori T, Hayat MQ, Imadi SR, Gul A, Altay V, Ozturk M (2019) Heavy metal stress and responses in plants. Int J Environ Sci Technol 16(3):1807–1828
Glick BR (2012) Plant growth-promoting bacteria: mechanisms and applications. Scientifica 2012: 963401
Goto H, Qadis AQ, Kim YH, Ikuta K, Ichijo T, Sato S (2016) Effects of a bacterial probiotic on ruminal pH and volatile fatty acids during subacute ruminal acidosis (SARA) in cattle. J Vet Med Sci 78(10):1595–1600
Govindasamy V, Senthilkumar M, Magheshwaran V, Kumar U, Bose P, Sharma V, Annapurna K (2010) Bacillus and Paenibacillus spp.: potential PGPR for sustainable agriculture. In: Plant growth and health promoting bacteria. Springer, Berlin, pp 333–364
Gowtham HG, Hariprasad P, Nayak SC, Niranjana SR (2016) Application of rhizobacteria antagonistic to Fusarium oxysporum f. sp. lycopersici for the management of Fusarium wilt in tomato. Rhizosphere 2:72–74
Gowtham HG, Singh B, Murali M, Shilpa N, Prasad M, Aiyaz M et al (2020) Induction of drought tolerance in tomato upon the application of ACC deaminase producing plant growth promoting rhizobacterium Bacillus subtilis Rhizo SF 48. Microbiol Res 234:126422
Granados-Amores A, Campa-Córdova ÁI, Araya R, Mazón-Suástegui JM, Saucedo PE (2012) Growth, survival and enzyme activity of lions-paw scallop (Nodipecten subnodosus) spat treated with probiotics at the hatchery. Aquac Res 43(9):1335–1343
Hameed A, Rasool S, Azooz MM, Hossain MA, Ahanger MA, Ahmad P (2016) Heavy metal stress: plant responses and signaling. In: Plant metal interaction. Elsevier, Amsterdam, pp 557–583
Hassan MM, Alam MZ, Anwar MN (2013) Biodegradation of textile azo dyes by bacteria isolated from dyeing industry effluent. Int Res J Biol Sci 2(8):27–31
He Z, Huang Y, Manohar AK, Mansfeld F (2008) Effect of electrolyte pH on the rate of the anodic and cathodic reactions in an air-cathode microbial fuel cell. Bioelectrochemistry 74(1):78–82
Heilmann J, Logan BE (2006) Production of electricity from proteins using a microbial fuel cell. Water Environ Res 78(5):531–537
Henkel M, Müller MM, Kügler JH, Lovaglio RB, Contiero J, Syldatk C, Hausmann R (2012) Rhamnolipids as biosurfactants from renewable resources: concepts for next-generation rhamnolipid production. Process Biochem 47(8):1207–1219
Hu J, Kim YH, Kim IH (2021) Effects of two bacillus strains probiotic supplement on reproduction performance, nutrient digestibility, blood profile, fecal score, excreta odor contents and fecal microflora in lactation sows, and growth performance in sucking piglets. Livest Sci 244:104293
Huang H, Ullah F, Zhou DX, Yi M, Zhao Y (2019) Mechanisms of ROS regulation of plant development and stress responses. Front Plant Sci 10:800
Ilangovan R, Daniel D, Krastanov A, Zachariah C, Elizabeth R (2006) Enzyme based biosensor for heavy metal ions determination. Biotechnol Biotechnol Equip 20(1):184–189
Ilyas N, Mumtaz K, Akhtar N, Yasmin H, Sayyed RZ, Khan W et al (2020) Exopolysaccharides producing bacteria for the amelioration of drought stress in wheat. Sustainability 12(21):8876

Imran M, Crowley DE, Khalid A, Hussain S, Mumtaz MW, Arshad M (2015) Microbial biotechnology for decolorization of textile wastewaters. Rev Environ Sci Biotechnol 14(1):73–92
Irianto A, Austin B (2002a) Probiotics in aquaculture. J Fish Dis 25(11):633–642
Irianto A, Austin B (2002b) Use of probiotics to control furunculosis in rainbow trout, Oncorhynchus mykiss (Walbaum). J Fish Dis 25(6):333–342
Ishaq SL, Kim CJ, Reis D, Wright ADG (2015) Fibrolytic bacteria isolated from the rumen of North American moose (Alces alces) and their use as a probiotic in neonatal lambs. PLoS One 10(12): e0144804
Islam S, Akanda AM, Prova A, Islam MT, Hossain MM (2016) Isolation and identification of plant growth promoting rhizobacteria from cucumber rhizosphere and their effect on plant growth promotion and disease suppression. Front Microbiol 6:1360
Jadhav SU, Ghodake GS, Telke AA, Tamboli DP, Govindwar SP (2009) Degradation and detoxification of disperse dye Scarlet RR by Galactomyces geotrichum MTCC 1360. J Microbiol Biotechnol 19(4):409–415
Jadhav JP, Kalyani DC, Telke AA, Phugare SS, Govindwar SP (2010) Evaluation of the efficacy of a bacterial consortium for the removal of color, reduction of heavy metals, and toxicity from textile dye effluent. Bioresour Technol 101(1):165–173
Jang JK, Chang IS, Kim BH (2004) Improvement of cathode reaction of a mediatorless microbial fuel cell. J Microbiol Biotechnol 14(2):324–329
Jang JH, Woo SY, Kim SH, Khaine I, Kwak MJ, Lee HK et al (2017) Effects of increased soil fertility and plant growth-promoting rhizobacteria inoculation on biomass yield, energy value, and physiological response of poplar in short-rotation coppices in a reclaimed tideland: a case study in the Saemangeum area of Korea. Biomass Bioenergy 107:29–38
Jasińska A, Różalska S, Bernat P, Paraszkiewicz K, Długoński J (2012) Malachite green decolorization by non-basidiomycete filamentous fungi of Penicillium pinophilum and Myrothecium roridum. Int Biodeterior Biodegradation 73:33–40
Jasińska A, Bernat P, Paraszkiewicz K (2013) Malachite green removal from aqueous solution using the system rapeseed press cake and fungus Myrothecium roridum. Desalin Water Treat 51(40–42):7663–7671
Jayaprakash NS, Pai SS, Anas A, Preetha R, Philip R, Singh IB (2005) A marine bacterium, Micrococcus MCCB 104, antagonistic to vibrios in prawn larval rearing systems. Dis Aquat Org 68(1):39–45
Jayaprakashvel M, Mathivanan N (2011) Management of plant diseases by microbial metabolites. In: Bacteria in agrobiology: plant nutrient management. Springer, Berlin, pp 237–265
Jha Y, Subramanian RB (2014) PGPR regulate caspase-like activity, programmed cell death, and antioxidant enzyme activity in paddy under salinity. Physiol Mol Biol Plants 20(2):201–207
Jiménez G (2010) Probiotics in animal nutrition—a century of research. All About Feed 1(5):14–16
Jin XC, Liu GQ, Xu ZH, Tao WY (2007) Decolorization of a dye industry effluent by Aspergillus fumigatus XC6. Appl Microbiol Biotechnol 74(1):239–243
Jones SW, Karpol A, Friedman S, Maru BT, Tracy BP (2020) Recent advances in single cell protein use as a feed ingredient in aquaculture. Curr Opin Biotechnol 61:189–197
Kadandale S, Marten R, Smith R (2019) The palm oil industry and noncommunicable diseases. Bull World Health Organ 97(2):118
Kalam S, Basu A, Podile AR (2020) Functional and molecular characterization of plant growth promoting Bacillus isolates from tomato rhizosphere. Heliyon 6(8):e04734
Kannan N, Rao AS, Nair A (2021) Microbial production of omega-3 fatty acids: an overview. J Appl Microbiol 131(5):2114–2130
Karim F, Fakhruddin ANM (2012) Recent advances in the development of biosensor for phenol: a review. Rev Environ Sci Biotechnol 11(3):261–274
Karim ME, Dhar K, Hossain MT (2018) Decolorization of textile reactive dyes by bacterial monoculture and consortium screened from textile dyeing effluent. J Genet Eng Biotechnol 16(2):375–380

Kaushik P, Malik A (2010) Effect of nutritional conditions on dye removal from textile effluent by Aspergillus lentulus. World J Microbiol Biotechnol 26(11):1957–1964
Kesarcodi-Watson A, Kaspar H, Lategan MJ, Gibson L (2010) Alteromonas macleodii 0444 and Neptunomonas sp. 0536, two novel probiotics for hatchery-reared Greenshell™ mussel larvae, Perna canaliculus. Aquaculture 309(1–4):49–55
Kesarcodi-Watson A, Miner P, Nicolas JL, Robert R (2012) Protective effect of four potential probiotics against pathogen-challenge of the larvae of three bivalves: Pacific oyster (Crassostrea gigas), flat oyster (Ostrea edulis) and scallop (Pecten maximus). Aquaculture 344:29–34
Khan N, Bano A (2016) Role of plant growth promoting rhizobacteria and Ag-nano particle in the bioremediation of heavy metals and maize growth under municipal wastewater irrigation. Int J Phytoremediation 18(3):211–221
Khan N, Mishra A, Nautiyal CS (2012) Paenibacillus lentimorbus B-30488r controls early blight disease in tomato by inducing host resistance associated gene expression and inhibiting Alternaria solani. Biol Control 62(2):65–74
Khan TA, Fariduddin Q, Yusuf M (2017) Low-temperature stress: is phytohormones application a remedy? Environ Sci Pollut Res 24(27):21574–21590
Kim BH, Chang IS, Cheol Gil G, Park HS, Kim HJ (2003) Novel BOD (biological oxygen demand) sensor using mediator-less microbial fuel cell. Biotechnol Lett 25(7):541–545
Kim BH, Park HS, Kim HJ, Kim GT, Chang IS, Lee J, Phung NT (2004) Enrichment of microbial community generating electricity using a fuel-cell-type electrochemical cell. Appl Microbiol Biotechnol 63(6):672–681
Kiran EU, Trzcinski AP, Ng WJ, Liu Y (2014) Bioconversion of food waste to energy: a review. Fuel 134:389–399
Knecht MR, Sethi M (2009) Bio-inspired colorimetric detection of Hg2+ and Pb2+ heavy metal ions using Au nanoparticles. Anal Bioanal Chem 394(1):33–46
Köhler HR, Triebskorn R (2013) Wildlife ecotoxicology of pesticides: can we track effects to the population level and beyond? Science 341(6147):759–765
Kumar S, Chaurasia P, Kumar A (2016) Isolation and characterization of microbial strains from textile industry effluents of Bhilwara, India: analysis with bioremediation. J Chem Pharm Res 8(4):143–150
Kumar N, Mukherjee I, Sarkar B, Paul RK (2017) Degradation of tricyclazole: effect of moisture, soil type, elevated carbon dioxide and Blue Green Algae (BGA). J Hazard Mater 321:517–527
Kumar A, Kumar D, George N, Sharma P, Gupta N (2018) A process for complete biodegradation of shrimp waste by a novel marine isolate Paenibacillus sp. AD with simultaneous production of chitinase and chitin oligosaccharides. Int J Biol Macromol 109:263–272
Kumar KK, Sridhar J, Murali-Baskaran RK, Senthil-Nathan S, Kaushal P, Dara SK, Arthurs S (2019) Microbial biopesticides for insect pest management in India: current status and future prospects. J Invertebr Pathol 165:74–81
Kumar P, Kamle M, Borah R, Mahato DK, Sharma B (2021) Bacillus thuringiensis as microbial biopesticide: uses and application for sustainable agriculture. Egypt J Biol Pest Control 31(1): 1–7
Lambo MT, Chang X, Liu D (2021) The recent trend in the use of multistrain probiotics in livestock production: an overview. Animals 11(10):2805
Lan RX, Lee SI, Kim IH (2016) Effects of multistrain probiotics on growth performance, nutrient digestibility, blood profiles, faecal microbial shedding, faecal score and noxious gas emission in weaning pigs. J Anim Physiol Anim Nutr 100(6):1130–1138
Lara-Flores M, Olvera-Novoa MA, Guzmán-Méndez BE, López-Madrid W (2003) Use of the bacteria Streptococcus faecium and Lactobacillus acidophilus, and the yeast Saccharomyces cerevisiae as growth promoters in Nile tilapia (Oreochromis niloticus). Aquaculture 216(1–4): 193–201
Lazado CC, Caipang CMA, Estante EG (2015) Prospects of host-associated microorganisms in fish and penaeids as probiotics with immunomodulatory functions. Fish Shellfish Immunol 45(1): 2–12

Leahy J, Mendelsohn M, Kough J, Jones R, Berckes N (2014) Biopesticide oversight and registration at the US Environmental Protection Agency. In: Biopesticides: state of the art and future opportunities. American Chemical Society, Washington, DC, pp 3–18
Lee JH (2013) Polyunsaturated fatty acids in children. Pediatr Gastroenterol Hepatol Nutr 16(3): 153–161
Liu H, Ramnarayanan R, Logan BE (2004) Production of electricity during wastewater treatment using a single chamber microbial fuel cell. Environ Sci Technol 38(7):2281–2285
Liu WC, Ye M, Liao JH, Zhao ZH, Kim IH, An LL (2018) Application of complex probiotics in swine nutrition—a review. Ann Anim Sci 18(2):335–350
Logan BE (2004) Peer reviewed: extracting hydrogen and electricity from renewable resources. Environ Sci Technol 38(9):160A–167A
Logan BE (2009) Exoelectrogenic bacteria that power microbial fuel cells. Nat Rev Microbiol 7(5): 375–381
Lovley DR (2008) The microbe electric: conversion of organic matter to electricity. Curr Opin Biotechnol 19(6):564–571
Maake TW, Adeleke M, Aiyegoro OA (2021) Effect of lactic acid bacteria administered as feed supplement on the weight gain and ruminal pH in two South African goat breeds. Trans R Soc South Africa 76(1):35–40
Mabood F, Zhou X, Smith DL (2014) Microbial signaling and plant growth promotion. Can J Plant Sci 94(6):1051–1063
Mahesh MS, Mohanta RK, Patra AK (2021) Probiotics in livestock and poultry nutrition and health. In: Advances in probiotics for sustainable food and medicine. Springer, Singapore, pp 149–179
Marques A, Thanh TH, Sorgeloos P, Bossier P (2006) Use of microalgae and bacteria to enhance protection of gnotobiotic Artemia against different pathogens. Aquaculture 258(1–4):116–126
Marulanda A, Azcón R, Chaumont F, Ruiz-Lozano JM, Aroca R (2010) Regulation of plasma membrane aquaporins by inoculation with a Bacillus megaterium strain in maize (Zea mays L.) plants under unstressed and salt-stressed conditions. Planta 232(2):533–543
Masurkar AAK, Datar AG, Vakil BV (2015) New microbial sources of PUFA from the Arabian sea and fresh water habitats of Indian territory. Int J Pharm Sci Res 6(1):349
Maszenan AM, Liu Y, Ng WJ (2011) Bioremediation of wastewater with recalcitrant organic compounds and metals by aerobic granules. Biotechnol Adv 29:111–123
Matassa S, Boon N, Pikaar I, Verstraete W (2016) Microbial protein: future sustainable food supply route with low environmental footprint. Microb Biotechnol 9(5):568–575
Matassa S, Papirio S, Pikaar I, Hülsen T, Leijenhorst E, Esposito G et al (2020) Upcycling of biowaste carbon and nutrients in line with consumer confidence: the "full gas" route to single cell protein. Green Chem 22(15):4912–4929
Matharu AS, de Melo EM, Houghton JA (2016) Opportunity for high value-added chemicals from food supply chain wastes. Bioresour Technol 215:123–130
Mayak S, Tirosh T, Glick BR (2004) Plant growth-promoting bacteria confer resistance in tomato plants to salt stress. Plant Physiol Biochem 42(6):565–572
Mazhandu ZS, Muzenda E, Mamvura TA, Belaid M, Nhubu T (2020) Integrated and consolidated review of plastic waste management and bio-based biodegradable plastics: challenges and opportunities. Sustainability 12(20):8360
Meena M, Swapnil P (2019) Regulation of WRKY genes in plant defence with beneficial fungus Trichoderma: current perspectives and future prospects. Arch Phytopathol Plant Protect 52(1–2):1–17
Merzaeva OV, Shirokikh IG (2006) Colonization of plant rhizosphere by actinomycetes of different genera. Microbiology 75(2):226–230
Min B, Kim J, Oh S, Regan JM, Logan BE (2005) Electricity generation from swine wastewater using microbial fuel cells. Water Res 39(20):4961–4968
Mohanan N, Montazer Z, Sharma PK, Levin DB (2020) Microbial and enzymatic degradation of synthetic plastics. Front Microbiol 11:580709

Mondal S, Palit D (2019) Effective role of microorganism in waste management and environmental sustainability. In: Sustainable agriculture, forest and environmental management. Springer, Singapore, pp 485–515

Moqsud MA, Omine K, Yasufuku N, Hyodo M, Nakata Y (2013) Microbial fuel cell (MFC) for bioelectricity generation from organic wastes. Waste Manag 33(11):2465–2469

Mostafa TH, Elsayed FA, Ahmed MA, Elkholany MA (2014) Effect of using some feed additives (TW-probiotics) in dairy cow rations on production and reproductive performance. Egypt J Anim Prod 51(1):1–11

Mundt CC (2014) Durable resistance: a key to sustainable management of pathogens and pests. Infect Genet Evol 27:446–455

Murthy PS, Naidu MM (2012) Sustainable management of coffee industry by-products and value addition—a review. Resour Conserv Recycl 66:45–58

Mustafa S, Kabir S, Shabbir U, Batool R (2019) Plant growth promoting rhizobacteria in sustainable agriculture: from theoretical to pragmatic approach. Symbiosis 78(2):115–123

Negi NP, Choephel T (2020) Biosensor: an approach towards a sustainable environment. In: Nanobiosensors for agricultural, medical and environmental applications. Springer, Singapore, pp 43–62

Nelson G, Bogard J, Lividini K, Arsenault J, Riley M, Sulser TB et al (2018) Income growth and climate change effects on global nutrition security to mid-century. Nat Sustain 1(12):773–781

Newaj-Fyzul A, Al-Harbi AH, Austin B (2014) Developments in the use of probiotics for disease control in aquaculture. Aquaculture 431:1–11

Nezarat S, Gholami A (2009) Screening plant growth promoting rhizobacteria for improving seed germination, seedling growth and yield of maize. Pak J Biol Sci 12(1):26

Ng HS, Kee PE, Yim HS, Chen PT, Wei YH, Lan JCW (2020) Recent advances on the sustainable approaches for conversion and reutilization of food wastes to valuable bioproducts. Bioresour Technol 302:122889

Ning Z, Zhang H, Li W, Zhang R, Liu G, Chen C (2018) Anaerobic digestion of lipid-rich swine slaughterhouse waste: methane production performance, long-chain fatty acids profile and predominant microorganisms. Bioresour Technol 269:426–433

Niu X, Song L, Xiao Y, Ge W (2018) Drought-tolerant plant growth-promoting rhizobacteria associated with foxtail millet in a semi-arid agroecosystem and their potential in alleviating drought stress. Front Microbiol 8:2580

Nnachi RC, Sui N, Ke B, Luo Z, Bhalla N, He D, Yang Z (2022) Recent progress on biosensors for rapid detection of bacterial pathogens in water, food and environment. Environ Int 166:107357

Oh R (2005) Practical applications of fish oil (Ω-3 fatty acids) in primary care. J Am Board Fam Pract 18(1):28–36

Oh S, Logan BE (2005) Hydrogen and electricity production from a food processing wastewater using fermentation and microbial fuel cell technologies. Water Res 39(19):4673–4682

Oldach L, Zhang J (2014) Genetically encoded fluorescent biosensors for live-cell visualization of protein phosphorylation. Chem Biol 21(2):186–197

Oteino N, Lally RD, Kiwanuka S, Lloyd A, Ryan D, Germaine KJ, Dowling DN (2015) Plant growth promotion induced by phosphate solubilizing endophytic Pseudomonas isolates. Front Microbiol 6:745

Pal P, Bhattacharyay D, Mukhopadhyay A, Sarkar P (2009) The detection of mercury, cadium, and arsenic by the deactivation of urease on rhodinized carbon. Environ Eng Sci 26(1):25–32

Pan X, Wu T, Zhang L, Song Z, Tang H, Zhao Z (2008) In vitro evaluation on adherence and antimicrobial properties of a candidate probiotic Clostridium butyricum CB2 for farmed fish. J Appl Microbiol 105(5):1623–1629

Pandey S, Ghosh PK, Ghosh S, De TK, Maiti TK (2013) Role of heavy metal resistant Ochrobactrum sp. and Bacillus spp. strains in bioremediation of a rice cultivar and their PGPR like activities. J Microbiol 51(1):11–17

Parashar A, Jin Y, Mason B, Chae M, Bressler DC (2016) Incorporation of whey permeate, a dairy effluent, in ethanol fermentation to provide a zero waste solution for the dairy industry. J Dairy Sci 99(3):1859–1867

Parate VR, Talib MI (2015) Characterization of tur dal (Cajanus cajan) husk carbon and its kinetics and isotherm study for removing Cu (II) ions. IOSR J Environ Sci 9:27–41

Patel Y, Gupte A (2016) Evaluation of bioremediation potential of isolated bacterial culture YPAG-9 (Pseudomonas aeruginosa) for decolorization of sulfonated di-azodye reactive red HE8B under optimized culture conditions. Int J Curr Microbiol Appl Sci 5(8):258–272

Patel PR, Shaikh SS, Sayyed RZ (2016) Dynamism of PGPR in bioremediation and plant growth promotion in heavy metal contaminated soil. Indian J Exp Biol 54(4):286–290

Perumal SM, Munuswamy D, Sellamuthu PS, Kandasamy M, Puthupalayam Thangavelu K (2007) Biosorption of textile dyes and effluents by Pleurotus florida and Trametes Hirsuta with evaluation of their laccase activity. Iran J Biotechnol 5(2):114–118

Phugare SS, Kalyani DC, Surwase SN, Jadhav JP (2011) Ecofriendly degradation, decolorization and detoxification of textile effluent by a developed bacterial consortium. Ecotoxicol Environ Saf 74(5):1288–1296

Phulpoto AH, Qazi MA, Mangi S, Ahmed S, Kanhar NA (2016) Biodegradation of oil-based paint by Bacillus species monocultures isolated from the paint warehouses. Int J Environ Sci Technol 13(1):125–134

Pieters N, Brunt J, Austin B, Lyndon AR (2008) Efficacy of in-feed probiotics against Aeromonas bestiarum and Ichthyophthirius multifiliis skin infections in rainbow trout (Oncorhynchus mykiss, Walbaum). J Appl Microbiol 105(3):723–732

Pii Y, Mimmo T, Tomasi N, Terzano R, Cesco S, Crecchio C (2015) Microbial interactions in the rhizosphere: beneficial influences of plant growth-promoting rhizobacteria on nutrient acquisition process. A review. Biol Fertil Soils 51(4):403–415

Pikaar I, De Vrieze J, Rabaey K, Herrero M, Smith P, Verstraete W (2018a) Carbon emission avoidance and capture by producing in-reactor microbial biomass based food, feed and slow release fertilizer: potentials and limitations. Sci Total Environ 644:1525–1530

Pikaar I, Matassa S, Bodirsky BL, Weindl I, Humpenöder F, Rabaey K et al (2018b) Decoupling livestock from land use through industrial feed production pathways. Environ Sci Technol 52(13):7351–7359

Podile AR, Kishore GK (2007) Plant growth-promoting rhizobacteria. In: Plant-associated bacteria. Springer, Dordrecht, pp 195–230

Pourbabaee AA, Malekzadeh F, Sarbolouki MN, Najafi F (2006) Aerobic decolorization and detoxification of a disperse dye in textile effluent by a new isolate of Bacillus sp. Biotechnol Bioeng 93(4):631–635

Prameela K, Venkatesh K, Immandi SB, Kasturi APK, Krishna CR, Mohan CM (2017) Next generation nutraceutical from shrimp waste: the convergence of applications with extraction methods. Food Chem 237:121–132

Rabaey K, Verstraete W (2005) Microbial fuel cells: novel biotechnology for energy generation. Trends Biotechnol 23(6):291–298

Rabaey K, Lissens G, Siciliano SD, Verstraete W (2003) A microbial fuel cell capable of converting glucose to electricity at high rate and efficiency. Biotechnol Lett 25(18):1531–1535

Rabaey K, Boon N, Siciliano SD, Verhaege M, Verstraete W (2004) Biofuel cells select for microbial consortia that self-mediate electron transfer. Appl Environ Microbiol 70(9): 5373–5382

Rajmohan KVS, Ramya C, Viswanathan MR, Varjani S (2019) Plastic pollutants: effective waste management for pollution control and abatement. Curr Opin Environ Sci Health 12:72–84

Ramlucken U, Ramchuran SO, Moonsamy G, Lalloo R, Thantsha MS, van Rensburg CJ (2020) A novel Bacillus based multi-strain probiotic improves growth performance and intestinal properties of Clostridium perfringens challenged broilers. Poult Sci 99(1):331–341

Rana A, Saharan B, Joshi M, Prasanna R, Kumar K, Nain L (2011) Identification of multi-trait PGPR isolates and evaluating their potential as inoculants for wheat. Ann Microbiol 61(4): 893–900

Rani N, Sangwan P, Joshi M, Sagar A, Bala K (2019) Microbes. In: Microbial wastewater treatment. Elsevier, Amsterdam, pp 83–102

Ravindra P (2000) Value-added food: Single cell protein. Biotechnol Adv 18(6):459–479

Ray SM, Shivanand Ghule S, Muthukumar AB, Maji C (2020) Effects of dietary supplementation of a single- and a multi-strain probiotic on growth performance and intestinal histomorphology of commercial broiler chickens. Int J Poult Sci 19:363–371

Reshma P, Naik MK, Aiyaz M, Niranjana SR, Chennappa G, Shaikh SS, Sayyed RZ (2018) Induced systemic resistance by 2,4-diacetylphloroglucinol positive fluorescent Pseudomonas strains against rice sheath blight. Indian J Exp Biol 56(3):207–212

Rhoads A, Beyenal H, Lewandowski Z (2005) Microbial fuel cell using anaerobic respiration as an anodic reaction and biomineralized manganese as a cathodic reactant. Environ Sci Technol 39(12):4666–4671

Rohm H, Brennan C, Turner C, Günther E, Campbell G, Hernando I et al (2015) Adding value to fruit processing waste: innovative ways to incorporate fibers from berry pomace in baked and extruded cereal-based foods—a SUSFOOD project. Foods 4(4):690–697

Sagar A, Sayyed RZ, Ramteke PW, Sharma S, Marraiki N, Elgorban AM, Syed A (2020) ACC deaminase and antioxidant enzymes producing halophilic Enterobacter sp. PR14 promotes the growth of rice and millets under salinity stress. Physiol Mol Biol Plants 26(9):1847–1854

Salvedia CB, Supangco EP, Vega RS, Elegado FB, Rayos AA (2015) Effect of probiotic feeding on milk yield and components of crossbred dairy goats. Philipp J Vet Anim Sci 41(1):21–30

Sangeetha T, Muthukumar M (2013) Influence of electrode material and electrode distance on bioelectricity production from sago-processing wastewater using microbial fuel cell. Environ Prog Sustain Energy 32(2):390–395

Saraf M, Pandya U, Thakkar A (2014) Role of allelochemicals in plant growth promoting rhizobacteria for biocontrol of phytopathogens. Microbiol Res 169(1):18–29

Saratale RG, Saratale GD, Chang JS, Govindwar SP (2010) Decolorization and biodegradation of reactive dyes and dye wastewater by a developed bacterial consortium. Biodegradation 21(6): 999–1015

Satti SM, Shah AA (2020) Polyester-based biodegradable plastics: an approach towards sustainable development. Lett Appl Microbiol 70(6):413–430

Sayyed RZ, Patel PR, Shaikh SS (2015) Plant growth promotion and root colonization by EPS producing Enterobacter sp. RZS5 under heavy metal contaminated soil. Indian J Exp Biol 53(2): 116–123

Schieber A (2017) Side streams of plant food processing as a source of valuable compounds: selected examples. Annu Rev Food Sci Technol 8:97–112

Sedighi M, Karimi A, Vahabzadeh F (2009) Involvement of ligninolytic enzymes of Phanerochaete chrysosporium in treating the textile effluent containing Astrazon Red FBL in a packed-bed bioreactor. J Hazard Mater 169(1–3):88–93

Seenivasan C, Bhavan PS, Radhakrishnan S, Shanthi R (2012) Enrichment of Artemia nauplii with Lactobacillus sporogenes for enhancing the survival, growth and levels of biochemical constituents in the post-larvae of the freshwater prawn Macrobrachium rosenbergii. Turk J Fish Aquat Sci 12(1):23–31

Seesuriyachan P, Takenaka S, Kuntiya A, Klayraung S, Murakami S, Aoki K (2007) Metabolism of azo dyes by Lactobacillus casei TISTR 1500 and effects of various factors on decolorization. Water Res 41(5):985–992

Selvam A, Wong J (2016) Waste management and sustainability an introduction. In: Sustainable solid waste management. Springer, Berlin, pp 1–6

Shapiro MJ, Downs SM, Swartz HJ, Parker M, Quelhas D, Kreis K et al (2019) A systematic review investigating the relation between animal-source food consumption and stunting in children aged 6–60 months in low and middle income countries. Adv Nutr 10(5):827–847

Sharifi R, Lee SM, Ryu CM (2018) Microbe-induced plant volatiles. New Phytol 220(3):684–691

Sharifuzzaman SM, Austin B (2010) Kocuria SM1 controls vibriosis in rainbow trout (Oncorhynchus mykiss, Walbaum). J Appl Microbiol 108(6):2162–2170

Sharma Y, Li B (2010) The variation of power generation with organic substrates in single-chamber microbial fuel cells (SCMFCs). Bioresour Technol 101(6):1844–1850

Sharma P, Jha AB, Dubey RS, Pessarakli M (2012) Reactive oxygen species, oxidative damage, and antioxidative defense mechanism in plants under stressful conditions. J Bot 2012:217037

Sharma S, Chatterjee S, Kataria S, Joshi J, Datta S, Vairale MG, Veer V (2017) A review on responses of plants to UV-B radiation related stress. In: UV-B radiation: from environmental stressor to regulator of plant growth. Wiley, New York, pp 75–97

Sharma P, Gaur VK, Kim SH, Pandey A (2020) Microbial strategies for bio-transforming food waste into resources. Bioresour Technol 299:122580

Sharma P, Gaur VK, Sirohi R, Varjani S, Kim SH, Wong JW (2021) Sustainable processing of food waste for production of bio-based products for circular bioeconomy. Bioresour Technol 325: 124684

Shedbalkar U, Jadhav JP (2011) Detoxification of malachite green and textile industrial effluent by Penicillium ochrochloron. Biotechnol Bioprocess Eng 16(1):196–204

Shin HJ (2011) Genetically engineered microbial biosensors for in situ monitoring of environmental pollution. Appl Microbiol Biotechnol 89(4):867–877

Shreedhar JN, Patil M, Kumar P (2016) Effect of probiotics supplementation on milk yield and its composition in lactating Holstein Friesian and Deoni cross bred cows. J Med Bioeng 5(1):19–23

Silva LMC, Salgado AM, Coelho MAZ (2010) Agaricus bisporus as a source of tyrosinase for phenol detection for future biosensor development. Environ Technol 31(6):611–616

Silva-Aciares FR, Carvajal PO, Mejías CA, Riquelme CE (2011) Use of macroalgae supplemented with probiotics in the Haliotis rufescens (Swainson, 1822) culture in Northern Chile. Aquac Res 42(7):953–961

Sorroza L, Padilla D, Acosta F, Román L, Grasso V, Vega J, Real F (2012) Characterization of the probiotic strain Vagococcus fluvialis in the protection of European sea bass (Dicentrarchus labrax) against vibriosis by Vibrio anguillarum. Vet Microbiol 155(2–4):369–373

Spector AA, Kim HY (2015) Discovery of essential fatty acids. J Lipid Res 56(1):11–21

Srivastava S, Bist V, Srivastava S, Singh PC, Trivedi PK, Asif MH et al (2016) Unraveling aspects of Bacillus amyloliquefaciens mediated enhanced production of rice under biotic stress of Rhizoctonia solani. Front Plant Sci 7:587

Standen BT, Rawling MD, Davies SJ, Castex M, Foey A, Gioacchini G et al (2013) Probiotic Pediococcus acidilactici modulates both localised intestinal- and peripheral-immunity in tilapia (Oreochromis niloticus). Fish Shellfish Immunol 35(4):1097–1104

Sui W, Xiao Y, Liu R, Wu T, Zhang M (2019) Steam explosion modification on tea waste to enhance bioactive compounds' extractability and antioxidant capacity of extracts. J Food Eng 261:51–59

Swain SM, Singh C, Arul V (2009) Inhibitory activity of probiotics Streptococcus phocae PI80 and Enterococcus faecium MC13 against vibriosis in shrimp Penaeus monodon. World J Microbiol Biotechnol 25(4):697–703

Tahir HA, Gu Q, Wu H, Raza W, Hanif A, Wu L et al (2017) Plant growth promotion by volatile organic compounds produced by Bacillus subtilis SYST2. Front Microbiol 8:171

Talebi A, Amirzadeh B, Mokhtari B, Gahri H (2008) Effects of a multi-strain probiotic (PrimaLac) on performance and antibody responses to Newcastle disease virus and infectious bursal disease virus vaccination in broiler chickens. Avian Pathol 37(5):509–512

Tamboli DP, Kurade MB, Waghmode TR, Joshi SM, Govindwar SP (2010) Exploring the ability of Sphingobacterium sp. ATM to degrade textile dye Direct Blue GLL, mixture of dyes and textile effluent and production of polyhydroxyhexadecanoic acid using waste biomass generated after dye degradation. J Hazard Mater 182(1–3):169–176

Telke A, Kalyani D, Jadhav J, Govindwar S (2008) Kinetics and mechanism of reactive red 141 degradation by a bacterial isolate Rhizobium radiobacter MTCC 8161. Acta Chim Slov 55(2):320–329
Telke AA, Joshi SM, Jadhav SU, Tamboli DP, Govindwar SP (2010) Decolorization and detoxification of Congo red and textile industry effluent by an isolated bacterium Pseudomonas sp. SU-EBT Biodegradation 21(2):283–296
Tevini M (2004) Plant responses to ultraviolet radiation stress. In: Chlorophyll a fluorescence. Springer, Dordrecht, pp 605–621
Tewari S, Mishra A (2018) Flooding stress in plants and approaches to overcome. In: Plant metabolites and regulation under environmental stress. Academic Press, New York, pp 355–366
Thanavel M, Kadam SK, Biradar SP, Govindwar SP, Jeon BH, Sadasivam SK (2019) Combined biological and advanced oxidation process for decolorization of textile dyes. SN Appl Sci 1(1): 1–16
Timmis K, De Vos WM, Ramos JL, Vlaeminck SE, Prieto A, Danchin A et al (2017) The contribution of microbial biotechnology to sustainable development goals. Microb Biotechnol 10(5):984–987
Tomaszewska E, Kwiecień M, Dobrowolski P, Klebaniuk R, Muszyński S, Olcha M et al (2018) Dose-dependent effects of probiotic supplementation on bone characteristics and mineralisation in meat-type female turkeys. Anim Prod Sci 58(3):507–516
Tsukanova KA, Meyer JJM, Bibikova TN (2017) Effect of plant growth-promoting rhizobacteria on plant hormone homeostasis. S Afr J Bot 113:91–102
Tuan TN, Duc PM, Hatai K (2013) Overview of the use of probiotics in aquaculture. Int J Res Fish Aquac 3(3):89–97
Urbanek AK, Rymowicz W, Mirończuk AM (2018) Degradation of plastics and plastic-degrading bacteria in cold marine habitats. Appl Microbiol Biotechnol 102(18):7669–7678
Vandamme EJ, Revuelta JL (2016) Industrial biotechnology of vitamins, biopigments, and antioxidants. Wiley, New York
Vejan P, Abdullah R, Khadiran T, Ismail S, Nasrulhaq Boyce A (2016) Role of plant growth promoting rhizobacteria in agricultural sustainability—a review. Molecules 21(5):573
Vendrell D, Balcázar JL, de Blas I, Ruiz-Zarzuela I, Gironés O, Muzquiz JL (2008) Protection of rainbow trout (Oncorhynchus mykiss) from lactococcosis by probiotic bacteria. Comp Immunol Microbiol Infect Dis 31(4):337–345
Verma N, Bhardwaj A (2015) Biosensor technology for pesticides—a review. Appl Biochem Biotechnol 175(6):3093–3119
Verma AK, Raghukumar C, Verma P, Shouche YS, Naik CG (2010) Four marine-derived fungi for bioremediation of raw textile mill effluents. Biodegradation 21(2):217–233
Verma P, Yadav AN, Khannam KS, Kumar S, Saxena AK, Suman A (2016) Molecular diversity and multifarious plant growth promoting attributes of Bacilli associated with wheat (Triticum aestivum L.) rhizosphere from six diverse agro-ecological zones of India. J Basic Microbiol 56(1):44–58
Vurukonda SSKP, Vardharajula S, Shrivastava M, SkZ A (2016) Enhancement of drought stress tolerance in crops by plant growth promoting rhizobacteria. Microbiol Res 184:13–24
Wang Y, Beydoun MA (2009) Meat consumption is associated with obesity and central obesity among US adults. Int J Obes 33(6):621–628
Wang W, Vinocur B, Altman A (2003) Plant responses to drought, salinity and extreme temperatures: towards genetic engineering for stress tolerance. Planta 218(1):1–14
Wasi S, Tabrez S, Ahmad M (2013) Toxicological effects of major environmental pollutants: an overview. Environ Monit Assess 185(3):2585–2593
Wei J, Liang P, Cao X, Huang X (2011) Use of inexpensive semicoke and activated carbon as biocathode in microbial fuel cells. Bioresour Technol 102(22):10431–10435
Wei L, Han H, Shen J (2012) Effects of cathodic electron acceptors and potassium ferricyanide concentrations on the performance of microbial fuel cell. Int J Hydrog Energy 37(17): 12980–12986

Weylandt KH, Serini S, Chen YQ, Su HM, Lim K, Cittadini A, Calviello G (2015) Omega-3 polyunsaturated fatty acids: the way forward in times of mixed evidence. Biomed Res Int 2015: 143109

Wiebe MG (2004) QuornTM myco-protein—overview of a successful fungal product. Mycologist 18(1):17–20

Xu H, Huang W, Hou Q, Kwok LY, Sun Z, Ma H et al (2017) The effects of probiotics administration on the milk production, milk components and fecal bacteria microbiota of dairy cows. Sci Bull 62(11):767–774

Yaakob MA, Mohamed RMSR, Al-Gheethi A, Tiey A, Kassim AHM (2019) Optimising of Scenedesmus sp. biomass production in chicken slaughterhouse wastewater using response surface methodology and potential utilisation as fish feeds. Environ Sci Pollut Res 26(12): 12089–12108

Yadav IC, Devi NL, Syed JH, Cheng Z, Li J, Zhang G, Jones KC (2015) Current status of persistent organic pesticides residues in air, water, and soil, and their possible effect on neighboring countries: a comprehensive review of India. Sci Total Environ 511:123–137

Yan N, Chen X (2015) Sustainability: don't waste seafood waste. Nature 524(7564):155–157

Yan J, Niu J, Chen D, Chen Y, Irbis C (2014) Screening of Trametes strains for efficient decolorization of malachite green at high temperatures and ionic concentrations. Int Biodeterior Biodegradation 87:109–115

Yang J, Kloepper JW, Ryu CM (2009) Rhizosphere bacteria help plants tolerate abiotic stress. Trends Plant Sci 14(1):1–4

Yang G, He Y, Cai Z, Zhao X, Wang L (2011) Isolation and characterization of Pseudomonas putida WLY for reactive brilliant red x-3b decolorization. Afr J Biotechnol 10(51): 10456–10464

Yang CM, Cao GT, Ferket PR, Liu TT, Zhou L, Zhang L et al (2012) Effects of probiotic, Clostridium butyricum, on growth performance, immune function, and cecal microflora in broiler chickens. Poult Sci 91(9):2121–2129

Yang SM, Jia B, Bai D, Zhang Y, Han Y (2014) The laboratory bioactivity of 8 fungicides against Sclerotinia sclerotiorum (Lib.) de Bary. J Shanxi Agr 1:17–19

You W, Henneberg M (2016) Meat consumption providing a surplus energy in modern diet contributes to obesity prevalence: an ecological analysis. BMC Nutr 2(1):1–11

Zahran HH (2001) Rhizobia from wild legumes: diversity, taxonomy, ecology, nitrogen fixation and biotechnology. J Biotechnol 91(2–3):143–153

Zhang YJ, Yu JJ, Zhang YN, Zhang X, Cheng CJ, Wang JX et al (2009) Effect of carbendazim resistance on trichothecene production and aggressiveness of Fusarium graminearum. Mol Plant-Microbe Interact 22(9):1143–1150

Zhang Y, Mo G, Li X, Zhang W, Zhang J, Ye J et al (2011) A graphene modified anode to improve the performance of microbial fuel cells. J Power Sources 196(13):5402–5407

Zhou D, Feng H, Schuelke T, De Santiago A, Zhang Q, Zhang J et al (2019) Rhizosphere microbiomes from root knot nematode non-infested plants suppress nematode infection. Microb Ecol 78(2):470–481

Zhuo R, Ma L, Fan F, Gong Y, Wan X, Jiang M et al (2011) Decolorization of different dyes by a newly isolated white-rot fungi strain Ganoderma sp. En3 and cloning and functional analysis of its laccase gene. J Hazard Mater 192(2):855–873

Zorriehzahra MJ, Delshad ST, Adel M, Tiwari R, Karthik K, Dhama K, Lazado CC (2016) Probiotics as beneficial microbes in aquaculture: an update on their multiple modes of action: a review. Vet Q 36(4):228–241

Part II

Emerging Technologies in Gut Microbiome Research

Emerging Technologies and Current Advances in Human Bacteriome Research

7

Achhada Ujalkaur Avatsingh and Nasib Singh

Abstract

The human body is recognized as one of the most densely populated microbial ecosystems. The human microbiome, a collection of microbial communities that occur in various anatomical sites, play a profound and indispensable role in promoting overall human health. Among these microorganisms, bacteria, representing the bacteriome, exhibit overwhelming predominance over other microbial groups as these extensively influence the physiological functions of the human body. Microbiome in general and bacteriome in particular, is recognized as a major influencer of health and disease in humans. In recent times, the relationship among human bacteriome, dysbiosis and metabolic disorders or diseases has gained considerable significance. Similarly, health interventions involving the modulation of the human bacteriome such as faecal microbiota transplantation, live biotherapeutics, bioengineered microbes and phage therapy have become attractive area of research with huge futuristic implications. Here, we present the current understanding of the human bacteriome, its compositional modulation for exerting beneficial health effects and the emerging technologies, viz. deep-sequencing, multiomics among others for deeper insights into dynamic interplay among bacterial communities in the human body.

Keywords

Microbiome · Dysbiosis · Bacteriome · Gastrointestinal tract · Next-generation sequencing · Multiomic technologies · Faecal microbiota transplantation · Live biotherapeutics

A. U. Avatsingh · N. Singh (✉)
Department of Microbiology, Akal College of Basic Sciences, Eternal University, Baru Sahib, Himachal Pradesh, India
e-mail: drnasibmicro@eternaluniversity.edu.in

R. Sobti et al. (eds.), *Role of Microbes in Sustainable Development*,
https://doi.org/10.1007/978-981-99-3126-2_7

7.1 Introduction

The human body is recognized as an ecosystem supporting vast communities of bacteria, archaea, fungi, viruses and protozoa (Locey and Lennon 2016; Singh et al. 2018; Gilbert et al. 2018; Aggarwal et al. 2022). These co-existing commensal and symbiotic microbial communities present on or inside the body represent the human microbiota and the microbiome. Precisely, the microbial cells of the body are referred to as microbiota and the microbiome represents the entire set of genes of these microbes, although these terms are often used interchangeably (Aggarwal et al. 2022; Mousa et al. 2022). Human bacteriome, on the other hand, specifically represents the bacterial inhabitants of the body. Recently, Berg et al. (2020) redefined the term microbiome as "a characteristic microbial community occupying a reasonably well-defined habitat which has distinct physio-chemical properties". The human microbiome is also considered our second genome and the human body inhabiting them is recognized as a superorganism (Gilbert et al. 2018; Aggarwal et al. 2022). The human microbiome has co-evolved for centuries with human beings and is now considered an essential organ for human health (Hooper and Gordon 2001; Gilbert and Neufeld 2014; Michels et al. 2022)

The human microbiota is composed of hundreds of species and strains of bacteria, archaea, fungi, protozoa and viruses. Their genomes containing thousands of genes offer much greater genetic diversity and versatility than our own genome (Locey and Lennon 2016; Perez-Carrasco et al. 2021). The human gut harbours more than 1000 bacterial species belonging to four phyla, namely Firmicutes, Bacteroidetes, Actinobacteria and Proteobacteria (Hooper and Gordon 2001; Fujisaka et al. 2022). In numerical terms, the estimated ratio of bacteria to human cells in the human body is 1.3:1, although at one time it was considered to be 10:1 (Sender et al. 2016; Locey and Lennon 2016). The Human Microbiome Project which was initiated in 2007 revealed interesting facts about the human body ecosystem and marks one of the most significant milestones in the history of biology (Gilbert et al. 2018). The human body carries more microbial cells than its own cells, whereas the number of genes expressed by the human microbiome is 50–100 folds higher than our genome (Hooper and Gordon 2001; Aggarwal et al. 2022). Microbiome-oriented studies have witnessed unprecedented impetus in the last two decades. Abdill et al. (2022) found that publicly available data on human microbiome samples has increased from 3 samples in 2010 to 123,302 in 2020. These microbiome samples are from 115 different countries where the USA represents 40.2% of the total samples. From India, 2997 microbiome samples have been studied and analysed as per available data (Abdill et al. 2022).

7.2 Human Bacteriome Diversity and its Importance

The microorganisms present in a normal human body are collectively known as commensal microbiota which has also been recognized as our forgotten organ. Human microbiota and microbiome have been extensively investigated in a plentitude of studies for their beneficial roles in several vital functions and overall health. Experimental studies revealed the influence of microbiota in host physiology, digestion, immunity (both innate and adaptive), homeostasis, neurodevelopment and metabolism (Gilbert et al. 2018; Liu et al. 2022a, b; Michels et al. 2022). The majority of human bacteriome is present in the gastrointestinal tract, mainly the large intestine, small intestine and colon (Table 7.1). However, other anatomical sites of the body such as skin, oral cavity, nose, respiratory tract, lungs, eyes, ears and urogenital tract harbour distinct microbial communities (Gilbert et al. 2018; Fujisaka et al. 2022; Aggarwal et al. 2022).

Microbial diversity, abundance and predominance in different sites of the human body are impacted either by environmental factors such as pH, temperature, osmolarity, salts, oxygen levels, sebum and nutrient concentration or by intrinsic factors such as ethnicity, genetics, age and gender (Aggarwal et al. 2022). Similarly, the human bacteriome is diverse and dynamic in nature and is influenced by diet, food types, age, habits, gender, body-mass-index, demographics, genetics, probiotics, prebiotics, antibiotics usage and other environmental factors (Girija and Ganesh 2022; Villemin et al. 2022). The human microbiome of different sites can be used as a reliable biomarker or microbial signature to assess the risk of occurrence of several metabolic conditions and disorders. Therefore, deeper insights into the microbiota of the human body under different physiological conditions are needed. These data may assist the scientific community and health professionals in designing and developing newer diagnostic tools and therapeutic interventions. Among different microbes present in the human body, the most prominent role is exhibited by gut microbiota or gut microbiome. It is essential for food digestion, vitamin production, protection against pathogens and immune system development (Arnold et al. 2016; Mirzaei and Maurice 2017; Fujisaka et al. 2022; Mousa et al. 2022). Gut microbiota is known to influence almost all physiological functions of the body including that of the brain, pancreas, heart, kidney, skin and eyes (Yan et al. 2017; Wang et al. 2021; Sani et al. 2021; Matsukawa et al. 2021; Sorboni et al. 2022; Wehedy et al. 2022; Zhou et al. 2022; Liu et al. 2022a, b). The gut bacteriome is represented by numerous bacterial species from mainly four phyla as shown in Table 7.1. In the Indian population, the gut microbiome exhibited the predominance of *Prevotella copri* and *Faecalibacterium prausnitzii* which can be attributed to the high content of resistant starch in Indian diets (Dubey et al. 2018).

Human bacteriome has been recognized as a vital player in systemic health. Dysbiosis refers to the lack of balance among microbial communities within certain areas of the body that may lead to the onset or progression of certain chronic diseases or disorders (Byrd et al. 2018; Gurung et al. 2020; Aggarwal et al. 2022; Carmona-Cruz et al. 2022). Bacteriome perturbations or dysbiosis is detrimental to the host health and can lead to chronic metabolic diseases such as obesity, type II diabetes mellitus, insulin resistance syndrome, cancers, allergies, gingivitis, periodontitis, endocarditis, asthma, osteoporosis, inflammatory bowel disease, ulcerative colitis, Crohn disease, rheumatoid arthritis, neurological disorders, Parkinson's disease, Alzheimer's disease or atopic eczema among others (Lamont et al. 2018; Perez-Carrasco et al. 2021; Aggarwal et al. 2022; Kenneally et al. 2022; Sorboni et al.

Table 7.1 Prominent bacterial phyla and/or genera present in different sites of the human body and dysbiosis-associated diseases

Organ/site	Predominant bacterial phyla[a]	Predominant bacterial genera/species	Disease/disorder due to dysbiosis	Reference
Gastrointestinal tract or gut	*Firmicutes*, *Bacteroidetes*, *Proteobacteria*, *Actinobacteria*, *Fusobacteria*, *Verrucomicrobia*	*Bacillus*, *Clostridium*, *Clostridioides difficile*, *E. coli*, *Collinsella*, *Eubacterium*, *Enterococcus*, *Enterobacter*, *Lactobacillus*, *Ruminococcus*, *Staphylococcus*, *Bacteroides*, *Prevotella*, *Bifidobacterium*, *Akkermansia muciniphila*,, *Faecalibacterium prausnitzii*, *Roseburia*, *Veillonella*	Inflammatory bowel disease, obesity, type II diabetes, ulcerative colitis, Crohn's disease, celiac disease	Fujisaka et al. (2022), Sorboni et al. (2022), Michels et al. (2022), Liu et al. (2022a, b), Mousa et al. (2022)
Skin	*Actinobacteria*, *Firmicutes*, *Proteobacteria*, *Bacteroidetes*	*Cutibacterium acnes*, *Staphylococcus epidermidis*, *Staphylococcus aureus*, *Corynebacterium*	Atopic dermatitis, psoriasis vulgaris, acne vulgaris	Byrd et al. (2018), Aggarwal et al. (2022), Carmona-Cruz et al. (2022)
Jejunum	*Actinobacteria*, *Firmicutes*, *Proteobacteria*, *Bacteroidetes*	*Streptococcus salivarius*, *S. mitis*, *Rothia mucilaginosa*, *Actinomyces odontolyticus*, *Granulicatella adiacens*, *Corynebacterium kroppenstedtii*, *Prevotella*, *Veillonella*, *Fusobacterium*	–	Villmones et al. (2022)
Oral cavity	*Actinobacteria*, *Bacteroidetes*, *Firmicutes*, *Fusobacteria*, *Proteobacteria*, *Spirochaetes*	*Actinomyces*, *Corynebacterium*, *Capnocytophaga*, *Eubacterium*, *Eikenella*, *Fusobacterium*, *Granulicatella*,	Dental caries, gingivitis and periodontitis	Kilian et al. (2016), Lamont et al. (2018), Aggarwal et al. (2022)

(continued)

Table 7.1 (continued)

Organ/site	Predominant bacterial phyla[a]	Predominant bacterial genera/ species	Disease/ disorder due to dysbiosis	Reference
		Gemella, Neisseria, Rothia, Prevotella, Porphyromonas, Treponema, Peptostreptococcus, Propionibacterium, Haemophilus, Leptotrichia, Lactobacillus, Staphylococcus, Streptococcus, Veillonella		
Stomach (gastric mucosa)	*Proteobacteria, Firmicutes, Bacteroidetes, Actinobacteria, Fusobacteria*	*Clostridium, Dialister, Helicobacter, Fusobacterium, Peptostreptococcus, Parvimonas, Streptococcus, Slackia, Veillonella*	Gastric cancer	Aggarwal et al. (2022), Mousa et al. (2022), Liu et al. (2022a, b)
Nose/nostrils	*Actinobacteria, Firmicutes, Proteobacteria*	*Staphylococcus aureus, Staphylococcus epidermidis, Cutibacterium acnes, Corynebacterium, Moraxella*	Chronic rhinosinusitis	Gilbert et al. (2018), Aggarwal et al. (2022)
Respiratory tract and lungs	Firmicutes, Proteobacteria, Bacteroidetes	*Prevotella, Corynebacterium, Haemophilus, Neisseria, Veillonella, Streptococcus*	Asthma, chronic obstructive pulmonary disease, pulmonary fibrosis, cystic fibrosis	Aggarwal et al. (2022), Mousa et al. (2022)
Esophagus	*Firmicutes, Bacteroides, Actinobacteria, Proteobacteria Fusobacteria*	*Streptococcus, Veillonella, Prevotella, Fusobacterium, Tannerella, Aggregatibacter, Porphyromonas*	Oesophageal cancer	Lv et al. (2019), Laserna-Mendieta et al. (2021), Muszyński et al. (2022)
Urogenital tract (bladder and vagina)	*Firmicutes, Bacteroidetes, Proteobacteria, Actinobacteria*	*Lactobacillus, Streptococcus, Corynebacterium, Prevotella,*	Urgency urinary incontinence, bladder	Perez-Carrasco et al. (2021),

(continued)

Table 7.1 (continued)

Organ/site	Predominant bacterial phyla[a]	Predominant bacterial genera/ species	Disease/ disorder due to dysbiosis	Reference
		Staphylococcus, *Escherichia*, *Enterococcus*, *Corynebacterium*, *Citrobacter*, *Gardnerella vaginalis*	cancer, bacterial vaginosis	Kenneally et al. (2022)

[a]Nomenclature of bacterial phyla has been changed recently (Oren and Garrity 2021). However, for the sake of clarity, old bacterial phyla names have been used here

2022; Michels et al. 2022; Table 7.2). It is now widely accepted that gut microbiota communicates with distant organs, viz. brain, lungs, kidney, liver and heart through an array of signalling molecules. Dysbiosis of gut microbiota composition has been implicated in the development and progression of several neurological disorders including anxiety, stress, depression, epilepsy, stroke, Parkinson's disease, multiple sclerosis, Alzheimer's disease and schizophrenia (Cekanaviciute et al. 2017; Sani et al. 2021; Sorboni et al. 2022). These pathophysiological outcomes of microbiome perturbations emphasized the significance of the microbiota–gut–organ axis (Liu et al. 2022a, b; Radjabzadeh et al. 2022).

7.3 Modulation of Human Bacteriome

The accumulated and ever-increasing microbiome knowledge, understandings and datasets as well as the outcomes of several clinical trials indicated that human microbiota is amenable to dynamic shifts in abundance and diversity by dietary modifications, fermented foods, probiotics, prebiotic polysaccharides, postbiotics, microbial infections and pre-existing metabolic conditions (Marco et al. 2017; Leeuwendaal et al. 2022; Fujisaka et al. 2022). The modulation of the human microbiome, particularly the gut microbiota, is thus considered an attractive strategy for creating better therapeutics, diagnostic tools and disease markers for health diseases (Arnold et al. 2016; Chua et al. 2017; O'Toole et al. 2017; Aggarwal et al. 2022; Callens et al. 2022; Villemin et al. 2022; Fig. 7.1). Faecal microbiota transplantation (FMT) was first performed in human subjects by van Nood et al. (2013) in which the duodenal infusion of donor faeces was carried out in recurrent *Clostridioides difficile* patients. This strategy witnessed a high success rate and paved the way for microbiome-mediated health interventions. Gut bacteriomes could potentially be modulated or reprogrammed by specific medicines or antibiotics, engineered microbes, live bacterial therapeutics, phage therapy and microbiome mimetics to alleviate the symptoms of several metabolic diseases or to minimize their occurrence (Mirzaei and Maurice 2017; El Haddad et al. 2022; Mousa et al. 2022; Milligan-McClellan et al. 2022; Aggarwal et al. 2022).

Table 7.2 Relationship between bacteriome dysbiosis and metabolic diseases in the human body

Disease/disorder	Key bacterial taxa/species/phyla/families[a]	Reference
Atherosclerosis	*Firmicutes, Proteobacteria, Lachnospiraceae, Erysipelotrichaceae, Pseudomonas luteola*	Koren et al. (2011)
Oral squamous cell carcinoma	*Porphyromonas gingivalis Fusobacterium nucleatum*	Irfan et al. (2020)
Esophageal cancer	*Tannerella forsythia, Streptococcus anginosus, Aggregatibacter actinomycetemcomitans, Porphyromonas gingivalis, Fusobacterium nucleatum*	Lv et al. (2019), Laserna-Mendieta et al. (2021), Muszyński et al. (2022)
Pancreatic ductal adeno carcinoma	*Clostridium bolteae, Acinetobacter, Klebsiella pneumoniae, Streptococcus mutans, Parabacteroides, Pseudomonas*	Matsukawa et al. (2021), Mousa et al. (2022)
Obesity	*Fusobacterium, Lactobacillus reuteri, Akkermansia muciniphila, Bifidobacterium animalis, Bacteroides fragilis, Lactobacillus plantarum, Staphylococcus aureus, Ruminococcus gnavus*	Michels et al. (2022), Mousa et al. (2022)
Type 2 diabetes	*Akkermansia muciniphila, Blautia, Bacteroides vulgatus, Clostridium hathewayi, C. symbiosum, C. coccoides and C. leptum, Escherichia coli, Lactobacillus, Veillonella denticariosi, Fusobacterium, Bifidobacterium, Faecalibacterium, Roseburia, Ruminococcus, Fusobacterium*	Zhou et al. (2022), Barlow and Mathur (2023)
Inflammatory bowel disease, Crohn's disease and ulcerative colitis	*Clostridium, Faecalibacterium prausnitzii, Escherichia coli, Ruminococcus, Actinomyces, Veillonella*	Caruso et al. (2020), Qiu et al. (2022)
Gingivitis and periodontitis	*Aggregatibacter actinomycetemcomitans, Porphyromonas gingivalis, Bacteroides, Catonella, Desulfobulbus, Eikenella, Peptostreptococcus*	Kilian et al. (2016), Lamont et al. (2018)
Coronary artery disease and hypertension	*Clostridium, Klebsiella, Faecalibacterium, Streptococcus, Parabacteroides, Roseburia, Lactobacillus*	Yan et al. (2017), Mousa et al. (2022)
Pulmonary disorders	*Alloprevotella, Dubosiella, Clostridium, Helicobacter, Lactobacillus, Olsenella, Parasutterella, Rikenella*	Mousa et al. (2022)

(continued)

Table 7.2 (continued)

Disease/disorder	Key bacterial taxa/species/phyla/families[a]	Reference
Parkinson's disease	*Prevotellaceae, Enterobacteriaceae*	Liu et al. (2022a, b)
Neurodegenerative diseases (Alzheimer's disease and dementia)	*Bacteroides, Bifidobacterium, Clostridium, Lactobacillus, Aggregatibacter actinomycetemcomitans, Porphyromonas gingivalis*	Mousa et al. (2022)
Multiple sclerosis	*Akkermansia muciniphila, Acinetobacter calcoaceticus, Fusobacterium*	Cekanaviciute et al. (2017)
Prostate cancer and bladder diseases	*Actinobaculum urinale, Anaerococcus, Propionimicrobium lymphophilum, Cutibacterium acnes, Streptococcus anginosus, Varibaculum cambriense*	Kenneally et al. (2022)
Urothelial carcinoma or transitional cell carcinoma	*Acinetobacter, Actinomyces, Anaerococcus, Fusobacterium, Sphingobacterium, Herbaspirillum, Porphyrobacter, Bacteroides*	Perez-Carrasco et al. (2021)
Bacterial vaginosis	*Lactobacillus, Gardnerella vaginalis*	
Depression, anxiety and other psychiatric disorders	*Bacteroides, Coprococcus, Eggerthella, Eubacterium, Hungatella, Subdoligranulum, Ruminococcus, Lachnoclostridium, Sellimonas*	Radjabzadeh et al. (2022), Mousa et al. (2022)
Skin-related diseases (atopic dermatitis, psoriasis vulgaris, acne vulgaris	*Corynebacterium mastitidis, C. bovis, Cutibacterium acnes, Staphylococcus aureus, Staphylococcus epidermidis, Streptococcus, Gemella, Dermacoccus, Coprobacillus, Ruminococcus, Streptococcus*	Byrd et al. (2018), Aggarwal et al. (2022), Carmona-Cruz et al. (2022)
Chronic kidney disease (CKD)	*Alistipes, Bifidobacterium, Collinsella, Eggerthella, Fusobacterium, Lactobacillus, Paraprevotella, Pseudobutyrivibrio*	Krukowski et al. (2022), Wehedy et al. (2022)

[a]Nomenclature of bacterial phyla has changed recently (Oren and Garrity 2021). However, for the sake of clarity, old bacterial phyla names have been used

Another microbiome-based approach for extracting health benefits is live bacterial therapeutics or biotherapeutics. In this microbiome-modulation strategy, well-defined single species-based or rationally selected consortium-based live bacterial therapeutics are derived from healthy subjects after extensive screening, optimization and safety assessment (Gulliver et al. 2022). Live biotherapeutics are non-vaccine biological products that contain live organisms for the prevention and treatment of a disease or condition in humans (Gulliver et al. 2022). Towards

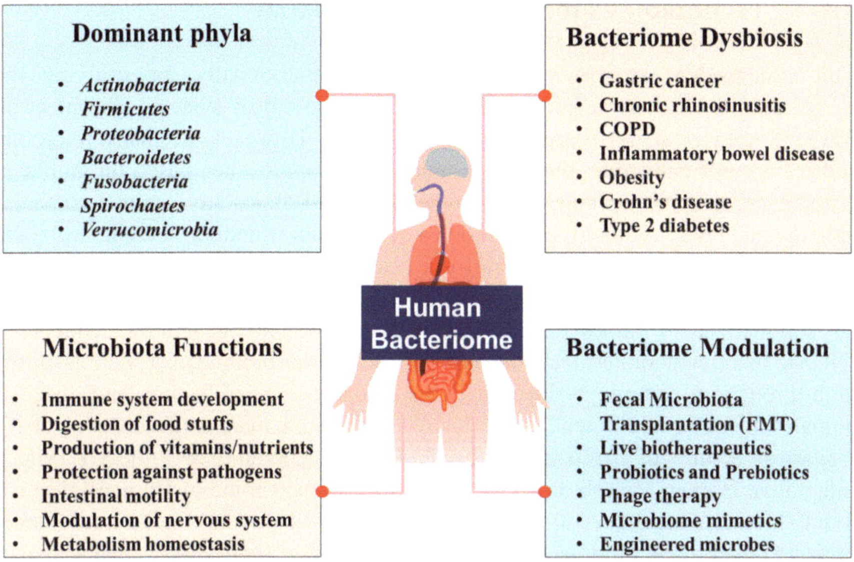

Fig. 7.1 Human bacteriome in human health, consequences of its dysbiosis and modulation interventions for preventive and therapeutic effects

microbiome-based biotherapeutics, several bacterial species have shown promising outcomes. As reviewed by Yadav and Chauhan (2022), *Christensenella* sp., *Akkermansia muciniphila*, *Lactobacillus johnsonii*, *Bifidobacterium longum*, *Oxalibacterium formigenes* and *Bacteroides* spp. are known to protect against depression, atherosclerosis, cancer, Crohn's disease, kidney stones adiposity, respectively. Microbiome modulation is also explored for the development of personalized therapeutics against hypotension by using gut bacteria and their metabolites (Sun et al. 2022). Phage-mediated bacteriome manipulations for managing gastrointestinal diseases are considered a better alternative than antibiotic therapy. By using exogenous phages, engineered phages or phage-containing faecal transplantation, the gut bacteriome balance can be restored with consequent control over gastrointestinal disease progression (El Haddad et al. 2022). Microbiome-based therapeutics are now considered a reliable option for the management of gut inflammatory conditions such as inflammatory bowel disease, Crohn's disease and ulcerative colitis. The administration of probiotic strains of *Lactobacillus* and *Bifidobacterium* has offered encouraging outcomes in people with gut-related disorders (Caruso et al. 2020; Qiu et al. 2022)

7.4 Technologies for Bacteriome Research

Culture-dependent approaches have been used traditionally for studying the microbiota of the human body and many other biological systems (Sirohi et al. 2012; Arnold et al. 2016; Aggarwal et al. 2022). However, as many microbial species are not cultivable, the culture-independent approaches based on amplicon sequencing, deep-sequencing technologies and multi-omics have gained considerable favour. These technologies afforded a deeper understanding of the microbiome and provided enormous sets of data for a detailed analysis and interpretation. The microbiome-based databases, projects and latest microbiome-oriented technologies are summarized in Tables 7.3 and 7.4, respectively. Next-generation sequencing (NGS), functional omics technologies and advanced bioinformatics tools allow high-throughput community-level analyses of bacteriome of healthy and diseased human subjects in an efficient, cost-effective and reliable manner (Zhang et al. 2019; Aggarwal et al. 2022; Yu et al. 2022). Human Microbiome Project (HMP1), Integrative Human Microbiome Project (iHMP), Metagenome of Human Intestinal Tract (MetaHIT) and several other public-funded and private-funded microbiome projects relied on an array of advanced next-generation sequencing, deep sequencing, "omics" technologies and sophisticated bioinformatics databases for microbiota identification, data generation, interpretation and analysis. To gain a better understanding of intricate interactions between diverse bacterial species in healthy persons and alteration in bacterial diversity and abundance under a dysbiosis state, high-throughput sequence-based studies are being performed for elucidating their genomic configurations and functional contributions using culture-independent next-generation sequencing technologies mainly 16S rRNA amplicon sequencing and shotgun metagenomic sequencing (Dubey et al. 2018; Zhang et al. 2019; Aggarwal et al. 2022; Yu et al. 2022; Table 7.4). Next-generation sequencing allows for high-volume studies of microbiome samples of different anatomical sites of the human body in a rapid, efficient and reliable manner. With the spectacular expansion of genomic information owing to technological advancements in the field of metagenomics, about 130,000 bacterial genomes have been sequenced and more than 20,000 metagenomic projects are accessible in the public domain (Aggarwal et al. 2022). Metagenomics, metatranscriptomics, metaproteomics and metabolomics further enhance our understanding of microbiota diversity, functional associations and kinetics of their abundance or decline (Mousa et al. 2022; Milligan-McClellan et al. 2022). Despite technical challenges and high-cost deterrents, high-throughput sequencing technologies are poised to further our knowledge of the microbiome and allow us for implementing microbiome–intervention based preventive and therapeutic strategies in the near future.

Table 7.3 Human microbiome projects, microbiome-related databases and microbiome-data analysis software/packages

Project/database	Description	Weblink
Human microbiome project (HMP)	First phase of NIH human microbiome project, established in 2007	https://hmpdacc.org/
The HMP data portal	Repository of human microbiome datasets from both HMP1 and iHMP	https://portal.hmpdacc.org/
The integrative human microbiome project (iHMP)	Second phase of NIH human microbiome project, established in 2014	https://hmpdacc.org/ihmp/overview/
Metagenome of human intestinal tract (MetaHIT)	Collaborative project funded by the European Commission for analysis of gut microbiota (2008–2012)	https://cordis.europa.eu/project/id/201052
Microbiome centers consortium (MCC)	Collaborative network of microbiome centers for sharing and developing resources and data	http://microbiomecenters.org/
GMrepo	Database of human gut metagenomes	https://gmrepo.humangut.info/
gutMDisorder	A comprehensive database for gut microbiota dysbiosis	http://bio-annotation.cn/gutMDisorder
GIMICA	Database for host genetic and immune factors shaping human microbiota	https://idrblab.org/gimica/
DISBIOME	Database for linking the microbiome to diseases	https://disbiome.ugent.be/
Amadis	Database for microbiota and disease association	http://gift2disease.net/GIFTED
MEtaGenome atlas (gutMEGA)	Database for published quantitative human gut microbiota datasets	http://gutmega.omicsbio.info
MicrobiomeDB	A data discovery and analysis platform for microbiome datasets	https://microbiomedb.org/mbio/app/
Forensics microbiome database (FMD)	Database of human microbiome obtained from multiple body sites for predicting the geographical location of subjects	http://fmd.jcvi.org/
Expanded human Oral microbiome database (eHOMD)	Database of bacteria in the human mouth and aerodigestive tract	https://www.homd.org/
Microbiome drug database	Global database of microbiome companies developing pharmaceuticals	https://www.microbiometimes.com/drug-database-2/
PharmacoMicrobiomics database	Database for microbiome-drug interactions	http://www.pharmacomicrobiomics.org
Microbiome search engine 2 (MSE 2)	Platform for the taxonomic and functional search of microbiomes	http://mse.ac.cn/

Table 7.4 Conventional and advanced technologies for human bacteriome analysis

S. No.	Current technologies for bacteriome analysis
1.	Culture-dependent approaches (a) in vitro microbiome culturing models (HuMiX, SHIME) (b) High-throughput platforms for drug–microbiome interactions (c) High throughput ex vivo microbiome RapidAIM assay
2.	Culture-independent approaches (a) Low-throughput sequencing (16S rRNA-based amplicon sequencing) (b) Next-generation sequencing (high-throughput genome-wide sequencing) • Shotgun metagenomics • Metatranscriptomics • Metaproteomics • Metabolomics (c) Whole-community functional assays (d) Whole-metagenome shotgun (WMS) sequencing (e) Integrative multi-omics for functional data analyses (f) HPLC-MS

Adapted from Arnold et al. (2016), Zhang et al. (2019), Yu et al. (2022) and Milligan-McClellan et al. (2022)

7.5 Conclusion and Future Directions

A plethora of evidences suggest the direct involvement of microbiota dysbiosis in the development and manifestation of chronic metabolic diseases. Our understanding of human microbiota and bacteriome has been revolutionized in the last decades by high-throughput next-generation sequencing technologies. These developments afforded compositional, abundance and functional analyses of the microbiome under varying conditions of host health. Microbiome-oriented research has emerged as a multidisciplinary field encompassing microbiology, biochemistry, molecular biology, neurology, clinical practice, biomaterials engineering, systems biology and synthetic biology. The significance of the microbiome in human health has received considerable impetus due to technological advancements, viz. deep-sequencing, microfluidics assays and high-throughput culturing and enhanced focus on microbiome-interventional strategies and personalized medicines. The rejuvenated emphasis on the microbiome is going to witness further attention from interdisciplinary scientific community in the near future. Unravelling the mechanistic details of microbiota-host cross-talk is desired in order to devise preventive and therapeutic strategies against various human chronic diseases.

Author Contributions AUA and NS wrote the manuscript and prepared the figures and tables. All authors contributed to the article and approved the submitted version.

Acknowledgements The authors are grateful to the Chancellor, Eternal University, for providing the necessary facilities and support.

References

Abdill RJ, Adamowicz EM, Blekhman R (2022) Public human microbiome data are dominated by highly developed countries. PLoS Biol 20(2):e3001536. https://doi.org/10.1371/journal.pbio.3001536

Aggarwal N, Kitano S, Puah GRY, Kittelmann S, Hwang IY, Chang MW (2022) Microbiome and human health: current understanding, engineering, and enabling technologies. Chem Rev 123:31. https://doi.org/10.1021/acs.chemrev.2c00431

Arnold JW, Roach J, Azcarate-Peril MA (2016) Emerging technologies for gut microbiome research. Trends Microbiol 24:887–901. https://doi.org/10.1016/j.tim.2016.06.008

Barlow GM, Mathur R (2023) Type 2 diabetes and the microbiome. J Endocr Soc 7(2):bvac184. https://doi.org/10.1210/jendso/bvac184

Berg G, Rybakova D, Fischer D, Cernava T, Vergès MC, Charles T, Chen X, Cocolin L, Eversole K, Corral GH, Kazou M, Kinkel L, Lange L, Lima N, Loy A, Macklin JA, Maguin E, Mauchline T, McClure R, Mitter B, Ryan M, Sarand I, Smidt H, Schelkle B, Roume H, Kiran GS, Selvin J, Souza RSC, van Overbeek L, Singh BK, Wagner M, Walsh A, Sessitsch A, Schloter M (2020) Microbiome definition re-visited: old concepts and new challenges. Microbiome 8(1):103. https://doi.org/10.1186/s40168-020-00875-0

Byrd AL, Belkaid Y, Segre JA (2018) The human skin microbiome. Nat Rev Microbiol 16(3): 143–155. https://doi.org/10.1038/nrmicro.2017.157. Epub 2018 Jan 15. PMID: 29332945.

Callens K, Fontaine F, Sanz Y, Bogdanski A, D'Hondt K, Lange L, Smidt H, van Overbeek L, Kostic T, Maguin E, Meisner A, Sarand I, Sessitsch A (2022) Microbiome-based solutions to address new and existing threats to food security, nutrition, health and agrifood systems' sustainability. Front Sustain Food Syst 6:1047765. https://doi.org/10.3389/fsufs.2022.1047765

Carmona-Cruz S, Orozco-Covarrubias L, Sáez-de-Ocariz M (2022) The human skin microbiome in selected cutaneous diseases. Front Cell Infect Microbiol 12:834135. https://doi.org/10.3389/fcimb.2022.834135

Caruso R, Lo BC, Núñez G (2020) Host-microbiota interactions in inflammatory bowel disease. Nat Rev Immunol 20(7):411–426. https://doi.org/10.1038/s41577-019-0268-7

Cekanaviciute E, Yoo BB, Runia TF, Debelius JW, Singh S, Nelson CA, Kanner R, Bencosme Y, Lee YK, Hauser SL, Crabtree-Hartman E, Sand IK, Gacias M, Zhu Y, Casaccia P, Cree BAC, Knight R, Mazmanian SK, Baranzini SE (2017) Gut bacteria from multiple sclerosis patients modulate human T cells and exacerbate symptoms in mouse models. Proc Natl Acad Sci U S A 114(40):10713–10718. https://doi.org/10.1073/pnas.1711235114

Chua KJ, Kwok WC, Aggarwal N, Sun T, Chang MW (2017) Designer probiotics for the prevention and treatment of human diseases. Curr Opin Chem Biol 40:8–16. https://doi.org/10.1016/j.cbpa.2017.04.011

Dubey AK, Uppadhyaya N, Nilawe P, Chauhan N, Kumar S, Gupta UA, Bhaduri A (2018) LogMPIE, pan-India profiling of the human gut microbiome using 16S rRNA sequencing. Sci Data 5:180232. https://doi.org/10.1038/sdata.2018.232

El Haddad L, Mendoza JF, Jobin C (2022) Bacteriophage-mediated manipulations of microbiota in gastrointestinal diseases. Front Microbiol 17(13):1055427. https://doi.org/10.3389/fmicb.2022.1055427

Fujisaka S, Watanabe Y, Tobe K (2022) The gut microbiome: a core regulator of metabolism. J Endocrinol 256(3):e220111. https://doi.org/10.1530/JOE-22-0111

Gilbert JA, Neufeld JD (2014) Life in a world without microbes. PLoS Biol 12:e1002020. https://doi.org/10.1371/journal.pbio.1002020

Gilbert JA, Blaser MJ, Caporaso JG, Jansson JK, Lynch SV, Knight R (2018) Current understanding of the human microbiome. Nat Med 24:392–400. https://doi.org/10.1038/nm.4517

Girija ASS, Ganesh PS (2022) Functional biomes beyond the bacteriome in the oral ecosystem. Jpn Dent Sci Rev 58:217–226. https://doi.org/10.1016/j.jdsr.2022.05.002

Gulliver EL, Young RB, Chonwerawong M, D'Adamo GL, Thomason T, Widdop JT, Rutten EL, Rossetto Marcelino V, Bryant RV, Costello SP, O'Brien CL, Hold GL, Giles EM, Forster SC

(2022) Review article: the future of microbiome-based therapeutics. Aliment Pharmacol Ther 56:192–208. https://doi.org/10.1111/apt.17049
Gurung M, Li Z, You H, Rodrigues R, Jump DB, Morgun A, Shulzhenko N (2020) Role of gut microbiota in type 2 diabetes pathophysiology. EBioMedicine 51:102590. https://doi.org/10.1016/j.ebiom.2019.11.051
Hooper LV, Gordon JI (2001) Commensal host-bacterial relationships in the gut. Science 292: 1115–1118. https://doi.org/10.1126/science.1058709
Irfan M, Delgado RZR, Frias-Lopez J (2020) The oral microbiome and cancer. Front Immunol 11: 591088. https://doi.org/10.3389/fimmu.2020.591088
Kenneally C, Murphy CP, Sleator RD, Culligan EP (2022) The urinary microbiome and biological therapeutics: novel therapies for urinary tract infections. Microbiol Res 259:127010. https://doi.org/10.1016/j.micres
Kilian M, Chapple IL, Hannig M, Marsh PD, Meuric V, Pedersen AM, Tonetti MS, Wade WG, Zaura E (2016) The oral microbiome - an update for oral healthcare professionals. Br Dent J 221(10):657–666. https://doi.org/10.1038/sj.bdj.2016.865
Koren O, Spor A, Felin J, Fåk F, Stombaugh J, Tremaroli V, Behre CJ, Knight R, Fagerberg B, Ley RE, Bäckhed F (2011) Human oral, gut, and plaque microbiota in patients with atherosclerosis. Proc Natl Acad Sci U S A 108(Suppl 1):4592–4598. https://doi.org/10.1073/pnas.1011383107
Krukowski H, Valkenburg S, Madella AM, Garssen J, van Bergenhenegouwen J, Overbeek SA, Huys GRB, Raes J, Glorieux G (2022) Gut microbiome studies in CKD: opportunities, pitfalls and therapeutic potential. Nat Rev Nephrol 19(2):87–101. https://doi.org/10.1038/s41581-022-00647-z
Lamont RJ, Koo H, Hajishengallis G (2018) The oral microbiota: dynamic communities and host interactions. Nat Rev Microbiol 16(12):745–759. https://doi.org/10.1038/s41579-018-0089-x
Laserna-Mendieta EJ, FitzGerald JA, Arias-Gonzalez L, Ollala JM, Bernardo D, Claesson MJ, Lucendo AJ (2021) Esophageal microbiome in active eosinophilic esophagitis and changes induced by different therapies. Sci Rep 11(1):7113. https://doi.org/10.1038/s41598-021-86464-z
Leeuwendaal NK, Stanton C, O'Toole PW, Beresford TP (2022) Fermented foods, health and the gut microbiome. Nutrients 14(7):1527. https://doi.org/10.3390/nu14071527
Liu L, Huh JR, Shah K (2022a) Microbiota and the gut-brain-axis: implications for new therapeutic design in the CNS. EBioMedicine 77:103908. https://doi.org/10.1016/j.ebiom.2022.103908
Liu D, Zhang R, Chen S, Sun B, Zhang K (2022b) Analysis of gastric microbiome reveals three distinctive microbial communities associated with the occurrence of gastric cancer. BMC Microbiol 22:184. https://doi.org/10.1186/s12866-022-02594-y
Locey KJ, Lennon JT (2016) Scaling laws predict global microbial diversity. Proc Natl Acad Sci U S A 113(21):5970–5975. https://doi.org/10.1073/pnas.1521291113
Lv J, Guo L, Liu JJ, Zhao HP, Zhang J, Wang JH (2019) Alteration of the esophageal microbiota in Barrett's esophagus and esophageal adenocarcinoma. World J Gastroenterol 25:2149–2161. https://doi.org/10.3748/wjg.v25.i18.2149
Marco ML, Heeney D, Binda S, Cifelli CJ, Cotter PD, Foligné B, Gänzle M, Kort R, Pasin G, Pihlanto A, Smid EJ (2017) Health benefits of fermented foods: microbiota and beyond. Curr Opin Biotechnol 44:94–102. https://doi.org/10.1016/j.copbio.2016.11.010
Matsukawa H, Iida N, Kitamura K, Terashima T, Seishima J, Makino I, Kannon T, Hosomichi K, Yamashita T, Sakai Y, Honda M, Yamashita T, Mizukoshi E, Kaneko S (2021) Dysbiotic gut microbiota in pancreatic cancer patients form correlation networks with the oral microbiota and prognostic factors. Am J Cancer Res 11:3163–3175
Michels N, Zouiouich S, Vanderbauwhede B, Vanacker J, Indave Ruiz BI, Huybrechts I (2022) Human microbiome and metabolic health: an overview of systematic reviews. Obes Rev 23(4): e13409. https://doi.org/10.1111/obr.13409
Milligan-McClellan KC, Dundore-Arias JP, Klassen JL, Shade A, Kinkel LL, Wolfe BE (2022) Deciphering the microbiome: integrating theory, new technologies, and inclusive science. mSystems 7:e0058322. https://doi.org/10.1128/msystems.00583-22

Mirzaei MK, Maurice CF (2017) Menage a trois in the human gut: interactions between host, bacteria and phages. Nat Rev Microbiol 15:397–408. https://doi.org/10.1038/nrmicro.2017.30
Mousa WK, Chehadeh F, Husband S (2022) Recent advances in understanding the structure and function of the human microbiome. Front Microbiol 13:825338. https://doi.org/10.3389/fmicb.2022.825338
Muszyński D, Kudra A, Sobocki BK, Folwarski M, Vitale E, Filetti V, Dudzic W, Kaźmierczak-Siedlecka K, Połom K (2022) Esophageal cancer and bacterial part of gut microbiota - a multidisciplinary point of view. Front Cell Infect Microbiol 12:1057668. https://doi.org/10.3389/fcimb.2022.1057668
O'Toole PW, Marchesi JR, Hill C (2017) Next-generation probiotics: the spectrum from probiotics to live biotherapeutics. Nat Microbiol 2:17057. https://doi.org/10.1038/nmicrobiol.2017.57
Oren A, Garrity GM (2021) Valid publication of the names of forty-two phyla of prokaryotes. Int J Syst Evol Microbiol 71(10). https://doi.org/10.1099/ijsem.0.005056
Perez-Carrasco V, Soriano-Lerma A, Soriano M, Gutiérrez-Fernández J, Garcia-Salcedo JA (2021) Urinary microbiome: yin and Yang of the urinary tract. Front Cell Infect Microbiol 11:617002. https://doi.org/10.3389/fcimb.2021.617002
Qiu P, Ishimoto T, Fu L, Zhang J, Zhang Z, Liu Y (2022) The gut microbiota in inflammatory bowel disease. Fron Cell Infect Microbiol 12:733992. https://doi.org/10.3389/fcimb.2022.733992
Radjabzadeh D, Bosch JA, Uitterlinden AG, Zwinderman AH, Ikram MA, van Meurs JBJ, Luik AI, Nieuwdorp M, Lok A, van Duijn CM, Kraaij R, Amin N (2022) Gut microbiome-wide association study of depressive symptoms. Nat Commun 13(1):7128. https://doi.org/10.1038/s41467-022-34502-3
Sani G, Manchia M, Simonetti A, Janiri D, Paribello P, Pinna F, Carpiniello B (2021) The role of gut microbiota in the high-risk construct of severe mental disorders: a mini review. Front Psychol 11:585769. https://doi.org/10.3389/fpsyt.2020.585769
Sender R, Fuchs S, Milo R (2016) Revised estimates for the number of human and bacteria cells in the body. PLoS Biol 14(8):e1002533. https://doi.org/10.1371/journal.pbio.1002533
Singh N, Singh J, Singh K (2018) Small at size, big at impact: microorganisms for sustainable development. In: Singh J, Sharma D, Kumar G, Sharma NR (eds) Microbial bioprospecting for sustainable development. Springer, pp 3–28. https://doi.org/10.1007/978-981-13-0053-0_1
Sirohi SK, Singh N, Dagar SS, Puniya AK (2012) Molecular tools for deciphering the microbial community structure and diversity in rumen ecosystem. App Microbio Biotechnol 95:1135–1154. https://doi.org/10.1007/s00253-012-4262-2
Sorboni SG, Moghaddam HS, Jafarzadeh-Esfehani R, Soleimanpour S (2022) A comprehensive review on the role of the gut microbiome in human neurological disorders. Clin Microbiol Rev 35(1):e0033820. https://doi.org/10.1128/CMR.00338-20
Sun D, Xiang H, Yan J, He L (2022) Intestinal microbiota: a promising therapeutic target for hypertension. Front Cardiovasc Med 9:970036. https://doi.org/10.3389/fcvm.2022.970036
van Nood E, Vrieze A, Nieuwdorp M, Fuentes S, Zoetendal EG, de Vos WM, Visser CE, Kuijper EJ, Bartelsman JF, Tijssen JG, Speelman P, Dijkgraaf MG, Keller JJ (2013) Duodenal infusion of donor feces for recurrent Clostridium difficile. N Engl J Med 368:407–415. https://doi.org/10.1056/NEJMoa1205037
Villemin C, Six A, Neville BA, Lawley TD, Robinson MJ, Bakdash G (2022) The heightened importance of the microbiome in cancer immunotherapy. Trends Immunol 44(1):44–59. https://doi.org/10.1016/j.it.2022.11.002
Villmones HC, Svanevik M, Ulvestad E, Stenstad T, Anthonisen IL, Nygaard RM, Dyrhovden R, Kommedal Ø (2022) Investigating the human jejunal microbiota. Sci Rep 12(1):1682. https://doi.org/10.1038/s41598-022-05723-9
Wang DD, Nguyen LH, Li Y, Yan Y, Ma W, Rinott E, Ivey KL, Shai I, Willett WC, Hu FB, Rimm EB, Stampfer MJ, Chan AT, Huttenhower C (2021) The gut microbiome modulates the protective association between a Mediterranean diet and cardiometabolic disease risk. Nat Med 27(2):333–343. https://doi.org/10.1038/s41591-020-01223-3

Wehedy E, Shatat IF, Al Khodor S (2022) The human microbiome in chronic kidney disease: a double-edged sword. Front Med 8:790783. https://doi.org/10.3389/fmed.2021.790783

Yadav M, Chauhan NS (2022) Microbiome therapeutics: exploring the present scenario and challenges. Gastroenterol Rep (Oxf) 10:goab046. https://doi.org/10.1093/gastro/goab046

Yan Q, Gu Y, Li X, Yang W, Jia L, Chen C, Han X, Huang Y, Zhao L, Li P, Fang Z, Zhou J, Guan X, Ding Y, Wang S, Khan M, Xin Y, Li S, Ma Y (2017) Alterations of the gut microbiome in hypertension. Front Cell Infect Microbiol 7:381. https://doi.org/10.3389/fcimb.2017.00381

Yu Y, Wen H, Li S, Cao H, Li X, Ma Z, She X, Zhou L, Huang S (2022) Emerging microfluidic technologies for microbiome research. Front Microbiol 13:906979. https://doi.org/10.3389/fmicb.2022.906979

Zhang X, Li L, Butcher J, Stintzi A, Figeys D (2019) Advancing functional and translational microbiome research using meta-omics approaches. Microbiome. 7(1):154. https://doi.org/10.1186/s40168-019-0767-6. PMID: 31810497; PMCID: PMC6898977.

Zhou Z, Sun B, Yu D, Zhu C (2022) Gut microbiota: an important player in type 2 diabetes mellitus. Front Cell Infect Microbiol 12:834485. https://doi.org/10.3389/fcimb.2022.834485

Emerging Microbial Technologies: Mitigating Challenges to Humans

8

Manish Thakur, Tejinder Kaur, and R. C. Sobti

Abstract

This book chapter discusses the applications of emerging microbial technologies in mitigating challenges in agriculture, environment, and health sectors. The chapter begins by highlighting the importance of microbes in addressing the challenges faced by humanity. It emphasizes the need for integrated strategies to control microorganisms for more efficient performance. The role of microbes in agriculture is then explored, including their contribution to organic matter breakdown, plant germination, nutrient utilization efficiency, and disease and pest management. The chapter also discusses the use of biofertilizers, which are mixtures of microbial cells that enhance nutrient availability and uptake in plants. Additionally, the potential of biopesticides as safe alternatives to synthetic insecticides is examined. The chapter highlights the use of specific preparations containing live microorganisms or botanical substances for pest control. Furthermore, recent advances in microbial technologies for soil health improvement are discussed, including the use of designer soil bacteria for crop protection, artificial intelligence platforms for analysing soil health, and the use of microalgae for fertilization. The chapter concludes by emphasizing the need for further research and collaboration between companies and research institutions to maximize the potential of microbial technologies in addressing the challenges faced by humanity. Overall, this chapter provides insights into the applications of microbial

M. Thakur
Department of Microbiology, School of Bio Engineering and Biosciences, Lovely Professional University, Phagwara, India

T. Kaur (✉)
Department of Zoology, DAV University, Jalandhar, India

R. C. Sobti
Panjab University, Chandigarh, Chandigarh, India

R. Sobti et al. (eds.), *Role of Microbes in Sustainable Development*,
https://doi.org/10.1007/978-981-99-3126-2_8

technologies in mitigating challenges in various sectors, highlighting their potential for sustainable and environmentally friendly solutions.

Keywords

Base editing · Artificial microbial consortia · Metagenomics · Bioremediation · CRISPR/Cas

8.1 Introduction

Currently, there are approximately seven billion people living on this planet and addressing a wide range of challenges in order to maintain or perhaps enhance the quality of life (Walker et al. 2014). It is apparent that microbes will play a critical role in addressing the issues of the next several decades. Controlling microorganisms using an integrated strategy is necessary for more efficient performance with greater efficiency (Verstraete and De Vrieze 2017). Plant growth promoters and phytopathogen controls are essential for a wide range of agricultural crops, and many species are employed as biofactories for significant pharmaceutical compounds. Using microorganisms to synthesize various chemicals, fuel molecules, and industrial polymers is not the end of the possibilities for biofactories. Strains that are environmentally significant due to their biodecomposition or biosorption capacity have piqued the interest of research laboratories and industry. Figure 8.1 depicts the various applications of microbial technology (Fig. 8.1).

Microbial technology is the name given to this new applied science, and we anticipate that the use of complicated methods will lead to new and better goods and services. Understanding the difficulties and identifying possible prospects for emerging microbial technologies within the perspective of environmental protection are significant. This article is written with the aim to shed light on some of the applications of microbial technology in mitigating challenges in agriculture, the environment, and the health sector.

8.2 Role of Microbes in the Agricultural Industry

Microorganisms play a vital function in agriculture. They play an important role in the breakdown of organic matter and the creation of humus. Plant germination and development may be sparked by the presence of microbes, which can counteract common illnesses, boost resistance to stress, increase general physical vigour and robustness, enhance nutrient utilization efficiency, and offer tolerance against abiotic stressors (Shinwari et al. 2019). Microorganisms also play a critical role in nitrogen fixation, phosphate solubilization, potassium mobilization, disease and pest management, and many other processes in the environment. Crop production, crop protection, soil health improvement, and compost preparation are all benefits of the use of

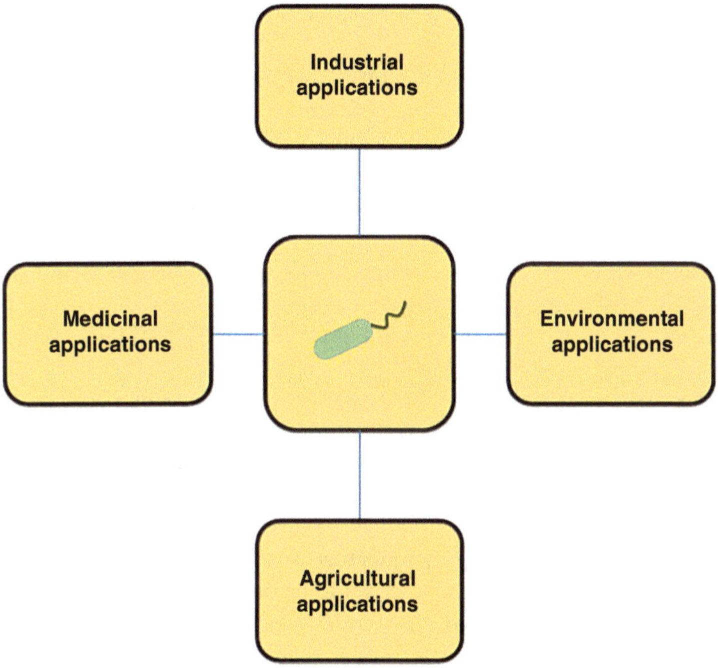

Fig. 8.1 Applications of microbial technology

many soil and non-soil microorganisms. As a result, microorganisms are critical in agricultural production.

Studies are now underway on a number of microbial species, such as *Micrococcus, Enterobacter,* and *Acinetobacter*, for their potential to create plant hormones and dissolve phosphates. Several of these microbes, such as *Pseudomonas, Azospirillum, Bacillus*, and *Burkholderia,* are related to the ability to withstand biotic and abiotic stress, *Azotobacter* and *Bacillus* for nutrient absorption. In addition, the biological control agents *Bacillus amyloliquefaciens* and *Microbacterium oleovorans* have proven that they can control *Fusarium verticillioides* by inhibiting the development of phytopathogens and reducing the prevalence of plant illnesses (Díaz-Rodríguez et al. 2021).

Microorganisms that cause disease in plants are also a major and long-term concern to the availability of food and ecological health. More and more farmers are using agrochemicals as a dependable way of crop protection as agricultural output has expanded over the last several decades, which eventually provides economic stability to their activities. There are, however, several downsides to using more chemical inputs. Integrating nutrition and disease and pest control into a single strategy is a need in today's world, but it must be done in accordance with agricultural and economic realities. Biological agents are used as part of the management strategy in this method.

8.2.1 Biofertilizers

Plant development may be encouraged by using bio-fertilizers, which are mixtures of microbial cells, either single or multiple strains that increase the availability and uptake of nutrients for plants (Riaz et al. 2020). Furthermore, it is important to note that biofertilizers may also give various direct and indirect advantages to plant development, such as phyto-stimulation, abiotic stress tolerance, and biocontrol. Biofertilizers are a kind of fertilizer that can be used to nourish plants (Shirmohammadi et al. 2020). A rising number of novel approaches are being employed to characterize soil microbial populations and their impact on plant nutrient uptake, as well as other plant growth-promoting (PGP) characteristics (Saad et al. 2020). In both symbiotic and non-symbiotic relationships, most plant species are associated with the PGPB (plant growth-promoting bacteria) and other plant growth-promoting organisms. When it comes to PGPBs, PGPRs (plant growth-promoting rhizobacteria) are the ones that have received the most consideration (Kloepper and Schroth 1978). Depending on the crop species or genotype that is used, biofertilizers may provide a variety of different results. A few new genetic markers have been discovered (e.g. quantitative trait loci) that are connected with this differential response, but many are yet to be discovered (Kaeppler et al. 2000). Currently, most commercial biofertilizers include N_2-fixing microbes (e.g. *Azospirillum* spp., *Azotobacter* spp., *Actinorhizobium* spp., and *Rhizobium* spp.) (Mitter et al. 2021). Figure 8.2 summarizes the role of microbes as biofertilizers. (Fig. 8.2).

8.2.2 Biopesticides

Crop output must expand dramatically to fulfil rising food demand and diminishing land availability. Pests like insects and pathogens affect plants at all phases of development, including harvest and storage, causing considerable damage and

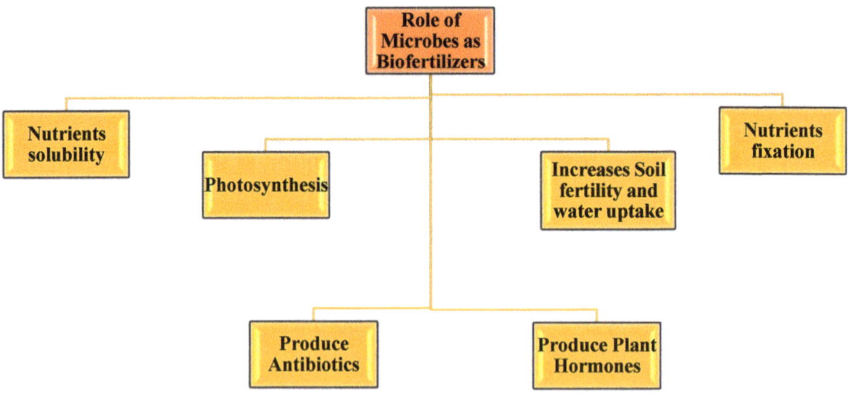

Fig. 8.2 Various roles of microbes to improve growth and development of plants as biofertilizers

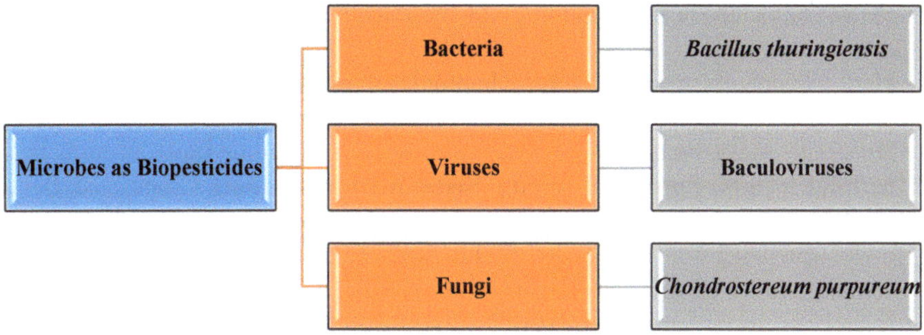

Fig. 8.3 Different microbes as bio pesticides

losses (Gasic and Tanovic 2013). Pest control without damaging the ecology is a big worldwide problem. Synthetic insecticides protect crops by killing pests. Contamination of food, the ability to change human biological systems, and long-term ecosystem persistence are serious threats to their long-term use. Alternative crop protection methods are urgently needed. Biopesticides may become more important crop protection solutions in the future since they are safe for people and non-target species in both individual and integrated pest control applications (IPM). For these reasons, biopesticides have gained popularity. Biopesticides are characterized as "specific preparations containing live microorganisms", e.g. *Actinomycete, B. thuringiensis*; "botanical substances", e.g. pheromones (Xiaoman et al. 2019); or viruses, e.g. Baculoviruses (Inceoglu et al. 2001) (Fig 8.3). For the time being, pesticides such as Cry and Cyt are the most well-known endotoxin proteins. These endotoxins are produced by the soil bacterium *Bacillus thuringiensis* (Bt), which has entomopathogenic effects and suppresses pests found in cabbage, potatoes, and cereals (Sarwar 2015). As a result of the efficacy of Bt protein crystals in inhibiting the spread of caterpillars, such as Lepidoptera, transgenic crops such as tomato, tobacco, and maize have been planted across the globe (Khan et al. 2016). *Spodoptera frugiperda*, for example, is a pest caterpillar that may be infected by the baculovirus thus lowering agricultural losses caused by this caterpillar, particularly in maize. Progress made in the genetic enhancement of this virus has also boosted the virus's efficacy as an insecticide (Popham et al. 2016).

Extensive research on biological control agents, such as biopesticides, will be necessary for the industry of biopesticides to be successful in the future. There are a limited number of available thorough studies despite the fact that intensive study is being conducted on the topic at a variety of academic institutes throughout the world. It would seem that a future in which biopesticides completely replace chemical pesticides would be impossible to achieve unless there is a maximal collaboration between companies and research institutions. In today's agricultural industry, it is necessary to use both biological and chemical methods of pest control. Increasing the rate at which laboratory data are used should be beneficial to the expansion of large-scale manufacturing (Kumar et al. 2021).

Recent advances in the technology of the microbiome of the soil are discussed below.

Here are three examples of new technologies that make our soil healthier.

1. AgBiome: With the help of designer soil bacteria, AgBiome protects crops from harmful insects, fungi, and weeds. Better Microbes, Better Crops, and Better World is the motto of this company, which was established in 2017. AgBiome is now sequencing a microbial library that has been generated from environmental samples collected from all around the world. More than 80,000 sequenced microbials have been developed by the North Carolina Company, 3500 of which are focused on controlling insects. Insect pests, fungus infections, and weeds may all be targeted using their technology, which can find and target specific bacteria and proteins. Soil microbe "cocktails" may then be created for use by farmers as a natural pest control approach.
2. Biome Makers: It is an artificial intelligence platform that analyses the state of a field by taking into account not only the farming methods currently in use but also the capacity of the soil to support any type of crop. Through collaboration with Bayer and over 70 other producers of agricultural inputs, they will assist farmers in gaining a better understanding of what works effectively and how it impacts the health of their land.
3. MyLand: It is a fertilizer that grows algae in the soil. Microalgae from the farm are harvested by MyLand for the purpose of boosting soil health and productivity, as well as carbon sequestration. It is exactly like our own microbiome, which is unique to each farm. Algae that are most suited for proliferation are sampled by MyLand technicians. The algae are grown in little containers with lights and a temperature that is just right. They produce millions of cells, and the farmer's irrigation system returns them to the soil. Farmers can now use 25% less fertilizer, 15% less water, and 40% less tillage as a consequence of this change. As a result, revenue soars by 40% and yield rises by 25%.
4. Biofloc technology: Fish and crustaceans may benefit from the application of the biofloc technology (BFT), which utilizes microbial biotechnology to boost the effectiveness and usage of fish meals by treating and converting harmful chemicals such as nitrogen components. Aquaculture systems with high fish stocking density, robust aeration, and biota might benefit greatly from the usage of biofloc technology because of this (Jamal et al. 2020). Hargreaves (2013) mentioned that "A combination of algae, bacteria, protozoans, and other sorts of particulate organic matter such as faeces and uneaten feed, in addition to some of zooplankton and nematodes, created together to be an integrated and interdependent ecosystem".

8.3 Emerging Technologies in the Environmental Industry

Throughout the globe, environmental contaminants in soil and water are a source of worry. Many harmful and carcinogenic elements are present in the environment and constitute a major hazard to human health. Microbes can be useful to make the environment free from contaminants. It is possible to treat contaminated water and wastewater by using chemical, physical, and biological methods in order to remove and/or detoxify the contaminants in it. As with contaminated soil, several treatment procedures such as thermal desorption and landfilling may be utilized to remediate the situation. However, these soil treatments are ineffective in restoring the indigenous flora and fauna of the area. Using microorganisms to eliminate hazardous pollutants from the environment is the most promising technology since it is eco-friendly, safe, and efficient even when the pollutants are present in low quantities.

8.3.1 Microbial Fuel Cell

Richard Smalley, a Nobel Laureate, once stated that "energy is the single greatest challenge facing humanity". Energy production and efficient use in industry and agriculture are critical for the prosperity of any nation. The generation of enough electricity in a developing country like India will be a major challenge in the coming years. To these issues, microbial fuel cells (MFCs) provide an answer. It is possible to use MFCs to both treat wastewater and generate power. It has been known since 1911 that bacteria are capable of generating electrical currents (Potter 1911). But it has come to light in recent years. To generate energy from microorganisms, an MFC uses an anaerobic compartment as the anode and an active microorganism as the catalyst. In an oxygen-free environment, microbial fuel cells get their power from the oxidation of an electron donor followed by the transmission of electrons to the anode. Electrons that are contributed to the anode are then transported to the cathode via a resistor or another form of electrical device. It is possible for the cathode to be open to the atmosphere or to be submerged in oxygenated water. The protons that are liberated during the oxidation of the organic matter make their way to the cathode, typically by passing through a cation-selective membrane that controls the amount of oxygen that can diffuse into the anode chamber. At the surface of the cathode, the elements electrons, protons, and oxygen mix to make water. Bacterial species identified till now for their use in microbial cells are *Rhodoferax ferrireducens* (Chaudhuri and Lovely 2003), *Escherichia coli*, *Pseudomonas aeruginosa* (Rabaey and Verstrate 2005), and *Clostridium butyricum* The efficiency and power of microbial fuel cells are improving all the time. There is no question that wastewater treatment can be used in the medium term at market value rates. It will be possible to transform carbs into energy in the future if one understands more about how biology works, enhance electrochemical technology, and lower the overall cost of electrodes (Rabaey and Verstrate 2005).

8.3.2 Microbial Concrete

In the building industry, concrete is used almost as frequently as water, and the need for both construction and construction supplies is on the rise throughout the world. Concrete is by far the most common building material. As a result, the demand for environmentally friendly, clean, and long-lasting concrete has increased (Kaur et al. 2022). Due to microbial concrete's ability to effectively mend cracks, several mechanical properties have improved, including compressive strength, water absorption, and permeability (Mondal and Ghosh 2019). Microbiologically induced calcite precipitation (MICP) is the process of making calcite precipitate by adding spore-forming bacteria in the concrete. These bacteria are able to produce calcite continuously (MICP). Ammonia and carbon dioxide are produced as urea is broken down by microbial urease. As a result of the ammonia released into the environment, the pH rises, resulting in a buildup of calcium carbonate that is insoluble. Previous studies found that the compressive strength of cement mortar was significantly increased about 18% by the addition of aerobic microorganisms such as *Bacillus pasteurii*, *Pseudomonas aeruginosa*, and *Bacillus subtilis* (Mondal and Ghosh 2019).

8.3.3 Role of Microbes as Bioremediators

Bioremediation is a technique that may be used to clean up contaminated sites. Bioremediation techniques might well be carried out by a diverse variety of microorganisms, and many of these microorganisms have already been deployed in sites that had previously been contaminated by polycyclic aromatic hydrocarbons (PAHs) and chlorinated organics (Vitorino and Bessa 2017). Bioremediation is the process of using organisms or derivatives of organisms to break down contaminants in a natural environment. When compared to traditional approaches, the primary benefit of bioremediation is the lower cost of the process itself. The use of microbe-based sorbents for the removal and recovery of strategic and valuable heavy metals from industrial wastewater has recently been developed in environmental microbial technology. Heavy metals may be removed from aqueous solutions by a variety of microorganisms, including bacteria, yeast, fungus, and algae. In recent years, researchers have been looking at ways to employ microbial fuel cells (MFCs) in the treatment of solid wastes or even wastewater, microbial fuel cells are those that utilize electrons from low-value organic substrates to produce energy (Xu et al. 2016). It is possible to use this alternative technology to treat wastewater and generate power at the same time because of the versatility of mixed MFC cultures (Pendyala et al. 2016). Microfuels for the generation of bioelectricity have recently been made from a variety of bacteria, including *E. coli* and *Geobacter* spp. (Rahimnejad et al. 2015).

8.3.4 Mitigating Climate Using Microorganisms

The climate system includes the atmosphere, land surface, snow and ice, and seas, as well as any organisms that live there. A thick blanket of gases envelops the globe, keeping it warm and encouraging life to flourish. If these gases were not there, the Earth would be 20–30 °C colder and less suitable for life (Kumar et al. 2020). A progressive increase in the temperature of the Earth's atmosphere is called global warming. These gases enable incoming sunlight to pass yet absorb heat reflected back from the Earth's surface. Because of human actions like burning fossil fuels, this blanketing effect is being increased (Olufemi et al. 2014). Human activities and natural phenomena like volcano eruptions have boosted greenhouse gas emissions in recent years. These gases build up in the atmosphere, increasing concentrations over time. During the industrial age, all of these gases increased significantly. Carbon dioxide, methane, nitrous oxide, and halocarbons are the principal greenhouse gases (Abatenh et al. 2018). Microbes are important contributors to global carbon dioxide, methane, and nitrous oxide fluxes, and they are expected to react efficiently to climate change. Microbes regulate the flow of greenhouse gases on Earth. This entails taking into account the intricate interactions that microbes have with various biotic and abiotic elements. The promise of lowering greenhouse gas emissions by controlling terrestrial microbial activities to combat climate change is a tempting option for the future. Microorganisms are generally acknowledged to have had a crucial role in determining greenhouse gas concentrations in the atmosphere (Zimmer 2010). Using nutrient cycle mechanisms and activating their functional genetic material for degrading and removing compounds or gases that cause global warming are the key feedback response mechanism for climate change (Zhou et al. 2011). A good method to tackle climate change is microbial communities and biogeochemical cycles. Cellular energy from greenhouse gases is crucial for microorganisms.

8.3.5 Metagenomics

Microorganisms boom all the way through the natural world, and microbes have adjusted to endure under a wide range of punitive or unaccommodating conditions, ensuing in adaptation by the microorganisms to specific niches (Edge et al. 2020; Handelsman et al. 1998). As a result of these microbe-based adaptations, noteworthy phenotypes have evolved that might be used for biotechnological purposes. For example, after the development of Taq DNA polymerase from *Thermus aquaticus*, a thermophile, and similar trends followed to identify key enzymes like pfu from *Pyrococcus furiosus*, a thermophile. There were two phospholipases from hot springs and a naphthalene catabolic gene found in oil-polluted areas that were identified using a metagenome technique. Furthermore, according to the most recent estimates, around 99.99% of the microorganisms that are present in many natural habitats are not easily culturable, and as a result, they are not accessible for either fundamental or technological study. Based on this analysis, it was determined that an alternative method of microbial biotechnology could be able to shed light on these specially modified exclusive bacteria, as well as their potentially valuable gene types or gene families. There is an abundance of microorganisms in the natural world, and through time, they have developed the ability to thrive in a wide range of harsh or inhospitable settings, which has led to their specialization in various niches (Kirubakaran et al. 2020). Metagenomics is a powerful tool that can access the abundant biodiversity of environmental samples, including uncultivable microbes. It can be used to determine microbial diversity and population structure, as well as the roles that microbes play in the ecosystem, and it can reveal novel genes of interest. The metagenomic method contributes to a better understanding of microbial ecology as well as the identification of beneficial microbial derivatives such as antibiotics, poisons, and enzymes with a variety of improved functions (Kirubakaran et al. 2020). The powerful method of metagenomics is the separation of genomic DNA directly from the surrounding environment. This eliminates the necessity for the cultivation of organisms in a laboratory setting in advance (Wang et al. 2021).

8.4 Role of Microbes in Health Sector

An important addition to clinical research in microbiology has been the study of ideal settings and test performance for a broad range of medical microbiological diagnostic and therapeutic tests. Thanks to the improved technology for studying complex microbial populations, the microbiome has gained increasing attention in recent years. Biocontrol of illnesses, manufacturing of vaccines, antibiotics, and biotherapeutics are all examples of microorganisms' involvement in the development of medical goods or services. Recombinant DNA technology's first commercially viable product, artificial insulin, was developed in the 1970s by genetic use of *Escherichia coli* bacteria (Walsh 2012). Vaccinations have had a huge influence on health at a relatively modest cost, thanks to the rapid development of these vaccines. In spite of this, the vast majority of vaccines now in use were created using procedures that were first used more than a century ago and do not make use of the field's full potential. Vaccine technology has advanced rapidly since the advent of genetic engineering, which is now paving the way for new goods to enter the market. Vaccines can be of a variety of types containing attenuated organisms known as attenuated or live vaccines. Whole or fractionated; subunit vaccines, which use proteins, peptides, or nucleic acids as antigens; carbohydrate vaccines, which are made up of polysaccharides, oligosaccharides, and glycans; toxoids, which use inactivated pathogen toxins; and conjugate vaccines, which include polysaccharides in combination with other antigenic components. Vaccines that have inactivated immunogenic antigen genes are supplied to the body by the use of DNA vaccines, which include plasmids carrying the antigenic genes. The development of vaccines is generated from recombinant viruses that carry antigens encoded by genes taken from other pathogen-causing viruses. Figure 8.4 depicts the role of microrganisms in health sector.

8.4.1 Probiotics

Pathogens such as yeasts, other bacteria, and viruses are crowded out, and a beneficial symbiosis develops between them and the human gastrointestinal system. Preventing pathogen adhesion to the intestinal epithelium and competing for nutrients necessary for pathogen survival are just two ways in which they have an anti-microbial effect. They also reverse some of the effects of infection on the epithelium, such as secretory changes and neutrophil migration, by altering the microflora. Lactose intolerance may be treated with probiotics by increasing the synthesis of β-galactosidase, an enzyme that breaks down the lactose into simpler sugars. This shows a possible benefit of probiotics in health industry.

8.4.2 Base Editing

CRISPR/Cas and nucleobase deaminase have been skilfully combined to provide a new method for genome editing that is both programmable and catalytic. Exogenous donors, homology-directed repair, or double-strand DNA breaks are not required to introduce point mutations at many target loci.

Base editing's rapid acceptance in the genetic alteration of several industrially and therapeutically significant microorganisms is made possible by its simplicity, precision, and multiplex editing capacity (Wang et al. 2021). Through either transient RNA or long-term DNA base changes, base editing has significant promise to cure many genetic illnesses and disorders. DNA and RNA base editors have recently improved in terms of selectivity, efficiency, accuracy, and delivery, opening up new therapeutic possibilities (Porto et al. 2020).

Clustered Regularly Interspaced Short Palindromic Repeat (CRISPR) PlatformsBacteria and archaeal CRISPR systems have been used to alter genomes and transcriptomes by causing double-stranded DNA breaks (DSBs) or RNA

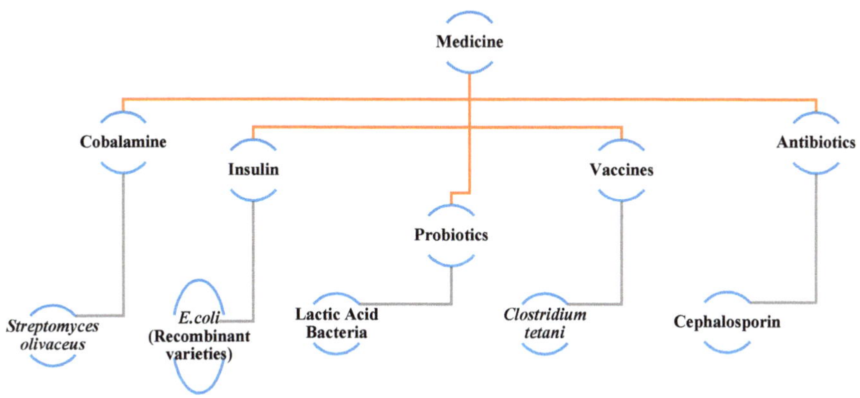

Fig. 8.4 Role of microbes in medicine

cleavage in live cells to target specific genes. Genome editing at distinct genomic loci may be reprogrammed using Watson–Crick–Franklin base pairing principles to change the sequence of a piece of RNA (referred to as a spacer, guide RNA (gRNA), or single guide RNA). One or both of the two competing mammalian endogenous repair mechanisms, NHEJ and Homologous End Joining (HNEJ), may be used to repair DSBs produced by the CRISPR-associated proteins, Cas9 and Cas12 (HDR). These non-specific indels (indels) occur at the location of NHEJ-resolved double-stranded breaks (DSBs), leading to frameshifts and gene knockouts. Exogenous DNA repair templates may be used by researchers to co-opt the HDR process and insert desired and precise sequence alterations into the genome.

Single-base editors are a unique technique for precise genetic modification that does not need the introduction of double-strand breaks (DSBs) or donor DNA templates. This was made possible by the discovery of single-base editors (Komor et al. 2016). The cytidine base editor (CBE) is made up of a catalytically impaired Cas nuclease in addition to a cytidine deaminase (such as APOBEC) and the uracil DNA glycosylase inhibitor UGI. Together, these three components catalyse the conversion of cytosine (C) to uracil (U) at targeted sites, which ultimately results in the substitution of C for thymine (T) (Komor et al. 2017). The application of base editing systems to microbes, such as *Escherichia coli,* has been made possible, *Staphylococcus aureus*, *Klebsiella pneumoniae*, and *Streptomyces* (Zhou et al. 2011), encourage the genetic modification of these microorganisms in a significant way. Infections caused by strains of *Mycobacterium tuberculosis* (Mtb) that are resistant to several drugs pose a significant risk to the health of people all over the world, generating an immediate need for innovative treatment approaches. Effective techniques for editing the genome may make it easier to locate critical genes and pathways that are involved in the physiology, pathogenesis, and drug resistance processes of bacteria. This can lead to the discovery of innovative therapies for drug-resistant TB. In this paper, we report on a two-plasmid system called MtbCBE that may be utilized to inactivate genes in Mtb and induce point mutations. In this system, the assistant plasmid pRecX-NucSE107A expresses RecX and NucSE107A in order to repress RecA-dependent and NucS-dependent DNA repair systems. Additionally, the base editor plasmid pCBE expresses a fusion protein that combines cytidine deaminase APOBEC1, Cas9 nickase (nCas9), and uracil DNA (UGI). The combination of the two plasmids allowed for the effective conversion of G:C base pairs to A:T base pairs at certain locations within the Mtb genome. The successful development of a base editing system will facilitate the elucidation of the molecular mechanisms underlying Mtb pathogenesis and drug resistance and provide critical inspiration for the development of base editing tools in other microbes (Ding et al. 2021).

8.4.3 Artificial Microbial Consortia

The metabolic variety of the organisms in natural microbial ecosystems facilitates interaction and syntrophic interactions among the members of the community. To boost bioprocesses' productivity and yields, artificial microbial consortia engineering is employed (Ergal et al. 2022). The development of artificially constructed communities, also known as "artificial microbial consortia" (AMC) or "synthetic communities", has been made possible by advances in culture collection and techniques (SynComs). Specifically, the structure and function of the microbiome are being reconstructed by using core microbiomes. In SynComs research, members may be added, removed, or replaced as required, allowing for a complete and accurate examination under controlled circumstances (Mitter et al. 2021). These can then be utilized in numerous applications.

To reduce our environmental impact while also creating a more circular economy, we may use these microorganisms to power our next generation of chemical production processes (NGCP).

8.5 Conclusion

Due to the fact that microorganisms are the most biodiverse class of creatures, we may expect the emergence and spread of new human and/or agricultural illnesses to become a recurrent problem in the future, which will be aggravated by globalization. Thus, we think that measures to manage epidemics, environmental stress, and agricultural pests should be addressed globally, to avoid uncertain repercussions on health and the economy. These initiatives must take into consideration the current technological framework, which includes transgenic and recombinant DNA, in order for microbiological research to be translated into commodities and services for the benefit of all people.

References

Abatenh E, Gizaw B, Tsegaye Z, Tefera G (2018) Microbial function on climate changes a review. Environ Pollut Clim Chang 3(1):001–007

Chaudhuri SK, Lovley DR (2003) Electricity generation by direct oxidation of glucose in mediatorless microbial fuel cells. Nat Biotechnol 21(10):1229–1232. https://doi.org/10.1038/nbt867. [Epub 2003 Sep 7. PMID: 12960964]

Díaz-Rodríguez AM, Lilian Alejandra SG, Carmen María FP, Parra-Cota FI et al (2021) The current and future role of microbial culture collections in food security worldwide. Front Sustain Food Syst 4:614739. https://doi.org/10.3389/fsufs.2020.614739

Ding X-Y, Li S-S, Geng Y-M, Yan M-Y, Li G-B, Zhang G-L, Sun Y-C (2021) Programmable base editing in mycobacterium tuberculosis using an engineered CRISPR RNA-guided cytidine deaminase. Front Genome Edit 3:734436. https://doi.org/10.3389/fgeed.2021.734436. ISSN=2673-3439.

Edge TA, Baird DJ, Bilodeau G et al (2020) The Ecobiomics project: advancing metagenomics assessment of soil health and freshwater quality in Canada. Sci Total Environ 710:135906. https://doi.org/10.1016/j.scitotenv.2019.135906

Ergal I, Buchman G, Fuchs W, Rittmann S (2022) Design and engineering of artificial microbial consortia for biohydrogen production. Curr Opin Biotechnol 73:74–80

Gasic S, Tanovic B (2013) Biopesticide formulations, possibility of application and future trends. Pestic Fitomed 28(2):97–102

Handelsman J, Rondon MR, Brady SF, Clardy J, Goodman RM (1998) Molecular biological access to the chemistry of unknown soil microbes: a new frontier for natural products. Chem Biol 5 (10):R245–R249. https://doi.org/10.1016/s1074-5521(98)90108-9. [PMID: 9818143]

Hargreaves JA (2013) Biofloc production systems for aquaculture. Southern Regional Aquaculture Center, Stoneville, NC, pp 1–11

Inceoglu AB, Kamita SG, Hinton AC, Huang Q, Seversson TF, Kyung-Don K, Hammock BD (2001) Recombinant baculoviruses for insect control. Pest Manag Sci 57(10):981–987

Jamal MT, Broom M, Al-Mur BA, Al Harbi M, Ghandourah M, Al Otaibi A, Haque MF (2020) Biofloc technology: emerging microbial biotechnology for the improvement of aquaculture productivity. Pol J Microbiol. 69(4):401–409. https://doi.org/10.33073/pjm-2020-049. Epub 2020 Dec 27. PMID: 33574868; PMCID: PMC7812359.

Kaeppler SM, Parke JL, Mueller SM, Senior L, Stuber C, Tracy WF (2000) Variation among maize inbred lines and detection of quantitative trait loci for growth at low phosphorus and responsiveness to arbuscular mycorrhizal fungi. Crop Sci 40:358–364. https://doi.org/10.2135/cropsci2000.402358x

Kaur P, Singh V, Arora A (2022) Microbial concrete—a sustainable solution for concrete construction. Appl Biochem Biotechnol 194:1401–1416. https://doi.org/10.1007/s12010-021-03604-x

Khan MA, Paul B, Ahmad W, Paul S, Aggarwal C, Khan Z et al (2016) Potential of Bacillus thuringiensis in the management of pernicious lepidopteran pests. In: Hakeem KR, Akhtar MS (eds) Plant, soil and microbes. Springer, New York City, NY, pp 277–301. https://doi.org/10.1007/978-3-319-29573-2_13

Kirubakaran R, ArulJothi KN, Revathi S, Shameem N, Parray JA (2020) Emerging priorities for microbial metagenome research. Bioresour Technol Rep 11:100485. https://doi.org/10.1016/j.biteb.2020.100485. ISSN 2589-014X.

Kloepper JW, Schroth MN (1978) Plant growth-promoting rhizobacteria on radishes. In: Proceedings of the 4th international conference on plant pathogenic bacteria. Gilbert-Clarey, Tours, pp 879–882

Komor AC, Kim YB, Packer MS, Zuris JA, Liu DR (2016) Programmable editing of a target base in genomic DNA without double-stranded DNA cleavage. Nature 533(7603):420–424. https://doi.org/10.1038/nature17946. [Epub 2016 Apr 20. PMID: 27096365; PMCID: PMC4873371]

Komor AC, Zhao KT, Packer MS, Gaudelli NM, Waterbury AL, Koblan LW, Kim YB, Badran AH, Liu DR (2017) Improved base excision repair inhibition and bacteriophage Mu Gam protein yields C:G-to-T:A base editors with higher efficiency and product purity. Sci Adv 3 (8):eaao4774. https://doi.org/10.1126/sciadv.aao4774. [PMID: 28875174; PMCID: PMC5576876]

Kumar BP, Reddy NS, Sparjanbabu DS (2020) Role of microbial communities to mitigate climate change in agriculture. Int J Curr Microbiol App Sci 9(10):1477–1483. https://doi.org/10.20546/ijcmas.2020.910.176

Kumar J, Ayyagari R, Mallick D, Mishra V (2021) An overview of some bio pesticides and their importance in plant protection for commercial acceptance. Plants 10(6):1185. https://doi.org/10.3390/plants10061185

Mitter EK, Tosi M, Obregón D, Dunfield KE, James J (2021) Germida. Rethinking crop nutrition in times of modern microbiology: innovative biofertilizer technologies. Front Sustain Food Syst 5: 606815. https://doi.org/10.3389/fsufs.2021.606815

Mondal S, Ghosh A (2019) Review on microbial induced calcite precipitation mechanisms leading to bacterial selection for microbial concrete. Constr Build Mater 225:67–75

Olufemi A, Okocha R, Olufemi O (2014) Global climate change. J Geosci Environ Protect 2:114–122

Pendyala B, Chaganti SR, Lalman JA, Heath DD (2016) Optimizing the performance of microbial fuel cells fed a combination of different synthetic organic fractions in municipal solid waste. Wast Manag 49:73–82

Popham HJ, Nusawardani T, Bonning BC (2016) Introduction to the use of Baculoviruses as biological insecticides. Methods Mol Biol 1350:383–392

Porto EM, Komor AC, Slaymaker IM, Yeo GW (2020) Base editing: advances and therapeutic opportunities. Nat Rev Drug Discov. 19(12):839–859. https://doi.org/10.1038/s41573-020-0084-6. Epub 2020 Oct 19. PMID: 33077937; PMCID: PMC7721651.

Potter MC (1911) Electrical effects accompanying the decomposition of organic compounds. Proc R Soc B Biol Sci 84(571):260–276. https://doi.org/10.1098/rspb.1911.0073. [Bibcode:1911RSPSB..84..260P. JSTOR 80609]

Rabaey K, Verstraete W (2005) Microbial fuel cells: novel biotechnology for energy generation. Trends Biotechnol 23(6):291–298. https://doi.org/10.1016/j.tibtech.2005.04.008. [PMID: 15922081]

Rahimnejad M, Adhami A, Darvari S, Zirepour A (2015) Microbial fuel cell as new technology for bioelectricity generation: a review. Alex Eng J 54(3):745–756

Riaz U, Mehdi SM, Iqbal S, Khalid HI, Qadir AA, Anum W et al (2020) Bio-fertilizers: eco-friendly approach for plant and soil environment. In: Hakeem KR, Bhat RA, Qadri H (eds) Bioremediation and biotechnology: sustainable approaches to pollution degradation. Springer, Cham, pp 188–214. https://doi.org/10.1007/978-3-030-35691-0_8

Saad MM, Eida AA, Hirt H (2020) Tailoring plant-associated microbial inoculants in agriculture: a roadmap for successful application. J Exp Bot 71:3878–3901. https://doi.org/10.1093/jxb/eraa111

Sarwar M (2015) Biopesticides: an effective and environmentally friendly insect-pests inhibitor line of action. Int J Eng Adv Res Technol 1:10–15

Shinwari ZK, Tanveer F, Iqrar I (2019) Role of microbes in plant health, disease management, and abiotic stress management. In: Kumar V et al (eds) Microbiome plant health dis, pp 231–250. https://doi.org/10.1007/978-981-13-8495-0_11

Shirmohammadi E, Alikhani HA, Pourbabaei AA, Etesami H (2020) Improved phosphorus (P) uptake and yield of rainfed wheat fed with p fertilizer by drought-tolerant phosphate-solubilizing fluorescent pseudomonads strains: a field study in drylands. J Soil Sci Plant Nutr 20:2195–2211. https://doi.org/10.1007/s42729-020-00287-x

Verstraete W, De Vrieze J (2017) Microbial technology with major potentials for the urgent environmental needs of the next decades. Microb Biotechnol 10(5):988–994. https://doi.org/10.1111/1751-7915.12779

Vitorino LC, Bessa LA (2017) Technological microbiology: development and applications. Front Microbiol 8:827. https://doi.org/10.3389/fmicb.2017.00827

Walker RV, Beck MB, Hall JW, Dawson RJ, Heidrich O (2014) The energy-water-food nexus: strategic analysis of technologies for transforming the urban metabolism. J Environ Manag 141: 104–115

Walsh G (2012) New biopharmaceuticals. Biopharm Int 25:34–38

Wang Y, Liu Y, Zheng P, Sun J, Wang M (2021) Microbial base editing: a powerful emerging technology for microbial genome engineering. Trends Biotechnol 39(2):165–180, . ISSN 0167-7799. https://doi.org/10.1016/j.tibtech.2020.06.010

Xiaoman L, Aocheng C, Dongdong Y, Canbin Q, Qiuxia W, Yuan L (2019) Overview of mechanisms and uses of biopesticides. Int J Pest Manage 67:65–72. https://doi.org/10.1080/09670874.2019.1664789

Xu L, Zhao Y, Doherty L, Hu Y, Hao X (2016) The integrated processes for wastewater treatment based on the principle of microbial fuel cells: a review. Crit Rev Environ Sci Technol 46:60–91. https://doi.org/10.1080/10643389.2015.1061884

Zhou J, Xue K, Xie J, Deng Y, Liyou W (2011) Microbial mediation of carbon-cycle feedbacks to climate warming. Nat Clim Chang 2:1–5

Zimmer C (2010) The microbe factor and its role in our climate future. Open J Environ Biol 1(4): 209–232

Modern Tools of Genome Engineering and Their Applications

9

Rajinder Kaur, Ashish Kumar Singh, Dinesh Kumar Singh, and Samer Singh

Abstract

Genome editing systems have emerged as an advanced bioengineering tool capable of targeting and editing the genomes of almost all organisms in a sequence-specific manner. This chapter presents an overview of the leading developments in the modern tool armory for genome editing that meet the high standards of efficiency, safety, and accessibility in genome engineering, i.e., ZFNs, TALENs, and CRISPR. These novel tools, primarily based on engineered nucleases, have proved to be one of the most effective and reliable tools for genome engineering. The engineered nucleases have enabled the alteration of targeted DNA sequences in a wide range of organisms and cell types. We will cover the mechanism and application of these methods for genome editing in current biology, functional genome screening, healthcare, agriculture, gene therapy, biological sciences, drug development, etc. General strategies used for designing specific ZFNs, TALENs, and CRISPR/Cas9 systems and analyzing their activity have been indicated. The therapeutic applications of these tools in controlling disease and their potential usage in the development of agricultural and industrial products, environmental protection, food development, immunotherapy, and treatment of genetic diseases, neurodegenerative diseases, and cancer are also briefly touched upon.

Keywords

ZFNs · TALEN · CRISPR/Cas9 · Genome editing · DNA repair

R. Kaur · A. K. Singh · S. Singh (✉)
Centre of Experimental Medicine and Surgery, Banaras Hindu University, Varanasi, India
e-mail: samer.singh10@bhu.ac.in

D. K. Singh
District Hospital, Chandauli, India

R. Sobti et al. (eds.), *Role of Microbes in Sustainable Development*,
https://doi.org/10.1007/978-981-99-3126-2_9

9.1 Introduction

The modifications (like insertions, deletions, and substitutions) in the genomes of organisms are commonly referred to as genome engineering (El-Mounadi et al. 2020). Genome or gene editing techniques are used for genome engineering to incorporate site-specific modifications into any genomic DNA, making use of different DNA repair mechanisms found endogenously. Gene editing usually deals with one target gene, i.e., a single gene is modified, whereas genome editing refers to large-scale modifications of the complete genomic DNA (Robb 2019). This technology has addressed the unmet need for the tools to introduce different types of genetic modifications that can cause a change in the physical as well as genomic traits of an individual or a population. Currently, scientists are making headway in developing gene-therapy treatment strategies by employing these advanced genome editing tools to prevent and treat various diseases in humans and animals. A breakthrough in the field occurred when Capecchi in 1989 for the first time demonstrated that the introduction of a segment of DNA, having homologous arms at both ends, into embryonic stem cells allowed its integration into the host genome via homologous recombination, causing inheritable changes in the cell (Capecchi 1989). Later, the discovery and development of methods to introduce an artificial DNA restriction enzyme into cells, which can cut genomic or dsDNA and generate a double-stranded break (DSB) at specific recognition sites, increased our ability to use cellular repair systems for genome engineering (Zhang et al. 2011). The mechanism of action for site-directed nucleases is based on the site-specific cleavage of the DNA or induction of a double-stranded break/nick (DSB) at targeted regions of DNA sequence by nucleases followed by triggering the two prominent DNA repair pathways of the cell, i.e., homology-directed repair (HDR) and nonhomologous end joining (NHEJ) (Fig. 9.1). The HDR repair mechanism uses homologous donor DNA to repair DNA damage, whereas NHEJ is an error-prone mechanism in which broken ends of DNA are joined together, often resulting in a heterogeneous pool of insertions or deletions. Though one of the efficient repair mechanisms active in the cells, NHEJ has a high rate of mutation and results in frequent nucleotide insertions or deletions (indels). HDR has low efficiency as it requires higher sequence similarity between the template and donor DNA strands. In the HDR mechanism, there is always a chance for reversion if the used DNA template is identical to the original undamaged DNA (Jasin and Rothstein 2013; Ghezraoui et al. 2014). The development of various techniques for genome engineering has focused on the use of different endonucleases (to create DSB with high precision) followed by employing these repair mechanisms to develop an engineered genome with new properties (Mandip and Steer 2019; Khalil 2020) (Fig. 9.1).

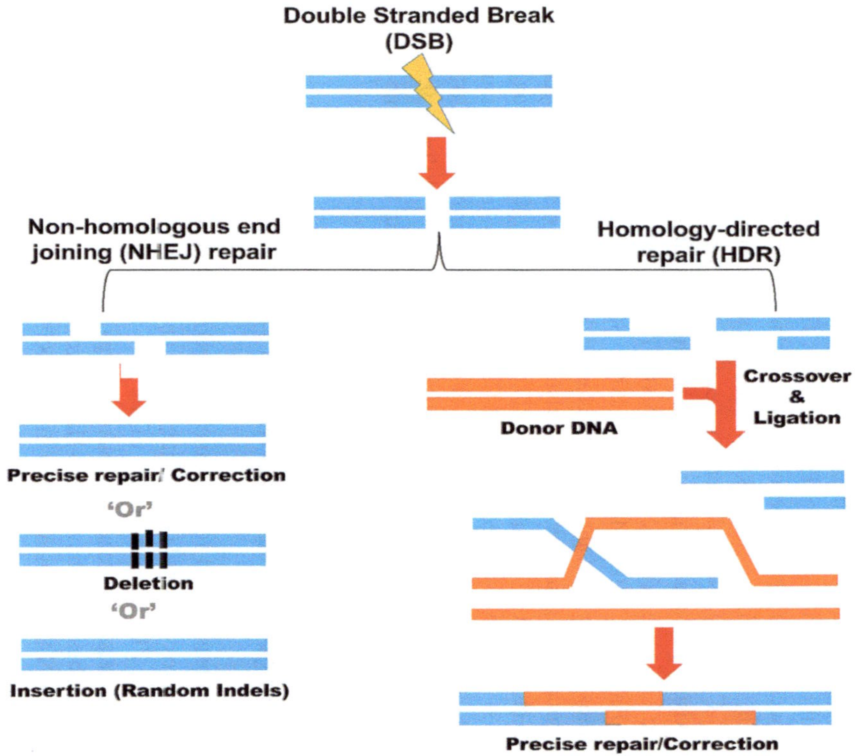

Fig. 9.1 An overview of the mode of action of ZFNs, TALENs, and CRISPR/Cas9 system: ZFNs TALENs, and CRISPR/cas9 nucleases will induce double-strand breaks (DSBs) in a gene at a targeted location and can be repaired by either NHEJ or HDR. NHEJ-mediated repair leads to the introduction of variable-length indel mutations. HDR with double-stranded DNA "donor templates" can lead to the introduction of precise nucleotide substitutions or insertions

9.2 Tools/Methods for Genome Engineering

The invention of genome editing tools opened up a whole set of opportunities for assisting the treatment of various diseases at the genome level (National Human Genome Research Institute 2019). Though various tools or methods for DNA modification have already existed for several decades, the development of more precise methods has made genome editing much cheaper, faster, and more efficient. Every genome editing tool being employed so far is based on one mechanism in which a targeted broken portion of DNA sequence in a gene or genome activates the cell repair mechanism that repairs the break in DNA sequence, i.e., HDR and NHEJ for DNA repair (National Institute of Health 2020). These tools and techniques allow efficient and accurate changes in genomic DNA by introducing DSB at a specific or targeted site in DNA followed by known modifications (i.e., insertion, deletion, indels, etc.). Currently, several genome engineering or genome/gene editing techniques exist which are primarily based on the following tools (nucleases) to target-specific sequences (molecular scissors): (1) zinc finger nucleases (ZFNs) (Porteus and Baltimore 2003; Miller et al. 2007; Sander et al. 2011; Wood et al. 2011), (2) transcription activator-like effector nucleases (TALENs) (Boch et al. 2009; Moscou and Bogdanove 2009; Christian et al. 2010; Hockemeyer et al. 2011; Wood et al. 2011; Zhang et al. 2011; Reyon et al. 2012; Sanjana et al. 2012), and (3) the RNA-guided CRISPR-Cas nuclease system (Deveau et al. 2010; Horvath and Barrangou 2010; Bhaya et al. 2011; Makarova et al. 2011; Cho et al. 2013; Cong et al. 2013; Jinek et al. 2013; Mali et al. 2013) (Table 9.1). The tools used for making sequence-specific cuts in the genome for the genome editing tool are briefly described below.

9.2.1 Zinc Finger Nucleases (ZFNs)

The first endonucleases used for genome engineering were Zinc finger nucleases (ZFNs), which were composed of zinc finger domains fused with FokI endonuclease (Kim et al. 1996). ZFNs are members of the zinc finger protein (ZFP) family, in which the zinc fingers (ZF) are novel DNA-binding domains that can bind to discrete base sequences. These ZFs have Cys2-His2 fingers and each ZF can recognize a triplet (3 bp) of DNA sequence (Miller et al. 1985; Wolfe et al. 2000). The ZFNs used for genomic engineering are comprised of a tandem array of ZFs, also known as the ZF array that confers unique nucleotide sequence-binding specificity. The dimerization of *Fok*I endonuclease of ZFNs on the binding of two ZFNs to the opposite DNA strands allows the cleavage of the dsDNA at the target sites (Fig. 9.2). For genome editing, two recombinant ZFNs recognizing two different (one each) closely located nucleotide sequences within the target DNA sequence are employed, which with the help of FokI, creates a double-strand break (DSB) at desired target DNA sequence. Since the series of linked ZF domains (ZF arrays) determine the specificity of the target nucleotide sequence, by changing the array of ZFs, any desired sequence may be targeted. A certain degree of off-target effects (nonspecific/

Table 9.1 Target specificity, mechanism of action, and experimental design of commonly used gene/genome editing nucleases

Feature	ZFNs	TALENs	CRISPR-Cas9
Recognition by	Zinc finger protein	RVD tandem repeat region of TALE protein	Single-strand guide RNA
Modification pattern	*Fok*1 nuclease	*Fok*1 nuclease	Cas9 nuclease
Length of recognized DNA target	18–36 bp/zinc finger pair; guanine-rich region	30–40 bp per TALEN	22 bp; followed immediately by 50-NGG-30 PAM sequence
Origin	Eukaryotes	Bacteria	Bacteria/Archaea
Structure (functional)	Dimer	Dimer	Monomer
Mechanism of target DNA recognition	DNA–protein interaction	DNA–protein interaction	DNA–RNA interaction via Watson-Crick base pairing
Mechanism of DNA cleavage and repair	Double-strand break-induced by *FokI*	Double-strand break-induced by *FokI*	Single- or double-strand break induced by Cas9
Design	Challenging. Available libraries of zinc finger motifs with predefined target specificity, but zinc finger motifs assembled in arrays can affect the specificity of neighboring zinc finger motifs, making the design challenge.	Easy. TALE motifs with target specificities are well defined	Easy. SgRNA design is based on complementarity with the target DNA
Cloning	Requires engineering linkages between zinc finger motifs	TALEs do not require linkages. Cloning of separate TALE motifs can be done using Golden Gate assembly	Expression vectors for Cas9 are available. SgRNA can be delivered to cells as a DNA expression vector or directly as an RNA molecule or preloaded Cas9-RNA complex
Specificity	Tolerating a small number of positional mismatches (Two-finger modules are stitched together to form a zinc finger protein, each with a specificity of ≥24 bp. b.)	Tolerating a small number of positional mismatches	Tolerating positional/multiple consecutive mismatches

(continued)

Table 9.1 (continued)

Feature	ZFNs	TALENs	CRISPR-Cas9
Targeting limitations	Difficult to target non-G-rich sites	5′-targeted base must be a T for each TALEN monomer	The targeted site must precede a PAM sequence
Difficulties of engineering	Requiring substantial protein engineering	Requiring complex molecular cloning methods	Using standard cloning procedures and oligo synthesis
Difficulties of delivering	Relatively easy as the small size of ZFN expression elements is suitable for a variety of viral vectors	Difficult due to the large size of functional components	Moderate as the commonly used SpCas9 is large and may cause packaging problems for viral vectors such as AAV, but smaller orthologs exist
Methods employed to deliver editing systems in vivo	Adeno-associated viruses (AAV)	AAV	AAV lentivirus
Multiplexing ability	No	No	Yes
Off-target mutagenesis	Unknown and hard to determine mutagenic sites due to many possible indiscriminate protein-DNA interactions that can occur (arrays can bind to homologous sites, leading to off-target DNA cleavages)	Unknown and hard to determine mutagenic sites due to many possible indiscriminate protein-DNA interactions that can occur (arrays can bind to homologous sites, leading to off-target DNA cleavages)	Easier to predict possible mutagenic sites by utilizing Watson–Crick base-pairing rules. (off-target mutations are caused by both Cas9 and gRNAs)
Delivery vehicle	Easy via electroporation and viral vectors transduction	Easy in vitro delivery; difficult in vivo due to the large size of TALEN DNA and their high probability of recombination.	Easy in vitro; the moderate difficulty of delivery in vivo due to poor packaging of the large Cas9 by viral vectors
Use as a gene activator	Yes; activation of endogenous genes; with minimal off-target effects; may require engineering to target particular sequences	Yes; activation of endogenous genes; minimal off-target effects; no sequence limitations	Yes; activation of endogenous genes; minimal off-target effects; requires "NGG" PAM next to the target sequence
Use as gene inhibitor	Yes; works by blocking transcription elongation via chromatin repression; minimal off-target effects; may require engineering to target particular sequences	Yes; works by blocking transcription elongation via chromatin repression; minimal off-target effects; no sequence limitations	Yes; works by blocking transcription elongation via chromatin repression; minimal off-target effects; requires "NGG" PAM next to the target sequence

Success rate[a]	Low (~24%)	High (>99%)	High (~90%)
Cytotoxicity	Variable to high	Low	Low
Advantages	Small protein size (<1 kb) allows packaging into a single AAV	High specificity with each module recognizing 1 bp; no need to engineer linkage between repeats	Enables multiplexing (targeting multiple genes)
Limitations	Length of target sequence confined to the multiples of three; cumbersome cloning methods that need additional linker sequences to fuse modules together	Large protein size makes it challenging to utilize viral system; repetitive sequences may induce undesirable recombination events within the TALE array	Limited PAM sequences in human genome; Cas9 nuclease (~4.2 kb) is large for packaging into AAV
Suppliers nonprofit organizations	Addgene (https://www.addgene.org/) *Sigma-Aldrich/ToolGen	Addgene/TALEN library resource *Cellectis Bioresearch/Life Technologies/ToolGen/Transposagen Biopharmaceuticals	Addgene *Life Technologies/ SigmaAldrich/System Biosciences/Tool Gen/TransposagenBiopharmaceuticals

[a]The success rate is defined as the proportion of nucleases that induce mutations at frequencies >0.5% in HEK293 cells; *companies

CRISPR-Cas9: clustered regularly interspaced short palindromic repeat-2 associated protein 9, *crRNA:* CRISPR RNA, *N:* any nucleotide, *PAM:* protospacer adjacent motif, *RGEN:* RNA-guided engineered nuclease, *sgRNA:* single-chain guide RNA, *TALEN:* transcription activator-like effector nuclease, *ZFN:* zinc finger nuclease. A wide range of success rates and mutation rates (which depend on factors such as the methods used to construct these nucleases, delivery methods, and cell lines or organisms) have been reported. Mutation frequencies are found to be higher in K562 cells and HeLa cells than in HEK293 cells (Gaj et al. 2016b; Khalil 2020)

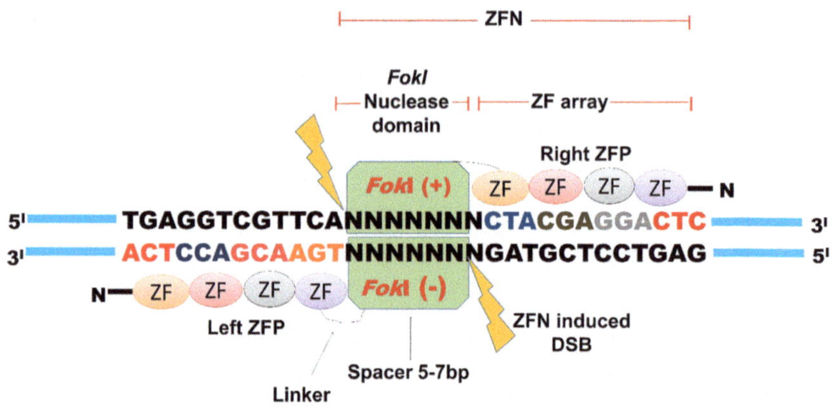

Fig. 9.2 Illustration of a pair of ZFNs bound to targeted nucleotide sequence: Zinc fingers are shown as ZF, with short circles indicating binding with the DNA base pairs. *Fok*I cleavage domains are shown as shaded boxes, with common cleavage sites, spaced by N bp, and indicated by vertical arrows as ZFN-induced DSB. Zinc fingers are numbered from the N-terminus. The linker between the binding and cleavage domains of one protein is labeled. The spacer between the zinc finger-binding sites is 5–7 bp in this case

desired sequence cleavage) sometimes occurs when the employed pair of ZFNs is not able to recognize the desired target sequence for cleavage. The addition of more fingers per ZFN is recommended to minimize off-target effects and successfully specify rarer and longer target cleavage sites.

The *FokI* domains of ZFNs are key to their successful application as they carry features that help in the cleavage of a complex genome at a specific target. *FokI* dimerization is crucial for the cleavage of the dsDNA. The lower strength of the interaction between *FokI* monomer domains causes the cleavage of DNA by *FokI* of ZFNs, requiring independent and appropriately placed two adjacent binding occurrences of ZFNs in correct orientations to allow catalytically active dimer formation (Miller et al. 1985; Vanamee et al. 2001; Szczepek et al. 2007) (Fig. 9.2). ZFNs-based genome editing is mainly dependent on the ability of endonuclease to create site-specific double-strand break (DSB) onto the locus of interest. In all eukaryotic cells, the DSBs generated by ZFNs are efficiently repaired by the NHEJ or HDR pathway (Szczepek et al. 2007; Lieber 2010; Moynahan and Jasin 2010) (Fig. 9.1).

Different strategies have been reported for the synthesis of ZFNs of desired DNA-binding specificity by "modular assembly" of different ZFs that have unique triplet base specificities (Segal et al. 2003; Sander et al. 2010; Thakore and Gersbach 2015). The ZFs developed for the modular assembly had been mostly for triplet sequences only (Choo and Klug 1994; Jamieson et al. 1994; Rebar and Pabo 1994; Segal et al. 1999, 2003; Dreier et al. 2001, 2005; Bae et al. 2003; Thakore and Gersbach 2015). The modular assembly of ZF components led to the generation of active ZFNs with specificity to a large number of endogenous sequences (Kim et al.

2009; Remy et al. 2010; Gaj et al. 2013b; Gupta and Musunuru 2014; Shiva and Suma 2019). Apart from the modular assembly approach, several other alternative strategies have also been developed for making ZFPs (Wu et al. 2007, 2013; Chandrasegaran and Carroll 2016; Paschon et al. 2019). These new approaches were focused on accommodating the deviation from strict functional modularity (like many natural and designed fingers can only contact with the adjacent ZF and to bases present outside of their proximal DNA triplet) which was observed for many of the ZF and making them specific (Fairall et al. 1993; Pavletich and Pabo 1993; Houbaviy et al. 1996; Nolte et al. 1998; Wolfe et al. 2001; Segal et al. 2006). These approaches could permit more selective binding and reduce the complications and wasted efforts that occur in modular designing for producing new ZFPs (Ramirez et al. 2008; Chandrasekharan et al. 2009; Chandrasegaran and Carroll 2016; Paschon et al. 2019).

Whatever may have been the methods used for designing ZFNs module, they were always first evaluated in vitro for their affinity and specificity toward the target DNA sequence followed by their application in vivo system. It is done as there is always a possibility that ZFNs/ZFPs which are validated in vitro could fail in performing the genome editing in vivo (Urnov et al. 2010; Wang et al. 2013a; Paschon et al. 2019). Many times, it may arise from the complexity of the genome which sometimes contains multiple copies of sequences that are identical or highly related (paralogues or pseudogenes) to the intended targeted sequences which can act as an additional target for ZFNs. The researchers have tried to address this problem by focusing on DNA-protein interactions and creating minor sequence divergence to reduce the chances of nonspecific targeting of related genomic regions (Carroll 2011; Urnov et al. 2010; Laoharawee et al. 2018). The specificity, recognition, and cleavage of desired sites by ZFNs are determined by the amino acid sequence of each ZF, nuclease (*FokI*) domain interactions, and quantity of the ZFs. The structure of both the functional domains of ZFNs, i.e., a catalytic domain and binding domains, can be optimized to increase specificity and enhance the affinity for the novel models developed by genome engineering (Jackson and Bartek 2009; Paschon et al. 2019). For improving the accuracy of targeting by ZFN, the "selection-based methods" have been also developed to optimize its cleavage specificity and reduce the nonspecific toxicity (Rahim et al. 2021).

9.2.2 Transcription Activator-Like Effector Nucleases (TALENs)

The second tool developed for genome editing or genome engineering is Transcription Activator-Like Effector Nucleases (TALENs) which display better specificity and functionality than ZFNs. Similar to ZFNs, TALENs also consist of an endonuclease, i.e., DNA cleavage domain, and a site-specific DNA-binding domain derived from transcription activator-like effectors (TALEs) proteins which together allow the creation of DSBs at specific sites. The DNA cleavage domain used for TALENs is primarily the *FokI* nuclease. The DNA-binding domains of TALENs, i.e., TALEs originated from a repeated sequence of highly conserved proteins of "phytopathogenic *Xanthomonas*" (Boch et al. 2009; Boch and Bonas 2010; Chandrasegaran and Carroll 2016). In *Xanthomonas*, the transcription activator-like effectors (TALEs) proteins are present in the cytoplasm where they promote the modification of genes that help in transcription. TALE proteins are capable of localization to the nucleus, DNA binding, and transcription activation of the target gene (Schornack et al. 2006). The studies conducted on the mechanism of action of these effector proteins showed that these proteins can mimic the functioning of eukaryotic transcription factors in binding with DNA and activating gene expression (Becker and Boch 2021).

Soon after the realization of the TALE domains simplicity, i.e., one monomer binds/recognizes one nucleotide, the first chimeric TALE domain-fused nuclease (TALEN) was constructed (Joung and Sander 2013; Gaj et al. 2013b; Nemudryi et al. 2014; Becker and Boch 2021). The chimera was developed by inserting the DNA-binding domain of TALE in a plasmid vector which was used for ZFNs (Christian et al. 2010). This ultimately leads to the formation of a genetic construct that has DNA binding and catalytic domain of restriction endonuclease, i.e., *FokI*. The DNA-binding domain (i.e., TALE) monomers that bind with the single nucleotide in the targeted DNA sequence are repeats of 34 amino acid residues in which amino acids at 12 and 13 positions are highly variable and known as repeat variable domain (RVD). The RVD region of TALE is responsible for the recognition of specific nucleotides. The variation in RVDs allows them to bind to different nucleotides with different efficiencies (Fig. 9.3). The TALEs with different RVDs were combined to form artificial nucleases (i.e., TALENs) which bind and cleave the targeted DNA sequences. The TALENs nucleases contain a half repeat (i.e., 20 amino acid residues of the last tandem repeat that bind to the nucleotide at the 3′ end of the recognition site), N-terminal domain, nuclear localization signals, and *FokI* catalytic domain (Fig. 9.3). The presence of thymidine at the 5′ end of the target sequence interacts with the N-terminal domain of the TALE and affects its overall binding efficiency (Lamb et al. 2013). TALENs always work in pair, their binding sites are located at the opposite site of DNA strands and are separated by small fragment (i.e., 12–25 bp) known as "spacer sequence." After the TALENs enter the nucleus, they bind with the targeted sequence and the *FokI* domains located at the C-terminal of TALE cause the DSBs (Fig. 9.1).

Despite simple designing codes as compared to ZFNs, there has been difficulty in the cloning of the designed TALE arrays comprised of large-scale repeats. To overcome this problem, different strategies have been developed such as High-Throughput Solid Phase Assembly, Golden Gate Cloning, and Connection-independent cloning techniques which help in assembling the desired TALE arrays (Schmid-Burgk et al. 2013). Several other modifications have also been made to TALENs to make them a better tool than the ZFNs such as (1) site selection enhancement by varying the length of the spacer sequence (Nemudryi et al. 2014); and (2) development of mutant variants of the TALE's N-terminal domains that could more specifically bind to A, G, and C nucleotide (Nemudryi et al. 2014; Mak et al. 2012; Lamb et al. 2013).

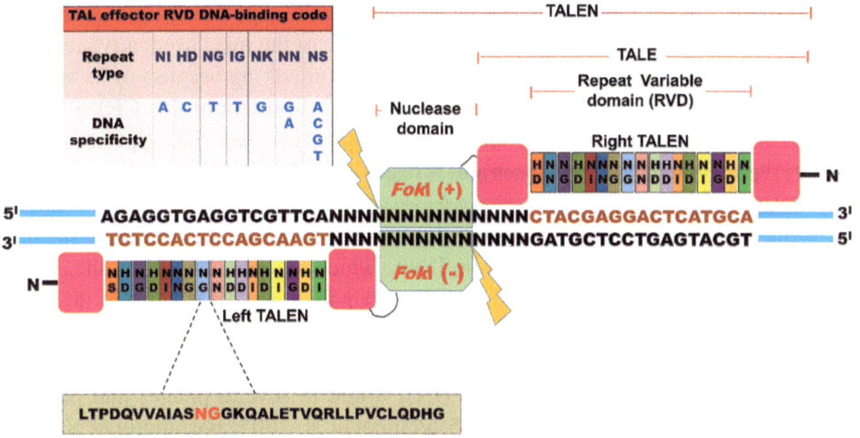

Fig. 9.3 Illustration of a pair of transcription activator-like effector nucleases (TALENs) bound to targeted nucleotide sequence: TALE repeats, i.e., RVD are shown as colored boxes that are responsible for the recognition of specific nucleotides. RVDs bind to different nucleotides with different efficiency. Letters inside each repeat represent the two hypervariable residues. TALE-derived amino (N domain) and carboxy-terminal domains required for DNA-binding activity are shown as pink boxes. The nonspecific nuclease domain from the *FokI* endonuclease is shown as a larger shaded green box. TALENs bind and cleave as dimers on a target DNA site. The TALE-derived amino- and carboxy-terminal domains flanking the repeats may make some contact with the DNA. Cleavage by the *FokI* domains occurs in the "spacer" sequence that lies between the two regions of the DNA bound by the two TALEN monomers. The amino acid sequence of a single TALE repeat is expanded below with the two hypervariable residues highlighted in red and bold text. TALE-derived DNA-binding domain aligned with its target DNA sequence is shown in the box indicated as TAL effector RVD DNA-binding codes

9.2.3 Clustered Regularly Interspaced Short Palindromic Repeats (CRISPR)

After two years of the discovery of TALENS, the discovery of Clustered Regularly Interspersed Short Palindromic Repeats (CRISPR) led to the development of a third genome engineering tool that has revolutionized the field of biotechnology and the health sector tremendously (Nemudryi et al. 2014; Lino et al. 2018; Kaminski et al. 2021). The CRISPR was first time discovered in *Escherichia coli* (*E. coli*) in 1987, and later on, found in many other prokaryotes too, e.g., 87% in archaea and 48% in eubacteria (Grissa et al. 2007). The CRISPR system has noncoding RNAs and CRISPR-associated (Cas) protein which has a nuclease activity (Ishino et al. 1987; Jore et al. 2012). In bacteria, the CRISPR-Cas system plays an important role in the adaptive immune response. It helps in protecting bacteria from phage infection by generating memory in the bacterial chromosomes against phage (Barrangou and Marraffini 2014; Renaud et al. 2016; Kim et al. 2021).

There are two types of CRISPR/Cas systems depending on the structural variation that existed in Cas genes (1) Class 1 systems contain multiple protein effectors complexes, and (2) Class 2 has one effector protein. To date, six types of CRISPR/

Cas systems and 29 subtypes of Cas-system have been reported (Moon et al. 2019; Liu et al. 2020). CRISPR-Cas9 type II system is one of the most used, advanced, and versatile CRISPR systems for genome engineering or editing because of its specificity which is stemming from the Cas protein. The Cas protein of this CRISPR-Cas9 type II system was extracted from *Streptococcus pyogenes* (i.e., SpCas9) which targets the specific DNA sequences and is responsible for the advanced specificity of the system (Jiang et al. 2013).

Sequencing of the various bacterial genomes revealed the presence of short unique DNA regions known as spacers DNA, which are separated from one another by short palindromic sequence repeats (Deshpande et al. 2015; Lino et al. 2018). These structures are found to be located in the proximity of Cas genes. The *cas* gene gives rise to protein products that have nuclease and helicase activity (Haft et al. 2005). Spacer DNA is a homologous DNA found in several phages and plasmids (Bolotin et al. 2005; Mojica et al. 2005; Pourcel et al. 2005; Barrangou et al. 2007) (Fig. 9.4). Cas9 protein is polyfunctional, it interferes with the foreign DNA and pre-crRNA processing (Sapranauskas et al. 2011). The processing of crRNA depends on small noncoding RNA known as transactivating RNA (tracrRNA). The tracrRNA forms a duplex after binding with the complementary repeat sequence present in the pre-rRNA. RNase III (present in the host cell) in the presence of Cas9 cleaves this duplex and leads to the formation of mature crRNA (Makarova et al. 2006; Marraffini and Sontheimer 2010). The CRISPR-Cas9 system employs two main components, i.e., Cas9 endonuclease and a single-stranded guide RNA (sgRNA) or tracrRNA-crRNA chimera (Cong et al. 2013). The sgRNA recognizes and binds with the targeted sequence, and Cas9 cleaves the DNA causing DSB (Fig. 9.4). The site of cleavage for Cas9 endonuclease is 3 bp upstream of an "NGG" PAM located on genomic DNA. This DSB generated gets repaired by NHEJ or HDR (Fig. 9.1) (Pawelczak et al. 2018).

The mechanism of genome editing with the help of the CRISPR system both inside the prokaryotic cells and in vitro is divided into three stages, i.e., adaptation, transcription, and intervention. *In adaptation*, a small fragment of foreign DNA entering the bacterial cell gets inserted into the CRISPR locus of the bacterial genome leading to the formation of the new spacer or protospacer (i.e., a viral genome fragment). Viral protospacer is complementary to the spacer present in the host cell and these protospacers are flanked by a short, conserved sequence (2–5 bp) which is known as a protospacer adjacent motif (PAM) (Mojica et al. 2009a). The PAM is inserted at the AT-rich side of the sequence that also has a promoter element and a landing site for regulatory proteins present just before the CRISPR cassette (Deltcheva et al. 2011; Jinek et al. 2012). *In the transcription* step, the complete CRISPR locus formed is transcribed into a long poly-spacer precursor crRNA (pre-crRNA) (Fig. 9.4). The Cas6 endonucleases are responsible for the formation of mature crRNA in most of the CRISPR/Cas systems (Carte et al. 2008; Lillestøl et al. 2009; Mojica et al. 2009b; Hale et al. 2012; Pawelczak et al. 2018). The short nucleotide CRISPR RNA (crRNAs) has one spacer sequence whose repeat ends are involved in the formation of a stem-like loop structure. The 5′ end with eight nucleotide repeats has an OH group and forms a stem whereas the 3′ end with

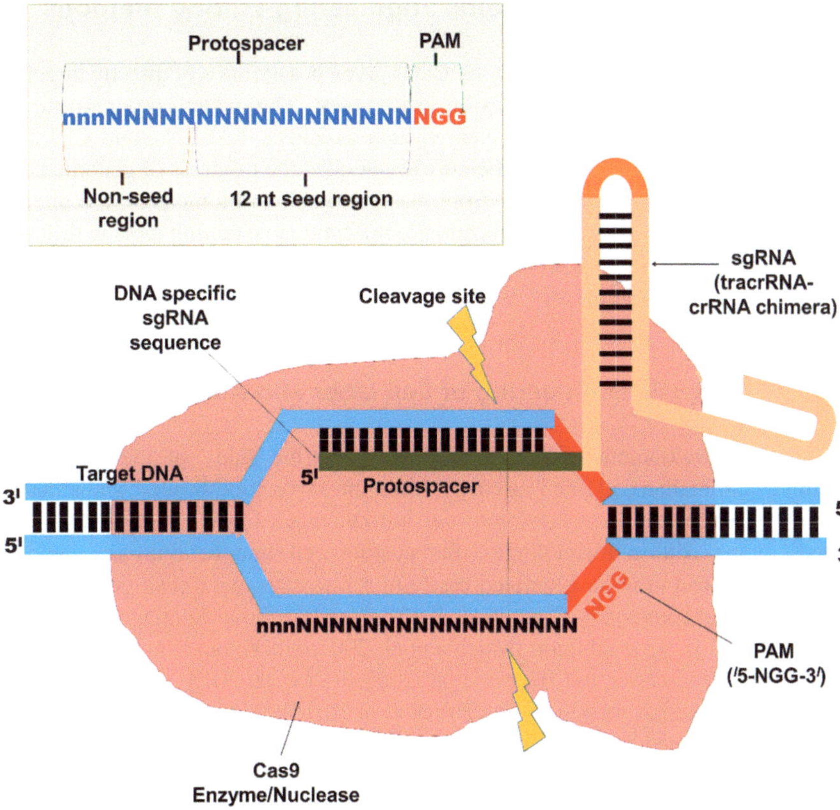

Fig. 9.4 Illustration of clustered regularly interspaced short palindromic repeats (CRISPR-Cas9 System) bound with targeted DNA sequence: CRISPR system has a single chimeric sgRNA (crRNA and tracrRNA) to which introduces a DSB into the target nucleotide sequence. A protospacer is a site that is recognized by the CRISPR/ Cas9 system. A spacer is a sequence in sgRNA that is responsible for complementary binding to the target site. PAM is a short motif (NGG in the case of CRISPR/Cas9) whose presence at the 3′-end of the protospacer is required for introducing a break. A Cas9 nuclease is capable of introducing DSB into selected DNA site

2′,3′-cyclic phosphate (hairpin structure) forms a loop (Haurwitz et al. 2010; Gesner et al. 2011).

During *the intervention step*, the viral DNA or RNA interacts with the crRNA and Cas proteins. The crRNA identifies the complementarily of the protospacer sequence whereas Cas protein leads to its degradation (Marraffini and Sontheimer 2010; Rath et al. 2015; Shabbir et al. 2019). The coevolution of viruses/phages with their host over time has led to the formation of a wide range of CRISPR/Cas system in prokaryotes (Hale et al. 2009; Sashital et al. 2011; Richter et al. 2012; Bondy-Denomy et al. 2013; Newsom et al. 2021).

9.3 Applications of Genome Engineering/Editing Methods

The development of genome editing tools has given possibilities of directly targeting and modifying genome sequences in eukaryotes. The recent progress in the development of programmable nucleases such as ZFNs, TALENs, and CRISPR-Cas-associated nucleases has significantly accelerated the progress of genome engineering in different fields ranging from basic research to biomedical and applied biotechnological research. The application of different gene editing tools in different fields of biological sciences and their future possibilities are briefly indicated below and summarized in Table 9.2.

9.3.1 In Genetic Engineering of Cell Lines and Animal Models

Before the development of engineered nucleases, the study of the genetically modified mammalian cell line was costly, labor-intensive, time-limited, and required specialized expertise. However, with the introduction of cost-effective and user-friendly gene editing technologies, the custom cell line bearing any genome modifications can now be generated easily in a few days, e.g., gene deletion (Lee et al. 2010), gene inversion (Xiao et al. 2013), gene knockout (Santiago et al. 2008; Mali et al. 2013), gene addition (Moehle et al. 2007; Hockemeyer et al. 2011; Hou et al. 2013), gene correction (Urnov et al. 2005; Ran et al. 2013), gene addition as well as chromosomal translocation (Torres et al. 2014). Along with cell line engineering, the targeted nucleases have also accelerated the generation of genetically modified organisms, such as the accelerated creation of transgenic zebrafish (Doyon et al. 2008; Sander et al. 2011; Hwang et al. 2013), livestock (Hauschild et al. 2011; Carlson et al. 2012), monkeys (Liu et al. 2014), mice (Cui et al. 2011; Wang et al. 2013a, b; Wu et al. 2013), rats (Geurts et al. 2009; Tesson et al. 2011; Li et al. 2013), etc.

9.3.2 In Genetic Engineering of Plant Cells

These engineered nucleases have also emerged as a dominant tool for plant engineering (Baltes and Voytas 2015). For example, both CRISPR-Cas9 and TALENs have been used for the modification of multiple alleles in the haploid breed of wheat to create resistance variety against powdery disease (Wang et al. 2014b). Moreover, TALENs were used for soybean to knock out the nonessential gene that is used for fatty acid metabolism and thus produce simple plant cells with reduced metabolic constituents (Haun et al. 2014). The purified proteins comprised of various genomic engineering tools can be directly injected into the plant protoplast to effect germline-transmissible changes which are almost indistinguishable from the natural variety (Luo et al. 2015; Woo et al. 2015). The technological advancement of these tools could be very much helpful to reduce some regulatory problems which are associated with the use of transgenic plants. The targeted nucleases have been also

Table 9.2 Application of genome editing tools in different fields of biological sciences

Application	Tools	Use	Reference
Genetic cloning of living organisms	ZFNs, TALENs, and CRISPR/Cas9	Introduction of Cas9 mRNA in the one-cell-stage zebrafish embryo. The first viable cloned cattle offspring were produced using somatic cell nuclear transfer (SCNT) in Russia. The CRISPR/Cas9 was used to knock out the beta-lacto-globulin gene as well as the beta-lacto-globulin-like protein gene in fibroblast cells. Transgenic-cloned buffalo embryos were generated using CRISPR/Cas9 facilitated targeted integration of enhanced green fluorescent protein (EGFP) in the Y chromosome	Charpentier and Doudna (2013), Zhao et al. (2019, 2020), Singina et al. (2021)
Establishment of animal models	ZFNs, TALENs, and CRISPR/Cas9	The CRISPR/Cas9 gene editing tool was used to generate a unique TshrDf/Df rat model that recapitulates the phenotype in the THSR (Thyroid-stimulating hormone receptor) Y444X patient. THSr was truncated at the second transmembrane domain in this rat model leading to congenital hypothyroidism (CH) phenotype which includes thyroid aplasia, infertility, dwarfism, low serum thyroid hormone level as well as TSH resistance. This phenotype can be partially reversed after weaning with the treatment of levothyroxine (L-T4). Cas9 and sgRNA were co-injected into the zygote to develop mutant ANO5 rabbits. The CRISPR-mediated small insertion in the mutant rabbit at the exon 12 or 13 leads to the display of	Yang et al. (2018), Zhao et al. (2019, 2020)

(continued)

Table 9.2 (continued)

Application	Tools	Use	Reference
		muscular dystrophy symptoms such as elevation of serum creatine kinase, muscle regeneration and necrosis, fibrosis, and fatty replacement. This type of novel ANO5 mutant rabbit is very much useful to study disease pathogenesis and treatment of ANO5 deficient muscular dystrophy. The CRISPR/Cas9-mediated genome editing was also used to develop a stable cell line expressing alpha-synuclein (SNCA) tagged with a nano-Luc luciferase reporter. The ability to monitor endogenous SNCA transcription can be used as an efficient drug screening tool for therapeutics identification for Parkinson's disease	
Discovery of drugs	CRISPR/Cas9	The CRISPR/ Cas9 mutagenesis was used to target exons that encode the functional protein domains. It generates a large number of null mutations which eventually increases the capability of negative selection. The inhibitors of receptor tyrosine kinase (RTK)/Ras/mitogen-activated protein kinase (MAPK) pathway are therapeutically used for the treatment of lung cancer as well as other cancers. The use of the CRISPR/Cas9 gene deletion tool in lung cancer showed that cell metabolism changes after deletion of KEAP1 in the presence of multiple inhibitors for the targeted RTK/Ras/MAPk pathway and allowing the cells to	Shi et al. (2015), Martinez-Lage et al. (2017)

(continued)

Table 9.2 (continued)

Application	Tools	Use	Reference
		multiply without the signaling of MAPK. CRISPR/CAs9 tool is also used to identify the essential genes such as Rosa26, Rpa3, Brd4, Smarca4, Eed, Suz12, and Rnf20 for murine MLL-AF9/ NrasG12D acute myeloid leukemia cell	
Agricultural products	ZFNs, TALENs, and CRISPR/ Cas9	Several applications of CRISPR/Cas9 have been reported for crop breeding. For example, targeting the ZmTMS5 gene of maize for traits of thermosensitive genic male sterility, the K1C gene of Sorghum to target trait of high lysine content and increased protein digestibility, TaEDR1 gene of wheat for target trait of powdery mildew resistance, OsRR22 gene of rice for target trait enhanced salinity tolerance, Gn1a, DEP1, GS3, and IPA1 gene of rice for target trait enhanced grain size and number, MS1 gene of wheat for target trait male sterility, PDS gene of banana for target trait albinism phenotype, etc. The CRISPR/Cas9 has been used to study mutation in rice for abscisic acid production which leads to cell deaths, altered seed development, and dormancy as well as enhanced disease resistance. The CRISPR/ Cas9 gave new insight into the role of abscisic acid in rice disease resistance. The CRISPR/Cas9 tool is also used for the production of genetically modified mushrooms. The CRISPR/	Shi et al. (2015), Huang et al. (2017), Maiti et al. (2017), Martinez-Lage et al. (2017), Liao et al. (2018), Sui et al. (2018), Lee et al. (2021), Chattopadhyay et al. (2022), Guo et al. (2022)

(continued)

Table 9.2 (continued)

Application	Tools	Use	Reference
		Cas9 system has been used to knock out the maize gene CCT transcription factor (ZmCCT9) which promotes early flowering under long days. The immunity of the plants has been enhanced after the expression of the pathogen-targeted CRISPR-Cas9 tool in plants	
Food	CRISPR/ Cas9	The CRISPR/Cas9 was used to make *Streptococcus thermophilus* (a thermophilic bacteria) as a bacteriophage-insensitive mutant to improve the product quality such as cheese and yogurt. This technology was also used to edit *Agaricus bisporus*, a white button mushroom that resists the browning of mushrooms. For this property, the gene knockout technique is used to target the gene encoding the polyphenol oxidase that causes the browning of mushrooms.	Barrangou et al. (2007), Waltz (2015)
Industrial products	CRISPRs/ Cas9	The CRISPR/Cas9 has been also used to make marine algae useful for industrial purposes such as the production of fuel, health foods, biomolecules, material for nanotechnology, pharmaceuticals, etc. They are also used for the bioremediation of contaminated water. The CRISPR/Cas9 is also used to capture and firmly store the many real data in the living cells' genome. The beta-carotene synthetic pathways were integrated into the genome of *Escherichia coli* by using the CRISPR-Cas system.	Waltz (2015), Nymark et al. (2016), Huang et al. (2017), Khalil et al. (2017), Su et al. (2018), Kachhawaha (2021)

(continued)

Table 9.2 (continued)

Application	Tools	Use	Reference
		This system has been also used for the modifications in the methylerythritol-phosphate (MEP) and central metabolic pathways for the overproduction of beta-carotene. The production of γ-aminobutyric acid has been enhanced within a few weeks after integrating synthetic single-stranded oligodeoxyribonucleotides with the help of recombinase RecT and knocking out the gene in *Corynebacterium*. The production of galactic acid, succinate, fatty acids, and citric acid has been increased using the CRISPR-cas9 gene knockout strategy to knock out several genes. CRISPR-Cas9 is also used for insertions and deletions of a few nucleotides in the filamentous fungus *Ashbya gossypii* to increase the production of folic acid, nucleosides, and biolipids	
Environmental protection	CRISPRs/ Cas9	CRISPR/Cas9 is used to modify the genome of diatoms and used for reducing global warming. CRISPR/Cas9 tool is also used to study the pathway involved in the effects of prenatal exposure to inorganic arsenic (iAs) that induces birth defects in animals and humans. The in silico study was performed to identify the glucocorticoid receptor pathway that acts as a key pathway for metal-induced prenatal toxicity. The CRISPR/Cas9 is also used to investigate the	Rager et al. (2011), Nymark et al. (2016)

(continued)

Table 9.2 (continued)

Application	Tools	Use	Reference
		mechanism of herbicide by which it potentially induces respiratory failure and Parkinson's disease. Different CRISPR/Cas nucleic acid detection tools viz., SHERLOCK (Cas13), Cas-12, and Cas-14 detectors were developed to detect the accurate, fast, and ultrasensitive diagnostic testing of pathogens from the patients and environmental samples. The CRISPR-Cas-12-based lateral flow assay was developed to detect Covid-19 from the respiratory swabs and monitor the presence of SARS-CoV-2 in the wastewater samples	
Restoration of extinct animals	CRISPRs/Cas9	The TRPV3 gene of mammoths which is responsible for thermal sensation and hair growth could be used for modifying the genes of Asian elephants. Thus transgenic "mammoth" can be restored through the CRISPR/Cas9 system on a successful transfer of the modified embryo into the uterus of living elephants	Lynch et al. (2015)
Medicine screening for humans	CRISPRs/Cas9	The novel method was developed using CRISPR/Cas9 to monitor the endogenous alpha-synuclein (α-SYN) transcription by inserting NanoLuc luciferase tag at the 3′ end and make it possible to screen strategy rapidly for Parkinson's disease (PD) therapy efficiently. CRISPR/Cas9 has been also used for cancer immunotherapy such as targeting chimeric antigen receptors T cells,	Maeder and Gersbach (2016), Porteus (2016), Limsirichai et al. (2016), Basu et al. (2017), Martinez-Lage et al. (2017), Raquel et al. (2019), Li et al. (2020), Qasim (2021)

(continued)

Table 9.2 (continued)

Application	Tools	Use	Reference
		inactivating HIV receptors such as CCR5, CXCR5, LEDGF/p75, inhibiting viral replication, inactivating essential viral genes, knocking out the endogenous T-cell receptors, correction of β-globin mutations in iPSCs, inactivation of the enhancer of BCL11A, gene correction in stem cells, gene correction in mouse lung epithelium, gene correction in human iPSCs and liver cells neuromuscular. The CRISPR/Cas9 is also used to remove β2M and CCR5 on CD34+ HSCs with retaining the ability of cells to undergo multidifferentiation that acts as a possible future treatment for ischemic heart disease. This tool is also used to remove cytochrome P450 (CYP) 2E1 in rat models and elucidate the role of the CYP2E1 gene in toxicology, biochemical metabolism, and diseases such as alcoholic cirrhosis and diabetes	
Preparation for cell therapy or immunotherapy for humans	CRISPRs/ Cas9	The engineered chimeric antigen receptor T (CART) and induced pluripotent stem (iPS) cells were developed. CRISPR/Cas9 has been also used to develop iPS cells to escape immune rejection in immunocompetent allogeneic recipients. Now CRISPR/Cas9 system may further improve the safety and efficiency of CART cells by altering therapeutic T cells. CRISPR/Cas9 has been also used in the	Deuse et al. (2019), Davis and Stokoe (2010), Hu (2016), Maeder and Gersbach (2016), Perales et al. (2018), Yang and Huang (2019), Universitesi (2020), Xu and Li et al. (2020), Qasim (2021)

(continued)

Table 9.2 (continued)

Application	Tools	Use	Reference
		application of hematopoietic cells mainly for the genes which are responsible for β-thalassemia, sickle cell disease (SCD), and HIV infection. This tool is used to disrupt the CXCR4 and CCR5 genes in T cells that act as a functional cure for HIV. Recently, CRISPR/Cas9 has been used for removing CXCR4 and CCR5 from the induced pluripotent stem cells (iPSCs) and targeting the HIC genome for treatment purposes. This tool is also used to produce the iPSC derived from patients with hemoglobinopathies by correcting the hemoglobin (HBB) mutation. The first gene therapy clinical trial using the CRISPR/Cas9 approach involves the treatment of children suffering from severe combined immunodeficiency symptoms by targeting the retroviral delivery of therapeutic adenosine deaminase (ADA). Also, gene editing in CD34+ HSCs, as well as pluripotent cells, provides new options for the treatment of hematologic disorders such as sickle cell disease caused by a point mutation in the β-globin gene and β-thalassemia which is caused by other types of mutations in β-globin	
Virus latent infection	TALENs and CRISPRs/Cas9	The anti-EBV activity of CRISPR/Cas9 that targets the Epstein-Barr virus genome in infected NPC cells has been reported. The	Barrangou et al. (2007), van Diemen et al. (2016), Cohen (2017), Imran et al. (2017), Zhang et al. (2017)

(continued)

Table 9.2 (continued)

Application	Tools	Use	Reference
		CRISPR/Cas9 has been also used in vivo and in vitro to battle etiological factors of cervical cancer such as HPV16 and HPV18. One of the studies explored the therapeutic value of CRISPR/Cas9 when combined with cisplatin for mutations in HPV16 and E7 oncogenes of the virus—showing inhibition of tumor growth	
Genetic disease	CRISPRs/Cas9	Different genetic diseases such as sickle cell anemia, β-thalassemia, α1-antitrypsin deficiency, muscular dystrophy, cystic fibrosis, etc. can be cured by gene therapies. The screening of transcriptional alteration due to CRISPR-Cas showed protective genes against the toxicity of alpha-synuclein. The gene editing facilitated by CRISPR/Cas9 has tremendously improved the prospects of gene therapy. Different ex vivo and in vivo experiments of gene editing technology showed positive results and some are under various stages of clinical trials	Cohen (2017)
Neurodegenerative disease	CRISPRs/Cas9	The perturbing regulatory interactions by synthetic modulators (PRISM) are applied to the yeast model of Parkinson's disease in which sgRNAs were identified that modify the transcriptional network and prevent the cells from experiencing α-Syn toxicity. Alzheimer's disease was caused by APPswe (Swedish) mutation in the amyloid precursor protein. The	Chen et al. (2017), György et al. (2018), Ricci and Colasanta (2021)

(continued)

Table 9.2 (continued)

Application	Tools	Use	Reference
		CRISPR/Cas9 tool can be also used to selectively disrupt the mutant APP^{sw} allele that decreases the pathogenic amyloid-β (Aβ)	
Cancer	CRISPRs/ Cas9	The hepatocellular carcinoma (HCC) cells were inhibited by using the CRISPR/Cas9 gene editing. The expression of SIRT6 can be suppressed by miR-125b which targets the seed-matching region of 3' UTR. After knocking out the expression of the SIRT6 gene through CRISPR/ Cas9, the invasiveness and viability of HCC cells were reduced which had a similar function after overexpression of miR-125b. Gene editing technology such as CRISPR/Cas9 was also tried to inhibit breast cancer. Presently, cyclin-dependent kinases (CDKs) are prominent anticancer drug targets and CDKs inhibitors give clinical benefits to cancer patients. The deactivated CRISPR/ Cas9 (dCRISPR) approach has been used for genetic manipulations of breast cancer cells	Barone et al. (2018), Song et al. (2018), Yang et al. (2021), Wang et al. (2022)
Development of testing tools and reagents	CRISPR/ Cas9	As compared to traditional methods CRISPR-Cas9 can be improved by bacterial genotyping to become more adaptive to the differences in food pathogens. The CRISPR locus present in different bacterial species shows high variances which could be developed as a highly sensitive basis for genotyping. The recently developed CRISPR/Cas12a (Cpf1) detector system can	Makarova et al. (2013), Chen et al. (2018)

(continued)

Table 9.2 (continued)

Application	Tools	Use	Reference
		be used for the diagnosis of cancers, microbial infections, gene mutations, and test antimicrobial resistance	

used for the inactivation of pathogenic genes that help in the prevention of parasitic or viral infections and knock out specific factors leading to the development of pathogens resistance varieties (Ghorbal et al. 2014; Lin et al. 2014; Wu et al. 2015).

9.3.3 In Genetic Engineering for Insect-Borne Disease

Interestingly, the targeted nuclease has been also used to limit mosquito or insect-borne diseases (Burt 2003; Sinkins and Gould 2006). Genome editing enables the introduction of a particular gene or mutation in the host that can also get transferred to its progeny (Windbichler et al. 2011). This gene editing technique has been used in the vector of malaria, i.e., *Aedes aegypti*, *Anopheles stephensi*, and *Anopheles gambiae* for disease control and prevention (Gantz and Bier 2015; Hammond et al. 2016). Countries like Saudi Arabia, Turkey, Korea, Philippines, India, USA, Europe, China, and Japan are using the CRISPR technique for combating vector-borne diseases (Mahto et al. 2022). Smidler et al. reported the targeted disruption of the thioester-containing protein1 (TEP1) gene using TALEN in *Anopheles gambiae* mosquitos, which transmit malaria. The TEP1 gene of *An. gambiae* has been identified as a key gene for immunity against plasmodium infection (Miller et al. 2011). Gene editing in *Ae. aegypti* and *An. stephensi* using ZFNs and TALENs was reported in 2013 (Degennaro et al. 2013). De Gennaro et al. investigated the involvement of the odorant receptor coreceptor (*orco*) gene and the odorant receptor pathway in host identification and susceptibility to the chemical repellent *N,N*-diethyl-meta-toluamide (DEET) in *Ae. aegypti* (Christian et al. 2010). The developed ZFN was injected into embryos of *Ae. aegypti* in this experiment with promising results.

9.3.4 In Genetic Engineering of Industrially Important Microorganisms

The targeted nucleases also offer a convenient means for developing modified bacterial and yeast strains for synthetic biology such as metabolic pathway engineering. For example, the bacterial species belonging to the order *Actinomycetales* are one of the key sources of industrially relevant secondary metabolites. However, the large numbers of *Actinomycetales* species are historically resistant to genetic manipulation and had severely hindered their use for metabolic engineering. Now, CRISPR-Cas9 has been used to deactivate several genes of actinomycetes (Tong et al. 2015). This indicates the ability of the CRISPR/Cas9 system to create designer bacteria with enhanced secondary metabolite production capabilities. The CRISPR has also helped in metabolic pathway engineering in yeast by creating random mutagenesis in yeast chromosomal DNA at high efficacy (Jakočiūnas et al. 2015), allowing rapid screening of the desired mutants (Ryan et al. 2014).

9.3.5 In Genetic Engineering for Functional Genomics

The CRISPR-based knockout strategy has been playing an important role in functional genomics (Hilton and Gersbach 2015), e.g., facilitated the discovery of genomic loci that make cells drug-resistant (Koike-Yusa et al. 2014; Shalem et al. 2014; Wang et al. 2014a; Zhou et al. 2014; Blancafort et al. 2008). The genome editing tools also uncovered how the cells can initiate host immune response (Parnas et al. 2015), as well as keep giving new insights into the genetic basis of cell fitness (Hart et al. 2015; Wang et al. 2015). The genome editing tools have also increased the understanding of how certain viruses affect cell death (Ma et al. 2015). The genome-wide application of the CRISPR strategy has helped in the discovery of functional noncoding elements (Kim et al. 2013; Korkmaz et al. 2016) and understanding of their role in the structure and evolution of the human genome (Findlay et al. 2014). The CRISPR has also helped in identifying the factors key to zebrafish development (Shah et al. 2015) as well as disease development in mice (Chen et al. 2015).

9.3.6 In Genetic Engineering for Therapeutics

Genome editing technologies have great potential to treat/cure various diseases at genetic levels (Cox et al. 2015; Porteus 2015; Maeder and Gersbach 2016). For example, the ZFN-mediated disruption of HIV co-receptor CCR5 allowed the development of resistance against HIV in both $CD4^+$ T cells (Perez et al. 2008) and CD34 hematopoietic stem/progenitor cells (HSPCs) (Holt et al. 2010; Tebas et al. 2014). Along with the introduction of gene modifications that enhance autologous cell therapies, targeted nucleases can also mediate genome editing in situ through combining viral vector, such as AAV (Gaj et al. 2016a). The delivery of an AAV vector designed to target a defective copy of the factor IX gene and provide a repair template had led to effective gene correction in mouse liver increasing factor IX protein production in both neonatal (Li et al. 2011) and adult (Anguela et al. 2013) mice. Recently, the in vivo gene editing tool has been used for the restoration of expression of the dystrophin gene allowing the rescue of muscle function in mouse models of Duchenne muscular dystrophy (Long et al. 2016; Nelson et al. 2016; Tabebordbar et al. 2016). A therapeutic gene editing tool has been successfully used in a mouse model of human hereditary tyrosinemia disease (Yin et al.

2014). This approach has been also used for the correction of disease-causing mutations in the ornithine transcarbamylase gene in the liver in a neonatal model of disease (Yang et al. 2016).

9.3.7 In Genetic Engineering: Epigenome Editing (Modulating Gene Expression)

Along with the DNA recognition ability of CRISPR-Cas9, the flexibility associated with constructing arrays of ZFs and TALEs proteins capable of binding to specific sequences allows their fusion with transcriptional activator and expression protein domains to modulate the expression of any gene from its promoter or enhancer sequences. The fusion of engineered zinc finger proteins either with the transcriptional domain derived from herpes-simplex or Kruppel-associated box (KRAB) repression protein had been used for the generation of the first fully synthetic transcriptional effector protein (Sadowski et al. 1988; Margolin et al. 1994; Beerli et al. 1998). Several other types of effector domains were extended and featured over the next 15 years using the zinc finger-based transcriptional modulators (Beerli and Barbas 2002). For example, modulation of transcription through targeted methylation or demethylation was done using the Dnmt3a methyltransferase domain (Rivenbark et al. 2012; Siddique et al. 2013) and the ten-eleven translocation methylcytosine dioxygenase 1 (TET1) (Chen et al. 2014). The TALE transcription factor has also emerged as an effective platform to achieve modulation of targeted transcription (Miller et al. 2011; Zhang et al. 2011). Similar to zinc finger, the TALE is also compatible with several modifiers such as the TET1 hydroxylase catalytic domain which is used for targeted CpG demethylase domains (Maeder et al. 2013), and the lysine-specific histone demethylase domain (LSD1) which has been used for targeted CpG histone demethylation (Beerli et al. 2000; Pollock et al. 2002; Magnenat et al. 2008; Polstein and Gersbach 2012; Maeder et al. 2013; Mendenhall et al. 2013; Perez-Pinera et al. 2013). TALE activators have also been effectively engineered to regulate gene expression in response to endogenous chemical stimuli (Li et al. 2012), proteolytic cues (Copeland et al. 2016; Lonzarić et al. 2016), external stimuli (Mercer et al. 2014), and optical signals (Konermann et al. 2013). The potential is immense.

9.3.8 Genome Engineering for Transcription Modulator

Because of the excellent ease, the CRISPR-Cas9 system has been also used for transcriptional modulation via fusion of a particular effector domain with the catalytically disabled variant of Cas9 protein (Qi et al. 2013). The mutant form is unable to cleave DNA and is referred to as dCas9 (Dead Cas9 Endonuclease) because of its ability to bind to the DNA in an RNA-directed manner. The carboxyl domain of dCas9 protein fused with the effector domain can modulate the gene expression from either strand of the targeted DNA sequences (Farzadfard et al. 2013;

Maeder et al. 2013; Perez-Pinera et al. 2013; Gilbert et al. 2014; Hu et al. 2014). Moreover, dCas9 can inhibit gene expression by simply blocking the transcription initiation or elongation via the process known as CRISPR interference (Qi et al. 2013), whereas the fusion of transcriptional repressor domains with dCas9 can also be used to effectively silence a gene from the promoter region (Gilbert et al. 2013; Balboa et al. 2015; Zalatan et al. 2015). The light-inducible dCAs9-based system has been shown to be capable of allowing optical control of gene expression or achieving altered conditional control of gene expression (Nihongaki et al. 2015; Polstein and Gersbach 2015). The first-generation dCas9 activators were found to display a sub-optimal level of activation (Karlson et al. 2021). The development of second-generation CRISPR activators has rapidly emerged and expanded as a hugely promising area of research (Vora et al. 2016; Chen and Qi 2017).

Even though a lot has been accomplished using these genetic engineering tools still many challenges remain to limit the realization of the full potential of the genome editing tool. Most importantly are the development of new techniques which are capable of introducing gene modifications without DNA breaks such as Oligonucleotide-Directed Mutagenesis (ODM) and Base Editing (Komor et al. 2018). These methods can convert one target base pair to a different base pair without requiring DSBs and in the future can be promising technologies for the study of potential treatments for genetic diseases (Komor et al. 2018). The targeted recombinases that can recognize specific DNA sequences and incorporate desired therapeutic factors into the human genome can be designed and developed (Akopian et al. 2003; Pruett-Miller et al. 2009; Mercer et al. 2012; Gaj et al. 2013a; Sirk et al. 2014; Wallen et al. 2015). This could very well herald the era of the beginning of the union between regenerative medicines and genome engineering (see Table 9.2). However, despite the existence of substantial knowledge gained from genome editing in immortalized cell lines, its application in regenerative medicine that requires genetic manipulation of the progenitor or stem cell populations is still in its infancy as their epigenome as well as the organization of the genome and its functional regulation is inherently different from the transformed cell lines. It is important to fully explore and understand the functional landscape of the potential role and usage of these technologies in progenitor cells and stem cells before their large-scale usage in designer therapeutic applications that could mean reprograming the cell fate and behavior for the next generation of advancement in gene therapy and synthetic biology.

References

Akopian A, He J, Boocock MR, Stark WM (2003) Chimeric recombinases with designed DNA sequence recognition. Proc Natl Acad Sci 100:8688–8691

Anguela XM, Sharma R, Doyon Y, Miller JC, Li H, Haurigot V et al (2013) Robust ZFN-mediated genome editing in adult hemophilic mice. Blood J Am Soc Hematol 122:3283–3287

Bae K-H, Do Kwon Y, Shin H-C, Hwang M-S, Ryu E-H, Park K-S et al (2003) Human zinc fingers as building blocks in the construction of artificial transcription factors. Nat Biotechnol 21:275–280

Balboa D, Weltner J, Eurola S, Trokovic R, Wartiovaara K, Otonkoski T (2015) Conditionally stabilized dCas9 activator for controlling gene expression in human cell reprogramming and differentiation. Stem Cell Rep 5:448–459
Baltes NJ, Voytas DF (2015) Enabling plant synthetic biology through genome engineering. Trends Biotechnol 33:120–131
Barone G, Arora A, Ganesh A, Abdel-fatah T, Moseley P, Ali R et al (2018) The relationship of CDK18 expression in breast cancer to clinicopathological parameters and therapeutic response. Oncotarget 9:29508–29524
Barrangou R, Marraffini LA (2014) CRISPR-Cas systems: prokaryotes upgrade to adaptive immunity. Mol Cell 54(2):234–244
Barrangou R, Fremaux C, Deveau H, Richards M, Boyaval P, Moineau S et al (2007) CRISPR provides acquired resistance against viruses in prokaryotes. Science 315:1709–1712. https://doi.org/10.1126/science.1138140
Basu S, Adams L, Guhathakurta S, Kim Y (2017) A novel tool for monitoring endogenous alpha-synuclein transcription by NanoLuciferase tag insertion at the 3′ end using CRISPR-Cas9 genome editing technique. Nat Publ Gr:1–11. https://doi.org/10.1038/srep45883
Becker S, Boch J (2021) TALE and TALEN genome editing technologies. Gene Genome Edit 2: 100007
Beerli RR, Barbas CF (2002) Engineering polydactyl zinc-finger transcription factors. Nat Biotechnol 20:135–141
Beerli RR, Segal DJ, Dreier B, Barbas CF (1998) Toward controlling gene expression at will: specific regulation of the erbB-2/HER-2 promoter by using polydactyl zinc finger proteins constructed from modular building blocks. Proc Natl Acad Sci 95:14628–14633
Beerli RR, Schopfer U, Dreier B, Barbas CF (2000) Chemically regulated zinc finger transcription factors. J Biol Chem 275:32617–32627
Bhaya D, Davison M, Barrangou R (2011) CRISPR-Cas systems in bacteria and archaea: versatile small RNAs for adaptive defense and regulation. Annu Rev Genet 45:273–297
Blancafort P, Tschan MP, Bergquist S, Guthy D, Brachat A, Sheeter DA et al (2008) Modulation of drug resistance by artificial transcription factors. Mol Cancer Ther 7:688–697
Boch J, Bonas U (2010) Xanthomonas AvrBs3 family-type III effectors: discovery and function. Annu Rev Phytopathol 48:419–436
Boch J, Scholze H, Schornack S, Landgraf A, Hahn S, Kay S et al (2009) Breaking the code of DNA binding specificity of TAL-type III effectors. Science 326:1509–1512
Bolotin A, Quinquis B, Sorokin A, Ehrlich SD (2005) Clustered regularly interspaced short palindrome repeats (CRISPRs) have spacers of extrachromosomal origin. Microbiology 151: 2551–2561
Bondy-Denomy J, Pawluk A, Maxwell KL, Davidson AR (2013) Bacteriophage genes that inactivate the CRISPR/Cas bacterial immune system. Nature 493:429–432
Burt A (2003) Site-specific selfish genes as tools for the control and genetic engineering of natural populations. Proc R Soc London Ser B Biol Sci 270:921–928
Capecchi MR (1989) Altering the genome by homologous recombination. Science 244:1288–1292
Carlson DF, Tan W, Lillico SG, Stverakova D, Proudfoot C, Christian M et al (2012) Efficient TALEN-mediated gene knockout in livestock. Proc Natl Acad Sci 109:17382–17387
Carroll D (2011) Genome engineering with zinc-finger nucleases. Genetics 188(4):773–782
Carte J, Wang R, Li H, Terns RM, Terns MP (2008) Cas6 is an endoribonuclease that generates guide RNAs for invader defense in prokaryotes. Genes Dev 22:3489–3496
Chandrasegaran S, Carroll D (2016) Origins of programmable nucleases for genome engineering. J Mol Biol 428:963–989
Chandrasekharan S, Kumar S, Valley CM, Rai A (2009) Proprietary science, open science and the role of patent disclosure: the case of zinc-finger proteins. Nat Biotechnol 27(2):140–144
Charpentier E, Doudna JA (2013) Rewriting a genome. Nature 495:50–51

Chattopadhyay A, Purohit J, Mehta S, Parmar H, Achary VMM, Reddy MK (2022) Precision genome editing toolbox: applications and approaches for improving rice's genetic resistance to pathogens. Agronomy 12:565
Chen M, Qi LS (2017) Repurposing CRISPR system for transcriptional activation. Adv Exp Med Biol 983:147–157
Chen H, Kazemier HG, de Groote ML, Ruiters MHJ, Xu G-L, Rots MG (2014) Induced DNA demethylation by targeting Ten-Eleven Translocation 2 to the human ICAM-1 promoter. Nucleic Acids Res 42:1563–1574
Chen S, Sanjana NE, Zheng K, Shalem O, Lee K, Shi X et al (2015) Genome-wide CRISPR screen in a mouse model of tumor growth and metastasis. Cell 160:1246–1260
Chen Y, Farzadfard F, Gharaei N, Chen WCW, Cao J, Lu TK et al (2017) Randomized CRISPR-Cas transcriptional against alpha-synuclein toxicity randomized CRISPR-Cas transcriptional perturbation screening reveals protective genes against alpha-synuclein toxicity. Mol Cell 68: 247–257.e5. https://doi.org/10.1016/j.molcel.2017.09.014
Chen JS, Chen JS, Ma E, Harrington LB, Da Costa M, Tian X et al (2018) CRISPR-Cas12a target binding unleashes indiscriminate single-stranded DNase activity. Science 6245:1–8
Cho SW, Kim S, Kim JM, Kim J-S (2013) Targeted genome engineering in human cells with the Cas9 RNA-guided endonuclease. Nat Biotechnol 31:230–232
Choo Y, Klug A (1994) Toward a code for the interactions of zinc fingers with DNA: selection of randomized fingers displayed on phage. Proc Natl Acad Sci 91:11163–11167
Christian M, Cermak T, Doyle EL, Schmidt C, Zhang F, Hummel A et al (2010) Targeting DNA double-strand breaks with TAL effector nucleases. Genetics 186:757–761
Cohen J (2017) The Birth of CRISPR Inc. Science 355:680–684
Cong L, Ran FA, Cox D, Lin S, Barretto R, Habib N et al (2013) Multiplex genome engineering using CRISPR/Cas systems. Science 339:819–823
Copeland MF, Politz MC, Johnson CB, Markley AL, Pfleger BF (2016) A transcription activator–like effector (TALE) induction system mediated by proteolysis. Nat Chem Biol 12:254
Cox DBT, Platt RJ, Zhang F (2015) Therapeutic genome editing: prospects and challenges. Nat Med 21:121–131
Cui X, Ji D, Fisher DA, Wu Y, Briner DM, Weinstein EJ (2011) Targeted integration in rat and mouse embryos with zinc-finger nucleases. Nat Biotechnol 29:64–67
Davis D, Stokoe D (2010) Zinc Finger Nucleases as tools to understand and treat human diseases. BMC Med 8:1–11
Degennaro M, McBride CS, Seeholzer L, Nakagawa T, Dennis EJ, Goldman C et al (2013) orco mutant mosquitoes lose strong preference for humans and are not repelled by volatile DEET. Nature 498:487–491
Deltcheva E, Chylinski K, Sharma CM, Gonzales K, Chao Y, Pirzada ZA et al (2011) CRISPR RNA maturation by trans-encoded small RNA and host factor RNase III. Nature 471:602–607
Deshpande K, Vyas A, Balakrishnan A, Vyas D (2015) Clustered regularly interspaced short palindromic repeats/Cas9 genetic engineering: robotic genetic surgery. Am J Robot Surg 2(1): 49–52
Deuse T, Hu X, Gravina A, Wang D, Tediashvili G, De C et al (2019) Hypoimmunogenic derivatives of induced pluripotent stem cells evade immune rejection in fully immunocompetent allogeneic recipients. Nat Biotechnol 37(3):252–258. https://doi.org/10.1038/s41587-019-0016-3
Deveau H, Garneau JE, Moineau S (2010) CRISPR/Cas system and its role in phage-bacteria interactions. Annu Rev Microbiol 64:475–493
Doyon Y, McCammon JM, Miller JC, Faraji F, Ngo C, Katibah GE et al (2008) Heritable targeted gene disruption in zebrafish using designed zinc-finger nucleases. Nat Biotechnol 26:702–708
Dreier B, Beerli RR, Segal DJ, Flippin JD, Barbas CF (2001) Development of zinc finger domains for recognition of the 5′-ANN-3′ family of DNA sequences and their use in the construction of artificial transcription factors. J Biol Chem 276:29466–29478

Dreier B, Fuller RP, Segal DJ, Lund CV, Blancafort P, Huber A et al (2005) Development of zinc finger domains for recognition of the 5′-CNN-3′ family DNA sequences and their use in the construction of artificial transcription factors. J Biol Chem 280:35588–35597

El-Mounadi K, Morales-Floriano ML, Garcia-Ruiz H (2020) Principles, applications, and biosafety of plant genome editing using CRISPR-Cas9. Front Plant Sci 11:56

Fairall L, Schwabe JWR, Chapman L, Finch JT, Rhodes D (1993) The crystal structure of a two zinc-finger peptide reveals an extension to the rules for zinc-finger/DNA recognition. Nature 366:483–487

Farzadfard F, Perli SD, Lu TK (2013) Tunable and multifunctional eukaryotic transcription factors based on CRISPR/Cas. ACS Synth Biol 2:604–613

Findlay GM, Boyle EA, Hause RJ, Klein JC, Shendure J (2014) Saturation editing of genomic regions by multiplex homology-directed repair. Nature 513:120–123

Gaj T, Mercer AC, Sirk SJ, Smith HL, Barbas CF III (2013a) A comprehensive approach to zinc-finger recombinase customization enables genomic targeting in human cells. Nucleic Acids Res 41:3937–3946

Gaj T, Gersbach CA, Barbas CF III (2013b) ZFN, TALEN, and CRISPR/Cas-based methods for genome engineering. Trends Biotechnol 31(7):397–405

Gaj T, Epstein BE, Schaffer DV (2016a) Genome engineering using adeno-associated virus: basic and clinical research applications. Mol Ther 24:458–464

Gaj T, Sirk SJ, Shui SL, Liu J (2016b) Genome-editing technologies: principles and applications. Cold Spring Harb Perspect Biol 8(12):a023754

Gantz VM, Bier E (2015) The mutagenic chain reaction: a method for converting heterozygous to homozygous mutations. Science 348:442–444

Gesner EM, Schellenberg MJ, Garside EL, George MM, MacMillan AM (2011) Recognition and maturation of effector RNAs in a CRISPR interference pathway. Nat Struct Mol Biol 18:688–692

Geurts AM, Cost GJ, Freyvert Y, Zeitler B, Miller JC, Choi VM et al (2009) Knockout rats via embryo microinjection of zinc-finger nucleases. Science 325:433

Ghezraoui H, Piganeau M, Renouf B, Renaud J-B, Sallmyr A, Ruis B et al (2014) Chromosomal translocations in human cells are generated by canonical nonhomologous end-joining. Mol Cell 55:829–842

Ghorbal M, Gorman M, Macpherson CR, Martins RM, Scherf A, Lopez-Rubio J-J (2014) Genome editing in the human malaria parasite Plasmodium falciparum using the CRISPR-Cas9 system. Nat Biotechnol 32:819–821

Gilbert LA, Larson MH, Morsut L, Liu Z, Brar GA, Torres SE et al (2013) CRISPR-mediated modular RNA-guided regulation of transcription in eukaryotes. Cell 154:442–451

Gilbert LA, Horlbeck MA, Adamson B, Villalta JE, Chen Y, Whitehead EH et al (2014) Genome-scale CRISPR-mediated control of gene repression and activation. Cell 159:647–661

Grissa I, Vergnaud G, Pourcel C (2007) The CRISPRdb database and tools to display CRISPRs and to generate dictionaries of spacers and repeats. BMC Bioinform 8:1–10

Guo M, Chen H, Dong S, Zhang Z, Luo H (2022) CRISPR - Cas gene editing technology and its application prospect in medicinal plants. Chin Med 17:1–19. https://doi.org/10.1186/s13020-022-00584-w

Gupta RM, Musunuru K (2014) Expanding the genetic editing tool kit: ZFNs, TALENs, and CRISPR-Cas9. J Clin Invest 124(10):4154–4161

György B, Lööv C, Zaborowski M, Takeda S, Kleinstiver B, Commins C et al (2018) CRISPR/Cas9 mediated disruption of the Swedish APP allele as a therapeutic approach for early-onset Alzheimer's disease. Mol Ther Nucleic Acid 1–31. https://doi.org/10.1016/j.omtn.2018.03.007

Haft DH, Selengut J, Mongodin EF, Nelson KE (2005) A guild of 45 CRISPR-associated (Cas) protein families and multiple CRISPR/Cas subtypes exist in prokaryotic genomes. PLoS Comput Biol 1:e60

Hale CR, Zhao P, Olson S, Duff MO, Graveley BR, Wells L et al (2009) RNA-guided RNA cleavage by a CRISPR RNA Cas protein complex. Cell 139:945–956

Hale CR, Majumdar S, Elmore J, Pfister N, Compton M, Olson S et al (2012) Essential features and rational design of CRISPR RNAs that function with the Cas RAMP module complex to cleave RNAs. Mol Cell 45:292–302

Hammond A, Galizi R, Kyrou K, Simoni A, Siniscalchi C, Katsanos D et al (2016) A CRISPR-Cas9 gene drive system targeting female reproduction in the malaria mosquito vector Anopheles gambiae. Nat Biotechnol 34:78–83

Hart T, Chandrashekhar M, Aregger M, Steinhart Z, Brown KR, MacLeod G et al (2015) High-resolution CRISPR screens reveal fitness genes and genotype-specific cancer liabilities. Cell 163:1515–1526

Haun W, Coffman A, Clasen BM, Demorest ZL, Lowy A, Ray E et al (2014) Improved soybean oil quality by targeted mutagenesis of the fatty acid desaturase 2 gene family. Plant Biotechnol J 12: 934–940

Haurwitz RE, Jinek M, Wiedenheft B, Zhou K, Doudna JA (2010) Sequence-and structure-specific RNA processing by a CRISPR endonuclease. Science 329:1355–1358

Hauschild J, Petersen B, Santiago Y, Queisser A-L, Carnwath JW, Lucas-Hahn A et al (2011) Efficient generation of a biallelic knockout in pigs using zinc-finger nucleases. Proc Natl Acad Sci 108:12013–12017

Hilton IB, Gersbach CA (2015) Enabling functional genomics with genome engineering. Genome Res 25:1442–1455

Hockemeyer D, Wang H, Kiani S, Lai CS, Gao Q, Cassady JP et al (2011) Genetic engineering of human pluripotent cells using TALE nucleases. Nat Biotechnol 29:731–734

Holt N, Wang J, Kim K, Friedman G, Wang X, Taupin V et al (2010) Human hematopoietic stem/ progenitor cells modified by zinc-finger nucleases targeted to CCR5 control HIV-1 in vivo. Nat Biotechnol 28:839–847

Horvath P, Barrangou R (2010) CRISPR/Cas, the immune system of bacteria and archaea. Science 327:167–171

Hou Z, Zhang Y, Propson NE, Howden SE, Chu L-F, Sontheimer EJ et al (2013) Efficient genome engineering in human pluripotent stem cells using Cas9 from Neisseria meningitidis. Proc Natl Acad Sci 110:15644–15649

Houbaviy HB, Usheva A, Shenk T, Burley SK (1996) Cocrystal structure of YY1 bound to the adeno-associated virus P5 initiator. Proc Natl Acad Sci 93:13577–13582

Hu X (2016) CRISPR/Cas9 system and its applications in human hematopoietic cells. Blood Cells. Mol Dis 62:6–12. https://doi.org/10.1016/j.bcmd.2016.09.003

Hu J, Lei Y, Wong W-K, Liu S, Lee K-C, He X et al (2014) Direct activation of human and mouse Oct4 genes using engineered TALE and Cas9 transcription factors. Nucleic Acids Res 42:4375–4390

Huang C, Sun H, Xu D, Chen Q, Liang Y, Wang X et al (2017) ZmCCT9 enhances maize adaptation to higher latitudes. PNAS 10:1–8. https://doi.org/10.1073/pnas.1718058115

Hwang WY, Fu Y, Reyon D, Maeder ML, Tsai SQ, Sander JD et al (2013) Efficient genome editing in zebrafish using a CRISPR-Cas system. Nat Biotechnol 31:227–229

Imran M, Waheed Y, Ghazal A, Ullah S, Safi SZ, Jamal M et al (2017) Modern biotechnology-based therapeutic approaches against HIV infection (Review). Biomed Rep 7:504–507. https://doi.org/10.3892/br.2017.1006

Ishino Y, Shinagawa H, Makino K, Amemura M, Nakata A (1987) Nucleotide sequence of the iap gene, responsible for alkaline phosphatase isozyme conversion in Escherichia coli, and identification of the gene product. J Bacteriol 169:5429–5433

Jackson SP, Bartek J (2009) The DNA-damage response in human biology and disease. Nature 461: 1071–1078

Jakočiūnas T, Bonde I, Herrgård M, Harrison SJ, Kristensen M, Pedersen LE et al (2015) Multiplex metabolic pathway engineering using CRISPR/Cas9 in Saccharomyces cerevisiae. Metab Eng 28:213–222

Jamieson AC, Kim S-H, Wells JA (1994) In vitro selection of zinc fingers with altered DNA-binding specificity. Biochemistry 33:5689–5695

Jasin M, Rothstein R (2013) Repair of strand breaks by homologous recombination. Cold Spring Harb Perspect Biol 5:a012740
Jiang W, Bikard D, Cox D, Zhang F, Marraffini LA (2013) RNA-guided editing of bacterial genomes using CRISPR-Cas systems. Nat Biotechnol 31:233–239
Jinek M, Chylinski K, Fonfara I, Hauer M, Doudna JA, Charpentier E (2012) A programmable dual-RNA–guided DNA endonuclease in adaptive bacterial immunity. Science 337:816–821
Jinek M, East A, Cheng A, Lin S, Ma E, Doudna J (2013) RNA-programmed genome editing in human cells. elife 2:e00471
Jore MM, Brouns SJJ, van der Oost J (2012) RNA in defense: CRISPRs protect prokaryotes against mobile genetic elements. Cold Spring Harb Perspect Biol 4:a003657
Joung JK, Sander JD (2013) TALENs: a widely applicable technology for targeted genome editing. Nat Rev Mol Cell Biol 14(1):49–55
Kachhawaha N (2021) CRISPR edited microbes and their industrial potential review. Res J Biotechnol 16:223–228. https://doi.org/10.6084/m9.figshare.14216363.v1
Kaminski MM, Abudayyeh OO, Gootenberg JS, Zhang F, Collins JJ (2021) CRISPR-based diagnostics. Nat Biomed Eng 5(7):643–656
Karlson CKS, Mohd-Noor SN, Nolte N, Tan BC (2021) CRISPR/dCas9-based systems: mechanisms and applications in plant sciences. Plan Theory 10(10):2055
Khalil AM (2020) The genome editing revolution. J Genet Eng Biotechnol 18:1–16
Khalil K, Elayat M, Khalifa E, Daghash S, Elaswad A, Miller M et al (2017) Generation of myostatin gene- edited channel catfish (Ictalurus punctatus) via zygote injection of CRISPR/Cas9 system. Sci Rep 7:7301. https://doi.org/10.1038/s41598-017-07223-7
Kim Y-G, Cha J, Chandrasegaran S (1996) Hybrid restriction enzymes: zinc finger fusions to Fok I cleavage domain. Proc Natl Acad Sci 93:1156–1160
Kim HJ, Lee HJ, Kim H, Cho SW, Kim J-S (2009) Targeted genome editing in human cells with zinc finger nucleases constructed via modular assembly. Genome Res 19:1279–1288
Kim Y-K, Wee G, Park J, Kim J, Baek D, Kim J-S et al (2013) TALEN-based knockout library for human microRNAs. Nat Struct Mol Biol 20:1458–1464
Kim S, Hupperetz C, Lim S, Kim CH (2021) Genome editing of immune cells using CRISPR/Cas9. BMB Rep 54(1):59
Koike-Yusa H, Li Y, Tan E-P, Velasco-Herrera MDC, Yusa K (2014) Genome-wide recessive genetic screening in mammalian cells with a lentiviral CRISPR-guide RNA library. Nat Biotechnol 32:267–273
Komor AC, Badran AH, Liu DR (2018) Editing the genome without double-stranded DNA breaks. ACS Chem Biol 13:383–388
Konermann S, Brigham MD, Trevino AE, Hsu PD, Heidenreich M, Cong L et al (2013) Optical control of mammalian endogenous transcription and epigenetic states. Nature 500:472–476
Korkmaz G, Lopes R, Ugalde AP, Nevedomskaya E, Han R, Myacheva K et al (2016) Functional genetic screens for enhancer elements in the human genome using CRISPR-Cas9. Nat Biotechnol 34:192–198
Lamb BM, Mercer AC, Barbas CF III (2013) Directed evolution of the TALE N-terminal domain for recognition of all 5′ bases. Nucleic Acids Res 41:9779–9785
Laoharawee K, DeKelver RC, Podetz-Pedersen KM, Rohde M, Sproul S, Nguyen HO et al (2018) Dose-dependent prevention of metabolic and neurologic disease in murine MPS II by ZFN-mediated in vivo genome editing. Mol Ther 26(4):1127–1136
Lee HJ, Kim E, Kim J-S (2010) Targeted chromosomal deletions in human cells using zinc finger nucleases. Genome Res 20:81–89
Lee HG, Kim DH, Choi Y-R, Yu J, Hong S-A, Seo PJ et al (2021) Enhancing plant immunity by expression of pathogen-targeted CRISPR-Cas9 in plants. Gene Genome Ed. 1:100001. https://doi.org/10.1016/j.ggedit.2021.100001
Li H, Haurigot V, Doyon Y, Li T, Wong SY, Bhagwat AS et al (2011) In vivo genome editing restores haemostasis in a mouse model of haemophilia. Nature 475:217–221

Li Y, Moore R, Guinn M, Bleris L (2012) Transcription activator-like effector hybrids for conditional control and rewiring of chromosomal transgene expression. Sci Rep 2:1–7

Li W, Teng F, Li T, Zhou Q (2013) Simultaneous generation and germline transmission of multiple gene mutations in rat using CRISPR-Cas systems. Nat Biotechnol 31:684–686

Li H, Yang Y, Hong W, Huang M, Wu M, Zhao X (2020) Applications of genome editing technology in the targeted therapy of human diseases: mechanisms, advances and prospects. Signal Transduct Target Ther 5:1–23. https://doi.org/10.1038/s41392-019-0089-y

Liao Y, Bai Q, Xu P, Wu T, Guo D, Peng Y et al (2018) Mutation in rice abscisic acid2 results in cell death, enhanced disease-resistance, altered seed dormancy and development. Front Plant Sci 9:405. https://doi.org/10.3389/fpls.2018.00405

Lieber MR (2010) The mechanism of double-strand DNA break repair by the nonhomologous DNA end-joining pathway. Annu Rev Biochem 79:181–211

Lillestøl RK, Shah SA, Brügger K, Redder P, Phan H, Christiansen J et al (2009) CRISPR families of the crenarchaeal genus Sulfolobus: bidirectional transcription and dynamic properties. Mol Microbiol 72:259–272

Limsirichai P, Gaj T, Schaffer DV (2016) CRISPR-mediated activation of latent HIV-1 expression. Mol Ther 24:499–507

Lin S-R, Yang H-C, Kuo Y-T, Liu C-J, Yang T-Y, Sung K-C et al (2014) The CRISPR/Cas9 system facilitates clearance of the intrahepatic HBV templates in vivo. Mol Ther Acids 3:e186

Lino CA, Harper JC, Carney JP, Timlin JA (2018) Delivering CRISPR: a review of the challenges and approaches. Drug Deliv 25(1):1234–1257

Liu Z, Zhou X, Zhu Y, Chen Z-F, Yu B, Wang Y et al (2014) Generation of a monkey with MECP2 mutations by TALEN-based gene targeting. Neurosci Bull 30:381–386

Liu Z, Dong H, Cui Y, Cong L, Zhang D (2020) Application of different types of CRISPR/Cas-based systems in bacteria. Microb Cell Factories 19(1):1–14

Long C, Amoasii L, Mireault AA, McAnally JR, Li H, Sanchez-Ortiz E et al (2016) Postnatal genome editing partially restores dystrophin expression in a mouse model of muscular dystrophy. Science 351:400–403

Lonzarić J, Lebar T, Majerle A, Manček-Keber M, Jerala R (2016) Locked and proteolysis-based transcription activator-like effector (TALE) regulation. Nucleic Acids Res 44:1471–1481

Luo S, Li J, Stoddard TJ, Baltes NJ, Demorest ZL, Clasen BM et al (2015) Non-transgenic plant genome editing using purified sequence-specific nucleases. Mol Plant 8:1425–1427

Lynch VJ, Bedoya-Reina OC, Ratan A, Perry GH, Miller W, Schuster SC (2015) Elephantid genomes reveal the molecular bases of woolly mammoth adaptations to the arctic. Cell Rep 12:217–228. https://doi.org/10.1016/j.celrep.2015.06.027

Ma H, Dang Y, Wu Y, Jia G, Anaya E, Zhang J et al (2015) A CRISPR-based screen identifies genes essential for West-Nile-virus-induced cell death. Cell Rep 12:673–683

Maeder ML, Gersbach CA (2016) Genome-editing technologies for gene and cell therapy. Mol Ther 24:430–446. https://doi.org/10.1038/mt.2016.10

Maeder ML, Linder SJ, Reyon D, Angstman JF, Fu Y, Sander JD et al (2013) Robust, synergistic regulation of human gene expression using TALE activators. Nat Methods 10:243–245

Magnenat L, Schwimmer LJ, Barbas CF (2008) Drug-inducible and simultaneous regulation of endogenous genes by single-chain nuclear receptor-based zinc-finger transcription factor gene switches. Gene Ther 15:1223–1232

Mahto KK, Prasad P, Kumar M, Dubey H, Ranjan A (2022) Role of CRISPR technology in gene editing of emerging and re-emerging vector borne disease. In: Puerta-Guardo H, Manrique-Saide DP (eds) Mosquito research - recent advances in pathogen interactions, immunity, and vector control strategies. IntechOpen Limited. https://doi.org/10.5772/intechopen.104100

Maiti P, Manna J, Dunbar GL (2017) Current understanding of the molecular mechanisms in Parkinson's disease: targets for potential treatments. Transl Neurodegener 6:1–35. https://doi.org/10.1186/s40035-017-0099-z

Mak AN-S, Bradley P, Cernadas RA, Bogdanove AJ, Stoddard BL (2012) The crystal structure of TAL effector PthXo1 bound to its DNA target. Science 335:716–719

Makarova KS, Grishin NV, Shabalina SA, Wolf YI, Koonin EV (2006) A putative RNA-interference-based immune system in prokaryotes: computational analysis of the predicted enzymatic machinery, functional analogies with eukaryotic RNAi, and hypothetical mechanisms of action. Biol Direct 1:1–26

Makarova KS, Haft DH, Barrangou R, Brouns SJJ, Charpentier E, Horvath P et al (2011) Evolution and classification of the CRISPR–Cas systems. Nat Rev Microbiol 9:467–477

Makarova KS, Wolf YI, Koonin EV (2013) The basic building blocks and evolution of CRISPR – Cas systems. Biochem Soc Trans 41:1392–1400. https://doi.org/10.1042/BST20130038

Mali P, Yang L, Esvelt KM, Aach J, Guell M, DiCarlo JE et al (2013) RNA-guided human genome engineering via Cas9. Science 339:823–826

Mandip KC, Steer CJ (2019) A new era of gene editing for the treatment of human diseases. Swiss Med Wkly

Margolin JF, Friedman JR, Meyer WK, Vissing H, Thiesen H-J, Rauscher FJ (1994) Krüppel-associated boxes are potent transcriptional repression domains. Proc Natl Acad Sci 91:4509–4513

Marraffini LA, Sontheimer EJ (2010) Self versus non-self discrimination during CRISPR RNA-directed immunity. Nature 463:568–571

Martinez-Lage M, Torres-Ruiz R, Rodriguez-Perales S (2017) CRISPR/Cas9 technology: applications and human disease modeling. Elsevier Inc. https://doi.org/10.1016/bs.pmbts.2017.09.002

Mendenhall EM, Williamson KE, Reyon D, Zou JY, Ram O, Joung JK et al (2013) Locus-specific editing of histone modifications at endogenous enhancers. Nat Biotechnol 31:1133–1136

Mercer AC, Gaj T, Fuller RP, Barbas CF III (2012) Chimeric TALE recombinases with programmable DNA sequence specificity. Nucleic Acids Res 40:11163–11172

Mercer AC, Gaj T, Sirk SJ, Lamb BM, Barbas CF III (2014) Regulation of endogenous human gene expression by ligand-inducible TALE transcription factors. ACS Synth Biol 3:723–730

Miller J, McLachlan AD, Klug A (1985) Repetitive zinc-binding domains in the protein transcription factor IIIA from Xenopus oocytes. EMBO J 4:1609–1614

Miller JC, Holmes MC, Wang J, Guschin DY, Lee Y-L, Rupniewski I et al (2007) An improved zinc-finger nuclease architecture for highly specific genome editing. Nat Biotechnol 25:778–785

Miller JC, Tan S, Qiao G, Barlow KA, Wang J, Xia DF et al (2011) A TALE nuclease architecture for efficient genome editing. Nat Biotechnol 29:143–148

Moehle EA, Rock JM, Lee Y-L, Jouvenot Y, DeKelver RC, Gregory PD et al (2007) Targeted gene addition into a specified location in the human genome using designed zinc finger nucleases. Proc Natl Acad Sci 104:3055–3060

Mojica FJM, Díez-Villaseñor C, García-Martínez J, Soria E (2005) Intervening sequences of regularly spaced prokaryotic repeats derive from foreign genetic elements. J Mol Evol 60: 174–182

Mojica FJM, Díez-Villaseñor C, García-Martínez J, Almendros C (2009a) Short motif sequences determine the targets of the prokaryotic CRISPR defence system. Microbiology 155:733–740

Mojica FJM, Díez-Villaseñor C, García-Martínez J, Almendros C (2009b) Short motif sequences determine the targets of the prokaryotic CRISPR defence system. Microbiology 155:733–740. https://doi.org/10.1099/mic.0.023960-0

Moon SB, Kim DY, Ko JH, Kim YS (2019) Recent advances in the CRISPR genome editing tool set. Exp Mol Med 51(11):1–11

Moscou MJ, Bogdanove AJ (2009) A simple cipher governs DNA recognition by TAL effectors. Science 326:1501

Moynahan ME, Jasin M (2010) Mitotic homologous recombination maintains genomic stability and suppresses tumorigenesis. Nat Rev Mol Cell Biol 11:196–207

National Human Genome Research Institute (2019) What is genome editing? US Dep Heal Hum Serv, 1. Available at: https://www.genome.gov/about-genomics/policy-issues/what-is-Genome-Editing

National Institute of Health (2020) Gene editing – digital media kit. US Dep Heal Hum Serv, 1. Available at: https://www.nih.gov/news-events/gene-editing-digital-press-kit
Nelson CE, Hakim CH, Ousterout DG, Thakore PI, Moreb EA, Rivera RMC et al (2016) In vivo genome editing improves muscle function in a mouse model of Duchenne muscular dystrophy. Science 351:403–407
Nemudryi AA, Valetdinova KR, Medvedev SP, Zakian SM (2014) TALEN and CRISPR/Cas genome editing systems: tools of discovery. Acta Nat 6(3 (22)):19–40
Newsom S, Parameshwaran HP, Martin L, Rajan R (2021) The CRISPR-Cas mechanism for adaptive immunity and alternate bacterial functions fuels diverse biotechnologies. Front Cell Infect Microbiol 10:619763
Nihongaki Y, Yamamoto S, Kawano F, Suzuki H, Sato M (2015) CRISPR-Cas9-based photoactivatable transcription system. Chem Biol 22:169–174
Nolte RT, Conlin RM, Harrison SC, Brown RS (1998) Differing roles for zinc fingers in DNA recognition: structure of a six-finger transcription factor IIIA complex. Proc Natl Acad Sci 95: 2938–2943
Nymark M, Sharma AK, Sparstad T, Bones AM, Winge P (2016) OPEN A CRISPR / Cas9 system adapted for gene editing in marine algae. Nat Publ Gr, 6–11. https://doi.org/10.1038/srep24951
Parnas O, Jovanovic M, Eisenhaure TM, Herbst RH, Dixit A, Ye CJ et al (2015) A genome-wide CRISPR screen in primary immune cells to dissect regulatory networks. Cell 162:675–686
Paschon DE, Lussier S, Wangzor T, Xia DF, Li PW, Hinkley SJ et al (2019) Diversifying the structure of zinc finger nucleases for high-precision genome editing. Nat Commun 10:1–12
Pavletich NP, Pabo CO (1993) Crystal structure of a five-finger GLI-DNA complex: new perspectives on zinc fingers. Science 261:1701–1707
Pawelczak KS, Gavande NS, VanderVere-Carozza PS, Turchi JJ (2018) Modulating DNA repair pathways to improve precision genome engineering. ACS Chem Biol 13:389–396
Perales M, Kebriaei P, Kean LS, Sadelain M (2018) Reprint of: building a safer and faster CAR: seatbelts, airbags, and CRISPR. Biol Blood Marrow Transplant 24:S15–S19. https://doi.org/10.1016/j.bbmt.2017.12.789
Perez EE, Wang J, Miller JC, Jouvenot Y, Kim KA, Liu O et al (2008) Establishment of HIV-1 resistance in CD4+ T cells by genome editing using zinc-finger nucleases. Nat Biotechnol 26: 808–816
Perez-Pinera P, Ousterout DG, Brunger JM, Farin AM, Glass KA, Guilak F et al (2013) Synergistic and tunable human gene activation by combinations of synthetic transcription factors. Nat Methods 10:239–242
Pollock R, Giel M, Linher K, Clackson T (2002) Regulation of endogenous gene expression with a small-molecule dimerizer. Nat Biotechnol 20:729–733
Polstein LR, Gersbach CA (2012) Light-inducible spatiotemporal control of gene activation by customizable zinc finger transcription factors. J Am Chem Soc 134:16480–16483
Polstein LR, Gersbach CA (2015) A light-inducible CRISPR-Cas9 system for control of endogenous gene activation. Nat Chem Biol 11:198–200
Porteus MH (2015) Towards a new era in medicine: therapeutic genome editing. Genome Biol 16: 1–12
Porteus M (2016) Genome editing: a new approach to human therapeutics. Annu Rev Pharmacol Toxicol 56:163–190. https://doi.org/10.1146/annurev-pharmtox-010814-124454
Porteus MH, Baltimore D (2003) Chimeric nucleases stimulate gene targeting in human cells. Science 300(5620):763
Pourcel C, Salvignol G, Vergnaud G (2005) CRISPR elements in Yersinia pestis acquire new repeats by preferential uptake of bacteriophage DNA, and provide additional tools for evolutionary studies. Microbiology 151:653–663
Pruett-Miller SM, Reading DW, Porter SN, Porteus MH (2009) Attenuation of zinc finger nuclease toxicity by small-molecule regulation of protein levels. PLoS Genet 5:e1000376
Qasim W (2021) Genome editing of therapeutic T cells. Gene Genome Ed 2:100010. https://doi.org/10.1016/j.ggedit.2021.100010

Qi LS, Larson MH, Gilbert LA, Doudna JA, Weissman JS, Arkin AP et al (2013) Repurposing CRISPR as an RNA-guided platform for sequence-specific control of gene expression. Cell 152: 1173–1183
Rager JE, Carberry C, Fry RC (2011) Use of genome editing tools in environmental health research. Physiol Behav 176:139–148. https://doi.org/10.1016/j.cotox.2019.02.007.Use
Rahim J, Gulzar S, Zahid R, Rahim KA (2021) Systematic review on the comparison of molecular gene editing tools. Int J Innov Sci Res Tech 6:1–8
Ramirez CL, Foley JE, Wright DA, Müller-Lerch F, Rahman SH, Cornu TI et al (2008) Unexpected failure rates for modular assembly of engineered zinc fingers. Nat Methods 5:374–375
Ran FA, Hsu PD, Lin C-Y, Gootenberg JS, Konermann S, Trevino AE et al (2013) Double nicking by RNA-guided CRISPR Cas9 for enhanced genome editing specificity. Cell 154:1380–1389
Raquel D, Rodríguez R, Solís RR, Alberto H, Saldaña B (2019) Genome editing: a perspective on the application of CRISPR/Cas9 to study human diseases (Review). Int J Mol Med 43:1559–1574. https://doi.org/10.3892/ijmm.2019.4112
Rath D, Amlinger L, Rath A, Lundgren M (2015) The CRISPR-Cas immune system: biology, mechanisms and applications. Biochimie 117:119–128
Rebar EJ, Pabo CO (1994) Zinc finger phage: affinity selection of fingers with new DNA-binding specificities. Science 263:671–673
Remy S, Tesson L, Ménoret S, Usal C, Scharenberg AM, Anegon I (2010) Zinc-finger nucleases: a powerful tool for genetic engineering of animals. Transgenic Res 19(3):363–371
Renaud J-B, Boix C, Charpentier M, De Cian A, Cochennec J, Duvernois-Berthet E et al (2016) Improved genome editing efficiency and flexibility using modified oligonucleotides with TALEN and CRISPR-Cas9 nucleases. Cell Rep 14:2263–2272
Reyon D, Tsai SQ, Khayter C, Foden JA, Sander JD, Joung JK (2012) FLASH assembly of TALENs for high-throughput genome editing. Nat Biotechnol 30:460–465
Ricci R, Colasante G (2021) CRISPR/dCas9 as a therapeutic approach for neurodevelopmental disorders: innovations and limitations compared to traditional strategies. Dev Neurosci 43(3–4): 253–261
Richter H, Zoephel J, Schermuly J, Maticzka D, Backofen R, Randau L (2012) Characterization of CRISPR RNA processing in Clostridium thermocellum and Methanococcus maripaludis. Nucleic Acids Res 40:9887–9896
Rivenbark AG, Stolzenburg S, Beltran AS, Yuan X, Rots MG, Strahl BD et al (2012) Epigenetic reprogramming of cancer cells via targeted DNA methylation. Epigenetics 7:350–360
Robb GB (2019) Genome editing with CRISPR-Cas: an overview. Curr Protoc Essent Lab Tech e36 19:1–20
Ryan OW, Skerker JM, Maurer MJ, Li X, Tsai JC, Poddar S et al (2014) Selection of chromosomal DNA libraries using a multiplex CRISPR system. elife 3
Sadowski I, Ma J, Triezenberg S, Ptashne M (1988) GAL4-VP16 is an unusually potent transcriptional activator. Nature 335:563–564
Sander JD, Maeder ML, Reyon D, Voytas DF, Joung JK, Dobbs D (2010) ZiFiT (Zinc Finger Targeter): an updated zinc finger engineering tool. Nucleic Acids Res 38(suppl_2):W462–W468
Sander JD, Dahlborg EJ, Goodwin MJ, Cade L, Zhang F, Cifuentes D et al (2011) Selection-free zinc-finger-nuclease engineering by context-dependent assembly (CoDA). Nat Methods 8:67–69
Sanjana NE, Cong L, Zhou Y, Cunniff MM, Feng G, Zhang F (2012) A transcription activator-like effector toolbox for genome engineering. Nat Protoc 7:171–192
Santiago Y, Chan E, Liu P-Q, Orlando S, Zhang L, Urnov FD et al (2008) Targeted gene knockout in mammalian cells by using engineered zinc-finger nucleases. Proc Natl Acad Sci 105:5809–5814
Sapranauskas R, Gasiunas G, Fremaux C, Barrangou R, Horvath P, Siksnys V (2011) The Streptococcus thermophilus CRISPR/Cas system provides immunity in Escherichia coli. Nucleic Acids Res 39:9275–9282

Sashital DG, Jinek M, Doudna JA (2011) An RNA-induced conformational change required for CRISPR RNA cleavage by the endoribonuclease Cse3. Nat Struct Mol Biol 18:680–687

Schmid-Burgk JL, Schmidt T, Kaiser V, Höning K, Hornung V (2013) A ligation-independent cloning technique for high-throughput assembly of transcription activator–like effector genes. Nat Biotechnol 31:76–81

Schornack S, Meyer A, Römer P, Jordan T, Lahaye T (2006) Gene-for-gene-mediated recognition of nuclear-targeted AvrBs3-like bacterial effector proteins. J Plant Physiol 163:256–272

Segal DJ, Dreier B, Beerli RR, Barbas CF (1999) Toward controlling gene expression at will: selection and design of zinc finger domains recognizing each of the 5′-GNN-3′ DNA target sequences. Proc Natl Acad Sci 96:2758–2763

Segal DJ, Beerli RR, Blancafort P, Dreier B, Effertz K, Huber A et al (2003) Evaluation of a modular strategy for the construction of novel polydactyl zinc finger DNA-binding proteins. Biochemistry 42:2137–2148

Segal DJ, Crotty JW, Bhakta MS, Barbas CF III, Horton NC (2006) Structure of Aart, a designed six-finger zinc finger peptide, bound to DNA. J Mol Biol 363:405–421

Shabbir MAB, Shabbir MZ, Wu Q, Mahmood S, Sajid A, Maan MK et al (2019) CRISPR-cas system: biological function in microbes and its use to treat antimicrobial resistant pathogens. Ann Clin Microbiol Antimicrob 18(1):1–9

Shah AN, Davey CF, Whitebirch AC, Miller AC, Moens CB (2015) Rapid reverse genetic screening using CRISPR in zebrafish. Nat Methods 12:535–540

Shalem O, Sanjana NE, Hartenian E, Shi X, Scott DA, Mikkelsen TS et al (2014) Genome-scale CRISPR-Cas9 knockout screening in human cells. Science 343:84–87

Shi J, Wang E, Milazzo JP, Wang Z, Kinney JB, Vakoc CR (2015) Discovery of cancer drug targets by CRISPR-Cas9 screening of protein domains. Nat Biotechnol Adv 1:1–9. https://doi.org/10.1038/nbt.3235

Shiva KG, Suma K (2019) Gene editing nucleases-ZFNs, TALENS and CRISPR: a review. Chettinad Health City Med J 8:130–135

Siddique AN, Nunna S, Rajavelu A, Zhang Y, Jurkowska RZ, Reinhardt R et al (2013) Targeted methylation and gene silencing of VEGF-A in human cells by using a designed Dnmt3a–Dnmt3L single-chain fusion protein with increased DNA methylation activity. J Mol Biol 425: 479–491

Singina GN, Sergiev PV, Lopukhov AV, Rubtsova MP, Taradajnic NP, Ravin NV et al (2021) Production of a cloned offspring and CRISPR/Cas9 genome editing of embryonic fibroblasts in cattle. Dokl Biochem Biophys 496:48–51. https://doi.org/10.1134/S1607672921010099

Sinkins SP, Gould F (2006) Gene drive systems for insect disease vectors. Nat Rev Genet 7:427–435

Sirk SJ, Gaj T, Jonsson A, Mercer AC, Barbas CF III (2014) Expanding the zinc-finger recombinase repertoire: directed evolution and mutational analysis of serine recombinase specificity determinants. Nucleic Acids Res 42:4755–4766

Song S, Yang Y, Liu M, Liu B, Yang X, Yu M et al (2018) MiR-125b attenuates human hepatocellular carcinoma malignancy through targeting SIRT6. Am J Cancer Res 8:993–1007

Su X, Cui K, Du S, Li H, Lu F, Shi D et al (2018) Efficient genome editing in cultured cells and embryos of Debao pig and swamp buffalo using the CRISPR/Cas9 system. Vitr Cell Dev Biol Anim:1–9

Sui T, Xu L, Lau YS, Liu D, Liu T, Gao Y et al (2018) Development of muscular dystrophy in a CRISPR-engineered mutant rabbit model with frame-disrupting ANO5 mutations. Cell Death Dis Artic 9:609. https://doi.org/10.1038/s41419-018-0674-y

Szczepek M, Brondani V, Büchel J, Serrano L, Segal DJ, Cathomen T (2007) Structure-based redesign of the dimerization interface reduces the toxicity of zinc-finger nucleases. Nat Biotechnol 25:786–793

Tabebordbar M, Zhu K, Cheng JKW, Chew WL, Widrick JJ, Yan WX et al (2016) In vivo gene editing in dystrophic mouse muscle and muscle stem cells. Science 351:407–411

Tebas P, Stein D, Tang WW, Frank I, Wang SQ, Lee G et al (2014) Gene editing of CCR5 in autologous CD4 T cells of persons infected with HIV. N Engl J Med 370:901–910
Tesson L, Usal C, Ménoret S, Leung E, Niles BJ, Remy S et al (2011) Knockout rats generated by embryo microinjection of TALENs. Nat Biotechnol 29:695–696
Thakore PI, Gersbach CA (2015) Genome engineering for therapeutic applications. In: Translating gene therapy to the clinic. Academic Press, pp 27–43
Tong Y, Charusanti P, Zhang L, Weber T, Lee SY (2015) CRISPR-Cas9 based engineering of actinomycetal genomes. ACS Synth Biol 4:1020–1029
Torres R, Martin MC, Garcia A, Cigudosa JC, Ramirez JC, Rodriguez-Perales S (2014) Engineering human tumour-associated chromosomal translocations with the RNA-guided CRISPR–Cas9 system. Nat Commun 5:1–8
Universitesi B (2020) Advance genome editing technologies in the treatment of human diseases: CRISPR therapy (Review). Int J Mol Med 46:521–534. https://doi.org/10.3892/ijmm.2020.4609
Urnov FD, Miller JC, Lee Y-L, Beausejour CM, Rock JM, Augustus S et al (2005) Highly efficient endogenous human gene correction using designed zinc-finger nucleases. Nature 435:646–651
Urnov FD, Rebar EJ, Holmes MC, Zhang HS, Gregory PD (2010) Genome editing with engineered zinc finger nucleases. Nat Rev Genet 11(9):636–646
van Diemen FR, Kruse EM, Hooykaas MJG (2016) CRISPR/Cas9-mediated genome editing of herpesviruses limits productive and latent infections. PLoS Pathog 12:e1005701. https://doi.org/10.1371/journal.ppat.1005701
Vanamee ÉS, Santagata S, Aggarwal AK (2001) FokI requires two specific DNA sites for cleavage. J Mol Biol 309:69–78
Vora S, Tuttle M, Cheng J, Church G (2016) Next stop for the CRISPR revolution: RNA-guided epigenetic regulators. FEBS J 283(17):3181–3193
Wallen MC, Gaj T, Barbas CF III (2015) Redesigning recombinase specificity for safe harbor sites in the human genome. PLoS One 10:e0139123
Waltz E (2015) Gene-edited CRISPR mushroom escapes US regulation. Nature 532:293
Wang L, Lin J, Zhang T, Xu K, Ren C, Zhang Z (2013a) Simultaneous screening and validation of effective zinc finger nucleases in yeast. PLoS One 8(5):e64687
Wang H, Hu Y-C, Markoulaki S, Welstead GG, Cheng AW, Shivalila CS et al (2013b) TALEN-mediated editing of the mouse Y chromosome. Nat Biotechnol 31:530–532
Wang T, Wei JJ, Sabatini DM, Lander ES (2014a) Genetic screens in human cells using the CRISPR-Cas9 system. Science 343:80–84
Wang Y, Cheng X, Shan Q, Zhang Y, Liu J, Gao C et al (2014b) Simultaneous editing of three homoeoalleles in hexaploid bread wheat confers heritable resistance to powdery mildew. Nat Biotechnol 32:947–951
Wang T, Birsoy K, Hughes NW, Krupczak KM, Post Y, Wei JJ et al (2015) Identification and characterization of essential genes in the human genome. Science 350:1096–1101
Wang SW, Gao C, Zheng YM, Yi L, Lu JC, Huang XY et al (2022) Current applications and future perspective of CRISPR/Cas9 gene editing in cancer. Mol Cancer 21:1–27. https://doi.org/10.1186/s12943-022-01518-8
Windbichler N, Menichelli M, Papathanos PA, Thyme SB, Li H, Ulge UY et al (2011) A synthetic homing endonuclease-based gene drive system in the human malaria mosquito. Nature 473:212–215
Wolfe SA, Nekludova L, Pabo CO (2000) DNA recognition by Cys2His2 zinc finger proteins. Annu Rev Biophys Biomol Struct 29:183–212
Wolfe SA, Grant RA, Elrod-Erickson M, Pabo CO (2001) Beyond the "recognition code": structures of two Cys2His2 zinc finger/TATA box complexes. Structure 9:717–723
Woo JW, Kim J, Kwon SI, Corvalán C, Cho SW, Kim H et al (2015) DNA-free genome editing in plants with preassembled CRISPR-Cas9 ribonucleoproteins. Nat Biotechnol 33:1162–1164
Wood AJ, Lo T-W, Zeitler B, Pickle CS, Ralston EJ, Lee AH et al (2011) Targeted genome editing across species using ZFNs and TALENs. Science 333:307

Wu J, Kandavelou K, Chandrasegaran S (2007) Custom-designed zinc finger nucleases: what is next? Cell Mol Life Sci 64(22):2933–2944

Wu Y, Liang D, Wang Y, Bai M, Tang W, Bao S et al (2013) Correction of a genetic disease in mouse via use of CRISPR-Cas9. Cell Stem Cell 13:659–662

Wu H, Wang Y, Zhang Y, Yang M, Lv J, Liu J et al (2015) TALE nickase-mediated SP110 knockin endows cattle with increased resistance to tuberculosis. Proc Natl Acad Sci 112:E1530–E1539

Xiao A, Wang Z, Hu Y, Wu Y, Luo Z, Yang Z et al (2013) Chromosomal deletions and inversions mediated by TALENs and CRISPR/Cas in zebrafish. Nucleic Acids Res 41:e141–e141

Yang G, Huang X (2019) Methods and applications of CRISPR/Cas system for genome editing in stem cells. Cell Regen 8:33–41. https://doi.org/10.1016/j.cr.2019.08.001

Yang Y, Wang L, Bell P, McMenamin D, He Z, White J et al (2016) A dual AAV system enables the Cas9-mediated correction of a metabolic liver disease in newborn mice. Nat Biotechnol 34: 334–338

Yang J, Yi N, Zhang J, He W, He D, Wu W et al (2018) Generation and characterization of a hypothyroidism rat model with truncated thyroid stimulating hormone receptor. Sci Rep 8:4004. https://doi.org/10.1038/s41598-018-22405-7

Yang Y, Xu J, Ge S, Lai L (2021) CRISPR/Cas: advances, limitations, and applications for precision cancer research. Front Med 8:1–11. https://doi.org/10.3389/fmed.2021.649896

Yin H, Xue W, Chen S, Bogorad RL, Benedetti E, Grompe M et al (2014) Genome editing with Cas9 in adult mice corrects a disease mutation and phenotype. Nat Biotechnol 32:551–553

Zalatan JG, Lee ME, Almeida R, Gilbert LA, Whitehead EH, La Russa M et al (2015) Engineering complex synthetic transcriptional programs with CRISPR RNA scaffolds. Cell 160:339–350

Zhang F, Cong L, Lodato S, Kosuri S, Church GM, Arlotta P (2011) Efficient construction of sequence-specific TAL effectors for modulating mammalian transcription. Nat Biotechnol 29: 149–153

Zhang Z, Cheng T, Lui W (2017) Suppression of Epstein-Barr virus DNA load in latently infected nasopharyngeal carcinoma cells by CRISPR/Cas9. Virus Res. https://doi.org/10.1016/j.virusres.2017.04.019

Zhao J, Lai L, Ji W, Zhou Q (2019) Genome editing in large animals: current status and future prospects. Natl Sci Rev 6:402–420. https://doi.org/10.1093/nsr/nwz013

Zhao X, Nie J, Tang Y, He W, Xiao K, Pang C et al (2020) Generation of transgenic cloned buffalo embryos harboring the EGFP gene in the Y chromosome using CRISPR/Cas9-mediated targeted integration. Front Vet Sci 7:1–14. https://doi.org/10.3389/fvets.2020.00199

Zhou Y, Zhu S, Cai C, Yuan P, Li C, Huang Y et al (2014) High-throughput screening of a CRISPR/Cas9 library for functional genomics in human cells. Nature 509:487–491

Emerging Technologies to Investigate the Potential of Gut Microbiota in Human Health

10

Sachin Kumar, Amey Jedhe, Avesh Choubey, Kalash Pandey, and Azhar Khan

Abstract

The human-gut-microbiome is the diverse microbial community consisting predominantly of bacteria although also includes fungus, viruses, protists and other organisms that have tightly coevolved with human genome and diet. Accordingly, these significant communities play a significant role in supporting human robustness as a result of coevolution of microbiome and the host. Understanding the relevance of gut microbiome in modulating host health has grabbed the interest of researchers from multiple fields. Microbiome research, which is inherently interdisciplinary, has benefitted from developments in the systems and the conventional microbiology, biomaterials engineering and synthetic biology. This chapter highlights and provides an update on various technologies in GIM research and their applications in the gastrointestinal microbiota therapy, such as NGS (Next-Generation Sequencing), Omics, Crisper, Microfluidics, Metabolomics, Metatranscriptomics, FMT (Faecal microbiota transplantation) and advanced culturing technology, with the goal of increasing interest in the validation, evaluation and eventual practices of these technologies in the diagnosis as well as therapy incorporation. Here, we will discuss the emerging technologies and their potent effects on gut microbiota analysis.

Keywords

Gut-microbiome · Human-health · Dysbiosis · Emerging technology · High-throughput sequencing · Next-generation technology · Crispr technology · Omics-technology

S. Kumar · A. Jedhe · A. Choubey · K. Pandey · A. Khan (✉)
Faculty of Applied Sciences and Biotechnology, Shoolini University of Biotechnology and Management Sciences, Solan, Himachal Pradesh, India

R. Sobti et al. (eds.), *Role of Microbes in Sustainable Development*,
https://doi.org/10.1007/978-981-99-3126-2_10

10.1 Introduction

The human gastrointestinal tract contains a staggering variety of fungus, viruses, bacteria and protists, as well as trillions of other creatures, making it one of the most complex and diverse ecosystems ever discovered. The phrase 'microbiota' refers to this group of commensals, which is often dominated by bacteria, while the term 'microbiome' refers to their collective genome. The gut microbiome plays a crucial part in the health of host, that includes but it is not restricted to the maturation of immune system and the alteration of intestinal-morphology and angiogenesis, the prevention of pathogenic infection, the fermentation of undigested polysaccharides and the synthesis and conversion of bioactive compounds (Matsuki and Tanaka 2014; Blaut 2018; Valdes et al. 2018; Pires et al. 2019).

Furthermore, the microbiota has found and is considered as a significant modulator of human health, even being proposed as an 'essential organ' of human body (Kashyap et al. 2017; Wang et al. 2017). Whereas significant alterations in microbiome composition have been observed in many disorders, identifying a distinctive makeup of a 'healthy' microbiome has been problematic with respect to inter-individual heterogeneity (Lloyd-Price et al. 2016).

Electrophoresis-based methods, such as denaturing gradient gel electrophoresis (DGGE) and temperature gradient gel electrophoresis (TGGE) and polymerase chain reaction (PCR)-based methods, such as terminal-restriction fragment length polymorphism (T-RFLP) and random amplified polymorphic DNA (RAPD), have traditionally been used for studying this microbiome independently of culture. The cytogenetic technique of fluorescence in situ hybridisation (FISH) has been used to study certain gut microbiota members, including the pathogens Listeria monocytogenes, Salmonella species, Helicobacter pylori and Yersinia enterocoliticai (Guimarães et al. 2007; Baysal 2014; Becattini et al. 2017; Prudent and Raoult 2019). Russmann et al. (2001) employed fluorescence in situ hybridisation (FISH) for the analysis of Helicobacter pylori strains isolated from patients. There are many issues with these approaches, such as a lack of resolution, specificity and sensitivity, as well as the need for highly targeted probes. Several studies have shown that the gut has a diverse and abundant microbiome, but until recently, interpreting the resulting large data was prohibitively expensive and time-consuming. However, recent improvements in sequencing and culture—or emerging technologies—have changed both that. Such technologies include NGS (Next-Generation Sequencing), Omics, Crisper, Microfluidics, Metabolomics, Metatranscriptomics, FMT (Faecal microbiota transplantation), and advanced culturing technology (Arnold et al. 2016). They have significant advantages over more traditional or older technology. Furthermore, this paper will primarily focus on emerging technology and its advantages over traditional technology and on comprehending the role of emerging technology in host-microbiome interactions (Fig. 10.1).

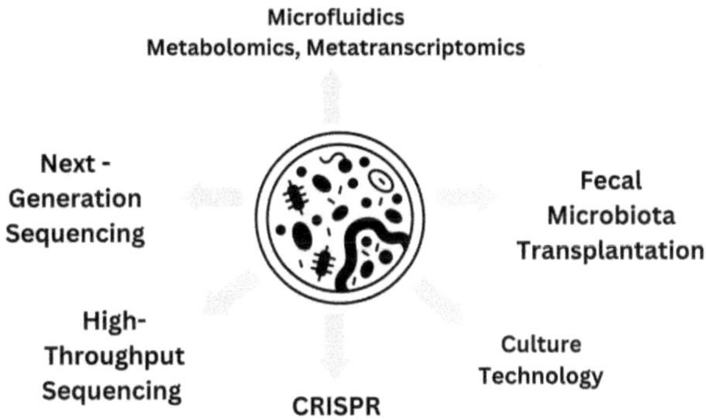

Fig. 10.1 Different Emerging Technologies in Exploring the Potential of Gut Microbiota in Human Health

10.2 The Gut Microbiome and Human Health

Humans are associated in a symbiotic-relationship with up to 10^{14} microorganisms (Savage 1977). Most of these host-specific bacteria are found in the gastrointestinal system, where they have incredible metabolic potential and are important for the maintenance of human health (De Vos and de Vos 2012). The total genetic repertory of all gut microbes represents one order of magnitude more than the genetic repertoire of the human genome (Fan and Pedersen 2020). It is also regarded as the 'essential organ' of the human body to certain extent (Ding et al. 2019). According to Kau et al. (2011), the gut microbiota influences host gene expression as well as immune response, which in turn affects general health. The gut microbiota also improves the host's response to pathogen invasion (Sobhani et al. 2011; Carding et al. 2015; Ramakrishna et al. 2015). Normal gastrointestinal tract residents aid in metabolism of the polysaccharides taken by host (Tremaroli and Bäckhed 2012), as well as interactions between bacteria within the microbiome improve this metabolic capacity, further increasing polysaccharide consumption (Gill et al. 2006).

Furthermore, particularly on association with the host, gut microbiota can produce a variety of metabolic products that can have an impact on human health, either positively or negatively. Indigestible carbohydrates such as hemi-cellulose, cellulose, pectin, resistant starch, lignin and oligosaccharides could be converted into short-chain fatty acids (SCFAs), i.e. propionic, butyric and acetic acids by these bacteria. These fatty acids enter the colon after escaping from the upper gastrointestinal tract during digestion (Lin and Zhang 2017; Thursby and Juge 2017). There are numerous pathogenic effects on the host whenever the production of short-chain fatty acids is disrupted (Perry et al. 2016). The gut microbiota plays a crucial part in manufacture of vitamins including thiamine, biotin, riboflavin, cobalamin, pantothenic acids and nicotine as well as vitamin B & K, thus can also have positive impacts on the host organism (LeBlanc et al. 2013). Additionally, the gut microbiota has peculiar ability to produce certain neurochemicals that can impact the peripheral and central neurological systems (Forsythe et al. 2010). In addition to promoting health, gut microbes can prevent disease by modulating the immune system (Medina

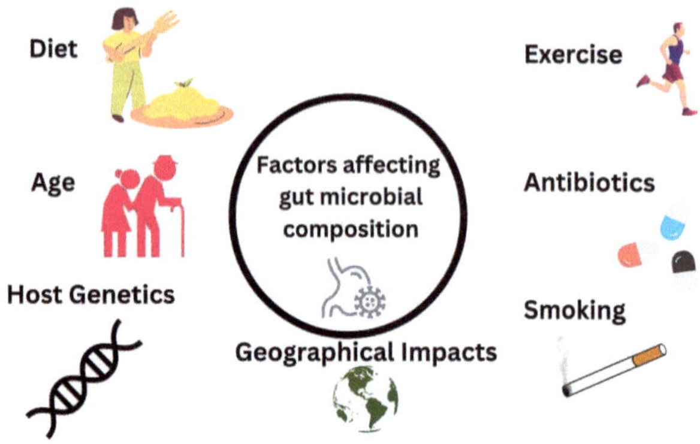

Fig. 10.2 Factors affecting gut microbial composition

et al. 2007). As demonstrated for Bifidobacterium longum, which significantly increases interleukin-10 and proinflammatory cytokines like TNF-production (Medina et al. 2007) that guards against tumour growth in the host (Lee et al. 2008). It is hypothesised that physiological changes in the colon and small intestine, such as nutritional and chemical gradients as well as isolated host immunological activity, impact the make-up of the bacterial communities (Donaldson et al. 2015). Further, a vital role of the interior environment of humans is played by the gut bacteria.

Several additional factors, such as nutrition, host genetics, age, exercise, use of antibiotics, smoking and geographic influences, all alter the composition of the gut microbes (Bäckhed et al. 2015; Schanche et al. 2015; Chen et al. 2018; Clarke et al. 2014; Biedermann et al. 2013; Ramnani et al. 2012). Additionally, similar compositional alterations in the microbiota (dysbiosis) have been linked to a number of illnesses, including obesity (Shen et al. 2013), diabetes (Naseer et al. 2014), colorectal cancer (Azcárate-Peril et al. 2011) and the allergies (Panzer and Lynch 2015) (Fig. 10.2).

10.3 NGS (Next-Generation Sequencing) Technology

Analysing the gut microbiota in the past relied on isolation and culture, but the accuracy of the research was severely hampered by the challenge of growing anaerobic bacteria, that are prevalent in the intestine. Research on the intestinal microbiome has recently been drawn to the development of next-generation sequencing (NGS), that could precisely assess microbial components without culture (Tang et al. 2020).

10.4 Fundamental Considerations in the Use of NGS

When examining the gut microbiota, one of the first questions to ask is which microorganisms are present in a particular sample. Finding the plethora and the functional profiles of the microorganisms present, as well as comprehending intra-species and population heterogeneity, are further questions that can be answered by NGS analysis (Durazzi et al. 2021). In order to answer these queries, NGS techniques directly sequence microbial DNA or RNA, for instance, from faecal, blood and/or tissue samples. Amplicon sequencing and shotgun metagenomic sequencing are the two main NGS methodologies currently in use, because of NGS's decreasing cost. However, because it enables the determination of the transcriptome, which is an additional step for characterising the function of the microbiota, RNA sequencing is also a valid and, in some respects, superior method for classifying microbes (Cottier et al. 2018; Clooney et al. 2016). NGS platforms are often used and come in a number of configurations. Among these are the Roche 454 GS FLX, Oxford Nanopore, Illumina (MiSeq and HiSeq), Ion Torrent/Ion Proton/Ion Proton and SOLiD 5500 series (Malla et al. 2019).

10.5 16S rRNA Gene Amplicon Sequencing

Through use of the metagenomic techniques and high-throughput sequencing technologies, the gut microbiome has been thoroughly studied. In order to analyse the diversity, community structure and functionality of microbial species, metagenomics involves the sequencing of the entire community's DNA (Albenberg and Kelsen 2016). To determine the microbial make-up of a community in an environment like the gut, bioinformatics and the hypervariable region of 16S rRNA gene sequencing have been extensively used. The gut microbiome of persons living in Amazon was characterised by 16S rDNA Illumina sequencing (Pires et al. 2019), which revealed a significant variance in composition when compared to individuals living in industrialised environments. Similar to this, Barone et al. (2019) used data from the 16S rRNA gene sequencing to understand gut microbiota response to a contemporary Palaeolithic diet in a setting of a Western lifestyle. The 16S rRNA gene is a perfect target because it is widely distributed and highly conserved among bacteria (without that, bacteria would not be able to translate mRNA in to the proteins and would therefore be non-functional), as well as because it has nine hypervariable regions (V1–V9) that vary within different bacterial species as well as genera. As a result, it is possible to construct PCR primers so that the forward and reverse primers bind to conserved areas but amplify an intervening variable region (Wensel et al. 2022). With the help of this technique, it is no longer necessary to cultivate individual bacteria, clone certain genes, or blot for a particular RNA in order to identify community members (Arnold et al. 2016). The biased nature of the databases used for comparisons is a significant flaw in this approach (Ajayi et al. 2020).

10.6 Whole-Genome Shotgun (WGS)

By enhancing the knowledge acquired by 16S/18S rDNA amplicon sequencing, whole-genome shotgun (WGS) sequencing can identify DNA viruses and reveal details about the composition of genes and metabolic pathways (Palmero et al. 2010). Bacteriophages, which are primarily bacterial DNA viruses, predominate in the gut virome, which also contains a varied population of eukaryotic viruses with both DNA and RNA encoding (Reyes et al. 2015, 2010). By influencing the bacterial ecology and interacting directly with host cells, the virome has a significant impact on host health (Reyes et al. 2015; Focà et al. 2015). However, as the majority of current findings which are based on the 16S/18S rRNA amplicon sequencing data, virome data are frequently left out of microbiome compositional investigations. By destroying their bacterial hosts while lytic growth or by changing gene expression while the lysogenic conversion (Mills et al. 2013), bacteriophages can modify the composition of the microbiome. The genes engaged in DNA replication, amino acid, carbohydrate, lipid metabolism, signal transduction as well as transcription control have also been found to be encoded by the eukaryotic viruses as bacteriophages in the gut (Reyes et al. 2015; Focà et al. 2015).

To obtain the right gut microbiome samples for NGS, though, is essential. The intestinal microbiome cannot be accurately represented by the sampling techniques currently used to gather samples from intestinal aspiration, faeces and mucosal biopsy, all of that which may have certain flaws (Tang et al. 2020).

10.7 Omics Technology in Gut Microbiota

'Meta-omics' approaches offer a way to investigate and comprehend the systems biology of the gut microbiome at many stages of expression (Lamendella et al. 2012). Here, we will talk about several cutting-edge meta-omics techniques used in the human digestive system. The study of complex human diseases has been revolutionised by the development of modern 'omics' methodologies and techniques that offer an unprecedented genome-wide perspective of genetic diversity, gene expression, interactions with microbes and environmentally responsive epigenetic alterations (Donlin et al. 2019; Nemtsova et al. 2019; Kishikawa et al. 2019).

10.8 Metagenomics

It refers to the environment's whole-community DNA being sequenced without being targeted (Escobar-Zepeda et al. 2015). Shotgun sequencing is frequently used to profile the taxonomic composition of a sample, such as faeces, with a diverse microbial community (down to the strain level) and to evaluate the functional potential of the sample. Large-scale studies of complex microbiomes have been made possible by metagenomics, which has also helped to clarify functional variations between the states of health and sickness. In addition to characterising non-bacterial microbial communities including fungi and viruses that have recently been revealed to possibly play a significant influence in host health, it enables strain-level resolution of gut bacteria (Gilbert and Dupont 2011; Oulas et al. 2015). By concurrently examining two facets of a microbial community—who is present and what they might be able to do—metagenomics offers the chance to learn more about both. Though effective, this method has a lot of drawbacks: In comparison, it is substantially more expensive than 16S rRNA gene sequencing. Additionally, there are numerous bacterial genomes that have not yet been fully annotated, and there are concerns about the correctness and even coverage of databases (Segal et al. 2019).

10.9 Metatranscriptomics

In metatranscriptomics, RNA sequencing is used to examine the transcriptional activity of microbiota (Aggarwal et al. 2022). While metagenomics outlines the community's microbiota's genetic potential, metatranscriptomics provides information about the actual genetic endeavour within a community phenotype and the genomic potential of a community's microbiota (Segal et al. 2019). Metatranscriptomics, as opposed to metagenomics, enables the detection of active microorganisms, genes and the associated pathways in microbial communities (Aggarwal et al. 2022). In the human microbiota, metatranscriptomics techniques have facilitated a deeper comprehension of host–microbiota interactions, active microbiota and their pathways and expression alterations in disease progression (Nowicki et al. 2018; Schirmer et al. 2018). As a result, the metatranscriptome provides dynamic, contextualises microbial functional activity to the human phenotype and, when combined with metagenomics, offers a profound understanding of the molecular pathways by which gut bacteria contribute to both health and illness (Bashiardes et al. 2016; Lavelle and Sokol 2018). It has tremendous utility in reorienting our knowledge of the descriptive gut microbiome towards a deeper comprehension of host-microbial causative pathways in causing disease and homeostasis (Segal et al. 2019). The field of metatranscriptomics has a number of significant limitations. Host contamination can be found in substantial amounts in tissues like colonic biopsies, when the biomass is composed almost entirely of host cells. Deep sequencing of the full mRNA is required in such scenarios to establish a representative insight into the mucosally adherent microbial pattern of gene expression. The translated protein or microbial transcriptome databases are incomplete and contain many genes that have not yet been assigned a recognised function. The microbial functional profile is frequently interpreted insufficiently and somewhat biasedly as a result of this information gap, although this is likely to alter as this field develops over time (Segal et al. 2019).

10.10 Metaproteomics

Alternately, metaproteomics can be used as substitutive method of observing gene activity in the microbiome. In a 2009 study, metaproteomics was first used to assess microbial function in ambient and gut microbiota samples from twins (Wilmes and Bond 2006; Verberkmoes et al. 2008). Numerous research have so far shown how human microbiome samples can be used for metaproteomics analysis (Issa Isaac et al. 2019; Long et al. 2020; Tanca et al. 2017). Since it has lower throughput than metatranscriptomics deep sequencer-based analysis, metaproteomics is not as popular. Metaproteomics can also provide information on post-translational modifications in proteins and the expression of proteins released from the host cell, even though metatranscriptomics is unable to explain these circumstances (Zhang et al. 2017). The inadequacy and a typical study's millions of tandem mass spectra (MS/MS) were produced with inadequate validation of the anticipated protein databases utilised for peptide matching (Lamendella et al. 2012).The ability of post-translational changes to affect microbiome function without changing their composition underlies the significance of metaproteomics in gaining mechanistic insights into the phenotypes connected to the microbiome. Multimeta-omics approaches have been used to thoroughly recognise the gene activity of the microbiota as well as interconnections between the microbiota and the host, taking into account the benefits and drawbacks of meta-omics approaches (Aggarwal et al. 2022).

10.11 CRISPR

According to Hille and Charpentier (2016), CRISPR-Cas is currently the sole adaptive immune system in prokaryotes. Although now it is generally acknowledged as a genetics tool, CRISPR was first identified in archaea as an immune system. Through the introduction of DNA breaks and homologous recombination that use donor DNAs, CRISPR is primarily used for gene editing (Aggarwal et al. 2022). Numerous organisms, even those whose genomes were before thought to be difficult to edit, have experienced an acceleration in genome engineering due to CRISPR-directed homologous recombination (Reardon 2019). In order to characterise the gene function of phenotypes connected to the microbiome, CRISPR has been further utilised to create the microbiome as well as commensal bacteria. Despite the widespread availability of CRISPR-driven gene editing for many organisms, most commensal bacteria with low homologous recombination activity experience cell death as a result of DNA breakage brought on by CRISPR/Cas9 rather than gene editing. Therefore, CRISPR/Cas9 cannot be applied to bulk of the commensal bacteria that are non-models. CRISPRi, CRISPRa, or base editors may be less harmful options for these microorganisms. CRISPRi and CRISPRa can be applied as customised transcription factors for building genomic circuits because of their great degree of programmability. Because of its lesser toxicity when compared to utilising bacteria, microbiome editing with base editors is likely to soon be used in therapies, unlike CRISPR/Cas9 gene editing, which introduces changes through

DNA strand breaks as well as subsequent homologous recombination. A variety of genomic DNA can be modified with CRISPR (Aggarwal et al. 2022). Despite the fact that a significant portion of commensal microorganisms are not cultivable, DNA delivery is nevertheless the very first step in experimental modification for downstream processes.

10.12 Faecal Microbiota Transplantation (FMT)

FMT is the most avant-garde therapy strategy (Quaranta et al. 2019). FMT involves injecting a healthy donor's faeces suspension into the patient's intestinal tract to cure a specific disorder linked to altered gut microbiota (Cammarota et al. 2017; Filip et al. 2018). Regardless of how FMT is administered, there is sufficient evidence to conclude that it is a highly effective treatment option for a number of intestinal illnesses, with the capacity to restore gut microbiota compositions and functions that are identical to those of recipients (Li et al. 2016). FMT can also be utilised to treat other extra-intestinal disorders caused by altered microbiota. Colorectal cancer, Parkinson's disease, atherosclerosis, coronary artery disease (CAD), rheumatoid arthritis, multiple sclerosis, irritable bowel syndrome, insulin resistance, obesity, autism, diarrhoea, allergic disorders, metabolic syndrome, colon cancer, anti-tumour immunity and neuropsychiatric disorders are a few clinical conditions for which FMT may be a potent therapeutic strategy (Holvoet et al. 2017; Johnsen et al. 2018; Aroniadis et al. 2018; Quaranta et al. 2019). In contrast to probiotics and prebiotics, whose colonisation appears to be temporary, it can be formed in a single-dose regimen, providing therapeutic potential, and it promotes microbial diversity without upsetting microbial gut ecology, which is employed in antibiotic treatment (Weingarden and Vaughn 2017). It is unclear exactly how FMT works to cure certain disorders. It may be caused by changes in the bacterial compositions, altered metabolic profiles of the hosts, the presence of donor-derived peptides that alter host immune responses and the participation of novel species of gut microbiota found in the healthy donor faeces (Gianotti and Moss 2017). Despite all of these benefits, there are still a lot of unfavourable side effects and challenges that this trend must overcome. It has been demonstrated that after therapy, the microbiota of the treated individuals resembles that of the donor. FMT's safety issue originates from the intricacy of the faeces microbial community, which is another drawback (Hansen and Sartor 2015). Numerous studies have demonstrated the potential for the transmission of microorganism-based infections or detrimental disease phenotypes such as metabolic syndromes, diabetes, obesity, and chronic cardiovascular diseases (Harsch and Konturek 2019).

10.13 Microfluidic

The gastrointestinal microbiota may be traced, examined and controlled at the single-cell level due to microfluidics technology (Ajayi et al. 2020). Organ Chips are basically microfluidic cell culture tools that were initially created utilising techniques modified from the production of computer microchips (e.g. soft lithography), they imitate tissue- and organ-level physiology by having constantly perfused chambers filled with living cells (Bhatia and Ingber 2014). Liu and Walther_Antonio discovered two potent microfluidic tool that could be harnessed in sorting of cells, cell screening, cell culture, metabolic screening/analysis, gene expression and genome applications. This approach made it possible to thoroughly examine particular bacterial species and determine how they assist keeping the integrity of the gastrointestinal tract (Ajayi et al. 2020). The establishment of organs on chip is yet another intriguing advancement in microfluidics for microbiome research. Additionally, the intestine is the location where majority of the commensal microbes in the gut microbiome reside and communicate with the host immune machinery and gut lymphoid tissues, which greatly aids in maintaining intestinal homeostasis (Garrett et al. 2010; Round and Mazmanian 2009).

10.14 Microfluidic Intestine Chip Models

The support for laminar fluid flow is provided by microfluidic devices with hollow micro-channels below 1 mm in width and fluid volume management from nanolitre to microlitre scales, thus making them feasible for use in living cell culture. The fluidic control also allows for a strictly regulated spatiotemporal allocation of nutrients, growth factors, drugs, or even sometime toxins to the intestinal epithelium developed over the microfluidic channels (Bein et al. 2018). A common porous polycarbonate or polyester membrane with ECM coating that has one of its surfaces cultivated with immortalised human intestinal epithelial cells separates two hollow channels that make up Intestine Chips (Gao et al. 2013). The solid polymer substance that obstructed the epithelium's abluminal surface prevented this design from allowing investigation of intestinal barrier function. Additionally, the HuMiX multichannel intestinal chip has been explained which uses nanoporous membranes to divide layers of intestinal Caco-2 epithelium from a luminal microbial compartment (Shah et al. 2016).

10.15 Mechanically Active Gut Chip Model

A more refined two-channel microfluidic Gut Chip prototype has been designed that permits human intestinal-epithelium to flourish and coexist along with immune cells, capillary endothelium and even the commensal microbial cells to develop, cohabit and communicate while in vitro undergoing peristalsis-like mechanical deformations and physiologically relevant fluid flow (Kim et al. 2015). Polydimethylsiloxane (PDMS), a gas-permeable, flexible, silicone polymer is used to design the Gut Chip which is crystal clear, so that it enables imaging at high-resolution using differential interference contrast, phase contrast or immunofluorescence-confocal-microscopy. It is bordered on either side by hollow, full-height side on chambers and has two parallel microchannels (<1 mm-wide) that are separated from one another by thin (w20 mm), flexible, porous, ECM-coated PDMS membrane (Kim et al. 2012, 2015; Kim and Ingber 2013; Huh et al. 2013). It is significant to note that coculture of living commensal microbes is also possible since the Gut Chip maintains continuous fluid flow, villi creation and mucus production (e.g. Lactobacillus rhamnosus GG) (Kim et al. 2012).

The fundamental application of microfluidics is to research bacterial cells' real-time susceptibility to antibiotics (Cama et al. 2020). Additionally, it has clinical applications for diagnosis, drug delivery, studying the pathophysiology of gastrointestinal illnesses and personalised or individualised medicine (Ajayi et al. 2020).

10.16 Advanced Culturing Techniques

With the development of various bacterial culture techniques over time, it is now feasible to cultivate a sizable number of hitherto uncultivated gut bacteria (Ajayi et al. 2020). Culturomics is one such cultural method. According to Lagier et al. (2016), culturomics is a method of culturing that uses various culture conditions along with 16S rRNA gene amplification/sequencing and matrix-assisted laser desorption/ionisation time of flight (MALDI-TOF) for identification. High-throughput culture techniques offer the clear advantage of improving the culturability of bacterial populations that would otherwise be 'non-culturable', allowing for a more thorough investigation of identified species. This method requires specialised laboratories, takes a lot of time and is highly complex. Additionally, this method can be helpful in the preparation and administration of probiotics. Only a small portion of the microorganisms that live in the gut can be cultured. Despite the fact that recent studies using gnotobiotic mice and anaerobic culturing methods were able to successfully culture 50% of the species of bacteria identified by 16S rDNA amplicon sequencing, covering nearly 70% known genera and >90% families (Goodman et al. 2011; Faith et al. 2010), the majority of the diversity present inside the gut microbiota is at the strain level, making identification and cultivation a challenging task. Additionally, individual microorganisms' morphology, physiology and biochemistry may be researched, and it is simple to assess how they react to or interact with medications. This makes it possible to treat gut disorders effectively. Traditional microbiology approaches have been advanced by advances in culturing technology, including the use of anaerobic environments and gnotobiotic animals. Since so many initially uncultivable bacteria may now be grown in environments created to mimic their natural growth circumstances, it is possible to isolate hitherto undescribed species (Connon and Giovannoni 2002). Additionally, improvements in culture control technology have made it possible to trigger gastrointestinal (GIT) parameters such as acidity and bile content (Adamberg et al. 2014). High-throughput culturing is now achievable due to new culturing technology and the knowledge offered by NGS (Connon and Giovannoni 2002).

10.17 Future and Conclusion

Our inability to culture majority of gut microorganisms, the fact that most of these bacteria are novel and lack any closely related previously cultivated strains and the lack of practical biomarkers of the microbiome functioning in body have all severely limited traditional studies on the exploration potential of the human gut microbiota. An emerging perspective is being created on the role that our gut microbiome plays in human systems biology thanks to recent developments in tools like NGS (next-generation sequencing), Omics, Crispr, Microfluidics, Metabolomics, Metatranscriptomics as well as FMT (faecal microbiota transplantation) and advanced culturing technology. To better understand these host-microbe interactions and to promote human health, these techniques give us better knowledge and information.

New gut microbiome research endeavours are made possible by these technical developments taken together. The gut microbiome is a key regulator of human health and will continue to draw researchers among a variety of scientific disciplines. At the same time, technological advances across many scientific disciplines will abide to give us tools we need to further unveil the potential of the gut microbiome as a target for personalised medicine. Additionally, this new technology will undoubtedly help us gain a clearer knowledge of gut microbial dysbiosis, which will help to lessen the burden this condition puts on human health.

References

Adamberg S, Sumeri I, Uusna R, Ambalam P, Kondepudi KK, Adamberg K, et al. Survival and synergistic growth of mixed cultures of bifidobacteria and lactobacilli combined with prebiotic oligosaccharides in a gastrointestinal tract simulator. Microb Ecol Health Dis 2014;25

Aggarwal N, Kitano S, Puah GR, Kittelmann S, Hwang IY, Chang MW (2022) Microbiome and human health: current understanding, engineering, and enabling technologies. Chem Rev 123:31

Ajayi A, Jolaiya T, Smith S (2020) Evolving technologies in gastrointestinal microbiome era and their potential clinical applications. J Clin Med 9(8):2565

Albenberg L, Kelsen J (2016) Advances in gut microbiome research and relevance to pediatric diseases. J Pediatr 178:16–23

Arnold JW, Roach J, Azcarate-Peril MA (2016) Emerging technologies for gut microbiome research. Trends Microbiol 24:887–901

Aroniadis O, Brandt L, Oneto C, Feuerstadt P, Sherman A, Wolkoff A, Downs I, Zanetti A, Ramos Y, Cotto C et al (2018) A double-blind, randomized, placebo-controlled trial of fecal microbiota transplantation capsules (FMTC) for the treatment of diarrhea-predominant irritable bowel syndrome (IBS-D). Gastroenterology 154(6):154–155

Azcárate-Peril MA, Sikes M, Bruno-Bárcena JM (2011) The intestinal microbiota, gastrointestinal environment and colorectal cancer: a putative role for probiotics in prevention of colorectal cancer? Am J Physiol Gastrointest Liver Physiol 301(3):G401

Bäckhed F, Roswall J, Peng Y, Feng Q, Jia H, Kovatcheva-Datchary P et al (2015) Dynamics and stabilization of the human gut microbiome during the first year of life. Cell Host Microbe 17(5):690–703

Barone M, Turroni S, Rampelli S, Soverini M, D'Amico F, Biagi E et al (2019) Gut microbiome response to a modern Paleolithic diet in a western lifestyle context. PLoS One 14(8):e0220619

Bashiardes S, Zilberman-Schapira G, Elinav E (2016) Use of metatranscriptomics in microbiome research. Bioinform Biol Insights 10:19–25
Baysal AH (2014) Comparison of conventional culture method and fluorescent in situ hybridization technique from detection of Listeria Spp. In ground beef, Turkey and chicken breast fillets in Izmir, Turkey. J Food Prot 77:2021–2030
Becattini S, Littmann ER, Carter RA, Kim SG, Morjaria SM, Ling L, Gyaltshen Y, Fontana E, Taur Y, Leiner IM et al (2017) Commensal microbes provide first line defence against listeria monocytogenes infection. J Exp Med 214:1973–1989
Bein A et al (2018) Microfluidic organ-on-a-chip models of human intestine. Cell Mol Gastroenterol Hepatol 5(4):659–668
Bhatia SN, Ingber DE (2014) Microfluidic organs-on-chips. Nat Biotechnol 32(8):760–772. https://doi.org/10.1038/nbt.2989
Biedermann L, Zeitz J, Mwinyi J, Sutter-Minder E, Rehman A, Ott SJ et al (2013) Smoking cessation induces profound changes in the composition of the intestinal microbiota in humans. PLoS One 8(3):e59260
Blaut M (2018) Composition and function of the gut microbiome. In: The gut microbiome in health and disease. Springer, Berlin, pp 5–30
Cama J et al (2020) Single-cell microfluidics facilitates the rapid quantification of antibiotic accumulation in gram-negative bacteria. Lab Chip 20(15):2765–2775
Cammarota G, Ianiro G, Tilg H, Rajilić-Stojanović M, Kump P, Satokari R et al (2017) European consensus conference on faecal microbiota transplantation in clinical practice. Gut 66(4): 569–580
Carding S, Verbeke K, Vipond DT, Corfe BM, Owen LJ. Dysbiosis of the gut microbiota in disease. Microb Ecol Health Dis 2015;26
Chen C, Huang X, Fang S, Yang H, He M, Zhao Y et al (2018) Contribution of host genetics to the variation of microbial composition of cecum lumen and feces in pigs. Front Microbiol 9:2626
Clarke SF, Murphy EF, O'Sullivan O, Lucey AJ, Humphreys M, Hogan A et al (2014) Exercise and associated dietary extremes impact on gut microbial diversity. Gut 63(12):1913–1920
Clooney AG, Fouhy F, Sleator RD, O'Driscoll A, Stanton C, Cotter PD et al (2016) Comparing apples and oranges?: next generation sequencing and its impact on microbiome analysis. PLoS One 11(2):e0148028
Connon SA, Giovannoni SJ (2002) High-throughput methods for culturing microorganisms in very-low-nutrient media yield diverse new marine isolates. Appl Environ Microbiol 68(8): 3878–3885
Cottier F, Srinivasan KG, Yurieva M, Liao W, Poidinger M, Zolezzi F, Pavelka N (2018) Advantages of meta-total RNA sequencing (MeTRS) over shotgun metagenomics and amplicon-based sequencing in the profiling of complex microbial communities. NPJ Biofilms Microbiomes. 4(1):1–7
De Vos W, de Vos E (2012) Role of the intestinal microbiome in health and disease: from correlation to causation. Nutr Rev 70:45–56
Ding R-X, Goh W-R, Wu R-N, Yue X-Q, Luo X, Khine WW et al (2019) Revisit gut microbiota and its impact on human health and disease. J Food Drug Anal 27(3):623–631
Donaldson GP, Lee SM, Mazmanian SK (2015) Gut biogeography of the bacterial microbiota. Nat Rev Microbiol 14(1):20–32
Donlin LT, Park S-H, Giannopoulou E, Ivovic A, Park-Min K-H, Siegel RM et al (2019) Insights into rheumatic diseases from next-generation sequencing. Nat Rev Rheumatol 15(6):327–339
Durazzi F, Sala C, Castellani G, Manfreda G, Remondini D, De Cesare A (2021) Comparison between 16S rRNA and shotgun sequencing data for the taxonomic characterization of the gut microbiota. Sci Rep 11(1):3030
Escobar-Zepeda A, de León AV-P, Sanchez-Flores A (2015) The road to metagenomics: from microbiology to DNA sequencing technologies and bioinformatics. Front Genet 6:6

Faith JJ, Rey FE, O'onnell D, Karlsson M, McNulty NP, Kallstrom G et al (2010) Creating and characterizing communities of human gut microbes in gnotobiotic mice. ISME J 4(9): 1094–1098

Fan Y, Pedersen O (2020) Gut microbiota in human metabolic health and disease. Nat Rev Microbiol 19(1):55–71

Filip M, Tzaneva V, Dumitrascu DL (2018) Fecal transplantation: digestive and extradigestive clinical applications. Med Pharm Rep 91(3):259–265

Focà A, Liberto MC, Quirino A, Marascio N, Zicca E, Pavia G (2015) Gut inflammation and immunity: what is the role of the human gut virome? Mediat Inflamm 2015:1–7

Forsythe P, Sudo N, Dinan T, Taylor VH, Bienenstock J (2010) Mood and gut feelings. Brain Behav Immun 24(1):9–16

Gao D et al (2013) Characterization of drug permeability in Caco-2 monolayers by mass spectrometry on a membrane-based microfluidic device. Lab Chip 13(5):978

Garrett WS, Gordon JI, Glimcher LH (2010) Homeostasis and inflammation in the intestine. Cell 140(6):859–870. https://doi.org/10.1016/j.cell.2010.01.023

Gianotti R, Moss A (2017) Fecal microbiota transplantation from Clostridium difficile to inflammatory bowel disease. Gastroenterol Hepatol 13:209–213

Gilbert JA, Dupont CL (2011) Microbial metagenomics: beyond the genome. Annu Rev Mar Sci 3(1):347–371

Gill SR, Pop M, DeBoy RT, Eckburg PB, Turnbaugh PJ, Samuel BS et al (2006) Metagenomic analysis of the human distal gut microbiome. Science 312(5778):1355–1359

Goodman AL, Kallstrom G, Faith JJ, Reyes A, Moore A, Dantas G et al (2011) Extensive personal human gut microbiota culture collections characterized and manipulated in gnotobiotic mice. Proc Natl Acad Sci 108(15):6252–6257

Guimarães N, Azevedo NF, Figueiredo C, Keevil CW, Vieira MJ (2007) Development and application of a novel peptide nucleic acid probe for the specific detection of *helicobacter pylori* in gastric biopsy specimens. J Clin Microbiol 45(9):3089–3094

Hansen J, Sartor R (2015) Therapeutic manipulation of the microbiome in IBD: current results and future approaches. Curr Treat Option Gastroenterol 13:105–120

Harsch I, Konturek P (2019) Adhesion ileus after fecal microbiota transplantation in longstanding radiation colitis. Case Rep Gastroint Med 6:1–4

Hille F, Charpentier E (2016) CRISPR-Cas: biology, mechanisms and relevance. Philos Trans R Soc Lond B Biol Sci 371(1707):20150496

Holvoet T, Joossens M, Wang J, Boelens J, Verhasselt B, Laukens D et al (2017) Assessment of faecal microbial transfer in irritable bowel syndrome with severe bloating. Gut 66(5):980–982

Huh D et al (2013) Microfabrication of human organs-on-chips. Nat Protoc 8(11):2135–2157. https://doi.org/10.1038/nprot.2013.137

Issa Isaac N, Philippe D, Nicholas A, Raoult D, Eric C (2019) Metaproteomics of the human gut microbiota: challenges and contributions to other omics. Clin Mass Spectrom 14:18–30

Johnsen PH, Hilpüsch F, Cavanagh JP, Leikanger IS, Kolstad C, Valle PC et al (2018) Faecal microbiota transplantation versus placebo for moderate-to-severe irritable bowel syndrome: a double-blind, randomised, placebo-controlled, parallel-group, single-Centre trial. Lancet Gastroenterol Hepatol 3(1):17–24

Kashyap PC, Chia N, Nelson H, Segal E, Elinav E (2017) Microbiome at the frontier of personalized medicine. Mayo Clin Proc 92(12):1855–1864

Kau AL, Ahern PP, Griffin NW, Goodman AL, Gordon JI (2011) Human nutrition, the gut microbiome and the immune system. Nature 474(7351):327–336

Kim HJ, Ingber DE (2013) Gut-on-a-chip microenvironment induces human intestinal cells to undergo villus differentiation. Integrat Biol 5(9):1130. https://doi.org/10.1039/c3ib40126j

Kim HJ et al (2012) Human gut-on-a-chip inhabited by microbial flora that experiences intestinal peristalsis-like motions and flow. Lab Chip 12(12):2165. https://doi.org/10.1039/c2lc40074j

Kim HJ et al (2015) Contributions of microbiome and mechanical deformation to intestinal bacterial overgrowth and inflammation in a human gut-on-a-chip. Proc Natl Acad Sci 113(1):E7. https://doi.org/10.1073/pnas.1522193112

Kishikawa T, Maeda Y, Nii T, Motooka D, Matsumoto Y, Matsushita M et al (2019) Metagenome-wide association study of gut microbiome revealed novel aetiology of rheumatoid arthritis in the Japanese population. Ann Rheum Dis 79(1):103–111

Lagier J-C, Khelaifia S, Alou MT, Ndongo S, Dione N, Hugon P et al (2016) Culture of previously uncultured members of the human Gut Microbiota by Culturomics. Nat Microbiol 1(12):16203

Lamendella R, VerBerkmoes N, Jansson JK (2012) 'Omics' of the mammalian gut – new insights into function. Curr Opin Biotechnol 23(3):491–500

Lavelle A, Sokol H (2018) Beyond metagenomics, metatranscriptomics illuminates microbiome functionality in IBD. Nat Rev Gastroenterol Hepatol 15(4):193–194

LeBlanc JG, Milani C, de Giori GS, Sesma F, van Sinderen D, Ventura M (2013) Bacteria as vitamin suppliers to their host: a gut microbiota perspective. Curr Opin Biotechnol 24(2): 160–168

Lee DK, Jang S, Kim MJ, Kim JH, Chung MJ, Kim KJ et al (2008) Anti-proliferative effects of Bifidobacterium adolescentis SPM0212 extract on human colon cancer cell lines. BMC Cancer 8(1):310

Li SS, Zhu A, Benes V, Costea PI, Hercog R, Hildebrand F et al (2016) Durable coexistence of donor and recipient strains after fecal microbiota transplantation. Science 352(6285):586–589

Lin L, Zhang J (2017) Role of intestinal microbiota and metabolites on gut homeostasis and human diseases. BMC Immunol 18:837–850

Lloyd-Price J, Abu-Ali G, Huttenhower C (2016) The healthy human microbiome. Genome Med 8(1):51

Long S, Yang Y, Shen C, Wang Y, Deng A, Qin Q et al (2020) Metaproteomics characterizes human gut microbiome function in colorectal cancer. NPJ Biofilms Microbiomes 6(1):14

Malla MA, Dubey A, Kumar A, Yadav S, Hashem A, Abd Allah EF (2019) Exploring the human microbiome: the potential future role of next-generation sequencing in disease diagnosis and treatment. Front Immunol 9:2868

Matsuki T, Tanaka R (2014) Function of the human gut microbiota. In: The human microbiota and microbiome. CABI Publishing, Cardi, pp 90–106

Medina M, Izquierdo E, Ennahar S, Sanz Y (2007) Differential immunomodulatory properties of *bifidobacteriumlogum* strains: relevance to probiotic selection and clinical applications. Clin Exp Immunol 150(3):531–538

Mills S et al (2013) Movers and shakers: influence of bacteriophages in shaping the mammalian gut microbiota. Gut Microbes 4(1):4–16

Naseer M, Bibi F, Alqahtani M, Chaudhary A, Azhar E, Kamal M et al (2014) Role of gut microbiota in obesity, type 2 diabetes and Alzheimer's disease. CNS Neurol Disord Drug Targets 13(2):305–311

Nemtsova MV, Zaletaev DV, Bure IV, Mikhaylenko DS, Kuznetsova EB, Alekseeva EA et al (2019) Epigenetic changes in the pathogenesis of rheumatoid arthritis. Front Genet 10:570

Nowicki EM, Shroff R, Singleton JA, Renaud DE, Wallace D, Drury J et al (2018) Microbiota and metatranscriptome changes accompanying the onset of gingivitis. MBio 9(2):e00575-18

Oulas A, Pavloudi C, Polymenakou P, Pavlopoulos GA, Papanikolaou N, Kotoulas G et al (2015) Metagenomics: tools and insights for analyzing next-generation sequencing data derived from biodiversity studies. Bioinform Biol Insights 9:75–88

Palmero D, Rodríguez JM, de Cara M, Camacho F, Iglesias C, Tello JC (2010) Fungal microbiota from rain water and pathogenicity of fusarium species isolated from atmospheric dust and rainfall dust. J Ind Microbiol Biotechnol 38(1):13–20

Panzer AR, Lynch SV (2015) Influence and effect of the human microbiome in allergy and asthma. Curr Opin Rheumatol 27(4):373–380

Perry R, Peng L, Barry N, Cline G, Zhang D, Cardone R, Petersen K, Kibbey R, Goodman A, Shulman G (2016) Acetate mediates a microbiome-brain-b-cell axis to promote metabolic syndrome. Nature 534:213–217

Pires ES, Hardoim CC, Miranda KR, Secco DA, Lobo LA, de Carvalho DP et al (2019) The gut microbiome and metabolome of two riparian communities in the Amazon. Front Microbiol 10: 2003

Prudent E, Raoult D (2019) Fluorescent in situ hybridization, a complementary molecular tool for the clinical diagnosis of infectious diseases by intracellular and fastidious bacteria. FEMS Microbiol Rev 43:88–107

Quaranta G, Sanguinetti M, Masucci L (2019) Fecal microbiota transplantation: a potential tool for treatment of human female reproductive tract diseases. Front Immunol 10:10

Ramakrishna BS, Jayakanthan P, Pugazhendhi S, Kabeerdoss J (2015) Alterations of mucosal microbiota in the colon of patients with inflammatory bowel disease revealed by real-time polymerase chain reaction amplification of 16S ribosomal ribonucleic acid. Indian J Med Res 142(1):23

Ramnani P, Chitarrari R, Tuohy K, Grant J, Hotchkiss S, Philp K et al (2012) In vitro fermentation and prebiotic potential of novel low molecular weight polysaccharides derived from agar and alginate seaweeds. Anaerobe 18(1):1–6

Reardon S (2019) CRISPR gene-editing creates wave of exotic model organisms. Nature 568(7753):441–442

Reyes A, Haynes M, Hanson N, Angly FE, Heath AC, Rohwer F et al (2010) Viruses in the faecal microbiota of monozygotic twins and their mothers. Nature 466(7304):334–338

Reyes A, Blanton LV, Cao S, Zhao G, Manary M, Trehan I et al (2015) Gut DNA viromes of malawian twins discordant for severe acute malnutrition. Proc Natl Acad Sci 112(38): 11941–11946

Round JL, Mazmanian SK (2009) The gut microbiota shapes intestinal immune responses during health and disease. Nat Rev Immunol 9(5):313–323. https://doi.org/10.1038/nri2515

Russmann H, Adler K, Haas R, Gebert B, Koletzko S, Heesemann J (2001) Rapid and accurate determination of genotypic clarithromycin resistance in cultured helicobacter pylori by fluorescent in situ hybridization. J Clin Microbial 39:4142–4144

Savage DC (1977) Microbial ecology of the gastrointestinal tract. Annu Rev Microbiol 31:107–133

Schanche M, Avershina E, Dotterud C, Øien T, Storrø O, Johnsen R et al (2015) High-resolution analyses of overlap in the microbiota between mothers and their children. Curr Microbiol 71(2): 283–290

Schirmer M, Franzosa EA, Lloyd-Price J, McIver LJ, Schwager R, Poon TW et al (2018) Dynamics of metatranscription in the inflammatory bowel disease gut microbiome. Nat Microbiol 3(3): 337–346

Segal JP, Mullish BH, Quraishi MN, Acharjee A, Williams HR, Iqbal T et al (2019) The application of omics techniques to understand the role of the gut microbiota in inflammatory bowel disease. Ther Adv Gastroenterol 12:175628481882225

Shah P et al (2016) A microfluidics-based in vitro model of the gastrointestinal human–microbe interface. Nat Commun 7(1):11535

Shen J, Obin MS, Zhao L (2013) The gut microbiota, obesity and insulin resistance. Mol Asp Med 34(1):39–58

Sobhani I, Tap J, Roudot-Thoraval F, Roperch JP, Letulle S, Langella P et al (2011) Microbial dysbiosis in colorectal cancer (CRC) patients. PLoS One 6(1):e16393

Tanca A, Abbondio M, Palomba A, Fraumene C, Manghina V, Cucca F et al (2017) Potential and active functions in the gut microbiota of a healthy human cohort. Microbiome 5(1):79

Tang Q, Jin G, Wang G, Liu T, Liu X, Wang B, Cao H (2020) Current sampling methods for gut microbiota: a call for more precise devices. Front Cell Infect Microbiol 10:151

Thursby E, Juge N (2017) Introduction to the human gut microbiota. Biochem J 474:1823–1836

Tremaroli V, Bäckhed F (2012) Functional interactions between the gut microbiota and host metabolism. Nature 489(7415):242–249

Valdes AM, Walter J, Segal E, Spector TD (2018) Role of the gut microbiota in nutrition and health. BMJ 361:k2179

Verberkmoes NC, Russell AL, Shah M, Godzik A, Rosenquist M, Halfvarson J et al (2008) Shotgun metaproteomics of the human distal gut microbiota. ISME J 3(2):179–189

Wang B, Yao M, Lv L, Ling Z, Li L (2017) The human microbiota in health and disease. Engineering 3(1):71–82

Weingarden AR, Vaughn BP (2017) Intestinal microbiota, fecal microbiota transplantation, and inflammatory bowel disease. Gut Microbes 8(3):238–252

Wensel CR, Pluznick JL, Salzberg SL, Sears CL (2022) Next-generation sequencing: insights to advance clinical investigations of the microbiome. J Clin Investig 132(7):e154944

Wilmes P, Bond PL (2006) Metaproteomics: studying functional gene expression in microbial ecosystems. Trends Microbiol 14(2):92–97

Zhang X, Chen W, Ning Z, Mayne J, Mack D, Stintzi A et al (2017) Deep metaproteomics approach for the study of human microbiomes. Anal Chem 89(17):9407–9415

Tools and Techniques for Exploring Hidden Microorganisms: A Potential Future of Human Health Diagnosis

11

Akanksha Vishwakarma, Ankita Srivastava, Vinay Kumar, and Digvijay Verma

Abstract

Microorganisms play a significant role in human health and the sustainability of the environment. The last two decades have revolutionized the world of microorganisms where one can explore those bacteria and viruses that were earlier inaccessible. Traditional tools and techniques contribute at the integral level to diagnose the various diseases of humans and animals. However, these techniques demand pure isolate as a prerequisite. Therefore, an alternate technique (metagenomics) was evolved that has replaced the cultivation glitches of microbiology and relies on direct cloning of the community DNA of a habitat. The functional and sequenced-based metagenomic approaches were well explored to retrieve the novel genes and explore the microbial community composition. Later, the microbiome-based studies were clubbed well with several other genomic and proteomic-based tools for a better understanding of the dynamics of the inhabitant microbiomes of a human host. With the decreasing cost of sequencing and generation of massive sequencing data, the microbiome-based investigation grabbed attention. The development of the fecal microbiota transplantation (FMT) technique for treating *Clostridium difficile* infection is one of the classic examples of the microbiome-based therapy. The assessment of the oral disease was also remarkably improved where most of the illnesses are polymicrobial. These tools exhibit the potential for developing a microbiome-based consortium to improve human and animal health. Besides, the diagnosis of infectious diseases also becomes more sensitive, accurate, rapid, and cost-effective with the aid of these tools.

A. Vishwakarma · A. Srivastava · V. Kumar · D. Verma (✉)
Department of Environmental Microbiology, School of Earth and Enviornmental Sciences, Babasaheb Bhimrao Ambedkar University, Lucknow, Uttar Pradesh, India

R. Sobti et al. (eds.), *Role of Microbes in Sustainable Development*,
https://doi.org/10.1007/978-981-99-3126-2_11

Keywords

Metagenomics · Microbiomics · Next-generation sequencing · Human diseases · Diagnosis · Health

11.1 Introduction

Microorganisms are ubiquitous in the environments including humans in and out of their surfaces. These unseen microorganisms share a similar count of the total number of cells in a human body (Sender et al. 2016). Therefore, this plethora of microorganisms can be similar to a multicelled organ (Plutchik 2001). The total count of microorganisms present in a habitat is defined as "Microbiome," which can be further specified on the basis of microbial types such as bacterial microbiome (bacteriome), viral microbiome (virome), and fungal microbiome (mycobiome). With the existence of one microorganism-one disease status, how a single microorganism exhibits the potential to affect the overall health of the human host? It also includes the nearby host cells as well as microbial cells. Most oral diseases are the consequence of polymicrobial associations that not only include various bacterial species but also microorganisms of other domains such as yeast and fungi (Yarieva 2022). Therefore, it is highly important to understand the overall microbiome dynamics. The count, as well as the diversity of the microorganism, differs in different niches of the human body. With a total count of 10–100 trillion bacterial cells, the gut is considered as the most abundant habitat for bacteria in the human body. It was followed by the oral cavity, where the count of the bacterial species is more than 700 (Verma et al. 2018). Besides, lung (Man et al. 2017), skin (Skowron et al. 2021), armpits (Akani et al. 2021), vagina (France et al. 2022), liver (Gola et al. 2021), kidneys (Mertowska et al. 2021), and other visceral organs are significantly high. The role of bacteria as commensals (Khan et al. 2019), symbionts (Henriquez et al. 2021), and another beneficial effects (Mohajeri et al. 2018) are well established in the literature to sustain the human well-being. However, in past few years, commendable research has been done to understand the microbiome dynamic in the context of human health. Due to several microhabitats in the human body and their interactions with inhabitant bacteria as well as external environments, the microbiota can be considered a dynamic ecological community (Khalighi et al. 2022). These microhabitats exhibit specific community composition that exclusively relies on the localized physiological environment of the organs (Foxman et al. 2008). Despite the several microbial habitats, the human body serves as an excellent ecosystem by maintaining a dynamic equilibrium. Moreover, it is now the prerequisite to this microbial structure unaltered to maintain the body healthy. This dynamic equilibrium can be altered due to exposure to environmental interferences (Dethlefsen et al. 2006). An altered microbiome (dysbiosis) led to the onset of several diseases (Fig. 11.1). Gut and oral habitats have been extensively studied for such dysbiotic states that resulted in diseased conditions. It has also been reported that most of the abundant bacteria in the gut have been associated with health status.

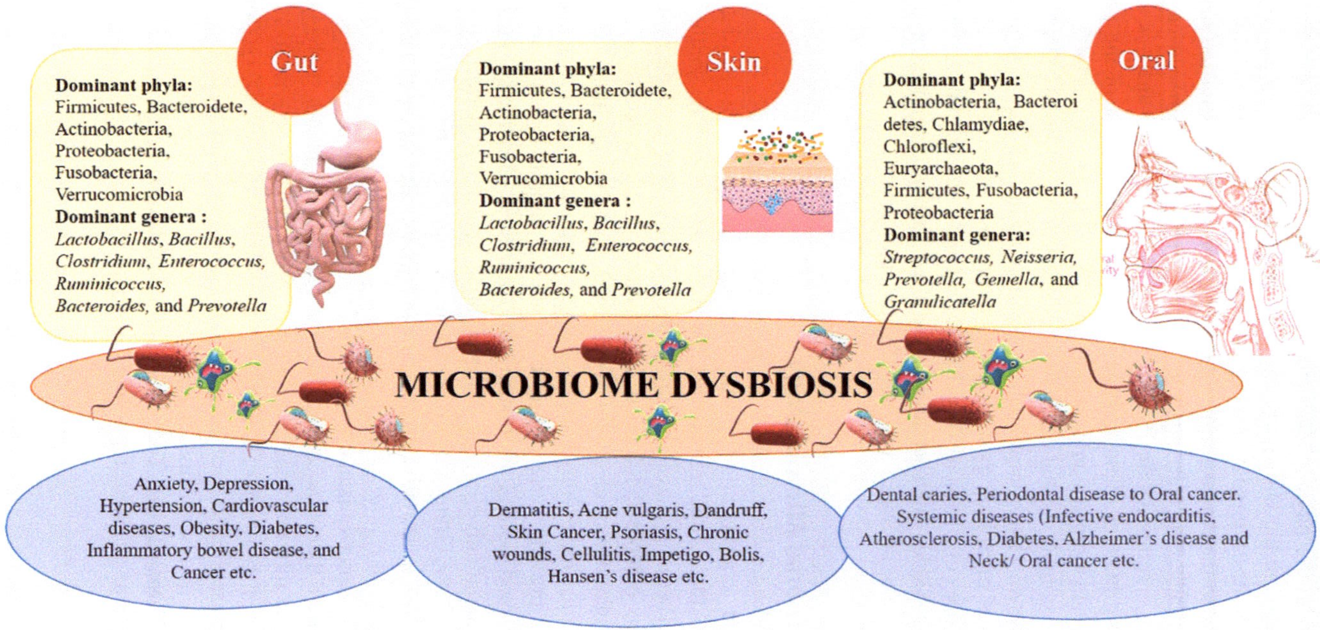

Fig. 11.1 The major habitats (gut, oral, and, skin) of bacteria in human hosts exhibit the dominant bacterial communities at the phyla and genera levels and the reported consequences of microbial dysbiosis on the development of diseases

Though it was oppositely reported in the case of the vaginal microbiome, where higher diversity is closer to the diseased state. It indicates that microbial profile can be correlated with the physiopathological states of the human body. It would be interesting to explore the knowledge of microbiome-based information for a better understanding of the diseased or healthy status of a host. However, the cultivation of microorganisms to their pure form is the biggest bottleneck. With the best of the traditional approaches of cultivation techniques, it is not possible to capture the entire microbial community on Petri-dishes for their extensive analysis from any habitat including humans. This way, microbiologists lose most of the portion of more than 99% and represent an incomplete analysis. Metagenomics has emerged to cope with these conditions that offer direct cloning of a community DNA for their extensive analysis. Thus, despite of the uncultivability of the microorganisms, advanced tools of microbiology has enabled to produce massive information of the inhabitant microbes of a habitat. This chapter describes various tools and techniques to explore such hidden microorganisms for their role in diagnosis of infectious diseases and developing rapid and cost-effective biomarkers.

11.2 Traditional Microbiology

With the great contribution of Antoine van Leeuwenhoek by introducing microscopes to the world, remarkable progress has been done in the field of microbiology. Thereafter, several microbiological techniques evolved with the passage of time to explore the limited information on microorganisms of various domains (Muthukumar et al. 2008). Traditional microbiology includes serology, antigen detection, microscopy, and isolations (Laupland and Valiquette 2013). These methods provide a rapid way of analyzing morphological parameters of bacteria, fungi, and several protozoan-based parasites. Traditional tests can be qualitative or quantitative; where the qualitative analysis is performed by using colonial morphology, Gram staining, endospore staining, and biochemical activities of the bacteria. The quantitative analysis includes the enumeration of the culture via different methods like pour plate, spread plate, surface drop, agar droplets, and microdilution. The prerequisite of traditional or conventional microbiological testing is an isolate/culture that is inoculated from the bacterial sample for studying various parameters. Even different media were used to inflate the bacterial identification essentially through biochemical testing (Gracias and McKillip 2004). One of the most significant applications of microbiology along with biotechnology is in the successful production of vaccines that prevent human as well as animals from many lethal diseases (Opal 2010). Today's pathological testing relies on pure culture-based investigations and their reports.

11.3 Problems Associated with Traditional Microbiology

The foremost problem of traditional microbiology is losing the major fraction of microorganisms in a habitat. The limitations rose in various facets that include (1) incomplete information on the nutritional parameters of the habitats required to cultivate the microbes. (2) Differences in the physiological parameters (pH, temperature, salt, metals, ionic state, etc.) of the habitats. (3) Loss of microbe-microbe and microbe–host interactions at the level of commensalism, symbiosis, protocorporation, and other mutual interactions (Pickup et al. 2003). The cultivation conditions in the laboratory are highly limited, where various microorganisms are cultivated in highly rich nutrient media (Keer and Birch 2003) in stringent parameters. With the best of efforts, environmental conditions cannot be simulated in the laboratory (Zhang et al. 2018). Besides, the traditional cultivation approaches are time-consuming and tedious to achieve the isolate in its pure state. Microbial diagnosis and therapies demand a high level of accuracy whereas traditional practices of microbiology are mostly manually conducted leading to lower accuracy of the test being conducted which is a major drawback to accurate diagnosis (Van Belkum et al. 2020). Contamination at any stage led to the failure of the entire diagnosis and enhances the possibility of relocation errors (Nathan et al. 2018).

Studying the obligate and even facultative anaerobic microorganisms is another big challenge where prior information on the source samples is a must to lose precious human samples for a diagnosis. Along with the mentioned restraints, another major limitation related to cell-based products includes the need for manual and visual examination of cultures to detect growth (Peris-Vicente et al. 2015). Hence, highly keen, and active observation is the demand for classical-based methods of microbiological testing which itself a laborious task for humans. Moreover, cultivation-based methods are unable to provide the community structure of a habitat (Rhoads et al. 2012). As most of the time, only dominating bacteria/microbes of that environment only appear during the defined cultivation conditions (Dawodu and Akanbi 2021). Another limitation is regarding the viability of the bacteria being cultured, i.e., viable bacteria are enumerated along with the nonviable colonies that provide the total count, and therefore, the proportion of viable bacteria is not known.

Traditional culturing methods are therefore often viewed as slow and outdated, although they still deliver an internationally accepted evidence-based analysis/diagnosis. In contrast, molecular tools have the potential for rapid analysis, and their operational utility and associated limitations and uncertainties should be assessed considering their use for regulatory monitoring (Oliver et al. 2014; Rhoads et al. 2012).

11.4 Molecular Techniques for Analyzing Bacteria for Human Diagnosis

Molecular techniques have come up with an immense contribution toward classification and identification of the bacteria in all the major fields like food, medicine, health diagnosis, etc. Since 1983, many molecular methods have been developed for the detection and genotyping of bacteria (Hallin et al. 2012). Several molecular typing methods are available such as ribotyping, restriction fragment length polymorphism (RFLP), Random amplified polymorphic DNA (RAPD), amplified fragment length polymorphism (AFLP), and amplified ribosomal DNA restriction analysis (ARDRA) find applications in identifying the bacterial types. However, these are time-consuming and tedious techniques that further demand highly trained manpower. Development of rapid molecular techniques has enhanced and specified the expeditious identification of the bacteria being considered.

11.4.1 PCR and Derived Approaches

The technique is well known for diagnosing infectious pathogens of all types that can amplify the millions of identical gene copies from a very less amount of a clinical sample. It is widely used in the diagnosis of viruses, bacteria, parasites, and fungi (Kurkela and Brown 2009). Several modifications of the technique empower it better for its utilization in the diagnosis of human diseases. For example, the quantitative-based PCR method depends upon the fluorescent probes reader which detects and quantifies the PCR product in real-time. This technique is more fruitful due to the deficiency of post-PCR processing. Where the amplified PCR product can be detected automatically by the fluorescence monitoring of real-time PCR. Nowadays, Light Cycler and and TaqMan are the commercially available advance real-time PCR versions. Light Cycler TM and smart cycler used SYBR-green dye are used to perform fluorescence monitoring. Whereas, TaqMan uses fluorescent probes, which specifically bind with the amplification- binding site of the target sequences (Dwivedi et al. 2017). Nummi and coworkers employed qPCR for the detection of *Mycoplasma pneumoniae* and *Chlamydia pneumoniae* infection as both cause pneumonia with similar expression (Nummi et al. 2015). Thus, qPCR enables to detection of the most common mutation associated with the 23S rRNA gene of macrolide resistance in *Mycoplasma pneumoniae*. The technique is widely used in respiratory pathogens and clinical specimens. The fungal disease has also been successfully screened by amplifying the signature sequences of the pathogenic strains such as *Aspergillus fumigatus* and *Aspergillus flavus*. Real-time PCR was widely used in the diagnosis of *Propionibacterium* spp., *Chlamydia* spp., *Legionella pneumophila*, and *Listeria monocytogenes* (Dwivedi et al. 2017; Mackay 2004). The application of PCR can be seen in detecting COVID-19 infection worldwide for detecting the coronavirus and its various mutants such as N501Y,69-70del, K417N, and E48K SARS-CoV2 (Vega-Magana et al. 2021). Multiplex PCR is another version of PCR that is widely used in the detection of multiple DNA sequences simultaneously in a single PCR reaction. Here, multiple sets of primer pairs along with multiple DNA templates are used in a single reaction. Where the primer pair binds with specific target DNA and is optimized at the different amplicon sizes. This technique generates information on multiple genes in a single run (Kurkela and Brown 2009). Dwivedi et al. (2017) use multiplex PCR for detecting infectious pathogenic bacterial strains of *Brucella abortus* and *Brucella melitend* in a very short time of 2–3 h with the genus-specific primer-probe sets. Loens et al. also used this technique for the detection of a respiratory specimens of *M. pneumoni*, *C. pneumonia*, and *Legionella* species with the use of molecular beacons (Loens et al. 2008).

Reverse transcription PCR (RT-PCR) is widely used in the detection of viruses. This technique synthesizes complementary DNA (cDNA) by using an mRNA template with the help of a reverse transcriptase enzyme. He et al. (2015) used this technique for the detection of porcine circovirus from domestic pigs in China. Reverse transcriptase quantitative PCR provides a major contribution to the investigation of viruses like *flaviviruses*, *hepadnaviruses*, *herpesviruses*, *orthomyxoviruses*, *parvoviruses*, *papovaviruses*, *paramyxoviruses*, *pestiviruses*, *picornaviruses*, *poxviruses*, *retroviruses*, *rhabdoviruses*, and *TT virus* (Bookout et al. 2006). Mackay (2004) is also used to detect the viral load to check the infectious interaction between the virus and the host.

11.4.2 Fluorescence In Situ Hybridization

Fluorescence in situ hybridization (FISH) is a rapid detection technique for the identification of pathogenic bacterial DNA sequences and the diagnosis of genetic diseases, gene mapping, and novel oncogenes which play role in various types of cancers. Besides, FISH is being utilized as a cytochemical technique for genetic detection and identifying the loci of nucleic acid-based probes (Cui et al. 2016). The technique has an advantage due to its accuracy, safety, and specificity toward the nucleic acids (Dwivedi et al. 2017). It is a low-cost molecular hybridization assay that was used to detect the DNA probe in *Chlamydia trachomatis* or *Neisseria gonorrhea*. This technique was also used in the detection of respiratory infections, gastrointestinal diseases, mycobacterial infections, and fastidious bacteria like spirochetes (Jensen et al. 2001). Prudent and Raoult reported the FISH technique for the detection of Q fever due to *Coxiella burnetii* infection which may cause serious complications in humans and animals. Similarly, Maiwald et al. (2003) used the FISH technique for the detection of two strains of *Tropheryma whipplei* from the cerebrospinal fluid of two patients of Whipple's disease. The FISH assay provides confirmation of toxin/antitoxin elements work as a pathogenic factor in bacterial cells (Audoly et al. 2011). Zhang et al. (2012) used peptide nucleic acid (PNA) probes for the rapid detection *of Listeria* spp., *L. monocytogenes*, and *L. ivanovii* within 1 h. Since early detection can decrease the infection potential of *L. monocytogenes* which causes meningoencephalitis. (Goulet et al. 2012). Few studies reported the PNA-FISH technique for uncovering tuberculosis causing Mycobacteria (MT) and nontuberculosis mycobacteria (NMT) strains (Soini and

Musser 2001). It was successfully used for the rapid detection of several *Mycobacterial* spp. such as *M. leprae*, *M. avium*, and *M. kansasii* within 3 h (Lefmann et al. 2006). The technique has been used for visualizing the microbes and their distribution in oral biofilm (Malic et al. 2009).

11.4.3 Microarray

Microarray is a DNA hybridization biochip-based technology that analyzes thousands of genes simultaneously and detects the specific gene of DNA and RNA. The technique is being widely used to detect single- nucleotide polymorphism and mutation in genomic DNA (You et al. 2008). Microarray technique has been employed for the identification of bacteria at species and subspecies level to check their pathogenicities such as *E. coli*, *Vibrio cholerae*, *Salmonella enterica*, *Campylobacter jejuni*, *Shigella* spp., *Yersinia enterocolitica*, and *Listeria monocytogenes* (You et al. 2008). Few studies also reported the microarray for the identification of yeast and molds by targeting the ITS region of the fungal 18S rRNA gene (Huang and Zheng 2006). Wang and colleagues use 16S-based microarray technique for the detection of intestinal bacteria from the fecal sample (Wang et al. 2002). From this technique, 20 predominant intestinal bacteria were directly detected using a microarray-based technique. Lee and coworkers detected 44 pathogenic bacterial strains using Pathochip DNA microarray from the clinical sample of blood, sputum, cerebral spinal fluid, pus, and urine. Bacterial-specific genes such as housekeeping genes, virulence factors, and antibiotic-resistant genes were also used in place of 16S rRNA sequence for the detection of *S. aureus*, *E. coli*, and *Pseudomonas aeruginosa-like* pathogens (Lee et al. 2003). DNA microarray was also implemented in the detection of pathogenic species in food samples (Cleven et al. 2006). Palka-Santini et al. (2009) employed DNA microarray for DNA-DNA hybridization with specific oligonucleotides for the identification of contaminating microorganisms. Besides, the bacterial species of some viruses and fungi are also detected by the microarray technique. Quantitative microarray was attempted for the early detection of the hepatitis virus that causes chronic liver disease and hepatocellular carcinoma-like diseases (Sakai et al. 2012).

11.4.4 Nucleic Acid Sequence-Based Amplification

In this technique, T7 RNA polymerase, RNaseH, and primer-mediated T7 promoters are used to convert the RNA template into double-stranded DNA. This technology is used in the detection of *Mycoplasma pneumoniae*, *Chlamydophila pneumoniae*, and *Legionella* spp. From the clinical diagnosis of respiratory specimens (Loens et al. 2008). Lau and Coworkers used two detection methods of nucleic acid sequence-based amplification (NASBA) technique: NASBA-electrochemiluminescence (NASBA-ECL) and enzyme-linked oligonucleotide for the rapid identification of foot and mouth disease virus (FMDV) (Lau et al. 2008). Prateek et al. (2010) used this technique for the detection of cytomegalovirus (CMV) infection. Similarly, Shan et al. (2003) also used the NASBA-ECL technique for the detection of avian influenza, a subtype H5 from the allantoic fluid harvested from inoculated chick embryos (Shan et al. 2003). Guoshuai and colleagues detect classical swine fever virus (CSFV) without interfering other viral RNA by using G4-THT-NASBA, since G4-THT-NASBA is a highly sensitive, easy-to-use, and rapid technique for RNA detection (Guoshuai et al. 2022; Jia et al. 2022).

11.5 Advanced Tools and Techniques of Microbiology

It took around 400 years to develop the techniques to explore the hidden and uncultivable microorganisms in the human body. Remarkable development has been done in the line of human microbiome-based research. However, prior to NGS, it was quite challenging due to the inability to capture the entire microbial community of a habitat. Several reasons have been discussed in Sect. 11.3 under the heading of "Problems associated with traditional microbiology." Research never waits for new techniques; it pursues the available facilities. Though, the human gut microbiome was studied by culturing the inhabitant gut microbiomes in large numbers on various media under varying physiological conditions. We had gathered huge information on the gut bacterial diversity prior to the advent of NGS (Guarner and Malagelada 2003). For example, significant research has been done on *Heliobacter pylori*; a cancer-causing bacterium of the gut. In 1984, Simon and Gorbach predicted that the gut microflora is more populated with anaerobes than the aerobic bacteria where the gut harbors more than 500 bacteria species (Simon and Gorbach 1984). *Firmicutes*, *Proteobacteria*, *Bacteroides*, *Clostridium*, *Fusobacterium*, *Eubacterium*, *Ruminococcus*, *Peptococcus*, *Peptostreptococcus*, and *Bifidobacterium* were already known dominant genera of the human gut microbiome (Harmsen et al. 2000). The admirable research on pro- and prebiotics in the context of gut health must be noted. Several Lactic acid bacteria (LAB) were established for their role in the betterment of gut health. Similarly, the salutary effect of various *Clostridium* spp. was well known for its role in attenuating inflammation and allergic diseases (Samarkos et al. 2018). Today's microbiome-based research on *Clostridium difficile* is highly influenced by prior research on its isolated cultures only (Britton and Young 2012). Several diseases have been diagnosed suing various NGS tools (Table 11.1). Though NGS-based investigations come out with enormous information, even the traditional cultivation approaches are irreplaceable and play an integral role to achieve a concrete solution (Vishwakarma and Verma 2021).

11.5.1 Metagenomics

Metagenomics is defined as the study of collective genomics analysis of microorganisms by direct extracting and cloning DNA from an assemblage of microorganisms present in a particular environment. The term metagenomics was

Table 11.1 Various NGS platforms used identification of human diseases

S. No.	NGS sequencing technology	Employed in human diseases	References
1.	Serial Sanger sequencing of multiple genes	Common variable immunodeficiency disorders (CVIDs)	Ameratunga et al. (2021)
2.	PacBio RSII sequencer	Epigenetic characterization	Pereira et al. (2020)
3.	Prometh ION	Characterization of structural variants (SVs) from a human genome	De Coster et al. (2019)
4.	ION Torrent (PGM and MiSeq)	Detection of disease-causing mutations among oculocutaneous albinism and congenital neutropenia patients	Cullinane et al. (2011)
5.	ION torrent	Identification uncertain genetic defects in SLC45A2 and G6PC3 genes using whole- exome sequencing	Grada and Weinbrecht (2013)
6.	Illumina (CMA-infinium iSelect HD and HTS)	Detection in copy number variation (CNV) using whole-exome sequencing	Yao et al. (2017)
7.	Ion torrent (IDP on PGM)	Detection in nonsense mutation in *DMD* gene of Duchenne muscular dystrophy patients	Niba et al. (2014)
8.	HiSeq 2500 sequencing system and *TBX1*-MLPA	Twenty-five pathogenic mutations were identified in five genes viz., *TBX1*, *AIRE*, *GATA3*, *FAM111A*, and *CASR* were identified in the patients of hypoparathyroidism patients	Wang et al. (2019)
9.	PGM sequencer	First report of *Clostridium hemolyticum; a causal bacterium of* bacillary hemoglobinuria in cattle, goat, sheep, and ruminants using	Saeb et al. (2017)
10.	NGS and real-time PCR	NGS employed for identifying various viruses and their correlation with ribosomal activity microbiome of gut parasites among nonspiked birds	Vibin et al. (2018)
11.	Illumina, ion torrent 454, pacific biosciences	Identification of ten genes involved in acute myeloid leukemia	Ley et al. (2008)
12.	VarScan	Exome- based sequencing revealed 29 large scale alterations, i.e., copy number alterations (CNAs) in tumor	Koboldt et al. (2012)
13.	HiSeq2500	Identification of significantly different genera among the oral microbiome of oral squamous cell carcinoma patients	Srivastava et al. (2022)
14.	Various gut-based NGS-based microbiome analysis	Developing fecal microbiota transplantation	Blaser (2019)
15.	Illumina	Determining shift in the oral microbiome during diseased stage	Tanner et al. (2018)

coined by Jo Handelsman in 1988 while doing her research on the discovery of natural products through biosynthetic gene clusters (BGC). Jo Handelsman recognized that entire sample DNA can be used for exploring Novel BGC loci (Handelsman et al. 1998). Unlike traditional microbiology, metagenomics has enabled the study of unculturable microorganisms in their native habitats by directly extracting DNA from the respective samples. Metagenomics explores the microbial genes and genomes either by functional-based or sequence-based approaches (Culligan et al. 2014). Metagenomic-derived studies are gaining interest in various fields to study taxonomic and functional annotation of the microbiomes of agricultural, environmental, human, and clinical samples (Zhou et al. 2020; Chiu and Miller 2019). Functional metagenomics is the activity-based screening of the metagenomic libraries for the desired bioactive molecules. Several successful achievements have been reported to retrieve the novel genes of various industrially relevant enzymes (Cui et al. 2019), antibiotics (De Coster et al. 2019), bacteriocin (Pal and Srivastava 2014), and other antimicrobial compounds (de Abreu et al. 2021). Of the sequences-based approaches, amplicon sequence metagenomics utilizes marker genes such as bacterial/archaeal- specific 16S rRNA, eukaryotic 18S rRNA, or ITS (internal transcribed spacers) regions for taxonomic and functional profiling of microbes. Whereas, shotgun-based metagenomics relies on whole metagenome or whole genome (unculturable) sequences (Pérez-Cobas et al. 2020). Thus, provides better resolution and reliable information about the taxonomic and functional characteristics of the inhabitant microorganisms of a particular environment or habitat. In most of the studies, amplicon sequencing is being used due to cost-effectiveness with high accuracy (Callahan et al. 2019). Various microbiome tools are available to study the microbiome and its functional activity in various environments (Galloway-Pena and Hanson 2020). For example, microbiome-based studies are the outcome of the sequence-based metagenomic approaches. Commendable research has been done in the last decade to understand the microbiome dynamics of humans (Sehli et al. 2021; Baker et al. 2021).

11.5.2 Microbiome-Based Tools

The term "Microbiome" can be defined as the total genome of all the microorganisms (commensals, parasitic, symbiotic, pathogenic, or nonpathogenic) present in a particular habitat or environment (Berg et al. 2021). Microbiome explores the entire microbial communities and has been successfully employed to study the microorganisms of various fields such as environment, soil, water, and air, including humans (Cullen et al. 2020). Several global projects on metagenomics and microbiome analysis have been successfully accomplished such as Earth Microbiome Projects (EMP) (Gilbert et al. 2010, 2014), Human Microbiome Project (HMP) (Turnbaugh et al. 2007), and European MetaHIT (Qin et al. 2010). Kho and Lal (2018) assume the gut microbiome is the controller between wellness and disease. The direct role of human microbiota in the immune system has been established that shapes the host immune system. The effect of environmental factors has also been studied to correlate the human microbiome dynamics. It includes geographical variations, antibiotic doses, temperature effects, vegetation, lifestyle, food habits, and mental status (Rodríguez et al. 2015; Biedermann et al. 2013; Tyakht et al. 2013). Maurice et al. (2013) reported the alteration in the physiology and gene expression of the human gut microbiome due to antibiotic doses. Disruption of short- and long-term microbial balance has been reported due to antibiotic treatment (Jernberg et al. 2007). In an interesting report, microbiota depletion was strongly correlated with serotonin and bile acid metabolism that consequently resulted in delayed GI motility (Ge et al. 2017). Besides, antibiotic-treated mice were also more susceptible to *S. Typhimurium* and *C. difficile*-like antibiotic-associated pathogens (Ng et al. 2013). Gut bacteria (LABs) also play a crucial role in the de novo synthesis of essential vitamins (LeBlanc et al. 2013). The LABs were exclusively known for the synthesis of vitamin B12 (LeBlanc et al. 2013). Besides, folate is constitutively produced from *Bifidobacteria* spp. (Pompei et al. 2007). Several other vitamins such as vitamin K, riboflavin, biotin, nicotinic acid, pantothenic acid, pyridoxine, and thiamine are chiefly produced by the gut microbiomes (Hill 1997). The role of gut bacteria in metabolizing the bile acid has also been reported that is hard to reabsorb. Alteration in such transformation in the respective bacteria causes human illnesses such as obesity and type 2 diabetes (Palau-Rodriguez et al. 2015). Therefore, microbiome-based investigations provide the potential to diagnose human diseases. To study the microbiome, we need certain tools and techniques. Marker Gene Analyses (Amplicon sequencing), shotgun metagenomics, metatranscriptomics, metaproteomics, and metabolomics are a few of them that can be used as well as clubbed with the microbiome-based analysis.

11.5.3 Marker Gene Analysis

Marker genes are defined as a specific region of DNA used to identify microbes in metagenomics samples. Commonly used methods are 16S ribosomal RNA gene sequencing for bacteria identification and internal transcribed spacer (ITS) region sequencing for fungal identification. Both markers possess hypervariable regions which participates in assigning genera and species. Bacterial or fungal-specific markers are usually amplified with their respective specific primers followed by NGS-based sequencing. The raw sequences undergo demultiplexing and quality filtering using various tools such as DADA2 (Callahan et al. 2016), Deblur (Amir et al. 2017), UNOISE3 (Edgar 2016), and FastQC (Andrews 2010) to achieve the quality reads. Processing the sequences is carried out either by picking OTUs (operational taxonomic units) with a similarity threshold of 97% or 99% or ASVs (Amplicon Sequence Variants). ASVs are more sensitive in detecting even single-nucleotide variations and thus provide more precise and detailed information on microbial diversity. These ASVs/OTUs are used for taxonomic assignment, diversity analysis, and functional profiling of the microbes present in the study samples (Hamady and Knight 2009). Several microbiome analysis interface packages or pipelines are available such as QIIME1 (Caporaso et al. 2010), QIIME2 (Bolyen et al. 2019), Mothur (Schloss et al. 2009), DADA2 (Callahan et al. 2016), and SILVAings (https://ngs.arb-silva.de/silvangs/). Taxonomy assignments are majorly done by using several classifiers and databases such as the RDP classifier (Wang et al. 2007), a naive Bayesian classifier such as Greengenes (McDonald et al. 2012) and SILVA (Yilmaz et al. 2014) and for fungi, UNITE (Kõjalg et al. 2005) database is used. The command-based tool Quantitative Insights into Microbial Ecology (QIIME) is a more widely used tool in recent times due to several advantages. It is noted that QIIME1.9.1 has been obsolete now and updated with its advanced version QIIME2 2020.1. The newer version of QIIME 2 has a diverse multiple-user interface and wraps many different tools required for the downstream analysis of sequences.

11.5.4 Shotgun Metagenomics

In contrast to amplicon sequencing which uses a short stretch of 16S rRNA or 18S rRNA, shotgun sequencing deals with the sequencing of entire metagenomic DNA in a sample. Thus, it comes out with massive information on bacteria, fungi, archaea, and other microorganisms. It parallelly depicts the information on its taxonomy as well as functionally active genes (Mukhopadhya et al. 2019; Moreno Gallego et al. 2019). The advantage of shotgun metagenomics is that it provides concrete information on functional profiling as compared to the marker-based analysis (16S and 18S). As it predicts the functionality of the microbial community using the available information in the available database. The PICRSUt tool is the highly recommended tool for predicting the function of 16S rRNA genes during marker gene analysis (Langille et al. 2013; Douglas et al. 2020). The shotgun-based analysis is still evolving and will be more reliable in the near future as the database will be evolved. In shotgun metagenomics, short reads are aligned for making contigs such as Layout or Consensus assembly (Ayling et al. 2019). These contigs and scaffolds are further used for retrieving the information based on the available databases. However, analysis of shotgun-based metagenome sequences requires more tedious steps over the QIIME platforms. It also requires complex bioinformatic tools and methods for analyzing reads, some bioinformatic pipelines and software used are Megahit (Li et al. 2015), StrainPhlAn (Truong et al. 2017), MetaPhlAn (Beghini et al. 2021), HUMAnN (Beghini et al. 2021), HOME-BIO (Ferravante et al. 2021), MetaVelvet (Namiki et al. 2012), IDBAUD (Peng et al. 2012), and metaSPAdes (Nurk et al. 2017). Profiling of marker or representative genes is mostly studied by read-based taxonomy assignment and gene annotation which is based on similar DNA constituents and patterns such as K-mers, gene homology, and GC content (Claesson et al. 2017). Kraken is frequently used for taxonomy assignment which is based on K-mer length (Wood and Salzberg 2014).

11.5.5 Metaproteomics

Metaproteomics is the study of microbial communities by analyzing total proteins, thus exploring the microbial community of a specific habitat at the molecular level. The term metaproteomics was first used by Rodriguez-Valera in 2004 to describe and identify proteins and related genes which are abundantly expressed in the environmental sample (Rodríguez-Valera 2004). The study relies on functional information rather than only gene-level information. The main advantage of this technology is that here the expressed protein is studied that finds a role in determining the overall physiology of the bacteria. It has now become an integral stream of proteomics which has enabled the identification of large-scale proteins in microbial populations/communities (Kleiner 2019). Several studies have used metaproteomics to explore the taxonomic and functional role of the complex microbiomes of specific habitats (Abram et al. 2011; Kan et al. 2005; Ram et al. 2005; Wilmes and Bond 2006). Metaproteomics correlates diseases or environmental parameters with the function and taxon of a specific environment (Erickson et al. 2012; Heyer et al. 2016; Kolmeder et al. 2016). Metaproteomics data analysis primarily identifies peptides through mass spectrometry and searches it against protein sequence databases. Further, these peptides are assigned to protein groups or proteins that might be unique or shared between several taxa followed by assignment of these groups to functional groups using several databases (Blank et al. 2018). Some algorithms/tools and databases used are eggNOG-mapper (Huerta-Cepas et al. 2017), MEGAN (Huson and Weber 2013), MetaGOmics (Riffle et al. 2017), MetaProteomeanalyzer (MPA) (Muth et al. 2018), ProPHAnE (https://www.prophane.de), and Unipept (Gurdeep Singh et al. 2019). Earlier, a few drawbacks/issues were associated with metaproteomics that was related to the bioinformatics evaluation of data (Muth et al. 2013). Firstly, it requires high computational efforts, large processors and efficient tools and algorithms, bigger hard drives, and memory. Secondly, the identification of redundant proteins also affects the accuracy of taxonomy and functional assignment (Herbst et al. 2016). Thirdly, for unknown taxonomic composition, it is difficult to identify their functional role in a particular environment. More advancement has been achieved in the field of metaproteomic in the context of microbiome-derived information.

11.5.6 Metabolomics

This is another interesting tool to explore the dynamics of microbiomes at the level of their total metabolites. Metabolites direct serve as a direct signature of the metabolic reaction and ongoing biochemical activity, thus metabolites can be employed as health indicators (Cavill et al. 2009; Tang et al. 2019). Recent developments uncovered the significant influence on the microbiome due to the metabolome of near and distant body sites. The commendable research on liver diseases and their association with gut microbiomes encourages to development of the technique by using the extended profile of metabolites as therapeutic targets or biomarkers (Del Chierico et al. 2017; Canfora et al. 2019). Metabolomics is the study of small molecules or metabolites (<1500 Da) such as smaller biochemical compounds, including simple amino acids and related amines, as well as lipids, sugars, nucleotides, and other intermediary metabolites within an organism, a cell, tissue or fluids. The interaction of these metabolites in a biological system is known as metabolome. Thus, unlike proteomics or genomics, it measures molecules having different properties like polarity, solubility, chirality, and other physicochemical properties (Kuehnbaum and Britz-McKibbin 2013). Metabolomics relies on various analytical techniques such as liquid or gas chromatography coupled to highresolution mass spectrometry (HRMS), Nuclear magnetic resonance (NMR) spectroscopy, mass spectrometry (MS), and, liquid chromatography–mass spectrometry (LC-MS). These techniques are frequently used in metabolomic-based research to analyze thousands of metabolites with high accuracy. Metabolomics studies have enabled researchers to know how food/diet and disease are related, the correlation of gut microbiome with cardiac diseases (Newgard et al. 2009; Koeth et al. 2013). Metabolomics based on mass spectrometry is most sensitive for analyzing various compounds; however, it has some problems with the standardization and quantitation information. NMR-based metabolomics exhibit less sensitivity as compared to mass spectrometry-based metabolomics; however, it provides an absolute concentration of detected compounds and is useful for elucidating molecular structure more accurately. Metabolomics can be categorized broadly into targeted and untargeted types, targeted metabolomics analyze a predefined set of compound or metabolites, however, untargeted or global metabolomics allows the estimation of extracted metabolites from respective samples and can be used for novel biological perturbations which work more effectively with a high-resolution mass spectrometer for better structural characterization of the compounds or metabolites (Johnson et al. 2016). Localization of specific metabolites within cells or tissue can be achieved with the use of imaging metabolomics consisting of imaging mass spectrometry techniques, such as MALDI (matrix-assisted laser desorption ionization), NIMS (Nanostructure-imaging Mass Spectrometry), DESI (Desorption Electrospray Ionization Mass Spectrometry), and SIMS (Secondary Ion Mass Spectrometry) (Palmer et al. 2016; Fletcher et al. 2013). Metabolomics not only provides biological information but also exhibits potential application on novel therapeutic molecules (Clish 2015). Advanced sequencing tools and extensive research on microbiomes have enabled the correlation of the metabolites between healthy and diseased states of the host.

11.5.7 Metatranscriptomics

Bashiardes et al. (2016) reviewed well the use of metatranscriptomics in microbiome research. In recent years, the development of advanced sequencing tools such as RNA-Seq showed a jump in transcriptome-based analysis providing deep insights into this line of research. Transcriptomics can be understood as the study of complex microbial community's gene expression within their natural environments/habitats.

The technique was first introduced in 2000, and now RNAs sequencing or meta transcriptomics has been significantly increased which enabled researchers to characterize microbial community and their interaction (Bashiardes et al. 2016; Bikel et al. 2015) enabling the detection of genes expression to understand the microbe–host relationship (Moniruzzaman et al. 2017). The major goal of a metatranscriptomics study is to explore the functional activity of the microbiome of a particular habitat. Functional annotation can be achieved either through reads or assembled contigs. Tools such as MetaCLADE (Ugarte et al. 2018), HMM-GRASPx (Zhong et al. 2016), and UProC (Meinicke 2015) are read based on functional profiler and take ORF as input which is used in metatranscriptomic-based investigations. Alternatively assembled contigs can also be used for functional annotation by using Prodigal (Hyatt et al. 2010) and FragGeneScan (Rho et al. 2010) like programs which are further followed by functional assignments primarily based on similarity searches approach by using tools such as DIAMOND (Buchfink et al. 2015) or functional databases like KEGG (Kyoto Encyclopedia of Genes and Genomes) (Kanehisa and Goto 2000), COG (Clusters of Orthologous Groups) (Tatusov et al. 2000), NCBI RefSeq (Oleary et al. 2016), and UniProt (Uniprot 2019), etc. Other tools are available to analyze the transcriptome data such as Prokka (Seemann 2014), EDGE Bioinformatics (Li et al. 2015), and MG-RAST (Wilke et al. 2016), which combine a few similarity searches against various databases pipelines and platforms. Further annotation and enzymatic functions are mapped using metabolic pathways tools such as MinPath (Ye and Doak 2009) or iPath (Yamada et al. 2011), MAPPS (Riaz et al. 2020) (https://mapps.lums.edu.pk), Metacyc (A multiorganism database of metabolic pathways and enzymes) (Caspi et al. 2010). Targeted approaches typically do not provide direct evidence of the functional potential of the microbial population of a specific habitat, tools such as PICRUSt (Phylogenetic Investigation of Communities by Reconstruction of Unobserved States) enable researchers to infer functional profiling of microbiome directly from marker gene (such as *16S rDNA) taxonomy profiling* (Aguiar-Pulido et al. 2016; Langille et al. 2013). Gosalbes and coworkers employed the 16S rRNA transcripts to determine the extensive bacterial profiling in the GI tract (Gosalbes et al. 2011). In another investigation, the transcriptome profiling of the gut microbiome revealed that the specific bacterial strain *Eggerthella lenta* exhibits cytochrome-encoding operon which is upregulated by digoxin that consequently inactivates the cardiac drug. Similarly, metatranscriptomics-microbiomics has been employed to assess the microbial community (Maurice et al. 2013), microbiome-immune interactions (Cullender et al. 2013), microbiome-antisense RNAs (Bao et al. 2015).

11.6 Development in Next-Generation Sequencing Platforms

It took approximately 400 years to understand that the entire microbial community of habitat cannot be analyzed due to several limitations of the traditional techniques of microbiology (Behjati and Tarpey 2013). The advancements in sequencing technology especially the emergence of NGS have immensely revolutionized the era of genomics (Buermans and Dunnen 2014). Several bacteria and archaea have been successfully sequenced and assembled using NGS-based technology (Goh et al. 2017; Panda et al. 2019). The technique extensively provides the comparative microbiome community structure of the healthy and diseased states (Malla et al. 2019). Significant development has been seen in the evolution of NGS- based platform. Earlier, Roche 454 GS FLX was exclusively used for NGS-based sequencing; however, due to low coverage and high cost, the technique has now been obsoleting today. At present, four NGS techniques are prevalent, which include Illumina, Ion Torrent technology, Pacific Biosciences technology, and Oxford nanopore sequencing. Illumina platforms dominate over the other sequencing platform for exploring human microbiomes using whole metagenome-based sequencing along with 16S rRNA amplicon-based sequencing (Chan et al. 2015; Saxena et al. 2017). The NGS has also made possible metagenome-based sequencing where a better resolution can be achieved to understand the functional and genetic diversity of unculturable communities (Lopez-Lopez et al. 2013; Sharon and Banfield 2013). Chan et al. (2015) added new information in shotgun-based metagenomics in terms of metabolic activity and dynamics of the inhabitant bacteria of the Sungai Klah (SK) hot spring that makes this hot spring unique. Expectedly, the bacterial diversity profile was almost similar in both amplicon as well as shotgun sequencing. NGS was the only solution to uncover the hidden microorganisms present in the human body with the successful completion of the human Microbiome project (Turnbaugh et al. 2007). Ion Torrent technology, an efficient sequencing platform has also been used for studying several environments and for understanding bacteria genomics (Bhalla et al. 2013; Mangrola et al. 2015; Pap et al. 2015). LeBlanc et al. (2013) studied the fecal microbiota composition using Ion torrent technology and suggested for a better DNA extraction strategy to avoid the biasness in bacterial community composition. Besides, a high error rate in Ion Torrent technology has been reported as compared to the Illumina Miseq platform which limits its employability. The Pacific biosciences single-molecule real-time (SMRT) sequencing technology is another major player in the sequencing world that is highly used for detecting specific DNA methylations (Ardui et al. 2018; Straub et al. 2018). Short-read massive parallel sequencing has emerged as a standard diagnostic tool in several medical diagnostic applications (Loomis et al. 2013; LeBlanc et al. 2013). However, the technique comes along with several limitations such as GC bias, challenges to map repetitive elements, and differentiating paralogous sequences. Therefore, it was gradually replaced with long-read single molecules that later clubbed with PacBio's single PacBio's single-molecule real-time (SMRT) sequencing technology (Ardui et al. 2018). Loomis et al. (2013) were the first to report the *FMR1* CGG repeat using SMRT sequencing technology to diagnose repeat biasing in Fragile X Syndrome. Oxford Nanopore Technologies (ONT) MinION long-read sequencer has emerged as an advanced sequencing tool (Bowden et al. 2019) that has been used for sequencing several microorganisms (Kato et al. 2020). Matsuo et al. (2021) utilized the technique to achieve the full-length 16S rRNA amplicon to confer the species-level classification. Though the technique is under evolution to minimize the high error rate, however, it can be adopted well for the diagnosis of infectious diseases in an efficient and time-saving mode. Several investigations have been carried out to diagnose the diseased state of the human microbiome. For example, pulmonary sepsis (Guillen-Guio et al. 2020). Dysbiotic stages of pre- and post-antibiotic-treated human microbiome (Leggett et al. 2020). Besides, the nanopore has been successfully utilized for assessing species engraftment after fecal microbiota transplantation (Benítez-Páez et al. 2020). At present, Illumina (HiSeq and MiSeq) technology has made incredible developments in data output and accuracy in a cost-effective manner (Dohm et al. 2008; Reuter et al. 2015). Therefore, the technology has dominated the NGS-based sequencing market. The technique has enormously been used to understand human microbiome alterations (Evans et al. 2014; Lambeth et al. 2015; Yasir et al. 2015). Illumina HiSeq technology was employed to identify the significantly different bacteria in the oral cavity of oral squamous cell carcinoma (Srivastava et al. 2022). Several studies have been done to understand the taxonomical and functional profiling of smokeless tobacco using Illumina-based sequencing technology (Sajid et al. 2021; Vishwakarma et al. 2023).

11.7 Conclusion and Prospects

With the remarkable development in omics technology, researchers are now able to explore microbial dynamics. Metagenomics has grabbed the attention and changed the perception of researchers. It is gaining huge interest in the field of biotechnology and substantially impacting and increasing industrial products (Lorenz and Eck 2005). Other than this, various novel bioactive molecules such as terragines, violacein, indirubin, cytarabine, and cephalosporins have been retrieved using metagenomics that finds applications in human wellness (Coughlan et al. 2015). Shotgun metagenomics gained interest in the field of human health (oral microbiome, gut microbiome) and the discovery of new drugs for the treatment of diseases and new genes and proteins from noncultivable microorganisms. Metabolomics is another noteworthy technique that has gained huge interest in the field of biomedical and pharmaceutical science and broadens microbiome-metabolome research. The technique has identified several metabolite biomarkers and caused diseases such as cancer, diabetes, and Alzheimer's disease which have no known therapeutic targets and strategies previously (Wishart 2016). Thus, it has bought more precision in medicine worldwide with the development of monitoring of each drug response, phenotyping of tumors, and targets for cancer therapies. MS-based metabolomics studies have recently been elevated and have been significantly used for the study of drug effects, toxins, and several diseases such as cancer of the kidney, bladder, breast, gastric, and other metabolic diseases and nutritional effects (Zhang et al. 2013a, b, 2014; Feng et al. 2018). In the future, metabolomics will help in elucidating various metabolic pathways and research related to metabolites with the help of combined chromatography. Metatranscriptomics provides the functional profiling of the microbial communities of an environment, thus giving information on the whole expressed genes in that community. The combined study of metagenomics and metatranscriptomics allows researchers to reveal the up-and-down expression of a particular function and role of the microbes present in an environment (Mason et al. 2012; Maurice et al. 2013; Duran-Pinedo et al. 2014). Therefore, at present, we are well equipped with various tools and techniques that assist in exploring the good as well as bad microorganisms at a broader scale. The taxonomic shift and their correlation with the hidden microorganisms can be explored for developing biomarkers in near future.

References

Abram F, Enright AM, O'Reilly J, Botting CH, Collins G, O'Flaherty V (2011) A metaproteomic approach gives functional insights into anaerobic digestion. J Appl Microbiol 110:1550–1560

Aguiar-Pulido V, Huang W, Suarez-Ulloa V, Cickovski T, Mathee K, Narasimhan G (2016) Metagenomics, metatranscriptomics, and metabolomics approaches for microbiome analysis. Evol Bioinformatics Online 12:5–16

Akani NP, Barika PN, Allison HB (2021) Prevalence and antibiogram of microbiome of selected body parts from students of Rivers State University, Nigeria. Nigerian J Microbiol 35:5804–5813

Ameratunga R, Allan C, Lehnert K, Woon ST (2021) Perspective: application of the American College of Medical Genetics variant interpretation criteria to common variable immunodeficiency disorders. Clin Rev Allergy Immunol 61:226–235

Amir A, McDonald D, Navas-Molina JA, Kopylova E, Morton JT et al (2017) Deblur rapidly resolves single-nucleotide community sequence patterns. mSystems 2:e00191–e00116

Andrews S (2010) FastQC: a quality control tool for high-throughput sequence data. Available at http://www.bioinformatics.babraham.ac.uk/projects/fastqc

Ardui S, Ameur A, Vermeesch JR, Hestand MS (2018) Single-molecule real-time (SMRT) sequencing comes of age: applications and utilities for medical diagnostics. Nucleic Acids Res 46:2159–2168

Audoly G, Vincentelli R, Edouard S et al (2011) Effect of rickettsial toxin VapC on its eukaryotic host. PLoS One 6:e26528

Ayling M, Clark MD, Leggett RM (2019) New approaches for metagenome assembly with short reads. Brief Bioinform 21:584–594

Baker JL, Morton JT, Dinis M, Alvarez R, Tran NC et al (2021) Deep metagenomics examines the oral microbiome during dental caries, revealing novel taxa and co-occurrences with host molecules. Genome Res 31:64–74

Bao G, Wang M, Doak TG, Ye Y (2015) Strand-specific community RNA-seq reveals prevalent and dynamic antisense transcription in human gut microbiota. Front Microbiol 6:896

Bashiardes S, Zilberman-Schapira G, Elinav E (2016) Use of metatranscriptomics in microbiome research. Bioinf Biol Insights 10:19–25

Beghini F, McIver LJ, Blanco-Míguez A, Dubois L, Asnicar F, Maharjan S, Mailyan A, Manghi P et al (2021) Integrating taxonomic, functional, and strain-level profiling of diverse microbial communities with bioBakery 3. elife 10:e65088

Behjati S, Tarpey PS (2013) What is next generation sequencing? Arch Dis Childhood Educ Pract 98:236–238

Benítez-Páez A, Hartstra AV, Nieuwdorp M, Sanz Y (2020) Strand-wise and bait-assisted assembly of nearly-full rrn operons applied to assess species engraftment after faecal microbiota transplantation. BioRxiv 2020 no. 11.292896:10-1101

Berg G, Rybakova D, Fischer D, Cernava T, Vergès MC et al (2021) Microbiome definition re-visited: old concepts and new challenges. Microbiome 8:103

Bhalla A, Kainth AS, Sani RK (2013) Draft genome sequence of lignocellulose-degrading thermophilic bacterium *Geobacillus* sp. strain WSUCF1. Genome Announc 1:10–1128

Biedermann L, Zeitz J, Mwinyi J, Sutter-Minder E et al (2013) Smoking cessation induces profound changes in the composition of the intestinal microbiota in humans. PLoS One 8:e59260

Bikel S, Valdez-Lara A, Cornejo-Granados F, Rico K, Canizales-Quinteros S, Soberón X et al (2015) Combining metagenomics, metatranscriptomics and viromics to explore novel microbial interactions: towards a systems-level understanding of human microbiome. Comput Struct Biotechnol J 13:390–401

Blank C, Easterly C, Gruening B, Johnson J, Kolmeder C, Kumar P et al (2018) Disseminating metaproteomic informatics capabilities and knowledge using the galaxy-P framework. Proteome 6:7

Blaser MJ (2019) Fecal microbiota transplantation for dysbiosis-predictable risks. N Engl J Med 381:2064–2066

Bolyen E, Rideout JR, Dillon MR, Bokulich NA, Abnet CC, Al-Ghalith GA, Alexander H, Alm EJ et al (2019) Reproducible, interactive, scalable, and extensible microbiome data science using QIIME 2. Nat Biotechnol 37:852–857

Bookout AL, Cummins CL, Mangelsdorf DJ, Pesola JM, Kramer MF (2006) High-throughput real-time quantitative reverse transcription PCR. Curr Protoc Mol Biol 73:15–18

Bowden R, Davies RW, Heger A et al (2019) Sequencing of human genomes with nanopore technology. Nat Commun 10:1869

Britton RA, Young VB (2012) Interaction between the intestinal microbiota and host in Clostridium difficile colonization resistance. Trends Microbiol 20:313–319

Buchfink B, Xie C, Huson DH (2015) Fast and sensitive protein alignment using DIAMOND. Nat Methods 12:59–60

Buermans HP, den Dunnen JT (2014) Next generation sequencing technology: advances and applications. Biochim Biophys Acta 1842:1932–1941

Callahan BJ, McMurdie PJ, Rosen MJ, Han AW, Johnson AJA, Holmes SP (2016) DADA2: high-resolution sample inference from Illumina amplicon data. Nat Methods 13:581–583

Callahan BJ, Wong J, Heiner C, Oh S, Theriot CM et al (2019) High-throughput amplicon sequencing of the full-length 16S rRNA gene with single-nucleotide resolution. Nucleic Acids Res 47:e103

Canfora EE, Meex RCR, Venema K, Blaak EE (2019) Gut microbial metabolites in obesity, NAFLD and T2DM. Nat Rev Endocrinol 15:261–273

Caporaso JG et al (2010) QIIME allows analysis of high-throughput community sequencing data. Nat Methods 7:335–336

Caspi R, Altman T, Dale JM, Dreher K, Fulcher CA, Gilham F, Kaipa P, Karthikeyan AS (2010) The MetaCyc database of metabolic pathways and enzymes and the BioCyc collection of pathway/genome databases. Nucleic Acids Res 38:D473–D479

Cavill R, Keun HC, Holmes E, Lindon JC, Nicholson JK, Ebbels TM (2009) Genetic algorithms for simultaneous variable and sample selection in metabonomics. Bioinformatics 25:112–118

Chan CS, Chan KG, Tay YL, Chua YH, Goh KM (2015) Diversity of thermophiles in a Malaysian hot spring determined using 16S rRNA and shotgun metagenome sequencing. Front Microbiol 6:177

Chiu CY, Miller SA (2019) Clinical metagenomics. Nat Rev Genet 20:341–355

Claesson MJ, Clooney AG, O'Toole PW (2017) A clinician's guide to microbiome analysis. Nat Rev Gastroenterol Hepatol 14:585–595

Cleven BE, Palka-Santini M, Gielen J, Meembor S, Krönke M, Krut O (2006) Identification and characterization of bacterial pathogens causing bloodstream infections by DNA microarray. J Clin Microbiol 44:2389–2397

Clish CB (2015) Metabolomics: an emerging but powerful tool for precision medicine. Cold Spring Harb Mol Case Stud 1:a000588

Coughlan LM, Cotter PD, Hill C et al (2015) Biotechnological applications of functional metagenomics in the food and pharmaceutical industries. Front Microbiol 6:672

Cui C, Shu W, Li P (2016) Fluorescence in situ hybridization: cell-based genetic diagnostic and research applications. Front Cell Dev Biol 4:89
Cui J, Mai G, Wang Z, Liu Q, Zhou Y, Ma Y (2019) Metagenomic insights into a cellulose-rich niche reveal microbial cooperation in cellulose degradation. Front Microbiol 10:1–12
Cullen CM, Aneja KK, Beyhan S, Cho CE, Woloszynek S, Convertino M, McCoy SJ (2020) Emerging priorities for microbiome research. Front Microbiol 11:136
Cullender TC, Chassaing B, Janzon A, Kumar K et al (2013) Innate and adaptive immunity interact to quench microbiome flagellar motility in the gut. Cell Host Microbe 14:571–581
Culligan EP, Sleator RD, Marchesi JR, Hill C (2014) Metagenomics and novel gene discovery: promise and potential for novel therapeutics. Virulence 5:399–412
Cullinane AR, Vilboux T, O'Brien K, Curry JA, Maynard DM, Carlson-Donohoe H, Ciccone C (2011) Homozygosity mapping and whole-exome sequencing to detect SLC45A2 and G6PC3 mutations in a single patient with oculocutaneous albinism and neutropenia. J Investig Dermatol 131:2017–2025
Dawodu OG, Akanbi RB (2021) Isolation and identification of microorganisms associated with automated teller machines on Federal Polytechnic Ede campus. PLoS One 16:e0254658
de Abreu VAC, Perdigão J, Almeida S (2021) Metagenomic approaches to analyze antimicrobial resistance: an overview. Front Genet 11:575592
De Coster W, De Rijk P, De Roeck A, De Pooter T, D'Hert S, Strazisar M, Van Sleegers K, Broeckhoven C (2019) Structural variants identified by Oxford nanopore PromethION sequencing of the human genome. Genome Res 29:1178–1187
Del Chierico F, Nobili V, Vernocchi P, Russo A, De Stefanis C, Gnani D et al (2017) Gut microbiota profiling of pediatric nonalcoholic fatty liver disease and obese patients unveiled by an integrated meta-omics-based approach. Hepatology 65:451–464
Dethlefsen L, Eckburg PB, Bik EM, Relman DA (2006) Assembly of the human intestinal microbiota. Trends Ecol Evol 21:517–523
Dohm JC, Lottaz C, Borodina T, Himmelbauer H (2008) Substantial biases in ultra-short read data sets from high-throughput DNA sequencing. Nucleic Acids Res 36:e105
Douglas GM, Maffei VJ, Zaneveld JR, Yurgel SN et al (2020) PICRUSt2 for prediction of metagenome functions. Nat Biotechnol 38:685–688
Duran-Pinedo AE, Chen T, Teles R et al (2014) Community-wide transcriptome of the oral microbiome in subjects with and without periodontitis. ISME J 8:1659–1672
Dwivedi S, Purohit P, Misra R, Pareek P, Goel A et al (2017) Diseases and molecular diagnostics: a step closer to precision medicine. Indian J Clin Biochem 32:374–398
Edgar RC (2016) UNOISE2: improved error-correction for Illumina 16S and ITS amplicon sequencing, abstract bioRxiv
Erickson AR, Cantarel BL, Lamendella R, Darzi Y, Mongodin EF, Pan C, Shah M (2012) Integrated metagenomics/metaproteomics reveals human host-microbiota signatures of Crohn's disease. PLoS One 7:e49138
Evans CC, LePard KJ, Kwak JW, Stancukas MC, Laskowski S, Dougherty J et al (2014) Exercise prevents weight gain and alters the gut microbiota in a mouse model of high fat diet-induced obesity. PLoS ONE 9:e92193
Feng CC, Zhang AH, Miao JH et al (2018) Recent advances in understanding cross-talk between bile acids and gut microbiota. Open J Proteom Genom 3:24–34
Ferravante C, Memoli D, Palumbo D et al (2021) HOME-BIO (sHOtgun MEtagenomic analysis of BIOlogical entities): a specific and comprehensive pipeline for metagenomic shotgun sequencing data analysis. BMC Bioinf 22:106
Fletcher JS, Kotze HL, Armitage EG, Lockyer NP, Vickerman JC (2013) Evaluating the challenges associated with time-of-fight secondary ion mass spectrometry for metabolomics using pure and mixed metabolites. Metabolomics 9:535–544
Foxman B, Goldberg D, Murdock C, Xi C, Gilsdorf JR (2008) Conceptualizing human microbiota: from multicelled organ to ecological community. Interdiscip Perspect Infect Dis 2008:613979

France M, Alizadeh M, Brown S, Ma B, Ravel J (2022) Towards a deeper understanding of the vaginal microbiota. Nat Microbiol 7:367–378

Galloway-Pena J, Hanson B (2020) Tools for analysis of the microbiome. Dig Dis Sci 65:674–685

Ge X, Ding C, Zhao W, Xu L, Tian H, Gong J, Zhu M, Li J, Li N (2017) Antibiotics-induced depletion of mice microbiota induces changes in host serotonin biosynthesis and intestinal motility. J Transl Med 15:13

Gilbert JA, Meyer F, Antonopoulos D, Balaji P, Brown CT et al (2010) Meeting reports: the terabase metagenomics workshop and the vision of the earth microbiome project. Standard Genom Sci 3:243–248

Gilbert JA, Jansson JK, Knight R (2014) The earth microbiome project: successed and aspiration. BMC Biol 12(1):69

Goh KM, Chan K-G, Sani RK, Donati ER, Reysenbach A-L (2017) Editorial: genetics, genomics and –omics of thermophiles. Front Microbiol 8:560

Gola A, Dorrington MG, Speranza E, Sala C, Shih RM, Radtke AJ, Wong HS (2021) Commensal-driven immune zonation of the liver promotes host defence. Nature 589:131–136

Gosalbes MJ, Durbán A, Pignatelli M, Abellan JJ, Jiménez-Hernández N, Pérez-Cobas AE, Latorre A, Moya A (2011) Meta-transcriptomic approach to analyze the functional human gut microbiota. PLoS One 6:e17447

Goulet V, Hebert M, Hedberg C et al (2012) Incidence of listeriosis and related mortality among groups at risk of acquiring listeriosis. Clin Infect Dis 54:652–660

Gracias KS, McKillip JL (2004) A review of conventional detection and enumeration methods for pathogenic bacteria in food. Can J Microbiol 50:883–890

Grada A, Weinbrecht K (2013) Next-generation sequencing: methodology and application. J Invest Dermatol 133:1–4

Guarner F, Malagelada JR (2003) Gut flora in health and disease. Lancet 361:512–519

Guillen-Guio B, Hernandez-Beeftink T, Ciuffreda L et al (2020) Could lung bacterial dysbiosis predict ICU mortality in patients with extra-pulmonary sepsis? A proof-of-concept study. Intensive Care Med 46:2118–2120

Guoshuai J, Xiaomeng X, Zemngdan G, Xingxing H, Qi P, Hanbing Z, Yi WA (2022) Rapid and high sensitivity RNA detection based on NASBA and G4-ThT fluorescent biosensor. Sci Rep 12:10076

Gurdeep Singh R, Tanca A, Palomba A, Van der Jeugt F et al (2019) Unipept 4.0: functional analysis of metaproteome data. J Proteome Res 18:606–615

Hallin M, Deplano A, Struelens MJ (2012) Molecular typing of bacterial pathogens: a tool for the epidemiological study and control of infectious diseases. In: New frontiers of molecular epidemiology of infectious diseases. Springer, Cham, pp 9–25

Hamady M, Knight R (2009) Microbial community profiling for human microbiome projects: tools, techniques, and challenges. Genome Res 19:1141–1152

Handelsman J, Rondon MR, Brady SF, Clardy J, Goodman RM (1998) Molecular biological access to the chemistry of unknown soil microbes: a new frontier for natural products. Chem Biol 5: R245–R249

Harmsen HJ, Wildeboer-Veloo AC, Raangs GC et al (2000) Analysis of intestinal flora development in breast-fed and formula-fed infants by using molecular identification and detection methods. J Paediatr Gastroenterol Nutr 30:61–67

He KW, Wen LB, Wang YS, Lu CP (2015) Development of real-time PCR assay for detection of porcine circovirus-like virus P1 in domestic pigs in China. BMC Vet Res 11:1–8

Henriquez FL, Mooney R, Bandel T, Giammarini E (2021) Paradigms of protist/bacteria symbioses affecting human health: *Acanthamoeba* species and *Trichomonas vaginalis*. Front Microbiol 11: 616213

Herbst FA, Lunsmann V, Kjeldal H, Jehmlich N, Tholey A, von Bergen M et al (2016) Enhancing meta-proteomics: the value of models and defined environmental microbial systems. Proteomics 16:783–798

Heyer R, Benndorf D, Kohrs F, De Vrieze J, Boon N, Hoffmann M, Rapp E, Schluter A, Sczyrba A, Reichl U (2016) Proteotyping of biogas plant microbiomes separates biogas plants according to process temperature and reactor type. Biotechnol Biofuels 9:155

Hill MJ (1997) Intestinal flora and endogenous vitamin synthesis. Eur J Cancer Prev 6:S43–S45

Huang DS, Zheng CH (2006) Independent component analysis-based penalized discriminant method for tumor classification using gene expression data. Bioinformatics 22:1855–1862

Huerta-Cepas J, Forslund K, Coelho LP, Szklarczyk D, Jensen LJ, von Mering C, Bork P (2017) Fast genome-wide functional annotation through orthology assignment by egg NOG-mapper. Mol Biol Evol 34:2115–2122

Huson DH, Weber N (2013) Microbial community analysis using MEGAN. Methods Enzymol 531:465–485

Hyatt D, Chen GL, LoCascio PF et al (2010) Prodigal: prokaryotic gene recognition and translation initiation site identification. BMC Bioinf 11:119

Jensen TK, Boye M, Ahrens P, Korsager B, Teglbjaerg PS, Lindboe CF, Møller K (2001) Diagnostic examination of human intestinal spirochetosis by fluorescent in situ hybridization for *Brachyspira aalborgi*, *Brachyspira pilosicoli*, and other species of the genus *Brachyspira* (Serpulina). J Clin Microbiol 39:4111–4118

Jernberg C, Löfmark S, Edlund C et al (2007) Long-term ecological impacts of antibiotic administration on the human intestinal microbiota. ISME J 1:56–66

Jia G, Xu X, Guo Z, Hu X, Pan Q, Zhu H, Yi W (2022) A rapid and high sensitivity RNA detection based on NASBA and G4-ThT fluorescent biosensor. Sci Rep 12:10076

Johnson CH, Julijana I, Gary S (2016) Metabolomics: beyond biomarkers and towards mechanisms. Nat Rev Mol Cell Biol 17:451–459

Kan J, Hanson TE, Ginter JM, Wang K, Chen F (2005) Metaproteomic analysis of Chesapeake Bay microbial communities. Saline Syst 1:7

Kanehisa M, Goto S (2000) KEGG: Kyoto encyclopaedia of genes and genomes. Nucleic Acids Res 28:27–30

Kato S, Itoh T, Ohkuma M (2020) Complete genome sequence of *Athalassotoga saccharophila* strain NAS-01, a deep-branching thermophilic lineage in the phylum Thermotogae. Microbiol Resour Announc 9:10–1128

Keer JT, Birch L (2003) Molecular methods for the assessment of bacterial viability. J Microbiol Methods 53:175–183

Khalighi M, Sommeria-Klein G, Gonze D et al (2022) Quantifying the impact of ecological memory on the dynamics of interacting communities. PLoS Comput Biol 18:e1009396

Khan R, Petersen FC, Shekhar S (2019) Commensal bacteria: an emerging player in defense against respiratory pathogens. Front Immunol 10:1203

Kho ZY, Lal SK (2018) The human gut microbiome – a potential controller of wellness and disease. Front Microbiol 9:1835

Kleiner M (2019) Meta-proteomics: much more than measuring gene expression in microbial communities. mSystems 4:e00115–e00119

Koboldt DC, Zhang Q, Larson DE, Shen D, McLellan MD et al (2012) VarScan 2: somatic mutation and copy number alteration discovery in cancer by exome sequencing. Genome Res 22:568–576

Koeth RA, Wang Z, Levison BS, Buffa JA, Org E et al (2013) Intestinal microbiota metabolism of L-carnitine, a nutrient in red meat, promotes atherosclerosis. Nat Med 19:576–585

Kõjalg U, Larsson K-H, Abarenkov K, Nilsson RH, Alexander IJ, Eberhardt U, Erland S (2005) UNITE: a database providing web-based methods for the molecular identification of *Ectomycorrhizal* fungi. New Phytol 166:1063–1068

Kolmeder CA, Salojarvi J, Ritari J, de Been M et al (2016) Faecal metaproteomic analysis reveals a personalized and stable functional microbiome and limited effects of a probiotic intervention in adults. PLoS One 11:e0153294

Kuehnbaum NL, Britz-McKibbin P (2013) New advances in separation science for metabolomics: resolving chemical diversity in a post-genomic era. Chem Rev 113:2437–2468

Kurkela S, Brown DW (2009) Molecular diagnostic techniques. Medicine 37:535–540

Lambeth SM, Carson T, Lowe J, Ramaraj T, Leff JW, Luo L et al (2015) Composition, diversity, and abundance of gut microbiome in prediabetes and type-2 diabetes. J Diabetes Obes 2:1

Langille MG, Zaneveld J, Caporaso JG, McDonald D, Knights D, Reyes JA et al (2013) Predictive functional profiling of microbial communities using 16S rRNA marker gene sequences. Nat Biotechnol 31:814–821

Lau LT, Reid SM, King DP, Lau AMF, Shaw AE, Ferris NP, Yu ACH (2008) Detection of foot-and-mouth disease virus by nucleic acid sequence-based amplification (NASBA). Vet Microbiol 126:101–110

Laupland KB, Valiquette L (2013) The changing culture of the microbiology laboratory. Can J Inf Dis Med Microbiol 24:125–128

LeBlanc JG, Milani C, de Giori GS, Sesma F, van Sinderen D, Ventura M (2013) Bacteria as vitamin suppliers to their host: a gut microbiota perspective. Curr Opin Biotechnol 24:160–168

Lee Y, Lee EK, Cho YW, Matsui T, Kang IC, Kim TS, Han MH (2003) Proteo chip: a highly sensitive protein microarray prepared by a novel method of protein immobilization for application of protein-protein interaction studies. Proteomics 3:2289–2304

Lefmann M, Schweickert B, Buchholz P et al (2006) Evaluation of peptide nucleic acid-fluorescence in situ hybridization for identification of clinically relevant mycobacteria in clinical specimens and tissue sections. J Clin Microbiol 44:3760–3767

Leggett RM, Alcon-Giner C, Heavens D, Caim S, Brook TC, Kujawska M et al (2020) Rapid MinION profiling of preterm microbiota and antimicrobial-resistant pathogens. Nat Microbiol 5:430–442

Ley TJ, Mardis ER, Ding L, Fulton B, McLellan MD, Chen K, Dooling D, Dunford-Shore BH, McGrath S, Hickenbotham M, Cook L (2008) DNA sequencing of a cytogenetically normal acute myeloid leukaemia genome. Nature 456:66–72

Li D et al (2015) MEGAHIT: an ultra-fast single-node solution for large and complex metagenomics assembly via succinct de Bruijn graph. Bioinformatics 31:1674–1676

Loomis EW, Eid JS, Peluso P, Yin J, Hickey L, Rank D, McCalmon S et al (2013) Sequencing the unquenchable: expanded CGG-repeat alleles of the fragile X gene. Genome Res 23:121–128

Loens K, Beck T, Ursi D, Overdijk M, Sillekens P, Goossens H, Ieven M (2008) Evaluation of different nucleic acid amplification techniques for the detection of *M. pneumoniae*, *C. pneumoniae* and *Legionella* spp. in respiratory specimens from patients with community-acquired pneumonia. J Microbiol Methods 73:257–262

Lopez-Lopez O, Cerdan ME, Gonzalez-Siso MI (2013) Hot spring metagenomics. Life 3:308–320

Lorenz P, Eck J (2005) Metagenomics and industrial applications. Nat Rev Microbiol 3:510

Mackay IM (2004) Real-time PCR in the microbiology laboratory. Clin Microbiol Infect 10:190–212

Maiwald M, von Herbay A, Fredricks DN et al (2003) Cultivation of *Tropheryma whipplei* from cerebrospinal fluid. J Infect Dis 188:801–808

Malic S, Hill KE, Hayes A, Percival SL, Thomas DW, Williams DW (2009) Detection and identification of specific bacteria in wound biofilms using peptide nucleic acid fluorescent in situ hybridization (PNAFISH). Microbiology 155:2603–2611

Malla MA, Dubey A, Kumar A, Yadav S, Hashem A, Abd Allah EF (2019) Exploring the human microbiome: the potential future role of next-generation sequencing in disease diagnosis and treatment. Front Immunol 9:2868

Man WH, de Steenhuijsen Piters WA, Bogaert D (2017) The microbiota of the respiratory tract: gatekeeper to respiratory health. Nat Rev Microbiol 15:259–270

Mangrola AV, Dudhagara P, Koringa P, Joshi CG, Patel RK (2015) Shotgun metagenomic sequencing-based microbial diversity assessment of Lasundra hot spring, India. Genomics data 4:73–75

Mason OU, Hazen TC, Borglin S et al (2012) Metagenome, metatranscriptome and single-cell sequencing reveal microbial response to deep-water horizon oil spill. ISME J 6:1715–1727

Matsuo Y, Komiya S, Yasumizu Y et al (2021) Full-length 16S rRNA gene amplicon analysis of human gut microbiota using MinION™ nanopore sequencing confers species-level resolution. BMC Microbiol 21:35

Maurice CF, Haiser HJ, Turnbaugh PJ (2013) Xenobiotics shape the physiology and gene expression of the active human gut microbiome. Cell 152:39–50

McDonald D et al (2012) An improved Greengenes taxonomy with explicit ranks for ecological and evolutionary analyses of bacteria and archaea. ISME J 6:610–618

Meinicke P (2015) UProC tools for ultra-fast protein domain classification. Bioinformatics 31: 1382–1388

Mertowska P, Mertowski S, Wojnicka J, Korona-Głowniak I, Grywalska E, Błażewicz A, Załuska W (2021) A link between chronic kidney disease and gut microbiota in immunological and nutritional aspects. Nutrients 13:3637

Mohajeri MH, Brummer RJ, Rastall RA, Weersma RK et al (2018) The role of the microbiome for human health: from basic science to clinical applications. Eur J Nutr 57:1–14

Moniruzzaman M, Wurch LL, Alexander H, Dyhrman ST, Gobler CJ, Wilhelm SW (2017) Virus-host relationships of marine single-celled eukaryotes resolved from metatranscriptomics. Nat Commun 8:16054

Moreno Gallego JL et al (2019) Virome diversity correlates with intestinal microbiome diversity in adult monozygotic twins. Cell Host Microbe 25:261–272.e5

Mukhopadhya I et al (2019) The gut virome: the 'missing link' between gut bacteria and host immunity? Ther Adv Gastroenterol 12:1756284819836620

Muth T, Benndorf D, Reichl U, Rapp E, Martens L (2013) Searching for a needle in a stack of needles: challenges in metaproteomics data analysis. Mol BioSyst 9:578–585

Muth T, Kohrs F, Heyer R, Benndorf D, Rapp E, Reichl U, Martens L, Renard BY (2018) MPA portable: a stand-alone software package for analyzing metaproteome samples on the go. Anal Chem 90:685–689

Muthukumar A, Zitterkopf NL, Payne D (2008) Molecular tools for the detection and characterization of bacterial infections: a review. Lab Med 39:430–436

Namiki T et al (2012) Meta velvet: an extension of velvet assembler to de novo metagenome assembly from short sequence reads. Nucleic Acids Res 40:155

Nathan S, Chieng S, Kingsley PV, Mohan A, Podin Y (2018) Melioidosis in Malaysia: incidence, clinical challenges, and advances in understanding pathogenesis. Trop Med Inf Dis 3:25

Newgard CB, An J, Bain JR, Muehlbauer MJ et al (2009) A branched-chain amino acid-related metabolic signature that differentiates obese and lean humans and contributes to insulin resistance. Cell Metab 9:311–326

Ng KM, Ferreyra JA, Higginbottom SK, Lynch JB et al (2013) Microbiota-liberated host sugars facilitate post-antibiotic expansion of enteric pathogens. Nature 502:96–99

Niba ETE, Vu DC, Nguyen NK, Tran TH, Lee T, Takeshima Y, Matsuo M (2014) Validation of ambiguous MLPA results by targeted next-generation sequencing discloses a nonsense mutation in the DMD gene. Clin Chim Acta 436:155–159

Nummi M, Mannonen L, Puolakkainen M (2015) Development of a multiplex real-time PCR assay for detection of *Mycoplasma pneumoniae*, *Chlamydia pneumoniae* and mutations associated with macrolide resistance in *Mycoplasma pneumoniae* from respiratory clinical specimens. SpringerPlus 4:1–8

Nurk S et al (2017) metaSPAdes: a new versatile metagenomic assembler. Genome Res 27:824–834

Oleary NA, Wright MW, Brister JR, Ciufo S, Haddad D, Mcveigh R et al (2016) Reference sequence (RefSeq) database at NCBI: current status, taxonomic expansion, and functional annotation. Nucleic Acids Res 44:733–745

Oliver DM, VanNiekerk M, Kay D, Heathwaite AL, Porter J et al (2014) Opportunities and limitations of molecular methods for quantifying microbial compliance parameters in EU bathing waters. Environ Int 64:124–128

Opal SM (2010) A brief history of microbiology and immunology. In: Artenstein A (ed) Vaccines: a biography. Springer, New York

Pal G, Srivastava S (2014) Cloning and heterologous expression of plnE, -F, -J and -K genes derived from soil metagenome and purification of active plantaricin peptides. Appl Microbiol Biotechnol 98:1441–1447

Palau-Rodriguez M, Tulipani S, Isabel Queipo-Ortuño M, Urpi-Sarda M, Tinahones FJ, Andres-Lacueva C (2015) Metabolomic insights into the intricate gut microbial-host interaction in the development of obesity and type 2 diabetes. Front Microbiol 27(6):1151

Palka-Santini M, Cleven BE, Eichinger L, Krönke M, Krut O (2009) Large- scale multiplex PCR improves pathogen detection by DNA microarrays. BMC Microbiol 9:1–14

Palmer A, Trede D, Alexandrov T (2016) Where imaging mass spectrometry stands: here are the numbers. Metabolomics 2:107

Panda AK, Bisht SS, De Mandal S, Kumar NS (2019) Microbial diversity of thermophiles through the lens of nextgeneration sequencing. In: Microbial diversity in the genomic era. Academic Press, pp 217–226

Pap B, Gyorkei Á, Boboescu IZ, Nagy IK, Biro T, Kondorosi E, Maroti G (2015) Temperature-dependent transformation of biogas-producing microbial communities points to the increased importance of hydrogenotrophic methanogenesis under thermophilic operation. Bioresour Technol 177:375–380

Peng Y et al (2012) IDBA-UD: a de novo assembler for single-cell and metagenomic sequencing data with highly uneven depth. Bioinformatics 28:1420–1428

Pereira R, Oliveira J, Sousa M (2020) Bioinformatics and computational tools for next-generation sequencing analysis in clinical genetics. J Clin Med 9:132

Pérez-Cobas AE, Gomez-Valero L, Buchrieser C (2020) Metagenomic approaches in microbial ecology: an update on whole-genome and marker gene sequencing analyses. Microb Genom 6: 000409

Peris-Vicente J, Carda-Broch S, Esteve-Romero J (2015) Validation of rapid microbiological methods. J Lab Autom 20:259–264

Pickup RW, Rhodes G, Hermon-Taylor J (2003) Monitoring bacterial pathogens in the environment: advantages of a multilayered approach. Curr Opin Biotechnol 14:319–325

Plutchik R (2001) The nature of emotions: human emotions have deep evolutionary roots, a fact that may explain their complexity and provide tools for clinical practice. Am Sci 89:344–350

Pompei A, Cordisco L, Amaretti A, Zanoni S, Matteuzzi D, Rossi M (2007) Folate production by bifidobacteria as a potential probiotic property. Appl Environ Microbiol 73:179–185

Prateek B, Narang A, Minz RW (2010) Neonatal cytomegalovirus infection: diagnostic modalities available for early disease detection. Indian J Pediatr 77:77–79

Qin J et al (2010) A human gut microbial gene catalogue established by metagenomic sequencing. Nature 464:59–65

Ram RJ, Verberkmoes NC, Thelen MP, Tyson GW et al (2005) Community proteomics of a natural microbial biofilm. Science 308:1915–1920

Reuter JA, Spacek DV, Snyder MP (2015) High-throughput sequencing technologies. Mol Cell 58:586–597

Rho M, Tang H, Ye Y (2010) FragGeneScan: predicting genes in short and error-prone reads. Nucleic Acids Res 38:191

Rhoads DD, Wolcott RD, Sun Y, Dowd SE (2012) Comparison of culture and molecular identification of bacteria in chronic wounds. Int J Mol Sci 13:2535–2550

Riaz MR, Preston GM, Mithani A (2020) MAPPS: a web-based tool for metabolic pathway prediction and network analysis in the postgenomic era. ACS Synth Biol 15:1069–1082

Riffle M, May DH, Timmins-Schiffman E, Mikan MP, Jaschob D, Noble WS et al (2017) MetaGOmics: a web-based tool for peptide-centric functional and taxonomic analysis of metaproteomics data. Proteome 6:2

Rodríguez JM, Murphy K, Stanton C, Ross RP, Kober OI, Juge N, Avershina E, Rudi K, Narbad A, Jenmalm MC, Marchesi JR, Collado MC (2015) The composition of the gut microbiota throughout life, with an emphasis on early life. Microb Ecol Health Dis 26:26050

Rodríguez-Valera F (2004) Environmental genomics, the big picture? FEMS Microbiol Lett 231: 153–158

Saeb AT, Abouelhoda M, Selvaraju M, Althawadi SI, Mutabagani M et al (2017) The use of next-generation sequencing in the identification of a fastidious pathogen: a lesson from a clinical setup. Evol Bioinforma 12:1176934316686072

Sakai T, Kohzaki K, Watanabe A, Tsuneoka H, Shimadzu M (2012) Use of DNA microarray analysis in diagnosis of bacterial and fungal endophthalmitis. Clin Ophthalmol 6:321

Samarkos M, Mastrogianni E, Kampouropoulou O (2018) The role of gut microbiota in *Clostridium difficile* infection. Eur J Intern Med 50:28–32

Sajid M, Srivastava S, Kumar A, Kumar A, Singh H, Bharadwaj M (2021) Bacteriome of moist smokeless tobacco products consumed in India with emphasis on the predictive functional potential. Front Microbiol 12:3908

Saxena R, Dhakan DB, Mittal P, Waiker P, Chowdhury A, Ghatak A, Sharma VK (2017) Metagenomic analysis of hot springs in Central India reveals hydrocarbon degrading thermophiles and pathways essential for survival in extreme environments. Front Microbiol 7:2123

Schloss PD et al (2009) Introducing Mothur: open-source, platform-independent, community-supported software for describing and comparing microbial communities. Appl Environ Microbiol 75:7537–7541

Seemann T (2014) Prokka: rapid prokaryotic genome annotation. Bioinformatics 30:2068–2069

Sehli S, Allali I, Chahboune R, Bakri Y, Al Idrissi N, Hamdi S, Nejjari C, Amzazi S, Ghazal H (2021) Metagenomics approaches to investigate the gut microbiome of COVID-19 patients. Bioinform Biol Insights 15:1177932221999428

Sender R, Fuchs S, Milo R (2016) Revised estimates for the number of human and bacteria cells in the body. PLoS Biol 14:e1002533

Shan S, Ko LS, Collins RA, Wu Z, Chen J et al (2003) Comparison of nucleic acid-based detection of avian influenza H5N1 with virus isolation. Biochem Biophys Res Commun 302:377–383

Sharon I, Banfield JF (2013) Microbiology. Genomes from metagenomics. Science 342:1057–1058

Simon GL, Gorbach SL (1984) Intestinal flora in health and disease. Gastroenterology 86:174–193

Skowron K, Bauza-Kaszewska J, Kraszewska Z, Wiktorczyk-Kapischke N, Grudlewska-Buda K (2021) Human skin microbiome: impact of intrinsic and extrinsic factors on skin microbiota. Microorganisms 9:543

Soini H, Musser JM (2001) Molecular diagnosis of mycobacteria. Clin Chem 47:809–814

Srivastava A, Mishra S, Garg PK, Dubey AK, Deo SVS, Verma D (2022) Comparative and analytical characterization of the oral bacteriome of smokeless tobacco users with oral squamous cell carcinoma. Appl Microbiol Biotechnol 106(11):4115–4128

Straub CT, Counts JA, Nguyen DMN, Wu CH, Zeldes BM, Crosby JR, Conway JM, Otten JK, Lipscomb GL et al (2018) Biotechnology of extremely thermophilic archaea. FEMS Microbiol Rev 42:543–578

Tang ZZ, Chen G, Hong Q, Huang S, Smith HM, Shah RD, Scholz M, Ferguson JF (2019) Multi-omic analysis of the microbiome and metabolome in healthy subjects reveals microbiome-dependent relationships between diet and metabolites. Front Genet 10:454

Tanner ACR, Kressirer CA, Rothmiller S, Johansson I, Chalmers NI (2018) The caries microbiome: implications for reversing dysbiosis. Adv Dent Res 29:78–85

Tatusov RL, Galperin MY, Natale DA, Koonin EV (2000) The COG database: a tool for genome-scale analysis of protein functions and evolution. Nucleic Acids Res 28:33–36

Truong DT, Tett A, Pasolli E, Huttenhower C, Segata N (2017) Microbial strain-level population structure and genetic diversity from metagenomes. Genome Res 27:626–638

Turnbaugh PJ et al (2007) The human microbiome projects. Nature 449:804–810

Tyakht A, Kostryukova E, Popenko A et al (2013) Human gut microbiota community structures in urban and rural populations in Russia. Nat Commun 4:2469

Ugarte A, Vicedomini R, Bernardes J, Carbone A (2018) A multi-source domain annotation pipeline for quantitative metagenomic and meta-transcriptomic functional profiling. Microbiome 6:149

Uniprot C (2019) UniProt: a worldwide hub of protein knowledge. Nucl Acids 47:515

Van Belkum A, Burnham CAD, Rossen JW, Mallard F, Rochas O, Dunne WM (2020) Innovative and rapid antimicrobial susceptibility testing systems. Nat Rev Microbiol 18:299–311

Vega-Magana N, Sánchez-Sánchez R, Hernández-Bello J, Venancio-Landeros AA et al (2021) RT-qPCR assays for rapid detection of the N501Y, 69-70del, K417N, and E484K SARS-CoV-2 mutations: a screening strategy to identify variants with clinical impact. Front Cell Infect Microbiol 11:672562

Verma D, Garg PK, Dubey AK (2018) Insights into the human oral microbiome. Arch Microbiol 200:525–540

Vibin J, Chamings A, Collier F, Klaassen M, Nelson TM et al (2018) Metagenomics detection and characterisation of viruses in faecal samples from Australian wild birds. Sci Rep 8:1–23

Vishwakarma A, Verma D (2021) Microorganisms: crucial players of smokeless tobacco for several health attributes. Appl Microbiol Biotechnol 105:6123–6132

Vishwakarma A, Srivastava A, Mishra S, Verma D (2023) Taxonomic and functional profiling of Indian smokeless tobacco bacteriome uncovers several bacterial-derived risks to human health. World J Microbiol Biotechnol 39:20

Wang RF, Beggs ML, Robertson LH, Cerniglia CE (2002) Design and evaluation of oligonucleotide-microarray method for the detection of human intestinal bacteria in fecal samples. FEMS Microbiol Lett 213:175–182

Wang Q et al (2007) Naive Bayesian classifier for rapid assignment of rRNA sequences into the new bacterial taxonomy. Appl Environ Microbiol 73:5261–5267

Wang Y, Nie M, Wang O, Li Y, Jiang Y, Li M, Xia W, Xing X (2019) Genetic screening in a large Chinese cohort of childhood onset hypoparathyroidism by next generation sequencing combined with TBX1MLPA. J Bone Miner Res 34:2254–2263

Wilke A, Bischof J, Gerlach W, Glass E, Harrison T, Keegan KP et al (2016) The MG-RAST metagenomics database and portal in 2015. Nucleic Acids Res 44:D590–D594

Wilmes P, Bond PL (2006) Meta-proteomics: studying functional gene expression in microbial ecosystems. Trends Microbiol 14:92–97

Wishart D (2016) Emerging applications of metabolomics in drug discovery and precision medicine. Nat Rev Drug Discov 15:473–484

Wood DE, Salzberg SL (2014) Kraken: ultrafast metagenomic sequence classification using exact alignments. Genome Biol 15:46

Yamada T, Letunic I, Okuda S, Kanehisa M, Bork P (2011) iPath2.0: interactive pathway explorer. Nucleic Acids Res 39:412–415

Yao R, Zhang C, Yu T, Li N, Hu X, Wang X, Wang J, Shen Y (2017) Evaluation of three read-depth based CNV detection tools using whole-exome sequencing data. Mol Cytogenet 10:1–7

Yarieva OO (2022) Criteria for early diagnosis and treatment of candidiasis of the oral mucosa. Middle Eur Sci Bull 24:122–124

Yasir M, Angelakis E, Bibi F, Azhar EI, Bachar D, Lagier JC et al (2015) Comparison of the gut microbiota of people in France and Saudi Arabia. Nutr Diabetes 5:e153

Ye Y, Doak TG (2009) A parsimony approach to biological pathway reconstruction/inference for genomes and metagenomes. PLoS Comput Biol 5:e1000465

Yilmaz P et al (2014) The SILVA and "all-species living tree project (LTP)" taxonomic frameworks. Nucleic Acids Res 42:643–648

You Y, Fu C, Zeng X, Fang D, Yan X, Sun B, Xiao D, Zhang J (2008) A novel DNA microarray for rapid diagnosis of enteropathogenic bacteria in stool specimens of patients with diarrhoea. J Microbiol Methods 75:566–571

Zhang X, Wu S, Li K, Shuai J, Dong Q, Fang W (2012) Peptide nucleic acid fluorescence in situ hybridization for identification of *Listeria* genus, *Listeria monocytogenes* and *Listeria ivanovii*. Int J Food Microbiol 157:309–313

Zhang A, Sun H, Qiu S et al (2013a) Metabolomics in non-invasive breast cancer. Clin Chim Acta 424:3–7

Zhang A, Sun H, Qiu S et al (2013b) NMR-based metabolomics coupled with pattern recognition methods in biomarker discovery and disease diagnosis. Magn Reson Chem 51:549–556

Zhang A, Qiu S, Xu H et al (2014) Metabolomics in diabetes. Clin Chim Acta 429:106–110

Zhang S, Hu Z, Wang H (2018) A retrospective review of microbiological methods applied in studies following the Deepwater Horizon oil spill. Front Microbiol 9:520

Zhong C, Edlund A, Yang Y, McLean JS, Yooseph S (2016) Metagenome and metatranscriptome analyses using protein family profiles. PLoS Comput Biol 12:e1004991

Zhou C, Zhao H, Xiao X, Guo R, Wang Q, Chen H et al (2020) Metagenomic profiling of the pro-inflammatory gut microbiota in ankylosing spondylitis. J Autoimmun 107:102360

CRISPR-Cas Fundamentals and Advancements in Translational Biotechnology

12

Deshraj Deepak Kapoor, Shilpi Yadav, and Ravi Kr. Gupta

Abstract

With the advent of genome editing technologies that allow direct targeting and editing of genome sequences across nearly every eukaryotic cell, has made it possible to uncover hidden facts and regulation of genetic diseases and many other diverse applications by developing more precise cellular models. In the past decade, genome editing technologies have advanced rapidly and emerged as highly useful technologies in various fields ranging from basic to applied research, including biomedicine. Since the development of CRISPR-Cas gene editing system, at least 45 Cas protein families now being recognised. The term CRISPR stands for **C**lustered **R**egularly **I**nterspaced **S**hort **P**alindromic **R**epeats. CRISPR-Cas system present in bacteria can neutralise the invasion of virus by destroying the viral genome. Thus, CRISPR works as an immune system of bacteria and responsible for protecting them.

CRISPR-Cas system has two essential components; the guide RNA, which needs to match the target sequence, as well as Cas (CRISPR-associated protein), an endonuclease responsible for DNA double-strand breaks that lead to genome editing.

Scientists are using CRISPR-Cas system to correct errors in genomes and turn on or off genes in cells and organisms rapidly and efficiently. In this chapter, we have elaborated the CRISPR-Cas system history, its working mechanism and applications in various fields. We have also included the ethical issues and limitations of this advance technique.

D. D. Kapoor · R. K. Gupta (✉)
Department of Environmental Microbiology, Babasaheb Bhimrao Ambedkar University, Lucknow, India

S. Yadav
Department of Biotechnology, Babasaheb Bhimrao Ambedkar University, Lucknow, India

R. Sobti et al. (eds.), *Role of Microbes in Sustainable Development*,
https://doi.org/10.1007/978-981-99-3126-2_12

Keywords

CRISPR · Cas9 · Gene editing · Genetic disorders · Genome modification

12.1 Introduction

12.1.1 History

During 1970s to 1980s, the first genetically modified mice were created (Jaenisch and Mintz 1974), and the first genetically modified bacteria were able to produce insulin and somatostatin (Goeddel et al. 1979). An impressive modification has been made to organisms that have been very difficult, costly and time-consuming to achieve. Scientists undertook more research into genetic engineering techniques in order to overcome these challenges. It led to the development of innovative tools such as zinc finger nucleases (ZFNs), TALENs (Transcriptional activator-like effector nucleases) and now CRISPR-Cas system. Combined with the cleavage domain of the restriction enzyme Fokl, zinc finger nucleases can produce DNA double-strand breaks by using their specificity to recognise DNA as well as their robust, yet controlled, activity. In several organisms, ZFNs have been used to create site-specific modifications; however, these have not been widely used in microbes (Urnov et al. 2010). ZFNs have several disadvantages, including context dependency, design difficulty and inefficiency in multiple gene targeting. A TALEN can produce double-strand breaks due to its DNA-binding specificity combined with its cleavage domain. Despite minimal off-target effects, they are not easy to clone and have a limited capability to target multiple genes.

CRISPR-Cas system simplicity and specificity are thought to make it superior to ZFNs and TALENs when compared. ZFN and TALEN proteins need to be produced separately for each DNA target, whereas CRISPR requires only matching a sequence of the guide RNA to a target region that will lead the Cas enzyme to a specific point where double-strand breaks can be introduced. It also has the advantage of being highly efficient since changes can be made directly through the system by inserting RNAs encoding the Cas protein and gRNA. Since, we can introduce multiple guide RNAs simultaneously, the CRISPR-Cas system can produce multiple gene modifications simultaneously (Jiang et al. 2015).

The CRISPR-Cas system was first described in 1987 when an unusual repetitive DNA sequence was observed in the *E. coli* genome during the analysis of genes involved in phosphate metabolism. Bacteria began inserting 32 nucleotide spacer sequences at regular intervals between the repeats whenever they encountered phage DNA (Ishino et al. 1987). The term CRISPR was coined in 2002 by Jansen et al. The repeat sequences were later found in 90% of archaea and 40% of sequenced bacterial genomes, although their functions were unclear (Horvath and Barrangou 2010). CRISPR-Cas history was made in 2005 when it was discovered that the spacer sequences actually originate from the phage genome (Mojica et al. 2005).

There were subsequent proposals that the CRISPR-Cas system could be used as a defence against phage attacks by bacteria and archaea. A spacer DNA similar to the DNA of phages can be added or deleted in *Streptococcus thermophilus* to increase or decrease its resistance to phage attack (Barrangou et al. 2007). Thus in 2007, CRISPR-Cas9 system was experimentally demonstrated as an acquired immune system in prokaryotes. Currently, researchers have revealed many Cas-proteins, CRISPR-associated genes, protospacer adjacent motif (PAM), CRISPR-RNA (crRNA) and transactivating crRNA (tracrRNA) which provide detail information about working mechanism of CRISPR-Cas system (Bolotin et al. 2005).

12.1.2 Mode of Action of CRISPR-Cas System

Researchers have used CRISPR-Cas system as a gene editing tool capable of making targeted genetic changes to any organism's DNA. In response to foreign DNA invasion by a phage or plasmid, CRISPR/Cas9 is most commonly used, which combines Cas9 endonuclease with a short guide RNA (gRNA) that contains two parts: a target-specific CRISPR RNA (crRNA) and a helper transactivating RNA (tracrRNA). Cas9 is guided by gRNA to specific genomic loci based on complementary nucleotides base pairing between the crRNA and target sequences (Pattanayak et al. 2013). The target DNA sequence containing the protospacer adjacent motif (PAM) on the 5′ end also have complementary sequences to the gRNA (Anders et al. 2014). A PAM sequence is necessary for the Cas enzyme to complement and distinguishes between the bacterial DNA from invaders DNA (Marraffini and Sontheimer 2010). In recent studies, researchers have shown that gRNAs derived from the fusion of guide sequence-containing crRNAs with tracrRNAs work as individual components (Jinek et al. 2012). Cas9 endonuclease binds specific sequences to induce specific double-strand breaks (DSBs), the repair of which is carried out by two distinct mechanisms:

(a) Non-Homologous End Joining (NHEJ): It is an error prone mechanism that results in random insertions or deletions during the repair procedure (Jeggo 1998).

(b) Homology-Directed Repair (HDR): A method that results in precise nucleotide edits but is less efficient, as DNA is repaired using either endogenous or exogenous templates (Komor et al. 2017; Hsu et al. 2014). For this reason, CRISPR/Cas9 can be employed to manipulate genetic sequences by inducing NHEJ or HDR. A CRISPR region is found in the bacterial genome; that helps them defend themselves against viruses.

Three main steps are involved in CRISPR-Cas system working mechanism (Fig. 12.1):

1. *Adaptation*-Short segments of viral DNA are inserted into the CRISPR sequence as new spacers after they have been processed.
2. *Processing and assembly*-In bacterial DNA, CRISPR repeats and spacers are transcribed, and this produces short RNA molecules known as CRISPR-RNAs.

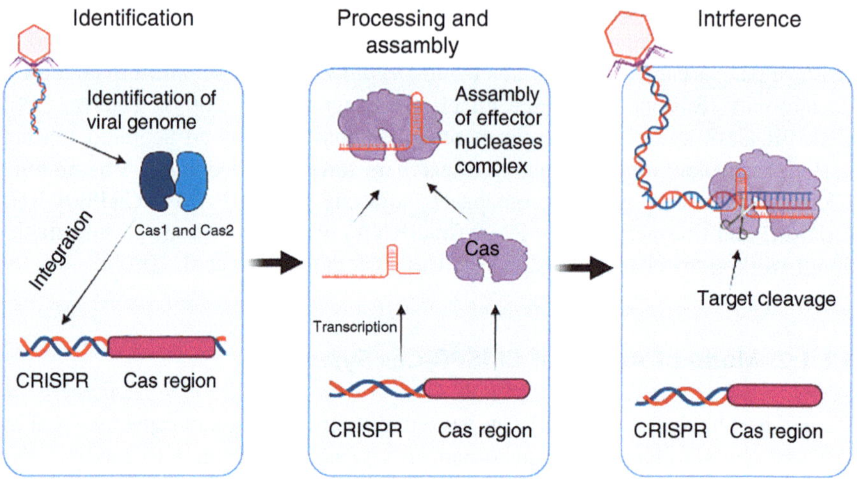

Fig. 12.1 Represents the working mechanism of CRISPR-Cas system in bacteria

3. *Targeting/Interference*-The machinery of bacteria is guided by CRISPR-RNAs to destroy the viral genetic material, since CRISPR-RNA sequences are copies of viral DNA acquired through adaptation; the sequences are exact matches of the viral genome, thus serving as an ideal guide.

12.2 Applications

The highly efficient and cost-effective CRISPR-Cas technology has many potential benefits and applications range vastly in translational biotechnology, from the introduction of point mutations to deletions, insertions, multiple-gene knock-downs and chromosomal rearrangements (Xue et al. 2014; Zhu et al. 2017). It has also potential applications in public health, species conservation, agriculture and basic research such as the ability to manipulate genetic sequences can be utilised to combat diseases such as malaria, dengue fever, Chagas and Lyme diseases. CRISPR could be used to analyse disease genes in viable human embryos and assist in immunotherapy, organoid engineering and development and identifying disease targets. In addition, it can be used to cure HIV, Haemophilia, Cancer, Duchenne muscular dystrophy, Amyotrophic lateral sclerosis, Sickle-cell anaemia, Cystic fibrosis and infertility. Some of the most potential applications of CRISPR-Cas system have been discussed below.

12.2.1 Genome Screening

Short hairpin RNAs (shRNAs) for RNA interference (RNAi) have been used in recent years to perturb transcript levels (Paddison et al. 2004). In this approach, the gene expression was incompletely abrogated and there were significant off-target effects which led to unexpected results in transcriptional analysis (Jackson and Linsley 2010). In some studies, Cas9, pooled guide RNA libraries and next-generation sequencing (NGS) have been used to adapt CRISPR for genome-scale screening (Schumann et al. 2015). CRISPR-Cas9 modified genomes can be used for genome-wide screening either by examining 20,000 genes or by studying one gene or signalling pathway in particular. Screening CRISPRs generally involves loss-of-function assays, which utilise indel-prone NHEJ repair or sequence repression. Certain applications also require the use of gain-of-function screens, which use endogenous HDR and CRISPR activation methods (Klann et al. 2017).

12.2.2 Cell Therapy

CRISPR-Cas technology has undergone a radical shift primarily linked to stem cells and immune cells (Chen et al. 2013). The treatment of cancer and autoimmune diseases using ex vivo gene-edited T cells has shown promising results (Bikard et al. 2013; Ren et al. 2017). An example is chimeric antigen receptor T Cells. By electroporation of this chimera with Cas9 ribonucleoproteins (RNPs), it is possible to target other receptors such as CXCR4, CCR5, PD-1 and CD7 on human cancer cells named as CAR-T cells therapy (Schumann et al. 2015). CRISPR-Cas9 technology has been approved for the treatment of muscle-invasive bladder cancer, castration-resistant prostate cancer, metastatic renal cancer and metastatic non-small cell lung cancer.

CRISPR/Cas9 was considered for the treatment of Cystic Fibrosis (CF) (Schwank et al. 2013). A successful correction of the most common mutation responsible for CF in intestinal organoids was achieved using adult intestinal stem cells. Study showed that the function of the CF transmembrane conductor receptor (CFTR) was restored once the mutation had been corrected.

12.2.3 HIV Treatment

HIV can also be treated with CRISPR/Cas9, though antiretroviral therapy is effective for treating HIV, there is no cure currently as the virus has been permanently incorporated into the host genome. It is possible to target HIV genome activity by using CRISPR/Cas9 technology. It inhibited the expression of the HIV gene and replication in the cells that are latently infected with HIV, without causing any toxic effects on the cells. Alternatively, cells can be immunised against HIV. As a result, this may prove to be a good therapeutic advancement in the quest to eradicate HIV. After further refinement, these findings may enable gene therapies or transplantation of genetically altered bone marrow stem cells or inducible pluripotent stem cells to eradicate HIV infection (Hu et al. 2014).

12.2.4 Editing of Human Zygotes

CRISPR/Cas9 for human germline editing appears to be at an infancy, based on the limited number of studies. However, human germline editing holds great promise for curing many genetic disorders which are lethal to human. The current accuracy of embryo mapping is limited due to off-target effects, embryo mosaicism and lack of access to the embryo. CRISPR/Cas9 germline studies in China revealed significant technical issues and made it apparent that further research on human zygotes is needed as well as comprehensive deliberation before any clinical use should be considered (Kang et al. 2016; Tang et al. 2017). CRISPR-Cas9 has achieved the highest success in human germline editing to date with intracytoplasmic sperm injection (ICSI) (Ma et al. 2017). CRISPR-Cas9 was further demonstrated to be effective in removing genetic mutations from embryos, but not in correcting the mutations in an established embryo (Tang et al. 2017).

12.2.5 Agriculture

Using CRISPR-Cas tools to edit the genomes of plant species has revolutionised agricultural science and provided new opportunities for crop improvement. Genetically engineered plants become more resistant to microbial pathogens (Dong and Ronald 2019). A biotechnological approach using CRISPR-Cas9 is able to alter the genetic code in a stable, permanent and heritable manner in order to reach a specific goal in agriculture.

The causative agent of citrus canker disease, *Xanthomonas citri* is one of the most important citrus pathogens, which severely reduces yields (Peng et al. 2017). One of the target genes present in plant was altered by the CRISPR-Cas9 system to provide resistance against *X. citri*. The promoter of *XLOB1* (lateral organ boundaries) gene has been mutated, which in turn results in the loss of the ability to acknowledge and respond to bacterial effectors hence showed increase resistance against infection (Jia et al. 2016) (Fig. 12.2).

CRISPR technology is not restricted to gene editing. It is possible to use CRISPR-Cas9 to activate (CRISPRa) or repress (CRISPRi) sequence-specific genes without altering the genome (Qi et al. 2013). In order to achieve this, deactivated Cas9 enzymes were developed (dCas9) that do not possess the catalytic domain for cleaving DNA but retain their ability to bind to target sequences (Gilbert et al. 2014). Cas9 has the ability to repress transcription by binding to target genes and preventing RNA polymerase activity. In prokaryotic genomes, CRISPRi works effectively, but is less effective when applied to eukaryotic genomes (Gilbert et al. 2013).

In a similar manner, CRISPRa has been made possible by pairing dCas9 with transcriptional activators, such as viral proteins (VP)16/VP64 or p65 and targeting them to gene promoters, resulting in gene transcription upregulation (Konermann et al. 2015; Perez-Pinera et al. 2013).

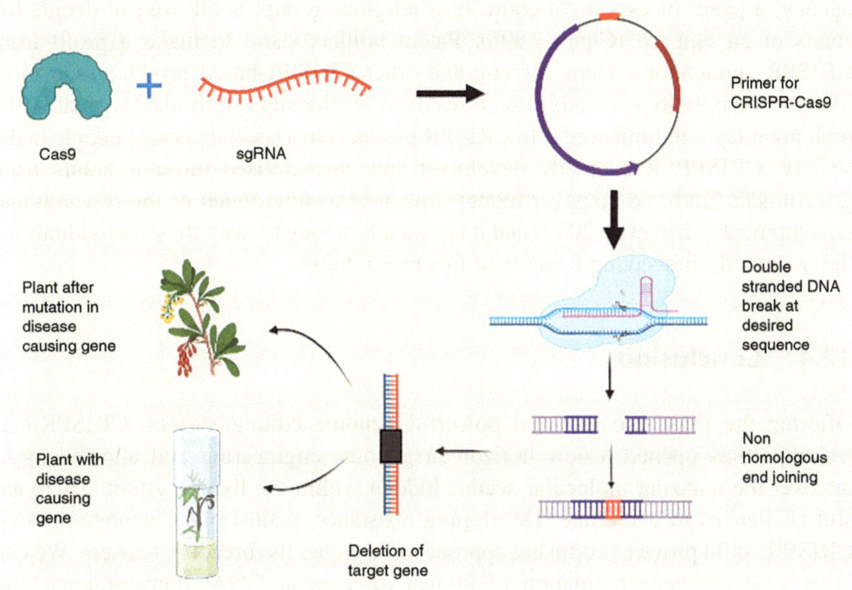

Fig. 12.2 Application of CRISPR-Cas9 system to edit the disease-causing gene in citrus plant

12.3 Limitations and Ethical Issues

There are incredible promising applications of CRISPR-Cas9 technology for the betterment of human life. But there are still some challenges to overcome. In vivo gene editing is highly challenging due to difficulties in the delivery of nuclease-encoding genes and guide RNAs to the appropriate cell types. As a means of safe delivery of cas9 nuclease genes and guide RNAs, a suitable vector must be used. In addition, there are potential off-target effects in the genome; non-intentional changes to the genome will have long-term effects on patients, including cancer.

CRISPR-Cas9 may have the greatest impact on human and its environment due to its potential applications and findings (Mulvihill et al. 2017). From an ethical perspective, use of CRISPR technology did not have the ethical issues pertaining to gene therapy and genetic engineering. In general, gene editing ethics can be divided into two groups: One which aims to correct defective genes (gene therapy) and the other one which aims to enhance physiologically normal genes (genetic enhancement). CRISPR technology does not violate the ethical issues if the gene editing is limited to the somatic cells. However, the genetic engineering of germline cells which could be inherited to the next generation should be carefully ethically reviewed (Sykora 2018). One of the major controversies about CRISPR technology emerges from its potential application in human embryos. Although group of scientists believe that experiments on human embryos after 14 days are ethically unacceptable, and no authority, whether it is the government, a law enforcement agency, a panel of experts, a court, or a religious group, is allowed to decide the status of an embryo (Charo 1995). Patent holders stand to make a profit from CRISPR applications. Gene therapy and other CRISPR-based products will most likely to be initially expensive. As a result, it is ethically questionable whether the high price-tag will limit access to CRISPR products to a special class of people in the society. CRISPR was mostly developed and characterised through grants from government funds, so taxpayer money was used to fund much of the research and development (Chen et al. 2015) and it is ethically wrong to deny these individuals for the potentially life-saving benefits of this technology.

12.4 Conclusions

Offering the most versatile and powerful genome editing system, CRISPR-Cas technique has opened a new horizon in genome engineering and allowed us to uncover the amazing molecular secrets hidden within the living system. There are still challenges to overcome. Developing resistance against plant pathogens using CRISPR could prove a promising approach to conquer the breeding barriers. We can investigate the gene regulation of human diseases at DNA, transcriptional and translational level by this revolutionary technology. As with every powerful tool, there are also potential risks involved. It is imperative that well-controlled, reproducible experiments and clinical trial research should be conducted in order to make truly informed decisions regarding ethically contentious areas. In the present, this is problematic since many international laws discourage research of this type or ban it outright; they also inhibit research from being funded. Due to this, it is difficult to determine the risks and benefits of a technology. Overall CRISPR-Cas system has been exploited for the benefit of human health in every aspect from curing of diseases to improvement of food and we hope in future it will also help in other untouched areas for the betterment of human and animal life.

References

Anders C, Niewoehner O, Duerst A, Jinek M (2014) Structural basis of PAM-dependent target DNA recognition by the Cas9 endonuclease. Nature. 513(7519):569–573. https://doi.org/10.1038/nature13579. Epub 2014 Jul 27. PMID: 25079318; PMCID: PMC4176945.

Barrangou R, Fremaux C, Deveau H, Richards M, Boyaval P, Moineau S, Romero DA, Horvath P (2007) CRISPR provides acquired resistance against viruses in prokaryotes. Science 315(5819): 1709–1712. https://doi.org/10.1126/science.1138140. PMID: 17379808.

Bikard D, Jiang W, Samai P, Hochschild A, Zhang F, Marraffini LA (2013) Programmable repression and activation of bacterial gene expression using an engineerined CRISPR-Cas system. Nucleic Acids Res 41:7429–7437

Bolotin A, Quinquis B, Sorokin A, Ehrlich SD (2005) Clustered regularly interspaced short palindrome repeats (CRISPRs) have spacers of extrachromosomal origin. Microbiology 151(8):2551–2561

Charo RA (1995) The hunting of the snark: the moral status of embryos, right-to-lifers, and Third World women. Stanford Law Pol Rev 6(2):11–37. PMID: 12645597.

Chen B, Gilbert LA, Cimini BA, Schnitzbauer J, Zhang W, Li G-W, Park J, Blackburn EH, Weissman JS, Qi LS, Huang B (2013. ISSN 0092-8674) Dynamic imaging of genomic loci in living human cells by an optimized CRISPR/Cas system. Cell 155(7):1479. https://doi.org/10.1016/j.cell.2013.12.001

Chen S, Sanjana NE, Zheng K, Shalem O, Lee K, Shi X et al (2015) Genome-wide CRISPR screen in a mouse model of tumor growth and metastasis. Cell 160:1246–1260

Dong OX, Ronald PC (2019) Genetic engineering for disease resistance in plants: recent progress and future perspectives. Plant Physiol 180:26–38

Gilbert LA, Larson MH, Morsut L, Liu Z, Brar GA, Torres SE, Stern-Ginossar N, Brandman O, Whitehead EH, Doudna JA, Lim WA, Weissman JS, Qi LS (2013) CRISPR-mediated modular RNA-guided regulation of transcription in eukaryotes. Cell. 154(2):442–451. https://doi.org/10.1016/j.cell.2013.06.044. Epub 2013 Jul 11. PMID: 23849981; PMCID: PMC3770145.

Gilbert LA, Horlbeck MA, Adamson B, Villalta JE, Chen Y, Whitehead EH, Guimaraes C, Panning B, Ploegh HL, Bassik MC, Qi LS, Kampmann M, Weissman JS (2014) Genome-scale CRISPR-mediated control of gene repression and activation. Cell. 159(3):647–661. https://doi.org/10.1016/j.cell.2014.09.029. Epub 2014 Oct 9. PMID: 25307932; PMCID: PMC4253859.

Goeddel DV, Kleid DG, Bolivar F, Heyneker HL, Yansura DG, Crea R, Hirose T, Kraszewski A, Itakura K, Riggs AD (1979) Expression in Escherichia coli of chemically synthesized genes for human insulin. Proc Natl Acad Sci U S A 76(1):106–110. https://doi.org/10.1073/pnas.76.1.106. PMID: 85300; PMCID: PMC382885.

Horvath P, Barrangou R (2010) CRISPR/Cas, the immune system of bacteria and archaea. Science 327(5962):167–170

Hsu PD, Lander ES, Zhang F (2014) Development and applications of CRISPR-Cas9 for genome engineering. Cell 157(6):1262–1278. https://doi.org/10.1016/j.cell.2014.05.010. PMID: 24906146; PMCID: PMC4343198.

Hu W, Kaminski R, Yang F et al (2014) RNA-directed gene editing specifically eradicates latent and prevents new HIV-1 infection. Proc Natl Acad Sci USA 111:11461–11466

Ishino Y, Shinagawa H, Makino K, Amemura M, Nakata A (1987) Nucleotide sequence of the iap gene, responsible for alkaline phosphatase isozyme conversion in Escherichia coli, and identification of the gene product. J Bacteriol 169:5429–5433

Jackson AL, Linsley PS (2010) Recognizing and avoiding siRNA off-target effects for target identification and therapeutic application. Nat Rev Drug Discov 9(1):57–67. https://doi.org/10.1038/nrd3010. PMID: 20043028.

Jaenisch R, Mintz B (1974) Simian virus 40 DNA sequences in DNA of healthy adult mice derived from preimplantation blastocysts injected with viral DNA. Proc Natl Acad Sci U S A 71(4):1250–1254. https://doi.org/10.1073/pnas.71.4.1250. PMID: 4364530; PMCID :PMC388203.

Jeggo PA (1998) DNA repair: PARP - another guardian angel? Curr Biol 8(2):R49–R51. https://doi.org/10.1016/s0960-9822(98)70032-6. PMID: 9427640.

Jia H, Orbovic V, Jones JB, Wang N (2016) Modification of the PthA4 effector binding elements in Type I CsLOB1 promoter using Cas9/sgRNA to produce transgenic Duncan grapefruit alleviating XccΔpthA4:dCsLOB1.3 infection. Plant Biotechnol J 14(5):1291–1301. https://doi.org/10.1111/pbi.12495. Epub 2015 Nov 2. PMID: 27071672.

Jiang Y, Chen B, Duan C, Sun B, Yang J, Yang S (2015) Multigene editing in the Escherichia coli genome via the CRISPR-Cas9 system. Appl Environ Microbiol. 81(7):2506–2514. https://doi.org/10.1128/AEM.04023-14. Epub 2015 Jan 30. Erratum in: Appl Environ Microbiol. 2016 Jun 15;82(12):3693. PMID: 25636838; PMCID: PMC4357945.

Jinek M, Chylinski K, Fonfara I, Hauer M, Doudna JA, Charpentier E (2012) A programmable dual-RNA-guided DNA endonuclease in adaptive bacterial immunity. Science 337(6096):816–821. https://doi.org/10.1126/science.1225829. Epub 2012 Jun 28. PMID: 22745249; PMCID: PMC6286148.

Kang X, He W, Huang Y et al (2016) Introducing precise genetic modifications into human 3PN embryos by CRISPR/Cas-mediated genome editing. J Assist Reprod Genet 33(5):581–588

Klann TS, Black JB, Chellappan M, Safi A, Song L, Hilton IB, Crawford GE, Reddy TE, Gersbach CA (2017) CRISPR-Cas9 epigenome editing enables high-throughput screening for functional regulatory elements in the human genome. Nat Biotechnol 35(6):561–568. https://doi.org/10.1038/nbt.3853. Epub 2017 Apr 3. PMID: 28369033; PMCID: PMC5462860.

Komor AC, Badran AH, Liu DR (2017) CRISPR-based technologies for the manipulation of eukaryotic genomes. Cell 168(1–2):20–36. https://doi.org/10.1016/j.cell.2016.10.044. Epub 2016 Nov 17. Erratum in: Cell. 2017 Apr 20;169(3):559. PMID: 27866654; PMCID: PMC5235943.

Konermann S, Brigham MD, Trevino AE, Joung J, Abudayyeh OO, Barcena C, Hsu PD, Habib N, Gootenberg JS, Nishimasu H, Nureki O, Zhang F (2015) Genome-scale transcriptional activation by an engineered CRISPR-Cas9 complex. Nature 517(7536):583–588. https://doi.org/10.1038/nature14136

Ma H, Marti-Gutierrez N, Park SW, Wu J, Lee Y, Suzuki K, Koski A, Ji D, Hayama T, Ahmed R, Darby H, Van Dyken C, Li Y, Kang E, Park AR, Kim D, Kim ST, Gong J, Gu Y, Xu X, Battaglia D, Krieg SA, Lee DM, Wu DH, Wolf DP, Heitner SB, Belmonte JCI, Amato P, Kim JS, Kaul S, Mitalipov S (2017) Correction of a pathogenic gene mutation in human embryos. Nature 548(7668):413–419. https://doi.org/10.1038/nature23305. Epub 2017 Aug 2. PMID: 28783728.

Marraffini LA, Sontheimer EJ (2010) CRISPR interference: RNA-directed adaptive immunity in bacteria and archaea. Nat Rev Genet 11(3):181–190. https://doi.org/10.1038/nrg2749. PMID: 20125085; PMCID: PMC2928866.

Mojica FJ, Díez-Villaseñor C, García-Martínez J, Soria E (2005) Intervening sequences of regularly spaced prokaryotic repeats derive from foreign genetic elements. J Mol Evol 60(2):174–182. https://doi.org/10.1007/s00239-004-0046-3. PMID: 15791728.

Mulvihill JJ, Capps B, Joly Y et al (2017) Ethical issues of CRISPR technology and gene editing through the lens of solidarity. Br Med Bull 122(1):17–29

Paddison PJ, Silva JM, Conklin DS, Schlabach M, Li M, Aruleba S, Balija V, O'Shaughnessy A, Gnoj L, Scobie K, Chang K, Westbrook T, Cleary M, Sachidanandam R, McCombie WR, Elledge SJ, Hannon GJ (2004) A resource for large-scale RNA-interference-based screens in mammals. Nature 428(6981):427–431. https://doi.org/10.1038/nature02370. PMID: 15042091.

Pattanayak V, Lin S, Guilinger JP, Ma E, Doudna JA, Liu DR (2013) High-throughput profiling of off-target DNA cleavage reveals RNA-programmed Cas9 nuclease specificity. Nat Biotechnol 31(9):839–843. https://doi.org/10.1038/nbt.2673. Epub 2013 Aug 11. PMID: 23934178; PMCID: PMC3782611.

Peng A, Chen S, Lei T, Xu L, He Y, Wu L, Yao L, Zou X (2017) Engineering canker-resistant plants through CRISPR/Cas9-targeted editing of the susceptibility gene CsLOB1 promoter in citrus. Plant Biotechnol J. 15(12):1509–1519. https://doi.org/10.1111/pbi.12733. Epub 2017 May 3. PMID: 28371200; PMCID: PMC5698050.

Perez-Pinera P, Kocak DD, Vockley CM, Adler AF, Kabadi AM, Polstein LR, Thakore PI, Glass KA, Ousterout DG, Leong KW, Guilak F, Crawford GE, Reddy TE, Gersbach CA (2013) RNA-guided gene activation by CRISPR-Cas9-based transcription factors. Nat Methods 10(10):973–976. https://doi.org/10.1038/nmeth.2600. Epub 2013 Jul 25. PMID: 23892895; PMCID: PMC3911785.

Qi LS, Larson MH, Gilbert LA, Doudna JA, Weissman JS, Arkin AP, Lim WA (2013) Repurposing CRISPR as an RNA-guided platform for sequence-specific control of gene expression. Cell 152(5):1173–1183. https://doi.org/10.1016/j.cell.2013.02.022. Erratum in: Cell 2021 Feb 4;184 (3):844. PMID: 23452860; PMCID: PMC3664290.

Ren J, Liu X, Fang C, Jiang S, June CH, Zhao Y (2017) Multiplex genome editing to generate universal CAR T cells resistant to PD1 inhibition. Clin Cancer Res 23(9):2255–2266. https://doi.org/10.1158/1078-0432.CCR-16-1300. Epub 2016 Nov 4. PMID: 27815355; PMCID: PMC5413401.

Schumann K, Lin S, Boyer E, Simeonov DR, Subramaniam M, Gate RE, Haliburton GE, Ye CJ, Bluestone JA, Doudna JA, Marson A (2015) Generation of knock-in primary human T cells

using Cas9 ribonucleoproteins. Proc Natl Acad Sci U S A 112(33):10437–10442. https://doi.org/10.1073/pnas.1512503112
Schwank G, Koo BK, Sasselli V et al (2013) Functional repair of CFTR by CRISPR/Cas9 in intestinal stem cell organoids of cystic fibrosis patients. Cell Stem Cell 13:653–658
Sykora P (2018) Germline gene therapy in the era of precise genome editing: how far should we go? In: Soniewicka M (ed) The ethics of reproductive genetics between utility, principles and virtues. Springer, pp 157–171
Tang L, Zeng Y, Du H et al (2017) CRISPR/Cas9-mediated gene editing in human zygotes using Cas9 protein. In: Molecular genetics and genomics, vol 292. Springer, Berlin Heidelberg, pp 525–533
Urnov FD, Rebar EJ, Holmes MC, Zhang HS, Gregory PD (2010) Genome editing with engineered zinc finger nucleases. Nat Rev Genet 11(9):636–646
Xue W, Chen S, Yin H, Tammela T, Papagiannakopoulos T, Joshi NS, Cai W, Yang G, Bronson R, Crowley DG, Zhang F, Anderson DG, Sharp PA, Jacks T (2014) CRISPR-mediated direct mutation of cancer genes in the mouse liver. Nature 514(7522):380–384. https://doi.org/10.1038/nature13589. Epub 2014 Aug 6. PMID: 25119044; PMCID: PMC4199937.
Zhu QM, Ko KA, Ture S, Mastrangelo MA, Chen MH, Johnson AD, O'Donnell CJ, Morrell CN, Miano JM, Lowenstein CJ (2017) Novel thrombotic function of a human SNP in STXBP5 revealed by CRISPR/Cas9 gene editing in mice. Arterioscler Thromb Vasc Biol 37(2):264–270. https://doi.org/10.1161/ATVBAHA.116.308614. Epub 2016 Dec 29. PMID: 28062498; PMCID: PMC5269475.

Part III

Gut Microbiome and Metabolic Disorders

Microbiome and Human Health: From Dysbiosis to Therapeutic Interventions

13

Gauri Garg Dhingra, Roshan Kumar, Utkarsh Sood, Princy Hira, Jasvinder Kaur, and Rup Lal

Abstract

Human body is a complex system housing multiple biomes such as virome, microbiome, and eukaryome which governs human health. Unicellular archaea, bacteria, virus, and multicellular eukaryotes together regulate the physiological functions along with the internal homeostatic mechanisms. With the next-generation sequencing technology, the composition of the "internal biome" and its role in health and disease has become much clearer. Joshua Lederberg coined the term "microbiome" in 2001, which defines the full complement of microbes (bacteria, viruses, fungi, and protozoa), their genes, and genomes in or on the human body. Human Microbiome Project was launched in 2007 through 2016 by the National Institutes of Health to characterize the genomic makeup of all microbes inhabiting the human body and analyze its role in health and disease. Bacteria outnumber the human cells tenfold and make up about 1–3% of body mass. The composition of the microbiome changes during the lifetime, impacting human physiology in healthy and diseased state by modulating the metabolic and immune functions. The current chapter provides an introduction to the "human

G. G. Dhingra · U. Sood
Department of Zoology, Kirori Mal College, University of Delhi, New Delhi, India

R. Kumar
Post-Graduate Department of Zoology, Magadh University, Bodh Gaya, Bihar, India

P. Hira
Department of Zoology, Maitreyi College, University of Delhi, New Delhi, India

J. Kaur
Department of Zoology, Gargi College, University of Delhi, New Delhi, India

R. Lal (✉)
Department of Zoology, Acharya Narendra Dev College, University of Delhi, New Delhi, Delhi, India

R. Sobti et al. (eds.), *Role of Microbes in Sustainable Development*,
https://doi.org/10.1007/978-981-99-3126-2_13

microbiome" and also compares the bacterial diversity in healthy and diseased state. A comparison of microbiome of Indian and Western population has also been added. Gut bacteria play a regulatory role in the metabolism and have a strong connection with brain, influencing behavior. Some bacterial species are found to be either abnormally abundant or reduced in certain diseases. Dysbiosis or imbalance in microbial flora has been recognized as a cause or consequence. Therefore there is a need to adopt certain therapeutic strategies for restoring the balance of microbiome.

Keywords

Microbiome · Dysbiosis · Gut microbiome · Homeostasis · Therapeutic strategies

13.1 Introduction

Microbes which include unicellular archaea, bacteria, virus, and multicellular eukaryotes are present in virtually every habitat on earth and in every organism. For example, a reference man weighing 70 kg would have approximately 3.8×10^{13} bacteria, which weigh about 0.2 kg and in the body, they work in a coordinated fashion performing various functions in metabolism (Sender et al. 2016). These trillions of microorganisms belong to thousands of different species (Kumar et al. 2022). The robustness in data generation and interpretation has shown the obvious role of these species in health homeostasis (Lynch and Pedersen 2016). Not only this, but also multiple reports suggest that human health is influenced by multiple microbiota interactions throughout their lives (Milani et al. 2017; Collado et al. 2012; Song et al. 2021). The multispecies microbial communities found in human hosts show a wide range of behavior depending on their microenvironment, which can be dynamic, interactive, commensal, or parasitic (Hou et al. 2022). Humans alone represent several ecological niches with different compositions and relative abundances of microbiota, such as the skin microbiome, oral microbiome, gut microbiome, urogenital microbiome (Hou et al. 2022). In order to maintain human health, the human gut microbiota play crucial roles in assisting in the breakdown of food substances and liberating nutrients inaccessible to the host otherwise (Kumar et al. 2020). Additionally, they modulate the immune system by triggering immune cell differentiation, protecting the host from pathogen colonization (Hou et al. 2022; Kumar et al. 2020). Microbes begin colonizing the neonatal gut immediately after birth. Early gut commensals are shown to be shaped by the immediate environment, especially the mother's microbiome (birth canal microbial interactions, followed by nursing) (Milani et al. 2017).

The gastrointestinal microbial composition majorly defines the state of health and disease of an individual through various metabolic processes. The recent scientific interpretations suggest that these microbial populations are not static and keep fluctuating based on environment and lifestyle (Flandroy et al. 2018; Nguyen et al. 2021). Coming on to the roles of microbes in the gut, they perform a vast range of functions such as degradation of undigested carbohydrates, production of vitamin B, antioxidants, protection against pathogens, etc. (Milani et al. 2017; Kumar et al. 2020). During the degradation of undigested carbohydrates in large intestine, these microbes produce the short-chain fatty acids (SCFA) through fermentation (den Besten et al. 2013). The food quantity and type determine the amount and type of SCFA production. Their concentration varies between 50 and 200 mM in the large intestine, mostly contributed by acetate (60%), propionate (25%), and butyrate (15%) (den Besten et al. 2013; McNabney and Henagan 2017). The *Bacteroidetes* phylum mainly produces acetate and propionate, while the butyrate is produced by the phylum *Firmicutes* (Magne et al. 2020). These SCFAs differ in their fate and tissue distribution as butyrate is the most preferred energy source by the gut mucosa, while the propionate mostly contributes to the gluconeogenesis in the liver and acetate is taken up for synthesis of essential molecules such as cholesterol, long-chain fatty acids, glutamine, and glutamate (Magne et al. 2020; Correa-Oliveira et al. 2016). These SCFAs are also known to regulate immune and inflammatory response and thus are considered to benefit health, including protection against colorectal cancer (Correa-Oliveira et al. 2016). Due to the abundance of these microbes in the lower gastrointestinal tract, this microbial mass is a significant contributor of fecal bulk and thus their presence and diversity can easily be detected with a scoop of fecal sample. Till now, there are numerous studies on human microbiome composition and it has been estimated that the collective human microbiota is composed of over 35000 bacterial species and over 10 million nonredundant genes (Jandhyala et al. 2015).

One of the major key players apart from environment and host genetics which defines the microbial relative abundance in a specific population is diet (Singh et al. 2017). The current paradigm shift in research mainly complies with the impact of different nutrients in shaping the gut microbial community. Nonetheless, defining the proper impact of a specific nutrient in a complex microbial niche still remains a challenge. Understanding this will be the main priority of the upcoming research.

13.2 Microbiome in Healthy Individuals and Its Impact

13.2.1 Gut Microbial Composition

The most abundant phyla in a healthy gut comprise *Firmicutes*, *Bacteroidetes*, *Actinobacteria*, and *Proteobacteria* (Xiao et al. 2021; Das and Nair 2019). The genus under firmicutes which are most abundant include *Clostridium*, *Lactobacillus*, *Streptococcus*, *Enterococcus*, *Ruminococcus*, *Faecalibacterium*, *Eubacterium*, *Catenibacterium*, *Megamonas*, *Veillonella*, etc. The two most abundant genera which belong to phylum *Bacteroidetes* include *Prevotella* and *Bacteroides*. *Bifidobacterium*, *Olsnella*, *Collinsella*, *Eggerthella* are the other dominant genera which belong to phylum *Actinobacteria*. *Eschericia* and *Shigella* are the abundant genera from phylum *Proteobacteria*. When the gut undergoes dysbiosis, the composition of microbes and their ecology shifts to an extent that it overcomes their resilience proficiencies. Thus, leading to the onset of variety of diseases namely inflammatory bowel disease (IBD) (Glassner et al. 2020), cardiovascular diseases (Witkowski et al. 2020), rheumatoid arthritis (Zaiss et al. 2021), colon cancer (Garrett 2019), depression (Lach et al. 2018), Parkinson's disease (Sampson et al. 2016), etc.

A healthy human gut microbiota predominantly differs between individuals based of factors such as diet, antibiotics, age, sex, host genetics, immunity (hyperimmunity or immunodeficiency), etc. and it keeps changing throughout the life (Das and Nair 2019). The massive data interpretation in the recent past focused on the microbe centric indicator of different populations. Based on the abundance of different taxa, the enterotypes are broadly divided in three categories: Enterotype 1: The best indicator of this enterotype is *Bacteroides* (a genus of Phylum *Bacteroidetes*); Enterotype 2: This enterotype is driven by genus *Prevotella* (another genus of Phylum *Bacteroidetes*) whose abundance is inversely proportional to *Bacteroides*, Enterotype 3: *Ruminococcus*, a genus of phylum *Firmicutes* distinguishes this enterotype (Costea et al. 2018). Countries like the USA, where people mostly consume animal-based diets and therefore their microbiota is mostly dominated by *Bacteroides*, while countries like India, where people prefer plant-based diets, the genus *Prevotella* dominates, suggesting the role of diet in shaping the microbial community (Costea et al. 2018). The third enterotype is common in European samples as the members of genus *Ruminococcus* are known to degrade mucins and subsequent hydrolysis of complex carbohydrates into simple sugars (Arumugam et al. 2011). The concept of enterotypes lies in the idea that they will be identifiable in the human microbiome of any cohort irrespective of age, gender, body weight, and national division.

Several studies have shown that microbial diversity in the form of a microbiome in the human body is extremely important for maintaining human health (Jandhyala et al. 2015; Das and Nair 2019; Clapp et al. 2017). As explained, dysbiosis of the microbiome has been linked to a large number of diseases (Glassner et al. 2020; Garrett 2019; El-Salhy et al. 2021; Horn et al. 2022). Therefore, it becomes extremely important to study what constitutes a healthy microbiome and what functions these microbes are performing. However, understanding what constitutes a healthy microbiome is challenging as the microbiome is affected by personal hygiene, genetics, gender, diet, lifestyle, and geography of stay for different individuals. Several populations' specific studies from all around the world have documented the diversity of microbes associated with different surfaces of the human body (Potbhare et al. 2022; Mahajan et al. 2022; Moffatt and Cookson 2017). Healthy individuals from different countries have been shown to have major differences in microbiome composition. In addition to this, the microbiome is reported to shift with the progression of the age of the individuals as well. Another limitation in deciphering what constitutes a healthy microbe is the presence of microbial dark matter, i.e., uncharacterized microbes.

The most widely studied microbiome from the human body is from the skin, gut, vagina, nasal cavity, oral cavity, conjunctiva, lung, urethra, bladder, uterus, placenta, and biliary tract (Table 13.1).

Table 13.1 The comparison of microbial diversity in Western and Indian populations

Microbiome	Western population	Indian population	References
Skin	Gram-positive genera: *Staphylococcus* spp., *Corynebacterium* spp., *Enhydrobacter* spp., *Micrococcus* spp., *Cutibacterium* spp., and *Veillonella* spp Gram-negative bacteria (GNB): *Roseomonas mucosa*, *Pseudomonas* spp., *Acinetobacter* spp., *Pantoea septica*, and *Moraxella osloensis* as commensal residents	*Staphylococcus*, *Bacillus*, *Corynebacterium*, and *Anaerococcus*, *Pseudomonas*, *Arthrobacter*, *Anaerococcus*, *Oceanobacillus*, *Cutibacterium*, *Acinetobacter*, *Salmonella*, *Moraxella*, *Pantoea*, *Enterobacter*, *Exiguobacterium*	Potbhare et al. (2022), Skowron et al. (2021)
Gut	*Lactobacillus*, *Bacillus*, *Clostridium*, *Enterococcus*, *Ruminicoccus*, *Bacteroides*, and *Prevotella*	*Prevotella*, *Dialister*, *Bacteroides*, *Megamonas*, and *Succinivibrio*	Rinninella et al. (2019), Chaudhari et al. (2020)
Vagina	*Lactobacillus* species such as *L. crispatus*, *L. gasseri*, and *L, jensenii*	In North East Indian Women: in addition to *Lactobacillus*, *Staphylococcus*, and rarely by *Propionibacterium avidum*, *Bacillus subtilis*, *Escherchia coli*, *Janthinobacterium lividum*, and *Kocuria kristinae*	Mahajan et al. (2022), Das Purkayastha et al. (2019)
Nasal cavity	*Firmicutes*, *Bacteroidetes*, *Proteobacteria*, *Actinobacteria*, *Cyanobacteria, and Fusobacteria*	No study	Chen et al. (2022)
Oral cavity	774 oral bacterial species identified and recorded in human oral microbiome database	*Neisseria*, *Streptococcus*, *Prevotella*, *Porphyromonas*, and *Haemophilus*	https://www.homd.org/ Chaudhari et al. (2020)
Conjunctiva	*Corynebacterium*, *Pseudomonas*, *Staphylococcus*, *Acinetobacter*, *Streptococcus*, *Millisia*, *Anaerococcus*, *Finegoldia*, *Simonsiella*, and *Veillonella*	Only fungal microbiome deciphered *Ascomycota*, *Basidiomycota*, *Zygomycota*	Huang et al. (2016), Prashanthi et al. (2019)
Lung	*Firmicutes*, *Bacteriodetes*, *Proteobacteria*, *Fusobacteria*, and *Actinobacteria* Genera: *Prevotella*, *Veillonella*, and *Streptococcus* spp	No study	Moffatt and Cookson (2017)

13.2.2 Brain-Gut-Microbiome (BGM)

The term "gut-microbiota-brain axis" describes the web of connections involving various biological systems that permit bidirectional communication between gut bacteria and the brain and is essential for preserving the homeostasis of the gastrointestinal, nervous, and microbial systems of animals (Morais et al. 2021).

Thus, a model of two-way communication between the gut, its microbiome, and the central nervous system (CNS), known as the BGM system, is being supported by new research. The three hubs in the wider BGM network are the microbiome, the gut connectome, and the brain connectome (Horn et al. 2022). The brain and gut microbiome thus communicate largely through the neuronal, endocrine, and immunoregulatory pathways (Osadchiy et al. 2019; Slyepchenko et al. 2017). Some of the molecules derived from microorganisms are ingested and travel to the brain via the vagus nerve and/or systemic circulation. Similar to this, the brain can also modify the microbiome indirectly through changes to the gut microbial environment or directly through the influence of neuroactive molecules released into the gut lumen, affecting gene expression and the behavior of microbes.

In continuation, tryptophan, an important amino acid, serves as a precursor for several neuroactive signaling molecules, such as serotonin, kynurenine, and indoles (Gao et al. 2020). Conversely, the synthesis of indoles is solely dependent on the metabolism of gut microbes, unlike the production of serotonin and kynurenine, which microorganisms just modulate (Yano et al. 2015; Tolhurst et al. 2012). The brain needs a variety of compounds to remain healthy and operate properly, and indoles are the precursor molecules for many of these neurotransmitters. In the GI tract, brain, and systemic circulation, these have been identified (Agus et al. 2018). Indole is further metabolized by the liver, and one of these by-products, indoxyl sulfate, is thought to have an impact on ASD, AD, and depression, among other brain illnesses (Osadchiy et al. 2019).

13.3 Disease, Dysbiosis, and Microbiome

Human microbiome has a great impact on health and diseases; however, to understand its role, it is important to know the microbial load as well as the diversity inhabiting different regions of the body (Das and Nair 2019). In diseases, the balance of diverse microorganisms is disturbed leading to a condition, termed dysbiosis (Jandhyala et al. 2015; Das and Nair 2019). In a healthy individual, the cross regulation and cross talk among microbes present in different parts of the body maintain the homeostasis balance. However, when the balance between the "beneficial" and "pathogenic" microbes is altered, the body shifts from an equilibrium state to a state of chaos, disturbing the underlying regulatory mechanisms of the healthy state. Dysbiosis can arise as a result in change of composition of microbial flora either by the loss of beneficial microbes or increase in pathogenic species, thus altering the entire microbial diversity of a homeostatic body (Glassner et al. 2020; Garrett 2019; El-Salhy et al. 2021; Horn et al. 2022; Chen et al. 2022).

Amongst all the systems of the human body, the gut is recognized as the most diverse region followed by skin and mouth. Gut microbiota is highly diverse probably due to different food habits of individuals and it varies from individual to individual (Li et al. 2012). Further, microbial communities in gut are mainly anaerobes outnumbering facultative anaerobes and aerobes. Human gut is found to harbor more than 50 bacterial genera to date but the dominant gut colonizers majorly belong to Bacteroidetes, Firmicutes, Proteobacteria, and Actinobacteria (Bull and Plummer 2014). It is reported that changes in the gut community structure have implications on human health, diseases, and treatment. Past studies were mainly focused to ascertain the gut microbiota changes between healthy individuals; however, with newer sequencing technologies, and improved computational tools, the exploration of normal healthy gut microbiota versus altered microbiome in diseased individuals has been an area of extensive research. There is a complex interplay between gut microbiome and neurodegenerative diseases and gastric inflammatory diseases (Table 13.2). Neurodegenerative diseases are regulated by the gut-brain axis that involves cross talk between gut microbiota and nervous system. Broadly they interact via three pathways, i.e., chemical signaling; neural pathways; and immune system (Fig. 13.1). There are a number of dysbiosis-related diseases which are mentioned in the Table 13.2 (Chen et al. 2021).

13.4 Therapeutic Strategies to Restore the Microbiome Balance

The recent microbiome research is expanding at a great pace. So far, this research has generated multiple strategies that are known to restore the microbial diversity within and help to control the various pathological conditions. Some are personalized dietary modulation, inclusion of probiotics, prebiotics, symbiotic, phage therapy, fecal transplantation, etc. which are discussed below.

13.4.1 Diet and Microbiome

The majority of the microbes in the gut are symbiotic (beneficial to both the human body and the microbiota), while a tiny number are harmful (promoting disease) as well. Pathogenic and symbiotic bacteria can coexist together in a healthy body. As mentioned earlier, the diet significantly influences the types of bacteria that exist in the colon, in addition to environmental factors, medication use, and genetics from the family (Flandroy et al. 2018; Singh et al. 2017; Chaudhari et al. 2020; Slyepchenko et al. 2017). Each individual's microbiome is distinct due to these characteristics. The diversity of bacteria in the gut can be influenced by dietary choices and certain foods. The addition of fruits and fiber-rich diets may elevate the production of SCFA, while in return helps in maintaining healthy gut (den Besten et al. 2013). In addition to having a substantial impact on the gut flora, diet can alter the anatomy and function of the brain (Horn et al. 2022). It is also well established that changes in the gut microbiota's composition or activity affect both health and

Table 13.2 Association of microbiomes with various diseases (↑ increased, ↓ decreased)

S. No.	Disease	Role of gut microbiome (GM)	References
1	Parkinson disease	*Bifidobacterium*, *Pasteurella*, and *Enterococcus*↑ *Brautella*, *Prevotella*, and *Faecoccus*↓	Hill-Burns et al. (2017), Hopfner et al. (2017), Peng et al. (2018), Scheperjans et al. (2015), Bedarf et al. (2017), Heintz-Buschart et al. (2018))
2	Alzheimer disease	Similar to Parkinson disease along with ↑ in proinflammatory bacteria *Escherichia* and *Shigella*	Hill-Burns et al. (2017), Hopfner et al. (2017), Peng et al. (2018), Scheperjans et al. (2015), Bedarf et al. (2017), Heintz-Buschart et al. (2018), Minter et al. (2017)
3	Cardiovascular diseases (CVD) like hypertension; atherosclerosis	Different metabolic products of GM like trimethylamine oxide (TMAO); SCFAs; bile acids interact and influence occurrence of CVD	Yan et al. (2017), Yang et al. (2015)
4	Obesity	*Fermicutes*↑ *Bacteroides*, *Akkermansia muciniphila*, *Faecalibacterium prausnitzii*↓	Turnbaugh et al. (2006), Vallianou et al. (2019)
5	Type I diabetes	*Clostridium*, *Prevotella*↓	Zhou et al. (2020)
6	Type II diabetes	*Dallella*↑ *Bifidobacteria*, *Akkermansia*↓	Li et al. (2020)
7	Gestational diabetes	*Desulfovibrio*, *Enterobacter*, *Prevotella*, *Ruminococcus*, *Bacteroides*↑ *Bifidobacterium*, *Fischeri*↓	Crusell et al. (2018)
8	Nonalcoholic fatty liver disease	*Lactobacillus*, *Streptococcus*, *Dorea*↑ *Prevotella*, *Ruminococcus*, *Flavobacterium*↓	Da Silva et al. (2018), Raman et al. (2013), Jiang et al. (2015)
9	Inflammatory bowel disease	*Enterobacter and Proteobacteria*↑ *Firmicutes*↓	Caruso et al. (2020), Li et al. (2014)
10	Colorectal cancers	*Escherichia coli*, *Bacteroides fragilis*, and *Fusobacterium nucleatum*↑ *Bifidobacteria*, *Lactobacillus*, and *Bacteroides*↓	Si et al. (2021), Tsoi et al. (2017)
11	Asthma and allergic rhinitis	*Rothia*, *Bacteriodes*, *Propionibacterium*, and *Corynebacterium*↑ *Sphingomonas*, *Halomonas*, and *Streptococcus* sp.↓	Chen et al. (2022), Sokolowska et al. (2018)
12	Anxiety and depression	*Actinomycineae*, *Coriobacterineae*,	Barandouzi et al. (2020)

(continued)

Table 13.2 (continued)

S. No.	Disease	Role of gut microbiome (GM)	References
		Bifidobacteriaceae, *Clostridiales incertae sedis XI*, *Porphyromonadaceae*, *Clostridiaceae*, *Lactobacillaceae*, *Streptococcaceae*, *Eubacteriaceae*, *Thermoanaerobacteriaceae*, *Fusobacteriaceae*, *Nocardiaceae*, *Streptomycetaceae*↑ *Veillonellaceae*, *Prevotellaceae*, *Bacteroidaceae*, *Sutterellaceae*, *Oscillospiraceae*, *Marniabilaceae*, and *Chitinophagaceae*↓	
13	Atopic dermatitis	*Staphylococcus aureus*, *Streptococcus*, *Gemella*, and *Haemophilus*↑ *Dermacoccus*↓	Kim and Kim (2019)
14	Psoriasis	*Firmicutes*, *Staphylococcus*↑ Actinobacteria, *Corynebacterium*↓	Chen et al. (2020)

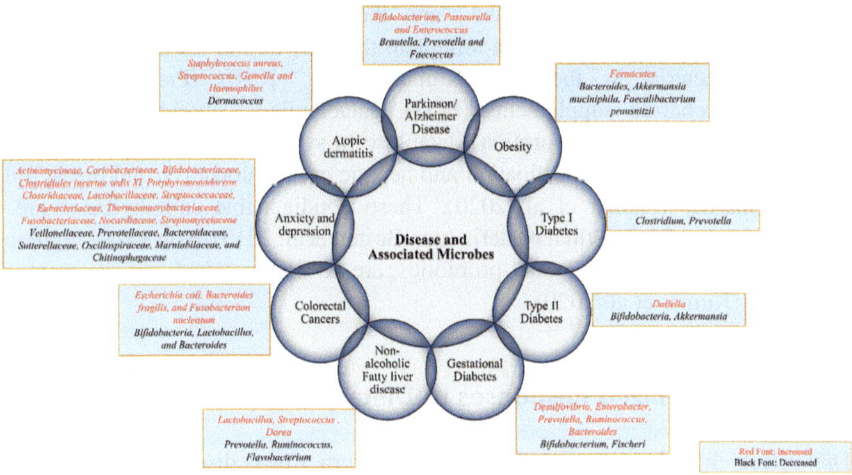

Fig. 13.1 Microbial dysbiosis and their association with disease (family/genus)

disease. The host bacterial species that are already present undergo predictable modifications as a result of the consumption of a particular diet (Singh et al. 2017), although these interactions are complex and might vary depending on the individual (Asnicar et al. 2021).

13.4.2 The Role of Probiotics as Therapeutic Agent

One of the earliest examples of use of probiotics for improving human health was in the early nineteenth century (Wieërs et al. 2020). Diarrhea and constipation among children is treated with *Bifidobacteria*. Incidentally *Bifidobacterium* is among the first bacterial species which colonize the intestine in newborns during the birth. Probiotics have been extensively used for diseases of the gut and skin (Kumar et al. 2020; Wieërs et al. 2020).

Probiotics help to replace the pathogenic bacteria and restore the eubiosis of the gut microbiome. These are defined as live microorganisms which when administered in adequate amounts can confer health benefits on the host. After prolonged antibiotic treatment, chronic diseases, and general health, dysbiosis can be reversed or prevented by the use of probiotics in the diet. Mostly *Lactobacilli*, *Bifidobacteria*, *Enterococci*, *Propionibacteria*, *Staphylococcus*, and some other species are used as probiotics (Kumar et al. 2020; Wieërs et al. 2020). These strains are regarded as safe for consumption as they form normal components of the gut microbiome. Probiotics provide both immunological and nonimmunological benefits including activation of local macrophages, modulation of cytokine profiles, production of immunoglobulins (Wieërs et al. 2020).

Prebiotics on the other hand, are the nonstarch polysaccharides and oligosaccharides in the diet which stimulate the growth and metabolic activity of beneficial bacteria in the gut (Kumar et al. 2020). Prebiotics include fructooligosaccharide supplements (FOS), galactooligosaccharides, inulin (also able to increase calcium absorption), lactulose (a synthetic disaccharide used as a drug for the treatment of constipation and hepatic encephalopathy), and breast milk oligosaccharides (Kumar et al. 2020). These nondigestible carbohydrates are a source of energy for intestinal bacteria (den Besten et al. 2013). Nowadays, common formulations of prebiotics and probiotics are available which are referred as synbiotics (Kumar et al. 2020).

13.4.3 Personalized Responses to Dietary Composition

Diet affects the microbiome and microbiota in turn affects disease risk. In order to predict a person's metabolic response to foods based on personal characteristics, the Personalized Responses to DIetary Composition Trial (PREDICT) used both standardized "test" meals and collected "free-living" nonstandardized food consumption (Berry et al. 2020). In this study, more than 1098 people were investigated by the researchers, and deep metagenomic sequencing was performed on 1203 of these gut microbiomes. In order to assess the dietary patterns of the participants, the researchers gathered comprehensive, long-term data on dietary intake from each of these individuals. This study found that the panel of intestinal species linked with good cardiometabolic and postprandial indicators overlapped with those related with healthy eating practices. The panel of intestinal species linked to good cardiometabolic and postprandial indicators coincided with those linked to healthy eating habits (Asnicar et al. 2021).

This study acquired the data on a wide range of factors, including pre- and postmeal assessments of blood sugar (glucose), cholesterol, and inflammation, which are known to affect metabolism and disease risk. Additionally, measurements were made of the study participants' age, weight, body fat, body mass index (BMI), and blood pressure (Asnicar et al. 2021). *Prevotella copri* and *Blastocystis* spp. were detected, indicating that postprandial glucose metabolism was favorable. These microorganisms were also connected with a wide range of cardiometabolic blood indicators, such as fasting and postprandial glycemic, lipemic, and inflammatory indices (Asnicar et al. 2021). Loss of keystone species, such *Faecalibacterium prausnitzii* (Banerjee et al. 2018), which control microbiome form and function, is unmistakably linked to a range of disease conditions.

Dietary habits that are less nutritious (such as dairy desserts, fatty meats, and processed foods) were seen to promote gut species that correlate with high blood sugar, cholesterol, and inflammatory conditions, all of which are strongly linked to a higher risk of cardiac events, strokes, and type 2 diabetes. On the contrary, a more varied gut microbiota was seen to be associated with good eating habits (high-fiber foods like spinach and broccoli, almonds, and healthy animal meals like fish and eggs), and polyunsaturated fats (fish, walnuts, flax and chia seeds, sunflower, etc.). All of which in turn correlated well with the metrics associated with a reduced risk of developing several chronic diseases (Asnicar et al. 2021). The gut microbiome can thus, flourish when minimally processed plant meals are chosen, it helps protect from and lowers the risk of chronic diseases like heart disease, diabetes, metabolic disease, and obesity (Asnicar et al. 2021).

Nevertheless, distinguishing correlation from causation is one of the most difficult tasks in microbiome research.

13.4.4 Precision Nutrition in Cancer

Dietary habits are also predicted in raising the risk of developing cancer (Zhu et al. 2013). Separately, many malignancies are influenced by microbial factors. More research is required, though, to fully comprehend how nutrition and the microbiome interact to affect how well cancer treatments work (Greathouse et al. 2022). Higher intakes of fiber, calcium, omega-3 fatty acids, and milk are specifically linked to a decreased risk of death from colorectal cancer (CRC), whereas whole-grain consumption is linked to a lower risk of death from CRC specifically (Song et al. 2017, 2018; Yang et al. 2019). On the other hand, a CRC diet high in processed meat is linked to a worse rate of disease-free survival (Zhu et al. 2013).

13.4.5 Antimicrobial Therapy

In sepsis patients, antimicrobial therapy with appropriate dosage and duration helps to provide protection against toxins and antigens of the harmful microflora (Niederman et al. 2021). During the sepsis disease, there is a dysregulation in the host response to infection. Bacteroidetes abundance in contrast to *Firmicutes* in stool samples of the deceased patients has been confirmed in case of sepsis death (Ojima et al. 2016). Surviving patients had higher ratios of *Pseudomonas aeruginosa, Bifidobacteria,* and other *Actinobacteria*. Antimicrobial therapy is provided to ill patients with β-lactamase inhibitor combination, cephalosporins, carbapenems, and fluoroquinolones (Bhalodi et al. 2019). Targeted antimicrobial therapy can be one of the ways to restore the original gut microbiota in the sepsis patients. However, there is a need for research to understand the effects of the dosage so that unnecessary exposure can be avoided.

13.4.6 Lifestyle Modifications

Overindulgence and consumption of processed food, fats, sweeteners, smoking, drinking, drugs, etc. has changed the normal microbiota with which a person is born. The gut dysbiosis is disturbed by an unbalanced diet, sleep deprivation, emotional stress or even genetics (Redondo-Useros et al. 2020). By changing sedentary lifestyle and including a high-fiber diet with water and high protein, a person can restore the balance of intestinal microflora. Exercise combined with a nourishing diet can help to improve cardiovascular health, weight loss, and improved insulin sensitivity.

13.4.7 Fecal Microbiota Transplantation

FMT is a method that involves administering the patient (i.e., the recipient) specially prepared stool material from healthy donors in an effort to treat certain medical disorders by reestablishing the balance of the gut microbial population (Ser et al. 2021). Currently, FMT is focused primarily on transferring bacteria from the donor to the recipient and these microbes are being isolated and purified to understand their precise roles (Aggarwala et al. 2021). However, evidence is pointing increasingly toward the possibility that bacteriophages are also contributing to FMT's efficacy in treating recurrent *Clostridium difficile* infections (Hanssen et al. 2021). Altogether, the microbiome alone can significantly modify human physiology (de Groot et al. 2017). The ability of FMT from lean donors to reorient host glucose metabolism is influenced by the recipient's initial microbial composition (Ser et al. 2021). This may be partially explained by the fact that following FMT, species from both donors and recipients continued to exist in the gut, illustrating the difficulties in effectively altering the microbiome's composition (Li et al. 2016). In spite of limited safety data and major safety concerns, the FMT seems to be very effective in the real world (Kelly et al. 2021).

13.4.8 The Challenges in Personalized Dietary Modulation of the Gut Microbiota

Whether it is chronic illnesses such as diabetes, inflammation, or disorders of the brain, the management of disease may depend on individualized dietary manipulation of the gut microbiota. But there is still a lack of comprehensive information about using diet as a tool. Because it is difficult to correctly manipulate the human microbiome, there is currently little proof of causation in people (Mendez-Garcia et al. 2018; Leeming et al. 2021).

New studies should consider improved dietary data collection, further characterization of metabolic interactions, and an increased focus on omic technologies like metabolomics in order to describe the bacterial and metabolic activity of food breakdown as well as its cross talk with the host. Furthermore, clinical evidence with regard to health outcomes is needed before therapeutic food plans for microbial rehabilitation may be developed.

13.5 Conclusions

The impact of microbes on human health is now universally recognized. This chapter discusses diversity-driven microbial composition, diet-driven modulation, and the consequences of microbial dysbiosis. We attempted to elucidate the healthy gut microbiota and its role in maintaining gut homeostasis. A detailed profile of microbial shifts in diseases such as neurodegenerative disease, cardiovascular disease, asthma, colorectal cancer, etc., has also been discussed along with the gut-microbiota-brain axis. Further, we have also emphasized the currently available/needed therapeutic strategies for restoring the microbial dysbiosis.

Acknowledgments RL thanks INSA for providing INSA Senior Scientist Position. RK acknowledges SERB International Research Experience (SIRE) for providing the fellowship (SIR/2022/000474).

References

Aggarwala V, Mogno I, Li Z, Yang C, Britton GJ, Chen-Liaw A et al (2021) Precise quantification of bacterial strains after fecal microbiota transplantation delineates long-term engraftment and explains outcomes. Nat Microbiol 6(10):1309–1318

Agus A, Planchais J, Sokol H (2018) Gut microbiota regulation of tryptophan metabolism in health and disease. Cell Host Microbe 23(6):716–724

Arumugam M, Raes J, Pelletier E, Le Paslier D, Yamada T, Mende DR et al (2011) Enterotypes of the human gut microbiome. Nature 473(7346):174–180

Asnicar F, Berry SE, Valdes AM, Nguyen LH, Piccinno G, Drew DA et al (2021) Microbiome connections with host metabolism and habitual diet from 1,098 deeply phenotyped individuals. Nat Med 27(2):321–332
Banerjee S, Schlaeppi K, van der Heijden MGA (2018) Keystone taxa as drivers of microbiome structure and functioning. Nat Rev Microbiol 16(9):567–576
Barandouzi ZA, Starkweather AR, Henderson WA, Gyamfi A, Cong XS (2020) Altered composition of gut microbiota in depression: a systematic review. Front Psych 11:541
Bedarf JR, Hildebrand F, Coelho LP, Sunagawa S, Bahram M, Goeser F et al (2017) Functional implications of microbial and viral gut metagenome changes in early stage L-DOPA-naive Parkinson's disease patients. Genome Med 9(1):39
Berry S, Drew D, Linenberg I, Hadjigeorgiou G, Davies R, Khatib H et al (2020) Personalised REsponses to DIetary Composition Trial (PREDICT): an intervention study to determine inter-individual differences in postprandial response to foods. Res Square. https://doi.org/10.21203/rs.2.20798/v1
Bhalodi AA, van Engelen TSR, Virk HS, Wiersinga WJ (2019) Impact of antimicrobial therapy on the gut microbiome. J Antimicrob Chemother 74(Suppl 1):i6–i15
Bull MJ, Plummer NT (2014) Part 1: the human gut microbiome in health and disease. Integr Med 13(6):17–22
Caruso R, Lo BC, Nunez G (2020) Host-microbiota interactions in inflammatory bowel disease. Nat Rev Immunol 20(7):411–426
Chaudhari DS, Dhotre DP, Agarwal DM, Gaike AH, Bhalerao D, Jadhav P et al (2020) Gut, oral and skin microbiome of Indian patrilineal families reveal perceptible association with age. Sci Rep 10(1):5685
Chen L, Li J, Zhu W, Kuang Y, Liu T, Zhang W et al (2020) Skin and gut microbiome in psoriasis: gaining insight into the pathophysiology of it and finding novel therapeutic strategies. Front Microbiol 11:589726
Chen Y, Zhou J, Wang L (2021) Role and mechanism of gut microbiota in human disease. Front Cell Infect Microbiol 11:625913
Chen M, He S, Miles P, Li C, Ge Y, Yu X et al (2022) Nasal bacterial microbiome differs between healthy controls and those with asthma and allergic rhinitis. Front Cell Infect Microbiol 12: 841995
Clapp M, Aurora N, Herrera L, Bhatia M, Wilen E, Wakefield S (2017) Gut microbiota's effect on mental health: the gut-brain axis. Clin Pract 7(4):987
Collado MC, Cernada M, Bauerl C, Vento M, Perez-Martinez G (2012) Microbial ecology and host-microbiota interactions during early life stages. Gut Microbes 3(4):352–365
Correa-Oliveira R, Fachi JL, Vieira A, Sato FT, Vinolo MA (2016) Regulation of immune cell function by short-chain fatty acids. Clin Transl Immunol 5(4):e73
Costea PI, Hildebrand F, Arumugam M, Backhed F, Blaser MJ, Bushman FD et al (2018) Enterotypes in the landscape of gut microbial community composition. Nat Microbiol 3(1):8–16
Crusell MKW, Hansen TH, Nielsen T, Allin KH, Ruhlemann MC, Damm P et al (2018) Gestational diabetes is associated with change in the gut microbiota composition in third trimester of pregnancy and postpartum. Microbiome 6(1):89
Da Silva HE, Teterina A, Comelli EM, Taibi A, Arendt BM, Fischer SE et al (2018) Nonalcoholic fatty liver disease is associated with dysbiosis independent of body mass index and insulin resistance. Sci Rep 8(1):1466
Das B, Nair GB (2019) Homeostasis and dysbiosis of the gut microbiome in health and disease. J Biosci 44(5):117
Das Purkayastha S, Bhattacharya MK, Prasad HK, Upadhyaya H, Lala SD, Pal K et al (2019) Contrasting diversity of vaginal lactobacilli among the females of Northeast India. BMC Microbiol 19(1):198
de Groot PF, Frissen MN, de Clercq NC, Nieuwdorp M (2017) Fecal microbiota transplantation in metabolic syndrome: history, present and future. Gut Microbes 8(3):253–267

den Besten G, van Eunen K, Groen AK, Venema K, Reijngoud DJ, Bakker BM (2013) The role of short-chain fatty acids in the interplay between diet, gut microbiota, and host energy metabolism. J Lipid Res 54(9):2325–2340

El-Salhy M, Patcharatrakul T, Gonlachanvit S (2021) Fecal microbiota transplantation for irritable bowel syndrome: an intervention for the 21(st) century. World J Gastroenterol 27(22): 2921–2943

Flandroy L, Poutahidis T, Berg G, Clarke G, Dao M-C, Decaestecker E et al (2018) The impact of human activities and lifestyles on the interlinked microbiota and health of humans and of ecosystems. Sci Total Environ 627:1018–1038

Gao K, Mu CL, Farzi A, Zhu WY (2020) Tryptophan metabolism: a link between the gut microbiota and brain. Adv Nutr 11(3):709–723

Garrett WS (2019) The gut microbiota and colon cancer. Science 364(6446):1133–1135

Glassner KL, Abraham BP, Quigley EMM (2020) The microbiome and inflammatory bowel disease. J Allergy Clin Immunol 145(1):16–27

Greathouse KL, Wyatt M, Johnson AJ, Toy EP, Khan JM, Dunn K et al (2022) Diet-microbiome interactions in cancer treatment: opportunities and challenges for precision nutrition in cancer. Neoplasia 29:100800

Hanssen NMJ, de Vos WM, Nieuwdorp M (2021) Fecal microbiota transplantation in human metabolic diseases: from a murky past to a bright future? Cell Metab 33(6):1098–1110

Heintz-Buschart A, Pandey U, Wicke T, Sixel-Döring F, Janzen A, Sittig-Wiegand E et al (2018) The nasal and gut microbiome in Parkinson's disease and idiopathic rapid eye movement sleep behavior disorder. Mov Disord 33(1):88–98

Hill-Burns EM, Debelius JW, Morton JT, Wissemann WT, Lewis MR, Wallen ZD et al (2017) Parkinson's disease and Parkinson's disease medications have distinct signatures of the gut microbiome. Mov Disord 32(5):739–749

Hopfner F, Kunstner A, Muller SH, Kunzel S, Zeuner KE, Margraf NG et al (2017) Gut microbiota in Parkinson disease in a northern German cohort. Brain Res 1667:41–45

Horn J, Mayer DE, Chen S, Mayer EA (2022) Role of diet and its effects on the gut microbiome in the pathophysiology of mental disorders. Transl Psychiatry 12(1):164

Hou K, Wu Z-X, Chen X-Y, Wang J-Q, Zhang D, Xiao C et al (2022) Microbiota in health and diseases. Signal Transduct Target Ther 7(1):135

Huang Y, Yang B, Li W (2016) Defining the normal core microbiome of conjunctival microbial communities. Clin Microbiol Infect 22(7):643:e7–e12

Jandhyala SM, Talukdar R, Subramanyam C, Vuyyuru H, Sasikala M, Nageshwar RD (2015) Role of the normal gut microbiota. World J Gastroenterol 21(29):8787–8803

Jiang C, Xie C, Li F, Zhang L, Nichols RG, Krausz KW et al (2015) Intestinal farnesoid X receptor signaling promotes nonalcoholic fatty liver disease. J Clin Invest 125(1):386–402

Kelly CR, Yen EF, Grinspan AM, Kahn SA, Atreja A, Lewis JD et al (2021) Fecal microbiota transplantation is highly effective in real-world practice: initial results from the FMT National Registry. Gastroenterology 160(1):183–192

Kim JE, Kim HS (2019) Microbiome of the skin and gut in atopic dermatitis (AD): understanding the pathophysiology and finding novel management strategies. J Clin Med 8(4):444

Kumar R, Sood U, Gupta V, Singh M, Scaria J, Lal R (2020) Recent advancements in the development of modern probiotics for restoring human gut microbiome dysbiosis. Indian J Microbiol 60(1):12–25

Kumar R, Sood U, Kaur J, Anand S, Gupta V, Patil KS et al (2022) The rising dominance of microbiology: what to expect in the next 15 years? Microb Biotechnol 15(1):110–128

Lach G, Schellekens H, Dinan TG, Cryan JF (2018) Anxiety, depression, and the microbiome: a role for gut peptides. Neurotherapeutics 15(1):36–59

Leeming ER, Louca P, Gibson R, Menni C, Spector TD, Le Roy CI (2021) The complexities of the diet-microbiome relationship: advances and perspectives. Genome Med 13(1):10

Li K, Bihan M, Yooseph S, Methe BA (2012) Analyses of the microbial diversity across the human microbiome. PLoS One 7(6):e32118

Li Q, Wang C, Tang C, He Q, Li N, Li J (2014) Dysbiosis of gut fungal microbiota is associated with mucosal inflammation in Crohn's disease. J Clin Gastroenterol 48(6):513–523
Li SS, Zhu A, Benes V, Costea PI, Hercog R, Hildebrand F et al (2016) Durable coexistence of donor and recipient strains after fecal microbiota transplantation. Science 352(6285):586–589
Li Q, Chang Y, Zhang K, Chen H, Tao S, Zhang Z (2020) Implication of the gut microbiome composition of type 2 diabetic patients from northern China. Sci Rep 10(1):5450
Lynch SV, Pedersen O (2016) The human intestinal microbiome in health and disease. N Engl J Med 375(24):2369–2379
Magne F, Gotteland M, Gauthier L, Zazueta A, Pesoa S, Navarrete P et al (2020) The firmicutes/bacteroidetes ratio: a relevant marker of gut dysbiosis in obese patients? Nutrients 12(5):1474
Mahajan G, Doherty E, To T, Sutherland A, Grant J, Junaid A et al (2022) Vaginal microbiome-host interactions modeled in a human vagina-on-a-chip. Microbiome 10(1):201
McNabney SM, Henagan TM (2017) Short-chain fatty acids in the colon and peripheral tissues: a focus on butyrate, colon cancer, obesity and insulin resistance. Nutrients 9(12):1348
Mendez-Garcia C, Barbas C, Ferrer M, Rojo D (2018) Complementary methodologies to investigate human gut microbiota in host health, working towards integrative systems biology. J Bacteriol 200:3
Milani C, Duranti S, Bottacini F, Casey E, Turroni F, Mahony J et al (2017) The first microbial colonizers of the human gut: composition, activities, and health implications of the infant gut microbiota. Microbiol Mol Biol Rev 81(4):e00036
Minter MR, Hinterleitner R, Meisel M, Zhang C, Leone V, Zhang X et al (2017) Antibiotic-induced perturbations in microbial diversity during post-natal development alters amyloid pathology in an aged APP(SWE)/PS1(DeltaE9) murine model of Alzheimer's disease. Sci Rep 7(1):10411
Moffatt MF, Cookson WO (2017) The lung microbiome in health and disease. Clin Med 17(6): 525–529
Morais LH, Schreiber HL, Mazmanian SK (2021) The gut microbiota-brain axis in behaviour and brain disorders. Nat Rev Microbiol 19(4):241–255
Nguyen J, Lara-Gutierrez J, Stocker R (2021) Environmental fluctuations and their effects on microbial communities, populations and individuals. FEMS Microbiol Rev 45(4):68
Niederman MS, Baron RM, Bouadma L, Calandra T, Daneman N, DeWaele J et al (2021) Initial antimicrobial management of sepsis. Crit Care 25(1):307
Ojima M, Motooka D, Shimizu K, Gotoh K, Shintani A, Yoshiya K et al (2016) Metagenomic analysis reveals dynamic changes of whole gut microbiota in the acute phase of intensive care unit patients. Dig Dis Sci 61(6):1628–1634
Osadchiy V, Martin CR, Mayer EA (2019) Gut microbiome and modulation of CNS function. Compr Physiol 10(1):57–72
Peng C, Gathagan RJ, Covell DJ, Medellin C, Stieber A, Robinson JL et al (2018) Cellular milieu imparts distinct pathological α-synuclein strains in α-synucleinopathies. Nature 557(7706): 558–563
Potbhare R, RaviKumar A, Munukka E, Lahti L, Ashma R (2022) Skin microbiota diversity among genetically unrelated individuals of Indian origin. PeerJ 10:e13075
Prashanthi GS, Jayasudha R, Chakravarthy SK, Padakandla SR, Sai Abhilash CR, Sharma S et al (2019) Alterations in the ocular surface fungal microbiome in fungal keratitis patients. Microorganisms 7(9):309
Raman M, Ahmed I, Gillevet PM, Probert CS, Ratcliffe NM, Smith S et al (2013) Fecal microbiome and volatile organic compound metabolome in obese humans with nonalcoholic fatty liver disease. Clin Gastroenterol Hepatol 11(7):868–875
Redondo-Useros N, Nova E, Gonzalez-Zancada N, Diaz LE, Gomez-Martinez S, Marcos A (2020) Microbiota and lifestyle: a special focus on diet. Nutrients 12(6):1776
Rinninella E, Raoul P, Cintoni M, Franceschi F, Miggiano GAD, Gasbarrini A et al (2019) What is the healthy gut microbiota composition? A changing ecosystem across age, environment, diet, and diseases. Microorganisms 7(1):14

Sampson TR, Debelius JW, Thron T, Janssen S, Shastri GG, Ilhan ZE et al (2016) Gut microbiota regulate motor deficits and neuroinflammation in a model of Parkinson's disease. Cell 167(6): 1469–1480
Scheperjans F, Aho V, Pereira PA, Koskinen K, Paulin L, Pekkonen E et al (2015) Gut microbiota are related to Parkinson's disease and clinical phenotype. Mov Disord 30(3):350–358
Sender R, Fuchs S, Milo R (2016) Revised estimates for the number of human and bacteria cells in the body. PLoS Biol 14(8):e1002533
Ser HL, Letchumanan V, Goh BH, Wong SH, Lee LH (2021) The use of fecal microbiome transplant in treating human diseases: too early for poop? Front Microbiol 12:519836
Si H, Yang Q, Hu H, Ding C, Wang H, Lin X (2021) Colorectal cancer occurrence and treatment based on changes in intestinal flora. Semin Cancer Biol 70:3–10
Singh RK, Chang HW, Yan D, Lee KM, Ucmak D, Wong K et al (2017) Influence of diet on the gut microbiome and implications for human health. J Transl Med 15(1):73
Skowron K, Bauza-Kaszewska J, Kraszewska Z, Wiktorczyk-Kapischke N, Grudlewska-Buda K, Kwiecinska-Pirog J et al (2021) Human skin microbiome: impact of intrinsic and extrinsic factors on skin microbiota. Microorganisms 9(3):543
Slyepchenko A, Maes M, Jacka FN, Kohler CA, Barichello T, McIntyre RS et al (2017) Gut microbiota, bacterial translocation, and interactions with diet: pathophysiological links between major depressive disorder and non-communicable medical comorbidities. Psychother Psychosom 86(1):31–46
Sokolowska M, Frei R, Lunjani N, Akdis CA, O'Mahony L (2018) Microbiome and asthma. Asthma Res Pract 4:1
Song M, Zhang X, Meyerhardt JA, Giovannucci EL, Ogino S, Fuchs CS et al (2017) Marine omega-3 polyunsaturated fatty acid intake and survival after colorectal cancer diagnosis. Gut 66(10):1790–1796
Song M, Wu K, Meyerhardt JA, Ogino S, Wang M, Fuchs CS et al (2018) Fiber intake and survival after colorectal cancer diagnosis. JAMA Oncol 4(1):71–79
Song X, Zhang J, Song J, Zhai Y (2021) Decisive effects of life stage on the gut microbiota discrepancy between two wild populations of hibernating asiatic toads (Bufo gargarizans). Front Microbiol 12:665849
Tolhurst G, Heffron H, Lam YS, Parker HE, Habib AM, Diakogiannaki E et al (2012) Short-chain fatty acids stimulate glucagon-like peptide-1 secretion via the G-protein-coupled receptor FFAR2. Diabetes 61(2):364–371
Tsoi H, Chu ESH, Zhang X, Sheng J, Nakatsu G, Ng SC et al (2017) Peptostreptococcus anaerobius induces intracellular cholesterol biosynthesis in colon cells to induce proliferation and causes dysplasia in mice. Gastroenterology 152(6):1419–1433
Turnbaugh PJ, Ley RE, Mahowald MA, Magrini V, Mardis ER, Gordon JI (2006) An obesity-associated gut microbiome with increased capacity for energy harvest. Nature 444(7122): 1027–1031
Vallianou N, Stratigou T, Christodoulatos GS, Dalamaga M (2019) Understanding the role of the gut microbiome and microbial metabolites in obesity and obesity-associated metabolic disorders: current evidence and perspectives. Curr Obes Rep 8(3):317–332
Wieërs G, Belkhir L, Enaud R, Leclercq S, Philippart de Foy J-M, Dequenne I et al (2020) How probiotics affect the microbiota. Front Cell Infect Microbiol 9:454
Witkowski M, Weeks TL, Hazen SL (2020) Gut microbiota and cardiovascular disease. Circ Res 127(4):553–570
Xiao S, Zhang G, Jiang C, Liu X, Wang X, Li Y et al (2021) Deciphering gut microbiota dysbiosis and corresponding genetic and metabolic dysregulation in psoriasis patients using metagenomics sequencing. Front Cell Infect Microbiol 11:605825
Yan Q, Gu Y, Li X, Yang W, Jia L, Chen C et al (2017) Alterations of the gut microbiome in hypertension. Front Cell Infect Microbiol 7:381
Yang T, Santisteban MM, Rodriguez V, Li E, Ahmari N, Carvajal JM et al (2015) Gut dysbiosis is linked to hypertension. Hypertension 65(6):1331–1340

Yang W, Ma Y, Smith-Warner S, Song M, Wu K, Wang M et al (2019) Calcium intake and survival after colorectal cancer diagnosis. Clin Cancer Res 25(6):1980–1988
Yano JM, Yu K, Donaldson GP, Shastri GG, Ann P, Ma L et al (2015) Indigenous bacteria from the gut microbiota regulate host serotonin biosynthesis. Cell 161(2):264–276
Zaiss MM, Joyce Wu HJ, Mauro D, Schett G, Ciccia F (2021) The gut-joint axis in rheumatoid arthritis. Nat Rev Rheumatol 17(4):224–237
Zhou H, Zhao X, Sun L, Liu Y, Lv Y, Gang X et al (2020) Gut microbiota profile in patients with type 1 diabetes based on 16S rRNA gene sequencing: a systematic review. Dis Markers 2020: 3936247
Zhu Y, Wu H, Wang PP, Savas S, Woodrow J, Wish T et al (2013) Dietary patterns and colorectal cancer recurrence and survival: a cohort study. BMJ Open 3(2):e002270

Gut Microbiota and Its Role in Human Metabolic Disorders

14

Asha Yadav, Shreya Vishwas Mohite, Arush Behl, Pratik Balwant Shinde, and Krishna Kant Sharma

Abstract

Gut microbiota is a collective population of all the microbial communities that live in the gastrointestinal tract of humans. It provides a range of structural, protective, and metabolic functions to the human host. The contribution of gut microbiota in human metabolic health is revealed by research findings based on omics, cellular, and observational studies. Microbial metabolites are important for the normal functioning of the host, and any disrupted production of metabolites leads to the emergence of common metabolic disorders like type 2 diabetes, obesity, malnutrition, non-alcoholic fatty liver disease, and cardio-metabolic disorder. Previously, gut microbiota studies were focused on the digestive and absorptive functions, but now the research is moving forward toward descriptive microbiome analysis that effect studies with new insights. The current study focuses on the role of gut microbes in host metabolism and the depiction of the role of microbial metabolites in host function, defining their targets. Here we highlighted a few microbiota-based emerging therapeutics for metabolic disorders that aim to enhance metabolic health. This knowledge will surely guide to define personalized, microbiota-based therapies for the improvement in human health.

Keywords

Gut microbiota · Metabolites · Metabolic disorders · SCFAs · Bile acids

A. Yadav · S. V. Mohite · A. Behl · P. B. Shinde · K. K. Sharma (✉)
Laboratory of Enzymology and Gut Microbiology, Department of Microbiology, Maharshi Dayanand University, Rohtak, Haryana, India
e-mail: kksharma.microbiology@mdurohtak.ac.in

R. Sobti et al. (eds.), *Role of Microbes in Sustainable Development*,
https://doi.org/10.1007/978-981-99-3126-2_14

14.1 Introduction

Recent advancement in the field of gut microbiome research has been majorly contributed by multiomics studies. It helps us to understand the role of gut microbiota (GM) in the maintenance of human health. Gut microbial communities live in mutualistic relationship with the host, maintaining the overall homeostasis (Liu et al. 2022). These commensal microbes include bacteria, archaea, viruses, and a few fungi collectively with its genetic factor termed as gut microbiome (Fan and Pedersen 2021). Arguably, GM is constituted by approximately 10^{15} microbial cells, and greater than 22 million microbial genes, both of them surpasses the human cells and genes, respectively (Qin et al. 2010). Every person has a unique gut microbial composition; however, healthy individuals share similar composition of gut microbiota. The microbial composition is shaped by different factors including mode of delivery, type of feeding, diet, environment, age, lifestyle, personal hygiene, use or abuse of drugs, and genetic factors (Lynch and Pedersen 2016; Ahlawat et al. 2021). The vaginally delivered infants have heavy colonization of *Bacteroides*, *Prevotella,* and *Lactobacillus*; whereas, caesarean born infants have *Escherichia coli*, *Clostridium difficile*, and *C. perfringens*. Children have less complex microbiota compared to adults with abundance of BPP (*Bacteroides-Porphyromonas-Prevotella*) group, *Bifidobacterium*, and *Enterobactericeae*. Gut microbes keep on changing with age and its composition is most stable during adulthood. The gut microbial composition of a healthy adult is denoted by the abundance of Bacteroidetes (*Bacteroides*), and Firmicutes (*Enterococcus*, *Clostridium*, and *Lactobacillus*) followed by Proteobacteria (*E. coli*), Actinobacteria (*Bifidobacterium*), *Verrucomicrobia*, and Cyanobacteria (Ahlawat et al. 2021). Gut with its microbes are considered to be "central body organ" that has a crucial role in various physiological functions like food digestion, drug metabolism, vitamin synthesis, production of signaling molecules (neurotransmitters and short-chain fatty acids (SCFAs)), intestinal barrier function, angiogenesis, colonization resistance to pathogens, and immune and neural modulation (Dey et al. 2021; Ahlawat et al. 2021). GM also regulates the functioning of other organs including liver, brain, kidney, heart, skin, and lungs tissues. Study of gut metabolomes by advanced metabolomic tools and techniques (UPLC/MS) help to identify the important metabolites like bile acids (BAs), SCFAs, amino acid derivatives, and choline, and their role in physiological and pathological functions of the host (Liu et al. 2022). In spite of major differences in the pathologies; common metabolic disorders such as type 2 diabetes (T2D), malnourishment (obesity and undernutrition), non-alcoholic fatty liver disease (NAFLD), and cardiovascular disease (CVD)/atherosclerosis are found to be associated with GM composition and functioning (Fan and Pedersen 2021). But the fact which initiates the disease pathology or the alteration in GM composition, is still a matter of dilemma (Hrncir 2022). The next-generation sequencing techniques and the development of advanced bioinformatics work platforms has lead to a bloom of knowledge about the role of microbes in energy metabolism, nutrient absorption, and digestion. In this chapter, we had discussed the role of gut microbes in host metabolism, with an explanation on the role of gut microbial metabolites as a messenger between host and microbes. Alteration in gut microbial composition, both at taxonomic and functional level has also been discussed with respect to major metabolic disorders, and finally concluded by providing new insights for the GM and microbial metabolite-based therapeutics for the treatment of metabolic disorders.

14.2 Gut as the Metabolic Activity Center of Human

The role of GM in food digestion, host metabolism, and nutrient absorption is well established. Gut microbes together with the intestinal enzymes help to digest the food components, and derive energy (Danneskiold-Samsøe et al. 2019). They help to metabolize various substrates such as dietary residues (amino acids, carbohydrates, phytochemicals, and certain lipids) endogenous metabolites, mucosal macromolecules (e.g., mucins), and xenobiotics (Krishnan et al. 2015). Based on their origination, gut microbial metabolites are categorized in three types: first, that directly produced by GM by acting on diet such as SCFAs and indole derivatives, second type are generated by host, and modified by microbes such as BAs, and third type are de novo produced by microbes like polysaccharide A, and vitamins (Liu et al. 2022) (Fig. 14.1). Microbial metabolites fulfill about 10% energy requirement of the host that is directly or indirectly produced by action on dietary constituents. Microbes like *Butyricicoccus pullicaecorum*, *Eubacterium rectale*, *Faecalibacterium prausnitzii*, *Ruminococcus bromii*, and *Roseburia* spp., that feed on plant-based diet result in production of SCFAs such as propionate, butyrate, and acetate. Whereas, microbes like *Bilophila*, *Alistipes*, and *Bacteroides* prefer animal-based food components. Thus, people with different food habits have an abundance of different types of microorganisms in their gut. *Prevotella*, a SCFAs producer, is found to be dominant in the gut of African children as their diet is completely plant based. Butyrate is specifically produced by bacterial genera of *Bifidobacterium*, *Eubacterium*, *Anerostipes*, and *Roseburia*. Lactic acid bacteria and *Bifidobacterium* particularly produce cobalamin and folate, respectively. These bacteria along with other microbial groups produce essential compounds such as riboflavin, nicotinic acid, biotin, pantothenic acid, thiamine, pyridoxine, and vitamin K. Microbes like *Propionibacterium*, *Streptococcus*, *Clostridium*, and *Bacteroides* help to digest the protein-rich diet, and release protein-derived metabolites such as tryptophan, polyamines, and fatty acids (Dey et al. 2021). The digestion of carbohydrate-rich diet is accompanied by the *Lactobacillus* and *Streptococcus* spp., with release of pyruvate, and lactate, whereas microorganisms like *Ruminococcus*, *Lachnospiraceae*, and *Bacteroides* spp. break down fibers into SCFAs, and succinate. Further, *Bilophila* together with *Bifidobacterium*, *Streptococcus*, *Stenotrophomonas* (different species and isolates have been reported in plants and animals), *Propionibacterium*, and *Lactobacillus* spp. are responsible for breakdown of fatty acids such as linoleic acids, hydroxy fatty acid, oxo-fatty acid, and 10-hydroxycis-12-octadeccenoid associated with maintenance of immunological functions including inhibition of inflammatory neutrophil recruitment, suppression

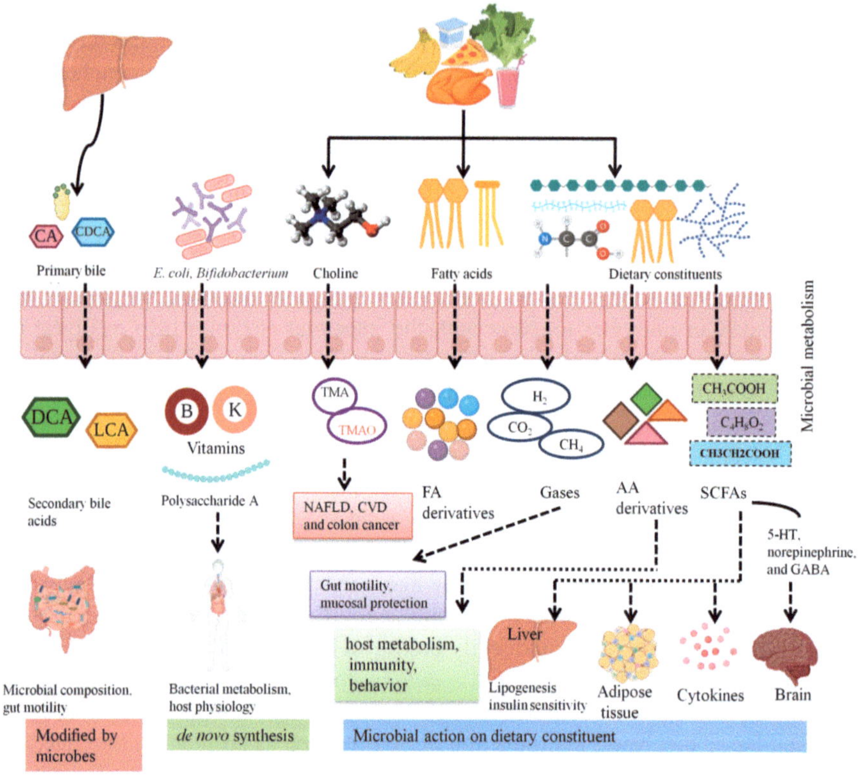

Fig. 14.1 Gut microbial metabolism corroborates with host metabolism via conversion of dietary components, de novo synthesis, or microbial modification of host generated metabolites (The figure created with BioRender.com)

of proinflammatory milieu, macrophage-mediated clearance of cell debris, and tissue remodeling (Blaut 2018; Dey et al. 2021). Moreover, products of fatty acid metabolism are utilized in microbial metabolic functions such as propionate metabolism, lactate utilization, sulfate reduction, succinate formation and decarboxylation, acetate utilization, and butyrate synthesis (Mirzaei et al. 2022). The phospholipase C activity of microbes (*Bacteroides* spp., *Gardnerella vaginalis*, and group B *Streptococcus*) is responsible for breakdown of phospholipids to release diacylglycerols (Dey et al. 2021). Certain bacterial species such as *Lactobacillus*, *Butyrivibrio*, and *Megasphaera* generate conjugate metabolites like linoleic acid that affect insulin sensitivity (Schoeler and Caesar 2019). Further, bacterial species from genera *Escherichia*, and *Bifidobacterium* can *de novo* synthesize vitamins (vitamin K and vitamin B group) (Liu et al. 2022). The gut microbial species *Lactobacillus casei*, *Lactobacillus acidophilus*, *Bifidobacterium adolescentis* or *Escherichia coli* Nissle 1917 influence intestinal enzyme cytochrome P450, a major enzyme involved in drug metabolism (Dey et al. 2021). Additionally, certain neurotransmitters such as serotonin (5-hydroxytryptamine (5-HT)), norepinephrine, and γ-aminobutyric acid (GABA) impacting the physiology of distant body organs are also released by gut microbial species like *Escherichia coli*, *Enterococcus faecium*, *Bifidobacterium brevis*, and *Streptococcus salivarius* (Liu et al. 2022) (Fig. 14.1). Experimental evidence suggests that microbes also play a role in the metabolism of neurotransmitters, a lower concentration of GABA was observed in the plasma and fecal samples of germ-free (GF) mice (Matsumoto et al. 2017). Further, antibiotic cocktail (vancomycin, ampicillin, metronidazole, neomycin) treatment results in lower GABA levels; whereas, a ketogenic diet results in increased levels of GABA (Olson et al. 2018). GF mice have lower serotonin levels compared with conventional mice (Matsumoto et al. 2017). Gut microbial metabolites alter host responses, thus, forming the tridirectional diet-microbiota-metabolites axis (Dey et al. 2021). A few important gut microbial metabolites with their functions are discussed below.

SCFAs are saturated aliphatic acids absorbed by intestinal colonocytes by hydrogen/sodium-dependent monocarboxylic transporters immediately after their production. A major part of SCFAs act as energy source for the colonocytes, and the remaining part enter the blood circulation targeted to act on distant body organs like lung, heart, and brain. They can be directly absorbed by the cells for their energy requirements or provide energy indirectly by the process of gluconeogenesis, and lipid biosynthesis. SCFAs such as acetate have role in improving lipid storage capacity of adipose tissue by inhibiting lipolysis and increasing adipogenesis (Liu et al. 2022; Rowland et al. 2018). The well-known receptors for SCFAs are G-protein-coupled receptors (GPR) including GPR 41 and 43 expressed on body tissues, and immune cells respectively. They bind to the GPR receptors on the intestinal cells and modulate certain host-specific metabolic events such as restriction of insulin and fatty acid secretion, reduction of bile acid synthesis, and cholesterol regulation (Krishnan et al. 2015). In the gastrointestinal epithelium, SCFAs regulate immune homeostasis by balancing the secretion of pro- and anti-inflammatory cytokines such as IL-1β, IL-10, IL-22, IL-6, and tumor necrosis factor α (TNF-α) (Liu et al. 2022). Butyrate, the most important SCFA, is essential for maintenance of intestinal barrier and prevents mucosal inflammation. It is the preferred energy source of colonocytes, and its major production occurs in colon and cecum (Dey et al. 2021; Krishnan et al. 2015). SCFAs also help to increase bioavailability of minerals by maintaining the gut pH balance (Whisner and Castillo 2018). SCFAs have a role in release of antimicrobial peptides, secretory immunoglobulin A, and mucins that prevent attachment and invasion of pathogenic microbes regulating intestinal barrier integrity (Liu et al. 2022).

Amino acid derivatives are the second major metabolic product of gut microbes produced from the amino acid, peptides, and proteins that escape digestion by host enzymes. These compounds include amines, p-cresol sulfate, sulfides, nitrogen compounds, and precursor to branched chain fatty acids, indole, and indole derivatives. Microbes act on tryptophan, the most studied amino acid, and release variety of compounds including skatole (3-methylindole), tryptamine, indole, and its derivatives (3-methyl-indole, indole-3-propionic acid, and indoxyl sulfate), that act on aryl hydrocarbon receptors (Arh) responsible for regulation of host metabolism, immunity, and behavior. Tryptophan induces secretion of interleukins in the intestinal cells, and promotes barrier function with expression of barrier protective genes (CLDN-1, OCC, ZO-1, MUC-2, and β-defensins) (Dey et al. 2021). Indole and indole derivatives play a role in maintenance of intestinal barrier by increasing the transepithelial resistance, expression of epithelial tight junction proteins, and reducing the expression of inflammatory cytokines (TNF-α, IL-1α, IL-1β, and IL-6). These metabolites act as ligands for epithelial nuclear receptors such as PXR, AhR, and retinoid-related orphan receptor gamma-t to regulate the gut barrier function. Another amino acid carnitine, abundant in red meat, is metabolized by gut microbes into trimethylamine (TMA), and TMA is further oxidized in liver by flavin-containing monooxygenase 1 and 3 into trimethylamine-N-oxide (TMAO). Higher TMAO levels in blood are responsible for increased risk of cardiovascular diseases (CVD). A few amino acid derivatives can also serve as neurotransmitters, for example, gut microbes help in release of phenylalanine and tyrosine derivative, dopamine by decarboxylation event of *L*-DOPA, which can be further converted into epinephrine and norepinephrine by the process of methylation and hydroxylation (Liu et al. 2022).

Bile acids such as cholate (CA), and chenodeoxycholate (CDCA) are synthesized in the liver and conjugated with glycine or taurine before entering into the intestine. In the intestine, microbial-deconjugation occurs that helps in the absorption of dietary lipids, fat-soluble vitamins, and maintains systemic cholesterol levels (Jiang et al. 2020). Bile salt hydrolases (BSHs) are the enzymes involved in deconjugation process in the large intestine released by the microbial genera such as Bacteroidetes (*Bacteroides*), Firmicutes (*Lactobacillus*, *Clostridium*), and Actinobacteria (*Bifidobacterium*). The primary bile acids such as CA and CDCA undergo dehydroxylation by the hydrolases and 7 α-dehydroxylases released by members of Firmicutes (Liu et al. 2022; Krishnan et al. 2015). Small shifts in BSH activity influence the intestinal bile acid profile and lipid metabolism of the host. There are two types of receptors present for bile acid action that is nuclear receptors (farnesoid X receptor (FXR), pregnane X receptor/steroid, constitutive androstane receptor, vitamin D3 receptor (VDR), and xenobiotic-sensing receptor (PXR/SXR)) and the membrane receptors (Takeda G-protein receptor 5 (TGR5), sphingosine 1-phosphate receptor 2, muscarinic acetylcholine, and formyl-peptide receptor). The balanced secretion of proinflammatory and anti-inflammatory cytokines is regulated by the action of bile acids targeted to TGR5. This interaction releases anti-inflammatory cytokines with growth factor β and IL-10, and reduces expression of proinflammatory cytokines (Liu et al. 2022). Bile acid signaling between liver and intestine is regulated by the farnesoid X receptor (FXR) present in both the organs. Synthesis of bile acid is under negative feedback control of intestinal FXR via fibroblast growth factor (FGF15)-dependent mechanism. The inhibition of FXR signaling corroborates with the reduced population of *Clostridium* and *Lactobacillus* species with lesser BSH and 7α-dehydroxylase activities. Additionally, bile acids also have a role in regulating gut microbial composition; certain secondary bile acids such as DCA are toxic (detergent property) to few microbes like *C. difficile* (Krishnan et al. 2015). The interaction between bile acids and FXR regulates the gut microbial composition by production of few antimicrobial compounds such as nitric oxide synthase and interleukin (IL)-18. Bile acids regulate appetite and gut motility function via release of gut hormones glucagon-like peptide-1 (GLP-1) and peptide YY (PYY), as they can bind to enteroendocrine cells to cause their release (Liu et al. 2022).

Choline is a necessary component for the phospholipids and bile acid biosynthesis. The direct target receptors for choline are unknown but it can bind to NF-κB, protein kinase C (PKC), and NLRP3 inflammasome, thus promote gut inflammation (Yang et al. 2019). Anaerobic choline metabolism by gut microbes results in production of acetate, trimethylamine (TMA), and ethanol that leads to increased systemic TMAO levels. Gut microbial dysbiosis resulting in altered choline metabolism is known to be a potential contributor to NAFLD, CVD, and colon cancer (Liu et al. 2022; Yang et al. 2019) (Fig. 14.1).

In addition to above described metabolites, gut microbes produce various gases such as hydrogen (H_2), carbon dioxide (CO_2), hydrogen sulfide (H_2S), methane (CH_4), and nitric oxide (NO) (Liu et al. 2022). These gases regulate gut functions such as gut motility (CH_4, H_2S), epithelial secretion, gut inflammation, susceptibility to infections (H_2S), gastric mucosal protection, and mucosal blood flow (NO) (Kalantar-Zadeh et al. 2019). The vitamins synthesized by gut microbes are required for basic bacterial metabolism and are also essential for metabolic and physiological functions of the host (Fig. 14.1). Further, gut microbes release various antibiotics both ribosomally synthesized, and post-translationally modified peptides such as microcins, lantibiotics, and bacteriocins. The antibiotics released by gut microbes draw the attention of researchers and pave the way to develop new therapeutic drugs in the era of global antibiotic abuse. The balanced production of microbial metabolites is crucial for healthy functioning of the host. Alteration in the production of metabolites disrupts the normal physiology of the human host and results in the diseased condition (Liu et al. 2022).

14.3 Dysbiotic Gut Microbiota: Cause or Consequence of Metabolic Disorders

14.3.1 Malnutrition

Malnutrition occurs when the balance of energy intake is altered, resulting from either excess or lower than required nutrient intake which includes both the conditions: undernutrition (stunting, underweight) and overnutrition (obesity).

14.3.1.1 Gut Microbiota and Obesity

Obesity is a multifactorial disease affecting nearly 33% of the world's population. It comprises multiple diseases such as CVD, diabetes mellitus, different types of cancers, muscle, and skeleton disorders impacting life quality and increasing health care cost (Chooi et al. 2019). Patients suffering from obesity have significantly different GM composition compared to healthy adults, demonstrating a higher ratio of Firmicutes to Bacteroides (Yuan et al. 2019; Du et al. 2022). Reduced fecal diversity in obese individuals is linked to adiposity, dyslipidemia, imbalanced glucose homeostasis, and high levels of low-grade inflammation (Chen et al. 2020). Microbial genera abundant in obese patients include *Turicibacter*, *Actinobacillus*, *Aggregatibacter*, *Campylobacter*, *Streptococcus* (SMB53), *Rothia*, *Granulicatella*, *Veillonella*, *Streptococcus*, *Megamonas*, *Phascolarctobacterium*, *Haemophilus*, *Fusobacterium*, *Lachnospira*, and *Sutterella* (Gérard 2016). The link of obesity at species level is denoted by the higher abundance of *Eubacterium ventriosum* and *Roseburia intestinalis* and reduced population of *Akkermansia muciniphila* (Han and Lin 2014; Vallianou et al. 2019) (Table 14.1). Colonization of germ-free (GF) mice with microbiota derived from conventional mice (CM) results in weight gain. The microbial species from CM are able to digest, and absorb previously undigestible food which result in hepatic lipogenesis, adipocyte hypertrophy, and elevated glucose levels (Bäckhed et al. 2004). In addition, microbial species also inhibit angiopoietin-like 4 (ANGPTL-4), which in turn inhibits lipoprotein lipases (LPL), thus leading to higher levels of fatty acid uptake by cells and triglyceride accumulation in adipocytes (Gérard 2016).

Microbial metabolites found to be associated with obesity are SCFAs and BAs (Pascale et al. 2019). SCFAs can inhibit inflammation associated with obesity by suppressing histone deacetylases (HDAC) which increases the transcription of noncoding region of Foxp3 locus that is involved in differentiation of Treg cells (Iddrisu et al. 2021) (Fig. 14.2). SCFAs enhance the release of anorexigenic peptide PYY as well as satietogenic hormone leptin. Obesity is characterized by increased bile acid production, linked to lower levels of FGF19 (Li et al. 2021; Castaner et al. 2018). The activation of FGF19 via FXR is involved in improved cell metabolism, and reduced production of triglycerides. BAs also maintain energy homeostasis in brown adipose tissues via TGR5 activation (Li et al. 2021) (Table 14.1).

14.3.1.2 Gut Microbiota and Undernutrition

Maternal and child undernutrition is a major problem in low-income and middle-income countries, and is a leading cause of death in early age (<5) children. Breastfeeding, unavailability of food, and good quality water are the major causes of malnourishment, and important factors for shaping gut microbial composition (Rokade et al. 2022; Ahlawat et al. 2021). Gut microbial composition typically found in undernourished populations consists of reduced abundance of *Bifidobacterium longum* and *B. pseudolongum*, and increased proportion of *Staphylococcus aureus* and *Escherichia coli* (Jensen et al. 2020) (Table 14.1). These bacteria reduce energy harvesting capacity, immune activity, vitamin synthesis, and increase presence of pathogenic factors (Du et al. 2022). In developing and underdeveloped countries, diarrhea is the primary reason for malnutrition in children linked to dysbiotic GM with increased abundance of Proteobacteria and reduced population of *Bifidobacterium* and *Lactobacillus* species (Jensen et al. 2020). Research studies on nutrient-depleted GF mice suggest that expression of ghrelin in liver, and insulin-like growth factor-1 (IGF-1) in both liver and serum are hampered, leading to stunted growth. SCFA and lactate produced by gut microbes

Table 14.1 Distribution of gut microbes and their metabolites in different metabolic disorders

Sr. No.	Microbe	Metabolite	Repercussion	References
Obesity				
1.	Decreased *Bacteroides thetaiotaomicron*, higher *Firmicutes/ Bacteroidetes* ratio	Increased serum glutamate levels	Glutamate fermentation	Liu et al. (2022)
2.	*Streptococcus*, *Campylobacter*, *Fusobacterium*, *Akkermansia muciniphila*, *Roseburia*	Decreased SCFAs, increased bile acid	Increased inflammation and triglyceride production, reduced cell metabolism, PYY, and leptin release	Gérard (2016), Vallianou et al. (2019), Sharma et al. (2020), Su et al. (2022)
Undernutrition				
3.	Reduced *Bifidobacterium longum*, *Lactobacillus*, increased *Staphylococcus aureus*, and *Escherichia coli*	Increased SCFAs	Decreased ghrelin levels, GH, and suppression of IGF-1	Jensen et al. (2020)
Type 2 diabetes mellitus				
4.	Reduction of *Bacteroides thetaiotamicron*, *Clostridium limosum*, *C. bifermentus*, *Enterococcus faecalis*, and *Ruminococcus gnavus*	Indole; tyramine reduction	Suppressed GLP-1 synthetic pathway, glucose uptake pathways, and glucose tolerance	Pascale et al. (2019), Zhai et al. (2021)
5.	Bacteroidetes, some Firmicutes, *Coprococcus catus* and *Megasphaera elsdenii*, *Roseburia inulinivorans*, *Ruminococcus obeum*, and *Salmonella enteric*	Propionate; acetate; butyrate reduction	Reduced GLP-1 synthesis and insulin secretion; increased lipid synthesis, fat deposition, β-cell apoptosis, and inflammation	Du et al. (2022)
6.	Decreased *Eubacterium rectale* and *Roseburia intestinalis*	Bile acid reduction	Decreased expression of GLP-1; reduced insulin secretion, glycogen synthesis, and energy expenditure	Du et al. (2022), González-Regueiro et al. (2018)
7.	Increased *Clostridium hathewayi*, *C. sporogenes*,	Increased systemic TMAO levels	Increased blood glucose levels, inflammatory	Du et al. (2022), Han and Lin (2014)

(continued)

Table 14.1 (continued)

Sr. No.	Microbe	Metabolite	Repercussion	References
	Escherichia fergusonii population			
Nonalcoholic fatty liver disease				
8.	High abundance of *Klebsiella pneumoniae*, *Escherichia*, and *Shigella* and fall in number of Firmicutes species	Increased LPS	Exacerbates inflammatory response and apoptosis in hepatocytes	Wang et al. (2020), Tokuhara (2021)
9.		Decreased SCFAs	Disruption of gut barrier, increased gluconeogenesis, and fat accumulation	Tokuhara (2021)
10.		Reduction in secondary bile acids and indole derivatives	Increased cholesterol synthesis, lipogenesis, and inflammation; reduced GLP-1 release	Jiang et al. (2020)
Cardiovascular disease				
11.	Abundance of *Helicobacteraceae*, *Thiotricaceae*, *Micrococcaceae*, *Streptococcaceae*	Increased TMAO	Inhibits cholesterol breakdown and bile acid transport, activates inflammatory pathways	Kazemian et al. (2020), Murphy et al. (2021)
12.		Decreased SCFAs	Higher cholesterol production, oxidative stress and endotoxemia, modulate vasodilation	Li et al. (2021), Kazemian et al. (2020), Papadopoulos et al. (2022)
13.		Reduced secondary BAs	Suppression of Cyp7a1 gene, increased cholesterol synthesis	Kazemian et al. (2020), Papadopoulos et al. (2022)

can reduce levels of ghrelin through their inhibitory action on growth hormone secretagogue receptor 1α, ultimately affecting growth hormone (GH) and IGF-1 resulting in disruption of host homeostasis, metabolism, and stunted growth (Fig. 14.2). SCFAs, primarily propionate, are also able to directly suppress GH synthesis via cAMP/PKA/CREB pathway (Hu et al. 2018).

14.3.2 Gut Microbiota and Type 2 Diabetes Mellitus (T2DM)

The most reported characteristic feature of T2DM is hyperglycemia, which is a result of inability of pancreatic β-cells to function properly, accompanied by insulin resistance, abnormal hyperglucagonemia, and altered GLP-1 secretion (Bellary

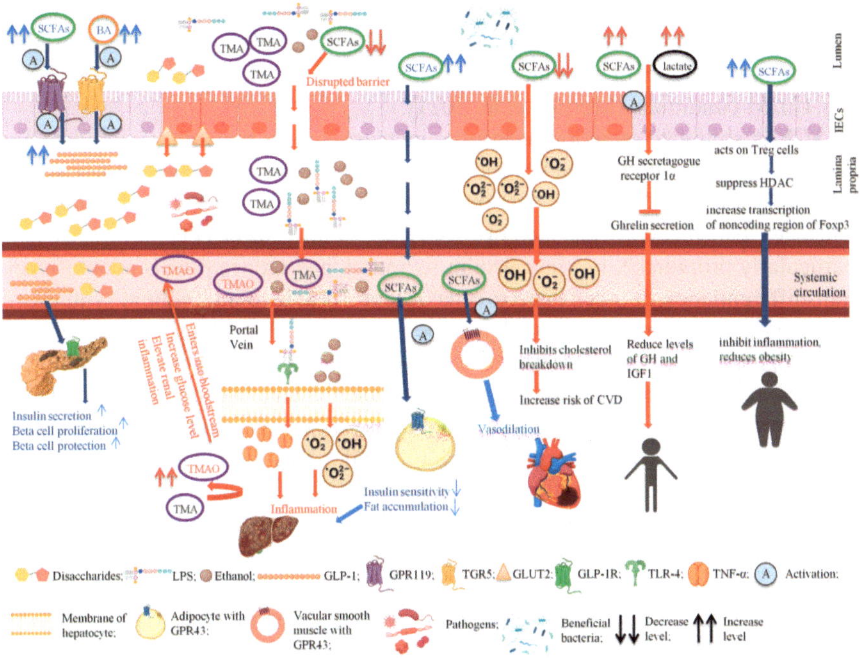

Fig. 14.2 Role of gut microbiota-derived metabolites in different metabolic disorders. Altered production of metabolites causes dysbiosis and affects the gut epithelium barrier resulting in translocation of pathogens and metabolites that induce immune response. These metabolites are transported systemically to distant body organs, alter organ functionality, and increase disease severity (red color). The gut microbiota-targeted therapies can be used to maintain gut microbial composition and their metabolites level which can decrease the risk of metabolic disorders (blue color)

et al. 2021). GM and its ability to secrete various gut hormones are involved in the regulation of host glucose metabolism. Experimental model studies revealed that hyperglycemia condition increases gut barrier permeability via a glucose transporter (GLUT) 2-dependent reprogramming of gut epithelial cells, and disrupts tight junction integrity that leads to a leaky gut mucosa (Fig. 14.1). However, whether the disrupted GM increases blood glucose level or it is a result of disease prognosis remains unclear (Fan and Pedersen 2021). Past reports suggest that depletion of *Akkermansia muciniphila* (Zhai et al. 2021) and butyrate-producing bacterial species (*Faecalibacterium prausnitzii* and *Roseburia intestinalis*) along with colonization of opportunistic pathogens (i.e., *Bacteroides caccae*, *Escherichia coli*, *Clostridium ramosum*, *Clostridium hathewayi*, *Clostridium symbiosum*, and *Eggerthella lenta*) in the gut are associated with T2DM (Vallianou et al. 2019). Microbes like *B. thetaiotaomicron*, *B. ovatus*, *C. limosum*, and *C. bifermentans* metabolize amino acids into their derivatives that stimulate enteroendocrine L cells to release GLP-1 which maintains glucose homeostasis (Fan and Pedersen 2021). Gut microbes *Enterococcus faecalis and Ruminococcus gnavus* can catabolize tyrosine to tyramine, which reduce the risk of diabetes by improving glucose tolerance ability of adipose tissues, cardiomyocytes, and skeletal muscle. Agmatine, a primary fermentation product of arginine, have insulin-like effects thus, beneficial in reducing T2DM conditions (Murphy et al. 2021).

SCFAs stimulate secretion of insulin, increase insulin sensitivity, stimulate gluconeogenesis by intestinal cells, and reduce fat deposition and inflammation (Fan and Pedersen 2021). SCFAs act on GPR119, inducing GLP-1 production that enhances pancreatic β-cells activity (Murphy et al. 2021) (Fig. 14.2). SCFAs, particularly propionate, increase insulin secretion, and reduce β-cells apoptosis and transdifferentiation of β-cells into α-cells. Acetate is able to inhibit lipid synthesis in the liver, and accumulation of lipid in adipose tissues. It also enhances the expression of myoglobin and GLUT-4 genes in muscle cells of the abdomen resulting in reduced serum glucose levels (Fan and Pedersen 2021). Acetate consumption at mM level stimulates browning of white adipose tissue (WAT) having anti-obesity and antidiabetic effects in murine models (González-Regueiro et al. 2018). Butyrate can cause a significant reduction in the levels of TNF-α, monocyte chemoattractant protein 1 (MCP-1) and IL-6, and in the activity of nuclear factor kappa B (NF-κB), reducing disease severity. Butyrate and propionate have been found to be associated with the activation of intestinal gluconeogenesis via a cAMP-dependent pathway and a gut-brain neural circuit, respectively, which suppress glucose production by hepatocytes. BAs act as a ligand for two primarily receptors, namely, Farnesoid X receptor (FXR) and Takeda G protein-coupled receptor 5 (TGR5) for regulating host glucose homeostasis and metabolizing lipids. BAs activate TGR5 present on intestinal L cells, stimulate release of GLP-1 that results in rise of ATP/ADP ratio ultimately inhibiting glycolysis (Fig. 14.1). In addition, activation of the TGR5 receptor of pancreatic α-cells can stimulate conversion of proglucagon to GLP-1, which stimulates pancreatic β-cells activity, thus, improving T2DM state. In case of FXR, contrasting results have been reported. Some studies have reported that inhibition of FXR exerts a positive effect on glucose metabolism by increasing the expression levels of GLP-1 gene (Prawitt et al. 2011). Some others have advocated that activation of FXR by BAs improves diabetic condition via release of Fibroblast growth factor (FGF) 15 or FGF19 (in case of humans), which inhibits gluconeogenesis, and enhances glycogen synthesis as well as energy expenditure (Massafra and van Mil 2018). TMAO, a metabolite that has been extensively studied in case of CVD is now considered a major marker for development of diabetes as well. TMAO may be responsible for promotion of metabolic dysfunction and increased blood glucose levels through direct binding and activation of RNA-like endoplasmic reticulum kinase (PERK). It elevates the level of transforming growth factor β1 (TGF-β1) and α-smooth muscle actin, promoting renal inflammation, indicating detrimental effects of TMAO in T2DM (Fan and Pedersen 2021) (Table 14.1; Fig. 14.2).

14.3.3 Gut Microbiota and Non-alcoholic Fatty Liver Disease (NAFLD)

NAFLD represents a series of liver conditions depending on severity of injured cells and fibrosis. Hepatic steatosis alone is NAFL and non-alcoholic steatohepatitis (NASH) is a worsen stage including inflammation as well as liver cell damage (Friedman et al. 2018). The incidence of NAFLD has dramatically increased; about 20–40% of the population of a country is suffering from this disease. It is considered as the hepatic manifestation of the altered host metabolism (Younossi et al. 2016). Reduced gut bacterial diversity is found to be associated with NAFLD. The members of the phylum Bacteroides are abundant in the NAFLD patients while Firmicutes species are reduced down. Further, research findings suggest presence of high-alcohol-producing microbes such as *Klebsiella pneumonia (HiAlc Kpn)*, *Escherichia*, and *Shigella* in the NAFLD patients (Yuan et al. 2019; Jiang et al. 2020). The altered gut microbial diversity causes disruption of intestinal barrier, leads to translocation of lipopolysaccharides (LPS), and ethanol to liver via portal vein, and reduced production of SCFAs. LPS can stimulate hepatocytes to produce TNF-α via Toll-like receptor-4 (TLR-4), resulting in hepatocyte apoptosis (Jiang et al. 2020) (Table 14.1; Fig. 14.1). Translocation of gut microbes through gut barrier causes activation of NF-κB pathway in hepatic cells leading to increased cytokine production, and inflammation that further exacerbates the disease progression. Also, NF-κB activation stimulates synthesis of TNFα, Fas ligand (FasL), and TGF-β, leading to development of NASH in severe cases (Davis 2016). Ethanol in the hepatic system is able to stimulate the process of oxidative stress and increase liver inflammation. SCFAs play a protective role against NAFLD. They activate GPR43 which inhibits the action of insulin in adipose cells leading to inhibition of fat accumulation and also aids in metabolism of un-utilized lipids and glucose moieties in the liver thus avoiding lipid accumulation (Fig. 14.2). Propionate can inhibit the expression of gene responsible for the synthesis of gluconeogenic enzymes via a h-pathway. Butyrate strengthens tight junctions, restricting the transport of LPS to the portal vein thus suppressing inflammation of hepatocytes (Jiang et al. 2020) (Table 14.1).

BAs show contradictory effects, i.e., both liver toxicity and liver protection (Jiang et al. 2020). Activation of TGR5 present on colonic L cells by secondary BAs acts as a stimulus to synthesize and release GLP-1, which inhibits both the glucose synthesis and fat deposition in hepatic cells (Fig. 14.2). Reduction of secondary bile acids consequently increases the proteins associated with cholesterol synthesis which in turn increase disease severity. Indole and its derivative like Indole-3-carbinol (I3C) show inhibitory effect on lipogenesis process in liver by reducing the mRNA copies of fatty acid synthase (FAS), SREBP1c, acetyl-CoA carboxylase 1 (ACC1) genes. In addition, they reduce the concentration of reactive oxygen species (ROS) and TNFα levels, promote the functioning of superoxide dismutase (SOD), and inhibit infiltration of macrophages in hepatocytes (Davis 2016).

14.3.4 Gut Microbiota and Cardiovascular Diseases (CVD)

Cardiovascular disease or atherosclerosis, a leading cause of deaths in the western world, is also a consequence of metabolic imbalance. Patients have primary symptoms related to host metabolism including increased levels of systemic glucose, lipids, and insulin. Therefore, there is a need to explore the changes in gut microbial composition and its impact on metabolic manifestations in early stages of disease (Fan and Pedersen 2021). Symptomatic plaques have a higher abundance of microbes from the family *Helicobacteracaea*, *Neisseriaceae*, and *Thiotrichacaea*; whereas asymptomatic atherosclerotic plaques have enhanced population of members of family *Porphyromonadacaea*, *Bacteroidaceae*, *Micrococcacaea*, and *Streprococcacaea* (Papadopoulos et al. 2022) (Table 14.1). Altered production of microbial metabolites is found to be associated with CVD prognosis. The most important gut microbial metabolites SCFAs suppress endotoxemia in mice (Sharma et al. 2020) and protect against the CVD. SCFAs exert their impact on renal olfactory receptor 78 and GPR41 situated on vascular smooth muscle cells, thus, modulating vasodilation. Further, reduced level of butyrate-producing bacterial species reportedly affects fatty acid breakdown, generates oxidative stress thus, promoting the occurrence of adverse cardiometabolic manifestations (Wang and Zhao 2018) (Fig. 14.2). BA-dependent FGF 19 activation induces FGF4 that inhibits the CYP7A1, thus, downregulating bile acid production (Papadopoulos et al. 2022). The altered gut microbial composition leads to reduced production of secondary BAs that result in accumulation of primary BAs, thus increasing cholesterol synthesis and risk of coronary artery disease (CAD) (Wang and Zhao 2018). An increased serum concentration of TMAO is linked to the onset of atherosclerosis, and higher risk of CVD-related death. TMAO can suppress the transcription of Cyp7a1 gene, which encodes for the primary enzyme that acts as a rate limiting step in cholesterol breakdown (Kazemian et al. 2020) (Table 14.1). Further, it also restricts activity of various bile acid transporters such as OAT4, OATP1, NTCP, and MRP2, reducing the pool of Bas, and decreasing reverse cholesterol efflux (Tokuhara 2021). TMAO activates TXNIP-NLRP3 inflammasomes, which increase the levels of TNF-α, IL-6, IL-18, and IL-1B, promoting the formation of cholesterol-packed foamy macrophages, finally leading to development of CAD (Papadopoulos et al. 2022).

These observations point out that a few specific gut microbial metabolites are associated with the progression or regression of major metabolic diseases. As discussed above, an increase of TMAO is generally associated with disease exacerbation, whereas increased concentration of SCFAs is usually linked to disease alleviation. But exceptions exist, such as, increased SCFAs level leads to a higher risk of undernourishment. In contrast to that, contradictory results are observed in studies focused on the role of BAs in metabolic disorders. Further, in case of cardiovascular diseases and NAFLD, low BAs concentration leads to disease development; however, obesity is positively linked with higher BAs level. Mostly, secondary BAs have been found to be usually positively associated with host health. Thus, it demands more GM-based research with a deep focus on the mining and association of microbial metabolites with disease progression. The deeper knowledge will prove beneficial for designing treatment therapies targeting GM and its metabolites.

14.4 Therapeutic Interventions for Metabolic Disorders

GM has a great impact on the functioning of other human organs, and ensures a healthy state of the body. The dysbiosis of microbial composition and their metabolites is correlated with the risk of various organ-related diseases. Thus, this association has led a path to find various GM-related therapies that can restore the gut microflora, and reestablish eubiosis.

14.4.1 Obesity and Undernutrition

Obesity is developed due to disbalance between energy intake and energy expenditure which ultimately results in fat accumulation. Over a decade, studies have been conducted to find effective probiotics like *Lactobacillus rhamnosus* GG (Vajro et al. 2011), *Bifidobacterium animalis* subsp. Lactis CECT 8145 (Pedret et al. 2019), *Lactobacillus gasseri* BNR17 (Kim et al. 2018), and *Bifidobacterium breve* B-3 (Minami et al. 2018). The probiotics mainly belonging to *Lactobacillus* are reported to modulate Firmicutes/Bacteroidetes ratio and improve the risk of obesity (Stojanov et al. 2020). The progression of obesity gains more research attention as it was found to be associated with other major metabolic diseases as well. Thus, the gut-targeted interventions to metabolic diseases contributed in reduction of body fat, improvement of lipid metabolism, and appetite control.

Undernutrition is a condition developed due to inability to provide sufficient nutrients and energy source. A number of pathogens residing in the gut demand for more nutrients and thus, induce an undernutrition state especially in children. The primary treatment to cope with this condition is intake of a nutrient-rich diet. The beneficial species like *Bifidobacterium*, *Butyrivibrio*, *Faecalibacterium*, *Lactobacillus*, and *Roseburia* can be reestablished by administration of probiotics or by including specific prebiotics in the diet. The yoghurt supplementation improves diarrhea and significant weight gain in children (He et al. 2005). The probiotic *L. plantarum* Dad-13 improved the nutritional status in infants, promoted the growth of butyric acid-producing bacteria and inhibited the growth of *Enterobacteriaceae* (Kamil et al. 2022). Thus, the undernutrition condition should be treated in early stage in order to prevent its long-term effect and onset of other associated metabolic diseases. Like obesity, the under nutrition associated with metabolic diseases can be significantly improved by administration of beneficial bacteria and their metabolites.

14.4.2 Type 2 Diabetes Mellitus

The traditional treatment for T2DM includes various antidiabetic drugs like Biguanides (Metformin), Alpha-glucosidase inhibitors (Acarbose, voglibose, miglitol), DPP-4 inhibitors (Vildagliptin, sitagliptin, saxagliptin), SGLT2 inhibitors (Dapagliflozin), and GLP-1 Receptor agonist (Liraglutide). These drugs successfully restore the community shift as well as functional shift occurred in gut microflora, and thereby improve the T2DM condition (Wang et al. 2016; Wu et al. 2017; Zhang et al. 2017a, b; Lee et al. 2018). The combinatorial drug therapy was introduced to achieve an intensified treatment in order to control the hyperglycemic condition. Recently, the combinations of drugs with prebiotics like sitagliptin/metformin + PolyGlycopleX (Reimer et al. 2014) and metformin + mannan-oligosaccharides (Zheng et al. 2018) are observed as effective multitarget treatment as it provides hypoglycemic condition along with modulation of gut microbiome. Probiotics like *L. casei* (Khalili et al. 2019), *L. reuteri ADR-1* (Hsieh et al. 2018), *B. lactis* (Bernini et al. 2016), and *L. acidophilus* (Ejtahed et al. 2011) have shown effective results in controlling hyperglycemic conditions and reduce the risk associated with T2DM. The administration of FMT in mice with high-fat-induced T2DM ensured uniform gut flora with improved glucose tolerance and insulin resistance (Wang et al. 2020). Also, it inhibits chronic inflammation, cell apoptosis, and promotes regeneration and repair of pancreatic cells. The combined treatment of specially designed diet, and FMT on T2DM patients has improved the blood glucose and lipid levels with significant decrease in sulfate-reducing bacteria and increase in abundance of *Bifidobacterium*, *Lactobacillus*, and *Prevotella* (Su et al. 2022).

14.4.3 Non-alcoholic Fatty Liver Disease

The disturbance in gut-liver axis affects the functioning of the liver, elicits immune responses, and triggers various risk factors involved in liver-related diseases and thus, various GM-targeted therapies are successfully used to control these factors. The probiotics *Lactobacillus plantarum* LC27 and *Bifidobacterium longum* LC67 are observed to reverse the AMP-activated protein kinase (AMPK) suppression leading to activation of PGC-1α (a regulator of lipolysis), and reduce LPS-induced inflammation via NF-κB inhibition (Kim et al. 2019). Similarly, a treatment of commercially available probiotics mainly consisting of *L. acidophilus*, *L. plantarum*, and *B. bifidum* showed improvement in lipid profile, liver functioning, histological features of liver, and inflammatory markers levels in rats with NAFLD (Al-Muzafar and Amin 2017).

Prebiotics have also shown significant impact on liver disease and obesity associated with it. The fructooligosaccharide (FOS) treatment prevents NAFLD by modulating the gut microflora, maintaining gut integrity, improving lipid content, and reducing hepatic inflammation (Chong et al. 2020; Bomhof et al. 2019; Wang and Wan 2019). The FMT treatment carried out for 8 weeks attenuated high-fat diet (HFD)-induced steatohepatitis in mice. This study showed decrease in intrahepatic triglyceride and cholesterol, and proinflammatory cytokines with increase in the abundance of *Christensenellaceae* and *Lactobacillus* (Zhou et al. 2017). The improvement in portal hypertension and liver condition was observed after FMT trial given to rat with artificial diet-induced nonalcoholic steatohepatitis (NASH) (García-Lezana et al. 2018).

14.4.4 Cardiovascular Disease

The interventions against CVD have gained more research interest due to its association with high mortality rate. As gut microflora governs the balance between healthy and diseased state, various types of diets are effective in reversing the symptoms of CVD. A healthy diet should be composed of dietary components necessary to maintain various factors like SCFAs levels, lipid metabolism, gut barrier integrity, and immune response. These factors are prerequisite to successfully modulate the gut microflora and improve human health. Recently, it has been found that the healthy Mediterranean diet is capable of enriching the human gut with beneficial bacteria which tries to maintain gut homeostasis, and improve cardiac risk factors (Wang et al. 2021). The strict vegetarian diet with high-dietary fibers helps to modulate SCFAs producers and promote weight loss which eventually reduces the symptoms of CVD (Kim et al. 2013). The other effective way to reestablish GM is to introduce probiotics in the gut environment. Over a decade, potent bacteria like *B. longum*, *B. lactis*, *L. sporogenes*, *L. bulgaricus*, *L. acidophilus*, and *L. reuteri* (Kumar et al. 2012) had successfully mediated cholesterol reduction.

Continuous research is still being carried out to explore the effect of various bacteria on CVD in order to achieve new probiotics which can target and improve multiple risk factors of CVD like inflammation, oxidative stress, impaired lipid metabolism, and high blood pressure. One such interesting study was carried out to replenish *Akkermansia muciniphila* which reduced the metabolic endotoxemia-induced inflammation and proved beneficial in improving atherosclerotic complications (Li et al. 2016). Similar results were found for *Bacteroides vulgatus* and *B. dorei* which significantly reduce lesions in atherosclerotic prone mice (Yoshida et al. 2018). The treatment of *L. reuteri* V3401 on patients diagnosed with metabolic syndrome shows decrease in soluble adhesion molecule sVCAM-1, a cardiovascular risk factor (Tenorio-Jiménez et al. 2019). Currently, FMT has been explored to find more effective treatment against CVD. The FMT from lean donor induces microflora shift in obese patients with metabolic syndrome but does not alter the pathway of TMAO conversion (Smits et al. 2018). The Th17/Treg balance and immune responses were restored in hypertensive rats after successful administration of FMT from wild rats (Toral et al. 2019). Further, FMT improved myocardial injuries in diseased mice by decreasing expression of IFN-γ gene in the heart tissue (Hu et al. 2019).

The antibiotics administration has been widely used for many decades to treat every type of metabolic disease. The inability of antibiotics to target the microbial metabolites and other risk factors was the major reason to develop multi-target interventions. A wide range of probiotics and prebiotics have been explored to effectively work against these diseases. Although the field of probiotics turned out to be a huge success, researchers found that combined therapy (antibiotics & probiotics) works more efficiently to improve diseased conditions. The current advancement in FMT has proved a boon for replenishing beneficial bacteria and trying to maintain the balance of GM. Researchers are working on FMT to make it a better therapy and minimize its side effects. The conventional as well as advanced treatments have proved their significant role in reducing the risk factors. Still, the need for more potent and multitarget treatments to combat the rigorously progressing metabolic diseases have led the path for in depth exploration of GM-targeted therapies. Thus, to achieve this goal, various gut microbial species, their combinations, and their effective administration methods have gained more research interest.

14.5 Conclusion

The microorganisms living in the human gut are very important as they are involved in different host reactions, and eventually affect overall host physiology. The diverse population of GM produces structurally different metabolites that interact with the host, both locally and systemically. Altered levels of microbial metabolites with microbial dysbiosis are associated with development of human disorders. The importance of gut microbial metabolites in human health is now well recognized, but the growing literature in this field suggests that it is in its juvenile phase. The next generation research develops various methods for the study of microbial metabolites, including sequence-based metagenomics, functional metagenomics, proteomics, and metabolomics. Therefore, research should be more focused to mine more microbial metabolites and explore their functions associated with human health. There is a need to develop more efficacious methods to study these metabolites as they can be helpful in developing precise and personalized medicine, and can also be used as diagnostic markers.

Conflict of Interest The authors do not have any potential competing interests.

References

Ahlawat S, Asha, Sharma KK (2021) Gut–organ axis: a microbial outreach and networking. Lett Appl Microbiol 72(6):636–668

Al-Muzafar HM, Amin KA (2017) Probiotic mixture improves fatty liver disease by virtue of its action on lipid profiles, leptin, and inflammatory biomarkers. BMC Complement Altern Med 17(1):1–9

Bäckhed F, Ding H, Wang T, Hooper LV, Koh GY, Nagy A, Gordon JI (2004) The gut microbiota as an environmental factor that regulates fat storage. Proc Natl Acad Sci 101(44):15718–15723

Bellary S, Kyrou I, Brown JE, Bailey CJ (2021) Type 2 diabetes mellitus in older adults: clinical considerations and management. Nat Rev Endocrinol 17(9):534–548

Bernini LJ, Simão ANC, Alfieri DF, Lozovoy MAB, Mari NL, de Souza CHB, Costa GN (2016) Beneficial effects of Bifidobacterium lactis on lipid profile and cytokines in patients with metabolic syndrome: a randomized trial. Effects of probiotics on metabolic syndrome. Nutrition 32(6):716–719

Blaut M (2018) Composition and function of the gut microbiome. In: The gut microbiome in health and disease. Springer, Cham, pp 5–30

Bomhof MR, Parnell JA, Ramay HR, Crotty P, Rioux KP, Probert CS, Reimer RA (2019) Histological improvement of non-alcoholic steatohepatitis with a prebiotic: a pilot clinical trial. Eur J Nutr 58(4):1735–1745

Castaner O, Goday A, Park YM, Lee SH, Magkos F, Shiow SATE, Schröder H (2018) The gut microbiome profile in obesity: a systematic review. Int J Endocrinol 2018(2018):4095789

Chen X, Sun H, Jiang F, Shen Y, Li X, Hu X, Wei P (2020) Alteration of the gut microbiota associated with childhood obesity by 16S rRNA gene sequencing. PeerJ 8:e8317

Chong CYL, Orr D, Plank LD, Vatanen T, O'Sullivan JM, Murphy R (2020) Randomised double-blind placebo-controlled trial of inulin with metronidazole in non-alcoholic fatty liver disease (NAFLD). Nutrients 12(4):937

Chooi YC, Ding C, Magkos F (2019) The epidemiology of obesity. Metabolism 92:6–10

Danneskiold-Samsøe NB, Barros HDDFQ, Santos R, Bicas JL, Cazarin CBB, Madsen L, Júnior MRM (2019) Interplay between food and gut microbiota in health and disease. Food Res Int 115:23–31

Davis CD (2016) The gut microbiome and its role in obesity. Nutr Today 51(4):167

Dey P, Chaudhuri SR, Efferth T, Pal S (2021) The intestinal 3M (microbiota, metabolism, metabolome) Zeitgeist–from fundamentals to future challenges. Free Radic Biol Med 176: 265–285

Du L, Li Q, Yi H, Kuang T, Tang Y, Fan G (2022) Gut microbiota-derived metabolites as key actors in type 2 diabetes mellitus. Biomed Pharmacother 149:112839

Ejtahed HS, Mohtadi-Nia J, Homayouni-Rad A, Niafar M, Asghari-Jafarabadi M, Mofid V, Akbarian-Moghari A (2011) Effect of probiotic yogurt containing Lactobacillus acidophilus and Bifidobacterium lactis on lipid profile in individuals with type 2 diabetes mellitus. J Dairy Sci 94(7):3288–3294

Fan Y, Pedersen O (2021) Gut microbiota in human metabolic health and disease. Nat Rev Microbiol 19(1):55–71

Friedman SL, Neuschwander-Tetri BA, Rinella M, Sanyal AJ (2018) Mechanisms of NAFLD development and therapeutic strategies. Nat Med 24(7):908–922

García-Lezana T, Raurell I, Bravo M, Torres-Arauz M, Salcedo MT, Santiago A, Augustin S (2018) Restoration of a healthy intestinal microbiota normalizes portal hypertension in a rat model of nonalcoholic steatohepatitis. Hepatology 67(4):1485–1498

Gérard P (2016) Gut microbiota and obesity. Cell Mol Life Sci 73(1):147–162

González-Regueiro JA, Moreno-Castañeda L, Uribe M, Chávez-Tapia NC (2018) The role of bile acids in glucose metabolism and their relation with diabetes. Ann Hepatol 16(1):15–20

Han JL, Lin HL (2014) Intestinal microbiota and type 2 diabetes: from mechanism insights to therapeutic perspective. World J Gastroenterol 20(47):17737

He M, Yang YX, Han H, Men JH, Bian LH, Wang GD (2005) Effects of yogurt supplementation on the growth of preschool children in Beijing suburbs. Biomed Environ Sci 18(3):192

Hrncir T (2022) Gut microbiota dysbiosis: triggers, consequences, diagnostic and therapeutic options. Microorganisms 10(3):578

Hsieh MC, Tsai WH, Jheng YP, Su SL, Wang SY, Lin CC, Chang WW (2018) The beneficial effects of Lactobacillus reuteri ADR-1 or ADR-3 consumption on type 2 diabetes mellitus: a randomized, double-blinded, placebo-controlled trial. Sci Rep 8(1):1–11

Hu J, Lin S, Zheng B, Cheung PC (2018) Short-chain fatty acids in control of energy metabolism. Crit Rev Food Sci Nutr 58(8):1243–1249

Hu XF, Zhang WY, Wen Q, Chen WJ, Wang ZM, Chen J, Shu YW (2019) Fecal microbiota transplantation alleviates myocardial damage in myocarditis by restoring the microbiota composition. Pharmacol Res 139:412–421

Iddrisu I, Monteagudo-Mera A, Poveda C, Pyle S, Shahzad M, Andrews S, Walton GE (2021) Malnutrition and gut microbiota in children. Nutrients 13(8):2727

Jensen EA, Young JA, Mathes SC, List EO, Carroll RK, Kuhn J, Berryman DE (2020) Crosstalk between the growth hormone/insulin-like growth factor-1 axis and the gut microbiome: a new frontier for microbial endocrinology. Growth Hormon IGF Res 53:101333

Jiang X, Zheng J, Zhang S, Wang B, Wu C, Guo X (2020) Advances in the involvement of gut microbiota in pathophysiology of NAFLD. Front Med 7:361

Kalantar-Zadeh K, Berean KJ, Burgell RE, Muir JG, Gibson PR (2019) Intestinal gases: influence on gut disorders and the role of dietary manipulations. Nat Rev Gastroenterol Hepatol 16(12): 733–747

Kamil RZ, Murdiati A, Juffrie M, Rahayu ES (2022) Gut microbiota modulation of moderate undernutrition in infants through gummy lactobacillus plantarum dad-13 consumption: a randomized double-blind controlled trial. Nutrients 14(5):1049

Kazemian N, Mahmoudi M, Halperin F, Wu JC, Pakpour S (2020) Gut microbiota and cardiovascular disease: opportunities and challenges. Microbiome 8(1):1–17

Khalili L, Alipour B, Jafar-Abadi MA, Faraji I, Hassanalilou T, Abbasi MM, Sani MA (2019) The effects of Lactobacillus casei on glycemic response, serum sirtuin1 and fetuin-a levels in patients with type 2 diabetes mellitus: a randomized controlled trial. Iran Biomed J 23(1):68

Kim MS, Hwang SS, Park EJ, Bae JW (2013) Strict vegetarian diet improves the risk factors associated with metabolic diseases by modulating gut microbiota and reducing intestinal inflammation. Environ Microbiol Rep 5(5):765–775

Kim J, Yun JM, Kim MK, Kwon O, Cho B (2018) Lactobacillus gasseri BNR17 supplementation reduces the visceral fat accumulation and waist circumference in obese adults: a randomized, double-blind, placebo-controlled trial. J Med Food 21(5):454–461

Kim HI, Kim JK, Kim JY, Jang SE, Han MJ, Kim DH (2019) Lactobacillus plantarum LC27 and Bifidobacterium longum LC67 simultaneously alleviate high-fat diet-induced colitis, endotoxemia, liver steatosis, and obesity in mice. Nutr Res 67:78–89

Krishnan S, Alden N, Lee K (2015) Pathways and functions of gut microbiota metabolism impacting host physiology. Curr Opin Biotechnol 36:137–145

Kumar M, Nagpal R, Kumar R, Hemalatha R, Verma V, Kumar A, Yadav H (2012) Cholesterol-lowering probiotics as potential biotherapeutics for metabolic diseases. Exp Diabetes Res 2012(2012):902917

Lee DM, Battson ML, Jarrell DK, Hou S, Ecton KE, Weir TL, Gentile CL (2018) SGLT2 inhibition via dapagliflozin improves generalized vascular dysfunction and alters the gut microbiota in type 2 diabetic mice. Cardiovasc Diabetol 17(1):1–14

Li J, Lin S, Vanhoutte PM, Woo CW, Xu A (2016) Akkermansia muciniphila protects against atherosclerosis by preventing metabolic endotoxemia-induced inflammation in Apoe$^{-/-}$ mice. Circulation 133(24):2434–2446

Li L, Zhang Y, Speakman JR, Hu S, Song Y, Qin S (2021) The gut microbiota and its products: establishing causal relationships with obesity-related outcomes. Obes Rev 22(12):e13341

Liu J, Tan Y, Cheng H, Zhang D, Feng W, Peng C (2022) Functions of gut microbiota metabolites, current status and future perspectives. Aging Dis 13(2):14336

Lynch SV, Pedersen O (2016) The human intestinal microbiome in health and disease. N Engl J Med 375:2369–2379

Massafra V, van Mil SW (2018) Farnesoid X receptor: a "homeostat" for hepatic nutrient metabolism. Biochim Biophys Acta 1864(1):45–59

Matsumoto M, Ooga T, Kibe R, Aiba Y, Koga Y, Benno Y (2017) Colonic absorption of low-molecular-weight metabolites influenced by the intestinal microbiome: a pilot study. PLoS One 12(1):e0169207

Minami J, Iwabuchi N, Tanaka M, Yamauchi K, Xiao JZ, Abe F, Sakane N (2018) Effects of Bifidobacterium breve B-3 on body fat reductions in pre-obese adults: a randomized, double-blind, placebo-controlled trial. Biosci Microbiota Food Health 37(3):67–75

Mirzaei R, Dehkhodaie E, Bouzari B, Rahimi M, Gholestani A, Hosseini-Fard SR, Karampoor S (2022) Dual role of microbiota-derived short-chain fatty acids on host and pathogen. Biomed Pharmacother 145:112352

Murphy K, O'Donovan AN, Caplice NM, Ross RP, Stanton C (2021) Exploring the gut microbiota and cardiovascular disease. Metabolites 11(8):493

Olson CA, Vuong HE, Yano JM, Liang QY, Nusbaum DJ, Hsiao EY (2018) The gut microbiota mediates the anti-seizure effects of the ketogenic diet. Cell 173(7):1728–1741

Papadopoulos PD, Tsigalou C, Valsamaki PN, Konstantinidis TG, Voidarou C, Bezirtzoglou E (2022) The emerging role of the gut microbiome in cardiovascular disease: current knowledge and perspectives. Biomedicine 10(5):948

Pascale A, Marchesi N, Govoni S, Coppola A, Gazzaruso C (2019) The role of gut microbiota in obesity, diabetes mellitus, and effect of metformin: new insights into old diseases. Curr Opin Pharmacol 49:1–5

Pedret A, Valls RM, Calderón-Pérez L, Llauradó E, Companys J, Pla-Pagà L, Solà R (2019) Effects of daily consumption of the probiotic Bifidobacterium animalis subsp. lactis CECT 8145 on anthropometric adiposity biomarkers in abdominally obese subjects: a randomized controlled trial. Int J Obes 43(9):1863–1868

Prawitt J, Abdelkarim M, Stroeve JH, Popescu I, Duez H, Velagapudi VR, Staels B (2011) Farnesoid X receptor deficiency improves glucose homeostasis in mouse models of obesity. Diabetes 60(7):1861–1871

Qin J, Li R, Raes J, Arumugam M, Burgdorf KS, Manichanh C et al (2010) A human gut microbial gene catalogue established by metagenomic sequencing. Nature 464:59–65

Reimer RA, Grover GJ, Koetzner L, Gahler RJ, Lyon MR, Wood S (2014) Combining sitagliptin/metformin with a functional fiber delays diabetes progression in Zucker rats. J Endocrinol 220(3):361–373

Rokade HG, Jidge AC, Nadgire SS (2022) Study of nutritional status and feeding practices in under five children from Solapur city. Int J Commun Med Public Health 9(8):3283

Rowland I, Gibson G, Heinken A, Scott K, Swann J, Thiele I, Tuohy K (2018) Gut microbiota functions: metabolism of nutrients and other food components. Eur J Nutr 57(1):1–24

Schoeler M, Caesar R (2019) Dietary lipids, gut microbiota and lipid metabolism. Rev Endocr Metab Disord 20(4):461–472

Sharma M, Li Y, Stoll ML, Tollefsbol TO (2020) The epigenetic connection between the gut microbiome in obesity and diabetes. Front Genet 10:1329

Smits LP, Kootte RS, Levin E, Prodan A, Fuentes S, Zoetendal EG, Nieuwdorp M (2018) Effect of vegan fecal microbiota transplantation on carnitine-and choline-derived trimethylamine-N-oxide production and vascular inflammation in patients with metabolic syndrome. J Am Heart Assoc 7(7):e008342

Stojanov S, Berlec A, Štrukelj B (2020) The influence of probiotics on the firmicutes/bacteroidetes ratio in the treatment of obesity and inflammatory bowel disease. Microorganisms 8(11):1715

Su L, Hong Z, Zhou T, Jian Y, Xu M, Zhang X, Wang J (2022) Health improvements of type 2 diabetic patients through diet and diet plus fecal microbiota transplantation. Sci Rep 12(1):1–12

Tenorio-Jiménez C, Martínez-Ramírez MJ, Del Castillo-Codes I, Arraiza-Irigoyen C, Tercero-Lozano M, Camacho J, Gómez-Llorente C (2019) Lactobacillus reuteri V3401 reduces inflammatory biomarkers and modifies the gastrointestinal microbiome in adults with metabolic syndrome: the PROSIR study. Nutrients 11(8):1761

Tokuhara D (2021) Role of the gut microbiota in regulating non-alcoholic fatty liver disease in children and adolescents. Front Nutr 8:700058

Toral M, Robles-Vera I, de la Visitación N, Romero M, Sanchez M, Gómez-Guzmán M, Duarte J (2019) Role of the immune system in vascular function and blood pressure control induced by faecal microbiota transplantation in rats. Acta Physiol 227(1):e13285

Vajro P, Mandato C, Licenziati MR, Franzese A, Vitale DF, Lenta S, Meli R (2011) Effects of Lactobacillus rhamnosus strain GG in pediatric obesity-related liver disease. J Pediatr Gastroenterol Nutr 52(6):740–743

Vallianou N, Stratigou T, Christodoulatos GS, Dalamaga M (2019) Understanding the role of the gut microbiome and microbial metabolites in obesity and obesity-associated metabolic disorders: current evidence and perspectives. Curr Obes Rep 8(3):317–332

Wang L, Wan YJY (2019) The role of gut microbiota in liver disease development and treatment. Liver Res 3(1):3–18

Wang Z, Zhao Y (2018) Gut microbiota-derived metabolites in cardiovascular health and disease. Protein Cell 9(5):416–431

Wang L, Li P, Tang Z, Yan X, Feng B (2016) Structural modulation of the gut microbiota and the relationship with body weight: compared evaluation of liraglutide and saxagliptin treatment. Sci Rep 6(1):1–10

Wang H, Lu Y, Yan Y, Tian S, Zheng D, Leng D, Bai Y (2020) Promising treatment for type 2 diabetes: fecal microbiota transplantation reverses insulin resistance and impaired islets. Front Cell Infect Microbiol 9:455

Wang DD, Nguyen LH, Li Y, Yan Y, Ma W, Rinott E, Huttenhower C (2021) The gut microbiome modulates the protective association between a Mediterranean diet and cardiometabolic disease risk. Nat Med 27(2):333–343

Whisner CM, Castillo LF (2018) Prebiotics, bone and mineral metabolism. Calcif Tissue Int 102: 443–479

Wu H, Esteve E, Tremaroli V, Khan MT, Caesar R, Mannerås-Holm L, Bäckhed F (2017) Metformin alters the gut microbiome of individuals with treatment-naive type 2 diabetes, contributing to the therapeutic effects of the drug. Nat Med 23(7):850–858

Yang S, Li X, Yang F, Zhao R, Pan X, Liang J, Wu M (2019) Gut microbiota-dependent marker TMAO in promoting cardiovascular disease: inflammation mechanism, clinical prognostic, and potential as a therapeutic target. Front Pharmacol 10:1360

Yoshida N, Emoto T, Yamashita T, Watanabe H, Hayashi T, Tabata T, Hirata KI (2018) Bacteroides vulgatus and Bacteroides dorei reduce gut microbial lipopolysaccharide production and inhibit atherosclerosis. Circulation 138(22):2486–2498

Younossi ZM, Koenig AB, Abdelatif D, Fazel Y, Henry L, Wymer M (2016) Global epidemiology of nonalcoholic fatty liver disease—meta-analytic assessment of prevalence, incidence, and outcomes. Hepatology 64(1):73–84

Yuan J, Chen C, Cui J, Lu J, Yan C, Wei X, Liu D (2019) Fatty liver disease caused by high-alcohol-producing Klebsiella pneumoniae. Cell Metab 30(4):675–688

Zhai L, Wu J, Lam YY, Kwan HY, Bian ZX, Wong HLX (2021) Gut-microbial metabolites, probiotics and their roles in type 2 diabetes. Int J Mol Sci 22(23):12846

Zhang X, Fang Z, Zhang C, Xia H, Jie Z, Han X, Ji L (2017a) Effects of acarbose on the gut microbiota of prediabetic patients: a randomized, double-blind, controlled crossover trial. Diabetes Ther 8(2):293–307

Zhang Q, Xiao X, Li M, Yu M, Ping F, Zheng J, Wang X (2017b) Vildagliptin increases butyrate-producing bacteria in the gut of diabetic rats. PLoS One 12(10):e0184735

Zheng J, Li H, Zhang X, Jiang M, Luo C, Lu Z, Shi J (2018) Prebiotic mannan-oligosaccharides augment the hypoglycemic effects of metformin in correlation with modulating gut microbiota. J Agric Food Chem 66(23):5821–5831

Zhou D, Pan Q, Shen F, Cao HX, Ding WJ, Chen YW, Fan JG (2017) Total fecal microbiota transplantation alleviates high-fat diet-induced steatohepatitis in mice via beneficial regulation of gut microbiota. Sci Rep 7(1):1–11

Influence of the Gut Microbiome on Cardiovascular Health and Hypertension 15

Sachin Kumar, Avesh Choubey, Anuj Sharma, Amey Jedhe, Kalash Pandey, Aman Prakash, and Azhar Khan

Abstract

Cardiovascular diseases are characterized by high rates of morbidity and mortality. Microbiota are closely associated with cardiovascular disease. There is much evidence that supports the aetiology of many cardiovascular diseases (CVD), and related risk states such as hypertension, atherosclerosis, coronary artery diseases, myocardial infarction, obesity or dyslipidaemia, heart failure, chronic kidney diseases, and diabetes mellitus may be influenced by gut microbial dysbiosis. In addition to dysbiosis, the metabolic potential of gut microbiota (producing bioactive metabolites) also has an effect on host physiology since it enters the systemic circulation and may amplify the inflammatory response. Moreover, it has been shown to be a risk factor for cardiovascular disorders. There are several mechanisms by which the microbiota communicates with the host, including the trimethylamine/trimethylamine N-oxide pathway, the short-chain fatty acid pathway, and the primary and secondary bile acid pathways. It has been hypothesized that these pathways may also play a role in the development of cardiovascular disease. This chapter is mostly about learning about the dynamic relationship between the gut microbiota and cardiovascular disease, with a focus on the pathogenic mechanisms and therapeutic implications of hypertension, atherosclerosis, coronary artery diseases, myocardial infarction, obesity or dyslipidaemia, heart failure, chronic kidney diseases, and diabetes mellitus.

S. Kumar · A. Choubey · A. Sharma · A. Jedhe · K. Pandey · A. Khan (✉)
Faculty of Applied Sciences and Biotechnology, Shoolini University of Biotechnology and Management Sciences, Solan, Himachal Pradesh, India

A. Prakash
Department of Molecular and Cellular Engineering, Jacob Institute of Biotechnology and Bioengineering, Sam Higginbottom University of Agriculture, Technology and Sciences, Naini, India

R. Sobti et al. (eds.), *Role of Microbes in Sustainable Development*,
https://doi.org/10.1007/978-981-99-3126-2_15

Keywords

Gut microbiomes · Secondary metabolites · Hormonal regulation · Cardiovascular disease · Hypertension

15.1 Introduction

Human microbiota is defined as a rigid group of organisms that inhabit and interact with the human body (Grice and Segre 2011). The diverse interactions can be communalistic, mutualistic, or pathogenic. The human microbiome is known as the genomic content material of organisms (microbiota) inhabiting a particular website within the human frame. Microorganisms colonize diverse anatomical frame websites along with the pores and skin, the mucosa, gastrointestinal tract, respiration tract, urogenital tract, and the mammary gland. They shape complicated and discrete surroundings that adapt to the environmental conditions of every area of interest (Whiteside et al. 2015). From childbirth, a steady interaction (symbiosis) between the human frame and its indigenous microbiota starts. Those interactions play crucial roles in preserving well-known health and wellness. In difficult coevolution, organisms make up the microbiota. They actively adjust to their specific habitats and live in their respective niches in the human body (Yilmaz et al. 2014; Reid and Schloss 2014; Hoeppli et al. 2015). Due to their biological sports, those organisms are identified as a part of the frame, leading to various adjustments from theory to loss of life. The human microbiome is constantly evolving in reaction to host factors. Elements together with age, nutrition, lifestyle, hormonal changes, inherited genes, and underlying ailment are major determinants of the human microbiome at any given point in time. However, an alteration in the makeup of the human (dysbiosis) microbiota can lead to life-threatening ailments (Whiteside et al. 2015). A balanced microbiota has been shown to play a crucial position in fitness sustenance (Whiteside et al. 2015). The largest concentration of the human microbiome is found within the gut. Those organisms are the predominant gamers in maintaining and sustaining the fitness of people. Previous research on the human microbiome venture has shown that changes within the immune environment can be directly related to dysbiotic plants of the gut. Furthermore, dysbiosis has been linked to life-threatening health conditions such as most cancers, cardiovascular disorders, bowel inflammatory diseases, and difficult-to-treat bacterial infections due to antibiotic resistance (Morgan and Huttenhower 2012; Pascal et al. 2018).

15.1.1 The Gut Microbiota

Human beings harbour an extremely complicated and plentiful aggregation of microbes (Norman et al. 2014). The microbes that inhabit the intestine are particularly divided into bacteria, archaea, bacteriophages, viruses, and meiofauna (e.g. fungus). This microbial network is called the microbiota. However, trans-nation interactions in the GI tract have been shown to have an impact on fitness and disease (Ursell et al. 2014). Microbes colonize the colon more densely than any other organ, and the colon is a site where microbes may also have a say in our biology. The human frame consists of approximately 100 trillion microorganisms in the intestines, which is about 10 times more than the full number of human cells inside the frame. This institution consists of more than 1000 exceptional species of bacteria, with about a hundred species comprising as much as 99% of the total populace (Guarner 2005). The number of microorganisms progressively increases alongside the small gut, from about 10^4 colony-forming devices (CFU) per gram of luminal content within the jejunum to 10^7 CFU at the ileum end, wherein gram-negative aerobes and a few obligate anaerobes are most important. Inside the colon, anaerobes are essential and bacterial counts attain around 10^{12} CFU per gram of luminal content. Microorganisms make up a contribution of 60% of the faecal mass. People showcase variants in the types and numbers of species inside their microbiota (Guarner and Malagelada 2003; Mahida and Rolfe 2004; Hollister et al. 2014). Traditional culturing strategies should locate the most effective 30% of overall bacteria within the gut (Guarner and Malagelada 2003). However, the use of 16S ribosomal RNA gene sequencing or entire-genome, series-based metagenomic analyses increases the capability to discover the numbers and variety of microbiota. Greater bacterial richness and diversity are associated with better nutritional fame, fewer comorbidities, and better overall health in an elderly population (Claesson et al. 2012).

15.1.2 Carbohydrate Metabolism

Carbohydrates and simple sugars are ample components of ingredients and substrates for microbial metabolism inside the human gut. The intake of carbohydrates impacts the composition and features of the human microbiome in healthy people (De Filippo et al. 2010). The intestine microbial shape of healthy children on plant carbohydrate-rich diets differs markedly from that of children on a typical Western diet (De Filippo et al. 2010). The intake of dietary carbohydrates results in a large percentage of the phylum Bacteroidetes; many of these bacteria are enriched with genes that encode carbohydrate-energetic enzymes (glycosyltransferases, glycoside hydrolases, and polysaccharide lyases) (Hollister et al. 2014; Cantarel et al. 2012). Regarding the phylum Bacteroidetes, folks who consume diets rich in plant carbohydrates have a greater proportion of *Prevotella* species (Wu et al. 2011), while folks that consume animal fats and protein have higher proportions of Bacteroides species. Hence, the intestine microbiome, in place of the human genome, plays a crucial function in the metabolism of dietary carbohydrates. The relative functions of the intestine microbiota within the metabolism of various meals can also provide strategies for nutritional modification and useful optimization of the microbiome (Wu et al. 2011).

15.1.3 Generation of Short-Chain Fatty Acids

Fermentation is the process by which anaerobic bacteria break down resistant starch or indigestible carbohydrates (dietary fibre) into short-chain fatty acids (SCFAs: C2-C6). This is an important function of the colon. Through collaboration with species specialized in oligosaccharide fermentation (e.g. *Bifidobacteria*) (Marchesi et al. 2016), the phyla Firmicutes and Bacteroidetes produce SCFAs from indigestible carbohydrates that escape digestion in the small intestine (Flint et al. 2012; Schwiertz et al. 2010). SCFAs constitute approximately two-thirds of the colonic anion concentration (70–130 mmol/l), mainly as acetate, propionate, and butyrate. Butyrate is a primary energy source for colonic epithelial cells (ECs), and it is estimated that SCFAs provide 10% of the total dietary energy supply in humans (McNeil 1984). Butyrate and propionate can control how the gut works and how the immune system works, while acetate helps lipogenesis and gluconeogenesis happen (Macfarlane and Macfarlane 2011). SCFA levels are associated with fermentation, mainly determined by microbial composition. A majority of fermentation occurs in the proximal colon, but the gut microbiota uses other substrates (e.g. protein or amino acids) in the distal colon due to the depletion of carbohydrates. So, the colonic pH is lower in the proximal colon where fermentation is highest (pH 5.5–6.5) compared to the pH in the distal colon (pH 6.5–7.0) (Simpson and Campbell 2015). Fermentation of amino acids leads to the generation of a range of potentially harmful compounds like ammonia, phenols, p-cresol, certain amines, and hydrogen sulphide. Some of these compounds may be involved in gut diseases such as colon cancer or IBD. In contrast, high luminal SCFAs inhibit the growth of gram-negative enterobacteriaceae, including the familiar pathogens *Salmonella* spp. and *Escherichia coli* (Simpson and Campbell 2015; Duncan et al. 2009) and contribute to the maintenance of a favourable environment in the colon.

15.2 Role of the Gut Microbiota in Immune and Inflammatory Responses

The gut microbiota regulates immune homeostasis and inflammatory response via the induction and development of the GI immune system (Kamada and Nunez 2014; Hooper and Macpherson 2010). On the other hand, the mucosal immune system also regulates the structure and function of the gut microbiota. This bidirectional interaction is finely tuned under healthy conditions, but its breakdown leads to GI or general disorders. Treg cells are a type of T cell that modulates the immune system, maintains tolerance to self-antigens, and prevents autoimmune disease (Kamada and Nunez 2014). These cells generally suppress or downregulate the induction and proliferation of effector T cells. Previous studies have demonstrated that the genus *Clostridium* is a strong inducer of Tregs through butyrate production (Kamada and Nunez 2014; Atarashi et al. 2011; Furusawa et al. 2013). In germ-free (GF) mice (deficient in the gut microbiota), a reduced luminal concentration of SCFAs is accompanied by an impaired development of intestinal Treg cells (Furusawa et al.

2013; Smith et al. 2013). In these GF mice, reconstitution with commensal bacteria or administration of SCFAs restores the number of Treg cells (Furusawa et al. 2013), supporting a role for bacterial metabolites in Treg development. Therefore, it was suggested that a decrease in the relative abundance of butyrate-producing microbes, such as *Faecalibacterium prausnitzii*, may lead to a disruption in mucosal homeostasis (Fujimoto et al. 2013). The gut microbiota also plays a crucial role in the induction of effector T cells in the GI tract. The Th17 cells are a novel class of helper CD4+ T cells characterized by the secretion of interleukin (IL)-17A, IL-17F, IL-21, and IL-22 (31). IL-17 is a potent pro-inflammatory cytokine that increases the expression of tumour necrosis factor (TNF)- and IL-1 in a variety of cell types (Weaver et al. 2007). Waite and Skokos (2012) found that inflammatory and autoimmune diseases are caused in part by the wrong way Th17 cells are controlled. In GF mice, the number of Th1 and Th17 cells was markedly reduced and then restored by reconstitution with conventional microbiota (Ivanov et al. 2008), indicating that the gut microbes play a role in Th1 and Th17 cell development. Recently, Atarashi et al. (2015) described the importance of the epithelial cells' adhesive character of microbes in the induction of Th17 cells in the mucosa. The EC-adhesive bacteria spp. triggered Th17 responses, whereas adhesion-defective mutants of these microbes failed to do so. Also, a mix of 20 bacterial strains that were taken from a person with ulcerative colitis triggered the Th17 response and had EC-adhesive properties (Atarashi et al. 2015).

15.3 Mechanism Associated with Gut Microbiota and Metabolites in Cardiovascular Diseases

The regulation of human physiology is greatly influenced by interactions between the intestinal microbiota and the host. Gut dysbiosis, or harmful changes in the gut microbiota, has been linked to the onset and progression of a number of diseases, including cardiovascular disease (CVD) (Novakovic et al. 2020). According to findings from genome sequencing and metagenomic studies (Karlsson et al. 2012; Vinje et al. 2014), the pathogenesis of CVD is thought to be influenced by changes in the constitution of gut flora and alterations in gut microbial metabolism (Xu et al. 2020). Trimethylamine/trimethylamine N-oxide, short-chain fatty acids, and bile acids are among the bioactive metabolites that are produced by the gut microbiota, which performs like an endocrine organ. Barrier functions are regulated by communication between the gut microbiota and intestinal cells, which continuously activate the immune system to fight pathogens (De Santis et al. 2015). Lipopolysaccharide (LPS) and peptidoglycan are structural elements of the microbiota that can interact with host intestinal cells directly through toll-like receptors (TLRs) (Larsson et al. 2012). A significant class of bacterial metabolites known as short-chain fatty acids is produced mostly in the colon from the bacterial fermentation of dietary fibres (Cummings et al. 1987). Firmicutes typically produce butyrate, whereas Bacteroidetes frequently produce acetate and propionate (Kasselman et al. 2018). Involvement with host G protein-coupled receptors (GPCRs), such as Gpr41 and

Olfr78, is one way that SCFAs have an impact on the host cells. When Olfr78 is stimulated, blood pressure rises while Gpr41 activation causes hypotension (Pluznick 2013). Alterations in the concentrations of bacterial metabolic products occur after changes in the bacteria's composition linked to hypertension (Kim et al. 2018). The production of different metabolites can be altered by gut dysbiosis, which can also have pro-atherosclerotic effects via metabolism-dependent pathways (Xu et al. 2020). Trimethylamine N-oxide (TMAO), one of the many metabolites produced by the gut microbiota, plays a significant role in the onset of atherosclerosis. Two different kinds of bacterial enzymes—carnitine-specific and choline-specific lyases—cleave particular carbon–nitrogen bonds to form TMA (trimethylamine), a TMAO precursor (Koeth et al. 2013). In the liver, the flavin-containing monooxygenase family of enzymes converts TMA into TMAO after it has been absorbed into the blood (Tang and Hazen 2014). Butyrate and BAs, two other metabolites produced by gut bacteria, have atheroprotective properties (Jones et al. 2012; Brown and Hazen 2015). The size of an aortic lesion is not significantly affected by N-acyl phosphatidylethanolamines (NAPEs), but the necrotic core area of atherosclerotic lesions can be significantly reduced (May-Zhang et al. 2019).

15.3.1 Hypertension

Over a billion individuals worldwide are affected by hypertension, a serious cardiovascular ailment that is relatively prevalent (Chockalingam 2007). Blood pressure levels rise with age in both males and females, especially in high-class civilizations, for a variety of reasons, with a high salt diet being the main cause of the disease's widespread prevalence (Grillo et al. 2019). In the past few years, a clear correlation between the gut microbiota and blood pressure has emerged (Marques et al. 2018). In the control of blood pressure, bacterial metabolites have a significant influence (Pevsner-Fischer et al. 2017). Unfolding literature on both human and animal trials suggests that disruption of the composition of the gut's microbial population elevates blood pressure (Al Khodor et al. 2017).

Diet inadequacies and lifestyle choices can influence how our cardiovascular system behaves and how it is regulated. Diets high in salt and cholesterol can make it more difficult to control cholesterol levels, which can result in hypertension and coronary artery disease (CAD), respectively. Many epidemiological studies have shown that a diet high in micronutrients, especially fibre, is important for maintaining BP homeostasis and a healthy gut. Typically associated with higher SCFA levels, fruits, vegetables, and legumes satisfy this criterion (Miura et al. 2004). More than 2900 commensal and opportunistic bacterial species dwell in the human gut, with 10^2 bacteria per gram found in the stomach and 10^{11} per gram in the colon (Rajilic-Stojanovic and de Vos 2014; Donaldson et al. 2016; Almeida et al. 2019). Shikata et al. (2019) found that using a combination of antibiotics got rid of all the gut microbiota, which greatly reduced the number of aneurysms caused by high blood pressure.

All of these studies point to a significant correlation between the composition of gut bacteria and hypertension. All of these studies show that there is a strong link between high blood pressure and the types of bacteria in the gut. SCFAs are produced as a result of the fermentation or breakdown of proteins and carbohydrates carried out by commensal gut bacteria. SCFAs are essential for preserving a healthy intestinal system. Nearly 20% of the tiny molecules circulating in human blood are metabolites generated by gut bacteria (Rook et al. 2017), which are responsible for maintaining lymphocyte infiltration into the intestinal tissues and stimulating the sympathetic nervous system (Diaz Heijtz et al. 2011; Straub et al. 2005). These SCFAs are either absorbed by the gut and contribute to a number of physiological functions, or they are expelled in faeces. Acetic, propionic, and butyric acids are these SCFAs' most prevalent types. The usage of the Mediterranean and DASH diets has been associated with a considerable reduction in blood pressure in hypertensive individuals and a reduction in inflammation (Al Khodor et al. 2017; Sureda et al. 2018; Saneei et al. 2014; Phillips et al. 2018).

Improvements in obesity, coronary artery disease, and other metabolic changes have also been observed, in addition to improvements in blood pressure and inflammation (Widmer et al. 2015; Krznaric et al. 2019). A few other studies, mostly in Chinese, American, and Brazilian populations, have investigated how the microbiome affects hypertension. These studies provide a variety of phylogenetic and functional signatures, suggesting that dysbiosis in the gut microbiota may play a role in the pathogenesis of hypertension (Silveira-Nunes et al. 2020; Sun et al. 2019; Li et al. 2019a, b). The gut microbiome's impact on hypertension sufferers has not been well studied. A number of population-based studies done on humans have shown a high correlation between blood pressure and bacterial metabolites (Holmes et al. 2008; Zheng et al. 2013). In comparison with subjects with normal blood pressure, hypertensive patients have a commensal microbial colony that is relatively less diverse. When compared to healthy controls, it was found that both prehypertensive and hypertension patients had lower levels of gene richness and microbial diversity but higher levels of the genus Prevotella of bacteria. These differences in microbiome clusters between disease groups suggest (Xu et al. 2020) that the metabolic profile of hypertension is linked to the number of gut microbes in the host.

15.3.2 Atherosclerosis

In the most extreme AS research, human GM has gotten a lot of attention. Based on associative information from metagenome-wide association studies, it is thought that disruption of the "normal" microbiome profile contributes to the development of atherosclerotic CVDs. According to histology, the atherosclerotic lesion is predominately composed of cellular proliferation and fibrosis of the intima of primarily medium-sized arteries, which results in a narrowing of the arterial lumen and subsequent organ ischaemia (Demetris et al. 1989). Localized lipid build-up in the artery wall causes the development of an atherosclerotic lesion, also known as a plaque. Low-density lipoprotein is the main source of this lipid build-up (LDL). Certain gut microorganisms, dietary components, platelet function, and the risk of thrombosis are related (Zhu et al. 2016). Dietary choline and high levels of the gut microbe-dependent metabolite TMAO have been shown to have a prothrombotic effect in both humans and animals (Organ et al. 2016; Zhu et al. 2016, 2017). Trimethylamine N-Oxide (TMAO) has been recognized as a significant human CVD and neurological comorbidity risk factor. The copious carnitine and choline found in meat, eggs, and other animal-derived meals are co-metabolized by the host and bacterium to produce the microbe-dependent metabolite TMAO. The proportion of Ruminococcus, Lachnospiraceae spp., Mollicutes spp., and Tenericutes spp. was shown to be enhanced in high carnitine consumers. However, the mechanisms driving the rise in these microbial taxa are still unclear (Wu et al. 2019a, b; Koeth et al. 2013, 2014). According to a recent study, the second phase (γ-butyrobetaine → TMA) in the two-phase metabolic process for carnitine (carnitine → γ-butyrobetaine → TMA) is a crucial stage that could establish if the gut bacteria is producing TMA from carnitine metabolism (Koeth et al. 2019). Koeth et al. identified a novel bacteria called Emergencia timonensis and discovered that it was exclusively in charge of the TMA step for γ-butyrobetaine, but more research is needed to confirm its effect on the disease's symptoms and the amount of TMAO in human plasma. Additionally, the composition of vegetarians' gut flora is different and it exhibits a reduced ability to manufacture the precursor to TMAO, trimethylamine (TMA) (Wu et al. 2019a, b; Koeth et al. 2013). As a result, the development of atherosclerosis linked to high TMAO levels is significantly influenced by gut flora. These results also suggest that the makeup of the microbiome could be an important therapeutic target for preventing and treating atherosclerosis.

The level of antioxidants, including polyphenols, which may directly affect the development of atherosclerosis, is significantly regulated by the gut flora. Quercetin has been demonstrated, for example, to reduce the production of ROS and NOS and nicotinamide-adenine dinucleotide phosphate (NADPH) oxidase in mouse models of atherosclerosis (Xiao et al. 2017; Loke et al. 2010). Matrix metalloproteinase-1 (MMP-1) is another matrix that quercetin targets since it is essential for the development of unstable atherosclerotic plaques that result in thrombogenesis (Song et al. 2001). In a recent study, animals with atherosclerosis caused by their diet were used to find out more about quercetin's anti-atherosclerosis effects. Without changing the phylum-level architecture of the gut microbiota, treatment with quercetin for 12 weeks improved the lipid profile and dramatically decreased the levels of the pro-inflammatory mediators TNF-α and IL-6. However, a considerable shift in the faecal metabolite profile showed that quercetin operated through modifying bile acid production (Wu et al. 2019a, b). The fatty stripe is the beginning of the atherosclerosis process, which leads to plaque rupture and acute atherothrombosis, which result in acute clinical events like stroke or myocardial infarction (Hansson 2005; Ross 1999). Even though inflammation is involved in all stages of atherosclerosis, most of the research on the microbiome and CAD has not made a clear distinction between chronic CAD and acute events. Cross-sectional research methods were used in all of the published studies about the involvement of the microbiome in CAD to examine patients with generally stable CAD (Jie et al. 2017; Zhu et al. 2018; Karlsson et al. 2012). Ross referred to atherosclerosis as an inflammatory illness almost 20 years ago (Ross 1999), The CANTOS trial was published by Ridker et al. (2017), demonstrating how the monoclonal antibody canakinumab can lower cardiovascular events by inhibiting IL-1b. Microbiota medication could speed up clinical translation, but there are still a lot of problems to solve, like safety concerns and the fact that each person's gut microbiota looks and works differently.

15.3.3 Coronary Artery Disease

Cardiovascular diseases are one of the leading causes of death worldwide. Even those with well-controlled hypertension who take medication run a high risk of developing a cardiovascular disorder. Therefore, it is crucial to pinpoint the various causes and mechanisms through which cardiovascular illnesses might develop and spread. Taxonomic changes in the gut may affect cardiovascular function, according to several research studies conducted on both animals and humans (Naqvi et al. 2021).

Recent studies have used different sequencing methods to look at the gut microbiota of people with coronary artery disease (CAD). According to Cui et al., individuals with coronary heart disease showed changes at the phylum level, with a lower proportion of Bacteroidetes and a higher proportion of firmicutes. Jie et al. reported reduced levels of Roseburia intestinalis and Faecalibacterium prausnitzii, known makers of the SCFA butyrate, and higher levels of various Streptococcus species and genera of the Enterobacteriaceae family in a much more thorough investigation.

Studies on mice have revealed that ω-3 PUFA obtained from flaxseed reduced the share of Bacteroidetes, while those derived from fish oil affected the number of firmicutes (Liu et al. 2012; Yu et al. 2014). The presence of Lactobacillus and Akkermansia Muciniphila was demonstrated to be enhanced in rats given either dietary fish oil or lard as a source of fat (Caesar et al. 2015). These two microorganisms in the human gut are well known for having anti-inflammatory properties and favourable effects on metabolism. Supplemental-3 PUFA may have an anti-inflammatory impact that is mediated by an increase in the synthesis of short-chain fatty acids (SCFA) (Santoru et al. 2017). The conclusion drawn is that supplementing with ω-3 PUFAs may have a positive impact on the diversity of bacterial phyla in the gut microbiota. In particular, these supplements seem to increase the number of anti-inflammatory bacteria like Lachnospiraceae and Bifidobacteria and bring the number of Firmicutes and Bacteroidetes back into balance. These findings imply that the prebiotic effect of ω-3 PUFA supplements on the composition of the gut flora is a key mechanism by which the positive effects of these supplements are partially mediated (Chockalingam 2007).

The microbial metabolism of dietary components like carnitine and choline is the most convincing proof of a connection between the gut microbiome and CAD (Koeth et al. 2013; Wang et al. 2011; Tang et al. 2013). In a seminal article from the Hazen group (Tang et al. 2019), the metabolite TMAO was discovered to be a powerful predictor of CAD. TMAO may be the cause of atherosclerosis rather than just a sign of the condition (Wang et al. 2011). Trimethylamine, the precursor to TMAO, is created by the gut microbiota from foods containing L-carnitine or phosphatidylcholine and then oxidized by hepatic monooxygenases that contain flavin (Tang et al. 2019; Wang et al. 2011). Animal models show that precursors of TMAO cause foam cell growth and atherosclerosis, but not when antibiotics are added to the water, which points to a mechanism that depends on the microbiota (Wang et al. 2011).

In a recent study, healthy individuals' TMAO levels rose after consuming red meat, compared to those who consumed a vegetarian or a diet heavy in white meat (Wang et al. 2019). Diet has a reversible impact on the levels of TMAO. Furthermore, the increased TMAO generation was shown to be caused by carnitine rather than choline by isotopic methods (Wang et al. 2019). However, TMAO is the microbiota-related metabolite that has been most extensively researched in terms of cardiovascular risk. Other metabolites found on the TMAO route may also be of relevance. Gamma-butyrobetaine (gBB) is a metabolite on the pathway from L-carnitine to TMAO that is partially related to the microbiota and has also been connected to CAD (Koeth et al. 2014). Trimethyl lysine (TML) is a precursor of both gBB and TMAO that is obtained both from dietary and endogenous sources. It has been connected to a higher risk of CAD, either by itself or in conjunction with TMAO (Li et al. 2018, 2019a, b). In individuals with carotid artery atherosclerosis, circulating levels of gBB and TML but not TMAO predicted cardiovascular mortality in recent investigations (Skagen et al. 2016). Because of this, future research must include TMAO precursors in order to find out how atherosclerosis and CAD happen.

15.3.4 Myocardial Infarction

Globally, atherothrombotic cardiovascular disorders (e.g. myocardial infarction (MI) and stroke) are the main offenders and a major cause of morbidity (Tilg 2016). The specific mechanisms by which a beneficial microbiota can lower MI risk factors or prevent post-MI events (Lam et al. 2012; Gan et al. 2014; McCafferty et al. 2012; Girard et al. 2009) are by modifications to the metabolism of cholesterol and lipogenesis, as well as the synthesis of antioxidants, which are all included. A greater trend is found in the gut microbiota in acute MI animal models, particularly in the Synergistetes, Spirochaetes, Lachnospiraceae, Syntrophomonadaceae, and Tissierella Soehngenia genera (Wu et al. 2017a, b). TMAO is linked to an increased risk of unfavourable cardiovascular events, including MI in stroke patients (Haghikia et al. 2018). It has the ability to foretell unfavourable outcomes, such as all-cause mortality or reinfarction, 2 years following MI (Suzuki et al. 2017). Studies conducted in living organisms have shown that dysbiosis in the gut microbiota can affect the efficiency of energy absorption from food, which in turn affects susceptibility to obesity and atherosclerosis through modifying inflammatory response and lipid metabolism (Zabell and Tang 2017; Caesar et al. 2010). Probiotic use was found to significantly lower cholesterol, LDL, and triglyceride levels, according to a meta-analysis (Guo et al. 2011).

Additionally, it can influence the host's physiology and gene expression and its cellular and immune responses (Rafter 2004). The role of toll-like receptors (TLRs) and the existence of bacterial effector protein(s) that may work to regulate the host signalling pathway are the most recent theories about the ways in which microbiota might alter the gene expression of the host. TLR receptors trigger downstream signalling cascades that regulate the functionality of NF-B, MAPK, AKT, or caspase-dependent signalling pathways.

The toll-like 13 receptors (TLRs) and NOD-like receptors (NLRs), which identify bacterial products such as lipopolysaccharides and peptidoglycans, play a key role in modulating the pro-inflammatory effects of microbiota. Numerous innate immune receptors, including TLR4, TLR2, NOD1, and NOD2, have been linked to a number of metabolic syndrome characteristics, such as obesity, insulin resistance, hepatic steatosis, and inflammation, in studies using mice models (Jin et al. 2013). Similar research on mice has directly linked TLR4, TLR3, and TLR2 to atherosclerosis (Ding et al. 2012). Additionally, a number of studies have connected inflammasome sensing to inflammation, fatty liver, and obesity (Jin et al. 2012; Henao-Mejia et al. 2012; Wlodarska et al. 2014). Positive in vitro, in vivo, and human investigations have clarified that various microbiota-derived factors modulate host signalling pathways and show advantageous heart-protective effects. Healthy gut microbiota can also regulate immunological responses, lower the degree of cardiac failure after an MI, and lower the size of the MI. Plasma levels of TMAO were used to predict myocardial infarction, stroke, and all-cause mortality in a number of independent cohorts from the USA and Europe (Tang et al. 2013, 2019; Wang et al. 2011; Koeth et al. 2013). The most common cause of death in developed cultures and a burgeoning epidemic in developing nations alike is acute myocardial infarction (AMI). The inflammation and cholesterol biosynthesis linked to hypertension, diabetes mellitus, dyslipidaemia, smoking, and the post-menopausal stage are currently the focus of research into the risk factors for AMI (Akbar et al. 2020). Protein disulphide isomerase activity, which involves the protein-folding procedure connected to inflammation, is influenced by the gut flora (Kiouptsi et al. 2019). Trimethylamine-N-oxide (TMAO) production from the gut microbiome is linked to changes in the phenotypes of endothelial and macrophage cells, stimulation of platelet hyperresponsiveness, coronary atherosclerotic burden, convolution of the plaque, acute coronary syndrome, and post-myocardial infarction sequelae (Jie et al. 2017; Li et al. 2017; Gao et al. 2020). Under conditions of ischaemia, the gut microbiota also controls the synthesis of adenosine triphosphate (ATP) and the metabolism of ketone bodies by anaerobic succinate generation (Lam et al. 2012; Pisarenko et al. 1986). Beginning with the astounding finding that gut microbiota silencing with an oral blend of poorly absorbed antibiotics significantly increased the rates of post-MI ventricular rupture and death in a murine model of chronic left anterior descending artery ligation, Tang and colleagues' studies go on to investigate this concept. They also demonstrated a substantial improvement in survival prior to MI following microbial reconstitution using faecal transplantation from untreated donors, indicating a role for the gut microbiota in early cardiac healing (Tang et al. 2019). In a series of add-back experiments, it is suggested that the beneficial effects in the post-MI setting may be partially mediated by gut microbiota-generated SCFAs, which induce recruitment of myeloid cells to the heart. This is a new mechanistic insight regarding the effect of gut microbiota on ventricular remodelling. The detrimental effects of antibiotics on mortality and ventricular rupture rates were considerably mitigated by either dietary SCFA supplementation or intravenous infusion of the monocytic cell line RAW264.7 given one day after MI (McMillan and Hazen 2019).

As a result, the gut microbiome affects immunological function, energy generation, and inflammation, and it may also play a role in AMI episodes.

15.3.5 Heart Disease

Heart failure (HF) is a disease that has a significant mortality rate (Kociol et al. 2010). About half of the patients die within 5 years of diagnosis (Suzuki et al. 2016). Reduced cardiac output and blood redistribution in heart failure patients lead to a reduction in intestinal perfusion and disruption of the intestinal barrier (Nagatomo and Tang 2015). Circulatory congestion can potentially cause intestinal hypoperfusion in heart failure patients with decreased cardiac output. Vasoconstriction is also hypothesized to contribute via increased sympathetic tone, which is initially compensatory but later becomes hyperactive. Preclinical investigations have shown downstream inflammatory effects that may amplify atherosclerosis and fibrosis and the course of heart failure (Pasini et al. 2016). Intestinal oedema has previously been linked to worse outcomes in hospitalized heart failure patients; moreover, bowel wall oedema is reported, with both characteristics likely to produce hypoxia in the intestine (Ikeda et al. 2018). Because of the lowered mucosal pH and associated effects on carrier-mediated transport, the gut becomes more permeable and its barrier functions are compromised. Following this, endotoxaemia and bacterial translocation into the circulation are possible, both of which might increase inflammation in heart failure patients (Chen et al. 2019). The intensity of heart failure is correlated with TMAO (trimethylamine N-oxide) and choline levels. When mice with transverse aortic constriction are fed choline or TMAO, pulmonary oedema and myocardial hypertrophy are significantly enhanced (Organ et al. 2016). A worse outcome is indicated by an elevated TMAO level, which is important to the pathophysiology of heart failure (Tang et al. 2014). The former hypothesis is supported mechanistically by recent preclinical data by Boccella et al., who used a mouse transverse aortic constriction (TAC) model of pressure-overload HF. Compared to previous preclinical models, the TAC model can simulate HF development over a more progressive time course. In contrast to sham-operated controls, it was demonstrated that TAC produced intestinal barrier disruption and dramatically elevated serum levels of cytokines and LPS (Boccella et al. 2021). The relationship between gut microbiota and chronic heart failure has been explained by a number of different pathways and mediators. When compared to individuals without oedema, heart failure patients with pedal oedema had greater levels of endotoxins and a lipopolysaccharide/lipopolysaccharide-binding proteins (LBP) ratio. Oedematous heart failure patients have been found to have greater plasma concentrations of C-reactive protein, TNF-a, soluble TNF receptor 1 and receptor 2, interleukin-6, and soluble CD14 than non-oedematous patients (Niebauer et al. 1999). Evidence suggests that these compounds may be absorbed into the intestinal circulation and then affect distal organs, such as the heart, thus acting as biomarkers and influencing the prognosis for HF (Chen et al. 2019). Probiotics and diuretics have both been investigated for their potential effects on microbiota related to chronic heart failure (CHF) patients. Niebauer et al. (1999) found that when people took diuretics for a short time, their endotoxin levels went down, but their cytokine levels did not change. These incidents might be avoided thanks to probiotic strains' capacity to strengthen the gut wall and reduce inflammation. Probiotic therapy with Lactobacillus rhamnosus has also been shown to lower levels of TMAO and systemic inflammation (Moludi et al. 2021), which may be helpful for CHF patients.

The bacteria in the phyla Bacteroidetes and Firmicutes make up the majority of the varied microbiota in the adult gut. Actinobacteria, Proteobacteria, Verrucomicrobia, and Fusobacteria are some of the other less prevalent phyla (Eckburg et al. 2005). From childhood to adulthood, the gut microbiota's makeup changes and may change depending on a number of circumstances and health conditions (Avershina et al. 2016; Rinninella et al. 2019). Various factors have the ability to change the gut's microbial community. The gut microbiota is strongly affected by the type of diet one consumes. Vegetarian diets have been linked to healthy, diverse gut microbiota, while non-vegetarian diets have been linked to a loss in overall gut microbiota and beneficial species (Hasan and Yang 2019). It has been discovered that the gut microbiota is crucial for the development and control of the host's immune system (Mazmanian et al. 2005). Additionally, it alters a number of host metabolic pathways. Human cardiovascular health has been discovered to be severely affected by changes in the composition of the gut microbiota. In the world, cardiovascular disease (CVD) is the leading cause of death.

15.3.6 Obesity and Dyslipidaemia

Low-density lipoprotein cholesterol (LDL-C), triglycerides, and high-density lipoprotein (HDL) are among the lipids that are imbalanced in dyslipidaemia. This disorder can occur from food, cigarette use, or genetics, and it can cause serious problems like cardiovascular disease. The relationship between dyslipidaemia, obesity, and CVD is well-established in the literature. The relationship between changes in lipid metabolism, obesity, and gut microbiota has been clarified by a number of studies. Low-density lipoprotein cholesterol (LDL-C), triglycerides, and high-density lipoprotein (HDL) are among the lipids that are imbalanced in dyslipidaemia. This disorder can occur from food, cigarette use, or genetics, and it can cause serious problems like cardiovascular disease. Plasma cholesterol levels have been revealed to be significantly influenced by gut flora. Bacteria belonging to the class

Betaproteobacteria, Bacteroidales, and phylum Firmicutes have been found to be significantly high in experimental high-cholesterol models (Le Roy et al. 2019). In a recent study, it was found that the gut microbiota of obese people is dominated by Firmicutes, while Bacteroidetes continue to be the most prevalent bacterial species in lean subjects. Additionally, this study demonstrated that diets low in fat and carbs increased the amount of the genus Bacteroidetes (Ley et al. 2006). It has been proposed that the production of SCFAs and the regulation of secondary bile acids, trimethylamine (TMA)/trimethylamine N-oxide (TMAO), and pro-inflammatory mediators such as lipopolysaccharides are some of the processes through which the gut microbiota affects lipid metabolism and obesity. It has been discovered that the gut microbiota affects the deconjugation and secretion of bile acids that regulate lipid metabolism via signalling from the farnesoid X receptor (FXR) and Takeda G protein (TGR5)-coupled bile acid receptor (Schoeler and Caesar 2019; Sayin et al. 2013). Through lipid metabolism and inflammatory pathways, TMA and TMAO made by certain bacteria may raise the risk of atherosclerosis and CVD. Certain gut bacteria may reduce gut permeability, which could stop endotoxins like lipopolysaccharides from entering the bloodstream and causing inflammation. It may be interesting to add that lipopolysaccharides have been linked to the onset of atherosclerosis and CVD in this context (Schoeler and Caesar 2019). The patho-physiological relationship between gut microbiota and lipid metabolism/obesity has been interpreted by a number of studies towards the development of treatment approaches treating these disease states. In a mouse model, it was discovered that co-supplementation with milk that had been fermented with Lactobacillus fermentum MTCC: 5898 improved the lipid profile, atherogenic index, coronary artery disease (CAD) risk index, and blood levels of a number of pro-inflammatory cytokines (Yadav et al. 2019). In a recent study, it was found that the probiotic Lactobacillus has a significant role in the reduction in total cholesterol and low-density lipoprotein cholesterol (Wu et al. 2017a, b). In a recent study, it was found that, in addition to probiotics, the efficiency of current therapy approaches for dyslipidaemia and obesity is related to mechanisms involving the gut flora. The fact that diet and exercise have positive effects on body weight control and are contagious through FMT points to a potential role for the gut microbiota in the effects of these interventions on obesity (Lai et al. 2018).

15.3.7 Chronic Kidney Disease

Chronic kidney disease occurs when a disease or condition impairs kidney function, causing kidney damage to worsen over several months or years. Diseases and conditions that cause chronic kidney disease include type 1 or type 2 diabetes and high blood pressure. Patients with CKD also have a changed gut microbiome. According to the evidence, the main causes of dysbiosis in CKD are reduced dietary fibre intake, constipation, poor protein metabolism, and medication (Ramezani and Raj 2014). Dysbiosis of the gut microbiome occurs in CKD as a result of structural and functional changes to the gut microbiota and reduced gut barrier function. Due to decreased urine excretion in CKD, this results in the creation of large amounts of uraemic toxins that may be retained. These toxins are produced as a result of an imbalanced nitrogen compound metabolism, primarily in conjunction with nondigestible carbohydrates like p-cresyl and indoxyl sulphates. They can exacerbate CKD by translocating into the systemic circulation through a compromised gut barrier. They are also linked to CVD development and increased mortality risk in CKD patients (Nallu et al. 2017; Armani et al. 2017). Tryptophan is converted by intestinal bacteria into indoles, which are then converted into indole indoxyl sulphate (IS) and indole-3-acetic acid (IAA). They cause tubulointerstitial fibrosis, aortic and vascular calcification, endothelial lining damage, and reduced erythropoietin production by activating nuclear factor (NF)-Kb and plasminogen activator inhibitor type 1. Tyrosine and phenylalanine are broken down by bacteria and converted into PCS, another uraemic toxin, which is then sulphonated in the liver. They frequently result in oxidative stress, inflammatory mediator production, and renal fibrosis. Most of the data (Nallu et al. 2017) show that high levels of IS, IAA, and PCS are linked to higher rates of death from all causes and cardiovascular events. It has also been observed that other uraemic toxins, including amines, polyamines, D-amino acids, endotoxins, blood urea nitrogen (BUN), urea, and their derivatives, are nephrotoxic and contribute to the development of CKD or CVD (Nallu et al. 2017). It has been demonstrated that consuming prebiotics and probiotics can reduce uraemic toxin levels, delay the onset of CKD, and lengthen life. Meijers et al. (2010) found that when haemodialysis patients consumed the prebiotic p-inulin orally, their serum concentrations of PCS and IS decreased (Meijers et al. 2010). In a recent study, it was found that healthy individuals' PCS levels decreased after receiving acarbose medication (Evenepoel et al. 2006). Treatment with oral Lactobacillus acidophilus in haemodialysis patients reduced levels of uraemic toxins like serum dimethylamine (Simenhoff et al. 1996).

In preclinical investigations, a number of new medicines have demonstrated efficacy, including the prostaglandin derivative lubiprostone and the trimethylamine (TMA) inhibitor 3,3-dimethyl-1-butanol (DMB). According to studies, lubiprostone lowers BUN levels and enhances the microbiota profile by rapidly increasing the species that break down sugars (Mishima et al. 2015). Also, it has been shown that 3, 3-dimethyl-1-butanol, which is found in several essential oils, stops the growth of atherosclerotic lesions and macrophage foam cells, as well as lowering TMAO levels (Wang et al. 2015).

15.3.8 Diabetes Mellitus

T2DM, or type 2 diabetes mellitus, is one of the leading causes of CVD. According to recent observational research and systematic reviews on the subject, there are significant changes in the composition of the gut microbiota between those with T2DM and healthy controls, according to recent observational research and systematic reviews on the subject. In contrast, it was discovered that people with T2DM had lower levels of bacteria from the genera Bifidobacterium, Bacteroides,

Faecalibacterium, Akkermansia, and Roseburia, which have protective effects on glucose metabolism and T2DM. T2DM was positively correlated with the phylum Firmicutes and the bacterial species Ruminococcus, Fusobacterium, and Blautia (Gurung et al. 2020; Li et al. 2020). Gurung et al. (2020) found that the genus Lactobacillus is the most diverse and has the most operational taxonomic units (OTUs) in the human gut. This is because its distribution patterns are not regular.

The potential connection between changed gut microbiota and T2DM has been explained by a number of mechanisms, including modulation of inflammation, gut permeability, glucose and fatty acid metabolism, and insulin sensitivity. The production of anti-inflammatory cytokines and chemokines like interleukin (IL)-10 and 22 and transforming growth factor (TGF)-b is induced by specific species of bacteria from the genera Roseburia, Bacteroides, Lactobacillus, and Akkermansia while reducing inflammation and improving insulin sensitivity by inhibiting pro-inflammatory cytokines and chemokines like IL-1b, monocyte chemoattractant protein-1, intercellular adhesion molecule-1, IL-8, IL-16, IL-17, CD36, nuclear factor-kappa B (NK-kB), interferon (IFN)-g, and C-reactive protein. On the other hand, it has been found that bacteria from the genera Ruminococcus and Fusobacterium increase the production of cytokines that cause inflammation. Scientists have found that the Bacteroides and Akkermansia genera can stop metabolic endotoxaemia by increasing the expression of tight junction genes by turning on adenosine monophosphate kinase (AMPK). This makes more tight junction genes.

Another crucial process is the modulation of fatty acid oxidation in adipose tissue. Short-chain fatty acids (SCFAs), which control a number of human endocrine pathways, are created by the activity of the gut bacteria on indigestible carbohydrates and proteins. They bind to the G protein-coupled receptors GPR41 and GPR43, which cause the release of postprandial plasma peptide YY (PYY) and glucagon-like peptide-1 (GLP-1) to occur (Samuel et al. 2008; Kimura et al. 2013). Moreover, SCFAs also stimulate AMPK, an important regulator of metabolic homeostasis (Den Besten et al. 2015).

As our knowledge of the involvement of gut microbiota dysbiosis in the pathogenesis of T2DM has grown, so have the potential treatment options. Faecal microbial transplantation (FMT) has recently been shown to help treat the symptoms of T2DM in mice. This course of treatment reduced the pancreatic inflammation and indicators of insulin resistance while reversing the elevated serum insulin levels linked to T2DM. With FMT, the pancreatic B-cell activity improved and the elevated levels of pro-inflammatory cytokines were returned to normal. Additionally, FMT reduced the expression of cleaved caspase-3 and Bax and the death of pancreatic cells (Wang et al. 2020). SCFA administration is another approach being developed for T2DM therapy. Rats receiving SCFAs like acetate experienced an increase in plasma GLP-1 and PYY levels as well as a decrease in inflammatory markers such as tumour necrosis factor-a (TNF-a) (Freeland and Wolever 2010). Additionally, it has been discovered that probiotics and prebiotics show promise for treating T2DM. Lactobacillus paracasei NL 41 has been proven to significantly increase insulin sensitivity and lower blood glucose levels in diabetic rats.

Additionally, the probiotic therapy decreased oxidative stress and provided B-cell protection (Zeng et al. 2019). In a recent study, it was found that patients with type 2 diabetes who took Lactobacillus casei supplements for eight weeks reported eating fewer carbs and fat and having better glycaemic control (Khalili et al. 2019).

15.4 Future Perspective and Conclusion

The strong link between gut–microbiota dysbiosis and CVDs and the risks associated with it is supported by a large body of evidence. Furthermore, a remarkable range of bacterial microbiomes have been found in numerous anatomical regions using new culture-independent genetic approaches for examining microbial diversity. This complexity was not fully understood in the past. A number of microbial pathways are now being explored as potential pharmacological targets and mediators in the treatment of cardiometabolic diseases as a result of the discovery of bacterial metabolites with the capacity to affect host physiological processes. In the treatment of cardiovascular diseases and the associated risk factors, it has been demonstrated that the interpretation of this important information into usable medications is both safe and effective. For the purpose of creating future targeted treatments that could prevent and treat cardiovascular illnesses (CVDs) and lessen the severity of their detrimental consequences, a deeper knowledge of this connection is essential. Blood levels of TMAOs are a key indication for predicting the outcome of specific cardiovascular events, such as a stroke, heart attack, or even death. Furthermore, this knowledge reduces the future risks and problems associated with the condition and its burden.

References

Akbar H, Foth C, Kahloon RA, Mountfort S (2020) Acute myocardial infarction ST elevation (STEMI). In: StatPearls. StatPearls Publishing LLC, Treasure Island

Al Khodor S, Reichert B, Shatat IF (2017) The microbiome and blood pressure: can microbes regulate our blood pressure? Front Pediatr 19(5):138. https://doi.org/10.3389/fped.2017.00138

Almeida A, Mitchell AL, Boland M, Forster SC, Gloor GB, Tarkowska A, Lawley TD, Finn RD (2019) A new genomic blueprint of the human gut microbiota. Nature 568(7753):499–504. https://doi.org/10.1038/s41586-019-0965-1

Armani RG, Ramezani A, Yasir A, Sharama S, Canziani ME, Raj DS (2017) Gut microbiome in chronic kidney disease. Curr Hypertens Rep 19(4):1–8

Atarashi K, Tanoue T, Shima T et al (2011) Induction of colonic regulatory T cells by indigenous Clostridium species. Science 331:337–341

Atarashi K, Tanoue T, Ando M et al (2015) Th17 cell induction by adhesion of microbes to intestinal epithelial cells. Cell 163:367–380

Avershina E, Lundgård K, Sekelja M, Dotterud C, Storrø O, Øien T, Johnsen R, Rudi K (2016) Transition from infant-to adult-like gut microbiota. Environ Microbiol 18(7):2226–2236

Boccella N, Paolillo R, Coretti L, D'Apice S, Lama A, Giugliano G, Schiattarella GG, Cuomo M, d'Aquino I, Cavaliere G, Paciello O (2021) Transverse aortic constriction induces gut barrier alterations, microbiota remodeling and systemic inflammation. Sci Rep 11(1):1

Brown JM, Hazen SL (2015) The gut microbial endocrine organ: bacterially derived signals driving cardiometabolic diseases. Annu Rev Med 66:343–359

Caesar R, Fak F, Backhed F (2010) Effects of gut microbiota on obesity and atherosclerosis via modulation of inflammation and lipid metabolism. J Intern Med 268(4):320–328

Caesar R, Tremaroli V, Kovatcheva-Datchary P, Cani PD, Bäckhed F (2015) Crosstalk between gut microbiota and dietary lipids aggravates WAT inflammation through TLR signaling. Cell Metab 22:658–668

Cantarel BL, Lombard V, Henrissat B (2012) Complex carbohydrate utilization by the healthy human microbiome. PLoS One 7:e28742

Chen X, Li HY, Hu XM, Zhang Y, Zhang SY (2019) Current understanding of gut microbiota alterations and related therapeutic intervention strategies in heart failure. Chin Med J 132(15): 1843–1855

Chockalingam A (2007) Impact of world hypertension day. Can J Cardiol 23(7):517–519. https://doi.org/10.1016/s0828-282x(07)70795-x

Claesson MJ, Jeffery IB, Conde S et al (2012) Gut microbiota composition correlates with diet and health in the elderly. Nature 488:178–184

Cummings JH, Pomare EW, Branch WJ, Naylor CP, Macfarlane GT (1987) Short chain fatty acids in human large intestine, portal, hepatic and venous blood. Gut 28:1221–1227

De Filippo C, Cavalieri D, Di Paola M et al (2010) Impact of diet in shaping gut microbiota revealed by a comparative study in children from Europe and rural Africa. Proc Natl Acad Sci U S A 107: 14691–14696

De Santis S, Cavalcanti E, Mastronardi M, Jirillo E, Chieppa M (2015) Nutritional keys for intestinal barrier modulation. Front Immunol 6:612

Demetris AJ, Zerbe T, Banner B (1989) Morphology of solid organ allograft arteriopathy: identification of proliferating intimal cell populations. Transplant Proc 1:3667–3669

Den Besten G, Bleeker A, Gerding A, van Eunen K, Havinga R, van Dijk TH, Oosterveer MH, Jonker JW, Groen AK, Reijngoud DJ, Bakker BM (2015) Short-chain fatty acids protect against high-fat diet–induced obesity via a PPARγ-dependent switch from lipogenesis to fat oxidation. Diabetes 64(7):2398–2408

Diaz Heijtz R, Wang S, Anuar F, Qian Y, Bjorkholm B, Samuelsson A, Hibberd ML, Forssberg H, Pettersson S (2011) Normal gut microbiota modulates brain development and behavior. Proc Natl Acad Sci U S A 108:3047–3052

Ding Y et al (2012) Toll-like receptor 4 deficiency decreases atherosclerosis but does not protect against inflammation in obese low-density lipoprotein receptor-deficient mice. Arterioscler Thromb Vasc Biol 32(7):1596–1604

Donaldson GP, Lee SM, Mazmanian SK (2016) Gut biogeography of the bacterial microbiota. Nat Rev Microbiol 14:20–32

Duncan SH, Louis P, Thomson JM et al (2009) The role of pH in determining the species composition of the human colonic microbiota. Environ Microbiol 11:2112–2122

Eckburg PB, Bik EM, Bernstein CN, Purdom E, Dethlefsen L, Sargent M, Gill SR, Nelson KE, Relman DA (2005) Diversity of the human intestinal microbial flora. Science 308(5728): 1635–1638

Evenepoel P, Bammens B, Verbeke K, Vanrenterghem Y (2006) Acarbose treatment lowers generation and serum concentrations of the protein-bound solute p-cresol: a pilot study. Kidney Int 70(1):192–198

Flint HJ, Scott KP, Louis P et al (2012) The role of the gut microbiota in nutrition and health. Nat Rev Gastroenterol Hepatol 9:577–589

Freeland KR, Wolever TM (2010) Acute effects of intravenous and rectal acetate on glucagon-like peptide-1, peptide YY, ghrelin, adiponectin and tumour necrosis factor-α. Br J Nutr 103(3): 460–466

Fujimoto T, Imaeda H, Takahashi K et al (2013) Decreased abundance of Faecalibacterium prausnitzii in the gut microbiota of Crohn’s disease. J Gastroenterol Hepatol 28:613–619

Furusawa Y, Obata Y, Fukuda S et al (2013) Commensal microbe-derived butyrate induces the differentiation of colonic regulatory T cells. Nature 504:446–450
Gan XT et al (2014) Probiotic administration attenuates myocardial hypertrophy and heart failure after myocardial infarction in the rat. Circ Heart Fail 7(3):491–499
Gao J, Yan KT, Wang JX, Dou J, Wang J, Ren M, Ma J, Zhang X, Liu Y (2020) Gut microbial taxa as potential predictive biomarkers for acute coronary syndrome and post-STEMI cardiovascular events. Sci Rep 10:2639. https://doi.org/10.1038/s41598-020-59235-5
Girard SA et al (2009) Lactobacillus helveticus and bifidobacterium longum taken in combination reduce the apoptosis propensity in the limbic system after myocardial infarction in a rat model. Br J Nutr 102(10):1420–1425
Grice EA, Segre JA (2011) E skin microbiome. Nat Rev Microbiol 9(4):244–253
Grillo A, Salvi L, Coruzzi P, Salvi P, Parati G (2019) Sodium intake and hypertension. Nutrients 11(9):1970. https://doi.org/10.3390/nu11091970
Guarner F (2005) The intestinal flora in inflammatory bowel disease: normal or abnormal? Curr Opin Gastroenterol 21:414–418
Guarner F, Malagelada JR (2003) Gut flora in health and disease. Lancet 361:512–519
Guo Z et al (2011) Influence of consumption of probiotics on the plasma lipid profile: a meta-analysis of randomised controlled trials. Nutr Metab Cardiovasc Dis 21(11):844–850
Gurung M, Li Z, You H, Rodrigues R, Jump DB, Morgun A, Shulzhenko N (2020) Role of gut microbiota in type 2 diabetes pathophysiology. EBioMedicine 51:102590
Haghikia A, Li XS, Liman TG, Bledau N, Schmidt D, Zimmermann F et al (2018) Gut microbiota-dependent trimethylamine N-oxide predicts risk of cardiovascular events in patients with stroke and is related to proinflammatory monocytes. Arterioscler Thromb Vasc Biol 38:2225–2235
Hansson GK (2005) Inflammation, atherosclerosis, and coronary artery disease. N Engl J Med 352(16):1685–1695
Hasan N, Yang H (2019) Factors affecting the composition of the gut microbiota, and its modulation. PeerJ 7:e7502
Henao-Mejia J et al (2012) Inflammasome-mediated dysbiosis regulates progression of NAFLD and obesity. Nature 482(7384):179–185
Hoeppli RE, Wu D, Cook L, Levings MK (2015) Environment of regulatory T cell biology: cytokines, metabolites, and the microbiome. Front Immunol 6:e1002808
Hollister EB, Gao C, Versalovic J (2014) Compositional and functional features of the gastrointestinal microbiome and their effects on human health. Gastroenterology 146:1449–1458
Holmes E, Loo R, Stamler J, Bictash M, Yap I, Chan Q et al (2008) Human metabolic phenotype diversity and its association with diet and blood pressure. Nature 453(7193):396–400. https://doi.org/10.1038/nature06882
Hooper LV, Macpherson AJ (2010) Immune adaptations that maintain homeostasis with the intestinal microbiota. Nat Rev Immunol 10:159–169
Ikeda Y, Ishii S, Yazaki M, Fujita T, Iida Y, Kaida T, Nabeta T, Nakatani E, Maekawa E, Yanagisawa T, Koitabashi T (2018) Portal congestion and intestinal edema in hospitalized patients with heart failure. Heart Vessel 33(7):740–751
Ivanov II, FrutosRde L, Manel N et al (2008) Specific microbiota direct the differentiation of IL-17-producing T-helper cells in the mucosa of the small intestine. Cell Host Microbe 4:337–349
Jie Z, Xia H, Zhong SL, Feng Q, Li S, Liang S, Zhong H, Liu Z, Gao Y, Zhao H et al (2017) The gut microbiome in atherosclerotic cardiovascular disease. Nat Commun 8:845. https://doi.org/10.1038/s41467-017-00900-1
Jin T et al (2012) Structures of the HIN domain: DNA complexes reveal ligand binding and activation mechanisms of the AIM2 inflammasome and IFI16 receptor. Immunity 36(4):561–571
Jin C, Henao-Mejia J, Flavell RA (2013) Innate immune receptors: key regulators of metabolic disease progression. Cell Metab 17(6):873–882

Jones ML, Martoni CJ, Parent M, Prakash S (2012) Cholesterol-lowering efficacy of a microencapsulated bile salt hydrolase-active *Lactobacillus reuteri* NCIMB 30242 yoghurt formulation in hypercholesterolaemic adults. Br J Nutr 107:1505–1513

Kamada N, Nunez G (2014) Regulation of the immune system by the resident intestinal bacteria. Gastroenterology 146:1477–1488

Karlsson FH, Fak F, Nookaew I, Tremaroli V, Fagerberg B, Petranovic D et al (2012) Symptomatic atherosclerosis is associated with an altered gut metagenome. Nat Commun 3:1245

Kasselman LJ, Vernice NA, DeLeon J, Reiss AB (2018) The gut microbiome and elevated cardiovascular risk in obesity and autoimmunity. Atherosclerosis 271:203–213

Khalili L, Alipour B, AsghariJafarabadi M, Hassanalilou T, Mesgari Abbasi M, Faraji I (2019) Probiotic assisted weight management as a main factor for glycemic control in patients with type 2 diabetes: a randomized controlled trial. Diabetol Metab Syndr 11(1):1–9

Kim S, Goel R, Kumar A, Qi Y, Lobaton G, Hosaka K et al (2018) Imbalance of gut microbiome and intestinal epithelial barrier dysfunction in patients with high blood pressure. Clin Sci 132: 701–718

Kimura I, Ozawa K, Inoue D, Imamura T, Kimura K, Maeda T, Terasawa K, Kashihara D, Hirano K, Tani T, Takahashi T (2013) The gut microbiota suppresses insulin mediated fat accumulation via short-chain fatty acid receptor GPR43. Nat Commun J 4:1829

Kiouptsi K, Finger S, Garlapati VS, Knorr M, Brandt M, Walter U, Wenzel P, Reinhardt C (2019) Hypoxia evokes increased PDI and PDIA6 expression in the infarcted myocardium of ex-germ-free and conventionally raised mice. Biol Open 8:bio038851. https://doi.org/10.1242/bio.038851

Kociol RD, Pang PS, Gheorghiade M, Fonarow GC, O'Connor CM, Felker GM (2010) Troponin elevation in heart failure: prevalence, mechanisms, and clinical implications. J Am Coll Cardiol 56(14):1071–1078

Koeth RA, Wang Z, Levison BS, Buffa JA, Org E, Sheehy BT et al (2013) Intestinal microbiota metabolism of L-carnitine, a nutrient in red meat, promotes atherosclerosis. Nat Med 19(5): 576–585

Koeth RA, Levison BS, Culley MK, Buffa JA, Wang Z, Gregory JC et al (2014) gamma-Butyrobetaine is a proatherogenic intermediate in gut microbial metabolism of L-carnitine to tmao. Cell Metab 20(5):799–812

Koeth RA, Lam-Galvez BR, Kirsop J, Wang Z, Levison BS, Gu X, Copeland MF, Bartlett D, Cody DB, Dai HJ, Culley MK, Li XS, Fu X, Wu Y, Li L, DiDonato JA, Tang WHW, Garcia-Garcia JC, Hazen SL (2019) L-Carnitine in omnivorous diets induces an atherogenic gut microbial pathway in humans. J Clin Invest 129:373–387

Krznaric Z, Vranesic Bender D, Mestrovic T (2019) The Mediterranean diet and its association with selected gut bacteria. Curr Opin Clin Nutr Metab Care 22:401–406

Lai ZL, Tseng CH, Ho HJ, Cheung CK, Lin JY, Chen YJ, Cheng FC, Hsu YC, Lin JT, El-Omar EM, Wu CY (2018) Fecal microbiota transplantation confers beneficial metabolic effects of diet and exercise on diet-induced obese mice. Sci Rep 8(1):1

Lam V, Su J, Koprowski S, Hsu A, Tweddell JS, Rafiee P, Gross GJ, Salzman NH, Baker JE (2012) Intestinal microbiota determine severity of myocardial infarction in rats. FASEB J 26:1727–1735. https://doi.org/10.1096/fj.11-197921

Larsson E, Tremaroli V, Lee YS, Koren O, Nookaew I, Fricker A et al (2012) Analysis of gut microbial regulation of host gene expression along the length of the gut and regulation of gut microbial ecology through MyD88. Gut 61:1124–1131

Le Roy T, Lécuyer E, Chassaing B, Rhimi M, Lhomme M, Boudebbouze S, Ichou F, Haro Barceló J, Huby T, Guerin M, Giral P (2019) The intestinal microbiota regulates host cholesterol homeostasis. BMC Biol 17(1):1–8

Ley RE, Turnbaugh PJ, Klein S, Gordon JI (2006) Human gut microbes associated with obesity. Nature 444(7122):1022–1023

Li XS, Obeid S, Klingenberg R, Gencer B, Mach F, Räber L, Windecker S, Rodondi N, Nanchen D, Muller O et al (2017) Gut microbiota-dependent trimethylamine N-oxide in acute coronary

syndromes: a prognostic marker for incident cardiovascular events beyond traditional risk factors. Eur Heart J 38:814–824. https://doi.org/10.1093/eurheartj/ehw582
Li XS, Wang Z, Cajka T, Buffa JA, Nemet I, Hurd AG et al (2018) Untargeted metabolomics identifies trimethyllysine, a TMAO-producing nutrient precursor, as a predictor of incident cardiovascular disease risk. JCI Insight 3(6):e99096
Li H, Liu B, Song J, An Z, Zeng X, Li J et al (2019a) Characteristics of gut microbiota in patients with hypertension and/or hyperlipidemia: a cross-sectional study on rural residents in Xinxiang County, Henan Province. Microorganisms 7:E399. https://doi.org/10.3390/microorganisms7100399
Li XS, Obeid S, Wang Z, Hazen BJ, Li L, Wu Y et al (2019b) Trimethyllysine, a trimethyl-amine N-oxide precursor, provides near- and long-term prognostic value in patients presenting with acute coronary syndromes. Eur Heart J 40(32):2700–2709
Li Q, Chang Y, Zhang K, Chen H, Tao S, Zhang Z (2020) Implication of the gut microbiome composition of type 2 diabetic patients from northern China. Sci Rep 10(1):1–8
Liu T, Hougen H, Vollmer AC, Hiebert SM (2012) Gut bacteria profiles of Mus musculus at the phylum and family levels are influenced by saturation of dietary fatty acids. Anaerobe 18:331–337
Loke WM, Proudfoot JM, Hodgson JM, McKinley AJ, Hime N, Magat M et al (2010) Specific dietary polyphenols attenuate atherosclerosis in apolipoprotein e-knockout mice by alleviating inflammation and endothelial dysfunction. Arterioscler Thromb Vasc Biol 30:749–757
Macfarlane GT, Macfarlane S (2011) Fermentation in the human large intestine: its physiologic consequences and the potential contribution of prebiotics. J Clin Gastroenterol 45(suppl):S120–S127
Mahida YR, Rolfe VE (2004) Host-bacterial interactions in inflammatory bowel disease. Clin Sci 107:331–341
Marchesi JR, Adams DH, Fava F et al (2016) The gut microbiota and host health: a new clinical frontier. Gut 65:330–339
Marques FZ et al (2018) Beyond gut feelings: how the gut microbiota regulates blood pressure. Nat Rev Cardiol 15:20–32. https://doi.org/10.1038/nrcardio.2017.120
May-Zhang LS, Chen Z, Dosoky NS, Yancey PG, Boyd KL, Hasty AH et al (2019) Administration of N-acyl-phosphatidylethanolamine expressing bacteria to low density lipoprotein receptor(-/-) mice improves indices of cardiometabolic disease. Sci Rep 9:420
Mazmanian SK, Liu CH, Tzianabos AO, Kasper DL (2005) An immunomodulatory molecule of symbiotic bacteria directs maturation of the host immune system. Cell 122(1):107–118
McCafferty K, Byrne C, Yaqoob M (2012) Intestinal microbiota determine severity of myocardial infarction in rats. FASEB J 26(11):4388. author reply 4388-9
McMillan A, Hazen S (2019) Gut microbiota involvement in ventricular remodeling post–myocardial infarction. Circulation 139(5):660–662
McNeil NI (1984) The contribution of the large intestine to energy supplies in man. Am J Clin Nutr 39:338–342
Meijers BK, De Preter V, Verbeke K, Vanrenterghem Y, Evenepoel P (2010) p-Cresylsulfate serum concentrations in haemodialysis patients are reduced by the prebiotic oligofructose-enriched inulin. Nephrol Dial Transplant 25(1):219–224
Mishima E, Fukuda S, Shima H, Hirayama A, Akiyama Y, Takeuchi Y, Fukuda NN, Suzuki T, Suzuki C, Yuri A, Kikuchi K (2015) Alteration of the intestinal environment by lubiprostone is associated with amelioration of adenine-induced CKD. J Am Soc Nephrol 26(8):1787–1794
Miura K, Greenland P, Stamler J, Liu K, Daviglus ML, Nakagawa H (2004) Relation of vegetable, fruit, and meat intake to 7-year blood pressure change in middle-aged men: the Chicago Western Electric Study. Am J Epidemiol 159:572–580
Moludi J, Kafil HS, Qaisar SA, Gholizadeh P, Alizadeh M, Vayghyan HJ (2021) Effect of probiotic supplementation along with calorie restriction on metabolic endotoxemia, and inflammation markers in coronary artery disease patients: a double blind placebo controlled randomized clinical trial. Nutr J 20(1):1

Morgan XC, Huttenhower C (2012) Chapter 12: human microbiome analysis. PLoS Comput Biol 8: e1002808

Nagatomo Y, Tang WW (2015) Intersections between microbiome and heart failure: revisiting the gut hypothesis. J Card Fail 21(12):973–980

Nallu A, Sharma S, Ramezani A, Muralidharan J, Raj D (2017) Gut microbiome in chronic kidney disease: challenges and opportunities. Transl Res 179:24–37

Naqvi S, Asar T, Kumar V, Al-Abbasi F, Alhayyani S, Kamal M, Anwar F (2021) A cross-talk between gut microbiome, salt and hypertension. Biomed Pharmacother 134:111156. https://doi.org/10.1016/j.biopha.2020.111156

Niebauer J, Volk HD, Kemp M, Dominguez M, Schumann RR, Rauchhaus M, Poole-Wilson PA, Coats AJ, Anker SD (1999) Endotoxin and immune activation in chronic heart failure: a prospective cohort study. Lancet 353(9167):1838–1842

Norman JM, Handley SA, Virgin HW (2014) Kingdom-agnostic metagenomics and the importance of complete characterization of enteric microbial communities. Gastroenterology 146:1459–1469

Novakovic M, Rout A, Kingsley T, Kirchoff R, Singh A, Verma V, Kant R, Chaudhary R (2020) Role of gut microbiota in cardiovascular diseases. World J Cardiol 12(4):110–122. https://doi.org/10.4330/wjc.v12.i4.110. PMID: 32431782; PMCID: PMC7215967.

Organ CL, Otsuka H, Bhushan S, Wang Z, Bradley J, Trivedi R, Polhemus DJ, Tang WW, Wu Y, Hazen SL, Lefer DJ (2016) Choline diet and its gut microbe–derived metabolite, trimethylamine N-oxide, exacerbate pressure overload–induced heart failure. Circulation 9(1):e002314

Pascal M, Perez-Gordo M, Caballero T et al (2018) Microbiome and allergic diseases. Front Immunol 9:1584

Pasini E, Aquilani R, Testa C, Baiardi P, Angioletti S, Boschi F, Verri M, Dioguardi F (2016) Pathogenic gut flora in patients with chronic heart failure. JACC Heart Fail 4(3):220–227

Pevsner-Fischer M, Blacher E, Tatirovsky E, Ben-Dov IZ, Elinav E (2017) The gut microbiome and hypertension. Curr Opin Nephrol Hypertens 26:1–8. https://doi.org/10.1097/MNH.0000000000000293

Phillips CM, Harrington JM, Perry IJ (2018) Relationship between dietary quality, determined by DASH score, and cardiometabolic health biomarkers: a cross-sectional analysis in adults. Clin Nutr 38(4):1620–1628. https://doi.org/10.1016/j.clnu.2018.08.028

Pisarenko OI, Khlopkov VN, Ruuge EK (1986) A 1H NMR study of succinate synthesis from exogenous precursors in oxygen-deprived rat heart mitochondria. Biochem Int 12:145–153

Pluznick JL (2013) Renal and cardiovascular sensory receptors and blood pressure regulation. Am J Physiol Renal Physiol 305:F439–F444

Rafter J (2004) The effects of probiotics on colon cancer development. Nutr Res Rev 17(2): 277–284

Rajilic-Stojanovic M, de Vos WM (2014) The first 1000 cultured species of the human gastrointestinal microbiota. FEMS Microbiol Rev 38:996–1047

Ramezani A, Raj DS (2014) The gut microbiome, kidney disease, and targeted interventions. J Am Soc Nephrol 25(4):657–670

Reid T, Schloss PD (2014) Dynamics and associations of microbial community types across the human body. Nature 509(7500):357–360

Ridker PM, Everett BM, Thuren T, MacFadyen JG, Chang WH, Ballantyne C et al (2017) Antiinflammatory therapy with canakinumab for atherosclerotic disease. N Engl J Med 377(12):1119–1131

Rinninella E, Raoul P, Cintoni M, Franceschi F, Miggiano GA, Gasbarrini A, Mele MC (2019) What is the healthy gut microbiota composition? A changing ecosystem across age, environment, diet, and diseases. Microorganisms 7(1):14

Rook G, Backhed F, Levin BR, McFall-Ngai MJ, McLean AR (2017) Evolution, human-microbe interactions, and life history plasticity. Lancet 390:521–530

Ross R (1999) Atherosclerosis-an inflammatory disease. N Engl J Med 340(2):115–126

Samuel BS, Shaito A, Motoike T, Rey FE, Backhed F, Manchester JK, Hammer RE, Williams SC, Crowley J, Yanagisawa M, Gordon JI (2008) Effects of the gut microbiota on host adiposity are modulated by the short-chain fatty-acid binding G protein-coupled receptor, Gpr41. Proc Natl Acad Sci 105(43):16767–16772

Saneei P, Hashemipour M, Kelishadi R, Esmaillzadeh A (2014) The dietary approaches to stop hypertension (DASH) diet affects inflammation in childhood metabolic syndrome: a randomized cross-over clinical trial. Ann Nutr Metab 64:20–27

Santoru ML, Piras C, Murgia A, Palmas V, Camboni T, Liggi S et al (2017) Cross sectional evaluation of the gut-microbiome metabolome axis in an Italian cohort of IBD patients. Sci Rep 7:9523

Sayin SI, Wahlström A, Felin J, Jäntti S, Marschall HU, Bamberg K, Angelin B, Hyötyläinen T, Orešič M, Bäckhed F (2013) Gut microbiota regulates bile acid metabolism by reducing the levels of tauro-beta-muricholic acid, a naturally occurring FXR antagonist. Cell Metab 17(2): 225–235

Schoeler M, Caesar R (2019) Dietary lipids, gut microbiota and lipid metabolism. Rev Endocr Metab Disord 20(4):461–472

Schwiertz A, Taras D, Schafer K et al (2010) Microbiota and SCFA in lean and overweight healthy subjects. Obesity 18:190–195

Shikata F, Shimada K, Sato H, Ikedo T, Kuwabara A, Furukawa H et al (2019) Potential influences of gut microbiota on the formation of intracranial aneurysm. Hypertension 73:491–496

Silveira-Nunes G, Durso DF Jr, Luiz Roberto AO, Cunha EHM, Maioli TU, Vieira AT, Speziali E, Corr Aaa-Oliveira R, Martins-Filho OA, Teixeira-Carvalho A, Franceschi C, Rampelli S, Turroni S, Brigidi P, Faria AMC (2020) Hypertension is associated with intestinal microbiota dysbiosis and inflammation in a Brazilian population. Front Pharmacol. https://doi.org/10.3389/fphar.2020.00258

Simenhoff ML, Dunn SR, Zollner GP, Fitzpatrick ME, Emery SM, Sandine WE, Ayres JW (1996) Biomodulation of the toxic and nutritional effects of small bowel bacterial overgrowth in end-stage kidney disease using freeze-dried Lactobacillus acidophilus. Miner Electrolyte Metab 22(1-3):92–96

Simpson HL, Campbell BJ (2015) Review article: dietary fibre-microbiota interactions. Aliment Pharmacol Ther 42:158–179

Skagen K, Troseid M, Ueland T, Holm S, Abbas A, Gregersen I et al (2016) The carnitine-butyrobetaine-trimethylamine-n-oxide pathway and its association with cardiovascular mortality in patients with carotid atherosclerosis. Atherosclerosis 247:64–69

Smith PM, Howitt MR, Panikov N et al (2013) The microbial metabolites, short-chain fatty acids, regulate colonic Treg cell homeostasis. Science 341:569–573

Song L, Xu M, Lopes-Virella MF, Huang Y (2001) Quercetin inhibits matrix metalloproteinase 1 expression in human vascular endothelial cells through extracellular signal-regulated kinase. Arch Biochem Biophys 391:72–78

Straub RH, Pongratz G, Weidler C, Linde HJ, Kirschning CJ, Gluck T, Scholmerich J, Falk W (2005) Ablation of the sympathetic nervous system decreases gram-negative and increases gram-positive bacterial dissemination: key roles for tumor necrosis factor/phagocytes and interleukin-4/lymphocytes. J Infect Dis 192:560–572

Sun S, Lulla A, Sioda M, Winglee K, Wu MC, Jacobs DR Jr, Shikany JM, Lloyd-Jones DM, Launer LJ, Fodor AA, Meyer KA (2019) Gut microbiota composition and blood pressure. Hypertension 73:998–1006. https://doi.org/10.1161/HYPERTENSIONAHA.118.12109

Sureda A, Bibiloni MDM, Julibert A, Bouzas C, Argelich E, Llompart I, Pons A, Tur JA (2018) Adherence to the Mediterranean diet and inflammatory markers. Nutrients 10:62

Suzuki T, Heaney LM, Bhandari SS, Jones DJ, Ng LL (2016) Trimethylamine N-oxide and prognosis in acute heart failure. Heart 102(11):841–848

Suzuki T, Heaney LM, Jones DJ, Ng LL (2017) Trimethylamine N-oxide and risk stratification after acute myocardial infarction. Clin Chem 63:420–428

Tang WH, Hazen SL (2014) The contributory role of gut microbiota in cardiovascular disease. J Clin Invest 124:4204–4211
Tang WH, Wang Z, Levison BS, Koeth RA, Britt EB, Fu X et al (2013) Intestinal microbial metabolism of phosphatidylcholine and cardiovascular risk. N Engl J Med 368(17):1575–1584
Tang WW, Wang Z, Fan Y, Levison B, Hazen JE, Donahue LM, Wu Y, Hazen SL (2014) Prognostic value of elevated levels of intestinal microbe-generated metabolite trimethylamine-N-oxide in patients with heart failure: refining the gut hypothesis. J Am Coll Cardiol 64(18): 1908–1914
Tang TVH, Chen H-C, Chen C-Y, Yen CYT, Lin C-J, Prajnamitra RP, Chen L-L, Ruan S-C, Lin J-H, Lin P-J, Lu H-H, Kuo CW, Chang CM, Hall AD, Vivas EI, Shui J-W, Chen P, Hacker TA, Rey FE, Kamp TJ, Hsieh PCH (2019) Loss of gut microbiota alters immune system composition and cripples postinfarction cardiac repair. Circulation 139:647–659. https://doi.org/10.1161/CIRCULATIONAHA.118.03523
Tilg H (2016) A gut feeling about thrombosis. N Engl J Med 374(25):2494–2496
Ursell LK, Haiser HJ, Van Treuren W et al (2014) The intestinal metabolome: an intersection between microbiota and host. Gastroenterology 146:1470–1476
Vinje S, Stroes E, Nieuwdorp M, Hazen SL (2014) The gut microbiome as novel cardio-metabolic target: the time has come! Eur Heart J 35:883–887
Waite JC, Skokos D (2012) Th17 response and inflammatory autoimmune diseases. Int J Inflamm 2012:819467
Wang Z, Klipfell E, Bennett BJ, Koeth R, Levison BS, Dugar B et al (2011) Gut flora metabolism of phosphatidylcholine promotes cardiovascular disease. Nature 472(7341):57–63
Wang Z, Roberts AB, Buffa JA, Levison BS, Zhu W, Org E, Gu X, Huang Y, Zamanian-Daryoush-M, Culley MK, DiDonato AJ (2015) Non-lethal inhibition of gut microbial trimethylamine production for the treatment of atherosclerosis. Cell 163(7):1585–1595
Wang Z, Bergeron N, Levison BS, Li XS, Chiu S, Jia X et al (2019) Impact of chronic dietary red meat, white meat, or non-meat protein on trimethylamine N-oxide metabolism and renal excretion in healthy men and women. Eur Heart J 40(7):583–594
Wang H, Lu Y, Yan Y, Tian S, Zheng D, Leng D, Wang C, Jiao J, Wang Z, Bai Y (2020) Promising treatment for type 2 diabetes: fecal microbiota transplantation reverses insulin resistance and impaired islets. Front Cell Infect Microbiol 9:455
Weaver CT, Hatton RD, Mangan PR et al (2007) IL-17 family cytokines and the expanding diversity of effector T cell lineages. Annu Rev Immunol 25:821–852
Whiteside SA, Razvi H, Dave S, Reid G, Burton JP (2015) E microbiome of the urinary tract—a role beyond infection. Nat Rev Urol 12(2):81–90
Widmer RJ, Flammer AJ, Lerman LO, Lerman A (2015) The Mediterranean diet, its components, and cardiovascular disease. Am J Med 128:229–238
Wlodarska M et al (2014) NLRP6 inflammasome orchestrates the colonic host-microbial interface by regulating goblet cell mucus secretion. Cell 156(5):1045–1059
Wu GD, Chen J, Hoffmann C et al (2011) Linking long-term dietary patterns with gut microbial enterotypes. Science 334:105–108
Wu ZX, Li SF, Chen H, Song JX, Gao YF, Zhang F et al (2017a) The changes of gut microbiota after acute myocardial infarction in rats. PLoS One 12:e0180717
Wu Y, Zhang Q, Ren Y, Ruan Z (2017b) Effect of probiotic Lactobacillus on lipid profile: a systematic review and meta-analysis of randomized, controlled trials. PLoS One 12(6): e0178868
Wu D, Guan L, Jiang YX, Ma SH, Sun YN, Lei HT et al (2019a) Microbiome and metabonomics study of quercetin for the treatment of atherosclerosis. Cardiovasc Diagn Ther 9:545–560
Wu W, Chen C, Liu P, Panyod S, Liao B, Chen P, Kao H, Kuo H, Kuo C, Chiu THT, Chen R, Chuang H, Huang Y, Zou H, Hsu C, Chang T, Lin C, Ho C, Yu H, Sheen L, Wu M (2019b) Identification of TMAO-producer phenotype and host–diet–gut dysbiosis by carnitine challenge test in human and germ-free mice. Gut 68:1439–1449

Xiao L, Liu L, Guo X, Zhang S, Wang J, Zhou F et al (2017) Quercetin attenuates high fat diet-induced atherosclerosis in apolipoprotein E knockout mice: a critical role of nadph oxidase. Food Chem Toxicol 105:22–33

Xu H, Wang X, Feng W, Liu Q, Zhou S, Liu Q, Cai L (2020) The gut microbiota and its interactions with cardiovascular disease. Microb Biotechnol 13(3):637–656. https://doi.org/10.1111/1751-7915.13524

Yadav R, Khan SH, Mada SB, Meena S, Kapila R, Kapila S (2019) Consumption of probiotic Lactobacillus fermentum MTCC: 5898-fermented milk attenuates dyslipidemia, oxidative stress, and inflammation in male rats fed on cholesterol-enriched diet. Probiot Antimicrob Prot 11(2):509–518

Yilmaz P, Parfrey LW, Yarza P et al (2014) e SILVA and "all species living tree project (LTP)" taxonomic frameworks. Nucleic Acids Res 42:D643–D648

Yu HN, Zhu J, Pan WS, Shen SR, Shan WG, Das UN (2014) Effects of fish oil with a high content of n-3 polyunsaturated fatty acids on mouse gut microbiota. Arch Med Res 45:195–202

Zabell A, Tang WH (2017) Targeting the microbiome in heart failure. Curr Treat Options Cardiovasc Med 19(4):27

Zeng Z, Yuan Q, Yu R, Zhang J, Ma H, Chen S (2019) Ameliorative effects of probiotic Lactobacillus paracasei NL41 on insulin sensitivity, oxidative stress, and beta-cell function in a type 2 diabetes mellitus rat model. Mol Nutr Food Res 63(22):1900457

Zheng Y, Yu B, Alexander D, Mosley T, Heiss G, Nettleton J, Boerwinkle E (2013) Metabolomics and incident hypertension among blacks. Hypertension 62(2):398–403. https://doi.org/10.1161/hypertensionaha.113.01166

Zhu W et al (2016) Gut microbial metabolite TMAO enhances platelet Hyperreactivity and thrombosis risk. Cell 165(1):111–124

Zhu W et al (2017) Gut microbe-generated trimethylamine N-oxide from dietary choline is prothrombotic in subjects. Circulation 135(17):1671–1673

Zhu Q, Gao R, Zhang Y, Pan D, Zhu Y, Zhang X et al (2018) Dysbiosis signatures of gut microbiota in coronary artery disease. Physiol Genomics 50(10):893–903

Role of Microbiome in Reproductive Health: An Expanding Dimension

16

Samridhi Pushkarna, Richa Bhatnager, Anil Kumar, Pooja Suneja, and Amita Suneja Dang

Abstract

Trillions of symbiotic microorganisms have evolved with and continue to live on and within human beings, serving an important role in regulating human health and disease. Their omnipresence has a great influence on almost all the physiological functions of the body, and female reproductive system is no exception. As per the Human Microbiome Project, the vaginal microbiota alone makes up about 9% of the total human microbiota. Any internal or external change in this resident microbial population might lead to dysbiosis, further causing disease pathologies. This undesirable shift in microbiota is also strongly associated with stress and lifestyle imbalance, which are directly linked to poor dietary habits and a sedentary routine. Such unfavourable outcomes of vast urbanisation and industrialisation have increased the incidence of major health issues. One such medical condition is infertility, which is currently perceived as a worldwide health problem. Pathophysiological conditions concerning the reproductive system like endometriosis, dysregulated ovarian functions, cervical factors, uterine complications and vaginal infections also affect female fertility to a great extent. Given that these conditions are usually associated with microbial dysbiosis, modulating the microbial population to reinstate eubiosis can certainly alleviate disease symptoms, thereby helping with disease management strategies. Gut microbiota also actively interacts with sex hormones (a concept called

S. Pushkarna · A. Kumar · A. S. Dang (✉)
Centre for Medical Biotechnology, Maharshi Dayanand University, Rohtak, India

R. Bhatnager
Royal Berkshire NHS Foundation Trust, Bracknell Lighthouse Laboratory, Bracknell Healthspace, Bracknell, UK

P. Suneja
Department of Microbiology, Maharshi Dayanand University, Rohtak, India

R. Sobti et al. (eds.), *Role of Microbes in Sustainable Development*,
https://doi.org/10.1007/978-981-99-3126-2_16

microgenderome) to regulate the circulating level of the latter and thus affect reproductive health in various ways. Therefore, many microorganisms are now being examined for their potential and promising use in developing various treatment approaches, including probiotics, synbiotics, faecal microbiota transfer (FMT) and vaginal microbiota transfer (VMT). The rationale behind all these options lies in rehabilitating the progenitive niche with a healthy microbial population in order to maintain and regulate the female reproductive system homeostasis.

16.1 Introduction

Humans are born in a virtually germ-free environment. Various microorganisms originating from the mother or other surroundings begin to settle in several body parts shortly after. This early colonisation of bacteria is essential for further development (Kataoka 2016). With time, the human body also referred to as a superorganism (Liang et al. 2018), fosters a tightly populated resident microbial community regulating the functions of their specific residential area. With the advent of high-throughput sequencing techniques and extensive molecular research potential, genital tract microbiota is now being explored in diverse arrays ranging from its identification of taxa to functional capabilities concerning reproductive health, especially in the case of females. The vagina of the female reproductive tract harbours roughly 9% of the human microbial load alone Instances of medical complications in pregnancy and fertility have been associated with a hampered endometrial and cervicovaginal bacterial population. Links have also been identified between infertility and bacterial vaginosis (BV) (Ravel et al. 2021). It is characterised by a transition from *Lactobacillus* spp. dominant bacterial community to a more heterogeneous population comprising mainly of anaerobic bacteria. Women suffering from PCOS, a complex reproductive, neuro-endocrinal and metabolic condition, exhibit a lower α-diversity correlated to higher testosterone concentration and hirsutism score (Thackray 2019). Impaired microbiota structure has also shown links with hyperandrogenism, ovulatory impairment, obesity, insulin resistance and cardiovascular diseases, hence majorly contributing to the pathogenesis of PCOS (He and Li 2020; Torres et al. 2018; Giampaolino et al. 2021). Reproductive health is not only affected by the tiny creatures residing at sites associated with the reproductive system but also by the distant ones present in the gut. Gut microbiota actively communicates with the sex hormones, thereby regulating their levels in circulation. This concept has been termed as "microgenderome" (Flak et al. 2013). Since the *Lactobacillus* species significantly dominate vaginal niche microflora, its role in maintaining reproductive homeostasis and health is also extensively studied. It is the major contributor to the production of antimicrobial factors like lactic acid, hydrogen peroxide, bacteriocins and other organic acids. An indispensable role of microbiota by virtue of its presence all over and inside the human body gives birth to a ray of the possibility of using these commensals for treatment purposes. Different approaches like probiotics, synbiotics, prebiotics and faecal microbiota transfer (FMT), or a combination of these, are employed to discover the probable hidden effects and establish additional treatment options for diseases in the near future.

16.2 Common Reproductive Problems and Their Microbial Association

16.2.1 Bacterial Vaginosis

The vaginal lumen is a nutrition-rich zone, acting as a favourable niche for a diverse array of microbiota. Though the microbial population in the vaginal milieu keeps fluctuating with different stages of a woman's life, a healthy vaginal microbiota (VMB) of most women is dominated by the Gram-positive, *Lactobacillus* spp capable of maintaining an acidic environment through glycogen fermentation, thereby preventing the overgrowth of exogenous bacteria and viruses. In addition to low pH resulting from the production of lactic acid, the synthesis of biosurfactants and bacteriocins also protects against urinary tract infections and pathogenic microorganisms. Disturbance in the VMB composition (also known as vaginal microbiota dysbiosis) can cause bacterial vaginosis, marked by the loss or a significant decline in the concentration of lactobacilli and an overgrowth of non-lactobacilli, primarily anaerobic bacterial population. The vaginal microbiota is classified into five different categories called as community state types (CSTs) (Ravel et al. 2011). CST I is dominated by *L. crispatus*, CST II by *L. gasseri,* CST III by *L. iners* and CST V by *L. jensenii*. CST IV comprises *Lactobacillus*-deficient type bacteria, most of which are anaerobic and are accountable for causing bacterial vaginosis. A small portion also belongs to partial aerobic bacteria, which are classified by aerobic vaginitis, AV (Gajer et al. 2012). Common microbial genera associated with BV include *Gardnerella*, *Mycoplasma*, *Roseburia*, *Mobiluncus*, *Dialister*, *Prevotella* and *Sneathia* (McMillan et al. 2015; Ceccarani et al. 2019). However, the distinction between a typical/healthy vaginal domain and BV cannot be just based on flora analysis since some bacterial strains are common to both environmental conditions and culturing of the bacterial load alone may result in ambiguous results. Thus, in addition to the vaginal microenvironment analysis, metabolomics also needs to be performed in order to properly characterise the vaginal dysbiosis resulting in diseased states (Vitali et al. 2015). BV most commonly occurs in women of childbearing age but in general, all women/females suffering from BV witness an increased probability of acquiring many gynaecological and obstetric complications like infertility, premature delivery/preterm birth, immature rupture of membranes and miscarriage (Workowski and Bolan 2015; Baqui et al. 2019; Peebles et al. 2019).

16.2.2 Polycystic Ovary Syndrome (PCOS)

PCOS is a highly complex heterogeneous endocrine disease and is one of the most common causes of female endocrine infertility. The worldwide prevalence of PCOS in women of childbearing age varies between 6% and 15%. Urban regions witness more PCOS patients because of improper dietary habits and lifestyle (Deswal et al. 2019). Timely diagnosis and prevention of PCOS are very important in order to combat its rising prevalence (Deswal et al. 2020). This syndrome is mainly characterised by chronic anovulation, hyperandrogenism, insulin resistance and obesity (Apridonidze et al. 2005; Bozdag et al. 2016; Bhatnager et al. 2018). Hyperandrogenism further leads to many dermatological issues like hirsutism, acne, and androgenic alopecia. Irregular ovarian folliculogenesis and alteration of the hypothalamus–pituitary–gonadal axis, on the other hand, cause severe menstrual abnormalities (Azziz et al. 2016). Although the exact aetiology of PCOS is still unclear, several factors have been identified that are suggested to disrupt the hormonal and metabolic balance and contribute to the pathogenesis of this syndrome. One of such important and critical factors is microbiota. A bidirectional relationship between gut microbiota and sex hormones is seen, which involves the modulation of enterohepatic androgen circulation. Elevated concentration of bacterial genera like *Escherichia*, *Shigella* and *Streptococcus* has also been correlated to decreased levels of gut hormones like PYY and Ghrelin (Batra et al. 2022). The widely accepted hypothesis states that this disease originates due to interactions between genetic and environmental factors. Genes regulating folliculogenesis, steroidogenesis, insulin resistance and adipocyte differentiation all might play a critical role in disease development (Franks et al. 2006). Apart from genetic, neuroendocrine, and metabolic factors, gut barrier integrity, endotoxemia and inflammation have been analysed as well. Lower diversity and an altered phylogenetic composition have been observed in the stool samples of PCOS patients. Differences in taxa abundance were also observed that varied with the metabolic factors and did not follow a single transition trend. For example, the concentration of some Gram-negative bacteria mainly belonging to the genera Bacteroides and Escherichia/Shigella significantly amplified in the gut of obese PCOS women. In such cases, lipopolysaccharides produced by these microorganisms induce obesity, insulin resistance and chronic inflammation (Liu et al. 2017), thereby worsening the symptoms of PCOS.

16.2.3 Endometrial Complications

Miscarriage or infertility is also associated with specific endometrial microbiota composition. With next-generation sequencing technologies and the possibility to culture even less than 1% of microorganisms (Wade 2002), the concept of the sterile womb hypothesis has been challenged. Primary studies concerning endometrial microbiota indicate its association with different gynaecological conditions like chronic endometritis (CE) (Moreno et al. 2018), dysfunctional endometrial bleeding

(Pelzer et al. 2018), endometriosis (Wee et al. 2018), endometrial polyps (Fang et al. 2016) and endometrial cancer (Walther-António et al. 2016). The primary cause of CE is a microbial infection in the uterine cavity. The microorganisms frequently detected in the endometrium with CE are *Escherichia coli*, *staphylococcus species Streptococcus species*, *Enterococcus faecalis*, *proteus species*, *Pseudomonas aeruginosa*, *Gardnerella vaginalis*, mycoplasma/ureaplasma species (*Mycoplasma hominis*, *Mycoplasma genitalium* and *Ureaplasma urealyticum*), *Corynebacterium*, *Klebsiella pneumoniae* and yeasts Cicinelli et al. 2008; Cicinelli et al. 2009; Kitaya et al. 2017). This microbiota structure is widely affected by factors like BMI, diet, physical activity, abnormal fluctuation in the levels of circulating hormones, and menstrual cycle irregularities (Dominianni et al. 2015; Claesson et al. 2012; Lynch and Pedersen 2016; Bracewell-Milnes et al. 2018; Wilson et al. 2007). Still, an active debate goes on about whether a unique microbiota resides in the uterine environment or not (Winters et al. 2019) since various technical challenges prevail that make it difficult to study samples with low biomass and to differentiate microorganisms between the ones that actually reside in small quantities from those that arise as a result of contamination/infection (Weyrich et al. 2019). Thus, more extensive research is needed to confirm whether dysbiosis in the uterus is a cause or consequence of disease pathology.

16.3 Pathophysiology

16.3.1 Microgenderome

Recently, the concept of "microgenderome" has come into play. It signifies the bidirectional interactions that take place between sex hormones and gut microbiota. Gut microbiota regulates sex hormone levels by means of interactions among its metabolites, the immune system and some nerve–endocrine axes, such as the gut–brain axis (He et al. 2021). The gut microbiome shows β-glucuronidase enzyme activity (Beaud et al. 2005). This enzyme prevents oestrogen from binding with glucuronic acid as a result of which, the oestrogen concentration increases in the body. Some studies also indicate the involvement of enterohepatic circulation (Liu et al. 2021). In this process, the β-glucuronidase activity shown by the bacterial species in the gut causes deconjugation of conjugated oestrogens that take place in the liver for excretion through bile. Such enhanced levels of oestrogen can further cause a surge in the migration and adhesion capabilities of the cells having oestrogen receptor, thereby contributing to the colonisation and metastasis of ovarian cancer (Wallace et al. 2010). Thus, it can be said that gut microbiota may play an indirect role in the development of ovarian cancer and other such oestrogen-driven diseases by regulating the amount of active oestrogen circulation (Shen et al. 2012). Microorganisms containing genes for β-glucuronidase also affect androgens in a similar way. These bacteria carry out side-chain cleavage of glucocorticoids, converting them into androgens, causing their elevation in the body and hence participating in the pathogenesis of PCOS. *Edwardsiella*, *Bacteroides*, *Collinsella*,

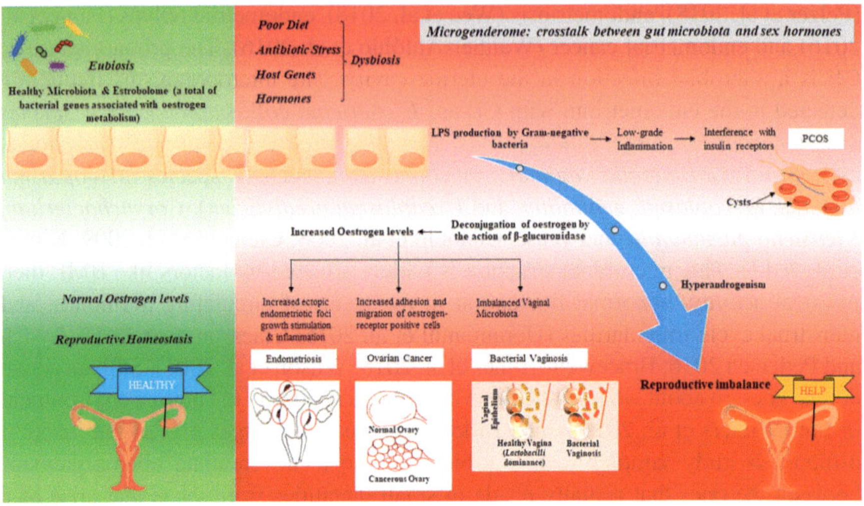

Fig. 16.1 Microgenderome: crosstalk between gut microbiota and sex hormones

Propionibacterium, *Bifidobacterium*, *Escherichia Clostridium*, *Faecalibacterium*, *Alistipes*, *Lactobacillus*, *Roseburia*, *Citrobacter*, etc., are some of the examples of such bacteria (Kwa et al. 2016). Figure 16.1 provides an overview of how microgenderome affects the reproductive health of a female.

16.3.2 Gut–Brain Axis

The gut and brain are very well interconnected through highly characterised channels of communication. Endocrine, neural, and inflammatory mechanisms are tightly integrated to make these channels work. Variations in the small intestinal permeability and blood–brain barrier may modulate this communication network. Another factor that tends to regulate this axis is brain–gut microbiome interaction (Osadchiy et al. 2019). The gut microbiome connects to the central nervous system via its metabolic intermediates such as short-chain fatty acids (SCFAs), secondary bile acids (2BAs), and tryptophan metabolites. 90–95% of colonic SCFAs include acetate, butyrate and propionate. They maintain a lower pH, favouring the growth of bacteria that promote a state of homeostasis, such as *Bifidobacteria* and *Lactobacillus,* and hinder the establishment of opportunistic pathogenic bacteria, including *Clostridium* and *E. coli*. SCFAs also aid in preserving the function of the gut barrier (Amabebe and Anumba 2020). These components stimulate the regeneration of epithelial cells and the production of mucus and antimicrobial peptides. This inhibits the translocation of toxins and bacteria into the bloodstream, preventing infections. The gut microbiome can regulate the activation of pathways leading to inflammation, brain–gut peptide secretion and proliferation of islet cell, which may result in abnormal or excessive fat accumulation, compensatory hyperinsulinemia and insulin resistance in case of dysbiosis (Vrieze et al. 2010; Barber et al. 2016). Food intake control is also modulated by these SCFAs via an increase in the production of hunger-suppressing hormones, like peptide YY, glucagon-like peptide-1 and leptin, eventually reducing excessive food intake. On the other hand, reduced production of SCFAs is associated with obesogenic and pro-inflammatory mechanisms (Larraufie et al. 2018). Obese PCOS women have an abnormal proportion of brain–gut axis intermediaries, which has been linked to PCOS-related clinical phenotypes. Gram-negative bacteria also produce lipopolysaccharides that pass through the gut wall, enter the circulation and cause low-grade inflammation. Immune system activation then interferes with the insulin receptors, elevating the insulin levels, which further increase testosterone production in the ovary, contributing to PCOS. Thus, it is important to avoid gut microbiota dysbiosis, which might take a toll on the reproductive health of women (Tremellen and Pearce 2012).

16.3.3 Prenatal Episodes of Microbiota Exposures

With the advent of new techniques, the placenta has now been shown to harbour microorganisms that affect the growth and development of the foetus (Cao et al. 2014). Several studies illustrate a unique microbial profile associated with the placenta, thereby opening the possibility for the maternal–foetal spread of commensal microorganisms, leading to colonisation of a microbial community at a prenatal stage. Through research done on samples from infant and maternal faeces, amniotic fluid, placenta, meconium and colostrum from mother–infant pairs, it was noted that a less abundant and less diverse microbiota with a predominance of Proteobacteria is present in placenta and amniotic fluid. Further, infant meconium displays shared features in the microbiota composition, thereby pointing towards a microbial transmission from mother to foetus. 3–4 days after birth, a newborn's gut microbiota structure resembles that of colostrum, thus outlining the stepwise establishment of the gut microbiome (Collado et al. 2016). Studies indicate a relationship between infections and inflammation with preterm births (PTBs). It has been assumed that infections that cause PTB initiate in the reproductive or genitourinary tract from where they rise upward via the cervix and breach the placental barrier. (Cao et al. 2014). Therefore, it is essential to shift our focus on the microbial ecology of the placenta and amniotic fluid, in particular, to avoid any malfunctioning of these sites and ensure proper growth and development of the foetus. Many eminent scientists favour the traditional belief and not the studies claiming microbiota colonisation of these parts of the human body. This is because most of these studies have identified the fragments of microbial DNA, their metabolites and some pathogenic strains, which might be translocated or leaked from maternal tissues into the so-called privileged sites, thus crossing the placental barrier and affecting the growth of the foetus. No clear evidence has been provided to assure the presence of live microbiota that has established a proper colony/niche with well-defined host–microbe interactions and their related functional outcomes. The presence of germ-free mice again points to sterility (Blaser et al. 2021). Thus, it can be said that actual microbiota is still in the process of being discovered in such germ-free habitats. New findings undoubtedly lay a foundational base for further research and help open up the gate to an altogether different direction that might bring out changes highly beneficial for society in the future. Hence, it can be said that there exists some association, direct or indirect, though way more exploration is required to understand the underlying interaction comprehensively and the magnitude of these connections affecting the in utero generations.

16.3.4 An Indispensable Role of Lactobacilli

In healthy reproductive-aged women, vaginal microenvironment and eubiosis are mainly characterised by the *Lactobacillus* species-dominated microbial population. These bacteria serve as active players in the game of reproductive homeostasis (Mendling 2016). Figure 16.2 provides a simple illustration of the establishment of Lactobacilli species in the vagina and the importance of lactic acid produced by them.

The rationale behind Lactobacillus dominance in VMB lies in the fact that these bacteria produce an ample amount of lactic acid as a result of glycogen fermentation, maintaining a pH < 4.5 and producing many other antimicrobial factors that help fight against infections. Some of these factors include the following:

1. *Lactic Acid:* The bactericidal action of this acid is facilitated in its protonated form (active at pH < 3.9) and not as a lactate anion (O'Hanlon et al. 2011). The latter anionic form is less membrane-permeant and requires the use of GPR81 receptor (Ahmed et al. 2009) or monocarboxylate transporters (Kuchiiwa et al.

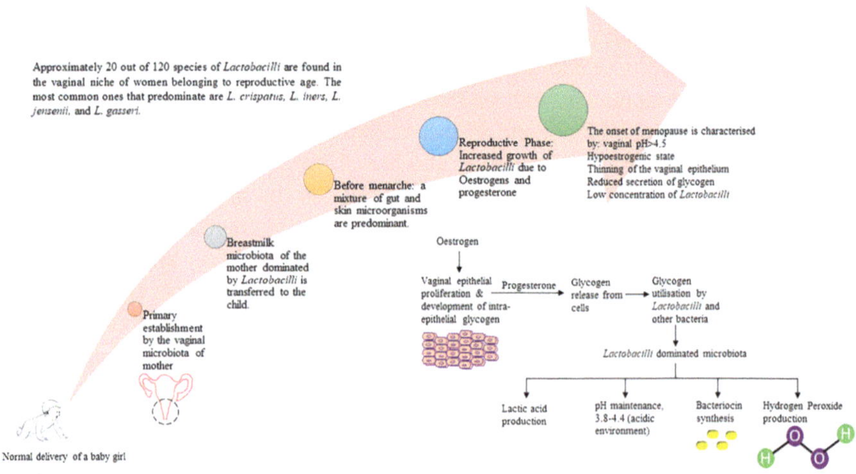

Fig. 16.2 Vaginal microbiota colonization and role of Lactobacilli

2011). This kind of complex lactate shuttle is not seen in the case of the protonated form of lactic acid, which, upon cell entry, acidifies the cytosol, interfering with the cellular metabolism and leading to cell death (Kashket 1987). Production of lactic acid alone by species like *L. gasseri* and *L. crispatus* can inhibit the infections caused by *Neisseria gonorrhoeae* (Graver and Wade 2011), *Chlamydia trachomatis* (Gong et al. 2014) and *Escherichia coli* (Valore et al. 2002). These pathogenic strains were seen to be inactivated in vitro. The acidic environment sustained in the vaginal niche also inhibits the growth and colonisation of pathogenic microorganisms. Lactic acid also exhibits virucidal activity. Reduced rate of viral shedding into the lower female reproductive tract has been seen in HIV-positive women harbouring Lactobacillus-enriched microbiota (Mitchell et al. 2013). The potential inactivation of HIV-1 in such cases is multifactorial that involves modulating the viral lipid envelop integrity, affecting the viral surface proteins and negating the viral core protein functionality (Tachedjian et al. 2017). This genus also makes women less likely to get infected with other viruses like herpes simplex virus 2 (HSV-2) (Borgdorff et al. 2014). Lactobacilli either indirectly prevent this disease by inhibiting viral entry, replication and reducing adhesion capacity or by directly exerting effects via lactic acid (Conti et al. 2009).

Apart from antimicrobial activities, lactic acid showcases *immunomodulatory properties* too. Studies have shown that treating cervicovaginal cells with the protonated form of lactic acid can result in a dampening effect on the production of pro-inflammatory cytokines and chemokines stimulated by Toll-like receptor 1/2 agonist Pam3CSK that mimics the bacterial and viral pathogen-associated molecular patterns (PAMPs) of BV-associated bacteria and HIV (i.e. gp120) (Hearps et al. 2014; Hearps et al. 2017). The mucus lining of the inner epithelium of the female genital tract acts as the first line of defence against pathogenic invasions. When any microorganism disrupts and makes its way across this layer, the epithelial cells identify the organism by using pattern recognition receptors and producing an inflammatory response. To maintain the immune equilibrium, lactic acid induces the production of anti-inflammatory cytokines like IL-10 to lower the production of pro-inflammatory cytokine IL-12 in dendritic cells (DCs) and reduces the cytotoxicity of natural killer cells. The presence of other organic acids produced by vaginal microbiota also favours the anti-inflammatory activity of lactic acid. A slight reduction in the production of other pro-inflammatory factors, for example, cytokine IL-6, IL-1 and macrophage inflammatory protein three alpha (MIP-3α), has also been documented (Li et al. 2020).

2. *Hydrogen Peroxide* (H_2O_2): It is another compound having antimicrobial properties. 94–95% of the strains of *L. crispatus* and *L. jensenii* produce H_2O_2 (Antonio et al. 1999). Hydrogen peroxide alone is microbicidal for many bacterial species. This microbicidal activity is 10- to 100-fold greater when combined with chloride anion and myeloperoxidase, both of which are found in the vagina (Ravel et al. 2011; Gajer et al. 2012). This vaginal antimicrobial defence system (H_2O_2, chloride anion and myeloperoxidase) has potent in vitro activity against *E. coli* and other microorganisms (Stapleton 2017). Using H_2O_2 resulted in

eliminating some major symptoms of BV like malodorous leucoxanthorrhoea in 89% of women under study. It also fostered the clue cells and anaerobic pathogenic flora to disappear from the vaginal smears and vaginal secretions in 100% of the cases (Cardone et al. 2003). Since the levels of dissolved O_2 are low in the vaginal compartment, high concentrations of (H_2O_2) cannot just be detrimental to the growth of lactobacillus species. Still, they might favour the growth of bacteria responsible for dysbiosis, causing gynaecological infections (O'Hanlon et al. 2011).

3. *Bacteriocins:* Bacteriocins are proteinaceous substances secreted by some bacteria and pose bactericidal activities to halt and prevent the growth and colonisation by microorganisms that are usually closely related to the microorganism producing the bacteriocins (Stoyancheva et al. 2014). Bacteriocins are mainly classified into two types. Class I constitutes lanthionine-containing bacteriocins (lantibiotics) and class II contains non-lanthionine-containing peptides. Class I elements cause pore formation, resulting in the uncontrolled efflux of small metabolites from the sensitive cells or inhibit enzyme activities, while class II types are active by inducing permeabilisation of membrane gains, causing the leakage of internal metabolites from target bacteria (Cotter et al. 2005). These are synthesised by *Lactobacillus* species, specifically *L. gasseri* and *L. crispatus*. These bacteria encode genes corresponding to antimicrobial bacteriocins like acidocin F221A, J46, IIa, IIc, type A lantibiotic and gassericin T (Stoyancheva et al. 2014). Bacteriocin production is one of the chief antagonistic mechanisms for avoiding undesirable microbial colonisation.
4. *Biosurfactants (BS):* Some microorganisms produce amphiphilic compounds called biosurfactants that either remain anchored to their surface or are secreted outside into the surrounding environment. These compounds are mostly secondary metabolites and are characterised by excellent emulsifying properties (Van Hamme et al. 2006). These molecules facilitate nutrient transportation to support the producer microorganism survival and actively fight against opportunistic and pathogenic microorganisms by serving as anti-adhesive, antimicrobial and anti-biofilm agents (Satpute et al. 2016; Sharma and Saharan 2016; Sambanthamoorthy et al. 2014). A non-homogenous lipopeptide molecule (BS) isolated from vaginal *L. crispatus* has been shown to reduce the growth of *Candida* spp. by interfering with its ability to adhere to the cervical epithelial wall and thus alleviating the mucosal damage caused during vulvovaginal candidiasis (VVC) (De Gregorio et al. 2020).

16.4 Microbiota in Therapeutics: An Evolving Concept!

An indispensable role of microbiota, by virtue of its presence all over and inside the human body, gives birth to a ray of the possibility of using these commensals for treatment purposes as well. Much research, including clinical and randomised controlled trials on various diseases, is being carried out to explore any possible therapeutic function of a microorganism if present. Different approaches like probiotics, faecal microbiota transfer (FMT), synbiotics, prebiotics or a combination of these are employed to discover the probable hidden effects and establish additional treatment options for diseases in the near future. Despite being characterised by differences in their constituents, nature, mode of administration and factors affecting the efficacy, the basic principle behind all these methods mentioned above lies in the ultimate aim to recover the typical microbial environment that has been imbalanced. These therapeutic options are usually used as adjuvant therapies and not straightaway as the first or only line of treatment.

Probiotics Probiotics are "live microorganisms, which when consumed in adequate amounts, confer a health benefit on the host" (FAO/WHO 2001). It is important to note that whether a probiotic contains a single microorganism or a consortium of microbial strains, and the composition is always defined/known with well-established viability and efficacy. Also, probiotics should not be confused with commensal microorganisms. In gynaecology and obstetrics, probiotics are living, beneficial microorganisms, mainly consisting of *Lactobacillus* species given in adequate quantity to restore physiological vaginal microflora to treat BV and other reproductive tract disorders (Buggio et al. 2019). *Lactobacillus* spp, owing to their capability of potentiating the effect of antibiotics, can act as effective antimicrobial adjuvants (Larsson et al. 2011). Different species of *Lactobacillus* like *Lactobacillus rhamnosus* (E21 and L3), *Lactobacillus salivarius* (N30) (Pino et al. 2019), *Lactobacillus* strain (SQ0048) (Niu et al. 2019), *Lactobacillus rhamnosus* GR1, *Lactobacillus buchneri* (DSM 32407) (Peter et al. 2018) and many other have remarkable properties that make them capable of regulating and maintain reproductive health. Some characteristics include tolerance to low pH/maintenance of low pH, ability to colonise both intestinal and vaginal epithelia, production of hydrogen peroxide, bacteriocins and organic acids and skill to co-aggregate and prevent the growth of pathogenic microorganisms. Some strains like *Lactobacillus rhamnosus* CECT8361 and *Bifidobacterium longum* CECT7347 also have anti-inflammatory and antioxidant activities (Valcarce et al. 2017). Other genera, for example, *Bifidobacterium lactis* V9 and Bacillus species (*B. clausii, B. subtilis and B. coagulans*), also showcase such probiotic effects. Further functional properties comprising tolerance to bile salts, acids and lysozyme are also essential for such microorganisms to get through the gastrointestinal tract and reach the target site of action (Hashem and Gonzalez-Bulnes 2022). Since glycogen is an energy source for lactobacilli, a combination (Gynoflor) of low-dose oestriol and live lactobacillus has been reported for glycogen production and re-establishment of normal vaginal microbiota (Ozkinay et al. 2005). Several studies have suggested that using probiotics helps decrease metabolic intestinal endotoxemia, insulin resistance and the production of inflammatory mediators and improves glycaemic control in women with PCOS (Akbari and Hendijani 2016; Ruan et al. 2015). Oral capsules containing *L. reuteri* (formerly *L. fermentum*) and *L. casei var. rhamnosus* significantly modulate the vaginal flora, resulting in a good percentage of treatment success rate (Reid et al. 2003). Oral feeding of *Bifidobacterium longum* NK49 and *Lactobacillus plantarum* NK3 alleviates *Gardnerella vaginalis* (GV)-induced vaginosis in female mice.

These bacteria exhibit anti-inflammatory activities by inhibiting the excessive expression of TNF-α and NF-κB activation in the vagina and uterus with a simultaneous decrease in the vaginal GV population (Kim et al. 2019). This example again reflects upon the importance of gut microbiota restoration in maintaining a homeostatic reproductive environment. However, the guidelines on using probiotic formulations for gynaecological concerns differ between countries since there is no universal background, and the presence of inconsistencies in results has led to inconclusiveness (Barrientos-Durán et al. 2020).

Prefilled vaginal applicators also use such species to encourage a favourable community state type. Post-antibiotic administration of *L crispatus* CTV-05 (Lactin-V, Osel) in women suffering from BV resulted in a lower rate of BV recurrence at 3 months compared to the placebo group (Cohen et al. 2020). Use of topical gels containing lactic acid has been assessed for tolerability and efficacy when administered as an adjuvant to metronidazole in a study concerning BV treatment. Gels like Lactacyd vaginal gel (LVG) contained lactic acid (225 mg) and prebiotic glycogen (5 g) to promote the growth of *Lactobacillus*. The combination was superior to antibiotic treatment alone (Decena et al. 2006).

Prebiotics Prebiotics can be defined as nutrients that promote the growth and activities of probiotic microorganisms, specifically favouring Lactobacilli and Bifidobacteria over potentially harmful bacteria, which may be proteolytic and putrefactive in nature. This way, prebiotics intend to deliver health benefits to the host and are gaining considerable interest in developing new therapeutic approaches. The most commonly known prebiotic nutrient components are inulin, fructooligosaccharides (FOS), lactulose and galactooligosaccharides (GOS). By favouring the growth of "good" bacteria, these prebiotics help reduce fasting glucose, total cholesterol, LDL cholesterol, HDL cholesterol and serum TG levels (Altun and Yıldız 2017). Some prebiotics also control hyperglycaemia and HOMA-IR (Fernandes et al. 2017). In women with PCOS, resident dextrin consumption may improve hirsutism, androgen levels and menstrual cycle irregularities (Gholizadeh Shamasbi et al. 2019). Considering the role of probiotics and prebiotics, if these components are combined, then what we would get will be a *synbiotic* featuring the advantages of both, resulting in coactive effects. Antibiotics also form a part of the strategic group, but it has been seen to reduce the bacterial diversity even of desired residents because of broad-spectrum activity, thereby disturbing communal harmony.

Faecal Microbiota Transfer (FMT) Lately, interest in using FMT to treat chronic gastrointestinal infections and inflammatory bowel diseases has been increasing. FMT involves the faecal microbiota transfer from a healthy donor into the recipient's (patient) intestinal tract (Smits et al. 2013). It helps transmit and increase the concentration of good bacteria at the disturbed site in the recipient and lower the severity of symptoms in ways such as restoring community structure and functions through engraftment or refurbishing the lost metabolome. However, the exact mechanisms are unclear (Libertucci and Young 2019). Recent findings suggest

FMT's promising role in managing female genital disorders associated with microbiota alterations. Crosstalk between the resident bacterial strains of the gut and vagina results in local and systemic immune regulation. Concerning this fact, genetic engineering of particular commensal bacterial species can be done to develop oral vectors proficient in eliciting immune responses to prevent or alleviate disease symptoms. These vectors can express specific antigenic proteins, thereby triggering a local immune response and the production of antibodies. Successful intravaginal immunization of mice against human papillomavirus infection suggests that mucosal colonisation with commensal recombinant bacteria can sure pave the way for developing new vaccines to fight sexually transmitted diseases (STDs) (Medaglini et al. 1997). In addition, rectal boosting, that is, administration of live attenuated strains, may also cause the cervicovaginal environment to become immunologically stimulated (Kutteh et al. 2001). Furthermore, faecal microbiota transfer poses an alternative to bacterial vectors and a better option since, with only one or a few more faecal microbiota infusions, an enormous number of bacteria along with their metabolites can be transferred, all of which are capable for volunteering to induce the local and systemic immune responses (Quaranta et al. 2019). An intervention of FMT from healthy rats into PCOS-induced rat models resulted in restored gut microbial profile, characterised by an increased *Clostridium* and *Lactobacillus* and a decreased *Prevotella* population indicating that the gut microbiota dysbiosis plays an important role in the pathogenesis of PCOS (Guo et al. 2016).

Vaginal Microbiota Transfer (VMT) It involves the transfer of healthy vaginal microbiota isolated from a healthy donor into the patient's (recipient's) vagina to retract dysbiosis and promote overall microbial diversity and stability. Since VMT has substantial uncertainty, whether it will always reinstate the beneficial microbiota or not still needs to be researched (DeLong et al. 2019). Table 16.1 lists different therapeutic interventions done using microbiota-targeted approaches to treat various female reproductive health-related conditions.

16.5 Conclusion

Human microbiome causes a significant impact on human health, regulating the metabolism at all levels and in all parts of the body. The past decade has provided evidence that the female reproductive system also harbours a specific microbial community and that a balance (eubiosis) among different populations is critical to reproductive fitness. Undesirable changes in the microbial community design can lead to dysbiosis and cause reproductive failure. Since these microorganisms form an essential part of the metabolism, immunity and many physiological interactions, manipulating these tiny residents can provide a great platform to manage disease pathologies and alleviate symptoms. Reinstating the favourable microbial population and establishing a state of eubiosis can lay the immense potential to not just serve as an adjunct therapy but also to provide a permanent solution to combat various female reproductive health conditions like infertility, PCOS and BV, which

Table 16.1 Therapeutic interventions using microbiota-targeted approaches to treat female reproductive health-related conditions

Study design	Location	Probiotic strains	Disease/ condition	Number of participants	Duration of intervention	Inference	Reference
Randomised double-blind Placebo-controlled trial	Auckland and Wellington, New Zealand	*Lactobacillus rhamnosus* HN001	Postpartum Symptoms of depression and anxiety	423 women (recruited at 14–16 weeks of gestation) Probiotic: 212 Placebo:211	From enrolment until 6 months postpartum if breastfeeding	Significant reduction in depression and anxiety scores	Slykerman et al. (2017)
Prospective double-blind randomised controlled trial	Brisbane, Australia	*Lactobacillus rhamnosus* and *Bifidobacterium animalis* subspecies *lactis*	Gestational diabetes mellitus (GDM)	411 overweight and obese women	From second trimester until 28 weeks gestation	Rates of GDM were not lowered in the probiotic group	Callaway et al. (2019)
Randomised, double-blind, placebo-controlled trial	Bangkok, Thailand	*Bifidobacterium* and *lactobacillus*	Insulin resistance in pregnant women (diet-controlled (GDM)	Women at 24–28 weeks of gestation Probiotic: 28 Placebo: 29	Four consecutive weeks	Glucose metabolism improved significantly	Kijmanawat et al. (2019)
Randomised, double-blinded, placebo-controlled clinical trial		*Lactobacillus acidophilus, lactobacillus reuteri, lactobacillus fermentum* and *Bifidobacterium bifidum* plus 200 µg/day selenium	Polycystic ovary syndrome	60 subjects (18–40 years old)	12 weeks	An improvement in mental health parameters, total serum testosterone and hirsutism was observed post-co-administration of the probiotic with selenium. MDA, GSH and TAC levels also improved	Jamilian et al. (2018)
Double-blind placebo-controlled	Iran	*Lactobacillus acidophilus* LA-5, *Bifidobacterium* BB-12, *streptococcus*	Inflammation and oxidative stress	64 pregnant women with GDM	Eight consecutive weeks	Statistically, significant improvement was observed in	Hajifaraji et al. (2018)

randomised clinical trial		*Thermophilus* STY-31 and *lactobacillus delbrueckii bulgaricus* LBY-27	biomarkers in GDM			C-reactive protein, tumour necrosis factor-α, erythrocyte glutathione peroxidase and malondialdehyde, glutathione reductase levels	
Two-Centre double-blind, randomised, placebo-controlled parallel	Auckland and Wellington, New Zealand	Lactobacillus rhamnosus HN001 (HN001)	GDM	Pregnant women (with a personal or partner history of atopic disease) randomised at 14–16 weeks' gestation. Participants (*n* 423) randomised to the HN001 (*n* 212) placebo group: (*n* 211)	From enrolment throughout pregnancy and until 6 months post birth if still breastfeeding	Probiotic supplementation might reduce the prevalence of GDM, especially in older women and women with a history of GDM	Wickens et al. (2017)
Randomised, blinded, placebo-controlled clinical trial	Rwanda	Lactobacillus rhamnosus GR-1 and lactobacillus reuteri RC-14	Microbiota and metabolomic analysis of vaginal samples	Healthy pregnant women (aged 18–55 years) with a gestational age <36 weeks	1 month	According to microbiota and metabolomics data, low diversity and lactobacillus dominance were seen to be associated with a healthy vagina, and these women were less likely to give preterm births	McMillan et al. (2018)
A randomised (1: 1) double-blind, placebo-controlled trial	Poznan University of Medical Sciences, Poland	2 strains of *Bifidobacterium lactis* (W51 and W52), *lactobacillus acidophilus* (W22),	Polycystic ovary syndrome	65 women with PCOS and BMI > 25	Evaluations first performed at baseline and then repeated after	8% decrease in the BMI of the symbiotic-supplemented group was noted	Chudzicka-Strugała et al. (2021)

(continued)

Table 16.1 (continued)

Study design	Location	Probiotic strains	Disease/ condition	Number of participants	Duration of intervention	Inference	Reference
		lactobacillus paracasei (W20), *lactobacillus plantarum* (W21), *lactobacillus salivarius* (W24) and *lactobacillus lactis* (W19) and the prebiotics fructooligosaccharides and inulin			3 months of intervention	with a reduction in the circumference of the waist, hip and thighs as well. Testosterone levels also lowered significantly post-symbiotic administration	
Randomised double-blind, placebo-controlled trial		*Lactobacillus acidophilus*, *lactobacillus casei* and *Bifidobacterium bifidum* plus 0.8 g inulin	Polycystic ovary syndrome	60 women diagnosed with PCOS Synbiotic: 30 Placebo: 30	12 weeks	A substantial increase in the plasma nitric oxide (NO) levels, serum sex hormone-binding globulin (SHBG) and a decrease in Ferriman Gallwey (mF-G) scores and serum high-sensitivity C-reactive protein (hs-CRP) was observed indicating beneficial effects on these parameters. No effect was seen on other oxidative biomarkers	Nasri et al. (2018)

Prospective, randomised, double-blind, placebo-controlled trial	Kashan, Iran	*Lactobacillus acidophilus* strain T16 (IBRC-M10785), *lactobacillus casei* strain T2 (IBRC-M10783) and *Bifidobacterium bifidum* strain T1 (IBRC-M10771) plus 800 mg inulin	Glycaemic control, lipid profile and atherogenic index (AIP) of women suffering from PCOS	60 patients with PCOS (aged 18–40 years) Synbiotic: 30 Placebo: 30	12 weeks	A substantial decrease in the following parameters was observed: Triglycerides, AIP and VLDL cholesterol levels. Lipid profiles were not much affected by the treatment	Samimi et al. (2019)
Randomised, controlled, triple-blinded, parallel trial study	Iran	SPJ: Inulin and lactobacillus per week, for 8 weeks PJ: Pomegranate juice per week, for 8 weeks SB: Synbiotic beverage per week, for 8 weeks, (each litre of the beverage contains 1 L of water +20 g of inulin $+2 \times 10^8$ CFU/g lactobacillus + pomegranate flavouring)	Sex hormone, glycaemic profiles and anthropometric indices in women with PCOS	92 PCOS patients (aged 15–48 years) 3 treatment groups (23 subjects each) Synbiotic pomegranate juice (SPJ) Pomegranate juice (PJ) Synbiotic beverage (SB) 1 control group (water + pomegranate flavouring)	2 L of beverage weekly	The new beverage SPJ can improve insulin, insulin resistance, BMI, weight and testosterone levels	Esmaeilinezhad et al. (2019)
Randomised double-blinded placebo-controlled clinical trial	Tabriz, Iran	Each synbiotic capsule contained seven strains (*lactobacillus casei*, *lactobacillus rhamnosus lactobacillus bulgaricus*, *lactobacillus acidophilus*, *Bifidobacterium longum* and *Streptococcus*	Metabolic factors and obesity values in women PCOS	68 overweight or obese women with PCOS (aged 20–44 years); 34 subjects in synbiotic and placebo groups each	8 weeks	Serum fasting glucose, insulin, BMI, waist and hip circumference decreased significantly. An increase in HDL cholesterol was also observed	Darvishi et al. (2021)

(continued)

Table 16.1 (continued)

Study design	Location	Probiotic strains	Disease/ condition	Number of participants	Duration of intervention	Inference	Reference
		thermophilus) and inulin-type prebiotics (Fructooligosaccharides (FOS))					
A pilot randomised controlled trial	Sydney, Australia	*Lactobacillus rhamnosus* GR-1 (GR-1) *and lactobacillus fermentum/reuteri* RC-14 (RC-14)	Group B streptococcal colonisation rates in vagina of pregnant women	34 women (36 weeks pregnant) positive for GBS Control group: 13 Intervention group: 21	Oral dose (daily for 3 weeks or until they gave birth)	More number of vaginal commensals found in probiotic group. Still, expected results not achieved Probable reasons: Problem might lie in the length of intervention, concentration of dosage, selection of probiotic strain, inadequate sample size or route of administration	Olsen et al. (2018)
Prospective pilot clinical assay	Spain	*L. salivarius* CECT 9145	Rectal and vaginal eradication of *Streptococcus agalactiae* (GBS)	57 pregnant women with age between 25 and 36 years (39 vaginal and rectal GBS positive and 18 rectal and vaginal GBS negative)	Week 26 to week 38 of pregnancy	Probiotic intervention caused 68% and 72% of the women to be GBS negative as analysed from in the vaginal and rectal samples, respectively	Martín et al. (2019)

Double-blind, placebo-controlled study	Finland	*Lactobacillus rhamnosus* GG and *Bifidobacterium lactis* Bb12	Pregnancy outcome; prenatal and postnatal growth	256 women at first trimester of pregnancy	First trimester of pregnancy to the end of exclusive breastfeeding	Frequency of GDM reduced post-probiotic treatment; no adverse events occurred in mother or child	Luoto et al. (2010)
Phase I placebo-controlled, randomised, double-blind study, dose-ranging safety trial	San Francisco (UCSF), USA	Lactobacillus crispatus CTV-05 (LACTIN-V) (vaginal applicator)	Safety trial for bacterial vaginosis	Twelve healthy volunteers	Product used for 5 consecutive days; follow-up on days 7 and 14. Phone interviews on days 2 and 35	31 mild and 4 moderate adverse events occurred out of which the moderate ones were unrelated to product use. Colposcopy findings and laboratory parameters were within normal limits. The product was well-tolerated and accepted	Hemmerling et al. (2009)
Phase 2a (randomised, double-blind, placebo-controlled trial)	San Francisco (UCSF), USA	Lactobacillus crispatus CTV-05 (LACTIN-V) administered by a vaginal applicator	Safety trial for bacterial vaginosis	24 women with BV Group receiving product: 18 Placebo: 6	Once daily for 5 days (days 1–5) followed by once weekly for 2 weeks (days 12 and 19)	Grade 3 or 4 adverse events not observed. No deep epithelial disruption seen during colposcopy. The product was well-tolerated, safe and accepted by women with BV	Hemmerling et al. (2010)

(continued)

Table 16.1 (continued)

Study design	Location	Probiotic strains	Disease/ condition	Number of participants	Duration of intervention	Inference	Reference
Prospective, randomised, single-blinded clinical trial	Kashan, Iran	Probiotic yoghurt containing *lactobacillus acidophilus* La5 and *Bifidobacterium animalis* BB12	Inflammatory factors in pregnant women	Primigravida and singleton pregnant women (18–30 years) Probiotic yoghurt group: 37 Conventional yoghurt group: 33	28–37 weeks of gestation	Probiotic yoghurt caused significant decrease in hs-CRP levels, but no effect was seen on TNF-alpha levels	Asemi et al. (2011)
Randomised double-blind, placebo-controlled trial	Arak, Iran	*Lactobacillus acidophilus*, *L. casei* and *Bifidobacterium bifidum*	Glycaemic control and lipid profiles in gestational diabetes	Sixty primigravida pregnant women (18–40 years of age) Probiotic group: 30 Placebo: 30	6 weeks	Serum insulin homoeostasis model assessment (HOMA) for insulin resistance and for β-cell function fasting plasma glucose levels significantly decreased. VLDL cholesterol and serum triglyceride levels also reduced. However, no changes were found in lipid profiles	Karamali et al. (2016)
Double-blind, placebo-controlled, randomised trial	Italy	4 strains of lactobacilli (*L. acidophilus* DSM 24735, *L. plantarum* DSM 24730, *L. paracasei* DSM 24733, *L. delbrueckii* subsp. bulgaricus DSM 24734), 3 strains of	Breast milk composition	66 women Probiotic: 33 Placebo: 33	36th week of pregnancy until 4 weeks after giving birth	The concentration of both lactobacilli and Bifidobacteria was higher in colostrum and mature milk of women taking probiotics. This	Mastromarino et al. (2015)

		Bifidobacteria (*B. longum* DSM 24736, *B. breve* DSM 24732, *B. infantis* DSM 24737) and 1 strain of *Streptococcus thermophilus* (DSM 24731)				study also indicates that modulation of breast milk composition probably occurs through a systemic effect	
Randomised placebo-controlled clinical trial	Iran	Probiotic *lactobacillus sporogenes* (1×10^7 CFU), 0.04 g inulin as prebiotic with 0.38 g isomalt, 0.36 g sorbitol and 0.05 g stevia as sweetener per 1 g	Glycaemic status and serum hs-CRP in pregnant women	52 pregnant women (primigravida), aged 18–35 years Control: 26 Synbiotic: 26	9 weeks	Substantial reduction in serum insulin levels with a significant rise in QUICKI score. No effect on FPG and serum hs-CRP levels	Taghizadeh and Asemi (2014)
Randomised, double-blind, placebo-controlled study	Malaysia	*L. Plantarum* LP115, *L. helveticus* LA25, *L. rhamnosus* LRH10, *L. paracasei* LPC12, *L. fermentum* LF26 and *L. delbrueckii* subsp. *lactis* LDL114 (mixture ratio of LP115: LA25: LRH10: LPC12: LF26: LDL114 = 3: 3: 3: 6: 1: 3)	Vaginal candidiasis in pregnant women	78 pregnant women with VC (lactobacilli, $n = 39$; placebo, $n = 39$)	8 weeks	Probiotic supplementation proved to be beneficial in alleviating the vulvovaginal symptoms (irritation, discharge and burning) and recurrences of VC. Social and emotional distress related to VC also reduced	Ang et al. (2022)
Prospective, double-blind randomised clinical trial	Taiwan	*Lactobacillus rhamnosus* GR-1 and *lactobacillus reuteri* RC-14	Group B streptococcus (GBS) colonisation in	10 group B streptococcus (GBS)-positive subjects (pregnant, at	From the time of recruitment until delivery	Probiotic administration could reduce the rate of vaginal and	Ho et al. (2016)

(continued)

Table 16.1 (continued)

Study design	Location	Probiotic strains	Disease/ condition	Number of participants	Duration of intervention	Inference	Reference
			pregnant women	35–37 weeks of gestation)		rectal GBS colonisation in pregnant women	
Randomised, double-blind, placebo-controlled trial	Iran	*Lactobacillus acidophilus*, *lactobacillus casei* and *Bifidobacterium bifidum* (2×10^9 colony-forming units/g each) plus 800 mg inulin	Insulin metabolisms and lipid profile in patients with GDM	70 patients with GDM (ages 18–40 years) Synbiotic group: 35 Placebo group: 35	6 weeks	Significant reduction serum TAG and VLDL cholesterol concentrations and serum insulin levels homoeostatic model assessment for insulin resistance and homoeostatic model assessment for β-cell function	Ahmadi et al. (2016)
Randomised single-blinded controlled clinical trial	Kashan, Iran	Commercially available yoghurt prepared with the starter cultures of *Streptococcus thermophilus* and *lactobacillus bulgaricus*, enriched with the probiotic culture of two strains of lactobacilli (*lactobacillus acidophilus* LA5) and bifidobacteria (*Bifidobacterium animalis* BB12) with a	Insulin resistance in pregnant women	Pregnant women, primigravida, aged 18–30 years old (singleton pregnancy at their third trimester) Probiotic yoghurt (n ¼ 37) Conventional yoghurt (n ¼ 33)	9 weeks	Daily consumption of probiotic yoghurt-maintained serum insulin levels. Therefore, it might help pregnant women prevent the development of insulin resistance	Asemi et al. (2013)

		total of min 1×10^7 colony-forming units					
Observational, prospective study	Belgrade	*Lactobacillus rhamnosus* BMX 54 (vaginal application)	Abnormal vaginal flora in pregnant women	60 pregnant women 30 in each group (treated and untreated)	12 weeks	Application of probiotic capsule prevented the development of an abnormal vaginal flora. It also lowered the chances of preterm delivery	Stojanović et al. (2012)
Pilot study (open-label, two-group quasi-experiment)	Midwest region, United States	Florajen3 ($>7.5 \times 10^9$ *L. acidophilus*, $> 6.0 \times 10^9$ *B. lactis* and $>1.5 \times 10^9$ *B. longum*)	GBS colonisation	10 pregnant women (at 28 ± 2-week gestation) received probiotic; 10 served as control	From the start of intervention till 36 ± 2 weeks gestation	No AEs or even minor side effects seen in the intervention group. One half of these participants presented improved gastrointestinal symptoms. Lower quantitative GBS colony counts found in the probiotic group	Hanson et al. (2014)
An open-label, double-blind, placebo-controlled, parallel-group randomised clinical trial	Tehran, Iran	Probiotic yoghurt contained *lactobacillus bulgaricus*, *Streptococcus thermophilus*, *probiotic lactobacillus* and *Bifidobacterium lactis*. And clindamycin	BV in pregnant women	310 pregnant women (third trimester) with BV	Twice a day for 1 week	Some women administered with probiotic yoghurt should pH decrease and those receiving both probiotic yoghurt and clindamycin, reported complete symptomatic cure	Hantoushzadeh et al. (2012)

(continued)

Table 16.1 (continued)

Study design	Location	Probiotic strains	Disease/ condition	Number of participants	Duration of intervention	Inference	Reference
Prospective, multicentre, double-blind, randomised phase III trial	Paris	*Lactobacillus crispatus* IP174178* (Lc)	Bacterial vaginosis	Probiotic group: 39 Placebo group: 39	Vaginal capsule of Lc or placebo: Once daily for 14 days over first two menstrual cycles and again the same treatment for 14 days for the following two menstrual cycles	Administration of Lc probiotic could lower the rate of BV recurrences and increase the time of recurrence	Bohbot et al. (2018)
Multicentre, randomised, double-blind, placebo-controlled, parallel–group study	Poland	*L. gasseri* 57C, *L. fermentum* 57A and *L. plantarum* 57B in a total number of $\geq 10^8$ c.f.u. with standard metronidazole treatment	Bacterial vaginosis and aerobic vaginitis	600 women with regular menstruation and histories of recurrent BV; 18–50-year-old (probiotic group, $n = 285$; placebo group, $n = 293$)	Antibiotic taken together with probiotic or placebo twice daily for 10 days	Oral probiotic treatment increased the remission time in patients with recurrent AV/BV. Clinical and microbiological parameters were also improved	Heczko et al. (2015)
Open-label, parallel-group, randomised, cross-over and controlled trial	Israel	5×10^9 *lactobacilli* (L.) *Rhamnosus* GR-1 and *L. reuteri* RC-14	Vaginal colonisation in pregnant women at high risk for preterm birth	40 participants (pregnant; normal vaginal flora) Probiotic: 20 No treatment: 20	Two oral capsules/ day for 2 months. Treatment crossed over for additional 2 months	No significant difference in the end points of the two groups. Oral administration of probiotics did not increase the vaginal lactobacilli colonisation rates	Yefet et al. (2020)
	Tehran, Iran	Symbiotic capsule (*lactobacillus*	PCOS	Women with PCOS (aged 19–37 years)	12 weeks	Beneficial effects on LDL and HDL	Karimi et al. (2020)

Double-blind placebo-controlled trial		*acidophilus* 3×10^{10} CFU/g, *lactobacillus casei* 3×10^{9} CFU/g, *lactobacillus bulgaricus* 5×10^{8} CFU/g, *lactobacillus rhamnosus* 7×10^{9} CFU/g, *Bifidobacterium longum* 1×10^{9} CFU/g, *Bifidobacterium breve* 2×10^{10} CFU/g, *Streptococcus thermophilus*) 3×10^{8} CFU/g) and prebiotic inulin (fructooligosaccharide)		Synbiotic supplement ($n = 50$) Placebo ($n = 49$)		cholesterol were observed. No differences in anthropometric indices between groups were found	
Phase 2 randomised (1: 1 allocation ratio) interventional parallel-group prospective placebo-controlled multicentre clinical study	Ukraine	1 capsule of the verum (5.4 billion lactobacillus crispatus LMG S-29995, lactobacillus brevis and lactobacillus acidophilus in proportion of 60%, 20% and 20%, respectively)	Symptomatic bacterial vaginosis	Women with recent symptomatic BV cured with metronidazole. 18 to 45 years old Supplement group: 82 Placebo: 82	1 capsule of the verum or the placebo two times daily for the first 7 days and one time daily for the next 8 to 120 days	Recurrences of BV significantly reduced in women treated with probiotic supplement	Reznichenko et al. (2020)
Randomised, double-blind, placebo-controlled clinical trial	Tehran, Iran	*Lactobacillus acidophilus* strain T16 (IBRC-M10785), *L. casei* strain T2 (IBRC-M10783) and *Bifidobacterium bifidum* strain T1 (IBRC-	Pregnancy outcomes in gestational diabetes	60 pregnant women (with GDM; aged 18–40 years) Synbiotic group: 30 Placebo group: 28	One capsule daily for 6 weeks	No effect on plasma nitric oxide levels observed post-symbiotic supplementation. However, symbiotic	Karamali et al. (2018)

(continued)

Table 16.1 (continued)

Study design	Location	Probiotic strains	Disease/ condition	Number of participants	Duration of intervention	Inference	Reference
		M10771) (2×10^9 CFU/ g each) plus 800 mg inulin				treatment showed beneficial effects on plasma TAC, GSH and MDA and serum hs-CRP levels	
Randomised, placebo-controlled, triple-blind, parallel-group trial	Germany	Oral supplementation with *lactobacillus rhamnosus* GR-1 and *L reuteri* RC-14 (10^9 colony-forming units)	Vaginal health in pregnancy	320 subjects enrolled with <12 completed weeks of pregnancy	8 weeks	No effect on vaginal health observed. Other routes of administration might be more effective in regulating the vaginal health	Gille et al. (2016)

are commonly seen to be associated with disturbed microbial taxa abundances. Hence, the microbiome indeed leads to an altogether different yet intricate dynamic horizon, expanding the dimensions of reproductive health and research.

References

Ahmadi S, Jamilian M, Tajabadi-Ebrahimi M, Jafari P, Asemi Z (2016) The effects of synbiotic supplementation on markers of insulin metabolism and lipid profiles in gestational diabetes: a randomised, double-blind, placebo-controlled trial. Br J Nutr 116(8):1394–1401

Ahmed K, Tunaru S, Offermanns S (2009) GPR109A, GPR109B and GPR81, a family of hydroxy-carboxylic acid receptors. Trends Pharmacol Sci 30(11):557–562

Akbari V, Hendijani F (2016) Effects of probiotic supplementation in patients with type 2 diabetes: systematic review and meta-analysis. Nutr Rev 74(12):774–784

Altun HK, Yıldız EA (2017) Prebiyotikler ve Probiyotiklerin Diyabet ile Iliskisi. Turk J Life Sci 11: 63–70

Amabebe E, Anumba DO (2020) Female gut and genital tract microbiota-induced crosstalk and differential effects of short-chain fatty acids on immune sequelae. Front Immunol 11:2184

Ang XY, Chung FYL, Lee BK, Azhar SNA, Sany S, Roslan NS et al (2022) Lactobacilli reduce recurrences of vaginal candidiasis in pregnant women: a randomized, double-blind, placebo-controlled study. J Appl Microbiol 132(4):3168–3180

Antonio MA, Hawes SE, Hillier SL (1999) The identification of vaginal lactobacillus species and the demographic and microbiologic characteristics of women colonized by these species. J Infect Dis 180(6):1950–1956

Apridonidze T, Essah PA, Iuorno MJ, Nestler JE (2005) Prevalence and characteristics of the metabolic syndrome in women with polycystic ovary syndrome. J Clin Endocrinol Metabol 90(4):1929–1935

Asemi Z, Jazayeri S, Najafi M, Samimi M, Mofid V, Shidfar F et al (2011) Effects of daily consumption of probiotic yoghurt on inflammatory factors in pregnant women: a randomized controlled trial. Pak J Biol Sci 14(8):476

Asemi Z, Samimi M, Tabassi Z, Naghibi Rad M, Rahimi Foroushani A, Khorammian H, Esmaillzadeh A (2013) Effect of daily consumption of probiotic yoghurt on insulin resistance in pregnant women: a randomized controlled trial. Eur J Clin Nutr 67(1):71–74

Azziz R, Carmina E, Chen Z, Dunaif A, Laven JS, Legro RS et al (2016) Polycystic ovary syndrome. Nat Rev Dis Primers 2(1):1–18

Baqui AH, Lee AC, Koffi AK, Khanam R, Mitra DK, Dasgupta SK et al (2019) Prevalence of and risk factors for abnormal vaginal flora and its association with adverse pregnancy outcomes in a rural district in north-East Bangladesh. Acta Obstet Gynecol Scand 98(3):309–319

Barber TM, Dimitriadis GK, Andreou A, Franks S (2016) Polycystic ovary syndrome: insight into pathogenesis and a common association with insulin resistance. Clin Med 16(3):262

Barrientos-Durán A, Fuentes-López A, de Salazar A, Plaza-Díaz J, García F (2020) Reviewing the composition of vaginal microbiota: inclusion of nutrition and probiotic factors in the maintenance of eubiosis. Nutrients 12(2):419

Batra M, Bhatnager R, Kumar A, Suneja P, Dang AS (2022) Interplay between PCOS and microbiome: the road less travelled. Am J Reprod Immunol 88:e13580

Beaud D, Tailliez P, Anba-Mondoloni J (2005) Genetic characterization of the β-glucuronidase enzyme from a human intestinal bacterium, Ruminococcus gnavus. Microbiology 151(7): 2323–2330

Bhatnager R, Nanda S, Dang AS (2018) Plasma prolidase levels as a biomarker for polycystic ovary syndrome. Biomark Med 12(6):597–606

Blaser MJ, Devkota S, McCoy KD, Relman DA, Yassour M, Young VB (2021) Lessons learned from the prenatal microbiome controversy. Microbiome 9(1):1–7

Bohbot JM, Daraï E, Bretelle F, Brami G, Daniel C, Cardot JM (2018) Efficacy and safety of vaginally administered lyophilized lactobacillus crispatus IP 174178 in the prevention of bacterial vaginosis recurrence. J Gynecol Obstetr Hum Reprod 47(2):81–86

Borgdorff H, Tsivtsivadze E, Verhelst R, Marzorati M, Jurriaans S, Ndayisaba GF et al (2014) Lactobacillus-dominated cervicovaginal microbiota associated with reduced HIV/STI prevalence and genital HIV viral load in African women. ISME J 8(9):1781–1793

Bozdag G, Mumusoglu S, Zengin D, Karabulut E, Yildiz BO (2016) The prevalence and phenotypic features of polycystic ovary syndrome: a systematic review and meta-analysis. Hum Reprod 31(12):2841–2855

Bracewell-Milnes T, Saso S, Nikolaou D, Norman-Taylor J, Johnson M, Thum MY (2018) Investigating the effect of an abnormal cervico-vaginal and endometrial microbiome on assisted reproductive technologies: a systematic review. Am J Reprod Immunol 80(5):e13037

Buggio L, Somigliana E, Borghi A, Vercellini P (2019) Probiotics and vaginal microecology: fact or fancy? BMC Womens Health 19(1):1–6

Callaway LK, McIntyre HD, Barrett HL, Foxcroft K, Tremellen A, Lingwood BE et al (2019) Probiotics for the prevention of gestational diabetes mellitus in overweight and obese women: findings from the SPRING double-blind randomized controlled trial. Diabetes Care 42(3): 364–371

Cao B, Stout J, Lee I, Mysorekar U (2014) Placental microbiome and its role in preterm birth. NeoReviews 15(12):e537–e545

Cardone A, Zarcone R, Borrelli A, Di Cunzolo A, Russo A, Tartaglia E (2003) Utilisation of hydrogen peroxide in the treatment of recurrent bacterial vaginosis. Minerva Ginecol 55(6): 483–492

Ceccarani C, Foschi C, Parolin C, D'Antuono A, Gaspari V, Consolandi C et al (2019) Diversity of vaginal microbiome and metabolome during genital infections. Sci Rep 9:14095

Chudzicka-Strugała I, Kubiak A, Banaszewska B, Zwozdziak B, Siakowska M, Pawelczyk L, Duleba AJ (2021) Effects of synbiotic supplementation and lifestyle modifications on women with polycystic ovary syndrome. J Clin Endocrinol Metabol 106(9):2566–2573

Cicinelli E, De Ziegler D, Nicoletti R, Colafiglio G, Saliani N, Resta L et al (2008) Chronic endometritis: correlation among hysteroscopic, histologic, and bacteriologic findings in a prospective trial with 2190 consecutive office hysteroscopies. Fertil Steril 89(3):677–684

Cicinelli E, De Ziegler D, Nicoletti R, Tinelli R, Saliani N, Resta L et al (2009) Poor reliability of vaginal and endocervical cultures for evaluating microbiology of endometrial cavity in women with chronic endometritis. Gynecol Obstet Investig 68(2):108–115

Claesson MJ, Jeffery IB, Conde S, Power SE, O'connor EM, Cusack S et al (2012) Gut microbiota composition correlates with diet and health in the elderly. Nature 488(7410):178–184

Cohen CR, Wierzbicki MR, French AL, Morris S, Newmann S, Reno H et al (2020) Randomized trial of Lactin-V to prevent recurrence of bacterial vaginosis. N Engl J Med 382(20):1906–1915

Collado MC, Rautava S, Aakko J, Isolauri E, Salminen S (2016) Human gut colonisation may be initiated in utero by distinct microbial communities in the placenta and amniotic fluid. Sci Rep 6(1):1–13

Conti C, Malacrino C, Mastromarino P (2009) Inhibition of herpes simplex virus type 2 by vaginal lactobacilli. J Physiol Pharmacol 60(Suppl 6):19–26

Cotter PD, Hill C, Ross RP (2005) Bacteriocins: developing innate immunity for food. Nat Rev Microbiol 3(10):777–788

Darvishi S, Rafraf M, Asghari-Jafarabadi M, Farzadi L (2021) Synbiotic supplementation improves metabolic factors and obesity values in women with polycystic ovary syndrome independent of affecting apelin levels: a randomized double-blind placebo-controlled clinical trial. Int J Fertil Steril 15(1):51

De Gregorio PR, Parolin C, Abruzzo A, Luppi B, Protti M, Mercolini L et al (2020) Biosurfactant from vaginal lactobacillus crispatus BC1 as a promising agent to interfere with Candida adhesion. Microb Cell Factories 19(1):1–16

Decena DCD, Co JT, Manalastas RM Jr, Palaypayon EP, Padolina CS, Sison JM et al (2006) Metronidazole with Lactacyd vaginal gel in bacterial vaginosis. J Obstet Gynaecol Res 32(2): 243–251

DeLong K, Zulfiqar F, Hoffmann DE, Tarzian AJ, Ensign LM (2019) Vaginal microbiota transplantation: the next frontier. J Law Med Ethics 47(4):555–567

Deswal R, Nanda S, Ghalaut VS, Roy PS, Dang AS (2019) Cross-sectional study of the prevalence of polycystic ovary syndrome in rural and urban populations. Int J Gynecol Obstet 146(3): 370–379

Deswal R, Narwal V, Dang A, Pundir CS (2020) The prevalence of polycystic ovary syndrome: a brief systematic review. J Hum Reproduct Sci 13(4):261

Dominianni C, Sinha R, Goedert JJ, Pei Z, Yang L, Hayes RB, Ahn J (2015) Sex, body mass index, and dietary fiber intake influence the human gut microbiome. PLoS One 10(4):e0124599

Esmaeilinezhad Z, Babajafari S, Sohrabi Z, Eskandari MH, Amooee S, Barati-Boldaji R (2019) Effect of synbiotic pomegranate juice on glycemic, sex hormone profile and anthropometric indices in PCOS: a randomized, triple blind, controlled trial. Nutr Metab Cardiovasc Dis 29(2): 201–208

Fang RL, Chen LX, Shu WS, Yao SZ, Wang SW, Chen YQ (2016) Barcoded sequencing reveals diverse intrauterine microbiomes in patients suffering with endometrial polyps. Am J Transl Res 8(3):1581

Food and Agricultural Organization of the United Nations and World Health Organization (2001). Health and nutritional properties of probiotics in food including powder milk with live lactic acid bacteria. World Health Organization

Fernandes R, do Rosario VA, Mocellin MC, Kuntz MG, Trindade EB (2017) Effects of inulin-type fructans, galacto-oligosaccharides and related synbiotics on inflammatory markers in adult patients with overweight or obesity: a systematic review. Clin Nutr 36(5):1197–1206

Flak MB, Neves JF, Blumberg RS (2013) Welcome to the microgenderome. Science 339(6123): 1044–1045

Franks S, McCarthy MI, Hardy K (2006) Development of polycystic ovary syndrome: involvement of genetic and environmental factors. Int J Androl 29(1):278–285

Gajer P, Brotman RM, Bai G, Sakamoto J, Schütte UM, Zhong X et al (2012) Temporal dynamics of the human vaginal microbiota. Sci Transl Med 4:132ra52

Gholizadeh Shamasbi S, Dehgan P, Mohammad-Alizadeh Charandabi S, Aliasgarzadeh A, Mirghafourvand M (2019) The effect of resistant dextrin as a prebiotic on metabolic parameters and androgen level in women with polycystic ovarian syndrome: a randomized, triple-blind, controlled, clinical trial. Eur J Nutr 58(2):629–640

Giampaolino P, Foreste V, Di Filippo C, Gallo A, Mercorio A, Serafino P et al (2021) Microbiome and PCOS: state-of-art and future aspects. Int J Mol Sci 22(4):2048

Gille C, Böer B, Marschal M, Urschitz MS, Heinecke V, Hund V, Poets CF (2016) Effect of probiotics on vaginal health in pregnancy. EFFPRO, a randomized controlled trial. Am J Obstet Gynecol. 215(5):608.e1–608.e7

Gong Z, Luna Y, Yu P, Fan H (2014) Lactobacilli inactivate chlamydia trachomatis through lactic acid but not H_2O_2. PLoS One 9(9):e107758

Graver MA, Wade JJ (2011) The role of acidification in the inhibition of Neisseria gonorrhoeae by vaginal lactobacilli during anaerobic growth. Ann Clin Microbiol Antimicrob 10(1):1–5

Guo Y, Qi Y, Yang X, Zhao L, Wen S, Liu Y, Tang L (2016) Association between polycystic ovary syndrome and gut microbiota. PLoS One 11(4):e0153196

Hajifaraji M, Jahanjou F, Abbasalizadeh F, Aghamohammadzadeh N, Abbasi MM, Dolatkhah N (2018) Effect of probiotic supplements in women with gestational diabetes mellitus on inflammation and oxidative stress biomarkers: a randomized clinical trial. Asia Pac J Clin Nutr 27(3): 581–591

Hanson L, Vusse LV, Duster M, Warrack S, Safdar N (2014) Feasibility of oral prenatal probiotics against maternal group B streptococcus vaginal and rectal colonization. J Obstet Gynecol Neonatal Nurs 43(3):294–304

Hantoushzadeh S, Golshahi F, Javadian P, Khazardoost S, Aram S, Hashemi S et al (2012) Comparative efficacy of probiotic yoghurt and clindamycin in treatment of bacterial vaginosis in pregnant women: a randomized clinical trial. J Matern Fetal Neonatal Med 25(7):1021–1024

Hashem NM, Gonzalez-Bulnes A (2022) The use of probiotics for management and improvement of reproductive Eubiosis and function. Nutrients 14(4):902

He FF, Li YM (2020) Role of gut microbiota in the development of insulin resistance and the mechanism underlying polycystic ovary syndrome: a review. J Ovarian Res 13(1):1–13

He S, Li H, Yu Z, Zhang F, Liang S, Liu H et al (2021) The gut microbiome and sex hormone-related diseases. Front Microbiol 2699:711137

Hearps A, Gugasyan R, Srbinovski D, Tyssen D, Aldunate M, Anderson DJ et al (2014) Lactic acid, a vaginal microbiota metabolite, elicits an anti-inflammatory response from vaginal and cervical epithelial cells. AIDS Res Hum Retrovir 30(S1):A238–A239

Hearps AC, Tyssen D, Srbinovski D, Bayigga L, Diaz DJD, Aldunate M et al (2017) Vaginal lactic acid elicits an anti-inflammatory response from human cervicovaginal epithelial cells and inhibits production of pro-inflammatory mediators associated with HIV acquisition. Mucosal Immunol 10(6):1480–1490

Heczko PB, Tomusiak A, Adamski P, Jakimiuk AJ, Stefański G, Mikołajczyk-Cichońska A et al (2015) Supplementation of standard antibiotic therapy with oral probiotics for bacterial vaginosis and aerobic vaginitis: a randomised, double-blind, placebo-controlled trial. BMC Womens Health 15(1):1–12

Hemmerling A, Harrison W, Schroeder A, Park J, Korn A, Shiboski S, Cohen CR (2009) Phase 1 dose-ranging safety trial of lactobacillus crispatus CTV-05 (LACTIN-V) for the prevention of bacterial vaginosis. Sex Transm Dis 36(9):564

Hemmerling A, Harrison W, Schroeder A, Park J, Korn A, Shiboski S et al (2010) Phase 2a study assessing colonization efficiency, safety, and acceptability of Lactobacillus crispatus CTV-05 in women with bacterial vaginosis. Sex Transm Dis 37:745–750

Ho M, Chang YY, Chang WC, Lin HC, Wang MH, Lin WC, Chiu TH (2016) Oral lactobacillus rhamnosus GR-1 and lactobacillus reuteri RC-14 to reduce group B streptococcus colonization in pregnant women: a randomized controlled trial. Taiwan J Obstet Gynecol 55(4):515–518

Jamilian M, Mansury S, Bahmani F, Heidar Z, Amirani E, Asemi Z (2018) The effects of probiotic and selenium co-supplementation on parameters of mental health, hormonal profiles, and biomarkers of inflammation and oxidative stress in women with polycystic ovary syndrome. J Ovarian Res 11(1):1–7

Karamali MARYAM, Dadkhah F, Sadrkhanlou M, Jamilian M, Ahmadi S, Tajabadi-Ebrahimi M et al (2016) Effects of probiotic supplementation on glycaemic control and lipid profiles in gestational diabetes: a randomized, double-blind, placebo-controlled trial. Diabetes Metab 42(4):234–241

Karamali M, Nasiri N, Taghavi Shavazi N, Jamilian M, Bahmani F, Tajabadi-Ebrahimi M, Asemi Z (2018) The effects of synbiotic supplementation on pregnancy outcomes in gestational diabetes. Probiotics Antimicrob Proteins 10(3):496–503

Karimi E, Heshmati J, Shirzad N, Vesali S, Hosseinzadeh-Attar MJ, Moini A, Sepidarkish M (2020) The effect of synbiotics supplementation on anthropometric indicators and lipid profiles in women with polycystic ovary syndrome: a randomized controlled trial. Lipids Health Dis 19(1):1–9

Kashket ER (1987) Bioenergetics of lactic acid bacteria: cytoplasmic pH and osmotolerance. FEMS Microbiol Rev 3(3):233–244

Kataoka K (2016) The intestinal microbiota and its role in human health and disease. J Med Investig 63(1.2):27–37

Kijmanawat A, Panburana P, Reutrakul S, Tangshewinsirikul C (2019) Effects of probiotic supplements on insulin resistance in gestational diabetes mellitus: a double-blind randomized controlled trial. J Diabetes Investig 10(1):163–170

Kim DE, Kim JK, Han SK, Jang SE, Han MJ, Kim DH (2019) Lactobacillus plantarum NK3 and Bifidobacterium longum NK49 alleviate bacterial vaginosis and osteoporosis in mice by suppressing NF-κ B-linked TNF-α expression. J Med Food 22(10):1022–1031

Kitaya K, Matsubayashi H, Takaya Y, Nishiyama R, Yamaguchi K, Takeuchi T, Ishikawa T (2017) Live birth rate following oral antibiotic treatment for chronic endometritis in infertile women with repeated implantation failure. Am J Reprod Immunol 78(5):e12719

Kuchiiwa T, Nio-Kobayashi J, Takahashi-Iwanaga H, Yajima T, Iwanaga T (2011) Cellular expression of monocarboxylate transporters in the female reproductive organ of mice: implications for the genital lactate shuttle. Histochem Cell Biol 135(4):351–360

Kutteh WH, Kantele A, Moldoveanu Z, Crowley-Nowick PA, Mestecky J (2001) Induction of specific immune responses in the genital tract of women after oral or rectal immunization and rectal boosting with salmonella typhi ty 21a vaccine. J Reprod Immunol 52(1–2):61–75

Kwa M, Plottel CS, Blaser MJ, Adams S (2016) The intestinal microbiome and estrogen receptor-positive female breast cancer. J Natl Cancer Inst 108(8):1–10

Larraufie P, Martin-Gallausiaux C, Lapaque N et al (2018) SCFAs strongly stimulate PYY production in human enteroendocrine cells. Sci Rep 8(1):1–9

Larsson PG, Brandsborg E, Forsum U, Pendharkar S, Andersen KK, Nasic S et al (2011) Extended antimicrobial treatment of bacterial vaginosis combined with human lactobacilli to find the best treatment and minimize the risk of relapses. BMC Infect Dis 11(1):1–14

Li H, Zang Y, Wang C, Li H, Fan A, Han C, Xue F (2020) The interaction between microorganisms, metabolites, and immune system in the female genital tract microenvironment. Front Cell Infect Microbiol 10:609488

Liang S, Wu X, Jin F (2018) Gut-brain psychology: rethinking psychology from the microbiota–gut–brain axis. Front Integr Neurosci 12:33

Libertucci J, Young VB (2019) The role of the microbiota in infectious diseases. Nat Microbiol 4 (January):35–45

Liu R, Zhang C, Shi Y, Zhang F, Li L, Wang X et al (2017) Dysbiosis of gut microbiota associated with clinical parameters in polycystic ovary syndrome. Front Microbiol 8:324

Liu S, Liu L, Luo D, Su Y, Yu C, Guan Q (2021) Gut microbiome: a potential controller of androgen- modulated disease. Online J Complement Altern Med 6(1):1–7

Luoto R, Laitinen K, Nermes M, Isolauri E (2010) Impact of maternal probiotic-supplemented dietary counselling on pregnancy outcome and prenatal and postnatal growth: a double-blind, placebo-controlled study. Br J Nutr 103(12):1792–1799

Lynch SV, Pedersen O (2016) The human intestinal microbiome in health and disease. N Engl J Med 375(24):2369–2379

Martín V, Cárdenas N, Ocaña S, Marín M, Arroyo R, Beltrán D et al (2019) Rectal and vaginal eradication of Streptococcus agalactiae (GBS) in pregnant women by using lactobacillus salivarius CECT 9145, a target-specific probiotic strain. Nutrients 11(4):810

Mastromarino P, Capobianco D, Miccheli A, Praticò G, Campagna G, Laforgia N et al (2015) Administration of a multistrain probiotic product (VSL# 3) to women in the perinatal period differentially affects breast milk beneficial microbiota in relation to mode of delivery. Pharmacol Res 95:63–70

McMillan A, Rulisa S, Sumarah M, Macklaim JM, Renaud J, Bisanz JE et al (2015) A multi-platform metabolomics approach identifies highly specific biomarkers of bacterial diversity in the vagina of pregnant and non-pregnant women. Sci Rep 5(1):1–14

McMillan A, Rulisa S, Gloor GB, Macklaim JM, Sumarah M, Reid G (2018) Pilot assessment of probiotics for pregnant women in Rwanda. PLoS One 13(6):e0195081

Medaglini D, Rush CM, Sestini P, Pozzi G (1997) Commensal bacteria as vectors for mucosal vaccines against sexually transmitted diseases: vaginal colonization with recombinant streptococci induces local and systemic antibodies in mice. Vaccine 15(12–13):1330–1337

Mendling W (2016) Vaginal microbiota. Adv Exp Med Biol 902:83–93

Mitchell C, Balkus JE, Fredricks D, Liu C, McKernan-Mullin J, Frenkel LM et al (2013) Interaction between lactobacilli, bacterial vaginosis-associated bacteria, and HIV type 1 RNA and DNA genital shedding in US and Kenyan women. AIDS Res Hum Retrovir 29(1):13–19

Moreno I, Cicinelli E, Garcia-Grau I, Gonzalez-Monfort M, Bau D, Vilella F et al (2018) The diagnosis of chronic endometritis in infertile asymptomatic women: a comparative study of histology, microbial cultures, hysteroscopy, and molecular microbiology. Am J Obstet Gynecol 218(6):602–6e1

Nasri K, Jamilian M, Rahmani E, Bahmani F, Tajabadi-Ebrahimi M, Asemi Z (2018) The effects of synbiotic supplementation on hormonal status, biomarkers of inflammation and oxidative stress in subjects with polycystic ovary syndrome: a randomized, double-blind, placebo-controlled trial. BMC Endocr Disord 18(1):1–8

Niu C, Cheng C, Liu Y, Huang S, Fu Y, Li P (2019) Transcriptome profiling analysis of bovine vaginal epithelial cell response to an isolated lactobacillus strain. Msystems 4(5):e00268–e00219

O'Hanlon DE, Moench TR, Cone RA (2011) In vaginal fluid, bacteria associated with bacterial vaginosis can be suppressed with lactic acid but not hydrogen peroxide. BMC Infect Dis 11(1): 1–8

Olsen P, Williamson M, Traynor V, Georgiou C (2018) The impact of oral probiotics on vaginal group B streptococcal colonisation rates in pregnant women: a pilot randomised control study. Women Birth 31(1):31–37

Osadchiy V, Martin CR, Mayer EA (2019) The gut–brain axis and the microbiome: mechanisms and clinical implications. Clin Gastroenterol Hepatol 17(2):322–332

Ozkinay E, Terek MC, Yayci M, Kaiser R, Grob P, Tuncay G (2005) The effectiveness of live lactobacilli in combination with low dose oestriol (Gynoflor) to restore the vaginal flora after treatment of vaginal infections. BJOG Int J Obstet Gynaecol 112(2):234–240

Peebles K, Velloza J, Balkus JE, McClelland RS, Barnabas RV (2019) High global burden and costs of bacterial vaginosis: a systematic review and meta-analysis. Sex Transm Dis 46(5): 304–311

Pelzer ES, Willner D, Buttini M, Huygens F (2018) A role for the endometrial microbiome in dysfunctional menstrual bleeding. Antonie Van Leeuwenhoek 111(6):933–943

Peter S, Gärtner MA, Michel G, Ibrahim M, Klopfleisch R, Lübke-Becker A et al (2018) Influence of intrauterine administration of lactobacillus buchneri on reproductive performance and pro-inflammatory endometrial mRNA expression of cows with subclinical endometritis. Sci Rep 8(1):1–13

Pino A, Bartolo E, Caggia C, Cianci A, Randazzo CL (2019) Detection of vaginal lactobacilli as probiotic candidates. Sci Rep 9(1):1–10

Quaranta G, Sanguinetti M, Masucci L (2019) Fecal microbiota transplantation: a potential tool for treatment of human female reproductive tract diseases. Front Immunol 10:2653

Ravel J, Gajer P, Abdo Z, Schneider GM, Koenig SS, McCulle SL et al (2011) Vaginal microbiome of reproductive-age women. Proc Natl Acad Sci 108(Suppl 1):4680–4687

Ravel J, Moreno I, Simón C (2021) Bacterial vaginosis and its association with infertility, endometritis, and pelvic inflammatory disease. Am J Obstet Gynecol 224(3):251–257

Reid G, Charbonneau D, Erb J, Kochanowski B, Beuerman D, Poehner R, Bruce AW (2003) Oral use of lactobacillus rhamnosus GR-1 and L. fermentum RC-14 significantly alters vaginal flora: randomized, placebo-controlled trial in 64 healthy women. FEMS Immunol Med Microbiol 35(2):131–134

Reznichenko H, Henyk N, Maliuk V, Khyzhnyak T, Tynna Y, Filipiuk I et al (2020) Oral intake of lactobacilli can be helpful in symptomatic bacterial vaginosis: a randomized clinical study. J Low Genit Tract Dis 24(3):284–289

Ruan Y, Sun J, He J, Chen F, Chen R, Chen H (2015) Effect of probiotics on glycemic control: a systematic review and meta-analysis of randomized, controlled trials. PLoS One 10(7): e0132121

Sambanthamoorthy K, Feng X, Patel R, Patel S, Paranavitana C (2014) Antimicrobial and antibiofilm potential of biosurfactants isolated from lactobacilli against multi-drug-resistant pathogens. BMC Microbiol 14(1):1–9

Samimi M, Dadkhah A, Haddad Kashani H, Tajabadi-Ebrahimi M, Seyed Hosseini E, Asemi Z (2019) The effects of synbiotic supplementation on metabolic status in women with polycystic ovary syndrome: a randomized double-blind clinical trial. Probiot Antimicrob Protein 11(4): 1355–1361

Satpute SK, Kulkarni GR, Banpurkar AG, Banat IM, Mone NS, Patil RH, Cameotra SS (2016) Biosurfactant/s from lactobacilli species: properties, challenges and potential biomedical applications. J Basic Microbiol 56(11):1140–1158

Sharma D, Saharan BS (2016) Functional characterization of biomedical potential of biosurfactant produced by lactobacillus helveticus. Biotechnol Rep 11:27–35

Shen RL, Dang XY, Dong JL, Hu XZ (2012) Effects of oat β-glucan and barley β-glucan on fecal characteristics, intestinal microflora, and intestinal bacterial metabolites in rats. J Agric Food Chem 60(45):11301–11308

Slykerman RF, Hood F, Wickens K, Thompson JMD, Barthow C, Murphy R et al (2017) Effect of lactobacillus rhamnosus HN001 in pregnancy on postpartum symptoms of depression and anxiety: a randomised double-blind placebo-controlled trial. EBioMedicine 24:159–165

Smits LP, Bouter KEC, De Vos WM, Borody TJ, Nieuwdorp M (2013) Therapeutic potential of fecal microbiota transplantation. In: Gastroenterology, vol 145, Issue 5. W.B. Saunders, pp 946–953

Stapleton AE (2017) The vaginal microbiota and urinary tract infection. In: Urinary tract infections: molecular pathogenesis and clinical management, pp 79–86

Stojanović N, Plećaš D, Plešinac S (2012) Normal vaginal flora, disorders and application of probiotics in pregnancy. Arch Gynecol Obstet 286(2):325–332

Stoyancheva G, Marzotto M, Dellaglio F, Torriani S (2014) Bacteriocin production and gene sequencing analysis from vaginal lactobacillus strains. Arch Microbiol 196(9):645–653

Tachedjian G, Aldunate M, Bradshaw CS, Cone RA (2017) The role of lactic acid production by probiotic lactobacillus species in vaginal health. Res Microbiol 168(9–10):782–792

Taghizadeh M, Asemi Z (2014) Effects of synbiotic food consumption on glycemic status and serum hs-CRP in pregnant women: a randomized controlled clinical trial. Hormones 13(3): 398–406

Thackray VG (2019) Sex, microbes, and polycystic ovary syndrome. Trends Endocrinol Metab 30(1):54–65

Torres PJ, Siakowska M, Banaszewska B, Pawelczyk L, Duleba AJ, Kelley ST, Thackray VG (2018) Gut microbial diversity in women with polycystic ovary syndrome correlates with hyperandrogenism. J Clin Endocrinol Metabol 103(4):1502–1511

Tremellen K, Pearce K (2012) Dysbiosis of gut microbiota (DOGMA)–a novel theory for the development of polycystic ovarian syndrome. Med Hypotheses 79(1):104–112

Valcarce DG, Genovés S, Riesco MF, Martorell P, Herráez MP, Ramón D, Robles V (2017) Probiotic administration improves sperm quality in asthenozoospermic human donors. Benefic Microbes 8(2):193–206

Valore EV, Park CH, Igreti SL, Ganz T (2002) Antimicrobial components of vaginal fluid. Am J Obstet Gynecol 187(3):561–568

Van Hamme JD, Singh A, Ward OP (2006) Physiological aspects: part 1 in a series of papers devoted to surfactants in microbiology and biotechnology. Biotechnol Adv 24(6):604–620

Vitali B, Cruciani F, Picone G, Parolin C, Donders G, Laghi L (2015) Vaginal microbiome and metabolome highlight specific signatures of bacterial vaginosis. Eur J Clin Microbiol Infect Dis 34(12):2367–2376

Vrieze A, Holleman F, Zoetendal EG, De Vos WM, Hoekstra JBL, Nieuwdorp M (2010) The environment within: how gut microbiota may influence metabolism and body composition. Diabetologia 53(4):606–613

Wade W (2002) Unculturable bacteria—the uncharacterized organisms that cause oral infections. J R Soc Med 95(2):81–83
Wallace BD, Wang H, Lane KT, Scott JE, Orans J, Koo JS et al (2010) Alleviating cancer drug toxicity by inhibiting a bacterial enzyme. Science 330(6005):831–835
Walther-António MR, Chen J, Multinu F, Hokenstad A, Distad TJ, Cheek EH et al (2016) Potential contribution of the uterine microbiome in the development of endometrial cancer. Genome Med 8(1):1–15
Wee BA, Thomas M, Sweeney EL, Frentiu FD, Samios M, Ravel J et al (2018) A retrospective pilot study to determine whether the reproductive tract microbiota differs between women with a history of infertility and fertile women. Aust N Z J Obstet Gynaecol 58(3):341–348
Weyrich LS, Farrer AG, Eisenhofer R, Arriola LA, Young J, Selway CA et al (2019) Laboratory contamination over time during low-biomass sample analysis. Mol Ecol Resour 19(4):982–996
Wickens KL, Barthow CA, Murphy R, Abels PR, Maude RM, Stone PR et al (2017) Early pregnancy probiotic supplementation with lactobacillus rhamnosus HN001 may reduce the prevalence of gestational diabetes mellitus: a randomised controlled trial. Br J Nutr 117(6): 804–813
Wilson JD, Lee RA, Balen AH, Rutherford AJ (2007) Bacterial vaginal flora in relation to changing oestrogen levels. Int J STD AIDS 18(5):308–311
Winters AD, Romero R, Gervasi MT, Gomez-Lopez N, Tran MR, Garcia-Flores V et al (2019) Does the endometrial cavity have a molecular microbial signature? Sci Rep 9(1):1–17
Workowski KA, Bolan GA (2015) Sexually transmitted diseases treatment guidelines, 2015. MMWR Recomm Rep 64(RR-03):1
Yefet E, Colodner R, Strauss M, Letova YGZ, Nachum Z (2020) A randomized controlled open label crossover trial to study vaginal colonization of orally administered lactobacillus reuteri RC-14 and rhamnosus GR-1 in pregnant women at high risk for preterm labor. Nutrients 12(4): 1141

Role of Bacteriocins in Modulation of Microbiome in Human Diseases

17

Pushpa Rani and Santosh Kumar Tiwari

Abstract

The inherent microbiome of humans provides protection against different pathogenic micro-organisms by producing antimicrobial products and nutrient competition. Changes in microbiome of humans cause disruption of barrier and colonization resistance against pathogens and ultimately cause infectious diseases. The change in microbiome can be caused by different antibiotics or exogenous pathogens like *Salmonella typhimurium, Escherichia coli, and Clostridioides difficile*. To overcome this problem, bacteriocins can be used to modulate the human microbiome and cure different diseases. Bacteriocins are small antimicrobial peptides produced by several bacteria as defense system and are ribosomally synthesized. Several bacteriocins like nisin, lacticin, and pediocin have been applied in food safety but also known to modulate microbiome by inhibiting growth of exogenous pathogens through pore formation, disturbance in proton motive force, and inhibition of cell wall biosynthesis. In this review, role of bacteriocins in possible modulation of microbiome leading to healthy microbiota has been explored. The infectious diseases caused by change in human microbiome and use of bacteriocins in modulation of human microbiome to cure infectious diseases are briefly described.

Keywords

Bacteriocins · Microbiome · Modulation · Human diseases · Dysbiosis

P. Rani · S. K. Tiwari (✉)
Department of Genetics, Maharshi Dayanand University, Rohtak, Haryana, India
e-mail: santoshgenetics@mdurohtak.ac.in

R. Sobti et al. (eds.), *Role of Microbes in Sustainable Development*,
https://doi.org/10.1007/978-981-99-3126-2_17

17.1 Introduction

Humans inhabit trillions of microbes forming a host-specific complex microbiota. Dysbiosis in such microbiota can lead to many diseases including allergies, diabetes (type I and II), cancer, and asthma (Lloyd-Price et al. 2016). Natural selection, mutation, and changes in environmental conditions for humans lead to evolutionary adaptations of microbiota. In addition, diet and scarcity of food can be major factor for the evolution (Dominguez-Bello et al. 2019). Over the past decade, many strategies like next-generation sequencing (NGS) and bioinformatics tools have been developed to know the role of different microbiota in case of human health and diseases.

The microbiome represents a genome of community of micro-organisms that resides on or within body. The term "microbiome" was coined by Laureate Joshua Lederberg in 2001. Dysbiosis in the microbiome can cause many diseases like inflammatory bowel disease, diabetes, cancer, and autism (Lloyd-Price et al. 2016). The microbiome associated with human body is very important for immunity and health of human. Human Microbiome Project (HMP) hypothesized that microbial communities might be universally the same across all healthy humans. Although few phyla of microbes that inhabit human body are the same, their relative abundance varies by ten-fold or more. Another hypothesis stated that fundamental microbiome of humans is defined by balance between metabolic and other functions within whole body or microhabitat. Fundamental microbiome of healthy human may have common gene families and pathways. Sequence-based techniques can be used to investigate microbes: (1) targeted amplicon-based techniques (e.g., 16S rRNA gene sequencing) and (2) shotgun metagenomic or metatranscriptomic approaches (Shafquat et al. 2014). The habitats within oral cavity of humans are supra-gingival, saliva, sub-gingival, tongue, and inter-dental area. Saliva provides hydration and nutrients to bacteria. Skin is interface between humans and external environment. Skin prevents loss of moisture and entry of pathogens within body. The skin of humans also has various niches like hairs, moist underarms, and skin. Many antibiotics and chemicals have been used to cure diseases caused from dysbiosis of human microbiota, but all these can be toxic for humans and have many side effects. Pathogens can also develop resistance against antibiotics. Thus, bacteriocins from lactic acid bacteria can be better alternate to modulate human microbiota to cure diseases (Lee et al. 2021; Grice et al. 2009).

17.2 Factors Affecting Human Microbiome

1. **Smoking:** Smoking is a very important factor affecting oral microbiome of human. Toxins of cigarettes reduce beneficial micro-organisms and their colonization and biofilm formation on human cells. In smokers, many microbes, for example, *Streptococcus sanguinis*, *S. parasanguinis*, *Fusobacterium nucleatum,* and *F. naviforme,* are high in number. Genera *Capnocytophaga*, *Peptostreptococcus,* and *Leptotrichia* were also depleted in smokers as compared

with never smokers, while *Atopobium* and *Streptococcus* were increased (Lee et al. 2021).

2. **Pregnancy:** Various complex and dramatic changes in hormones take place during the period of pregnancy, which affect the microbiota. *Porphyromonas gingivalis* in periodontal pockets is associated with the formation of amniotic cavity in preterm labor. Genera *Neisseria, Treponema,* and *Porphyromonas* were abundant, while *Streptococcus* and *Veillonella* were less in number during pregnancy (Lee et al. 2021).
3. **Antibiotics:** Amoxicillin affected diversity of approx. 35 taxa like increased in *Actinobacteria* and *Proteobacteria* in saliva in 3 weeks of treatment. Amoxicillin and azithromycin amoxicillin-clavulanate or phenoxymethylpenicillin also affected saliva microbiota (Lee et al. 2021).
4. **Diet:** Diet affects microbiota; e.g., frequent intake of sugar increases the production of acid, which dissolves tooth surfaces and also increases the risk of dental caries. Intake of high sugar diet leads to reduction in pH and buffering capacity of saliva, which results in change in oral microbiota, i.e., increase in *Streptococcus mutans* and *Lactobacilli.* Acidic pH also altered gene expression in some bacteria, resulting in growth of pathogens like *P. gingivalis* (Lee et al. 2021).
5. **Host Factors:** Host factors like age, location, and sex affect microbiota of gut, skin, mouth, etc. Age is very important factor for change in colonization of bacteria on skin by the involvement of some succession mechanisms. *In uterus*, fetal skin is sterile, but during birth, many microbes are provided by mother through birth canal. Babies show immunological tolerance induced by mother through regulation of T-lymphocytes. These microbes colonize as first inoculums on the skin of babies (Grice and Segre 2011; Dominguez-Bello et al. 2010).

17.3 Bacteriocins

Bacteriocins are small antimicrobial peptides, which are ribosomally synthesized, cationic, and thermostable in nature. They show bacteriolytic or bacteriostatic activity against narrow or broad range of bacteria but protect themselves by the synthesis of specific immunity proteins (Yang et al. 2014). They have been used for human health in recent years because of their effective nature against pathogens. Nisin produced by several strains of *Lactococcus lactis* is one of the most studied bacteriocins for food safety and human health (O'sullivan et al. 2002).

17.4 Bacteriocins from Gram-Positive Bacteria

Class I: These bacteriocins are called as lantibiotics, which are modified, heat-stable, globular or linear peptides, and have molecular weight < 5 kDa and < 28 amino acids. The lantibiotics have lanthionine, methyllanthionine and dehydrated amino acids generated by post-translation modifications (Yang et al. 2014; Ghodhbane et al. 2015; Bali et al. 2016; Kumariya et al. 2019). Class I is further subdivided into six subclasses: (1) class Ia: lanthipeptides—in this group, nisin is the most studied example; (2) class Ib: head-to-tail cyclized peptides formed by linkage of N- and C-terminal by peptide bond. These bacteriocins consist of only α-helices, e.g., enterocin AS-48; (3) class Ic: sactibiotics. These peptides contain sulfur to alpha carbon, e.g., subtilosin A; (4) Class Id: It contains linear azol(in)e and possesses cysteine, threonine, and serine residues derived from heterocyclic rings of (methyl)-oxazole and thiazole, e.g., streptolysin S. (5), Class Ie contains glycosylated residues, e.g., glycocin F having two α-helices linked with disulfide bonds. They have N-acetylhexosamine linked with C-terminal cysteine. (6) Class If is lasso peptides in which first amino acid is linked with amide bond and a ring is formed at +7 to +9 positions (Alvarez-Sieiro et al. 2016).

Class II: These are non-lantibiotics, unmodified, small (30–60 A.A.), thermostable, and positively charged (Yang et al. 2014; Bali et al. 2016) bacteriocins. These bacteriocins have been further subdivided into four subclasses: (1) Class IIa: This class consists of pediocin-like bacteriocins. In these bacteriocins, two cysteine residues are linked by disulfide bonds and the motif consists of conserved sequence YGNGVXC at N-terminal region, e.g., sakacin, pediocin PA-1, and leucocin A. (2) Class IIb: This class consists of two-peptide bacteriocins, which have two-component system to form an active complex of pore formation, e.g., lactococcin and plantaricin. (3) Class IIc: leaderless bacteriocins—these bacteriocins lack leader peptide at N-terminal end, which generally helps in recognition sequence for modification, secretion, and maintaining bacteriocins inactive into the producer cells, e.g., enterocin; and (4) class IId: non-pediocin-like bacteriocins, e.g., bactofencin A (Alvarez-Sieiro et al. 2016; Kumariya et al. 2019).

Class III: These bacteriocins are large molecular weight > 30 kDa and heat-labile (Bali et al. 2016). This group is further divided into two groups: One is bacteriolytic that lyses the cell wall of target cells, e.g., enterolysin. Other is non-lytic bacteriocins that have bactericidal action without causing cell lysis, e.g., helveticin (Yang et al. 2014; Alvarez-Sieiro et al. 2016).

17.5 Bacteriocins from Gram-Negative Bacteria

Bacteriocins from Gram-negative bacteria can be divided into four main classes as follows: **Class I: Colicins**—these bacteriocins are sensitive to proteases and heat. They are high molecular weight (30–80 kDa).

Class II: Colicin-Like Bacteriocins—These bacteriocins are structurally and functionally similar to colicins. They have one R-domain at central position, which is responsible for binding of bacteriocins with receptor, one T-domain at N-terminal position, which takes place in transportation and penetration of bacteriocins into the cell, and one C-domain at C-terminal, which acts as active center for bacteriocins and cause toxicity (Zimina et al. 2020).

Class III: Tailocins—These are phage tail-like bacteriocins having molecular weight (20–100 kDa). These bacteriocins consist of 8–14 different polypeptides similar to bacteriophages tail nodules. The genomes size that encode for tailocins in bacteria is less than 40 kbp and having genes for structural proteins, assembly enzymes, lysis cassettes, regulatory genes, and chaperones for synthesis and release of bacteriocins. They are subdivided into two groups—R and F. R-type tailocins form a long tube encircled with shell having receptor-binding proteins (RBP) at one end. They are evolutionary related to tails of phages in *Myoviridae* family. F-type does not consist any tube and related to tails of phages related to *Siphoviridae* family. Most studied examples of this class are R- and F-type pyocins (Zimina et al. 2020).

Class IV: Microcins—The bacteriocins of this class are stable and low molecular weight (<10 kDa) peptides, i.e., resistant to proteases and intense range of pH and temperature. They are subdivided into two classes. Subclass I: The bacteriocins of this class have molecular weight < 5 kDa and are post-translational modified, for example, microcins B17, C-7-C-51, and J-25. They show bactericidal activity by inhibiting bacterial enzymes like DNA gyrases I and II, 68S, and RNA polymerases. They also inhibit respiratory chain. Subclass II: These bacteriocins have molecular weight (5–10 kDa) and are linear and not modified post-translationally (MccL, Mcc24, and MccV). The genes encoded for linear microcins present on chromosome and carry a siderophore at C-terminal position (microcins E492, M, M47, I47, and G47). They inhibit bacterial growth through pore formation and breakage of plasma membrane of targeted bacteria (Zimina et al. 2020).

17.6 Mode of Action of Bacteriocins

The nisin belongs to class I bacteriocin (lantibiotics) is one of the well-known bacteriocins till date, which has been thoroughly characterized for antimicrobial activity. Lantibiotics generally cause disruption of cell wall and pore formation in target cells. The cell wall of bacteria is made up of peptidoglycans, which consist of lipid II molecules, which are anionic in nature. Therefore, lipid II acts as docking molecule for cationic lantibiotics (Moll et al. 1999; Gharsallaoui et al. 2016; Ramos et al. 2020). The electrostatic interaction between lipid II molecule and lantibiotics causes adsorption of lantibiotics on cell membrane. The hydrophobic part of lantibiotics inserts into membrane and causes conformational changes and pore formation (Islam et al. 2012). Pore formation causes increase in free energy and inhibits biosynthesis of cell wall by blocking intake of glucose and D-alanine into cell wall material. They also inhibit the biosynthesis of peptidoglycan through formation of a complex with lipid II molecule at the time of trans-glycosylation. Bacteriocins cause autolysis of target bacteria by activation of enzymes, which are involved in lysis of cell. For example, nisin and Pep5 bind with teichoic and lipoteichoic acids, which cause changes in enzymes involved in hydrolyzing of cell wall and with N-acetylmuramoyl-L-alanine amidase and N-acetylglucosaminidase, leading to activation of series of enzymes for cell lysis. Bacteriocins have ability to form complex with phosphatidylethanol-amine that cause increase in membrane permeability and inhibition of ATP-dependent protein

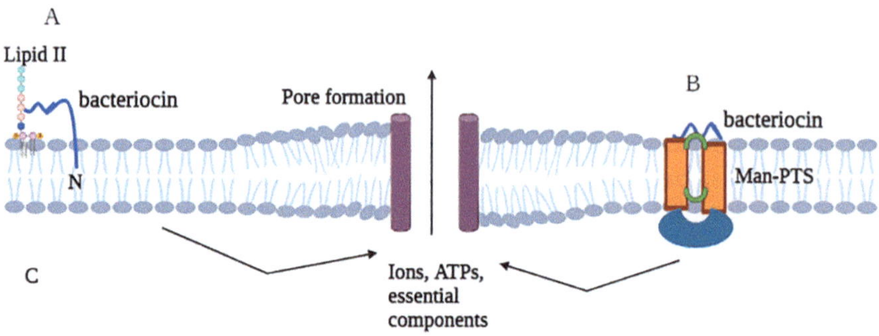

Fig. 17.1 Action mechanism of bacteriocins on cell membrane of targeted bacteria. There are two mode of action of bacteriocins. (**a**) Binding of bacteriocins to lipid II molecule, e.g., nisin. (**b**) Binding of bacteriocins to Man-PTS, e.g., pediocin PA-1. (**c**) Binding of bacteriocins with cell membrane of targeted bacteria leads to pore formation and leakage of intracellular matrix and disruption of plasma membrane of bacteria

translocation, uptake of calcium ions, and number of metabolic processes (Ramos et al. 2020).

Mainly class II bacteriocins (pediocin PA-1) show high activity against *Listeria monocytogenes* through mannose phosphotransferase system (Man-PTS). Man-PTS is a complex of proteins responsible for phosphorylation and transport of different molecules. This system has three enzyme components: enzyme I (E1); histidine-containing phosphor carrier protein (HPr); and enzyme II (EII), which consists of either three (EIIA, EIIB, and EIIC) or four (EIIA, EIIB, EIIC, and EIID) sub-components. EI, HPr, EIIA, and B reside in cytoplasm, and EIIC and D penetrate the cell membrane (Balandin et al. 2019). The Man-PTS acts as docking molecule for bacteriocin, which allows the insertion and oligomerization of bacteriocin in membrane to form channel (Colombo et al. 2018). Then after, bacteriocin causes disturbance in proton motive force (combination of membrane potential and pH gradient) resulting in cell growth inhibition and loss of intracellular ATP, nutrients, and ions (K^+ and Mg^{2+}) ultimately cell death. Enterolysin-like bacteriocins follows cell wall-degrading mechanism in which lysis of cell takes place (Braïek and Smaoui 2019). The protein EntA of enterolysin produces some protease enzymes like lysostaphin, LytM, and ALE-1, which contain metal ions like Zn^{2+} to hydrolyze the peptide bond of peptidoglycan, leading to degradation of the cell wall and cell lysis (Nilsen et al. 2003) (Fig. 17.1).

17.7 Bacteriocins in Modulation of Human Microbiome in Different Diseases

Bacteriocins play an important role in maintaining balance of micro-organisms because of interaction between bacteriocin-producing, sensitive, and resistant bacteria in the same environment. In some cases, resistant bacteria can be higher than bacteriocin-producing bacterial communities, but growth rate can also affect number of bacterial communities. The bioinformatic analysis of bacteriocins showed that approximately 317 microbes of human intestine had 175 bacteriocins in Firmicutes, 34 in Bacteroidetes, 79 in Proteobacteria, and 25 in Actinobacteria. The analysis showed that bacteriocins vary in size and amino acid compositions. These bacteriocins have less leucine, arginine, aspartic acid, and glutamic acid but more methionine and lysine. The gut microbiota interacts with signals from brain, endocrine system, immune system, and gut microbiota itself for metabolism and health. Therefore, depending upon the response of different micro-organisms to different signaling molecules the micro-organisms of host exert effect health and disease. For example, plantaricin P1053 from *Lactobacillus plantarum* strain PBS067 increases viability of healthy cells and reduces carcinogenic epithelial cells in human. Bacteriocin serpin from *Bifidobacterium longum* subsp. *longum* NCC2705 inhibits pancreatic and neutrophil elastases through modulating anti-inflammatory response. The bacteriocins from human gut microbiota affect many metabolic processes they can either improve health or can cause dysbiosis and disease. Some disease caused from dysbiosis in human microbiota have been discussed (Cesa-Luna et al. 2021). Role of few bacteriocins in modulation of gut microbiota is depicted in Table 17.1.

1. **Acne:** Acne is a chronic skin disease, which is caused by an increase in sebum production and affects pilosebaceous unit, altered keratinization, inflammation, and colonization of bacteria in hair follicles. Mainly *Propionibacterium acnes* and fungus *Malassezia* spp. are involved in acne formation. Species of *Streptococcus salivarius* and *Enterococcus faecalis* inhibit growth of *P. acnes* by producing a bacteriocin-like inhibitory substance (BLIS). These species also have immunomodulatory effects in epithelial cells and keratinocytes for acne treatment (Mottin and Suyenaga 2018). Live fermenting micro-organisms inhibit *Staphylococcus epidermidis* and *P. acnes*. A BLIS produced by *Streptococcus salivarius* inhibited *P. acnes,* which is the main factor for pathogenesis of disease acne vulgaris (Bowe et al. 2006). Bacteriocin produced from *Lactococcus* sp. HY 499 showed inhibition against *Staphylococcus epidermidis*, *Staphylococcus aureus*, *Streptococcus pyogenes,* and *P. acnes* (Oh et al. 2006). *Lactobacillus paracasei* CNCM I-2116 reduced skin inflammation by decreasing vasodilation, mast cell degranulation, edema, and release of TNF-α. *L. paracasei* CNCM I-2116 co-culture with Caco-2/PMBC was tested ex vivo on human abdominal plastic skin explants. The extract of *Lactobacillus* was found effective in reduction in skin erythema, repair of skin barrier, and reduction in skin microflora and then exhibited reduction in size of acne (Muizzuddin et al. 2012).
2. **Atopic Dermatitis:** Atopic dermatitis (AD) is multifactorial chronic inflammatory disease, which is characterized by pruritic, erythematous, and scaly lesions. Prevalence of this disease in children and adult is 1–20% over the world. It is caused by genetic defect in filaggrin that leads to rupturing of epidermis, which results in contact of dermis directly with pathogens or environmental antigens. This also causes itchiness, which leads to rupturing of skin's epidermis barrier (Mottin and Suyenaga 2018). Chronic atopic dermatitis is also caused by

Table 17.1 Bacteriocins of lactic acid bacteria in restoring of microbiota in different diseases

S. No.	Bacteriocins	Mechanism of bacteriocin activity	Diseases	References
1	Bacteriocin from *Lactococcus* sp. HY 499	Inhibited several pathogens causing acne like *Staphylococcus epidermidis*, *S. aureus*, *streptococcus pyogenes*, and *P. acnes*	Acne	Oh et al. (2006)
	Bacteriocin -like inhibitory substance (BLIS) by *Streptococcus salivarius*	Inhibited *Propionibacterium acne* which is main pathogen of acne vulgaris	Acne	Bowe et al. (2006)
2	Nisin, bacteriocin of *Staphylococcus epidermidis*, plantaricin EF and JK	Showed antistaphylococcal activity Inhibited the growth of *S. aureus*	Atopic dermatitis	Valenta et al. (1996) and Jang et al. (2020)
3	Reuterin 6 and/or gassericin	Inhibited the formation of biofilm of *S. mutans* and reduced dental lesions	Dental caries	Liang et al. (2021)
	Nisin and thermophilin 110	Inhibited oral pathogens like *Streptococcus mutans*, *Lactobacillus fermenti*, *L. acidophilus*, and *S. sanguinis*	Dental caries	Renye and Steinberg (2021); Tong et al. (2010)
4	Nisin, two-peptide bacteriocin from *L. plantarum,and* bacteriocin from *Actinobacillus actinomycetemcomitans*	Inhibited biofilm of oral pathogens like *Porphyromonas gingivalis*, *Prevotella intermedia*, *Treponema denticola*, *enterococcus faecalis*, and *acidophilus actinomycetemcomitans*	Periodontitis	Nguyen et al. (2020) and Khalaf et al. (2016)
5	Nisin ZP	Inhibited cell migration, invasion, and formation of tumorsphere by *Treponema denticola*, *Porphyromonas gingivalis*, and *Fusobacterium nucleatum* in oral squamous cell carcinoma	Oral cancer	Kamarajan et al. (2015)
	Nisin	Upregulated the expression of CHAC1 gene that promotes apoptosis and induction of cell cycle arrest	Oral cancer	Joo et al. (2012)

dysbiosis in gut microbiota mainly *Faecalibacterium prausnitzii* resulting in reduction in short chain fatty acids and dysregulation of gut epithelial inflamma- tion. Then damage of epithelial cells has been associated with atopic dermatitis because it increased the permeability of epithelium to toxins and pathogens into blood stream and skin (Song et al. 2016). Atopic dermatitis pathogenesis is caused mainly by increase in *S. aureus* count during flare-ups. *Corynebacterium mastitidis* and *Corynebacterium bovis* inhibited growth of *S. aureus* and reverse dysbiosis (Mottin and Suyenaga 2018). Probiotic *Lactobacillus salivarius* and placebo were given to 38 patients aged from 18 to 46 with atopic dermatitis. After treatment, the probiotic reduced in Th2 production and maintained Th1 cytokine stable. Nisin showed significant results against *S. aureus* infection in AD (Valenta et al. 1996). A cytoplasmic protein was extracted from *Staphylococcus epidermidis* by TCA/acetone precipitation method. This cytoplasmic bacteriocin inhibited growth of *S. aureus* on agar plate (Jang et al. 2020) from human swab and was tested against *S. aureus*. It was found that *L. plantarum* LB244R and LB356R had shown highest antimicrobial activity. Genome analysis of these genes revealed that LB244R and LB356R genes encode for bacteriocin plantaricin EF and JK complex (Christensen et al. 2021).

3. **Dental Caries:** Dental caries is the most common oral disease, which causes severe pain and tooth loss. It is generally caused by demineralization of hard tissue of teeth by acid-producing bacteria from fermentation of carbohydrate that leads to formation of cavity. Dysbiosis of natural microflora is caused due to this acidic environment during dental caries. Secretory IgA from saliva and serum IgG from gingival crevicular may also influence dysbiosis of microbiota in infection of dental caries. Dental caries is caused by a group of pathogenic micro-organisms like *Streptococcus mutans*, *Veillonella*, *Bifidobacterium*, *Propionibacterium*, *Lactobacillus*, and *Neisseria* (Zhang et al. 2018). Pit and tissue structures or space between teeth can be seen in dental caries. Difference in velocity and clearance of salivary film in different sites of mouth might affect the microbiome and make the surface prone to demineralization (Lee et al. 2021). Reutericin 6 and/or gassericin A named LN-7 controlled the growth of *S. mutans* for cure of dental caries. Effect of LN-7 biofilm of *S. mutans* was quantified by crystal violet scanning. Also morphological changes in cells of *S. mutans* after treatment with LN-7 were observed by confocal laser scanning microscopy and scanning electron microscopy. In vivo, LN-7 suppressed dental caries by reducing dental lesions (Liang et al. 2021). Nisin from *Lactococcus lactis* was found effective against different oral pathogens like *S. mutans*, *Lactobacillus fermenti*, *L. acidophilus*, and *Streptococcus sanguinis*. The activity was measured by spot on lawn assay. A thermophilin 110 (from *Streptococcus thermophilus* B59671) concentration $\geq$ 80 AU ml^{-1} inhibited *S. mutans* in batch culture, while $\geq$160 AU ml^{-1} concentration was effective against biofilm formation by *S. mutans*. At higher concentration (640–1280 AU ml^{-1}), thermophilin 110 also inhibited other commensal oral *Streptococci* strains for dental caries (Renye and Steinberg 2021).
4. **Periodontitis:** Periodontal disease is the most common oral disease in the world and characterized by increase of inflammation into teeth and losing of attachment and bone (Lee et al. 2021). Periodontitis is a progressed form of gingivitis. Gingivitis is reversible inflammatory disease, but periodontitis is chronic and

irreversible inflammatory disease in which infiltrate of immune cells causes disruption of connective tissue, destruction of alveolar bone, and proliferation of vascular tissue. This all shifting causes dysbiosis of microbiota of oral cavity and leads to inflammation and bone loss (Zhang et al. 2018). Many pathogens like *Porphyromonas gingivalis* and *Treponema denticola*, *Tannerella forsythia, Filifactor alocis*, *Peptoanaerobacter stomatis*, *Megasphaera* sp., and *Selenomonas* sp. are associated with periodontitis disease. Inflammation in periodontitis is derived from increase in micro-organisms that release many nutrients like collagen, heme, amino acids, and iron. Addition of these nutrients to biofilm of oral pathogens acts as pathobionts and causes inflammation by production of pro-inflammatory cytokines (Lamont et al. 2018). Nisin-producing *Lactococcus lactis* showed its potential to prevent accumulation of plaque and gingivitis as compared to placebo and 0.12% chlorhexidine. Nisin with concentration of 100 or 300 μg/mL was used for mouthrinse twice in a daily, which lowered the number of bleeding sites. Nisin also controlled growth of many oral salivary-derived biofilms or oral pathogens including *Porphyromonas gingivalis*, *Prevotella intermedia*, *Treponema denticola*, *Enterococcus faecalis,* and *Aggregatibacter actinomycetemcomitans* (Nguyen et al. 2020). *Weissella cibaria* was tested in vivo on 72 volunteers, which showed reduction in plaque up to 20% and inhibition of biofilm formation (Fusco et al. 2015). Fifty-nine patients with gingivitis were formulated with different *Lactobacillus reuteri* (LR-1 or LR-2) at dose of 2×10^8 CFU/day or placebo. *L. reuteri* was found effective in reducing gingivitis and plaque in patients (Krasse et al. 2006). *Lactobacillus* and *Bifidobacterium* spp. controlled the growth of oral pathogens and cariogenic Streptococci associated with diseases (Allaker and Douglas 2009). A two-peptide bacteriocin produced from *L. plantarum* NC 8 (PLNC 8 αβ) was effective against *Porphyromonas gingivalis* through permeabilization of membranes (Khalaf et al. 2016). Bacteriocin from *Actinobacillus actinomycetemcomitans* showed antagonist activity against various species of both Gram-positive and Gram-negative bacteria from oral microbiome (Lúcia et al. 2002).

5. **Oral Cancer:** Oral cancer is a cancer of oropharynx, lips, or mouth. The term oral cancer can be represented as oral squamous cell carcinoma (OSCC). Incidence of OSCC among young aged 18–44 years is increasing. Common symptoms of OSCC are ulcers with fissuring or exophytic margins (Markopoulos 2012). Culture-dependent methods evaluated that oral carcinoma surfaces have increased number of many aerobic and anaerobic micro-organisms like *Veillonella*, *Prevotella*, *Actinomyces*, *Haemophilus*, *Porphyromonas*, *Fusobacterium*, *Streptococcus* sp., and *Candida albicans* (Nagy et al. 1998). Also, firmicutes especially *Streptococcus* and *Actinobacteria* (especially *Rothia*) were found decreased in number at different tumor sites. Probiotic *Streptococcus salivarius* K12 was found effective against radiation-induced oral mucositis (RIOM), which is oral complication caused by radiotherapy and/or chemotherapy in patients with oral cancer. *S. salivarius* K12 reduced size of ulcers, increased cellularity of basal layer epithelial and thickness of mucosal layer, elevated

proliferation, and attenuated apoptosis in mice model (Wang et al. 2021). Nisin ZP inhibited cell migration and invasion induced by *T. denticola*, *P. gingivalis*, and *Fusobacterium nucleatum*. Nisin ZP also inhibited formation of tumorsphere in OSCC cells. Cell migration and tumorsphere formation are caused by upregulation of integrin alpha V and phosphorylation of FAK, TLR2, and MyD88 in host cells. Nisin downregulated the expression of these genes and also altered influx of ions specially calcium ions (Kamarajan et al. 2015). Head and neck squamous cell carcinoma (HNSCC) is caused due to increase in calcium ions level, induction of arrestation of cell cycle, and CHAC1 gene activation in cell. Nisin treatment showed alteration in genes responsible for function of plasma membrane and transportation of calcium ions. Nisin upregulated the expression of CHAC1 gene (encodes for a cation transport protein), which promotes apoptosis (Joo et al. 2012).

17.8 Conclusions and Future Perspectives

The microbiota of an ecological niche consists of several micro-organisms including diverse bacteria, and some of them are bacteriocin producer and other non-producer. Bacteriocins generally kill related strains but broad-range bacteriocins are also reported recently. The production of bacteriocin by a strain is a defense mechanism and helps in the access of nutrition and space by inhibiting micro-organisms present in the environment. Thus, type of bacteriocin producer determines the shape of microbiota of a particular niche. Though several bacteriocins have been reported for their application in food safety, their microbiome-shaping role is least explored. Therefore, there is need for further study to explore this area in human health and disease conditions.

References

Allaker RP, Douglas CW (2009) Novel anti-microbial therapies for dental plaque-related diseases. Int J Antimicrob Agents 33:8–13. https://doi.org/10.1016/j.ijantimicag.2008.07.014

Alvarez-Sieiro P, Montalbán-López M, Mu D, Kuipers OP (2016) Bacteriocins of lactic acid bacteria: extending the family. Appl Microbiol Biotechnol 100:2939–2951. https://doi.org/10.1007/s00253-016-7343-9

Balandin SV, Sheremeteva EV, Ovchinnikova TV (2019) Pediocin-like antimicrobial peptides of bacteria. Biochemistry (Moscow) 84:464–478. https://doi.org/10.1134/S000629791905002X

Bali V, Panesar PS, Bera MB, Kennedy JF (2016) Bacteriocins: recent trends and potential applications. Crit Rev Food Sci Nutr 56:817–834. https://doi.org/10.1080/10408398.2012.729231

Bowe WP, Filip JC, DiRienzo JM, Volgina A, Margolis DJ (2006) Inhibition of *propionibacterium acnes* by bacteriocin-like inhibitory substances (BLIS) produced by *streptococcus salivarius*. J Drugs Dermatol 5:868–870

Braïek BO, Smaoui S (2019) *Enterococci*: between emerging pathogens and potential probiotics. Biomed Res Int 2019:5938210. https://doi.org/10.1155/2019/5938210

Cesa-Luna C, Alatorre-Cruz JM, Carreño-López R, Quintero-Hernández V, Baez A (2021) Emerging applications of Bacteriocins as antimicrobials, anticancer drugs, and modulators of the gastrointestinal microbiota. Pol J Microbiol 70:143–159. https://doi.org/10.33073/pjm-2021-020

Christensen IB, Vedel C, Clausen ML, Kjærulff S, Agner T, Nielsen DS (2021) Targeted screening of lactic acid bacteria with antibacterial activity toward *Staphylococcus aureus* clonal complex type 1 associated with atopic dermatitis. Front Microbiol 12:733847. https://doi.org/10.3389/fmicb.2021.733847

Colombo NS, Chalón MC, Navarro SA, Bellomio A (2018) Pediocin-like bacteriocins: new perspectives on mechanism of action and immunity. Curr Genet 64:345–351. https://doi.org/10.1007/s00294-017-0757-9

Dominguez-Bello MG, Costello EK, Contreras M, Magris M, Hidalgo G, Fierer N, Knight R (2010) Delivery mode shapes the acquisition and structure of the initial microbiota across multiple body habitats in newborns. Proc Natl Acad Sci U S A 107:11971–11975. https://doi.org/10.1073/pnas.1002601107

Dominguez-Bello MG, Godoy-Vitorino F, Knight R, Blaser MJ (2019) Role of the microbiome in human development. Gut 68:1108–1114. https://doi.org/10.1136/gutjnl-2018-317503

Fusco V, Quero GM, Cho GS, Kabisch J, Meske D, Neve H, Bockelmann W, Franz CM (2015) The genus *Weissella* taxonomy, ecology and biotechnological potential. Front Microbiol 6:155. https://doi.org/10.3389/fmicb.2015.00155

Gharsallaoui A, Oulahal N, Joly C, Degraeve P (2016) Nisin as a food preservative: part 1: physicochemical properties, antimicrobial activity, and main uses. Crit Rev Food Sci Nutr 56:1262–1274. https://doi.org/10.1080/10408398.2013.763765

Ghodhbane H, Elaidi S, Sabatier JM, Achour S, Benhmida J, Regaya I (2015) Bacteriocins active against multi-resistant gram negative bacteria implicated in nosocomial infections. Infect Disord Drug Targets 15:2–12

Grice EA, Kong HH, Conlan S, Deming CB, Davis J, Young AC, Comparative Sequencing Program NISC, Bouffard GG, Blakesley RW, Murray PR, Green ED, Turner ML, Segre JA (2009) Topographical and temporal diversity of the human skin microbiome. Science 324:1190–1192. https://doi.org/10.1126/science.1171700

Grice EA, Segre JA (2011) The skin microbiome. Nat Rev Microbiol 9:244–253. https://doi.org/10.1038/nrmicro2537

Islam MR, Nagao J, Zendo T, Sonomoto K (2012) Antimicrobial mechanism of lantibiotics. Biochem Soc Trans 40:1528–1533. https://doi.org/10.1042/BST20120190

Jang IT, Yang M, Kim HJ, Park JK (2020) Novel cytoplasmic Bacteriocin compounds derived from *Staphylococcus epidermidis* selectively kill *Staphylococcus aureus*, including methicillin-resistant *Staphylococcus aureus* (MRSA). Pathogens (Basel, Switzerland) 9:87. https://doi.org/10.3390/pathogens9020087

Joo NE, Ritchie K, Kamarajan P, Miao D, Kapila YL (2012) Nisin, an apoptogenic bacteriocin and food preservative, attenuates HNSCC tumorigenesis via CHAC1. Cancer Med 1:295–305. https://doi.org/10.1002/cam4.35

Kamarajan P, Hayami T, Matte B, Liu Y, Danciu T, Ramamoorthy A, Worden F, Kapila S, Kapila Y (2015) Nisin ZP, a Bacteriocin and food preservative, inhibits head and neck cancer tumorigenesis and prolongs survival. PLoS One 10:e0131008. https://doi.org/10.1371/journal.pone.0131008

Khalaf H, Nakka SS, Sandén C, Svärd A, Hultenby K, Scherbak N, Aili D, Bengtsson T (2016) Antibacterial effects of *lactobacillus* and bacteriocin PLNC8 αβ on the periodontal pathogen *Porphyromonas gingivalis*. BMC Microbiol 16:188. https://doi.org/10.1186/s12866-016-0810-8

Krasse P, Carlsson B, Dahl C, Paulsson A, Nilsson A, Sinkiewicz G (2006) Decreased gum bleeding and reduced gingivitis by the probiotic *lactobacillus reuteri*. Swed Dent J 30:55–60

Kumariya R, Garsa AK, Rajput YS, Sood SK, Akhtar N, Patel S (2019) Bacteriocins: classification, synthesis, mechanism of action and resistance development in food spoilage causing bacteria. Microb Pathog 128:171–177. https://doi.org/10.1016/j.micpath.2019.01.002

Lamont RJ, Koo H, Hajishengallis G (2018) The oral microbiota: dynamic communities and host interactions. Nat Rev Microbiol 16:745–759. https://doi.org/10.1038/s41579-018-0089-x

Lee YH, Chung SW, Auh QS, Hong SJ, Lee YA, Jung J, Lee GJ, Park HJ, Shin SI, Hong JY (2021) Progress in Oral microbiome related to Oral and systemic diseases: an update. Diagnostics (Basel, Switzerland) 11:1283. https://doi.org/10.3390/diagnostics11071283

Liang J, Liang D, Liang Y, He J, Zuo S, Zhao W (2021) Effects of a derivative of reutericin 6 and gassericin a on the biofilm of Streptococcus mutans in vitro and caries prevention in vivo. Odontology 109:53–66. https://doi.org/10.1007/s10266-020-00529-5

Lloyd-Price J, Abu-Ali G, Huttenhower C (2016) The healthy human microbiome. Genome Med 8: 51. https://doi.org/10.1186/s13073-016-0307-y

Lúcia LF, Farias FF, Eustáquio CJ, Auxiliadora M, Carvalho R, Alviano CS, Farias LM (2002) Bacteriocin production by *Actinobacillus actinomycetemcomitans* isolated from the oral cavity of humans with periodontal disease, periodontally healthy subjects and marmosets. Res Microbiol 153:45–52. https://doi.org/10.1016/s0923-2508(01)01285-2

Markopoulos AK (2012) Role of human papillomavirus in the pathogenesis of oral squamous cell carcinoma. World J Exp Med 2:65–69. https://doi.org/10.5493/wjem.v2.i4.65

Moll GN, Konings WN, Driessen AJ (1999) Bacteriocins: mechanism of membrane insertion and pore formation. Antonie Van Leeuwenhoek 76:185–198. https://doi.org/10.1023/A:1002002718501

Mottin V, Suyenaga ES (2018) An approach on the potential use of probiotics in the treatment of skin conditions: acne and atopic dermatitis. Int J Dermatol 57:1425–1432. https://doi.org/10.1111/ijd.13972

Muizzuddin N, Maher W, Sullivan M, Schnittger S, Mammone T (2012) Physiological effect of a probiotic on skin. J Cosmet Sci 63:385–395

Nagy KN, Sonkodi I, Szöke I, Nagy E, Newman HN (1998) The microflora associated with human oral carcinomas. Oral Oncol 34:304–308

Nguyen T, Brody H, Lin GH, Rangé H, Kuraji R, Ye C, Kamarajan P, Radaic A, Gao L, Kapila Y (2020) Probiotics, including nisin-based probiotics, improve clinical and microbial outcomes relevant to oral and systemic diseases. Periodontology 2000(82):173–185. https://doi.org/10.1111/prd.12324

Nilsen T, Nes IF, Holo H (2003) Enterolysin a, a cell wall-degrading bacteriocin from *enterococcus faecalis* LMG 2333. Appl Environ Microbiol 69:2975–2984. https://doi.org/10.1128/AEM.69.5.2975-2984.2003

O'sullivan L, Ross RP, Hill C (2002) Potential of bacteriocin-producing lactic acid bacteria for improvements in food safety and quality. Biochimie 84:593–604. https://doi.org/10.1016/S0300-9084(02)01457-8

Oh S, Kim SH, Ko Y, Sim JH, Kim KS, Lee SH, Park S, Kim YJ (2006) Effect of bacteriocin produced by *Lactococcus* sp. HY 449 on skin-inflammatory bacteria. Food Chem Toxicol 44: 552–559. https://doi.org/10.1016/j.fct.2005.08.030

Ramos B, Brandão TR, Teixeira P, Silva CL (2020) Biopreservation approaches to reduce *listeria monocytogenes* in fresh vegetables. Food Microbiol 85:103282. https://doi.org/10.1016/j.fm.2019.103282

Renye JA, Steinberg DH (2021) Thermophilin 110 inhibits growth and biofilm formation of *Streptococcus mutans*. Biotechnol Rep (Amst) 31:e00647. https://doi.org/10.1016/j.btre.2021.e00647

Shafquat A, Joice R, Simmons SL, Huttenhower C (2014) Functional and phylogenetic assembly of microbial communities in the human microbiome. Trends Microbiol 22:261–266. https://doi.org/10.1016/j.tim.2014.01.011

Song H, Yoo Y, Hwang J, Na YC, Kim HS (2016) *Faecalibacterium prausnitzii* subspecies-level dysbiosis in the human gut microbiome underlying atopic dermatitis. J Allergy Clin Immunol 137:852–860. https://doi.org/10.1016/j.jaci.2015.08.021

Tong Z, Dong L, Zhou L, Tao R, Ni L (2010) Nisin inhibits dental caries-associated microorganism in vitro. Peptides 31:2003–2008. https://doi.org/10.1016/j.peptides.2010.07.016

Valenta C, Bernkop-Schnürch A, Rigler HP (1996) The antistaphylococcal effect of nisin in a suitable vehicle: a potential therapy for atopic dermatitis in man. J Pharm Pharmacol 48:988–991. https://doi.org/10.1111/j.2042-7158.1996.tb06019.x

Wang Y, Li J, Zhang H, Zheng X, Wang J, Jia X, Peng X, Xie Q, Zou J, Zheng L, Li J, Zhou X, Xu X (2021) Probiotic *streptococcus salivarius* K12 alleviates radiation-induced Oral mucositis in mice. Front Immunol 12:684824. https://doi.org/10.3389/fimmu.2021.684824

Yang SC, Lin CH, Sung CT, Fang JY (2014) Antibacterial activities of bacteriocins: application in foods and pharmaceuticals. Front Microbiol 5:241. https://doi.org/10.3389/fmicb.2014.00241

Zhang Y, Wang X, Li H, Ni C, Du Z, Yan F (2018) Human oral microbiota and its modulation for oral health. Biomed Pharmacother 99:883–893. https://doi.org/10.1016/j.biopha.2018.01.146

Zimina M, Babich O, Prosekov A, Sukhikh S, Ivanova S, Shevchenko M, Noskova S (2020) Overview of global trends in classification, methods of preparation and application of Bacteriocins. Antibiotics (Basel, Switzerland) 9:553. https://doi.org/10.3390/antibiotics9090553

18 Emerging Role of Gut Microbiome in Cancer Immunotherapy

Meghali Bharti, Sonakshi Modeel, Sheetal Yadav, Pankaj Yadav, Sneha Siwach, Padma Dolkar, Shekhar Nagar, Tarana Negi, and Ram Krishan Negi

Abstract

Current scientific advances have considerably added to our understanding of the complex association between the microbiome and cancer. Host and microbiota have co-evolved into a "super-organism," and several physiological processes and multifactorial disease conditions are influenced by host–microbiome interactions. In the past, microbial communities have been suggested to influence the development, progression, metastasis formation, and treatment response of multiple cancer types. However, a better molecular understanding of cancer-modulating interactions and influences on cancer treatment is considered of major scientific relevance and clinical importance. Here, we discuss the scientific evidence on the role of gut microbiota in cancer progression and its treatment and highlight the latest knowledge leveraged to target specific microbes contributing to tumorigenesis.

Keywords

Microbiome · Carcinogenesis · Therapy · Immunity

M. Bharti · S. Modeel · S. Yadav · P. Yadav · S. Siwach · P. Dolkar · R. K. Negi (✉)
Fish Molecular Biology Laboratory, Department of Zoology, University of Delhi, Delhi, India

S. Nagar
Fish Molecular Biology Laboratory, Department of Zoology, University of Delhi, Delhi, India

Present Address: Deshbandhu college, University of Delhi, Delhi, India

T. Negi
Department of Zoology, Government College, Dujana, India

R. Sobti et al. (eds.), *Role of Microbes in Sustainable Development*,
https://doi.org/10.1007/978-981-99-3126-2_18

18.1 Introduction

Microbes represent wide ecological adaptations dominating all four spheres of the earth and in that way captivate microbiologists to explore their niches. A narrative of the terminology has been proposed explaining that microbial taxa linked to a host organism or a dominant environment refer to "microbiota," whereas the "microbiome" is the exclusive assemblage of the microbes and their associated genes (Ursell et al. 2012). Next-generation sequencing (NGS) techniques have garnered ample attention for providing a comprehensive view of human microbiome. The human microbiome is a multi-kingdom repertoire of bacteria, archaea, fungi, protists, and viruses residing in and on the human body (Ursell et al. 2012). Microbiome performs vital functions such as regulating barrier, maintaining homeostasis, preventing pathogenic infection, and regulating metabolism and vitamin synthesis (Li et al. 2019). Oral, gut, and skin microbiomes comprise greatly enriched and diverse microbial consortia, whereas lungs, bladder, prostate, liver, pancreas, and vagina harbor low diverse microbial populations (Cho and Blaser 2012). The host and its microbiome exist in symbiosis as a super-organism by offering a nutrient-rich microenvironment (Schwabe and Jobin 2013). Despite the presence of trillions of beneficial microbes inhabiting human body, dysbiosis can still occur that might lead to the development of cancers and inflammatory diseases. Microbial diversity and abundance differ in different organs; thus, many diseases, including cancer, occur in specific locations within an organ. Chances of cancer are high in those locations where microbial densities are high (Human Microbiome Project Consortium 2012; Cullin et al. 2021).

Investigation of evidence of microbial influence on biology of cancer is in its infancy, and a better comprehensive view of cancer-modeling interfaces and its influence on cancer treatment are reflected as of great scientific relevance. Composition and functional repertoire of microbial communities can be characterized and used for defining pathology and physiology of human cancers (Cullin et al. 2021). Human bodies are continuously exposed to variety of microbes and their byproducts including some tumor-promoting metabolites such as high levels of polyamines, sulfides, and *N*-nitroso compounds (Louis et al. 2014). These metabolites while circulating in the body may lead to cancer progression at locations distinct from the specific microbial residence (Rajagopala et al. 2017). Microorganisms can also migrate to different locations and get associated with the development of tumors. However, the microbes impact the process of carcinogenesis by inflammation and immune system-independent mechanisms and the most decipherable link is via the immune system as the microbes themselves play a significant role in activating and regulation of host immunity (Rajagopala et al. 2017). Interaction of immune system and microorganisms can occur at (1) mucosal layers (via microbial metabolites) or (2) locally at lymphoid organs. Remote/local microbial signals may impact both innate and adaptive immune responses, leading to systemic immunity modulation and anti-tumor innate immune responses (Cullin et al. 2021). Microorganisms can modulate metabolism, inflammation, carcinogenesis, and genotoxicity through multiple mechanisms, and targeting these mechanisms could envision cancer prevention strategies. Genetic modification of microbiota producing/lacking particular enzymes could be utilized for expressing tumor-reducing phytochemicals or reducing tumor-promoting elements (Shwabe and Jobin 2013).

By means of proliferation, escaping cell growth suppression, activation of metastatic pathways, angiogenesis induction, and autophagy resistance cancer cells evade the immune system. All these processes have been explored in detail for decades, but role of microbiome in both cancer progression and treatment is still partly unknown. Microbiome data analysis may assist the advancement of novel cancer diagnostic strategies encompassing cancer detection (identification of microbial DNA/RNA in peripheral blood), surveying metastatic cancer progression, assessing prognosis, and applying artificial intelligence algorithms in foreseeing patient treatment responses. The understanding of the microbiome and cancer needs to be broadened with enhanced feasibility of cancer diagnosis based on microbial profile. Exploring various effects of the microbiome on carcinogenesis will provide new opportunities for diagnostic, preventive, and therapeutic strategies and would represent the next frontier of medical research.

18.2 Cancer Triggering Microbes and Their Cancer-Promoting Mechanisms

Hepatitis B (HBV), Epstein–Barr (EBV), hepatitis C (HCV), Kaposi sarcoma herpesvirus (KSV), human immunodeficiency virus-1 (HIV), human papillomaviruses (HPV), human T cell lymphotropic virus type 1 (HTLV), *Opisthorchis viverrini, Clonorchis sinensis, Schistosoma haematobium*, and *Helicobacter pylori* have been recognized as causes of distinct human cancer (IARC 2009). They participate in cancer progression through different mechanisms, such as B-cell differentiation, cell-cycle disruption, immune hyperactivation (HBV, EBV, HIV, and HCV), dysregulation of T cell (HTLV, EBV), and direct oncogenesis (HCV). Merkel cell polyomavirus (MCV) and Simian virus 40 (SV40) are involved in Merkel cell carcinoma (MCC) and mesothelioma, respectively (Pagano et al. 2004; Weitzman and Weitzman 2014).

In the case of *H. pylori*, CagA dissociates E-cadherin/b-catenin complex, leading to accumulation of b-catenin in both nucleoplasm and cytoplasm (Fig. 18.1). Further, this b-catenin forms complex with transcription factors (TCF/LEF) to activate target gene expression (Hamway et al. 2020). It binds to gastric epithelial cells by HopQ binding to CEACAM, whereby virulence factor CagA is directly injected into the epithelial cells via the T4SS. CagA activates Wnt/β-catenin pathways resulting in dysregulated cellular turnover and apoptosis (Cullin et al. 2021). On the other hand, *F. nucleatum* secretes Fap2 protein that interacts with TIGIT and hinders natural killer cell-facilitated immunosurveillance of cancer (Fig. 18.1). Another important adhesion, FadA, allows cellular internalization and induction of proinflammatory cascades mediated by NF-κB and IL-6. Fap2 interacts with D-galactose-β (1–3)-N-acetyl-D-galactosamine (Gal-GalNAc) carbohydrate moieties at the tumor surface to enhance cellular proliferation via Wnt/β-catenin pathway and increase

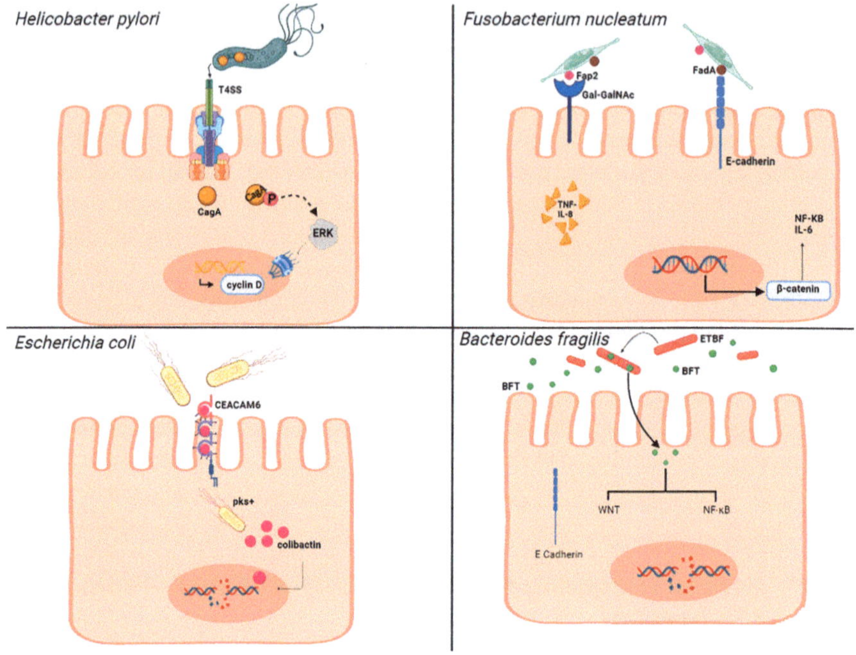

Fig. 18.1 Impact of microbes and their cancer-triggering mechanisms

proinflammatory cytokine production, leading to cancer cell invasion and therapy resistance (Zhang et al. 2020; Cullin et al. 2021).

The intestinal dysbiosis favors growth of adherent-invasive *Escherichia coli* (AIEC) and activation of pks island during inflammation, which increases IL-6-inducing CEACAM6 expression, hence increasing invasiveness of AIEC (Cullin et al. 2021). Once internalized, it secretes genotoxin colibactin, which induces inter-strand crosslink and double-stranded breaks with pro-tumoral cellular transformation (Fig. 18.1). Likewise, the enterotoxigenic *Bacteroides fragilis* (ETBF) produce BFT that can disrupt the intestinal environment by causing inflammation and increased permeability. BFT targets intestinal cell tight junction and causes cleavage of E-cadherin, which causes increased intestinal permeability and induces chronic intestinal inflammation via NF_KB signaling, leading to colorectal tumorigenesis (Fig. 18.1) (Pleguezuelos-Manzano et al. 2020).

18.3 Mechanisms of Microbial Carcinogenesis

Mechanism of microbial carcinogenesis involves (a) inflammation: Bacterial translocation may increase due to change in microbiome and host defense system, which leads to inflammation mediated by microorganism-associated molecular patterns (MAMPs) that activate Toll-like receptors (TLRs) in various cells like macrophage,

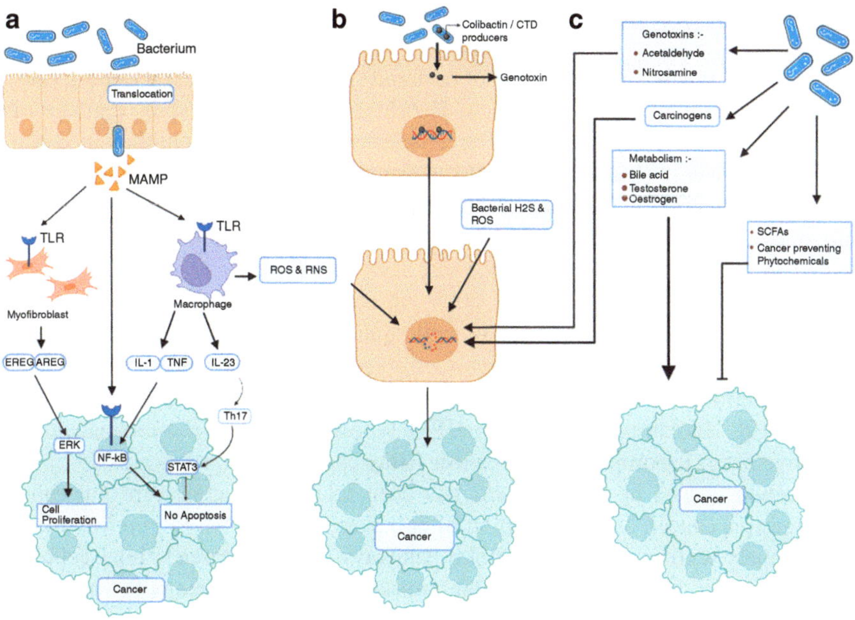

Fig. 18.2 Mechanism of bacterial microbiome-mediated carcinogenesis

myofibroblasts, and tumor cells; (b) genotoxin effect: Bacterial genotoxins like colibactin and cytolethal distending toxin (CTD), which when delivered to host cell nucleus cause DNA damage in various organs (Cullin et al. 2021). Reactive oxygen species (ROS) and reactive nitrogen species (RNS) released from inflammatory cells like macrophage and hydrogen sulfide (H_2S) from bacterial microbiota may also be genotoxic, thus triggering carcinogenesis; (c) metabolic effect: Genotoxins like acetaldehyde, dietary nitrosamine in metabolism of bile acids, and hormones like estrogen and testosterone may activate due to metabolic actions of microbiome. The microbiota also mediates tumor suppressive effect by inactivation of carcinogenesis via generation of short-chain fatty acids and activation of cancer-preventing phytochemicals (Schwabe and Jobin 2013) (Fig. 18.2).

18.4 Risk Factors of Specific Cancers

Numerous propensities especially lifestyle and diet stand as the risk factors for the changes in the microbial diversity of the gut, thus affecting the microenvironment of host's cells (Moskal et al. 2016). It is still not clear whether the shift in population pattern causes carcinogenesis or is an outcome of the emergence of tumors. Hence, this important variation will be the epicenter of research in cancer–microbiome area in the upcoming future. Different types of cancer and their association with microbiome are discussed below.

18.4.1 Oral and Gastric Cancer

Oral cavity is inhabited by a variety of bacterial species, which play a leading role in the development of oral diseases (Al-Hebshi et al. 2017). Dysbiosis in the oral microflora destabilizes the defense mechanisms of the host, resulting in chronic periodontal disease (Bullon et al. 2014; Johannsen et al. 2014), which is related to changes in the oral microflora that is caused by the outgrowth of certain pathogenic microbes. The two notable pathogenic members of the oral microbiome that are known to induce tumorogenesis in the oral cavity are *Fusobacterium nucleatum* and *Prevotella gingivalis* (Mager et al. 2005). Repeated periodontitis is considered to increase risk for the development of oral squamous cell carcinoma (OSCC) and *Prevotella intermedia* and *Porphyromonas gingivalis* are associated with the occurrence of periodontitis (Mysak et al. 2014; Zhang et al. 2017; Hsiao et al. 2018, Li et al. 2019). Human papillomavirus (HPV) type 16 has also been recognized as a causative agent for oropharyngeal cancer (Bray et al. 2018).

Infection caused by *P. gingivalis* has been related to increase in oral cancer, oro-digestive cancer, and propagation of oral cancer stem cells. *P. gingivalis* infection tempers with many anti-apoptotic pathways and also spreads inter-cellularly with the help of actin-based membrane protrusions and interferes with various cell signaling pathways (Mao et al. 2007; Yilmaz et al. 2004). Firstly, *P. gingivalis* activates Jak1/Akt/Stat3 signaling, which controls intrinsic mitochondrial apoptosis pathways (Mao et al. 2007; Yilmaz et al. 2004). It inhibits gingival epithelial cell apoptosis induced by ATP ligation of P2X7 receptors (Yilmaz et al. 2008). *P. gingivalis* also causes reduction in amount of p53 and accelerates the progression through the S-phase of the cell cycle (Kuboniwa et al. 2008). It promotes the expression of the B7-H1, which accelerates regulatory T-cell production and B7-DC receptors in primary gingival epithelial cells (Groeger et al. 2011). This may be the possible reason that B7-H1 expression contributes to immune evasion by oral cancers. AR2/NF-KB pathways are activated by *P. gingivalis* infection, which in turn induces expression of promatrix metalloproteinase (proMMP-9) (Inaba et al. 2014). *P. gingivalis* produces gingipains and cysteine proteinases, which play a key role in engaging the PAR2 receptor and also cleave the MMP-9 pro-enzyme into an active form. The active form of MMP-9 along with extracellular matrix degrades the basement membrane. This promotes carcinoma, cell migration, and invasion. The *P. gingivalis* might contribute to oral squamous cell carcinoma metastasis. The another pathogenic member *Fusobacterium nucleatum* works by activating inflammatory cytokines like TNF-α, IL-1β, and IL-6. Several antiapoptotic pathways are modulated by *F. nucleatum,* and it also activates p38, which results in the secretion of MMP-9 and MMP-13 (collagenase 3).

Role of human papilloma virus in carcinogenesis has been confirmed since the discovery of HPV16. HPV16 expresses E6 and E7 proteins, which further lead to the inactivation of tumor suppressor proteins p53 and Rb (Wiest et al. 2002). Inactivation of p53 and Rb causes dysregulation of host DNA synthesis, as a consequence cell cycle control is lost (D'Souza et al. 2007). This leads to destabilization of the genome, chromosomal aberrations, and abnormal centrosome numbers. HPV acts as an independent risk factor for the development of oral squamous cell carcinoma (Duensing et al. 2000).

18.4.1.1 GI Tract Cancer

Gastric cancer has been acknowledged as the model to study bacterial cancers. *Helicobacter pylori* in the gastrointestinal tract has been classified as a class I carcinogen by the International Agency for Research on Cancer (International Agency for Research on Cancer 1994). Strong host immune response is generated due to infection with *H. pylori,* which results in various gastric problems such as gastric inflammation, dysplasia, achlorhydria, and epithelial atrophy (Blaser and Atherton 2004). The key hallmarks of cancer driven by *H. pylori* include inflammation that promotes tumor, downregulation of antitumor immune destruction, and increase in proliferative signaling (Asano et al. 2015). *H. pylori* is able to adhere to gastric epithelial cells with the help of multiple adhesins, including SabA and BabA (Ilver et al. 1998). Once it firmly associates with the gastric epithelium, it deposits CagA and other virulence factors on human cells by utilizing its type IV secretion system. SHP2, a host oncoprotein, gets dysregulated by CagA after entering the cell leading to uncontrolled cell growth and motility (Murata-Kamiya et al. 2010). Other virulence factors such as VacA are responsible for making pores in host membranes, causing cell death and higher rates of cell turnover (Cover and Blanke 2005). Neutrophils and macrophages start producing reactive oxygen species (ROS) due to *H. pylori* infection (Tsugawa et al. 2012; Sung et al. 2022; Kuo et al. 2022). Inflammatory cytokines, including IL-1β, tumor necrosis factor-α, and IL-8, are also produced by *H. pylori* infection, ultimately leading to induction of Th1 immune behavior in the gut (Handa et al. 2011; Wei et al. 2010). NF-κB levels rise in patients with infection, which further drive inflammatory mechanisms of tumor initiation and succession. *H. pylori* is protected from ROS due to the presence of catalase and superoxide dismutase proteins but bystander effect damages host tissues (Chaturvedi et al. 2011). VacA, virulence factor of the pathogen, obstructs nuclear factor of activated T-cell (NFAT) activation in T cells. Blocking of NFAT leads to a deficiency of IL-2-driven T-cell proliferation, which would gradually result in expulsion of *H. pylori* (Jain et al. 2011). γ-glutamyltranspeptidase (GGT) produced by *H. pylori* also has the ability to block T-cell proliferation. Hence, a perfect storm is generated by the combination of cellular damage, innate procarcinogenic signals, and reduced immune surveillance in patients, which results in cancer.

18.4.2 Hepatic and Pancreatic Cancer

Hepatocellular carcinoma (HCC) is responsible for 80–90% of liver cancers and stands as the third leading cause of cancer-related deaths (Mima et al. 2017; Tong et al. 2011). Fox et al. (2010) observed the presence of *Helicobacter* spp. in gastric mucosa, leading to an increase in the chances of tumor progression in the hepatobiliary tract. *H. pylori* seizes in the host's cells by attaching to gastric epithelial cells by HopQ protein, which binds to carcinoembryonic antigen-related cell adhesion molecules (CEACAM) and destructs the antitumor immune system, further inducing tumorigenic inflammation (Schwabe and Jobin 2013). Murphy et al. (2014) studied the association of fifteen *H. pylori* proteins with hepatobiliary carcinoma using a multiplex serology panel and found an increase in antibodies against these proteins, predicting an increase in HCC and biliary tract cancer. *Helicobacter hepaticus* also promotes HCC by producing toxins that promote anti-apoptotic factors and activate nuclear factor kappa B (NF- κB) and WNT signaling pathways, thus promoting tumor-inducing metabolites and suppressing antitumor immunity (Fox et al. 2010; Beyoğlu and Idle 2022). Other experimental studies have shown the enrichment of *Methylophilaceae*, *Fusobacterium*, *Prevotella*, *Actinomyces*, and *Novosphingobium* along with *Helicobacter pylori* in extrahepatic carcinoma tissue specimens of 100 patients (Avilés-Jiménez et al. 2016).

Similarly, highly lethal pancreatic cancer is also influenced by gut microbiota, which induces oncogenic metabolites for tumor development. Pancreatic cancer has been found to be associated with periodontal diseases and gum inflammation (Michaud et al. 2007; Stolzenberg-Solomon et al. 2003). In a comparative study of salivary microbiome of pancreatic cancer patients and healthy people, *Neisseria elongata* and *Streptococcus mitis* were found to be associated with cancer (Farrell et al. 2012; Herremans et al. 2022). Another oral bacteria, *Fusobacterium* spp., was found to be present in the tumor tissues of 283 pancreatic ductal adenocarcinoma patients (Mitsuhashi et al. 2015). *Fusobacterium* increases the inflammatory cytokines and reactive oxygen species (ROS), which leads to epigenetic alteration of mismatch proteins such as mutL homolog 1 and CpG island methylator phenotype (tumor suppressor gene), thus promoting carcinogenesis (Kostic et al. 2013; Schetter et al. 2010). *Helicobacter* spp. has also been found to be associated with pancreatic cancer (Nilsson et al. 2002; Trikudanathan et al. 2011). Evidence from different studies suggests that the accumulation of specific microbes can lead to the activation of carcinogenic factors producing hepatocellular and pancreatic carcinoma. *Prevotella* sp., *Bacteroides, Ruminococcus, Faecalibacterium,* and *Ruminiclostridium* are involved in progression of hepatic cancer (Hu et al. 2020; Yu et al. 2017). The gut microbiota may impact not only the formation of tumor but also the adequacy of chemotherapies and immunotherapies for HCC and pancreatic medical procedure, leading to low survival rates of cancer patients (Mima et al. 2017).

18.4.3 Colorectal Cancer

Colorectal carcinoma (CRC) being the third most malignant tumor with high occurrence in Western countries stands as the fourth most common cause for cancer-related deaths with mortality rate of 9.2% (Mármol et al. 2017; Zhou et al. 2020b, b). 70% of the human microbiome is present in the colon, which makes it the most vigorously colonized part of the gastrointestinal system (Saus et al. 2019). The modulation of colonic flora can be responsible for the dysplasia and can cause colonic inflammation and biosynthesis of carcinogenic molecules, leading to CRC development (Arthur et al. 2012; Rubinstein et al. 2013). *Fusobacterium nucleatum, Peptostreptococcus stomatis, Peptostreptococcus anaerobius,* and *Bacteroides fragilis* along with twenty other microbial gene markers were found to be associated with CRC (Yu et al. 2017; Zhou et al. 2020b, b). On the other hand, the commonly found *Escherichia coli* can have the pathogenicity island (*pks*), which can synthesize colibactin, a genotoxin causing oncogenic mutations and DNA damage (Pleguezuelos-Manzano et al. 2020) (Fig. 18.1c). A polyamine catabolic enzyme, spermine, is highly inducible by inflammatory stimuli of enterotoxigenic *Bacteroides fragilis,* which results in DNA damage and increase in ROS, leading to colorectal carcinoma (Goodwin et al. 2011) (Fig. 18.1d). It is also known to produce another inflammatory toxin that causes diarrhea and colonizes in host, which can further induce CRC (Wu et al. 2009). *Enterococcus faecalis* also have capability to produce enterotoxins and ROS, causing inflammation and epithelial damage (Saus et al. 2019). The microbes alter the signaling pathway by producing toxic proteins or biochemicals, which creates unfavorable microenvironment for cells and promotes carcinogenesis. For instance, FasA surface protein produced by *Fusobacterium nucleatum* adheres to the epithelial walls of colon and interacts with E-cadherin (Zhou et al. 2022). The complex then alters the β-catenin and Wnt signaling pathways. Increase in FadA protein leads to increase in the inflammatory and oncogenic genes (Kostic et al. 2013; Rubinstein et al. 2013). Other bacterial species such as *Bacteroides, Parabacteroides, Lachnospiraceae bacterium,* and *Alistipes* spp. has high prevalence in CRC patients and are found to play role in development of colon carcinoma (Feng et al. 2015).

18.5 Role of Microbiome in Treatment of Cancer and Future Applications

Since certain microbial signatures are known to promote the development of cancer and affect the safety, tolerability, and effectiveness of treatments, there is growing evidence that the gut microbiome is connected to cancer in a variety of ways. From over past few years, tremendous progress has been made in the field of cancer treatment, but somehow the tumor microenvironment has a significant impact on how well tumor immunotherapy works. Studies have demonstrated that a variety of tumor microenvironment cells, including T cells, fibroblasts, natural killer (NK) cells, dendritic cells (DCs), and others, play a key role in tumor immunotherapy (Ferlazzo et al. 2004). NK cells additionally induce cDC1 to enter the tumor microenvironment (TME) and assist tumor immune control (Poutahidis and Erdman 2016; Zhang et al. 2018; Böttcher et al. 2018; Zhou et al. 2020b, b). Chemotherapy or immune checkpoint inhibitor resistance is linked to altered gut flora immune checkpoint inhibitors (ICIs). The antitumor effects of chemotherapy medicines or ICIs may be enhanced by altering the microbiota through the use of antibiotics, probiotics, fecal microbiota transplants, or nanotechnologies (Cheng et al. 2020).

18.5.1 Immunotherapy

The term "immunotherapy or immuno-anticancer treatments" refers to a variety of therapeutic modalities intended to stimulate a patient's immune system or recruit external immunological cells to combat cancer. Immunotherapy achieves the goal of eradicating cancers by inhibiting negative immune regulatory factors, stimulating the immune system, and improving immune cell identification, which results in the death of immune cells to tumors (Beatty and Gladney 2015).

In comparison with conventional cancer treatments, the gut microbiota has a stronger anti-cancer effect and further activates the host immune system. Additionally, there is growing acknowledgment for the role that the interaction between the gut microbiota and cancer ICIs plays in antitumor immune therapy, which is a form of targeted therapy for cancer immunotherapy (Qiu et al. 2021). Numerous species, including *Bifidobacterium breve* and *Bifidobacterium longum*, have been shown to improve dendritic cell activity and therefore trigger CD8+ T-cell priming and accumulation in tumor microenvironments (Tanoue et al. 2019; Sivan et al. 2015). In the tumor microenvironment, molecular cell refinement and immunological control of therapeutic targets are growing, and clinical application is expanding as well, such would be the resistance to the programmed cell death ligand 1, which plays a very crucial role in anti-tumor immunotherapy. The reduced programmed cell death ligand 1 activity that is resistant to the tumor can have favorable therapeutic effects and may be used for modifying the inert tumor microenvironment in future too (Qiu et al. 2021). Resistance to PD-1/PD-L1 plays many roles in tumor immunotherapy. Using an acceptable and selective combination of immunotherapy in a constrained tumor microenvironment reactivates the anti-tumor immune response in the host. This refers to the possibility of immune toxicity and immunotherapy to increase antitumor immunity (Ngiow and Young 2020). Antibodies that are able to act against the ligands of the PD-1 and CTLA-4 may inhibit affinity of T lymphocytes with their suppressive two or more ligands, which will act on the tumor cells, which in turn will activate the anti-tumor response in the immune system against the tumor cells (Cullin et al. 2021) (Fig. 18.3).

18.5.2 Chemotherapy

Chemotherapy refers to the treatment that requires anti-cancer drugs with high and powerful chemical content to treat any form of cancer by targeting the fast multiplying or growing cancer cells in the body. Chemotherapy causes DNA and non-DNA damage that is ROS-mediated, which allows bacteria to pass the intestinal epithelium. In turn, this causes an inflammatory reaction and may result in systemic infections (Kalasabail et al. 2021). The gut microbiota can influence how the body reacts to chemotherapy by enhancing medication efficacy, encouraging chemoresistance, and/or mediating adverse effects and toxicity.

Gut microbiota uses various mechanisms to regulate or modify the potency of the anti-cancer drugs. Many species of bacteria that are present in the gut impact the

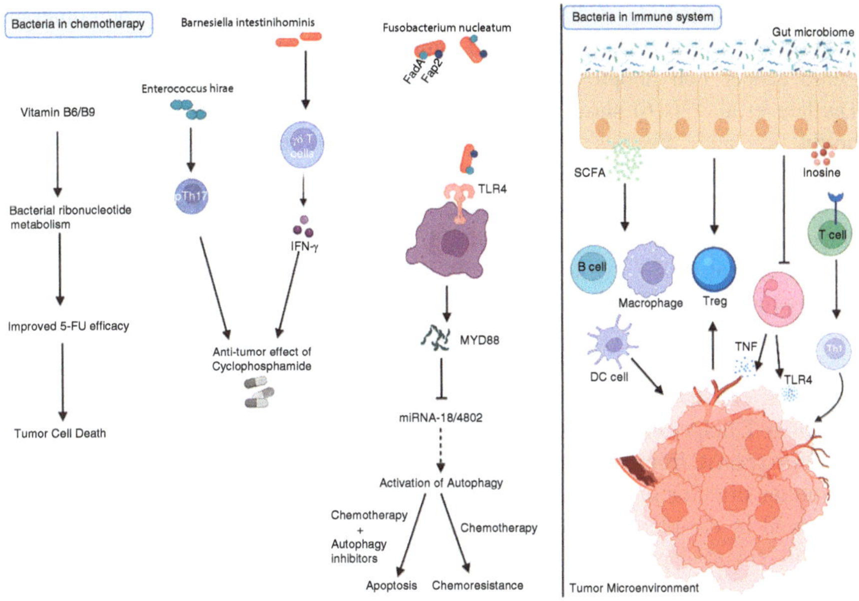

Fig. 18.3 Role of gut microbiota in modulating efficacy of anticancer drug in cancer therapy

competency of the drugs that are used in chemotherapy and immunotherapy, which also includes immune checkpoint inhibitors by various mechanisms (Cullin et al. 2021).

Fusobacterium nucleatum is one such bacterium found in the gut microbiota and with the use of antibiotics/drugs can also reduce the intensity of cancer. *F. nucleatum* functions via myeloid differentiation primary response 88(MYD88) and Toll-like receptors (TLR4), which further results in the selective deprivation of miR-18a and miR-4802, and this in turn initiates autophagy. This process can further assist in chemoresistance in cancer patients (Fig. 18.3).

An earlier research demonstrated that cyclophosphamide can alter the composition of the gut microbiota by causing some gram-positive bacteria to translocate into the secondary lymphoid organs, which in turn triggers the production of "pathogenic" T helper 17 (pTh17) cells and improves the host immune response brought on by memory T helper 1 (Th1) cells. Researchers using *Caenorhabditis elegans* models have also identified bacteria that play role in chemotherapeutic effectiveness, particularly in the metabolism of ribonucleotides and vitamins B6 and B9 (Scott et al. 2017; García-González et al. 2017), which further increase the efficacy of fluoropyrimidine, an antimetabolite by inhibiting bacterial deoxynucleotide metabolism (Cheng et al. 2020). Efficacy of chemotherapy drugs can be modulated by bacteria via different mechanisms. The efficiency of 5-FU can be modulated via B6, B9, and ribonucleotide metabolism. The efficiency of 5-FU was promoted by inhibiting bacterial deoxynucleotide metabolism (Fig. 18.3).

Cyclophosphamide (CTX), one of the most often prescribed chemotherapy medications for the treatment of solid tumors and lymphomas, stimulates immunogenic cancer cell apoptosis and immunomodulatory effects (Qiu et al. 2021). Daillère et al. (2016) showed the incorporation of *E. hirae* in mice treated with antibiotics and reverses cyclophosphamide chemoresistance, which promotes pTh-17 production and T-helper (Th)-1 cell differentiation by increasing CD8+ T cells and CD4+ T regulatory cell (Treg) in the intratumoral region (Fig. 18.3). As *Barnesiella intestinihominis* builds up in the colon, it activates Th1 and polyfunctional CD8+ cytotoxic T cells in the body, which encourages interferon (IFN)-producing γδT cells to invade tumors (Daillère et al. 2016; Cheng et al. 2020) on treatment with cyclophosphamide (Cheng et al. 2020) (Fig. 18.3). Non-enterotoxigenic *B. fragilis* and *Erysipelotrichaceae* are examples of immunogenic bacteria that promote migratory dendritic cells (DCs), which then promote follicular T helper (TFH) cells through interleukin (IL) 1 and IL-12. The increased IgG2b response from stimulated TFH cells then interacts with B cells to boost the antitumor effector or the memory CD8+ T-cell activity. The response is further enhanced in the presence of gut commensal and boosts antineoplastic drugs like oxaliplatin and cisplatin (Roberti et al. 2020).

18.5.3 Radiotherapy

Radiation therapy is a part of the treatment plans for more than 50% of patients with cancer, and it is thought to make up around 40% of therapeutic protocols with efficacy in more than 90% of cases particularly those in their first stage of diseases (Poonacha et al. 2022). The predominantly most significant impediment to malignant cancer cure in patients is radiation enteropathy and radiation toxicity (Hauer-Jensen et al. 2014). RT also affects the gut microbiota; however, only a few studies have attempted to examine the connection between the microbiota and radiotherapy response (Tonneau et al. 2021). In melanoma, lung, and cervical cancer models, an oral vancomycin-induced decrease in gram-positive gut commensals mediated by IFNg and CD8 T-cell-dependent pathways was linked to improved radiation efficiency (Uribe-Herranz et al. 2020). Cui et al. (2017) showed the benefits of fecal microbiota transplantation (FMT) against total body irradiation-induced acute radiation enteritis in the mouse model with an increase in microbiota diversity. Irradiated mice that received (FMT) survived longer and had enhanced digestive system performance and increased levels of peripheral white blood cells. An evaluation of FMT in the treatment of chronic radiation enteritis showed a radical shift in diversifying the microbiome composition and reducing gastrointestinal symptoms (Ding et al. 2020).

18.5.4 Targeting Microbiome for Therapeutic Modulation of Carcinogenesis

Gut microbiota influences the shape of the tumor microenvironment by acting on the host immune system. TLR4 signaling in tumor cells recruit neutrophils, and they release tumor necrosis factor (TNF), which causes metastasis. Gut microbiota reduces the number of neutrophils, and its metabolites inosine promotes differentiation of Th1 cells in the presence of exogenous interferon-γ. An additional therapeutic option might be the introduction of anaerobic bacteria since they colonize all tumor sites regardless of oxygenation level and can even eradicate hypoxic cancers (Riehl et al. 2019; Poonacha et al. 2022). Through anaerobic action, butyrate-producing bacteria can degrade polysaccharides to create short-chain fatty acids (SCFAs), which are crucial for reducing the proliferation of cancer cells (Wang et al. 2019; Wagner et al. 2018). SCFAs like butyrate along with metabolites of tryptophan lessen pro-inflammatory cytokines and encourage the release of anti-inflammatory cytokines and affect the class conversion of B-cells, activating dendritic cells and macrophages, which affect differentiation of memory T cells, which helps limit radiation-related toxicity (Fish et al. 2016; Badgeley et al. 2021; Cheng et al. 2020) (Fig. 18.3). An introduction of *Bifidobacterium infantis*, a commensal bacterium to mice, and *B. infantis* with monoclonal antibody along with RT shows improvement in tumor cell proliferation, decreases blood supply, and enhanced cell apoptosis of the tumor. Furthermore, mice that received the combination therapy showed a higher survival rate than mice that received either the RT or a bacterial antibody alone (Du et al. 2018). Similarly, the symbiotic introduction of *Lactobacillus rhamnosus* (ATCC 7469) exopolysaccharides (EPS) in rat models with a dose of ionizing radiation was tested to slow colorectal development, enhanced the modulation of signaling growth factors involved in inflammation, and also reduced the progression of colorectal carcinoma in mice when compared to untreated control or those given *L. rhamnosus* or irradiation alone (Ruiz-Ruiz et al. 2017). These findings regarding the significance of the unique characteristics of the patient's pre-existing microbiota suggest the possibility of incorporating microbiota analysis into personalized treatment protocols to predict how therapy will affect the patient and how the gut microbiota can be used as a biodosimeter to check the biological response for treatment planning (González-Mercado et al. 2021; Shi et al. 2020; Ding et al. 2020).

References

Al-Hebshi NN, Nasher AT, Maryoud MY, Homeida HE, Chen T, Idris AM, Johnson NW (2017) Inflammatory bacteriome featuring fusobacterium nucleatum and Pseudomonas aeruginosa identified in association with oral squamous cell carcinoma. Sci Rep 7(1):1–10. https://doi.org/10.1038/s41598-017-02079-3

Arthur JC, Perez-Chanona E, Mühlbauer M, Tomkovich S, Uronis JM, Fan TJ et al (2012) Intestinal inflammation targets cancer-inducing activity of the microbiota. Science 338(6103):120–123

Asano N, Iijima K, Koike T, Imatani A, Shimosegawa T (2015) Helicobacter pylori-negative gastric mucosa-associated lymphoid tissue lymphomas: a review. World J Gastroenterol: WJG 21(26): 8014. https://doi.org/10.3748/wjg.v21.i26.8014

Avilés-Jiménez F, Guitron A, Segura-Lopez F, Mendez-Tenorio A, Iwai S, Hernández-Guerrero A, Torres J (2016) Microbiota studies in the bile duct strongly suggest a role for helicobacter pylori in extrahepatic cholangiocarcinoma. Clin Microbiol Infect 22(2):178–e11. https://doi.org/10.1016/j.cmi.2015.10.008

Badgeley A, Anwar H, Modi K, Murphy P, Lakshmikuttyamma A (2021) Effect of probiotics and gut microbiota on anti-cancer drugs: mechanistic perspectives. Biochim Biophys Acta Rev Cancer 1875(1):188494. https://doi.org/10.1016/j.bbcan.2020.188494

Beatty GL, Gladney WL (2015) Immune escape mechanisms as a guide for cancer immunotherapy tailoring cancer immunotherapy. Clin Cancer Res 21(4):687–692. https://doi.org/10.1158/1078-0432.CCR-14-1860

Beyoğlu D, Idle JR (2022) The gut microbiota–a vehicle for the prevention and treatment of hepatocellular carcinoma. Biochem Pharmacol 115225:115225

Blaser MJ, Atherton JC (2004) Helicobacter pylori persistence: biology and disease. J Clin Invest 113(3):321–333. https://doi.org/10.1172/JCI20925

Böttcher JP, Bonavita E, Chakravarty P, Blees H, Cabeza-Cabrerizo M, Sammicheli S et al (2018) NK cells stimulate recruitment of cDC1 into the tumor microenvironment promoting cancer immune control. Cell 172(5):1022–1037. https://doi.org/10.1016/j.cell.2018.01.004

Braakhuis BJ, Snijders PJ, Keune WJH, Meijer CJ, Ruijter-Schippers HJ, Leemans CR, Brakenhoff RH (2004) Genetic patterns in head and neck cancers that contain or lack transcriptionally active human papillomavirus. J Natl Cancer Inst 96(13):998–1006. https://doi.org/10.1093/jnci/djh183

Bray F, Ferlay J, Soerjomataram I, Siegel RL, Torre LA, Jemal A (2018) Global cancer statistics 2018: GLOBOCAN estimates of incidence and mortality worldwide for 36 cancers in 185 countries. CA Cancer J Clin 68(6):394–424. https://doi.org/10.3322/caac.21492

Bullon P, Newman HN, Battino M (2014) Obesity, diabetes mellitus, atherosclerosis and chronic periodontitis: a shared pathology via oxidative stress and mitochondrial dysfunction? Periodontology 64(1):139–153. https://doi.org/10.1111/j.1600-0757.2012.00455.x

Chaturvedi R, Asim M, Romero Gallo J, Barry DP, Hoge Sablet TD, Delgado AG, Wroblewski LE, Piazuelo MB, Yan F, Israel DA, Casero RA, Correa P, Gobert AP, Polk DB, Peek RM, Wilson KT (2011) Spermine oxidase mediates the gastric cancer risk associated with helicobacter pylori CagA. Gastroenterology 141(5):1696–1708. https://doi.org/10.1053/j.gastro.2011.07.045

Cheng WY, Wu CY, Yu J (2020) The role of gut microbiota in cancer treatment: friend or foe? Gut 69(10):1867–1876. https://doi.org/10.1136/gutjnl-2020-321153

Cho I, Blaser MJ (2012) The human microbiome: at the interface of health and disease. Nat Rev Genet 13(4):260–270

Cover TL, Blanke SR (2005) Helicobacter pylori VacA, a paradigm for toxin multi functionality. Nat Rev Microbiol 3(4):320–332. https://doi.org/10.1038/nrmicro1095

Cui M, Xiao H, Li Y, Zhou L, Zhao S, Luo D, Fan S (2017) Faecal microbiota transplantation protects against radiation-induced toxicity. EMBO Mol Med 9(4):448–461. https://doi.org/10.15252/emmm.201606932

Cullin N, Antunes CA, Straussman R, Stein-Thoeringer CK, Elinav E (2021) Microbiome and cancer. Cancer Cell 39(10):1317–1341. https://doi.org/10.1016/j.ccell.2021.08.006

Daillère R, Vétizou M, Waldschmitt N, Yamazaki T, Isnard C, Poirier-Colame V et al (2016) Enterococcus hirae and Barnesiella intestinihominis facilitate cyclophosphamide-induced therapeutic immunomodulatory effects. Immunity 45(4):931–943. https://doi.org/10.1016/j.immuni.2016.09.009

Ding X, Li Q, Li P, Chen X, Xiang L, Bi L et al (2020) Fecal microbiota transplantation: a promising treatment for radiation enteritis? Radiother Oncol 143:12–18. https://doi.org/10.1016/j.radonc.2020.01.011

D'Souza G, Kreimer AR, Viscidi R, Pawlita M, Fakhry C, Koch WM, Gillison ML (2007) Case–control study of human papillomavirus and oropharyngeal cancer. N Engl J Med 356(19): 1944–1956. https://doi.org/10.1056/NEJMoa065497

Du SX, Jia YR, Ren SQ, Gong XJ, Tang H, Wan-Shui W, Li-Ming S (2018) The protective effects of bacillus licheniformis preparation on gastrointestinal disorders and inflammation induced by radiotherapy in pediatric patients with central nervous system tumor. Adv Med Sci 63(1): 134–139. https://doi.org/10.1016/j.advms.2017.09.005

Duensing S, Lee LY, Duensing A, Basile J, Piboonniyom SO, Gonzalez S et al (2000) The human papillomavirus type 16 E6 and E7 oncoproteins cooperate to induce mitotic defects and genomic instability by uncoupling centrosome duplication from the cell division cycle. Proc Natl Acad Sci 97(18):10002–10007. https://doi.org/10.1073/pnas.170093297

Farrell JJ, Zhang L, Zhou H, Chia D, Elashoff D, Akin D et al (2012) Variations of oral microbiota are associated with pancreatic diseases including pancreatic cancer. Gut 61(4):582–588. https://doi.org/10.1136/gutjnl-2012-302312

Feng Q, Liang S, Jia H, Stadlmayr A, Tang L, Lan Z et al (2015) Gut microbiome development along the colorectal adenoma–carcinoma sequence. Nat Commun 6(1):1–13. https://doi.org/10.1038/ncomms7528

Ferlazzo G, Pack M, Thomas D, Paludan C, Schmid D, Strowig T, Bougras G, Muller W, Moretta L, Münz C (2004) Distinct roles of IL-12 and IL-15 in human natural killer cell activation by dendritic cells from secondary lymphoid organs. Proc Natl Acad Sci 101(47): 16606–16611. https://doi.org/10.1073/pnas.0407522101

Fish BL, Gao F, Narayanan J, Bergom C, Jacobs ER, Cohen EP et al (2016) Combined hydration and antibiotics with lisinopril to mitigate acute and delayed high-dose radiation injuries to multiple organs. Health Phys 111(5):410. https://doi.org/10.1097/HP.0000000000000554

Fox JG, Feng Y, Theve EJ, Raczynski AR, Fiala JL, Doernte AL et al (2010) Gut microbes define liver cancer risk in mice exposed to chemical and viral transgenic hepatocarcinogens. Gut 59(01):88–97. https://doi.org/10.1136/gut.2009.204115

García-González AP, Ritter AD, Shrestha S, Andersen EC, Yilmaz LS, Walhout AJ (2017) Bacterial metabolism affects the C. elegans response to cancer chemotherapeutics. Cell 169(3):431–441. https://doi.org/10.1016/j.cell.2017.03.046

González-Mercado VJ, Henderson WA, Sarkar A, Lim J, Saligan LN, Berk L et al (2021) Changes in gut microbiome associated with co-occurring symptoms development during chemo-radiation for rectal cancer: a proof of concept study. Biol Res Nurs 23(1):31–41. https://doi.org/10.1177/1099800420942830

Goodwin AC, Shields CED, Wu S, Huso DL, Wu X, Murray-Stewart TR et al (2011) Polyamine catabolism contributes to enterotoxigenic Bacteroides fragilis-induced colon tumorigenesis. Proc Natl Acad Sci 108(37):15354–15359. https://doi.org/10.1073/pnas.1010203108

Groeger S, Domann E, Gonzales JR, Chakraborty T, Meyle J (2011) B7-H1 and B7-DC receptors of oral squamous carcinoma cells are upregulated by Porphyromonas gingivalis. Immunobiology 216(12):1302–1310. https://doi.org/10.1016/j.imbio.2011.05.005

Hamway Y, Taxauer K, Moonens K, Neumeyer V, Fischer W, Schmitt V et al (2020) Cysteine residues in helicobacter pylori Adhesin HopQ are required for CEACAM–HopQ interaction and subsequent CagA translocation. Microorganisms 8(4):465

Handa O, Naito Y, Yoshikawa T (2011) Redox biology and gastric carcinogenesis: the role of helicobacter pylori. Redox Rep 16(1):1–7. https://doi.org/10.1179/174329211X12968219310756

Hauer-Jensen M, Denham JW, Andreyev H (2014) Radiation enteropathy—pathogenesis, treatment and prevention. Nat Rev Gastroenterol Hepatol 11(8):470–479. https://doi.org/10.1038/nrgastro.2014.46

Herremans KM, Riner AN, Cameron ME, McKinley KL, Triplett EW, Hughes SJ, Trevino JG (2022) The oral microbiome, pancreatic cancer and human diversity in the age of precision medicine. Microbiome 10(1):1–14

Hsiao JR, Chang CC, Lee WT, Huang CC, Ou CY, Tsai ST et al (2018) The interplay between oral microbiome, lifestyle factors and genetic polymorphisms in the risk of oral squamous cell carcinoma. Carcinogenesis 39(6):778–787. https://doi.org/10.1093/carcin/bgy053

Hu H, Lin A, Kong M, Yao X, Yin M, Xia H et al (2020) Intestinal microbiome and NAFLD: molecular insights and therapeutic perspectives. J Gastroenterol 55(2):142–158

Human Microbiome Project Consortium (2012) Structure, function and diversity of the healthy human microbiome. Nature 486:207–214

IARC (2009) IARC monographs on the evaluation of carcinogenic risks to humans 100, B. https://publications.iarc.fr/Book-And-Report-Series/IarcMonographs-On-The-Identification-Of-Carcinogenic-Hazards-To-Humans/biological-Agents-2012

IARC, Working group on the evaluation of carcinogenic risks to humans, International Agency for Research on Cancer, World Health Organization (1994) Schistosomes, liver flukes and helicobacter pylori, vol 61. International Agency for Research on Cancer, Lyon

Ilver D, Arnqvist A, Ogren J, Frick IM, Kersulyte D, Incecik ET et al (1998) Helicobacter pylori adhesin binding fucosylated histo-blood group antigens revealed by retagging. Science 279(5349):373–377. https://doi.org/10.1126/science.279.5349.373

Inaba, H., Sugita, H., Kuboniwa, M., Iwai, S., Hamada, M., Noda, T. et al. (2014). Porphyromonas gingivalis promotes invasion of oral squamous cell carcinoma through induction of pro MMP 9 and its activation. Cell Microbiol, 16(1), 131–145. ://doi.org/https://doi.org/10.1111/cmi.12211

Jain P, Luo ZQ, Blanke SR (2011) Helicobacter pylori vacuolating cytotoxin A (VacA) engages the mitochondrial fission machinery to induce host cell death. Proc Natl Acad Sci 108(38):16032–16037. https://doi.org/10.1073/pnas.1105175108

Johannsen A, Susin C, Gustafsson A (2014) Smoking and inflammation: evidence for a synergistic role in chronic disease. Periodontol 64(1):111–126. https://doi.org/10.1111/j.1600-0757.2012.00456.x

Kalasabail S, Engelman J, Zhang LY, El-Omar E, Yim HCH (2021) A perspective on the role of microbiome for colorectal cancer treatment. Cancers 13(18):4623. https://doi.org/10.3390/cancers13184623

Kostic AD, Chun E, Robertson L, Glickman JN, Gallini CA, Michaud M et al (2013) Fusobacterium nucleatum potentiates intestinal tumorigenesis and modulates the tumor-immune microenvironment. Cell Host Microbe 14(2):207–215. https://doi.org/10.1016/j.chom.2013.07.007

Kuboniwa M, Hasegawa Y, Mao S, Shizukuishi S, Amano A, Lamont RJ, Yilmaz Ö (2008) *P. gingivalis* accelerates gingival epithelial cell progression through the cell cycle. Microbes Infect 10(2):122–128. https://doi.org/10.1016/j.micinf.2007.10.011

Kuo YC, Yu LY, Wang HY, Chen MJ, Wu MS, Liu CJ, Lin YC, Shih SC, Hu KC (2022) Effects of *helicobacter pylori* infection in gastrointestinal tract malignant diseases: from the oral cavity to rectum. World J Gastrointest Oncol 14(1):55–74. https://doi.org/10.4251/wjgo.v14.i1.55

Li W, Deng Y, Chu Q, Zhang P (2019) Gut microbiome and cancer immunotherapy. Cancer Lett 447:41–47

Louis P, Hold GL, Flint HJ (2014) The gut microbiota, bacterial metabolites and colorectal cancer. Nat Rev Microbiol 12(10):661–672

Mager DL, Haffajee AD, Devlin PM, Norris CM, Posner MR, Goodson JM (2005) The salivary microbiota as a diagnostic indicator of oral cancer: a descriptive, non-randomized study of cancer-free and oral squamous cell carcinoma subjects. J Transl Med 3(1):1–8. https://doi.org/10.1186/1479-5876-3-27

Mao S, Park Y, Hasegawa Y, Tribble GD, James CE, Handfield M, Stavropoulos MF, Yilmaz Ö, Lamont RJ (2007) Intrinsic apoptotic pathways of gingival epithelial cells modulated by *Porphyromonas gingivalis*. Cell Microbiol 9:1997–2007. https://doi.org/10.1111/j.1462-5822.2007.00931.x

Mármol I, Sánchez-de-Diego C, Pradilla Dieste A, Cerrada E, Rodriguez Yoldi MJ (2017) Colorectal carcinoma: a general overview and future perspectives in colorectal cancer. Int J Mol Sci 18(1):197. https://doi.org/10.3390/ijms18010197
Michaud DS, Joshipura K, Giovannucci E, Fuchs CS (2007) A prospective study of periodontal disease and pancreatic cancer in US male health professionals. J Natl Cancer Inst 99(2): 171–175. https://doi.org/10.1093/jnci/djk021
Mima K, Nakagawa S, Sawayama H, Ishimoto T, Imai K, Iwatsuki M et al (2017) The microbiome and hepatobiliary-pancreatic cancers. Cancer Lett 402:9–15. https://doi.org/10.1016/j.canlet.2017.05.001
Mitsuhashi K, Nosho K, Sukawa Y, Matsunaga Y, Ito M, Kurihara H et al (2015) Association of Fusobacterium species in pancreatic cancer tissues with molecular features and prognosis. Oncotarget 6(9):7209. https://doi.org/10.18632/oncotarget.3109
Moskal A, Freisling H, Byrnes G, Assi N, Fahey MT, Jenab M et al (2016) Main nutrient patterns and colorectal cancer risk in the European prospective investigation into cancer and nutrition study. Br J Cancer 115(11):1430–1440. https://doi.org/10.1038/bjc.2016.334
Murata-Kamiya N, Kikuchi K, Hayashi T, Higashi H, Hatakeyama M (2010) Helicobacter pylori exploits host membrane phosphatidylserine for delivery, localization, and pathophysiological action of the CagA oncoprotein. Cell Host Microbe 7(5):399–411. https://doi.org/10.1016/j.chom.2010.04.005
Murphy G, Michel A, Taylor PR, Albanes D, Weinstein SJ, Virtamo J et al (2014) Association of seropositivity to helicobacter species and biliary tract cancer in the ATBC study. Hepatology 60(6):1963–1971. https://doi.org/10.1002/hep.27193
Mysak J, Podzimek S, Sommerova P, Lyuya-Mi Y, Bartova J, Janatova T et al (2014) Porphyromonas gingivalis: major periodontopathic pathogen overview. J Immunol Res 2014: 1. https://doi.org/10.1155/2014/476068
Ngiow SF, Young A (2020) Re-education of the tumor microenvironment with targeted therapies and immunotherapies. Front Immunol 11:1633. https://doi.org/10.3389/fimmu.2020.01633
Nilsson HO, Stenram U, Ihse I, Wadström T (2002) Re: helicobacter pylori seropositivity as a risk factor for pancreatic cancer. J Natl Cancer Inst 94(8):632–633. https://doi.org/10.1016/S1473-3099(08)70066-5
Pagano JS, Blaser M, Buendia MA, Damania B, Khalili K, Raab-Traub N, Roizman B (2004) Infectious agents and cancer: criteria for a causal relation. In: Seminars in cancer biology, vol 14. Academic, Cambridge, MA, pp 453–471
Pleguezuelos-Manzano C, Puschhof J, Rosendahl Huber A, van Hoeck A, Wood HM, Nomburg J et al (2020) Mutational signature in colorectal cancer caused by genotoxic pks+ E. coli. Nature 580(7802):269–273. https://doi.org/10.1038/s41586-020-2080-8
Poonacha KNT, Villa TG, Notario V (2022) The interplay among radiation therapy, antibiotics and the microbiota: impact on cancer treatment outcomes. Antibiotics 11(3):331. https://doi.org/10.3390/antibiotics11030331
Poutahidis T, Erdman SE (2016) Commensal bacteria modulate the tumor microenvironment. Cancer Lett 380(1):356–358. https://doi.org/10.1016/j.canlet.2015.12.028
Qiu Q, Lin Y, Ma Y, Li X, Liang J, Chen Z et al (2021) Exploring the emerging role of the gut microbiota and tumor microenvironment in cancer immunotherapy. Front Immunol 11:612202. https://doi.org/10.3389/fimmu.2020.612202
Rajagopala SV, Vashee S, Oldfield LM, Suzuki Y, Venter JC, Telenti A, Nelson KE (2017) The human microbiome and cancer. The human microbiome and cancer. Cancer Prev Res 10(4): 226–234
Riehl TE, Alvarado D, Ee X, Zuckerman A, Foster L, Kapoor V et al (2019) Lactobacillus rhamnosus GG protects the intestinal epithelium from radiation injury through release of lipoteichoic acid, macrophage activation and the migration of mesenchymal stem cells. Gut 68(6):1003–1013. https://doi.org/10.1136/gutjnl-2018-316226
Roberti MP, Yonekura S, Duong CP, Picard M, Ferrere G, Tidjani Alou M et al (2020) Chemotherapy-induced ileal crypt apoptosis and the ileal microbiome shape

immunosurveillance and prognosis of proximal colon cancer. Nat Med 26(6):919–931. https://doi.org/10.1038/s41591-020-0882-8
Rubinstein MR, Wang X, Liu W, Hao Y, Cai G, Han YW (2013) Fusobacterium nucleatum promotes colorectal carcinogenesis by modulating E-cadherin/β-catenin signaling via its FadA adhesin. Cell Host Microbe 14(2):195–206. https://doi.org/10.1016/j.chom.2013.07.012
Ruiz-Ruiz JC, Matus-Basto AJ, Acereto-Escoffié P, Segura-Campos MR (2017) Antioxidant and anti-inflammatory activities of phenolic compounds isolated from Melipona beecheii honey. Food Agric Immunol 28(6):1424–1437. https://doi.org/10.1080/09540105.2017.1347148
Saus E, Iraola-Guzmán S, Willis JR, Brunet-Vega A, Gabaldón T (2019) Microbiome and colorectal cancer: roles in carcinogenesis and clinical potential. Mol Asp Med 69:93–106. https://doi.org/10.1016/j.mam.2019.05.001
Schetter AJ, Heegaard NH, Harris CC (2010) Inflammation and cancer: interweaving microRNA, free radical, cytokine and p53 pathways. Carcinogenesis 31(1):37–49. https://doi.org/10.1093/carcin/bgp272
Schwabe RF, Jobin C (2013) The microbiome and cancer. Nat Rev Cancer 13(11):800–812. https://doi.org/10.1038/nrc3610
Scott TA, Quintaneiro LM, Norvaisas P, Lui PP, Wilson MP, Leung KY et al (2017) Host-microbe co-metabolism dictates cancer drug efficacy in C. elegans. Cell 169(3):442–456. https://doi.org/10.1016/j.cell.2017.03.040
Shi W, Shen L, Zou W, Wang J, Yang J, Wang Y et al (2020) The gut microbiome is associated with therapeutic responses and toxicities of neoadjuvant chemoradiotherapy in rectal cancer patients—a pilot study. Front Cell Infect Microbiol 10:562463. https://doi.org/10.3389/fcimb.2020.562463
Sivan A, Corrales L, Hubert N, Williams JB, Aquino-Michaels K, Earley ZM et al (2015) Commensal Bifidobacterium promotes antitumor immunity and facilitates anti–PD-L1 efficacy. Science 350(6264):1084–1089. https://doi.org/10.1126/science.aac4255
Stolzenberg-Solomon RZ, Dodd KW, Blaser MJ, Virtamo J, Taylor PR, Albanes D (2003) Tooth loss, pancreatic cancer, and helicobacter pylori. Am J Clin Nutr 78(1):176–181. https://doi.org/10.1093/ajcn/78.1.176
Sung C-E, Lin F-G, Huang R-Y, Fang W-H, Cheng W-C, Tsai Y-WC, Chen W-L (2022) Periodontitis, *helicobacter pylori* infection, and gastrointestinal tract cancer mortality. J Clin Periodontol 49(3):210–220. https://doi.org/10.1111/jcpe.13590
Tanoue T, Morita S, Plichta DR, Skelly AN, Suda W, Sugiura Y et al (2019) A defined commensal consortium elicits CD8 T cells and anti-cancer immunity. Nature 565(7741):600–605. https://doi.org/10.1038/s41586-019-0878-z
Tong CM, Ma S, Guan XY (2011) Biology of hepatic cancer stem cells. J Gastroenterol Hepatol 26(8):1229–1237. https://doi.org/10.1093/ajcn/78.1.176
Tonneau M, Elkrief A, Pasquier D, Del Socorro TP, Chamaillard M, Bahig H, Routy B (2021) The role of the gut microbiome on radiation therapy efficacy and gastrointestinal complications: a systematic review. Radiother Oncol 156:1–9
Trikudanathan G, Philip A, Dasanu CA, Baker WL (2011) Association between helicobacter pylori infection and pancreatic cancer. A cumulative meta-analysis. JOP 12(1):26–31. https://doi.org/10.6092/1590-8577/3379
Tsugawa H, Suzuki H, Saya H, Hatakeyama M, Hirayama T, Hirata K et al (2012) Reactive oxygen species-induced autophagic degradation of helicobacter pylori CagA is specifically suppressed in cancer stem-like cells. Cell Host Microbe 12(6):764–777. https://doi.org/10.1016/j.chom.2012.10.014
Uribe-Herranz M, Rafail S, Beghi S, Gil-de-Gómez L, Verginadis I, Bittinger K et al (2020) Gut microbiota modulate dendritic cell antigen presentation and radiotherapy-induced antitumor immune response. J Clin Invest 130(1):466–479. https://doi.org/10.1172/JCI124332
Ursell LK, Metcalf JL, Parfrey LW, Knight R (2012) Defining the human microbiome. Nutr Rev 70(suppl_1):S38–S44

Wagner BD, Grunwald GK, Zerbe GO, Mikulich-Gilbertson SK, Robertson CE, Zemanick ET, Harris JK (2018) On the use of diversity measures in longitudinal sequencing studies of microbial communities. Front Microbiol 9:1037. https://doi.org/10.3389/fmicb.2018.01037

Wang Z, Wang Q, Wang X, Zhu L, Chen J, Zhang B et al (2019) Gut microbial dysbiosis is associated with development and progression of radiation enteritis during pelvic radiotherapy. J Cell Mol Med 23(5):3747–3756. https://doi.org/10.1111/jcmm.14289

Wei J, Nagy TA, Vilgelm A, Zaika E, Ogden SR, Romero Gallo J et al (2010) Regulation of p53 tumor suppressor by helicobacter pylori in gastric epithelial cells. Gastroenterology 139(4): 1333–1343. https://doi.org/10.1053/j.gastro.2010.06.018

Weitzman MD, Weitzman JB (2014) What's the damage? The impact of pathogens on pathways that maintain host genome integrity. Cell Host Microbe 15(3):283–294

Wiest T, Schwarz E, Enders C, Flechtenmacher C, Bosch FX (2002) Involvement of intact HPV16 E6/E7 gene expression in head and neck cancers with unaltered p53 status and perturbed pRb cell cycle control. Oncogene 21(10):1510–1517. https://doi.org/10.1038/sj.onc.1205214

Wu S, Rhee KJ, Albesiano E, Rabizadeh S, Wu X, Yen HR et al (2009) A human colonic commensal promotes colon tumorigenesis via activation of T helper type 17 T cell responses. Nat Med 15(9):1016–1022. https://doi.org/10.1038/nm.2015

Yilmaz O, Jungas T, Verbeke P, Ojcius DM (2004) Activation of the phosphatidylinositol 3-kinase/ Akt pathway contributes to survival of primary epithelial cells infected with the periodontal pathogen Porphyromonas gingivalis. Infect Immun 72(7):3743–3751. https://doi.org/10.1128/IAI.72.7.3743-3751.2004

Yilmaz O, Yao L, Maeda K, Rose TM, Lewis EL, Duman M, Lamont RJ, Ojcius DM (2008) ATP scavenging by the intracellular pathogen Porphyromonas gingivalis inhibits P2X7-mediated host-cell apoptosis. Cell Microbiol 10(4):863–875. https://doi.org/10.1111/j.1462-5822.2007.01089.x

Yu J, Feng Q, Wong SH, Zhang D, Yi Liang Q, Qin Y et al (2017) Metagenomic analysis of faecal microbiome as a tool towards targeted non-invasive biomarkers for colorectal cancer. Gut 66(1): 70–78. https://doi.org/10.1136/gutjnl-2015-309800

Zhang S, Li C, Liu J, Geng F, Shi X, Li Q et al (2020) Fusobacterium nucleatum promotes epithelial-mesenchymal transition through regulation of the lncRNA MIR4435-2HG/miR-296-5p/Akt2/SNAI1 signaling pathway. FEBS J 287(18):4032–4047

Zhang Y, Yu G, Chu H, Wang X, Xiong L, Cai G et al (2018) Macrophage-associated PGK1 phosphorylation promotes aerobic glycolysis and tumorigenesis. Mol Cell 71(2):201–215. https://doi.org/10.1016/j.molcel.2018.06.023

Zhang Y, Zhen M, Zhan Y, Song Y, Zhang Q, Wang J (2017) Population-genomic insights into variation in Prevotella intermedia and Prevotella nigrescens isolates and its association with periodontal disease. Front Cell Infect Microbiol 7:409. https://doi.org/10.3389/fcimb.2017.00409

Zhou Y, Fei M, Zhang G, Liang WC, Lin W, Wu Y et al (2020a) Blockade of the phagocytic receptor MerTK on tumor-associated macrophages enhances P2X7R-dependent STING activation by tumor-derived cGAMP. Immunity 52(2):357–373

Zhou Z, Ge S, Li Y, Ma W, Liu Y, Hu S et al (2020b) Human gut microbiome-based knowledge base as a biomarker screening tool to improve the predicted probability for colorectal cancer. Front Microbiol 11:596027. https://doi.org/10.3389/fmicb.2020.596027

Zhou Z, Wang Y, Ji R, Zhang D, Ma C, Ma W et al (2022) Vanillin derivatives reverse fusobacterium nucleatum-induced proliferation and migration of colorectal cancer through E-cadherin/β-catenin pathway. Front Pharmacol 13:841918

Microbial Secondary Metabolites: Targeting Tumors and Associated Challenges 19

Simran Rani, Pradeep Kumar, Priyanka Dahiya, Amita Suneja Dang, and Pooja Suneja

Abstract

Secondary metabolites derived from microbes have been utilized for a number of applications in the medical sector for a considerable time. Microbial secondary metabolites (MSMs) are utilized as enzyme inhibitors, immunosuppressants, hypocholesterolemic agents, etc., in the healthcare sector. Owing to their high efficacy and biocompatibility, several MSMs have also been applied for cancer treatment, responsible for major mortality in the world. Plenty of these metabolites are still under clinical trials awaiting large-scale application. This chapter summarizes the current status of tumor-targeting MSMs, recent innovations to elucidate the antitumor mechanism of these metabolites, and exploring potential microbes for the production of novel antitumor metabolites using co-cultivation and omics approaches.

Keywords

Metabolites · MSMs · Cancer · Co-cultivation · Omics

19.1 Introduction

Cancer is one of the leading causes of death worldwide among the Non-communicable diseases (NCDs) with breast cancer, lung cancer, colon cancer, rectum cancer, and prostate cancer being the most common (Siegel et al. 2019).

S. Rani · P. Kumar · P. Dahiya · P. Suneja (✉)
Department of Microbiology, M. D. University, Rohtak, India

A. S. Dang
Centre for Medical Biotechnology, M. D. University, Rohtak, India

R. Sobti et al. (eds.), *Role of Microbes in Sustainable Development*,
https://doi.org/10.1007/978-981-99-3126-2_19

Chemotherapy is the key treatment available for targeting tumors, immunotherapy, hormone therapy, nanovaccines, photocatalytic therapy, and ferroptosis (Son et al. 2020). A plethora of research in this particular field has facilitated the introduction of novel antibodies, drugs, and similar products for targeting several types of tumors, the result of this advancement has not been witnessed entirely (Hu et al. 2021).

Microorganisms generate a variety of metabolites in their entire lifecycle. Despite their formation in all living organisms, microbes remain the preferred source of metabolites owing to their accessibility, effectiveness, and environment friendliness (Singh et al. 2017). Microbial secondary metabolites (MSMs) are formed near or at the end of the stationary phase playing an indirect role in the growth, development, and reproduction of organisms. These are a heterogeneous group of natural products that include terpenoids and steroids, alkaloids, fatty acid-derived substances and polyketides, non-ribosomal polypeptides, and enzyme cofactors (Dowd and Kelley 2011; Thirumurugan et al. 2018). The word "secondary" may give us a sense of these metabolites being less important as compared to the central metabolites, but the research on genetics concerning their production has facilitated their innumerable applications from agriculture to biomedical science. A significant portion of genome is involved in the production of microbial secondary metabolites (MSMs) (Sharrar et al. 2020). Earlier, the use of MSMs in health care was restricted to discovering novel metabolites having antibiotic activities, but it has witnessed a gradual shift in the recent past covering an even wider range of research from modifying current antibiotics to isolating and identifying metabolites having activities other than antibioses like anti-neoplastic agents, immunosuppressive drugs, and enzyme inhibitors.

MSMs have been serving as anticancer agents for a very long time since the discovery of actinomycin from *Streptomyces antibioticus* way back in 1940. A significant proportion of antitumor drugs under clinical use are of microbial origin (more than 60%) (Ramírez-Rendon et al. 2022). Nevertheless, challenges such as the never-ending demand for novel and efficient anticancer compounds, nontargeted delivery, and adverse effects on tissues are major hindrances in the large-scale application of the same (Mohan et al. 2021). However, exploring the hitherto untouched microflora, co-culture engineering, and "omics" led revolution is improving the manipulation of microbial metabolites for the welfare of mankind. This chapter discusses the recent development associated with the use of MSMs to fight cancer along with their challenges and the future of research in this regard.

19.2 Antitumor MSMs: A Bibliometric Analysis

A bibliometric analysis was done to assess the current status of research on the use of MSMs to treat the four most common types of cancers according to WHO (used in the keywords). The keywords, "Microbe" AND "Metabolites" AND "Colorectal cancer" AND "Breast cancer" AND "Lung cancer" AND "Prostate cancer," were searched in the Scopus literature database and filtered in period between 2012 and 2022. According to the Scopus database, the maximum no. of results was produced

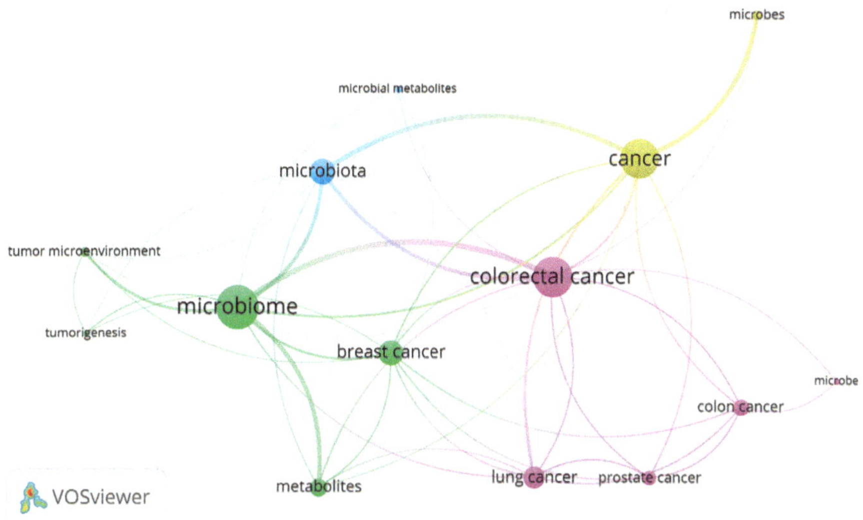

Fig. 19.1 Bibliometric analysis of the keywords of the antitumor activity of microbial metabolites

for colorectal cancer (147) followed by breast cancer (62), lung cancer (54), and prostate cancer (22). The United States tops the list of countries contributing to studying the potential microbial metabolites as antitumor agents against these particular cancers. China, India, Australia, and Germany were also found to be at the top of the list. The maximum proportion of literature belonged to the subject area of "Biochemistry, Genetics & Molecular biology" (in the case of colorectal and prostate cancer) and "Medicine" (breast cancer and lung cancer). "Immunology and microbiology" remained another significant subject area for all these cancers. All the papers obtained (285) were also extracted from Scopus and downloaded as CSV files that were imported to the VOSviewer software, Leiden University, The Netherland, Year (version 1.6.18), for the bibliographic data analysis based on the keywords. Owing to the good amount of literature produced, the highest number of occurrences, links, and total link strength (TLS), the circle with the largest diameter is occupied by colorectal cancer. The other three circles representing breast, lung, and prostate cancer are comparatively small and have a lesser number of occurrence links, and TLS with prostate cancer has the least amount of literature owned by it (Fig. 19.1).

19.3 Tumor-Targeting MSMs: Clinical Trials, Approvals, and Mechanism

Since the discovery of penicillin by Flemming in 1929 from *Penicillium notatum,* we have witnessed a long period of using microbial metabolites in the pharmaceutical sector. However, the focus has shifted toward newer applications in the recent past employing MSMs as immunosuppressants, enzyme inhibitors, motilides, hypocholesterolemic, and antitumor agents, etc. (Ramírez-Rendon et al. 2022).

The mechanism of some of the most effective anticancer drugs derived from microbes has been discussed below. Anthracyclines are one of the most effective antitumor agents against a variety of cancers (leukemia, breast, lymphomas, uterine, ovarian, and lung). They intercalate into the DNA strand inhibiting their synthesis. Daunomycin produced by *Streptomyces peucetius* was the first anthracycline to be discovered. A number of its analogs are under clinical trials (Kadurin et al. 2017; Ramírez-Rendon et al. 2022). Romidepsin approved against cutaneous T-cell lymphoma (CTCL) acts by inhibiting the histone deacetylase. It is produced by *Chromobacterium violaceum* and has shown good activity against testicular cancer among three other cancer cell lines (Xiong et al. 2019). Actinobacteria-derived carfilzomib has been approved by FDA against multiple myeloma. It leads to the accumulation of poly-ubiquitinated proteins and cell cycle arrest by binding to the 20S proteasome and inhibiting its chymotrypsin-like activity (Park et al. 2019). Camptothecin is a monoterpene indole alkaloid isolated initially from the plants but reported to be produced by fungal endophyte, *Entrophospora infrequens*. It has yielded promising results against lung, ovarian, and uterus cancer (Li et al. 2017). Taxol (steroidal alkaloid diterpenoid), a leading cancer-fighting molecule, initially isolated from *Taxus brevifolia*, is also produced by a number of fungal genera like *Alternaria, Aspergillus, Phoma, and Penicillium* (Shankar Naik 2019). Actinomycin D/dactinomycin (*Streptomyces parvulus, Streptomyces antibioticus*), bleomycin (*Streptomyces verticillus*), epothilones/ixabepilones (*Sorangium cellulosum*), mitomycin C (*Streptomyces caespitosus*), pentostatin (*Streptomyces antibioticus*), and rapamycin/sirolimus (*Streptomyces hygroscopicus*) are the other clinically used anticancer drugs derived from microbes especially actinobacteria (Mohan et al. 2021) (Table 19.1).

19.4 Advances Driving the Utilization of MSMs to Treat Cancer

19.4.1 Exploring the Unexplored

With huge changes in lifestyle and a significant increase in population, there is a need for a constant search for new metabolites with higher antitumor potential and lesser toxicity. Exploring the unexplored diversity of world microflora and improved computational tools is a way forward to this. Despite the enormous diversity of microorganisms, the proportion of culturable microbes is very less. It is essential to make dedicated efforts toward the ways to culture and utilize these uncultivable microorganisms. Besides this, microbes adapted to extreme environments such as desert soils, deep-sea sediments, highly acidic habitats, saline and hypersaline habitats, and high-temperature environment are a rich source of new specialized metabolites and can serve as a treasure of anticancer MSMs. However, these habitats are falling prey to anthropogenic factors like mining and climate change, for which microbiologists need to take a strong stand to ensure their conservation by the

Table 19.1 Current studies employing MSMs as antitumor agents

Microbial class	Metabolite	Microbial source	Antitumor activity	Mechanism of action	Reference
Bacteria	Anthrone derivatives	*Actinomadura* sp. BCC47066	KB, MCF-7, NCI-H187	Cytotoxicity	Bunbamrung et al. (2018)
	Pradimicin-IRD	*Amycolatopsis* sp. IRD-009	Colon cancer	Inducing DNA damage, apoptosis, and cell cycle arrest	Almeida et al. (2019)
	Thiocoraline	*Micromonospora marina*	MCF-7	Increase in Akt phosphorylation and BCR proteins	Jin et al. (2019)
	CFZ-albN	Actinomycetes	4 T1, MDA-MB-231, MCF-7, HCC1640, HCC1937, U87MG	Proteasome inhibition	Park et al. (2019)
	Crude methanol extract	*Streptomyces* sp. MUM 265	HT-29, Caco-2	Depolarization of mitochondrial membrane potential and accumulation of subG1 cells inducing apoptosis	Tan et al. (2019)
	Romipeptides A and B	*Chromobacterium violaceum* no. 968	SW620, HL60, A549	Cytotoxicity	Xiong et al. (2019)
	Butyrate	Gut microbiota	Promotes antitumor immune response	Modulation of CD8+ T-cell response	He et al. (2021)
Diatoms	Stigmasterol	*Navicula incerta*	HepG2	Upregulation of Bax, p53 (pro-apoptotic genes), and downregulation of Bcl-2 (anti-apoptotic gene)	Kim et al. (2014)
	Polyunsaturated aldehydes	Diatoms	A549, COLO205	Activation of an extrinsic apoptotic machinery	Sansone et al. (2014)
	Monoacylglycerides	*Skeletonema marinoi*	U-937, HCT-116	Caspase 3/7 activation	Miceli et al. (2019)
Yeast	Supernatant	*Saccharomyces boulardii*	MCF-7 and MCF-7/MX	Suppression of surviving gene expression-inducing apoptosis	Pakbin et al. (2022)
Fungi			A549, HepG2	–	

(continued)

Table 19.1 (continued)

Microbial class	Metabolite	Microbial source	Antitumor activity	Mechanism of action	Reference
	Ergosterimide B, demethylincisterol A5	*Aspergillus tubingensis* YP-2			Yu et al. (2021)
	Epicorazine A	*Epicoccum nigrum*	L5178Y, Ramos and Jurkat J16	Inducing cell death	Harwoko et al. (2021)

Cell lines—A549: lung adenocarcinoma; HepG2: hepatocarcinoma; L5178Y: mouse lymphoma, Ramos: human lymphoma, Jurkat J16: human leukemia; COLO205: colon adenocarcinoma; U-937: hematological cancer, HCT-116: colon cancer; MCF-7, MCF-7/MX: breast carcinoma; HT-29, Caco-2: colon cancer; 4 T1: murine breast cancer, MDA-MB-231, MCF-7, HCC1640, HCC1937, U87MG: human breast cancer; HL60: acute promyelocytic leukemia, SW620: colonic carcinoma policymakers (Sayed et al. 2020). Endophytes are also a reliable and non-exhaustible source of secondary metabolites having enormous potential to be utilized not just in the arena of cancer treatment but in the whole pharma industry (Bhutani et al. 2018; Rani et al. 2021). Concentrated crude extracts of three endophyte belonging to *Pantoea* genus residing in the *Solanum mauritianum* shoot were tested against two human cancer cell lines, A549 lung carcinoma and UMG87 glioblastoma cell lines (Uche-Okereafor et al. 2019). Wang et al. (2022) isolated and identified several trichothecenes from *Fusarium sporotrichioides* (host—*Rauvolfia yunnanensis* Tsiang) out of which 8-n-butyrylneosolaniol displayed a significant level of antitumor activity by inducing cell apoptosis in Huh-7 cells via the mitochondria-mediated apoptotic signaling pathway. Further analysis revealed that the cell cycle arrest at G2/M phase leads to cell proliferation inhibition and pro-apoptotic activity. Similar studies have demonstrated the potential of other fungal metabolites, jammosporin A, vinblastine, and paclitaxel, derived from *Rosellinia sanctae-cruciana* (*Albizia lebbeck*), *Curvularia verruculosa* (*Catharanthus roseus*), and *Aspergillus fumigatus* (*Taxus* sp.), respectively (Sharma et al. 2018; Parthasarathy et al. 2020; Kumar et al. 2019).

19.4.2 Co-Culturing

Several microbes produce an insignificant yield of secondary metabolites that makes them unsuitable for large-scale production and discourage capitalists and firms due to economic factors. Co-culture involves culturing more than one type of microbes together and exploiting the intermicrobial communications for the enhanced production of bioactive metabolites (Yuan and Alper 2019). Other than resulting in the production of novel metabolites, it has also been reported to enhance the production of already-known metabolites. A fungal endophyte of *Taxus*, *Paraconiothyrium* SSMoo1, has shown a small increase in the production of taxol after co-culturing it with *Alternaria* and a further increase after adding *Phomopsis* (Soliman and Raizada 2013). Shin et al. (2018) reported the production of an anticancer bioactive compound, Dentigerumycin E by *Streptomyces albogriseolus* B24 in the presence of an inducer, *Bacillus cereus*. However, compatibility issues, competition for substrates, and data acquisition problems exist for which there is a need to study the metabolic pathways in more detail and develop proper databases to make it popular for bioproduction (Jawed et al. 2019; Padmaperuma et al. 2018).

19.4.3 "OMICS" Approaches: A Leap Forward in the Search for MSMs

Biosynthetic gene clusters (BGCs) are equipped with enormous potential to yield valuable secondary metabolites but remain either silent or poorly expressed due to complex regulation involved at transcriptional, translational, and post-translational levels. Some techniques have been developed in the recent past to trigger their expression. Genome mining with tools such as antiSMASH, PRISM3, and comparative genomics estimates the genetic potential of a microbe by scanning the genome of interest and identifying BGCs. The complete genome sequencing and comparative analysis enable the reconstruction of both primary and secondary metabolic pathways suggesting key metabolic genes for metabolic engineering (Ziemert et al. 2016; Palazzotto and Weber 2018). Comparing protein expression levels as in proteomics provides information on different pathways regulation highlighting significant players in natural product biosynthesis that can be used to target rational engineering. Metagenomics is a unique approach that can explore the huge proportion of uncultivable microorganisms, which otherwise remains largely unexplored by analyzing the DNA isolated from the environment possible (Bodor et al. 2020). Metabolomics enables the metabolic comparison of different biological samples by nuclear magnetic resonance (NMR) and mass spectrometry (MS) leading to the identification of secondary metabolites from BGCs. Combined use of multi-meta-omics approaches like metabologenomics provides us with a complete picture of microbial metabolism throwing light on silent BGCs and the role of natural products encoded by the same (Palazzotto and Weber 2018).

19.5 Conclusion

Every new development introduced in biomedical science to cope with the existing challenges is associated with advantages and limitations. There is no doubt about the potential of microbial secondary metabolites targeting cancer, one of the deadliest diseases, causing 1 out of every 6 deaths globally (WHO 2018). Therefore, we require a continuous emphasis on searching metabolites with similar/higher scales of antitumor activity, addressing the glitches at the level of commercial production, and elucidating the molecular basis of the anticancer mechanism. Natural product-based antitumor drugs have an edge over those of synthetic origin with regard to the toxicity caused. Nonetheless, they are not completely devoid of the same (Mohan et al. 2021). Recent progress made in the field of nanotechnology is facilitating the resolution of this issue by using nano-carriers equipped with targeted delivery and improved toxicity profile of the product. Nanomedicines improve the time period for which the drug stays in the blood circulation and along with enhancing its stability. Nano-bacterial therapy utilizes nanomolecular properties of microbe-derived natural products for increasing permeability and retention of drugs, besides specific targeting (Cao et al. 2020; Xie et al. 2022). Chemical modification of these metabolites and improving the deliverability of metabolites are also required to unleash the potential of these compounds in a better way. Employing MSMs in combination with other antitumor therapies available also has the potential to serve this purpose.

References

Almeida LCD, Bauermeister A, Rezende-Teixeira P, Santos EAD, Machado-Neto JA, Costa-Lotufo LV (2019) Pradimicin-IRD exhibits antineoplastic effects by inducing DNA damage in colon cancer cells. Biochem Pharmacol 168:38–47

Bhutani N, Maheshwari R, Negi M, Suneja P (2018) Optimization of IAA production by endophytic bacillus spp. from Vigna radiata for their potential use as plant growth promoters. Isr J Plant Sci 65(1–2):83–96

Bodor A, Bounedjoum N, Vincze GE, Erdeiné Kis Á, Laczi K, Bende G, Szilágyi Á, Kovács T, Perei K, Rákhely G (2020) Challenges of unculturable bacteria: environmental perspectives. Rev Environ Sci Biotechnol 19(1):1–22

Bunbamrung N, Supong K, Intaraudom C, Dramae A, Auncharoen P, Pittayakhajonwut P (2018) Anthrone derivatives from the terrestrial actinomycete, Actinomadura sp. BCC47066. Phytochem Lett 25:109–117

Cao J, Huang D, Peppas NA (2020) Advanced engineered nanoparticulate platforms to address key biological barriers for delivering chemotherapeutic agents to target sites. Adv Drug Deliv Rev 167:170–188

Dowd CJ, Kelley B (2011) Purification process design and the influence of product and technology platforms. Elsevier, Amsterdam

Harwoko H, Daletos G, Stuhldreier F, Lee J, Wesselborg S, Feldbrügge M, Müller WE, Kalscheuer R, Ancheeva E, Proksch P (2021) Dithiodiketopiperazine derivatives from endophytic fungi Trichoderma harzianum and Epicoccum nigrum. Nat Prod Res 35(2):257–265

He Y, Fu L, Li Y, Wang W, Gong M, Zhang J, Dong X, Huang J, Wang Q, Mackay CR, Fu YX (2021) Gut microbial metabolites facilitate anticancer therapy efficacy by modulating cytotoxic CD8+ T cell immunity. Cell Metab 33(5):988–1000

Hu J, Yuan X, Wang F, Gao H, Liu X, Zhang W (2021) The progress and perspective of strategies to improve tumor penetration of nanomedicines. Chin Chem Lett 32(4):1341–1347

Jawed K, Yazdani SS, Koffas MA (2019) Advances in the development and application of microbial consortia for metabolic engineering. Metab Eng Commun 9:e00095

Jin J, Zhao Y, Guo W, Wang B, Wang Y, Liu X, Xu C (2019) Thiocoraline mediates drug resistance in MCF-7 cells via PI3K/Akt/BCRP signaling pathway. Cytotechnology 71(1):401–409

Kadurin A, Aliper A, Kazennov A, Mamoshina P, Vanhaelen Q, Khrabrov K, Zhavoronkov A (2017) The cornucopia of meaningful leads: applying deep adversarial autoencoders for new molecule development in oncology. Oncotarget 8(7):10883

Kim YS, Li XF, Kang KH, Ryu B, Kim SK (2014) Stigmasterol isolated from marine microalgae Navicula incerta induces apoptosis in human hepatoma HepG2 cells. BMB Rep 47(8):433

Kumar P, Singh B, Thakur V, Thakur A, Thakur N, Pandey D, Chand D (2019) Hyper-production of taxol from aspergillus fumigatus, an endophytic fungus isolated from Taxus sp of the Northern Himalayan region. Biotechnol Rep 24:e00395

Li F, Jiang T, Li Q, Ling X (2017) Camptothecin (CPT) and its derivatives are known to target topoisomerase I (Top1) as their mechanism of action: did we miss something in CPT analogue molecular targets for treating human disease such as cancer? Am J Cancer Res 7(12):2350

Miceli M, Cutignano A, Conte M, Ummarino R, Romanelli A, Ruvo M, Leone M, Mercurio FA, Doti N, Manzo E, Romano G (2019) Monoacylglycerides from the diatom Skeletonema marinoi induce selective cell death in cancer cells. Mar Drugs 17(11):625

Mohan CD, Rangappa S, Nayak SC, Jadimurthy R, Wang L, Sethi G, Garg M, Rangappa KS (2021) May. Bacteria as a treasure house of secondary metabolites with anticancer potential. In: Seminars in cancer biology, vol 86. Academic, Cambridge, MA, p 998

Padmaperuma G, Kapoore RV, Gilmour DJ, Vaidyanathan S (2018) Microbial consortia: a critical look at microalgae co-cultures for enhanced biomanufacturing. Crit Rev Biotechnol 38(5): 690–703

Pakbin B, Dibazar SP, Allahyari S, Javadi M, Amani Z, Farasat A, Darzi S (2022) Anticancer properties of probiotic saccharomyces boulardii supernatant on human breast cancer cells. Probiotics Antimicrob Proteins 14:1130–1138

Palazzotto E, Weber T (2018) Omics and multi-omics approaches to study the biosynthesis of secondary metabolites in microorganisms. Curr Opin Microbiol 45:109–116

Park JE, Park J, Jun Y, Oh Y, Ryoo G, Jeong YS, Gadalla HH, Min JS, Jo JH, Song MG, Kang KW (2019) Expanding therapeutic utility of carfilzomib for breast cancer therapy by novel albumin-coated nanocrystal formulation. J Control Release 302:148–159

Parthasarathy R, Shanmuganathan R, Pugazhendhi A (2020) Vinblastine production by the endophytic fungus Curvularia verruculosa from the leaves of Catharanthus roseus and its in vitro cytotoxicity against HeLa cell line. Anal Biochem 593:113530

Ramírez-Rendon D, Passari AK, Ruiz-Villafán B, Rodríguez-Sanoja R, Sánchez S, Demain AL (2022) Impact of novel microbial secondary metabolites on the pharma industry. Appl Microbiol Biotechnol 106:1–24

Rani S, Kumar P, Suneja P (2021) Biotechnological interventions for inducing abiotic stress tolerance in crops. Plant Gene 27:100315

Sansone C, Braca A, Ercolesi E, Romano G, Palumbo A, Casotti R, Francone M, Ianora A (2014) Diatom-derived polyunsaturated aldehydes activate cell death in human cancer cell lines but not normal cells. PLoS One 9(7):e101220

Sayed AM, Hassan MH, Alhadrami HA, Hassan HM, Goodfellow M, Rateb ME (2020) Extreme environments: microbiology leading to specialized metabolites. J Appl Microbiol 128(3): 630–657

Shankar Naik B (2019) Developments in taxol production through endophytic fungal biotechnology: a review. Orient Pharm Exp Med 19(1):1–13

Sharma N, Kushwaha M, Arora D, Jain S, Singamaneni V, Sharma S, Shankar R, Bhushan S, Gupta P, Jaglan S (2018) New cytochalasin from Rosellinia sanctae-cruciana, an endophytic fungus of Albizia lebbeck. J Appl Microbiol 125(1):111–120

Sharrar AM, Crits-Christoph A, Méheust R, Diamond S, Starr EP, Banfield JF (2020) Bacterial secondary metabolite biosynthetic potential in soil varies with phylum, depth, and vegetation type. MBio 11(3):e00416–e00420

Shin D, Byun WS, Moon K, Kwon Y, Bae M, Um S, Lee SK, Oh DC (2018) Coculture of marine Streptomyces sp. with bacillus sp. produces a new piperazic acid-bearing cyclic peptide. Front Chem 6:498

Siegel RL, Miller KD, Jemal A (2019) Cancer statistics, 2019. CA Cancer J Clin 69(1):7–34

Singh R, Kumar M, Mittal A, Mehta PK (2017) Microbial metabolites in nutrition, healthcare and agriculture. 3 Biotech 7(1):15

Soliman SS, Raizada MN (2013) Interactions between co-habitating fungi elicit synthesis of Taxol from an endophytic fungus in host Taxus plants. Front Microbiol 4:3

Son S, Kim JH, Wang X, Zhang C, Yoon SA, Shin J, Sharma A, Lee MH, Cheng L, Wu J, Kim JS (2020) Multifunctional sonosensitizers in sonodynamic cancer therapy. Chem Soc Rev 49(11): 3244–3261

Tan LTH, Chan KG, Pusparajah P, Yin WF, Khan TM, Lee LH, Goh BH (2019) Mangrove derived Streptomyces sp. MUM265 as a potential source of antioxidant and anticolon-cancer agents. BMC Microbiol 19(1):1–16

Thirumurugan D, Cholarajan A, Raja SS, Vijayakumar R (2018) An introductory chapter: secondary metabolites. In: Vijayakumar R, Raja SS (eds) Secondary metabolites: sources and applications BoD–books on demand. IntechOpen, London, pp 1–21

Uche-Okereafor N, Sebola T, Tapfuma K, Mekuto L, Green E, Mavumengwana V (2019) Antibacterial activities of crude secondary metabolite extracts from Pantoea species obtained from the stem of Solanum mauritianum and their effects on two cancer cell lines. Int J Environ Res Public Health 16(4):602

Wang YJ, Ma N, Lu YF, Dai SY, Song X, Li C, Sun Y, Pei YH (2022) Structure elucidation and anti-tumor activities of Trichothecenes from endophytic fungus fusarium sporotrichioides. Biomol Ther 12(6):778
Xie YJ, Huang M, Li D, Hou JC, Liang HH, Nasim AA, Huang JM, Xie C, Leung ELH, Fan XX (2022) Bacteria-based nanodrug for anticancer therapy. Pharmacol Res 182:106282
Xiong L, Chang-Fa CHEN, Tao-Ling MIN, Hai-Feng HU (2019) Romipeptides a and B, two new romidepsin derivatives isolated from Chromobacterium violaceum no. 968 and their antitumor activities in vitro. Chin J Nat Med 17(2):155–160
Yu S, Zhu YX, Peng C, Li J (2021) Two new sterol derivatives isolated from the endophytic fungus aspergillus tubingensis YP-2. Nat Prod Res 35(19):3277–3284
Yuan SF, Alper HS (2019) Metabolic engineering of microbial cell factories for production of nutraceuticals. Microb Cell Factories 18(1):1–11
Ziemert N, Alanjary M, Weber T (2016) The evolution of genome mining in microbes–a review. Nat Prod Rep 33:988–1005. https://doi.org/10.1039/C6NP00025H

Bacteria and Bacteria-Based Products in Cancer Therapy: Current Status and Future Advances

20

Nikita Jangra, Anubhuti Kawatra, Rakhi Dhankhar, and Pooja Gulati

Abstract

Cancer is one of the most frequent causes of global mortality with about ten million deaths reported in 2020, according to the World Health Organization (WHO). The present-day cancer armamentarium typically possesses limitations viz. lack of specificity, high toxicity, multidrug resistance, and other threatening side effects. Thus, there remains a need for novel antitumor drugs with high efficacy and minimum side effects. The use of live oncolytic bacteria has emerged as a plausible approach to meet these challenges. These tumor-targeted bacteria can proliferate in the cancer cells, initiate the immune response, and suppress the cancers. In addition, microbes also produce a variety of peptides, enzymes, toxins, other proteins, and several secondary metabolites, which have emerged as promising candidates for cancer therapy. The present chapter elaborates upon the various attributes of these bio-drugs such as sources, mode of action, targeted cancers, toxicity issues, and clinical status. In addition, the recent approaches to improve the efficacy and safety of therapeutic bacteria and their products are briefly discussed.

Keywords

Anticancer · Amino acid deprivation therapy · Prodrugs · Bacterial products · Antibiotics · Bacteriocins · Cyclopeptides

N. Jangra · A. Kawatra · R. Dhankhar · P. Gulati (✉)
Department of Microbiology, Maharshi Dayanand University, Rohtak, India

R. Sobti et al. (eds.), *Role of Microbes in Sustainable Development*,
https://doi.org/10.1007/978-981-99-3126-2_20

20.1 Introduction

The multifaceted physiology of cancer is a major public health concern, with an ever-increasing global mortality rate each year. The incidence rate was estimated to be 1.9 million newly diagnosed cases, and around 609,360 cancer deaths are expected by 2022 in the United States. The annual global mortality rate is expected to reach ~17 million by 2030 accounting for approximately 1 in 6 deaths per day from cancer alone (https://www.cancer.org/latest-news/facts-and-figures-2022.html). The existing anticancer modalities including radiotherapy, cytotoxic chemotherapy, surgical therapy, oncolytic virotherapy, and immunotherapy possess several limitations like toxicity, multidrug resistance, lack of specificity, and adverse side effects (Sedighi et al. 2019). The establishment of an efficacious therapeutic armamentarium against cancer poses a formidable challenge for researchers. The WHO statistical estimation and the bottlenecks of conventional therapies emphasize the imperative need to find effective and novel cancer therapeutics.

Over recent decades, the ameliorating interest in harnessing bacteria to eradicate cancer has been accompanied by their ability to target the hypoxic tumor microenvironment (Azevedo et al. 2020). The live oncolytic bacteria can selectively penetrate the cancer cells and possess colonization ability in both aerobic and anaerobic conditions. Various facultative and obligate anaerobes, including *Escherichia coli*, *Listeria*, *Bifidobacterium*, and *Salmonella* species, possess inherent anticancer properties. Recently, many other live-attenuated bacteria including *Clostridium* have been implemented in the different phases of clinical studies. The clinical trial studies for the treatment of metastatic cancer via intravenous administration of genetically engineered *Salmonella typhimurium* strain VNP20009 have completed the phase 1 trials (Toso et al. 2002).

Moreover, bacterial derivatives like toxins, peptides, enzymes, and several secondary metabolites have been used significantly as anticancer agents over the last few decades (Sugumaran et al. 2022; Vaishnav and Demain 2011; Zhang et al. 2021). The therapeutic bacteria allow secretion of tumoricidal proteins/secondary metabolites including bacteriocins, cyclopeptides, and antibiotics that can be harnessed as potential candidates for the management and treatment of cancer constantly (Azevedo et al. 2020; Karpiński and Adamczak 2018; Narsing Rao et al. 2017; Singh et al. 2019). The bacteria-derived product inhibits macrophage migration and cellular cytokine production. They also catalyze the chemical modifications of specific proteins and disrupt the epithelial barrier (Sedighi et al. 2019). Term “bacteriotherapy” involving the use of genetically modified, live-attenuated bacteria and their products for cancer therapy has been proposed to overcome the drawbacks of traditional cancer treatments.

The chapter comprehensively covers the role of live-attenuated bacteria and bacteria-based products in cancer therapeutics. In addition, we have elaborated upon various aspects like source, mechanistic aspects to target cancer, and clinical status of these bacteriotherapeutic agents. Further, the challenges faced in the establishment of bacteriotherapy and their future prospects in cancer treatment are also reviewed.

20.2 Live-Oncolytic Bacteria in the Cancer Therapy

Conventional therapies like viz. radiation and chemotherapy have a major disadvantage of low specificity, which results in collateral damage to the surrounding tissue (Wang et al. 2019). The major challenge to the oncologists and cancer researchers is to develop a strategy that can reduce this off-target toxicity. In this pursuit, the use of live bacteria as antitumor agents has emerged as a promising biological therapy. Some commensal bacteria have the ability to colonize the tumor microenvironment (TME) (Nallar et al. 2017). The oncolytic action of bacteria was first observed in 1813 by Vautier, when he found that the tumor was regressed in patients with gas gangrene (caused by infection of *Clostridium perfringens*). The other historical landmark in the field dates back to 1893, when New York-based surgeon William B Coley observed regression of sarcoma in patients with erysipelas infection. He treated hundreds of sarcoma patients by injecting the causal agent of erysipelas, i.e., *Streptococci,* and developed what we now call as Coley's toxin (heat-killed *Serratia marcescens* and *Streptococcus*) to treat inoperable sarcoma (Coley 1991). After >200 years from the first observation, we presently know that many facultative and obligate anaerobes like *Bifidobacterium*, *Clostridium*, *Escherichia coli*, *Listeria*, and *Salmonella* species have the inherent oncolytic properties (Duong et al. 2019).

The angiogenesis in cancer cells is very slow as compared to the fast growth of the tumor, and due to poor vascularization, the oxygen supply is low; thus, TME is the region of hypoxia and necrosis (Zu and Wang 2014). These conditions are ideal for the colonization of anaerobes, and these bacteria exhibit chemotaxis toward the dying cancerous cells and low O_2 concentration. The motile bacteria have the ability to penetrate deeper in tumors. A study revealed that *S. typhimurium* cells spread across the entire tumor tissue within 3 days of injection (Ganai et al. 2011). Another study reported that a bioengineered *S. typhimurium* strain VNP20009 proliferates faster in tumors at the ratio more than 1000:1 in comparison with the normal cells (Clairmont et al. 2000). This is in sharp contrast to the chemotherapy agents, which exhibit limited penetration and passive distribution in tumor cells.

Once the bacteria colonize the tumors, they exhibit different and unique mechanisms for tumor regression. For example, *Clostridium* secrete toxins like hemolysins and phospholipases, which impair various intracellular processes in tumors and disrupt the cellular membrane (Duong et al. 2019). *Listeria* suppress tumors via increasing the ROS levels by activating NADPH oxidase and increasing calcium levels (Kim et al. 2009). The antitumor mechanism of *Salmonella* is diverse, and its infection can induce apoptosis by production of toxins or by depriving cancerous cells of nutrients (Wang et al. 2021). It also promotes the formation of gap junction between dendritic and cancerous cells by upregulating the levels of connexin 43 (Chang et al. 2013). Besides, as a general mechanism, the presence of large number of bacteria and consequently various bacterial components such as LPS and flagellin in tumor cells elicit an immune response and macrophages, dendritic cells, and neutrophils target the tumor cells (Duong et al. 2019). It is crucial to remember that some of these bacteria are pathogenic so the virulence of these bacteria is attenuated to reduce their action against host immune system. Also sometimes, the virulence contributes to the intrinsic tumor suppression activity of the bacteria, and hence, attenuation is carried out without the loss of antitumor activity (Ozdemir et al. 2018).

The various oncolytic bacteria *Salmonella typhimurium* (VNP20009), *S. typhimurium* (LH430), *E. coli* K-12, and *Clostridium novyi* (NT) have shown remarkable antitumor activity in experimental tumors in mouse transplant models (Dang et al. 2004; Jiang et al. 2010; Luo et al. 2001; Nallar et al. 2017); however, for most of them, antitumor potential has not been validated in the human clinical trials. The first clinical trial was carried out with attenuated *S. typhimurium* strain VNP20009 (designed by Vion Pharmaceuticals) in 1999. The trial enrolled one patient with carcinoma and 24 patients of metastatic melanoma. Significant tumor colonization and enhanced levels of cytokines were reported, but the regression of tumors was not observed (Toso et al. 2002). Many other clinical trials were carried out using this strain and other cancers (adenocarcinoma, pancreatic cancer, etc.), but none of the clinical trials has entered phase II (Duong et al. 2019). Only *Listeria monocytogenes* has shown most promising results so far and have been tested in phase II and phase III clinical trials. Advaxis Inc., an American biotechnology company, have funded these trails, and encouraging results have been obtained for the treatment of cervical cancer (Basu et al. 2018).

The oncolytic bacteria also offer various challenges such as genetic instability, toxicity, and inconsistent performance as it is subjected to availability of hypoxia, which varies in different patients. Also, the engineering of "smart bacteria" is very tricky as we want the optimum toxicity to kill tumor cells without systemic toxicity to the surrounding healthy tissue (Wang et al. 2019). Thus, in place of relying on inherent oncolytic activity, the bacteria are often used as "delivery vehicles" for delivery of various antitumor agents like immunomodulators [interleukins, IL-18, IL-12, flagellin flab, and prostate-specific antigen (PSA)], prodrug-converting enzymes, and siRNA. The tumor colonization property of bacteria is also exploited to deliver conventional chemotherapeutic agents in the TME. This strategy is referred as combination bacteriolytic therapy (COBALT) (Zu and Wang 2014).

20.3 Bacterial-Derived Products as Anticancer Agents

Bacteria serve as repository of biologically active molecules endowed with a variety of therapeutic potential including anticancer activities. They can efficiently target the malignant tissues by employing various strategies such as secretion of enzymes, toxins, secondary metabolites, proteins, and peptides (Karpiński and Adamczak 2018; Kaur and Kaur 2015; Singh et al. 2019). Moreover, the anticancer effect of these bacterial-derived natural compounds could control immune response, restrain cell growth, and trigger apoptosis. The bacteria-derived products are extremely potent and exhibit anticancer activities even at nano-concentrations (Law and Zhang 2007). The application of these bacterial products in cancer therapy is discussed in depth below.

20.3.1 Enzymes in the Treatment of Auxotrophic Cancers

Targeting the metabolism of cancerous cells has emerged as a promising strategy for precise treatment in oncology. In order to maintain their high prolificacy and redox homeostasis, cancerous cells undergo significant metabolic changes including plasticity, upregulation of oncogenes, and elevated uptake of certain nutrients, mainly amino acids (Dhankhar et al. 2020; Kawatra et al. 2022). The increased nutritional demand causes dependency on exogenous amino acids (from normal cells), ultimately leading to auxotrophy for them (Kawatra et al. 2020, 2022). These auxotrophic cancerous cells are thus targeted specifically by the amino acid-degrading enzymes. The latter interferes with the availability of amino acids to auxotrophic tumors and suppresses them without causing any damage to normal healthy cells. Microbial-derived amino acid-degrading enzymes, in this context, have attained immense importance in "amino acid deprivation therapy" (AADT) to target tumors due to their cost-effectiveness, high productivity, and mass availability. These enzymes are discussed in detail as follows:

20.3.1.1 L-Asparaginase

L-Asparaginase (EC 3.5.1.1, ASNase) catalyzing the hydrolytic bioconversion of plasmatic L-asparagine into aspartate and ammonia was the earliest microbial enzyme approved for the targeted treatment of pediatric blood cancer (ALL, "acute lymphoblastic leukemia") (Brumano et al. 2019). The enzyme has been characterized from various microbes like bacteria, filamentous fungi, yeast, and algae but is absent in mammals. L-asparagine is involved in the synthesis of proteins/glycoproteins, bioregulation of various oncogenic transcription factors, and functioning of nervous and immune systems (Avramis and Tiwari 2006; Wu 2013). It also enhances the metastatic potency of cancerous cells by promoting epithelium-to-mesenchymal transition (Knott et al. 2018). Normal cells synthesize L-asparagine de novo using asparagine synthetase; however, some cancerous cells including leukemic cells have a poor expression for this enzyme and are more dependent on the exogenous supply (Fernandes et al. 2017). ASNase specifically depletes the plasmatic asparagine, thereby inducing autophagy/apoptosis, leading to regression of asparagine–auxotrophic cancer. ASNase isolated from *Escherichia coli* and *Erwinia chrysanthemi* is currently available under different brands namely Elspar (native *E. coli* ASNase), Erwinase (native *E. chrysanthemi* ASNase), Oncaspar (pegylated ASNase), and Spectrila (recombinant *E. coli* ASNase) for the biological treatment of ALL worldwide. The pharmacological aspects of these different approved ASNase formulations viz. dosage, serum half-life, and route of administration are described in Table 20.1.

Recently, various research groups have focused on overcoming the limitations of these commercially available asparaginases viz. immunogenicity, neurotoxicity, hepatotoxicity, treatment relapse, and improving its catalytic efficacy by using mutagenesis, PEGylation, and immobilization (Table 20.2).

Table 20.1 Various approved ASNase formulations and their pharmacological aspects

Formulation	Brand name	Composition	Route of administration	Serum half-life	Dosage	Data source
E. coli ASNase	Elspar	L-ASNase (10,000 U), mannitol (80 mg)	Intravenous/ intramuscular	26–30 h	6000 IU/m^2 three doses per week	Merck & Co, Inc.
E. chrysanthemi ASNase	Erwinase	L-ASNase (10,000 U), glucose (5 mg), sodium chloride (0.5 mg)	Intravenous/ intramuscular	16 h	6000 IU/m^2 three doses per week	Speywood pharmaceuticals, Inc.
Pegylated ASNase	Oncaspar	L-ASNase (750 U), dibasic sodium phosphate, USP (5.58 mg), monobasic sodium phosphate, USP (1.20 mg), sodium chloride (8.5 mg)	Intravenous/ intramuscular	5.5–7 days	25,000 IU/m^2 per 2 or 4 weeks	Enzon Pharmaceuticals
Recombinant *E. coli* ASNase	Spectrila	L-ASNase (10,000 U), sucrose	Intravenous	17.3 h	5000 IU/m^2 every 3 day	Medac GmbH

Table 20.2 Improvement strategies of bacterial enzymes employed in the treatment of auxotrophic cancers

Enzyme	Experimental study			References
	Improvement method	Microbial source	Outcome	
L-Asparaginase	Mutagenesis	*Erwinia chrysanthemi*	Mutant N133V with high thermal stability	Kotzia and Labrou (2009)
			Mutant A31I/E63Q/S254Q with improved anticancer efficacy	Nguyen et al. (2018)
		Escherichia coli	Mutant R195A/K1961/H197A with decreased immunogenicity	Jianhua et al. (2006)
			Mutant N24S with enhanced thermal stability and protease resistance	Maggi et al. (2017)
			Mutant L23G/K129L/S263C/R291F with reduced toxicity and improved pharmacokinetics	Mahboobi et al. (2017)
	PEGylation	*E. coli*	MonoPEGylated ASNase with 42% yield and enhanced stability	Meneguetti et al. (2019)
	Encapsulation	*E. coli*	RBCs encapsulated ASNase with enhanced half-life, stability, and lower side effects	Bailly et al. (2011)
			ASNase polymersomes with enhanced proteolytic stability and reduced immunogenicity	Blackman et al. (2018)
Arginine deiminase	Mutagenesis	*Pseudomonas plecoglossicida*	Mutant K5T/D44E/H404R with enhanced activity at physiological conditions	Zhu et al. (2010)
		P. plecoglossicida	Different mutants [K5T/D44E/H404R + A128T, K5T/D44E/H404R+A128T+E296K, K5T/D44E/H404R+A128T+ E296K +D38H] with enhanced activity at physiological conditions	Zhu et al. (2010)
		P. aeruginosa	Mutant H405R with improved kinetics and anticancer efficacy	Ding et al. (2012)
	PEGylation	*Mycoplasma hominis*	PEGylated recombinant enzyme with enhanced anticancer efficacy	Holtsberg et al. (2002)
	PEGylation and mutagenesis	*P. plecoglossicida*	Formulation displayed enhanced half-life and anticancer efficacy	Zhang et al. (2015)

(continued)

Table 20.2 (continued)

	Experimental study			
Enzyme	Improvement method	Microbial source	Outcome	References
Methioninase	Mutagenesis	*Citrobacter freundii*	Mutant V358Y with 1.9-fold high catalytic activity than native enzyme	Raboni et al. (2018)
	PEGylation	*P. putida*	PEGylated recombinant methioninase with low toxicity and high pharmacokinetics	Hoffman et al. (2019)
	Immobilization	*P. putida*	Cross-linked aggregates of methioninase with enhanced stability and cytotoxicity against lung carcinoma	Kannan and Marudhamuthu (2019))
Phenylalanine ammonia lyase	Immobilization	*Anabaena variabilis*	Thermosensitive micelles of phenylalanine ammonia lyase with improved efficacy against colorectal cancer	Yang et al. (2019)

20.3.1.2 Arginine Deiminase

Arginine deiminase (EC 3.5.3.6, ADI) is another widely employed enzyme in AADT. It causes the bioconversion of L-arginine into citrulline and ammonium ion. The semi-essential amino acid, L-arginine, is involved mainly in the metabolism of nitric oxide, polyamines, proteins, glutamate, proline, and nucleotides (Dhankhar et al. 2019; Kawatra et al. 2022). All these metabolites are associated with proliferation and metastasis of cancerous cells. Contrary to normal cells, cancerous cells are unable to synthesize arginine endogenously due to downregulated activity of enzymes argininosuccinate synthetase and arginosuccinate lyase and show arginine auxotrophy (Dhankhar et al. 2020; Kawatra et al. 2022). ADI can be thereby used as a cancer treatment modality for the regression of arginine auxotrophic tumors as it lowers plasmatic levels of arginine and suppresses protein synthesis, angiogenesis, and cellular proliferation along with the induction of apoptosis in cancerous cells.

Microbes like bacteria, fungi, and a few protists exclusively produce ADI as it is associated with survival benefits in extremely acidic conditions (Dhankhar et al. 2020; Kawatra et al. 2022). The therapeutic potential of ADI sourced from *Mycoplasma* has been proven in various cancer cell lines (Kawatra et al. 2022). Currently, *Mycoplasma* ADI conjugated to polyethylene glycol moieties (ADI-PEG20) is under clinical evaluation for melanoma (NCT00450372, phase II), malignant pleural mesothelioma (NCT02709512, phase II), and hepatocellular carcinoma (NCT01287585, advanced phase) (Dhankhar et al. 2018). The results are encouraging and well tolerated. Furthermore, ADI has been also manipulated genetically or modified chemically from various bacteria including *Pseudomonas plecoglossicida, P. aeruginosa*, and *Mycoplasma hominis* for improved kinetics and pharmacodynamics (Table 20.2).

20.3.1.3 Methioninase

Methioninase (EC 4.4.1.11, METase) catalyzes the deamination of L-methionine to alpha-ketobutyrate and ethanethiol. The enzyme belongs to PLP-dependent (pyridoxal L-phosphate) family of enzymes (El-Sayed 2010). Considered an essential amino acid, L-methionine plays an indispensable role in the synthesis of polyamines, glutathione, and proteins. While polyamines are required for cell division, glutathione aids in protecting cells from oxidative stress (Hens et al. 2016). Besides, it is also important for the biosynthesis of S-adenosylmethionine, which in turn is responsible for methylation of RNA, histones, DNA, and proteins. Even though L-methionine is one of the essential amino acids, its deprivation is not considered fatal as it is recycled in vivo by the activity of betaine–homocysteine methyltransferase, L-methionine synthase, and homocysteine re-methylation, the mechanism lacking in cancerous cells (Cellarier et al. 2003). Therefore, methioninase-mediated methionine depletion can alter the epigenetic landscape of these auxotrophic cancerous cells, aiding in their regression. To date, the enzyme has been obtained from several bacteria like *Pseudomonas putida, Aeromonas* sp., *Brevibacterium linens*, archaebacteria, fungi, few protists, and plants (Suganya et al. 2017). METase has been found to effectively suppress various cancers such as glioma, glioblastoma, leukemia, melanoma, non-small lung cancer and primary ductal carcinoma, and osteosarcoma (Cavuoto and Fenech 2012). At present, recombinant and PEGylated formulations of *P. putida* METase are under clinical trials for lymphoma breast cancer, lung cancer, and renal cancer (Hoffman et al. 2019). Promising results were attained at pilot phase with significant dropping of plasmatic L-methionine. In addition, no case of toxicity was observed upon administering METase formulation (Hoffman et al. 2019). Furthermore, various research groups have employed strategies like genetic engineering, PEGylation, and immobilization to improve the pharmacological properties of METase. These are discussed in detail in Table 20.2.

20.3.1.4 Phenylalanine Ammonia Lyase

Phenylalanine ammonia lyase (EC 4.3.1.24, PAL) catalyzing phenylalanine-to-trans-cinnamate bioconversion has been also implicated to effectively regress tumors. L-phenylalanine has been associated with the biosynthesis of neurotransmitters dopamine, nor-epinephrine, and epinephrine (Kawatra et al. 2020). Furthermore, it is important for tumorigenesis, and therefore, its depletion can aid in inhibiting cancer progression. This enzyme has been mainly isolated from yeast *Rhodosporidium toruloides* and *Rhodotorula glutinis* for therapeutic applications (Kawatra et al. 2020). Recently, Yang et al. (2019) screened the potency of *Anabaena variabilis* immobilized PAL against colorectal cancer. The bacterial-based formulation (9 mg/ml concentration) induced apoptosis in 35.75% of cells (Table 20.2). Although results are encouraging, but further studies are required before it could be employed clinically in AADT.

20.3.2 Microbial Enzymes in Prodrug Activation Therapy

In addition to the abovementioned anticancer enzymes, there are some other microbial enzymes that do not possess the anticancer activity themselves but can be used to activate a prodrug into an active anticancer drug. The prodrug activation therapy is a two-step process; in the first step, the microbial enzyme is expressed in tumor cells, and in the second step, the inactive prodrug (which is substrate to the enzyme) is administered in the tumors. The enzyme then converts the nontoxic prodrug into active anticancer drug (Xu and McLeod 2001). The process exhibits a bystander effect; i.e., the active drug is not limited to the transfected cells, rather it is transported to neighboring tumor cells by gap junctions, by simple diffusion or due the transfer of apoptosis factors from dying transfected cells (kiss of death) (Dachs et al. 2009). This targeted approach is highly preferred because of its high therapeutic index (benefits versus toxic side effects) (Dhankhar et al. 2021).

The enzyme prodrug activation therapy (EPT) involves three components: microbial enzyme, inactive prodrug, and enzyme/gene delivery vector. The active enzymatic protein is either directly administered in the tumor cells or in other approach, the gene coding for the enzyme is delivered in the cancer cells and the enzyme is expressed within the tumor itself (gene-directed enzyme prodrug therapy, GDPET) (Williams et al. 2015). On the basis of vectors used, EPT can be classified into "viral directed enzyme prodrug therapy (VDEPT) and bacterial directed enzyme prodrug therapy (BDEPT)" (Greco and Dachs 2001). The oncolytic bacteria (*Clostridium, E. coli, Bifidobacterium, and Salmonella*) discussed in Sect. 20.2 are used in the BDEPT approach. Retroviruses and adenoviruses are commonly used viral vectors. Besides virus and bacteria some other vectors like liposomes, stem cells, mesenchymal cells, and lectins are also used as vectors (Mooney et al. 2017).

The various microbial enzymes that are used in prodrug activation therapy are "nitro reductase, cytosine deaminase (CD), purine nucleoside phosphorylase (PNP), carboxypeptidase (CPG), and also methioninase, xanthine/guanine phosphoribosyl transferase (XG-PRT), and β-glucuronidase" (Dhankhar et al. 2021; Malekshah et al. 2016). These enzymes are used in combination with prodrugs 5-fluorocytosine (5-FC,CD/5-FCsystem), CMDA(4-[(2-chloroethyl)(2-mesyloxyethyl)amino]benzoyl-L-glutamic acid) (CPG2/CDMA system), CB1954 (NTR/CB154 system), 6-methylpurine deoxyriboside, selenomethionine (SeMET, Met/SeMET system), 9ACG (9-aminocamptothecin glucuronide) (βG-9ACG system), and 6-thioxanthine (6-TX)(XG-PRT/ 6-TX system) (Dhankhar et al. 2021). Various aspects namely the cytotoxin produced, mode of action, and types of targeted tumors of the important enzyme prodrug system are described in Table 20.3.

Among all these systems, CD/5-FC is the most widely studied EPT system. CD is an amino hydrolase that catalyzes the conversion of prodrug 5-FC into 5-FU (5-fluorouracil). 5-FU is then converted into various metabolites by cellular enzymes that inhibit DNA and RNA synthesis in tumors. The enzyme is obtained from *E. coli* or yeast *Saccharomyces cerevisiae* (Kaliberov et al. 2007). This enzyme prodrug system has completed phase I trial for the treatment of recurrent or progressive grade III or grade IV gliomas (NCT01985256 and NCT01470794). The phase II/III clinical trials are also carried out using retroviral replicating Vector Toca 511 (NCT02414165). The CD/5-FC system has shown promising results against different tumors like bladder cancer, gastrointestinal cancer, brain tumor, and glioma. Ntr/CB1954 system is another popular system; the prodrug CB1954 (5-(aziridin-1-yl)-2,4-dinitrobenzamide) is named so because this prodrug was discovered at Chester Beatty Laboratories (London) (Williams et al. 2015). Ntr is an oxidoreductase enzyme, and type I NTR from *E. coli* is used in EPT. This system inhibits the tumor growth by DNA cross-linking. Phase I clinical trials using this enzyme prodrug system against various cancers such as liver cancer, gastrointestinal cancer, and prostate cancer has been completed (Dhankhar et al. 2021). The enzyme prodrug system employing carboxypeptidase G2 and PNP is also tested in clinical trials (NCT00625430, NCT03754933). Antitumor potential of other microbial enzymes (methioninase, xanthine/guanine phosphoribosyl transferase, and β-glucuronidase) in prodrug therapy has not been analyzed in trials. The detail of various experimental tumors targeted by these enzymes in mouse xenograft models has been provided in Table 20.3.

Although the enzyme prodrug therapy has gained momentum in past few years and some of the EP systems have reached clinical trials, there is a long way until these therapies are used in the standard cancer armamentarium. There are severe challenges like immunogenicity (due to foreign enzyme) and low bystander effect. These problems are addressed by selection and improvement of prodrug and

Table 20.3 Details of the enzyme prodrug therapy (EPT) systems, mode of action of the enzymes, and target tumors (adapted from Dhankhar et al. 2020)

Enzyme	Class	Producer	Prodrug	Cytotoxin	Action	Type of EPT	Target Tumor
Cytosine deaminase (CD)	Pentosyl-Transferase	*E. coli*	5-fluorocytosine	5-fluorouracil (5-FU)	Inhibitition of DNA and RNA Synthesis	GDEPT	Colorectal and renal cancer, Breast cancer, Prostate cancer, Glioma, Pancreatic cancer, Head and neck cancer
Carboxypeptidase G2 (CPG2)	Oxido-reductase	*E. coli*	CMDA	Nitrogen ustards 4 [(2-chloroethyl) (2 mesyloxyethyl) amino] benzoic acid	DNA interstrand Crosslinking	ADEPT	Murine melanoma, Breast cancer, Colon cancer
Nitroreductase (NTR)	Metallo-carboxypeptidase	*Variovorax paradoxus (Pseudomonas* sp. strain RS-16)	CB1954	5-(aziridin-1-yl)-4-hydroxylamino-2-nitrobenzamide)	DNA interstrand crosslinking	GDEPT	Prostate cancer, Liver cancer, Walker carcinoma
Purine nucleoside phosphorylase (PNP)	Hydrolase	*E. coli, S. cerevisiae*	6-methylpurine deoxyriboside	6-methylpurine	Inhibition of RNA, protein and DNA synthesis	VDEPT	Chronic lymphocytic leukemia, Glioma, Prostate cancer, Bladder cancer, Head and neck cancer

Methioninase	Lyase	*Pseudomonas putida*	Selenomethionine (SeMET)	Methylselenol	Generation of superoxides	GDEPT	Murine melanoma B16F10, Fi-brosarcoma cells HT1080, Colon cancer derived HCT-116, Human prostate cancer cells LNCaP
β-glucuronidase	Hydrolase	*E. coli*	9ACG (9-aminocamptothecin glucuronide)	9AC (9-aminocamptothecin)	Inhibition of topoisomerase I	GDEPT	Cancer lines like LS174T for colon cancer, AGS for stomach cancer, MCF-7 for breast cancer, OVCAR-3 for ovarian cancer
Xanthine/guanine phosphoribosyltransferase	Transferases	*E. coli*	6-thioxanthine (6-TX)	6-thioxanthine monophosphate (6-XMP)	DNA cross linking	GDEPT	Intracerebral C6 glioma xenografts Model

transport vectors and the enzymes are pegylated to reduce the immunogenicity (AlQahtani et al. 2019).

20.3.3 Bacterial Toxins as Therapeutic Cancer Modality

Bacterial toxins are virulence proteins that manipulate vital biological processes and take control over host cellular machinery in order to promote bacterial infections (do Vale et al. 2016). Several bacterial toxins offer considerable potential for their use as anticancer therapeutics including diphtheria toxin, alpha-toxin, and Shiga toxin (Khoshnood et al. 2022). Bacterial toxins can precisely bind to the ligands specific for tumor surface receptors. Ligand–receptor complex internalization releases the active toxin inside the cell and causes cell death. Bacterial toxins can destroy cancer cells by triggering various apoptotic pathways or by regulating vital processes viz. differentiation, proliferation, and apoptosis. A variety of bacterial toxins can interfere with cell cycle progression. Here, we described the most common toxins that are used as cancer therapeutics.

20.3.3.1 Shiga Toxin

Shiga toxin (Stx) or verotoxin is primarily produced by *Shigella dysenteriae* serotype 1 and *E. coli* (Shiga toxin-producing *E.* coli, STEC). Stx consists of two non-covalently attached moieties: StxA (enzymatically active) and StxB (nontoxic binding moiety). Stx causes tumor regression due to the specific binding of StxB moiety to the glycosphingolipid cell surface receptor, globotriaosylceramide (Gb3), which is highly expressed by the cancer cells like breast, ovarian, testicular tumors, and Burkitt's lymphoma (BL) cells but having a restricted profile in normal human cells and tissues (Engedal et al. 2011; Robert and Wiels 2021). The Stx/Gb3 complex eventually results in membrane invagination and is further internalized via endocytosis (Fig. 20.1). The Stx is then transported in a retrogressive manner to the Golgi apparatus and then to the endoplasmic reticulum (ER) such that only StxA is transported to the cytosol (Robert and Wiels 2021). During this transport, StxA is cleaved into two disulfide-linked fragments: A1 (large, catalytic active fragment) and A2 (small, binding fragment). This disulfide bond is intact during transport from Golgi to ER and is reduced only inside the ER. StxA1 fragment is then translocated to the cytosol, while both StxA2 and StxB remain in the ER (Melton-Celsa 2014). The catalytically active StxA1 fragment halts protein synthesis by removing an adenine from 28S ribosomal subunit that ultimately leads to the tumor regression. The toxin also triggers apoptosis, cytotoxicity, and irreversible cell cycle arrest at G_2/M phase (Fig. 20.2), which has been explored as a potent treatment for malignancies (Oloomi et al. 2018). Recently in a study, the Shiga toxin from *E. coli* O157:H7 showed anti-carcinogenic effects in T47 breast cancer cell lines by inducing apoptosis (Pinatih et al. 2021). In addition, combinational therapy with various chemotherapeutic drugs including doxorubicin (an intercalating agent), SN38 (topoisomerase I inhibitor), and auristatin F (an antimitotic drug) demonstrates the more effective cancer treatments. The experimental evaluation shows the that

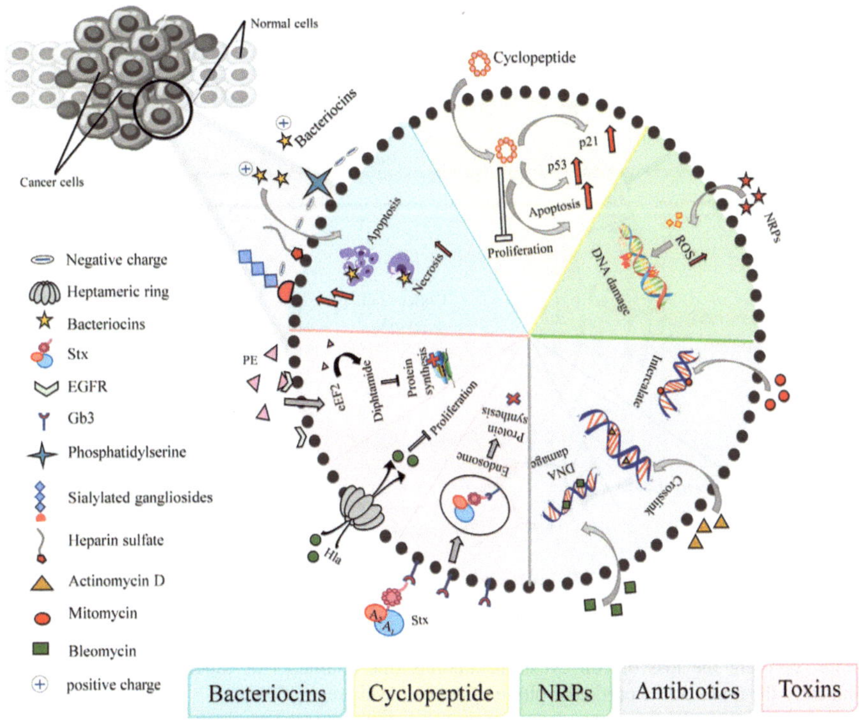

Fig. 20.1 Overview of the various mechanisms underlying the use of different bacterial products as anticancer agents (Stx, Shiga toxin; Hla, alpha-hemolysin toxin; PE, *Pseudomonas* exotoxin A; Gb3, globotriaosylceramide; EGFR, epidermal growth factor receptor; eEF2, elongation factor 2; ROS, reactive oxygen species)

StxB-doxorubicin and StxB-auristatin F combinational approach is highly selective against the HT29 (human colon adenocarcinoma cell lines). The STxB-SN38 has been potent against gastric and pancreatic cell lines (Geyer et al. 2016; Kostova et al. 2015; Maak et al. 2011).

20.3.3.2 Diphtheria Toxin

Diphtheria toxin (DT) is an extremely efficient bacterial exotoxin that was first detected in 1888 by Roux and Yersin in *Corynebacterium diphtheriae* (Murphy 1996). DT exhibits extremely high toxicity that even a single molecule in cell cytosol is enough to cause cell death. DT has been modified for targeted cancer therapy by removing the cell receptor-binding moiety from the toxin and ligating its catalytic active moiety with the specific proteins, which show selective binding to the tumor cell receptors (Y. Zhang et al. 2010). As a result, receptor-less DT cannot able to bind the cell surface on its own. DT has been explored as a potent toxic agent against a wide range of cancer cell lines. Also, cancer cells exhibit overexpression of heparin-binding epidermal growth factor (HB-EGF), which is a specific cell surface receptor for DT. Experimental evaluation shows that toxin binding with HB-EGF

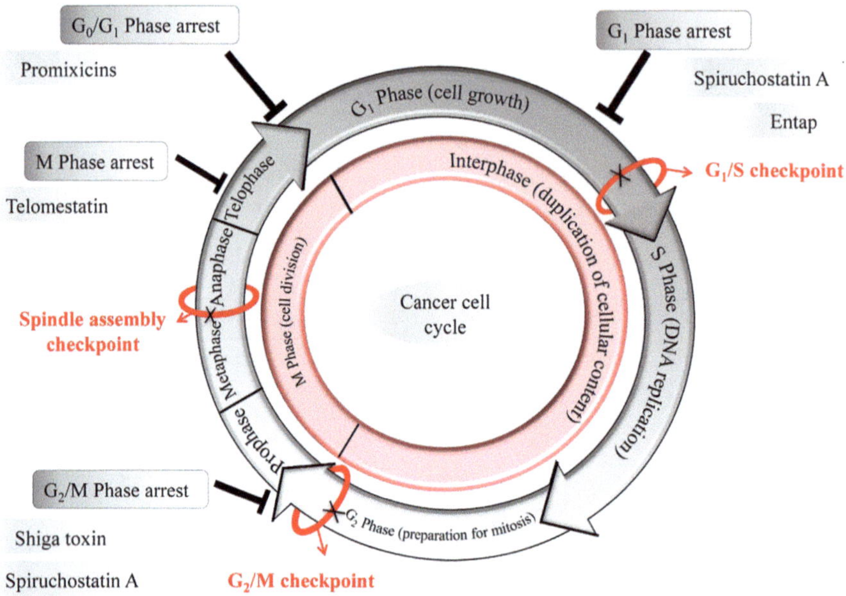

Fig. 20.2 Bacterial products mediated arrest at different phases of cancer cell cycle. Crossed red rings represent the loss of cell cycle checkpoints (the hallmarks of cancer). The blocked sign indicates the arrest at the particular phase of the cell cycle regulation

suppresses angiogenesis and proliferation of cancer cells and promotes apoptosis (Weerakkody and Witharana 2019).

In the 1970s, the new concept of immunotoxin was introduced to enhance toxin specificity by linking the enzymatic active subunit of the toxin to the specific antibody or cytokines. The U.S. Food and Drug Administration (FDA) approved the first DT-based immunotoxin (IT), Denileukin diftitox (Ontak®), administered for the treatment of recurrent cutaneous T-cell lymphoma (Hamamichi et al. 2020). Later, in 2018, Tagraxofusp-erzs (Elzonris®) immunotoxin was approved by FDA for the treatment of blastic plasmacytoid dendritic cell neoplasms (BPDCNs). The Elzonris® is a recombinant immunotoxin that consists of human IL-3 and the enzymatic active moiety of DT, which targets IL-3-alpha (CD123) receptors (Jen et al. 2020).

20.3.3.3 Staphylococcal Alpha-Toxin

Alpha-toxin or alpha-hemolysin toxin (Hla) is the pore-forming bacterial exotoxins (PFT) secreted by *Staphylococcus aureus*. It belongs to the beta-barrel pore-forming toxin (β-PFT) family that forms a heptameric ring-like transmembrane pore in the plasma membrane upon binding to A-disintegrin and metalloprotease 10 (ADAM10) receptor (Iacovache et al. 2010) (Fig. 20.1). In addition, Hla toxin also interfered with caveolin-1 and integrin in cell membranes (Liang and Ji 2006; Vijayvargia et al. 2004). It was demonstrated that Hla toxin exhibits higher toxicity to cancer cells and could induce the apoptosis in wide variety human cancer cell lines including human Jurkat T lymphocytes (via the intrinsic pathway by upregulating the activity of caspases 3 and 9) and ECV304 bladder cancer cell line (through the extrinsic pathway) (Bantel et al. 2001; Yu et al. 2013). It was also reported that caspase-2 was the first caspase to be involved in Hla-induced apoptosis in HeLa cell lines, followed by caspases 8 and 9 where the activation of caspase 2 was regulated by the potassium efflux caused by Hla toxin (Imre et al. 2012). Thus, different cell types have distinct mechanisms for Hla-induced apoptosis, including intrinsic pathways, extrinsic pathways, and caspase 2-initiated pathways. In addition, the interaction of α5β1-integrin with Hla-induced apoptosis in the A549 lung epithelial cell line (Liang and Ji 2007). Administration of *E.coli* engineered with staphylococcal alpha-toxin into murine tumors resulting in induced tumor regression and necrosis (St. Jean et al. 2014).

20.3.3.4 *Pseudomonas* Exotoxin A

Like diphtheria toxin, a typical AB toxin *Pseudomonas* exotoxin A (PE) primarily secreted by *Pseudomonas aeruginosa* consists of the A domain (catalytic active) and B domain (binding domain). PE is a highly toxic ADP-ribosyltransferase toxin that blocks cellular proliferation and protein synthesis by conversion of elongation factor eEF2 to diphthamide (Wolf and Elsässer-Beile 2009). The PE38 (38 kDa truncated fragment of PE toxin) possesses highly toxic effects on EGFR-expressing cancer cells (Fig. 20.1). Increased DNA fragmentation, inactivation of poly ADP-ribose polymerase (PARP), endoplasmic reticulum stress, and elevated intracellular Ca^{2+} levels have all been linked to exotoxin A-based cytotoxicity in tumor cells (Risberg et al. 2011). Various genetically engineered immunotoxin has been developed by the combination of truncated PE38 and receptor-specific monoclonal antibodies or cytokines. The deimmunized exotoxin fragment (dEGF4KDEL)-based bispecific ligand-directed toxin (BLT) specifically suppresses metastasis and shows cytotoxicity against various pancreatic cancer lines (SW1990, S2VP10, MIA PaCa-2) (Oh et al. 2012). Exotoxin T also exhibits cytotoxicity against a number of cancer cell lines, such as EMT6, HeLa, 4 T1, MDA-MB-231 (mammary cancer), SK-OV-3 (human ovarian cancer), MCA-205 (mouse fibrosarcoma), Calu-3 (human lung cancer), and B16 (murine tumor) (Goldufsky et al. 2015).

20.3.4 Other Anticancer Bacterial Proteins and Peptides

In addition to the bacterial enzymes and toxins, other bacterial-derived proteins and peptides like bacteriocins and cyclopeptides possess the potential to serve as cancer therapeutic agents. Mechanisms underlying the use of the bacterial products in anticancer therapy are summarized in Fig. 20.1. Some of these are already a part of the cancer armamentarium, while others are under clinical evaluation.

20.3.4.1 Bacteriocins

Bacteriocins are non-immunogenic, biodegradable, cationic, ribosomal synthesized peptides, which are secreted by mostly all bacterial species. Bacteriocins possess cancer cell-specific mechanism that could be explained by generalized differences between the membranes of cancer cells and normal cells (Kaur and Kaur 2015; Villarante et al. 2011). The cancer cell contains an increased level of anionic O-glycosylated mucins, phosphatidylserine, heparin sulfates, and sialylated gangliosides due to which the cell membrane carries a negative charge on its outer surface (Riedl et al. 2011; Schweizer 2009). The cationic bacteriocins preferentially bind to the anionic cancer cells as compared to the neutral normal cell membranes (Hoskin and Ramamoorthy 2008) (Fig. 20.1). Colicin, a bacteriocin produced by *Escherichia coli,* shows anticancer properties against human cervical adenocarcinoma (HeLa cell line), breast cancer, and bone cancer (Braun et al. 1994; Chumchalová and Šmarda 2003). Another bacteriocin such as bovicin HC5 secreted from *Streptococcus bovis* was reported to have cytotoxic activities against MCF-7 (human breast can cell lines) and HepG2 (liver cancer cell lines) (Mantovani et al. 2002; Paiva et al. 2012). Microcin E492 from *Klebsiella pneumoniae* was found to enhance apoptosis in HeLa, RJ2.25 (Burkitt's lymphoma cell line), and Jurkat (immortalized T lymphocyte cell line) (Hetz et al. 2002). Laterosporulin 10 (LS10) isolated from *Brevibacillus* sp. cause apoptotic and necrotic death of cancer cells. LS10 is known to target various human cancer cell lines including HEK293T, MCF-7, H1299, HT1080, and HeLa (Baindara et al. 2017).

20.3.4.2 Cyclopeptides

Cyclopeptides are cyclic polypeptides, generally formed of 4–10 amino acids arranged in a highly rigid cyclic ring structure. These are extremely potent biological entities that possess high binding affinity, specificity, stability, and rigidity in their confirmation (Zhang et al. 2021). Thus, cyclopeptides are considered as promising modalities for the development of novel therapeutic drugs like antibiotics, immune boosters, protease inhibitors, and anticancer agents. The cyclic ring structure of the peptide enhances membrane permeability due to the absence of an inflexible backbone of amino/carbonyl termini. As a result, cyclopeptides can be used as anticancer agents with their exceptional ability to penetrate tumors along with their anti-proliferative properties (Rezai et al. 2006) (Fig. 20.1). In past years, various anticancer cyclopeptides have been extracted from bacterial species. Mixirins A-C are three cyclic lipopeptides produced by marine *Bacillus* species. Mixirins exhibit anti-proliferative activity against HCT116 (human colorectal carcinoma cell line) with IC_{50} value of 0.65 μM, 1.6 μM, and 1.26 μM, respectively (Zhang et al. 2004). Dolyemycins A and B and cyclopeptides obtained from *Streptomyces griseus* showed significant anti-proliferative properties against MCF7, SF-268 (glioblastoma cell line), and NCL-H460 (lung carcinoma cell line) (Liu et al. 2018). Telomestatin, a macrocyclic peptide, was first extracted from *Streptomyces anulatus* and act as a telomerase inhibitor via binding to the 3′ telomere of the G-quadruplex structure. It shows significant cell cycle arrest, cytotoxicity (Fig. 20.2), apoptosis, and telomere shortening in neuroblastoma cell lines (Binz et al. 2005; Kim et al. 2002; Shin-ya et al. 2001). A cyclopeptide with a disulfide bond, YM753 (spiruchostatin A), was purified from *Pseudomonas* species. YM753 induces the G_1 and G_2/M cell cycle arrest-mediated histone acetylation (Fig. 20.2) and enhances apoptosis and p21 expression in WiDr (human colon carcinoma cell line) (Shindoh et al. 2008).

20.3.4.3 Azurin

Azurin is a 16 kDa anticancer redox protein that was isolated from *Pseudomonas aeruginosa*. It was reported that p28, a peptide fragment of azurin, mediates its preferential penetration into the cancer cell (Yaghoubi et al. 2020). Several mechanisms have been proposed for the anticancer activity of azurin including induction of apoptosis via binding of azurin to the DNA binding domain of p53 and inhibition of tumor angiogenesis by inhibiting the VEGFR2 phosphorylation, pro-apoptotic, and anti-proliferative activity of azurin (Karpiński and Adamczak 2018). At present, phase 1 clinical studies of p28 have been completed and findings have shown that p28 is safe as an anticancer drug (Yaghoubi et al. 2020). In addition, p28 exhibits anti-proliferative and cytotoxic activity against various cell lines including MDA-MB-231 (human breast cancer cell lines) and MCF7 (Noei et al. 2019).

Other anticancer peptides like enterococcal peptide, Entap, induce G_1 phase arrest and apoptosis (Fig. 20.2). Entap shows anti-proliferative activity against colorectal adenocarcinoma (HT-29), human gastric adenocarcinoma (AGS), uterine cervix adenocarcinoma (HeLa), and also against prostatic carcinoma (22Rv1) (Karpiński and Szkaradkiewicz 2013). Pep27 anal2, a signal peptide derived from *Streptococcus pneumoniae,* also exhibits properties like apoptosis and proliferation inhibition against MCF7, Jurkat, HL-60, AML-2, and SNU-601 cancer cell lines (Lee et al. 2005; Sung et al. 2007).

20.3.5 Secondary Metabolites for Cancer Chemotherapy

Secondary metabolites are the naturally occurring substances produced by an organism as a part of its metabolic activity. These are not required for an organism's growth or reproduction, but are produced to provide a specific advantage to the organism. Secondary metabolites enable an organism to engage with its neighbors, overcome stress, nutritional requirements, and compete with its commensals. Bacterial cells are considered as an irreplaceable source of secondary metabolites with considerable therapeutic potential including immunosuppressive, antineoplastic, and anticancer properties (Vaishnav and Demain 2011). The current knowledge and potential of bacteria for the production of secondary metabolites involving antibiotics and non-ribosomal peptides have been elucidated.

20.3.5.1 Non-Ribosomal Peptides

Non-ribosomal peptides (NRPs) are peptide secondary metabolites that are synthesized independently of ribosomal machinery, unlike other proteins. NRPs exhibit a variety of biological and pharmacological properties such as immunosuppressant, cytostatic, and anticancer activity. These are primarily produced by bacteria, fungi, soil-inhabiting microorganisms, and marine microorganisms (Dincer et al. 2022). Bleomycin is a NRP-based commercial antitumor drug typically produced by *Streptomyces verticillus* (Matsuo et al. 1997). Bleomycin promotes the reactive oxygen species-mediated oxidative DNA damage to the cancer cells (Fig. 20.1) and is generally used for the treatment of testicular germ cell tumors and Hodgkin's lymphoma (Llop-Hernández et al. 2022). Thiocoraline is another recent antitumor NRP typically produced by marine bacteria, *Micromonospora marina*. Thiocoraline possesses two extremely rare S-methylated L-cysteine residue peptidic backbone (Lombó et al. 2006). In addition, proximicins A-C are novel anticancer NRPs isolated from marine *Verrucosispora* sp. that show cytotoxic effects against HepG2 (hepatocellular carcinoma cell line), AGS (gastric adenocarcinoma cell line), and MCF7 (breast cancer cell line). Proximicins involves in the cell cycle arrest at G_0/G_1 phase in the AGC cell lines after 24-h incubation period (Fig. 20.2). An increase in the population of apoptotic cells in the sub-G_1 phase after incubation for 40 h was also observed. Additionally, it induces the upregulation of cyclin-dependent kinase inhibitor p21 and the p53 tumor suppressor gene in AGS cells (Fiedler et al. 2008).

20.3.5.2 Antibiotics

Antibiotics are the antimicrobial medications acting against bacteria by either inhibiting or eradicating them. Bacteria produce majority of antibiotics that are active against other members of their own genus. Among all, the genome of *Streptomyces* spp. possess more than 20 gene clusters for antibiotic production. Some antibiotics have also been demonstrated as potent anticancer therapeutics. Anticancer antibiotics have been shown to have a deleterious impact on cancer growth by preventing malignant cells from spreading throughout the body (metastasis), inducing cancer cell death, enhancing the immune system, and increasing the effectiveness of therapies like radiotherapy (Mohan et al. 2021). Some of these anticancer antibiotics are DNA damagers, while some are potent cross-linking and intercalating agents. Various cytotoxic antibiotics are currently used as cancer therapeutics including DNA intercalating (actinomycin D, doxorubicin, mitoxantrone, and daunorubicin), DNA cross-linker (mitomycin C), and DNA damager (bleomycin) (Fig. 20.1). Actinomycin D was isolated from *Streptomyces parvulus* and *Streptomyces antibioticus,* and it is the first antibiotic that was approved by FDA for the treatment of malignancies involving Wilms tumor, testicular cancer, choriocarcinoma, and rhabdomyosarcoma by intercalating in the minor grooves of DNA (Falzone et al. 2018; Gallego et al. 1997; Mohan et al. 2021). Other important DNA intercalating antibiotics obtained from *Streptomyces* species are phleomycin, bleomycin (from *Streptomyces verticillus*), zobramycin (from *Streptomyces flavoviridis*), doxorubicin (from *Streptomyces peucetius*), epirubicin, daunorubicin, and idarubicin are used for the treatment of melanoma, leukemia, ovarian cancer, lymphomas, and endometrium cancers. Mitomycin C isolated from *Streptomyces caespitosus* was a potent DNA cross-linker and alkylation-inhibiting anticancer antibiotic. Antineoplastic mitomycin C is a commercial antibiotic for the treatment of lung carcinoma, colorectal, pancreatic cancer, and others (Law et al. 2020; Mohan et al. 2021).

20.4 Challenges Associated with Bacteria-Based Cancer Therapy

Bacteria and their bioactive products have been identified as a possible alternative for the treatment of cancer. Tumor-targeting bacteria are a perfect vehicle for delivering therapeutic cargo selectively targeted at cancers of different origins. The oncolytic bacteria possess unique characteristic features such as tumor selectivity, targeting the hypoxic tumor microenvironment, and unique genome packaging (Gulati et al. 2022). However, there is still a long way to go before cancer treatment using bacteria reaches the level of acceptance as conventional cancer therapies (Zhou et al. 2018). The main issue with employing bacteria as anticancer drugs is their toxicity at the dose needed for therapeutic efficacy, and lowering the dose has an adverse influence on its efficacy (Gardlík et al. 2005). Secondly, live bacteria can proliferate in foreign bodies such as joint replacements, artificial heart valves, and implanted medical devices, which could act as infection reservoirs.

Thirdly, bacteria-based cancer treatment is not appropriate for patients who have undergone some types of chemotherapy since these may cause immunosuppression to the point where it is unable to respond to bacterial colonization. In addition, the unstable DNA and short half-life of the bacterial proteins/peptides present yet another major challenge in the therapeutic field (Gupta et al. 2021). Apart from these, their potential to develop antibiotic resistance or mutations that would reverse the bacteria's attenuated phenotype is of serious concern. Production of live bacteria is more challenging than producing anticancer medicines where real-time monitoring of the cultivation, purification, and harvesting of live bacteria under stringent aseptic conditions are practical techniques to assure the GMP grade test quality of the final products (Mayakrishnan et al. 2022). The chemotherapeutic drug gemcitabine, for instance, has been shown to be converted into its inactive form by *Mycoplasma hyorhinis* and species of *Shigella*, *Escherichia*, *Klebsiella*, *Salmonella*, *Citrobacter*, and *Serratia* that express cytidine deaminase (CDDL) (Geller et al. 2017). This results in the development of resistance to cancer treatments. Nonetheless, gut bacteria can influence tumor pathogenesis, chemotherapy resistance, anticancer immune responses, and health consequences of chemotherapy (Sivan et al. 2015; Vétizou et al. 2015).

20.5 Conclusion and Future Directions

In the evolutionary timeline, bacteria have evolved their habitats, ranging from abiotic components (soil, water, and air) to biotic components (human skin, anaerobic niches viz. human gut) of the ecosystem. The adaptation to these different conditions has enabled some bacteria to grow in hypoxic cellular conditions along with the production of diverse bioactive secondary metabolites viz. enzymes, toxins, peptides, and antibiotics aiding in their survival and nutrition. Scientific advancement in the recent past has enabled the exploration of bacteria and their bioactive compounds for therapeutic purposes including cancers. The latter has been a leading cause of increased rate of mortality since past decades. The molecular and histological heterogeneity of cancerous cells has also posed a major challenge for their regression with monotherapies like chemotherapy and radiotherapy and lower side effects. Thus, employment of bacteriotherapy as alternative cancer treatment modality is taken into immense considerations in recent times. Bacteriotherapy viz. live oncolytic bacteria and its secondary bioactive metabolites have been reported to target cancerous cells in vivo with high efficacy and specificity. Interestingly, many of them including genetically engineered bacteria and bacterial enzymes have been under clinical evaluation for their antitumor potency. In addition, alteration of cellular fluidity and phospholipid expression makes cancerous cells more susceptible toward bacteriotherapy. However, this novel biological approach has certain risks and challenges that need to be circumvented for clinical progression and establishment of bacteriotherapy. Also, considering the in vivo safety concerns, the main trends of bacteriotherapy are foreseen to be shifting toward personalized cancer treatment employing commensal bacteria. Furthermore, employment of novel approaches like immunoinformatics, smart nano-carriers, and conjugation of bacterial peptides with tumor-targeting peptides could enable researchers to design customized, controllable bacterial cells/communities, targeting this multifactorial disorder on a molecular level. In addition, a combinatorial approach harnessing the pharmacological potential of bacteria along with the currently employed monotherapies can aid in the complete remission of cancer.

References

AlQahtani AD, Al-Mansoori L, Bashraheel SS, Rashidi FB, Al-Yafei A, Elsinga P, Domling A, Goda SK (2019) Production of "biobetter" glucarpidase variants to improve drug detoxification and antibody directed enzyme prodrug therapy for cancer treatment. Eur J Pharm Sci 127:79–91. https://doi.org/10.1016/j.ejps.2018.10.014

Avramis VI, Tiwari PN (2006) Asparaginase (native ASNase or pegylated ASNase) in the treatment of acute lymphoblastic leukemia. Int J Nanomedicine 1(3):241–254

Azevedo MM, Pina-Vaz C, Baltazar F (2020) Microbes and cancer: friends or faux? Int J Mol Sci 21(9):3115. https://doi.org/10.3390/ijms21093115

Bailly J, Sezanne V, Godfrin Y (2011) L-asparaginase loaded red blood cells: from prescription to injection. Blood 118(21):4252. https://doi.org/10.1182/blood.V118.21.4252.4252

Baindara P, Gautam A, Raghava GPS, Korpole S (2017) Anticancer properties of a defensin like class IId bacteriocin Laterosporulin 10. Sci Rep 7:46541. https://doi.org/10.1038/srep46541

Bantel H, Sinha B, Domschke W, Peters G, Schulze-Osthoff K, Jänicke RU (2001) α-Toxin is a mediator of *Staphylococcus aureus*–induced cell death and activates caspases via the intrinsic death pathway independently of death receptor signaling. J Cell Biol 155(4):637–648. https://doi.org/10.1083/jcb.200105081

Basu P, Mehta A, Jain M, Gupta S, Nagarkar RV, John S, Petit R (2018) A randomized phase 2 study of ADXS11-001 *listeria monocytogenes*-Listeriolysin O immunotherapy with or

without cisplatin in treatment of advanced cervical cancer. Int J Gynecol Cancer 28(4):764–772. https://doi.org/10.1097/IGC.0000000000001235
Binz N, Shalaby T, Rivera P, Shin-ya K, Grotzer MA (2005) Telomerase inhibition, telomere shortening, cell growth suppression and induction of apoptosis by telomestatin in childhood neuroblastoma cells. Eur J Cancer 41(18):2873–2881. https://doi.org/10.1016/j.ejca.2005.08.025
Blackman LD, Varlas S, Arno MC, Houston ZH, Fletcher NL, Thurecht KJ, Hasan M, Gibson MI, O'Reilly RK (2018) Confinement of therapeutic enzymes in selectively permeable polymer vesicles by polymerization-induced self-assembly (Pisa) reduces antibody binding and proteolytic susceptibility. ACS Cent Sci 4(6):718–723. https://doi.org/10.1021/acscentsci.8b00168
Braun V, Pilsl H, Groß P (1994) Colicins: structures, modes of action, transfer through membranes, and evolution. Arch Microbiol 161(3):199–206. https://doi.org/10.1007/BF00248693
Brumano LP, da Silva FVS, Costa-Silva TA, Apolinário AC, Santos JHPM, Kleingesinds EK, Monteiro G, Rangel-Yagui C et al (2019) Development of L-Asparaginase biobetters: current research status and review of the desirable quality profiles. Front Bioeng Biotechnol 6:212. https://doi.org/10.3389/fbioe.2018.00212
Cavuoto P, Fenech MF (2012) A review of methionine dependency and the role of methionine restriction in cancer growth control and life-span extension. Cancer Treat Rev 38(6):726–736. https://doi.org/10.1016/j.ctrv.2012.01.004
Cellarier E, Durando X, Vasson MP, Farges MC, Demiden A, Maurizis JC, Madelmont JC, Chollet P (2003) Methionine dependency and cancer treatment. Cancer Treat Rev 29(6):489–499. https://doi.org/10.1016/S0305-7372(03)00118-X
Chang W-W, Lai C-H, Chen M-C, Liu C-F, Kuan Y-D, Lin S-T, Lee C-H (2013) *Salmonella* enhance chemosensitivity in tumor through connexin 43 upregulation. Int J Cancer 133(8): 1926–1935. https://doi.org/10.1002/ijc.28155
Chumchalová J, Šmarda J (2003) Human tumor cells are selectively inhibited by colicins. Folia Microbiol 48(1):111–115. https://doi.org/10.1007/BF02931286
Clairmont C, Lee KC, Pike J, Ittensohn M, Low KB, Pawelek J, Bermudes D, Brecher SM, Margitich D, Turnier J, Li Z, Luo X, King I, Zheng LM (2000) Biodistribution and genetic stability of the novel antitumor agent VNP20009, a genetically modified strain of *salmonella typhimurium*. J Infect Dis 181(6):1996–2002. https://doi.org/10.1086/315497
Coley WB (1991) The treatment of malignant tumors by repeated inoculations of erysipelas. With a report of ten original cases. 1893. Clin Orthop Relat Res 262:3–11
Dachs GU, Hunt MA, Syddall S, Singleton DC, Patterson AV (2009) Bystander or no bystander for gene directed enzyme prodrug therapy. Molecules (Basel, Switzerland) 14(11):4517–4545. https://doi.org/10.3390/molecules14114517
Dang LH, Bettegowda C, Agrawal N, Cheong I, Huso D, Frost P, Loganzo F, Greenberger L, Barkoczy J, Pettit GR, Smith AB, Gurulingappa H, Khan S, Parmigiani G, Kinzler KW, Zhou S, Vogelstein B (2004) Targeting vascular and avascular compartments of tumors with *C. novyi*-NT and anti-microtubule agents. Cancer Biol Ther 3(3):326–337. https://doi.org/10.4161/cbt.3.3.704
Dhankhar R, Gulati P, Kumar S, Kapoor RK (2018) Arginine-lowering enzymes against cancer: a technocommercial analysis through patent landscape. Expert Opin Ther Pat 28(8):603–614. https://doi.org/10.1080/13543776.2018.1508452
Dhankhar R, Gupta V, Kumar S, Kapoor RK, Gulati P (2020) Microbial enzymes for deprivation of amino acid metabolism in malignant cells: biological strategy for cancer treatment. Appl Microbiol Biotechnol 104(7):2857–2869. https://doi.org/10.1007/s00253-020-10432-2
Dhankhar R, Kawatra A, Mohanty A, Gulati P (2021) Microbial enzymes used in prodrug activation for cancer therapy: insights and future perspectives. Curr Protein Pept Sci 22(7):514–525. https://doi.org/10.2174/1389203721666201207231932
Dhankhar R, Kumar A, Kumar S, Chhabra D, Shukla P, Gulati P (2019) Multilevel algorithms and evolutionary hybrid tools for enhanced production of arginine deiminase from pseudomonas

furukawaii RS3. Bioresour Technol 290:121789. https://doi.org/10.1016/j.biortech.2019.121789
Dincer S, Takci HAM, Ozdenefe MS, Dincer S, Takci HAM, Ozdenefe MS (2022) Nonribosomal peptide synthesis. In: Molecular cloning. IntechOpen, London. https://doi.org/10.5772/intechopen.104722
Ding H, Liu H, Yin Y, Ding Y, Jia Y, Chen Q, Zou G, Zheng Z (2012) Insights into the modulation of optimum pH by a single histidine residue in arginine deiminase from *Pseudomonas aeruginosa*. Biol Chem 393(9):1013–1024. https://doi.org/10.1515/hsz-2012-0183
do Vale A, Cabanes D, Sousa S (2016) Bacterial toxins as pathogen weapons against phagocytes. Front Microbiol 7:42. https://doi.org/10.3389/fmicb.2016.00042
Duong MT-Q, Qin Y, You S-H, Min J-J (2019) Bacteria-cancer interactions: Bacteria-based cancer therapy. Exp Mol Med 51(12):1–15. https://doi.org/10.1038/s12276-019-0297-0
El-Sayed AS (2010) Microbial L-methioninase: production, molecular characterization, and therapeutic applications. Appl Microbiol Biotechnol 86(2):445–467. https://doi.org/10.1007/s00253-009-2303-2
Engedal N, Skotland T, Torgersen ML, Sandvig K (2011) Shiga toxin and its use in targeted cancer therapy and imaging. Microb Biotechnol 4(1):32–46. https://doi.org/10.1111/j.1751-7915.2010.00180.x
Falzone L, Salomone S, Libra M (2018) Evolution of cancer pharmacological treatments at the turn of the third millennium. Front Pharmacol 9:1300. https://doi.org/10.3389/fphar.2018.01300
Fernandes HS, Silva Teixeira CS, Fernandes PA, Ramos MJ, Cerqueira NMFSA (2017) Amino acid deprivation using enzymes as a targeted therapy for cancer and viral infections. Expert Opin Ther Pat 27(3):283–297. https://doi.org/10.1080/13543776.2017.1254194
Fiedler H-P, Bruntner C, Riedlinger J, Bull AT, Knutsen G, Goodfellow M, Jones A, Maldonado L, Pathom-aree W, Beil W, Schneider K, Keller S, Sussmuth RD (2008) Proximicin A, B and C, novel aminofuran antibiotic and anticancer compounds isolated from marine strains of the actinomycete *Verrucosispora*. J Antibiot 61(3):158–163. https://doi.org/10.1038/ja.2008.125
Gallego J, Ortiz AR, de Pascual-Teresa B, Gago F (1997) Structure–affinity relationships for the binding of actinomycin D to DNA. J Comput Aided Mol Des 11(2):114–128. https://doi.org/10.1023/A:1008018106064
Ganai S, Arenas RB, Sauer JP, Bentley B, Forbes NS (2011) In tumors *salmonella* migrate away from vasculature toward the transition zone and induce apoptosis. Cancer Gene Ther 18(7):457–466. https://doi.org/10.1038/cgt.2011.10
Gardlík R, Pálffy R, Hodosy J, Lukács J, Turna J, Celec P (2005) Vectors and delivery systems in gene therapy. Med Sci Monit 11(4):RA110-121
Geller LT, Barzily-Rokni M, Danino T, Jonas OH, Shental N, Nejman D, Gavert N, Zwang Y, Cooper ZA, Shee K, Thaiss CA, Reuben A, Livny J, Avraham R, Frederick DT, Ligorio M, Chatman K, Johnston SE, Mosher CM, Straussman R (2017) Potential role of intratumor bacteria in mediating tumor resistance to the chemotherapeutic drug gemcitabine. Science 357(6356):1156–1160. https://doi.org/10.1126/science.aah5043
Geyer PE, Maak M, Nitsche U, Perl M, Novotny A, Slotta-Huspenina J, Dransart E, Holtorf A, Johannes L, Janssen K-P (2016) Gastric adenocarcinomas express the glycosphingolipid Gb3/CD77: targeting of gastric cancer cells with Shiga toxin b-subunit. Mol Cancer Ther 15(5):1008–1017. https://doi.org/10.1158/1535-7163.MCT-15-0633
Goldufsky J, Wood S, Hajihossainlou B, Rehman T, Majdobeh O, Kaufman HL, Ruby CE, Shafikhani SH (2015) *Pseudomonas aeruginosa* exotoxin T induces potent cytotoxicity against a variety of murine and human cancer cell lines. J Med Microbiol 64(Pt 2):164–173. https://doi.org/10.1099/jmm.0.000003-0
Greco O, Dachs GU (2001) Gene directed enzyme/prodrug therapy of cancer: historical appraisal and future prospectives. J Cell Physiol 187(1):22–36. https://doi.org/10.1002/1097-4652(2001)9999:9999<::AID-JCP1060>3.0.CO;2-H
Gulati P, Kawatra A, Dhillon S, Dhankhar R (2022) Bacterial drug delivery vehicles for targeted treatment of tumors. In: Microbial products. CRC, Boca Raton, FL

Gupta KH, Nowicki C, Giurini EF, Marzo AL, Zloza A (2021) Bacterial-based cancer therapy (BBCT): recent advances, current challenges, and future prospects for cancer immunotherapy. Vaccine 9(12):1497. https://doi.org/10.3390/vaccines9121497

Hamamichi S, Fukuhara T, Hattori N (2020) Immunotoxin screening system: a rapid and direct approach to obtain functional antibodies with internalization capacities. Toxins 12:658. https://doi.org/10.3390/toxins12100658

Hens JR, Sinha I, Perodin F, Cooper T, Sinha R, Plummer J, Perrone CE, Orentreich D (2016) Methionine-restricted diet inhibits growth of MCF10AT1-derived mammary tumors by increasing cell cycle inhibitors in athymic nude mice. BMC Cancer 16(1):349. https://doi.org/10.1186/s12885-016-2367-1

Hetz C, Bono MR, Barros LF, Lagos R (2002) Microcin E492, a channel-forming bacteriocin from *Klebsiella pneumoniae*, induces apoptosis in some human cell lines. Proc Natl Acad Sci U S A 99(5):2696–2701. https://doi.org/10.1073/pnas.052709699

Hoffman RM, Tan Y, Li S, Han Q, Zavala J, Zavala J (2019) Pilot phase I clinical trial of methioninase on high-stage cancer patients: rapid depletion of circulating methionine. In: Hoffman RM (ed) Methionine dependence of cancer and aging: methods and protocols. Springer, Cham, pp 231–242. https://doi.org/10.1007/978-1-4939-8796-2_17

Holtsberg FW, Ensor CM, Steiner MR, Bomalaski JS, Clark MA (2002) Poly(ethylene glycol) (PEG) conjugated arginine deiminase: effects of PEG formulations on its pharmacological properties. J Control Release 80(1–3):259–271. https://doi.org/10.1016/s0168-3659(02)00042-1

Hoskin DW, Ramamoorthy A (2008) Studies on anticancer activities of antimicrobial peptides. Biochim Biophys Acta Biomembr 1778(2):357–375. https://doi.org/10.1016/j.bbamem.2007.11.008

Iacovache I, Bischofberger M, van der Goot FG (2010) Structure and assembly of pore-forming proteins. Curr Opin Struct Biol 20(2):241–246. https://doi.org/10.1016/j.sbi.2010.01.013

Imre G, Heering J, Takeda A-N, Husmann M, Thiede B, Zu Heringdorf DM, Green DR, van der Goot FG, Sinha B, Dötsch V, Rajalingam K (2012) Caspase-2 is an initiator caspase responsible for pore-forming toxin-mediated apoptosis. EMBO J 31(11):2615–2628. https://doi.org/10.1038/emboj.2012.93

Jen EY, Gao X, Li L, Zhuang L, Simpson NE, Aryal B, Wang R, Przepiorka D, Shen YL, Leong R, Liu C, Sheth CM, Bowen S, Goldberg KB, Farrell AT, Blumenthal GM, Pazdur R (2020) FDA approval summary: Tagraxofusp-erzs for treatment of blastic plasmacytoid dendritic cell neoplasm. Clin Cancer Res 26(3):532–536. https://doi.org/10.1158/1078-0432.CCR-19-2329

Jiang S-N, Phan TX, Nam T-K, Nguyen VH, Kim H-S, Bom H-S, Choy HE, Hong Y, Min J-J (2010) Inhibition of tumor growth and metastasis by a combination of *Escherichia coli*–mediated cytolytic therapy and radiotherapy. Mol Ther 18(3):635–642. https://doi.org/10.1038/mt.2009.295

Jianhua C, Yujun W, Ruibo J, Min W, Wutong W (2006) Probing the antigenicity of *E. coli* L-asparaginase by mutational analysis. Mol Biotechnol 33(1):57–65. https://doi.org/10.1385/MB:33:1:57

Kaliberov SA, Market JM, Gillespie GY, Krendelchtchikova V, Della Manna D, Sellers JC, Kaliberova LN, Black ME, Buchsbaum DJ (2007) Mutation of *Escherichia coli* cytosine deaminase significantly enhances molecular chemotherapy of human glioma. Gene Ther 14(14):1111–1119. https://doi.org/10.1038/sj.gt.3302965

Kannan S, Marudhamuthu M (2019) Development of chitin cross-linked enzyme aggregates of L-methioninase for upgraded activity, permanence and application as efficient therapeutic formulations. Int J Biol Macromol 141:218–231. https://doi.org/10.1016/j.ijbiomac.2019.08.246

Karpiński TM, Adamczak A (2018) Anticancer activity of bacterial proteins and peptides. Pharmaceutics 10(2):54. https://doi.org/10.3390/pharmaceutics10020054

Karpiński TM, Szkaradkiewicz AK (2013) Anti-cancer peptides from bacteria. Bangladesh J Pharmacol 8(3):Article 3

Kaur S, Kaur S (2015) Bacteriocins as potential anticancer agents. Front Pharmacol 6:272. https://doi.org/10.3389/fphar.2015.00272
Kawatra A, Dhankhar R, Gulati P (2022) Microbial arginine deiminase: a multifaceted green catalyst in biomedical sciences. Int J Biol Macromol 196:151–162. https://doi.org/10.1016/j.ijbiomac.2021.12.015
Kawatra A, Dhankhar R, Mohanty A, Gulati P (2020) Biomedical applications of microbial phenylalanine ammonia lyase: current status and future prospects. Biochimie 177:142–152. https://doi.org/10.1016/j.biochi.2020.08.009
Khoshnood S, Fathizadeh H, Neamati F, Negahdari B, Baindara P, Abdullah MA, Haddadi MH (2022) Bacteria-derived chimeric toxins as potential anticancer agents. Front Oncol 12:953678. https://doi.org/10.3389/fonc.2022.953678
Kim SH, Castro F, Paterson Y, Gravekamp C (2009) High efficacy of a *listeria*-based vaccine against metastatic breast cancer reveals a dual mode of action. Cancer Res 69(14):5860–5866. https://doi.org/10.1158/0008-5472.CAN-08-4855
Kim M-Y, Vankayalapati H, Shin-ya K, Wierzba K, Hurley LH (2002) Telomestatin, a potent telomerase inhibitor that interacts quite specifically with the human telomeric intramolecular G-quadruplex. J Am Chem Soc 124(10):2098–2099. https://doi.org/10.1021/ja017308q
Knott SRV, Wagenblast E, Khan S, Kim SY, Soto M, Wagner M, Turgeon M-O, Fish L, Erard N, Gable AL, Maceli AR, Dickopf S, Papachristou EK, D'Santos CS, Carey LA, Wilkinson JE, Harrell JC, Perou CM, Goodarzi H, Hannon GJ (2018) Asparagine bioavailability governs metastasis in a model of breast cancer. Nature 554(7692):Article 7692. https://doi.org/10.1038/nature25465
Kostova V, Dransart E, Azoulay M, Brulle L, Bai S-K, Florent J-C, Johannes L, Schmidt F (2015) Targeted Shiga toxin–drug conjugates prepared via cu-free click chemistry. Bioorg Med Chem 23(22):7150–7157. https://doi.org/10.1016/j.bmc.2015.10.010
Kotzia GA, Labrou NE (2009) Engineering thermal stability of L-asparaginase by in vitro directed evolution. FEBS J 276(6):1750–1761. https://doi.org/10.1111/j.1742-4658.2009.06910.x
Law JW-F, Law LN-S, Letchumanan V, Tan LT-H, Wong SH, Chan K-G, Ab Mutalib N-S, Lee L-H (2020) Anticancer drug discovery from microbial sources: the unique mangrove Streptomycetes. Molecules 25(22):E5365. https://doi.org/10.3390/molecules25225365
Law D, Zhang Z (2007) Stabilization and target delivery of nattokinase using compression coating. Drug Dev Ind Pharm 33(5):495–503. https://doi.org/10.1080/03639040601050247
Lee DG, Hahm K-S, Park Y, Kim H-Y, Lee W, Lim S-C, Seo Y-K, Choi C-H (2005) Functional and structural characteristics of anticancer peptide pep 27 analogues. Cancer Cell Int 5(1):21. https://doi.org/10.1186/1475-2867-5-21
Liang X, Ji Y (2006) Alpha-toxin interferes with integrin-mediated adhesion and internalization of *Staphylococcus aureus* by epithelial cells. Cell Microbiol 8(10):1656–1668. https://doi.org/10.1111/j.1462-5822.2006.00740.x
Liang X, Ji Y (2007) Involvement of α5β1-integrin and TNF-α in *Staphylococcus aureus* α-toxin-induced death of epithelial cells. Cell Microbiol 9(7):1809–1821. https://doi.org/10.1111/j.1462-5822.2007.00917.x
Liu X, Gu K, Xia S-S, Zhang D-J, Li Y-G (2018) Dolyemycins A and B, two novel cyclopeptides isolated from *Streptomyces griseus* subsp. *griseus* HYS31. J Antibiot 71(10):838. https://doi.org/10.1038/s41429-018-0071-x
Llop-Hernández À, Verdura S, Cuyàs E, Menendez JA (2022) Nutritional niches of cancer therapy-induced senescent cells. Nutrients 14(17):Article 17. https://doi.org/10.3390/nu14173636
Lombó F, Velasco A, Castro A, de la Calle F, Braña AF, Sánchez-Puelles JM, Méndez C, Salas JA (2006) Deciphering the biosynthesis pathway of the antitumor thiocoraline from a marine actinomycete and its expression in two *Streptomyces* species. ChemBioChem 7(2):366–376. https://doi.org/10.1002/cbic.200500325
Luo X, Li Z, Lin S, Le T, Ittensohn M, Bermudes D, Runyab JD, Shen SY, Chen J, King IC, Zheng LM (2001) Antitumor effect of VNP20009, an attenuated *salmonella*, in murine tumor models. Oncol Res 12(11–12):501–508. https://doi.org/10.3727/096504001108747512

Maak M, Nitsche U, Keller L, Wolf P, Sarr M, Thiebaud M, Rosenberg R, Langer R, Kleeff J, Friess H, Johannes L, Janssen K-P (2011) Tumor-specific targeting of pancreatic cancer with Shiga toxin b-subunit. Mol Cancer Ther 10(10):1918–1928. https://doi.org/10.1158/1535-7163.MCT-11-0006

Maggi M, Mittelman SD, Parmentier JH, Colombo G, Meli M, Whitmire JM, Merrell DS, Whitelegge J, Scotti C (2017) A protease-resistant *Escherichia coli* asparaginase with outstanding stability and enhanced anti-leukaemic activity in vitro. Sci Rep 7(1):Article 1. https://doi.org/10.1038/s41598-017-15075-4

Mahboobi M, Sedighian H, Ch MH, Bambai B, Soofian SE, Amani J (2017) Applying bioinformatic tools for modeling and modifying type II *E. coli* L-asparginase to present a better therapeutic agent/drug for acute lymphoblastic leukemia. Int J Cancer Manag 10(3):Article 3. https://doi.org/10.5812/ijcm.5785

Malekshah OM, Chen X, Nomani A, Sarkar S, Hatefi A (2016) Enzyme/prodrug systems for cancer gene therapy. Curr Pharmacol Rep 2(6):299–308. https://doi.org/10.1007/s40495-016-0073-y

Mantovani HC, Hu H, Worobo RW, Russell JB (2002) Bovicin HC5, a bacteriocin from *Streptococcus bovis* HC5. Microbiology 148(Pt 11):3347–3352. https://doi.org/10.1099/00221287-148-11-3347

Matsuo H, Mochizuki H, Davies J, Sugiyama M (1997) Production of bleomycin N-acetyltransferase in *Escherichia coli* and *Streptomyces verticillus*. FEMS Microbiol Lett 153(1):83–88. https://doi.org/10.1111/j.1574-6968.1997.tb10467.x

Mayakrishnan V, Kannappan P, Tharmalingam N, Bose RJC, Madheswaran T, Ramasamy M (2022) Bacterial cancer therapy: a turning point for new paradigms. Drug Discov Today 27(8):2043–2050. https://doi.org/10.1016/j.drudis.2022.03.007

Melton-Celsa AR (2014) Shiga toxin (Stx) classification, structure, and function. Microbiol Spectr 2(2):EHEC-0024-2013. https://doi.org/10.1128/microbiolspec.EHEC-0024-2013

Meneguetti GP, Santos JHPM, Obreque KMT, Barbosa CMV, Monteiro G, Farsky SHP, Marim de Oliveira A, Angeli CB, Palmisano G, Ventura SPM, Pessoa-Junior A, de Oliveira Rangel-Yagui C (2019) Novel site-specific PEGylated L-asparaginase. PLoS One 14(2):e0211951. https://doi.org/10.1371/journal.pone.0211951

Mohan CD, Rangappa S, Nayak SC, Jadimurthy R, Wang L, Sethi G, Garg M, Rangappa KS (2021) Bacteria as a treasure house of secondary metabolites with anticancer potential. Semin Cancer Biol 86:998. https://doi.org/10.1016/j.semcancer.2021.05.006

Mooney R, Abdul Majid A, Batalla J, Annala AJ, Aboody KS (2017) Cell-mediated enzyme prodrug cancer therapies. Adv Drug Deliv Rev 118:35–51. https://doi.org/10.1016/j.addr.2017.09.003

Murphy JR (1996) Corynebacterium diphtheriae. In: Baron S (ed) Medical microbiology, 4th edn. University of Texas Medical Branch, Galveston. http://www.ncbi.nlm.nih.gov/books/NBK7971/

Nallar SC, Xu D-Q, Kalvakolanu DV (2017) Bacteria and genetically modified bacteria as cancer therapeutics: current advances and challenges. Cytokine 89:160–172. https://doi.org/10.1016/j.cyto.2016.01.002

Narsing Rao MP, Xiao M, Li W-J (2017) Fungal and bacterial pigments: secondary metabolites with wide applications. Front Microbiol 8:1113. https://doi.org/10.3389/fmicb.2017.01113

Nguyen HA, Su Y, Zhang JY, Antanasijevic A, Caffrey M, Schalk AM, Liu L, Rondelli D, Oh A, Mahmud DL, Bosland MC, Kajdacsy-Balla A, Peirs S, Lammens T, Mondelaers V, De Moerloose B, Goossens S, Schlicht MJ, Kabirov KK, Lavie A (2018) A novel l-asparaginase with low L-glutaminase coactivity is highly efficacious against both T- and B-cell acute lymphoblastic leukemias in vivo. Cancer Res 78(6):1549–1560. https://doi.org/10.1158/0008-5472.CAN-17-2106

Noei A, Nili-Ahmadabadi A, Soleimani M (2019) The enhanced cytotoxic effects of the p - 28-apoptin chimeric protein as a novel anti-cancer agent on breast cancer cell lines. Drug Res 69(3):144–150. https://doi.org/10.1055/a-0654-4952

Oh S, Todhunter DA, Panoskaltsis-Mortari A, Buchsbaum DJ, Toma S, Vallera DA (2012) A deimmunized bispecific ligand-directed toxin that shows an impressive anti–pancreatic cancer effect in a systemic nude mouse orthotopic model. Pancreas 41(5):789–796. https://doi.org/10.1097/MPA.0b013e31823b5f2e

Oloomi M, Imani M, Behzadi R, Asori M, Bouzari S, Mokhlesi B (2018) Anti-tumor activity of *Escherichia coli* Shiga toxin a subunit delivered by SF9 insect cells. J Pharmacol Sci 138(1): 71–75. https://doi.org/10.1016/j.jphs.2018.09.003

Ozdemir T, Fedorec AJH, Danino T, Barnes CP (2018) Synthetic biology and engineered live biotherapeutics: toward increasing system complexity. Cell Syst 7(1):5–16. https://doi.org/10.1016/j.cels.2018.06.008

Paiva AD, de Oliveira MD, de Paula SO, Baracat-Pereira MC, Breukink E, Mantovani HC (2012) Toxicity of bovicin HC5 against mammalian cell lines and the role of cholesterol in bacteriocin activity. Microbiology 158(Pt 11):2851–2858. https://doi.org/10.1099/mic.0.062190-0

Pinatih KJP, Suardana IW, Widiasih DA, Suharsono H (2021) Shiga-like toxin produced by local isolates of *Escherichia coli* O157:H7 induces apoptosis of the t47 breast cancer cell line. Breast Cancer 15:11782234211010120. https://doi.org/10.1177/11782234211010120

Raboni S, Revtovich S, Demitri N, Giabbai B, Storici P, Cocconcelli C, Faggiano S, Rosini E, Pollegioni L, Galati S, Buschini A, Morozova E, Kulikova V, Nikulin A, Gabellieri E, Cioni P, Demidkina T, Mozzarelli A (2018) Engineering methionine γ-lyase from *Citrobacter freundii* for anticancer activity. BBA-Proteins Proteom 1866(12):1260–1270. https://doi.org/10.1016/j.bbapap.2018.09.011

Rezai T, Yu B, Millhauser GL, Jacobson MP, Lokey RS (2006) Testing the conformational hypothesis of passive membrane permeability using synthetic cyclic peptide diastereomers. J Am Chem Soc 128(8):2510–2511. https://doi.org/10.1021/ja0563455

Riedl S, Rinner B, Asslaber M, Schaider H, Walzer S, Novak A, Lohner K, Zweytick D (2011) In search of a novel target—phosphatidylserine exposed by non-apoptotic tumor cells and metastases of malignancies with poor treatment efficacy. Biochim Biophys Acta Biomembr 1808(11):2638–2645. https://doi.org/10.1016/j.bbamem.2011.07.026

Risberg K, Fodstad Ø, Andersson Y (2011) Synergistic anticancer effects of the 9.2.27PE immunotoxin and ABT-737 in melanoma. PLoS One 6(9):e24012. https://doi.org/10.1371/journal.pone.0024012

Robert A, Wiels J (2021) Shiga toxins as antitumor tools. Toxins 13(10):690. https://doi.org/10.3390/toxins13100690

Schweizer F (2009) Cationic amphiphilic peptides with cancer-selective toxicity. Eur J Pharmacol 625(1):190–194. https://doi.org/10.1016/j.ejphar.2009.08.043

Sedighi M, Zahedi Bialvaei A, Hamblin MR, Ohadi E, Asadi A, Halajzadeh M, Lohrasbi V, Mohammadzadeh N, Amiriani T, Krutova M, Amini A, Kouhsari E (2019) Therapeutic bacteria to combat cancer; current advances, challenges, and opportunities. Cancer Med 8(6): 3167–3181. https://doi.org/10.1002/cam4.2148

Shindoh N, Mori M, Terada Y, Oda K, Amino N, Kita A, Taniguchi M, Sohda K-Y, Nagai K, Sowa Y, Masuoka Y, Orita M, Sasamata M, Matsushime H, Furuichi K, Sakai T (2008) YM753, a novel histone deacetylase inhibitor, exhibits antitumor activity with selective, sustained accumulation of acetylated histones in tumors in the WiDr xenograft model. Int J Oncol 32(3):545–555. https://doi.org/10.3892/ijo.32.3.545

Shin-ya K, Wierzba K, Matsuo K, Ohtani T, Yamada Y, Furihata K, Hayakawa Y, Seto H (2001) Telomestatin, a novel telomerase inhibitor from *Streptomyces anulatus*. J Am Chem Soc 123(6): 1262–1263. https://doi.org/10.1021/ja005780q

Singh BP, Rateb ME, Rodriguez-Couto S et al (2019) Front Microbiol 10:914. https://doi.org/10.3389/fmicb.2019.00914

Sivan A, Corrales L, Hubert N, Williams JB, Aquino-Michaels K, Earley ZM, Benyamin FW, Man Lei Y, Jabri B, Alegre M-L, Chang EB, Gajewski TF (2015) Commensal *Bifidobacterium* promotes antitumor immunity and facilitates anti–PD-L1 efficacy. Science 350(6264): 1084–1089. https://doi.org/10.1126/science.aac4255

St. Jean AT, Swofford CA, Panteli JT, Brentzel ZJ, Forbes NS (2014) Bacterial delivery of *Staphylococcus aureus* α-hemolysin causes regression and necrosis in murine tumors. Mol Ther 22(7):1266–1274. https://doi.org/10.1038/mt.2014.36

Suganya K, Govindan K, Prabha P, Murugan M (2017) An extensive review on L-methioninase and its potential applications. Biocatal Agric Biotechnol 12:104–115. https://doi.org/10.1016/j.bcab.2017.09.009

Sugumaran A, Pandiyan R, Kandasamy P, Antoniraj MG, Navabshan I, Sakthivel B, Dharmaraj S, Chinnaiyan SK, Ashokkumar V, Ngamcharussrivichai C (2022) Marine biome-derived secondary metabolites, a class of promising antineoplastic agents: a systematic review on their classification, mechanism of action and future perspectives. Sci Total Environ 836:155445. https://doi.org/10.1016/j.scitotenv.2022.155445

Sung WS, Park Y, Choi C-H, Hahm K-S, Lee DG (2007) Mode of antibacterial action of a signal peptide, pep 27 from *Streptococcus pneumoniae*. Biochem Biophys Res Commun 363(3): 806–810. https://doi.org/10.1016/j.bbrc.2007.09.041

Toso JF, Gill VJ, Hwu P, Marincola FM, Restifo NP, Schwartzentruber DJ, Sherry RM, Topalian SL, Yang JC, Stock F, Freezer LJ, Morton KE, Seipp C, Haworth L, Mavroukakis S, White D, Mac Donald S, Mao J, Sznol M, Rosenberg SA (2002) Phase I study of the intravenous administration of attenuated *salmonella typhimurium* to patients with metastatic melanoma. J Clin Oncol 20(1):142–152

Vaishnav P, Demain AL (2011) Unexpected applications of secondary metabolites. Biotechnol Adv 29(2):223–229. https://doi.org/10.1016/j.biotechadv.2010.11.006

Vétizou M, Pitt JM, Daillère R, Lepage P, Waldschmitt N, Flament C, Rusakiewicz S, Routy B, Roberti MP, Duong CPM, Poirier-Colame V, Roux A, Becharef S, Formenti S, Golden E, Cording S, Eberl G, Schlitzer A, Ginhoux F, Zitvogel L (2015) Anticancer immunotherapy by CTLA-4 blockade relies on the gut microbiota. Science 350(6264):1079–1084. https://doi.org/10.1126/science.aad1329

Vijayvargia R, Suresh CG, Krishnasastry MV (2004) Functional form of Caveolin-1 is necessary for the assembly of α-hemolysin. Biochem Biophys Res Commun 324(3):1130–1136. https://doi.org/10.1016/j.bbrc.2004.09.164

Villarante KI, Elegado FB, Iwatani S, Zendo T, Sonomoto K, de Guzman EE (2011) Purification, characterization and in vitro cytotoxicity of the bacteriocin from *Pediococcus acidilactici* K2a2-3 against human colon adenocarcinoma (HT29) and human cervical carcinoma (HeLa) cells. World J Microbiol Biotechnol 27(4):975–980. https://doi.org/10.1007/s11274-010-0541-1

Wang Y, Guo W, Wu X, Zhang Y, Mannion C, Brouchkov A, Man Y-G, Chen T (2019) Oncolytic bacteria and their potential role in bacterium-mediated tumour therapy: a conceptual analysis. J Cancer 10(19):4442–4454. https://doi.org/10.7150/jca.35648

Wang D, Wei X, Kalvakolanu DV, Guo B, Zhang L (2021) Perspectives on oncolytic *salmonella* in cancer immunotherapy—A promising strategy. Front Immunol 12:615930. https://doi.org/10.3389/fimmu.2021.615930

Weerakkody LR, Witharana C (2019) The role of bacterial toxins and spores in cancer therapy. Life Sci 235:116839. https://doi.org/10.1016/j.lfs.2019.116839

Williams EM, Little RF, Mowday AM, Rich MH, Chan-Hyams JVE, Copp JN, Smaill JB, Patterson AV, Ackerley DF (2015) Nitroreductase gene-directed enzyme prodrug therapy: insights and advances toward clinical utility. Biochem J 471(2):131–153. https://doi.org/10.1042/BJ20150650

Wolf P, Elsässer-Beile U (2009) *Pseudomonas* exotoxin A: from virulence factor to anti-cancer agent. Int J Med Microbiol 299(3):161–176. https://doi.org/10.1016/j.ijmm.2008.08.003

Wu G (2013) Functional amino acids in nutrition and health. Amino Acids 45(3):407–411. https://doi.org/10.1007/s00726-013-1500-6

Xu G, McLeod HL (2001) Strategies for enzyme/prodrug cancer therapy. Clin Cancer Res 7(11): 3314–3324

Yaghoubi A, Khazaei M, Avan A, Hasanian SM, Cho WC, Soleimanpour S (2020) P 28 bacterial peptide, as an anticancer agent. Front Oncol 10:1303. https://doi.org/10.3389/fonc.2020.01303

Yang J, Tao R, Wang L, Song L, Wang Y, Gong C, Yao S, Wu Q (2019) Thermosensitive micelles encapsulating phenylalanine ammonia lyase act as a sustained and efficacious therapy against colorectal cancer. J Biomed Nanotechnol 15(4):717–727. https://doi.org/10.1166/jbn.2019.2734

Yu F-L, Liu T-T, Zhu X, Yang W-X, Zhang T, Lin N, Liu Y, Liu C-S, Jiang J, Guan J-C (2013) Staphylococcal enterotoxin B and α-toxin induce the apoptosis of ECV304 cells via similar mechanisms. Mol Med Rep 8(2):591–596. https://doi.org/10.3892/mmr.2013.1550

Zhang HL, Hua HM, Pei YH, Yao XS (2004) Three new cytotoxic cyclic acylpeptides from marine *bacillus* sp. Chem Pharm Bull 52(8):1029–1030. https://doi.org/10.1248/cpb.52.1029

Zhang L, Liu M, Jamil S, Han R, Xu G, Ni Y (2015) PEGylation and pharmacological characterization of a potential anti-tumor drug, an engineered arginine deiminase originated from *pseudomonas plecoglossicida*. Cancer Lett 357(1):346–354. https://doi.org/10.1016/j.canlet.2014.11.042

Zhang Y, Schulte W, Pink D, Phipps K, Zijlstra A, Lewis JD, Waisman DM (2010) Sensitivity of cancer cells to truncated diphtheria toxin. PLoS One 5(5):e10498. https://doi.org/10.1371/journal.pone.0010498

Zhang J-N, Xia Y-X, Zhang H-J (2021) Natural cyclopeptides as anticancer agents in the last 20 years. Int J Mol Sci 22(8):3973. https://doi.org/10.3390/ijms22083973

Zhou S, Gravekamp C, Bermudes D, Liu K (2018) Tumour-targeting bacteria engineered to fight cancer. Nat Rev Cancer 18(12):Article 12. https://doi.org/10.1038/s41568-018-0070-z

Zhu L, Tee KL, Roccatano D, Sonmez B, Ni Y, Sun Z-H, Schwaneberg U (2010) Directed evolution of an antitumor drug (arginine deiminase PpADI) for increased activity at physiological pH. ChemBioChem 11(5):691–697. https://doi.org/10.1002/cbic.200900717

Zhu L, Verma R, Roccatano D, Ni Y, Sun Z-H, Schwaneberg U (2010) A potential antitumor drug (arginine deiminase) reengineered for efficient operation under physiological conditions. ChemBioChem 11(16):2294–2301. https://doi.org/10.1002/cbic.201000458

Zu C, Wang J (2014) Tumor-colonizing bacteria: a potential tumor targeting therapy. Crit Rev Microbiol 40(3):225–235. https://doi.org/10.3109/1040841X.2013.776511

Communication with Gut Microbiota: An Emerging Strategy to Predict and Prevent Cancer

21

S. Ramadevi and Shanmugaraja Meenakshi

Abstract

Spectrum of microorganisms that live in and on different parts of the human body is collectively called human microbiome. Human microbiome is remarkably diverse comprising of bacteria, archea, fungi, protozoa, and non-living viruses. These microbes are further classified based on their location into oral, vaginal, gastrointestinal, and nasal microbiomes. Based on Human Microbiome Project data, human gut harbours the most complex and diverse microbial community having approximately 100 trillion bacterial species. Gut microbes play a key role in regulating almost all biological functions of the host, in particular the immunological, neurological, and metabolic activities. This countless microbial community exist in a delicate equilibrium which is very critical for the normal well-being of the host. Any disturbance to this equilibrium causes the state called Gut dysbiosis. Gut dysbiosis has a strong association with the development of several disorders including cancer. Among all other diseases, cancer is the highly investigated one for its association with gut dysbiosis as it has been shown to influence gastrointestinal cancer as well as distal cancer. It can either suppress or promote cancer and there exists a bidirectional connection between cancer and gut microbiome. These observations have accelerated the research on microbiome, especially in the aspect of therapeutic target.

S. Ramadevi
Faculty of Allied Health Sciences, Chettinad Hospital and Research Institute, Chettinad Academy of Research and Education, Chennai, Tamil Nadu, India

S. Meenakshi (✉)
Faculty of Allied Health Sciences, Chettinad Hospital and Research Institute, Chettinad Academy of Research and Education, Chennai, Tamil Nadu, India

Present Address: Department of Biotechnology, Mepco Schlenk Engineering College, Sivakasi, India

R. Sobti et al. (eds.), *Role of Microbes in Sustainable Development*,
https://doi.org/10.1007/978-981-99-3126-2_21

21.1 Introduction

Bacteria, archaea, viruses, and eukaryotic organisms live in and on our body and constitute the human microbiome (Fig. 21.1). These bacteria have enormous capability to affect human physiology in both good and bad health. They support metabolism, protect against various infections, stimulate the immune system, and influence most of our physiologic functions directly or indirectly through these basic roles (Varian et al. 2016). Eran Segal, a computational biologist, claims that gathering microbiome data would enable a 'deep phenotyping' method that will revolutionize drug discovery (Segal 2015). Numerous biological processes are now known to be controlled by gut microbiome. Intestinal disorders and non-intestinal disorders such as metabolic disorders, cancer, autoimmune diseases including multiple sclerosis, and autistic spectrum disorder are among them. Emerging studies report that microbiome can even influence the pharmacodynamic of most of the drugs, including several psychological medicines (Brody 2020). The microbiome is increasingly recognized as a significant human cell collaborator, interacting with almost all human cells (Cani 2018). Even though the majority of human intestine microbial genes are conserved amid individuals, the human gut microbiome encloses a few hundred bacterial communities and is very diverse in terms of bacterial

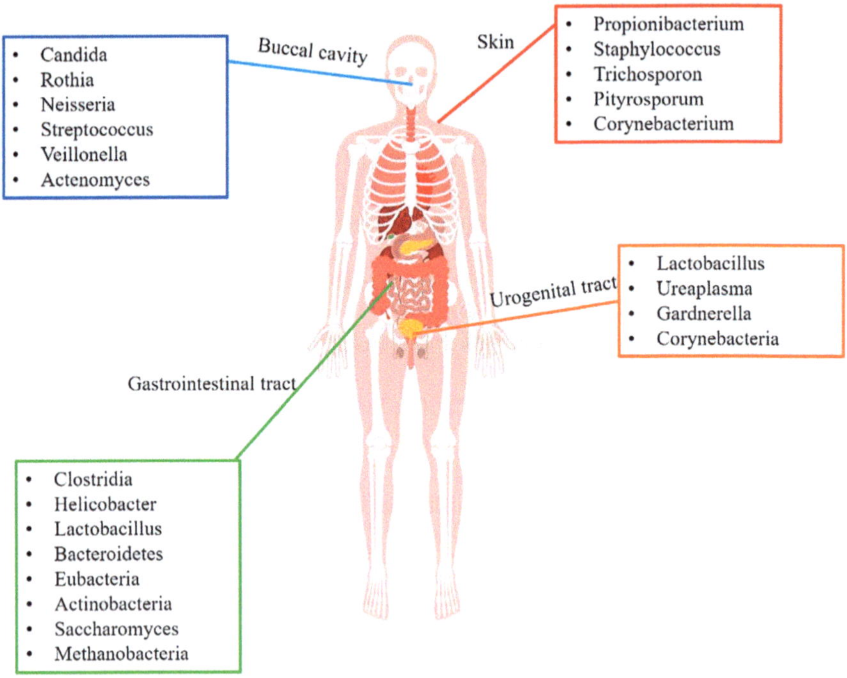

Fig. 21.1 Enlisting the microbial taxa found in different anatomical sites of human body (Source: Montalban-Arques and Scharl 2019; Jain et al. 2021)

composition and abundance. The Human Microbiome Project's (HMP) metagenomic data sets have created a strong platform for metaproteomic research of the human microbiome (Pan and Chen 2020). The development of current molecular tools enabled the delineation of the constitution as well as certain activities of the intestinal flora, which was aided by a boom in microbiota research during the last decade (Tilg et al. 2018). Behavioural characteristics, such as varied diet, restricted use of refined carbohydrates, sufficient nutritional fibre intake, sports, as well as body weight, all culminate in an altered constitution of the gut microbiota (Kyrgiou et al. 2017). Increasing evidence reporting the significance of intestinal microbiota in the normal biological functions of the host influenced the scientists to investigate the association of intestinal microbiota with colorectal cancer (CRC) in humans. Especially the methodical epidemiologic comparisons of CRC patients and control subjects revealed the following findings; reduced bacterial diversity in faeces, depletion of Gram-positive, fibre-fermenting Clostridia, and increased abundance of Gram-negative, pro-inflammatory genera *Fusobacterium* and *Porphyromonas*. All these findings were linked to an increased risk of CRC (Ahn et al. 2013). Inflammation is considered to be accountable for 20–30% of CRC cases. Inflammatory processes are well-known carcinogenesis promoters, as evidenced by the high risk of CRC in patients with IBD (Brennan and Garrett 2016). IBD patients are more likely to develop colorectal cancer as well as small intestinal adenocarcinoma. Dysbiosis toward selected microorganisms and lowered sophistication of commensal bacteria have been observed in patients with Crohn's disease (CD) and ulcerative colitis (UC), but it is unclear as to if dysbiosis helps in the advancement of IBD or is a consequence of the disease, as in the case of colon cancer patients. Patients with IBD have fewer bacteria with anti-inflammatory capabilities or more bacteria with pro-inflammatory characteristics than healthy controls (Raisch et al. 2014).

21.2 Microbiome and Cancer

Microbes in the human gut have a substantial influence on human wellness as well as immune system function located in close vicinity to the immunological response within the intestinal mucosa, which has been termed "the last undiscovered human organ." Dysbiosis, a rather inadequate state characterized by a less diversified and much less stable microbiota, with a prospective predominance of opportunistic harmful bacteria is found when the fine balance of commensal bacteria is disrupted (Lee et al. 2021). *Helicobacter pylori*, a class I carcinogen as per International Agency for Research on Cancer (IARC), is a causative agent of gastric cancer and MALT lymphoma, contributing to almost 5.5% of cancer cases globally. Chronic carriage of *Salmonella typhi,* the leading cause of typhoid fever, is speculated to be interrelated with gallbladder cancer. *Streptococcus bovis* (also known as *S. gallolyticus*), *Helicobacter pylori*, *Fusobacterium nucleatum, Coriobacterialies,* and enterotoxigenic *Bacteroides fragilis* (ETFB) have already been identified as probable candidates related to colorectal cancer (Hullar et al. 2014). It is well recognized that certain compounds are transformed into procarcinogens or carcinogens by indigenous gut microbiota (Berg 1996). Cancer, according to researchers, can occur as a consequence of a loss of immunosurveillance and the acquisition of immune tolerance to tumour-derived antigens. Several studies have demonstrated that the gut microbiota has a strong influence over host immune system by means of many secretory bioactive molecules that get transported even to the distant tissues and directly interact with the host cells. These properties make the bacteria a promising new cancer treatment option (Patyar et al. 2010).

21.3 Complex Bacteria as Cancer Therapy

Several facultative or essential anaerobic bacteria, notably *Escherichia coli*, *Clostridium*, *Listeria*, *Bifidobacterium*, and also *Salmonella* species, have tumour-identifying and tumour-killing abilities. It has been demonstrated that in rodent tumour models, live tumour-targeting bacteria can colonize tumours or tumour-impelled lymph nodes, suppress tumourigenesis, and enhance endurance following systemic infection (Duong et al. 2019). One technique that could be capable of overcoming many of the boundaries of conventional cancer treatment is the use of therapeutic microorganisms. Bacteria are potent antitumour agents on their own. Bacteria's flexibility to be genetically altered to change their ability to synthesize and release certain substances, and also customize their biochemical functions, is another remarkable characteristic that can be exploited for the treatment of various diseases in particular the cancer (Song et al., 2018; Sedighi et al. 2019).

21.4 The Gut Microbiome: Pro and Anti-Tumourigenic Agent

Recent reports convince the fact that change in the gut microbiota has a direct association with a higher risk of developing cancer in a variety of ways. Bacteria and their secretory by-products were also documented in several studies which not only contribute to the formation of cancer, but also influence the pharmacodynamics of anti-cancer medications. When the state of equilibrium in the gut microbiota is disturbed, commensal bacteria can infiltrate the gut lining and adjacent tissues, causing inflammation. Chronic inflammation has been demonstrated to be a trigger in the progression of malignancies in this regard (Akbar et al. 2022). The implementation of cancer-related lifestyle decisions such as smoking, 'westernized' diets, and lack of physical activity is continuing to increase cancer incidence in many developing countries as a result of demographics (Bultman 2014). Microbial-induced inflammation can also enhance the number of cytokines and also chemokines, which stimulate cellular propagation and prevent cell death, and may lead to cancer. The inflammatory process has a role in the progression of cancer in a variety of organs, such as the colon, abdomen, liver, and lungs (Dapito et al. 2012). Chronic inflammation is also involved in the presentation of an epigenetic environment for cancer development by inducing aberrant DNA methylation (Hattori and Ushijima 2016).

In the colons of IL-10 knockout mice with colitis, members of the Enterobacteriaceae family are elevated by up to 100-fold compared to wild-type control mice without colitis, according to a recent study (Arthur et al. 2012). The microbiome's metabolic potential is so enormous that it's been labeled as "second liver" and "second brain" of the human system and it's been correlated with predisposition to various disorders. This pertains to cancers, from preventing cancer in the onset of disease through wherewith people retort to chemo in the later phases (Nicholson et al. 2012). The impact of the microbiota in boosting or preventing tumour commencement and recurrence is not constrained to direct oncogenic effects. Increased oestrogen levels are a key risk factor for endometrial and breast cancers, and gut microbes play a significant role in oestrogen metabolism (Plottel and Blaser 2011). Clostridium was disproportionately portrayed in particular, and they can switch primary bile acids into a subsidiary bile acid known as deoxycholic acid (DCA). DCA is a toxin that can interact with DNA by inducing oxidative stress, and it has been linked to malignancies of the liver and colon. Colonic bacteria ferment heterocyclic amines from roasted meat, and some of the metabolites are susceptible to oxidation and thought to harm DNA and lead to colorectal cancer (Huycke and Rex Gaskins 2004). Numerous genes in the bacterium disrupt normal cell equilibrium, contributing to elevated levels of cytokines and other cancer-causing signaling biochemical compounds in *H. pylori*-infected people and possibly drive to gastric cancer (Moyat and Velin 2014). *Bacteroidetes*, *Firmicutes*, and *Proteobacteria* are the most common phyla that cause oesophageal malignancies among other gut flora (Nasrollahzadeh et al. 2015). The modification of bile acid (BA) conversion, generation of persistent low-grade inflammation, and alteration of the gut microbiota have all been linked to a high-fat diet (HFD) as a key independent predictor for CRC. Dysbiosis of the gut microbiota has been linked to CRC in several studies (Chen et al. 2020). Dysbiotic gut microbiome in mice has been linked to cancer initiation in trials. This is predicated on a mouse model that showed inflammation of the colon's inner lining and tumour formation, which was linked to a shift in the intestinal flora (Couturier-Maillard et al. 2013). *Helicobacter pylori* illness has been linked to stomach cancer and also mucosal-associated lymphoid tissue lymphoma (MALT). The prevalence of pathogenic *Escherichia coli* with *pks* toxicity genes has also been connected to local tissue inflammation and colon carcinogenesis, according to reports (Arthur et al. 2012). Prior studies into the microbiological etiology of breast cancer have focused on distinct viruses and their possible roles in the disease. While some studies have indicated HPV infection to breast cancer, others have not (Kroupis et al. 2006). Liver cancer is the most common primary hepatic tumour and the leading cause of death in persons with cirrhosis (Balogh et al. 2016). *Clostridium hylemonae*, *Clostridium hiranonis*, *Clostridium sordellii*, and *Clostridium scindens* convert the primary BA (7-hydroxyl BAs, chenodeoxycholic acid, and cholic acid) to the secondary BA: deoxycholic acid (DCA) through 7-dehydroxylation. High levels of secondary BA have been shown to produce reactive nitrogen (RNS) and reactive oxygen species (ROS), which affect the cellular membranes as well as mitochondria, causing cell death and DNA damage, ultimately leading to colon cancer (Ridlon et al. 2006). Inadequate eating routines, nicotine use,

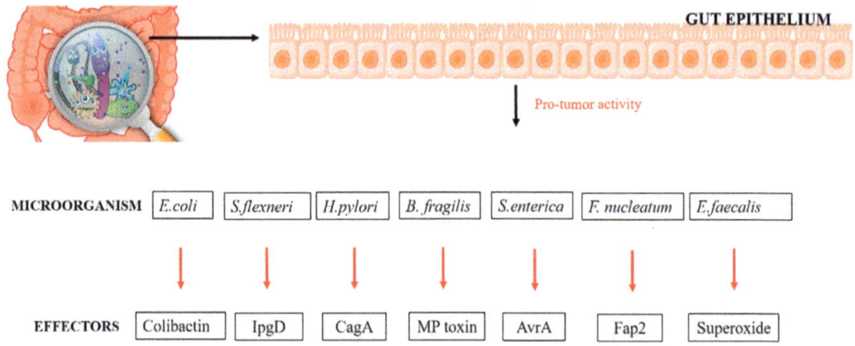

Fig. 21.2 Various bacterial species of the gut microbiota and their effector molecules that have been shown to be associated with pro-tumourigenic activity

poor oral hygiene, and excessive consumption of alcohol are all factors that contribute to oral cancer. Microbial infections (bacterial, fungal, and viral) have been linked to the advancement of mouth cancer (Arzmi et al. 2019). Cervical cancer is the fourth greatest inducer of cancer-related deaths among women worldwide, and the most common form of gynaecological tumours in developing countries (Arbyn et al. 2020). Proteobacteria were shown to be much more abundant in cervical cancer, as were the frequency of seven different taxa, including, *Roseburia, Lachnospiracea*, *Pseudomonas, Escherichia–Shigella, Lachnoclostridium, Succinivibrio,* and *Dorea.* Proteobacteria, the genera *Parabacteroides, Roseburia,* and *Escherichia,* according to the scientists, could be effective biomarkers for cervical cancer (Wang et al. 2019). Globally, ovarian cancer is the most lethal of all gynecological cancers. With 0.29 million cases and 0.18 million cancer deaths globally in 2018, ovarian cancer was the top cause of death among gynecologic malignancies in 2018 (Bray et al. 2018). Weight reduction and cardiac failure have been linked to changes in the gut microbiota, notably a reduction in *Firmicutes. Lactobacillus* supplementation may also aid to prevent cardiotoxicity caused by cisplatin by lowering inflammation (Zhao et al. 2018), (Fig. 21.2).

21.5 Bacterial Biomarkers and Cancer

21.5.1 Colorectal Cancer

Colorectal cancer (CRC) is a huge worldwide major health concern and among the most prevalent malignancies. CRC is the third most common cancer worldwide and the second most deadly, with 1.8 million new cases and 881,000 deaths in 2018 (Cheng et al. 2020). The gut microbiome is a non-invasive indicator for metabolic illnesses and malignancies that are found in the stool (Wu et al. 2021). Several bacteria, including *Bacteroides fragilis* and an *Escherichia coli* strain, have been linked to colorectal cancer in investigations (Yu et al. 2017). Fusobacterium, particularly *Fusobacterium nucleatum* (Fn), has been linked to colorectal cancer in recent investigations. Colorectal cancer patients' faeces and colonic mucosa are both high in Fn (McCoy et al. 2013). Bc is a rod-shaped gram-negative, facultatively anaerobic, non-spore-forming bacteria that was discovered in human faeces in 2010 (Watanabe et al. 2010). Ch is a gram-positive, completely anoxic, spore-forming, rod-shaped bacterium that participates in glucose metabolism, producing ethanol, acetate, hydrogen, carbon dioxide, and carbohydrates as fermentable substrates (Steer et al. 2001). Despite the widely known Fn, which is notorious to enhance colorectal carcinogenesis, it is also shown that the modified amplitudes of Ri, Bc, or m7 are a trigger of colorectal cancer formation or an outcome of colorectal cancer growth. Fn measurement alone has a moderate sensitivity and specificity as a noninvasive diagnostic technique for colorectal cancer. For colorectal cancer, combining four bacterial markers (Fn, Bc, Ch, and m7) increased the diagnostic capabilities of Fn alone (Liang et al. 2017).

21.5.2 Gall Bladder Cancer

As per autopsy surveys, gallbladder cancer is the most prevalent bile system cancer, accounting for 80 percent to 95 percent of all biliary tract malignancies globally. Gallstones are a strong predictor for gallbladder cancer, as they are found in the majority of patients (85 percent). Chronic bacterial cholangitis is a known risk factor for biliary tract cancer. *Salmonella* (e.g. *S. typhi* and *S. paratyphi*) and *Helicobacter* (e.g. *H. bilis*) spp. have been implicated the most. Gallbladder cancer affects about 6% of typhoid carriers, a 12-fold boost in risk (Hundal and Shaffer 2014). According to Xu et al. *Streptococcus* spp. are more common in urine from bladder cancer patients ($n = 8$) than healthy people ($n = 6$) (Xu et al. 2014). Bucevic et al. discovered a predominance of certain bacteria (*Facklamia*, *Fusobacterium*, *Actinobaculum*, *Subdoligranulum*, and *Campylobacter* as well as the family *Ruminococcaceae*) in urine from bladder cancer patients ($n = 12$) compared to healthy people ($n = 11$) (Popović et al. 2018). In urine from bladder cancer patients ($n = 31$) matched to nonneoplastic controls ($n = 18$), Wu et al. discovered an abundance of certain bacterial species (e.g. *Anaerococcus, Sphingobacterium*, and *Acinetobacter*) (Wu et al. 2018). Certain commonly agreed bacteria, particularly *Acinetobacter,* have been linked to bladder cancer, according to a study by Mai et al., but the underlying association is yet unknown. A deeper knowledge of the role of urine microfora in the initiation and progression of bladder cancer could lead to new diagnostic and prognostic indicators and additional microbial-targeted therapy options (Mai et al. 2019). Infection with *Helicobacter pylori* (H. pylori) is a well-known cause of stomach cancer. *H. hepaticus* is a known cause of biliary canaliculi infection, which can develop into liver cancer. The probability of biliary tract cancer was six times higher in *H. bilis*. Colon cancer can be caused by long-term infection with *Citrobacter rodentium,* a mouse pathogen that is genetically identical to enteropathogenic *Escherichia coli* (Nath et al. 2010).

21.5.3 Cervical Cancer

Cervical cancer is the fourth most frequent cancer in women, behind breast cancer and lung cancer (Buskwofie et al. 2020). Cervical intraepithelial neoplasia is caused by the human papillomavirus (HPV) (Fang et al. 2014). Depending on the bacterial species represented, the cervicovaginal microbiota can be divided into five groups using 16S rRNA high-throughput sequencing (16S-HTS) data. The Lactobacillus genus, which was initially discovered in 1892 by Döderlein, is observed in significant frequency in the cervicovaginal microbiota. *L. gasseri*, *Lactobacillus crispatus*, and *L. jensenii* create lactic acid and hydrogen peroxide (H_2O_2), which impede the development of pathogenic microorganisms and viruses. *L. iners*, on the other hand, is regarded as a species in the dysbiosis transition. Several studies have already shown a cervicovaginal microbiome shift associated with CIN and cancer, with increasing proportions of bacteria such as *Prevotella*, *Gardnerella,* and *Atopobium* and lower abundance of *Lactobacillus spp* (Curty et al. 2019). *Sneathia* spp. and *Fusobacterium* spp. were prevalent in squamous intraepithelial lesions and cervical cancer, respectively, for HPV- and HPV+ women without cytologic alterations, and *Sneathia* spp. and *Fusobacterium* spp. were dominant in squamous intraepithelial lesions as well as cervical cancer, respectively (Audirac-Chalifour et al. 2016). A majority of specific Lactobacillus species, such as *L. jensenii, L. gasseri,* and *L. crispatus,* are linked to a cervicovaginal environment that is relatively non-inflammatory. These pro-inflammatory states cause tissue damage, which may increase the carcinogenic potential of HPV (Kyrgiou et al. 2017). *Gardnerella vaginallis, Aerococcus, Schlegelella thermodepolymerans, Moryella,* and *Bifidobacterium bifidium,* all present in increased frequency in lesions, were found in the LEfSe analysis comparing the microbiota of normal and lesion samples (Curty et al. 2017), (Fig. 21.3 and Tables 21.1 and 21.2).

21.6 Crosstalk Between Microbiome and Host Immune System

The established interplay between the innate and adaptive immune systems and the tens of trillions of microorganisms that dwell in our gastrointestinal tracts have been documented in a number of recent reviews (Kau et al. 2011). Diversity, stability and resistance, and resilience are characteristics of a healthy intestinal microbial community, which are described as the ecosystem's richness, amenability to disruption, and ability to revert to its pre-perturbation form, respectively. The immune system is thought to be among the most powerful mechanisms at work in shaping the normal and dysbiotic microbiome configurations. As a result, characterizing the indirect impact of host immunity on dysbiosis-driven illnesses requires an understanding of immune system–microbiome interaction (Levy et al. 2017). Innate immune cells' transcriptional programming is heavily influenced by the microbiome. It's been demonstrated in the case of innate lymphoid cells (ILCs), where microbiota ablation has a genome-wide epigenetic and transcriptional effect (Gury-BenAri et al. 2016). Specific bacterial species have been found to have a direct effect on immune system

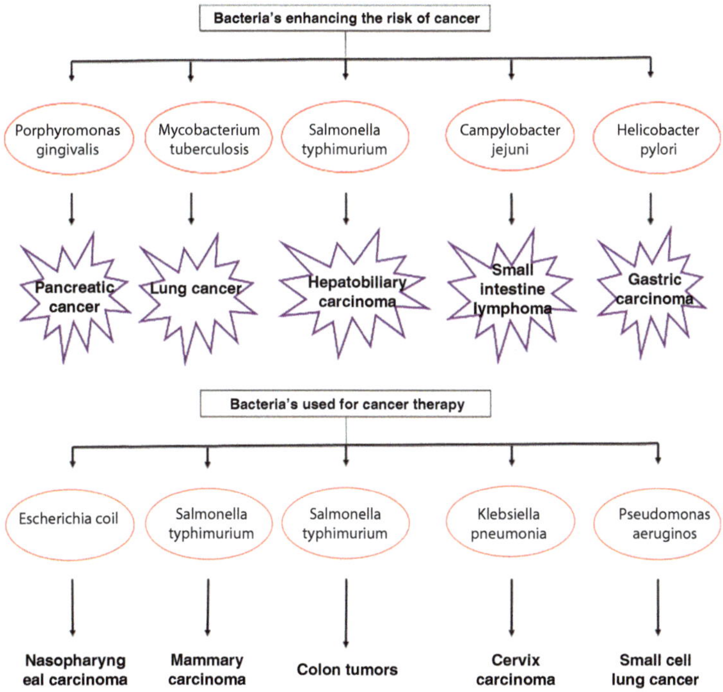

Fig. 21.3 Schematic illustration depicting the dual role of bacteria in promoting as well as treating the cancer

growth and differentiation in adaptive immune cells. Interleukin 23 (IL 23), IL 22, and serum amyloid A (SAA) protein are released when segmented filamentous bacteria (SFB) adhere to the intestinal epithelium, causing antigen-specific T helper 17 (TH17) cells to operate (Atarashi et al. 2015). The maturation of the immune system, as well as the establishment of resistance and suppression of microbial diseases, is influenced by the microbial flora at birth. The NF-B, c-Jun/JNK, and JAK/STAT3 pathways, which have roles in cell proliferation and immunosuppression, can be activated by upregulation of TLRs by LPS and other microbial metabolites (Bose and Mukherjee 2019). Many anti-cancer effects were later discovered to be due to the following bacterial components, which include Toll-like receptor (TLR) and NOD-like receptor (NLR) agonists. This led to the hypothesis that innate immunity could be used to convert tumour tolerance into anti-cancer immune responses (Garaude et al. 2012). Local and systemic inflammatory responses, oncogenic signalling, and tumour growth are all clearly modulated by interaction between the natural host microbiota and immune response. Immunotherapy efficacy is influenced by microbiome-induced innate and adaptive immune responses. Discovering the underlying immunological mechanisms and finding targetable molecules associated with the host's specific microbiota that regulate immune function are therefore critical (Fig. 21.4).

Table 21.1 (A): Treatment strategies: Bacteria that have reputed anti-cancer properties in human

Bacterial species	Cancer type	Interventions
Streptococcus pyogenes and *Serratia marcescens*	Osteosarcoma	Coley's toxins: Injection of *S. pyogenes* and *S. marcescens* in patients with sarcoma, with some evidence of objective response
Mycobacterium bovis BCG	Urothelial superficial cancers	Intravesical treatment of a live attenuated form of *M. bovis* reduces the risk of short- and long-term relapse
Lactobacillus casei str. Shirota (found in the fermented milk product Yakult)	Superficial bladder cancer	Immune-mediated effects (by NK cells and macrophages) and decreased tumour recurrence (except with multiple secondary tumours)
IMM-101 (heat-killed *Mycobacterium obuense*; NCTC 13365) with gemcitabine	Melanoma and advanced pancreatic ductal adenocarcinoma	Activation of APCs, granulocytes, and γδ T cells. Increased survival in metastatic disease in a randomized phase II trial
Live-attenuated *listeria monocytogenes* expressing mesothelin (CRS-207) with GVAX-cyclophosphamide	Advanced pancreatic ductal adenocarcinoma	Priming of mesothelin-specific CTLs, loss of regulatory T cells and tertiary lymphoid organ formation, and increased overall survival
Attenuated strain of *salmonella enterica* subsp. *enterica* serovar typhimurium: VNP20009	Metastatic melanoma and refractory solid tumours	Phase I trial of intravenous infusion of *S.* typhimurium led to inflammation, DC and T cell activation, and evidence of bacterial tumour colonization; however, there was no tumour regression
TAPET-CD: An attenuated *salmonella* bacterium that expresses the *Escherichia coli* cytosine deaminase gene	Head and neck squamous cell carcinoma or adenocarcinoma of the oesophagus	Evidence of bacterial colonization and confirmation of the conversion of 5-FC to 5-FU in 2 out of 3 tumours
Genetically modified *Corynebacterium diphtheriae*: Tf-CRM107 is a conjugate of transferrin and a point mutant of diphtheria toxin	Malignant brain tumour	MRI scans showed regression of tumour volume in 9 out of 15 patients with no evidence of severe local or systemic complications at low dose

21.7 Probiotics and Prebiotics in the Prevention of Cancer

Because of their capacity to control cancer cell proliferation and death in vitro and in vivo, probiotics have gotten a lot of interest. Several following researches in humans and rats confirmed the beneficial effect of probiotic strains on the activity of

Table 21.2 (B): Bacteria that have reputed anti-cancer properties in experimental models

Bacterial species	Cancer type	Interventions
Lactobacillus casei	Orthotopic and transplantable bladder tumours and their metastases	Oral or intravesical injection of dead or alive bacteria increased the levels of IFNγ and the recruitment of neutrophils
Lactobacillus rhamnosus GG	Bladder tumours	Weekly intravesical instillations directed chemokine and/or cytokine release, recruitment of NK cells and direct cytotoxic effects on cell lines ex vivo
Bacteroides fragilis and *Burkholderia cepacia*	MCA205 sarcomas and MC38 and CT26 colon cancers	Oral gavage of *B. fragilis* stimulated the production of IL-12 by bone marrow-derived DCs in vitro. The mechanism of *B. cepacia* remains unknown
Prevotella spp. and *Oscillibacter* spp.	Subcutaneous hepatocellular carcinoma	Oral administration of Prohep, a probiotic mixture, altered the microbiota and reduced tumour growth
Enterococcus hirae and *Barnesiella intestinihominis*	Sarcoma	Bacterial translocation: Induction of TH1 cells and pathogenic TH17 cells, intratumoural regulation of Treg cells, and IFNγ-producing γδ T cells, respectively
Bifidobacterium longum and *Bifidobacterium breve*	Melanoma	Oral gavage led to the activation of DCs and an increased frequency of tumour-specific CTLs
Lactobacillus casei str. Shirota	MCA-induced cancer	*L. casei* str. Shirota mixed into mouse diet delayed carcinogenesis through enhancement of NK cell cytotoxicity
Lactobacillus casei ATCC334	Colon cancer SW620 cells (Caco2 in vitro)	Secretion of ferrichrome, which induces JNK-associated induction of DNA damage-inducible transcript 3. Enhanced apoptosis of colon cancer cells
Lactobacillus casei BL23	DMH-associated colorectal cancer	Oral administration of *L. casei* BL23 led to differentiation of T cells towards a TH17-biased immune response
Bacillus polyfermenticus and its culture medium	HT-29, DLD-1, Caco2 human colon cancer in mice	Cyclin D1 expression required for ErbB-dependent cell transformation was decreased by culture medium injections near the tumour sites

bacterial enzymes implicated in tumour development. The binding and breakdown of possible carcinogens could be linked to a cancer-prevention method employing probiotic bacteria, particularly *Lactobacillus* and *Bifdobacillus* strains (Górska et al. 2019). Based on a recently reported successful strategy with Bifidobacterium expressing Wilms' tumour 1 (WT1) protein, new evidence suggests using probiotics in the delivery of tumour-associated antigens (TAAs) as an orally administered vaccine (Kitagawa et al. 2017). The use of *Lactococcus lactis* producing catalase

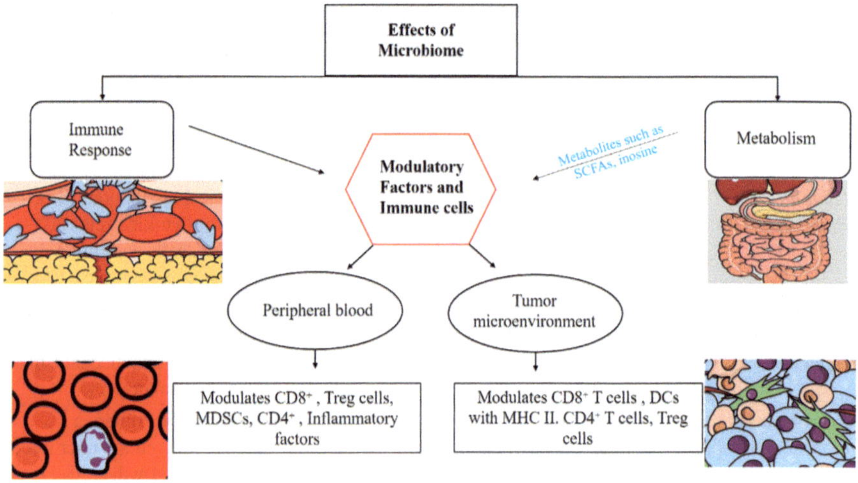

Fig. 21.4 Illustration describing the mechanism of crosstalk between microbiome and the host immune system which plays a determining factor for cancer

has been shown to reduce the formation of reactive oxygen species (ROS) such as H_2O_2, hence lowering colonic damage and inflammation and, as a result, distant metastasis and propagation (Moreno de LeBlanc et al. 2008). The importance of intestinal microbiota (*Bacteroides* and *Bifdobacterium*) in anti-PD-L1 (Programmed death-ligand 1) and anti-CTLA-4 (cytotoxic T lymphocyte-associated protein 4) therapy has been demonstrated in recent mouse trials (Vétizou et al. 2015). Probiotics may have an effect on other carcinogenic and mutagenic variables, assisting in cancer prevention. They have the ability to alter the activity of several enzymes involved in cellular detoxification, preventing free radicals and carcinogenic chemicals from causing harm. Some probiotic bacteria's cell walls may attach to carcinogenic substances. This is thought to be linked to cationic exchange between carcinogenic chemicals and peptidoglycan found in the cell surfaces of some probiotic microbes. As a result, carcinogenic chemicals, along with microorganisms, would be removed through the faeces (Śliżewska et al. 2020). When compared to the probiotic reference strain *L. casei ATCC 393*, the adhesion properties and antiproliferative effects of two lactobacillus strains, *L. pentosus B281* and *L. plantarum B282*, originally isolated from industrially fermented table olives, showed significantly higher adhesion rates to Caco-2 colon cancer cells (Saxami et al. 2016). Prebiotics are nondigestible microbial fermentation sugars that drive the development of probiotic organisms that have an immunomodulatory effect on malignant growth. Bacterial oncostatic capabilities are mediated in the tumour microenvironment via the recruitment of cytotoxic T lymphocytes, natural killer cells, and oxidative stress-induced apoptosis. Furthermore, efforts have been made to employ probiotics as an additive in cancer treatment (Samanta 2020). Prebiotics should boost the proliferation of one or a small number of bacteria in the colon, as well as their particular metabolites, which could help anti-cancer therapy (Gibson et al. 2017). Prebiotics are a potential therapeutic technique that is safe to utilize in a variety of clinical contexts. High fibre supplementation with prebiotics changes the microbiota community dramatically, increasing SCFA-producing bacteria, amplifying SCFA-related functional pathways, and increasing SCFA metabolite levels. Notably, the majority of research has linked an increase in SCFAs to a considerable reduction in tumour burdens (Mahdavi et al. 2021).

21.8 Conclusion

Gut microbiome is a highly explored area of research in recent days. Almost all human disorders have been associated with gut microbiome either directly or indirectly. Quantitative analysis of the gut microbiota clearly indicates that there exists a very delicate equilibrium among the microbial community; a small disturbance in this state has been found to be affecting the normal metabolic activities of the host. Gut dysbiosis can offer the opportunistic pathogens a chance to flourish and outcompete the friendly microbes, thereby resulting in various diseases. Qualitative and quantitative assessment of the gut microbiome is being largely explored in the aspect of exploiting them as a non-invasive marker for the diagnosis of various cancers. Moreover, gut microbiome plays a significant role in influencing the treatment efficiency of many drugs given for cancer, and therefore, microbiome therapy as a supplement might improve the treatment efficacy also. Imbalance in the gut microbiota could become the cause for the origin/progress of cancer at the same time; restoration of gut microbiome might cure the disease. This chapter emphasizes the role of gut microbiota as an eminent approach to predict and prevent cancer.

References

Ahn J et al (2013) Human gut microbiome and risk for colorectal cancer. J Natl Cancer Inst 105(24): 1907–1911

Akbar N et al (2022) The role of gut microbiome in cancer genesis and cancer prevention. Health Sci Rev 2:100010

Arbyn M et al (2020) Estimates of incidence and mortality of cervical cancer in 2018: a worldwide analysis. Lancet Glob Health 8(2):e191–e203

Arthur JC et al (2012) Intestinal inflammation targets cancer-inducing activity of the microbiota. Science 338(6103):120–123

Arzmi MH, Dashper S, McCullough M (2019) Polymicrobial interactions of Candida albicans and its role in oral carcinogenesis. J Oral Pathol Med 48(7):546–551

Atarashi K et al (2015) Th17 cell induction by adhesion of microbes to intestinal epithelial cells. Cell 163(2):367–380

Audirac-Chalifour A et al (2016) Cervical microbiome and cytokine profile at various stages of cervical cancer: a pilot study. PLoS One 11(4):e0153274

Balogh J et al (2016) Hepatocellular carcinoma: a review. J Hepatocell Carcinoma 3:41

Berg RD (1996) The indigenous gastrointestinal microflora. Trends Microbiol 4(11):430–435

Bose M, Mukherjee P (2019) Role of microbiome in modulating immune responses in cancer. Mediat Inflamm 2019:1

Bray F et al (2018) Global cancer statistics 2018: GLOBOCAN estimates of incidence and mortality worldwide for 36 cancers in 185 countries. CA Cancer J Clin 68(6):394–424
Brennan CA, Garrett WS (2016) Gut microbiota, inflammation, and colorectal cancer. Annu Rev Microbiol 70:395–411
Brody H (2020) The gut microbiome. Nature 577(7792):S5–S5
Bultman SJ (2014) Emerging roles of the microbiome in cancer. Carcinogenesis 35(2):249–255
Buskwofie A, David-West G, Clare CA (2020) A review of cervical cancer: incidence and disparities. J Natl Med Assoc 112(2):229–232
Cani PD (2018) Human gut microbiome: hopes, threats and promises. Gut 67(9):1716–1725
Chen D et al (2020) Clostridium butyricum, a butyrate-producing probiotic, inhibits intestinal tumor development through modulating Wnt signaling and gut microbiota. Cancer Lett 469:456–467
Cheng Y, Ling Z, Li L (2020) The intestinal microbiota and colorectal cancer. Front Immunol 11: 615056
Couturier-Maillard A et al (2013) NOD2-mediated dysbiosis predisposes mice to transmissible colitis and colorectal cancer. J Clin Invest 123:2
Curty G, de Carvalho PS, Soares MA (2019) The role of the cervicovaginal microbiome on the genesis and as a biomarker of premalignant cervical intraepithelial neoplasia and invasive cervical cancer. Int J Mol Sci 21(1):222
Curty G et al (2017) Analysis of the cervical microbiome and potential biomarkers from postpartum HIV-positive women displaying cervical intraepithelial lesions. Sci Rep 7(1):1–10
Dapito DH et al (2012) Promotion of hepatocellular carcinoma by the intestinal microbiota and TLR4. Cancer Cell 21(4):504–516
Duong MT-Q et al (2019) Bacteria-cancer interactions: bacteria-based cancer therapy. Exp Mol Med 51(12):1–15
Fang J, Zhang H, Jin S (2014) Epigenetics and cervical cancer: from pathogenesis to therapy. Tumor Biol 35(6):5083–5093
Garaude J et al (2012) Simultaneous targeting of toll-and nod-like receptors induces effective tumor-specific immune responses. Sci Transl Med 4(120):120ra16
Gibson Glenn R et al (2017) "The International Scientific Association for Probiotics and Prebiotics (ISAPP) consensus statement on the definition and scope of prebiotics"
Górska A et al (2019) Probiotic bacteria: a promising tool in cancer prevention and therapy. Curr Microbiol 76(8):939–949
Gury-BenAri M et al (2016) The spectrum and regulatory landscape of intestinal innate lymphoid cells are shaped by the microbiome. Cell 166(5):1231–1246
Hattori N, Ushijima T (2016) Epigenetic impact of infection on carcinogenesis: mechanisms and applications. Genome Med 8(1):1–13
Hullar MAJ, Burnett-Hartman AN, Lampe JW (2014) Gut microbes, diet, and cancer. Adv Nutr Cancer 2014:377–399
Hundal R, Shaffer EA (2014) Gallbladder cancer: epidemiology and outcome. Clin Epidemiol 6:99
Huycke MM, Rex Gaskins H (2004) Commensal bacteria, redox stress, and colorectal cancer: mechanisms and models. Exp Biol Med 229(7):586–597
Jain T et al (2021) New insights into the cancer–microbiome–immune axis: decrypting a decade of discoveries. Front Immunol 12:622064
Kau AL et al (2011) Human nutrition, the gut microbiome and the immune system. Nature 474(7351):327–336
Kitagawa K et al (2017) Development of oral cancer vaccine using recombinant Bifidobacterium displaying Wilms' tumor 1 protein. Cancer Immunol Immunother 66(6):787–798
Kroupis C et al (2006) Presence of high-risk human papillomavirus sequences in breast cancer tissues and association with histopathological characteristics. Clin Biochem 39(7):727–731
Kyrgiou M, Mitra A, Moscicki A-B (2017) Does the vaginal microbiota play a role in the development of cervical cancer? Transl Res 179:168–182
Kyrgiou M et al (2017) Adiposity and cancer at major anatomical sites: umbrella review of the literature. BMJ 356:j477

Lee KA et al (2021) The gut microbiome: what the oncologist ought to know. Br J Cancer 125(9): 1197–1209
Levy M et al (2017) Dysbiosis and the immune system. Nat Rev Immunol 17(4):219–232
Liang Q et al (2017) Fecal bacteria act as novel biomarkers for noninvasive diagnosis of colorectal cancer. Clin Cancer Res 23(8):2061–2070
Mahdavi M, Laforest-Lapointe I, Massé E (2021) Preventing colorectal cancer through prebiotics. Microorganisms 9(6):1325
Mai G et al (2019) Common core bacterial biomarkers of bladder cancer based on multiple datasets. Biomed Res Int 2019:1
McCoy AN et al (2013) Fusobacterium is associated with colorectal adenomas. PLoS One 8(1): e53653
Montalban-Arques A, Scharl M (2019) Intestinal microbiota and colorectal carcinoma: implications for pathogenesis, diagnosis, and therapy. EBioMedicine 48:648–655
Alejandra de Moreno de LeBlanc, et al. "Oral administration of a catalase-producing Lactococcus lactis can prevent a chemically induced colon cancer in mice." J Med Microbiol 57.1 (2008): 100–105
Moyat M, Velin D (2014) Immune responses to helicobacter pylori infection. World J Gastroenterol: WJG 20(19):5583
Nasrollahzadeh D et al (2015) Variations of gastric corpus microbiota are associated with early esophageal squamous cell carcinoma and squamous dysplasia. Sci Rep 5(1):1–6
Nath G, Gulati AK, Shukla VK (2010) Role of bacteria in carcinogenesis, with special reference to carcinoma of the gallbladder. World J Gastroenterol: WJG 16(43):5395
Nicholson JK et al (2012) Host-gut microbiota metabolic interactions. Science 336(6086): 1262–1267
Pan S, Chen R (2020) Metaproteomic analysis of human gut microbiome in digestive and metabolic diseases. Adv Clin Chem 97:1–12
Patyar S et al (2010) Bacteria in cancer therapy: a novel experimental strategy. J Biomed Sci 17(1): 1–9
Plottel CS, Blaser MJ (2011) Microbiome and malignancy. Cell Host Microbe 10(4):324–335
Popović B, Viljemka et al (2018) The urinary microbiome associated with bladder cancer. Sci Rep 8(1):1–8
Raisch J et al (2014) Colon cancer-associated B2 Escherichia coli colonize gut mucosa and promote cell proliferation. World J Gastroenterol: WJG 20(21):6560
Ridlon JM, Kang D-J, Hylemon PB (2006) Bile salt biotransformations by human intestinal bacteria. J Lipid Res 47(2):241–259
Samanta S (2020) Potential impacts of prebiotics and probiotics in cancer prevention. Anti Cancer Agents Med Chem 22:605–628
Saxami G et al (2016) Two potential probiotic lactobacillus strains isolated from olive microbiota exhibit adhesion and anti-proliferative effects in cancer cell lines. J Funct Foods 24:461–471
Sedighi M et al (2019) Therapeutic bacteria to combat cancer; current advances, challenges, and opportunities. Cancer Med 8(6):3167–3181
Segal E (2015) Perspective: another dimension for drug discovery. Nat Genet 47:856–860
Śliżewska K, Markowiak-Kopeć P, Śliżewska W (2020) The role of probiotics in cancer prevention. Cancers 13(1):20
Song S, Vuai MS, Zhong M (2018) The role of bacteria in cancer therapy–enemies in the past, but allies at present. Infect Agents Cancer 13(1):1–7
Steer T et al (2001) Clostridium hathewayi sp. nov., from human faeces. Syst Appl Microbiol 24(3): 353–357
Tilg H et al (2018) The intestinal microbiota in colorectal cancer. Cancer Cell 33(6):954–964
Varian BJ et al (2016) Beneficial bacteria inhibit cachexia. Oncotarget 7(11):11803
Vétizou M et al (2015) Anticancer immunotherapy by CTLA-4 blockade relies on the gut microbiota. Science 350(6264):1079–1084

Wang Z et al (2019) Altered diversity and composition of the gut microbiome in patients with cervical cancer. AMB Express 9(1):1–9

Watanabe Y et al (2010) Bacteroides clarus sp. nov., Bacteroides fluxus sp. nov. and Bacteroides oleiciplenus sp. nov., isolated from human faeces. Int J Syst Evol Microbiol 60(8):1864–1869

Wu P et al (2018) Profiling the urinary microbiota in male patients with bladder cancer in China. Front Cell Infect Microbiol 8:167

Wu Y et al (2021) Identification of microbial markers across populations in early detection of colorectal cancer. Nat Commun 12(1):1–13

Xu W et al (2014) Mini-review: perspective of the microbiome in the pathogenesis of urothelial carcinoma. Am J Clin Exp Urol 2(1):57

Yu J et al (2017) Metagenomic analysis of faecal microbiome as a tool towards targeted non-invasive biomarkers for colorectal cancer. Gut 66(1):70–78

Zhao L et al (2018) Lactobacillus supplementation prevents cisplatin-induced cardiotoxicity possibly by inflammation inhibition. Cancer Chemother Pharmacol 82(6):999–1008

Insights in the Cross-Talk Between Microbiota-Gut-Brain Axis: A Focus on Alzheimer's Disease

22

Thomson Soni, Ishwerpreet Kaur Jawanda, Seema Kumari, and Vijay Prabha

Abstract

The gut microbiota (GM) signifies a broad and dynamic group of microorganisms that impact the host's health as well as metabolic processes and defensive mechanisms. Gut dysbiosis can affect brain immune homeostasis through the microbiota–gut–brain axis and can play a key role in the pathogenesis of neurodegenerative diseases including Alzheimer's disease (AD). The complexity of these interactions is enclosed in the denomination of gut-brain axis (GBA) which integrates the gut functions as well as links the emotional and cognitive centers of the brain with peripheral intestinal functions. Although the mechanism behind neurological dysfunction and intestinal dysbiosis still remains vague, emerging evidence suggests that it can enhance the secretion of lipopolysaccharides and amyloids that may disrupt the blood–brain barrier and intestinal permeability. In addition, it can promote the hallmarks of AD, such as oxidative stress, neuroinflammation, amyloid-beta formation, insulin resistance, and ultimately the causation of neural death. Poor dietary lifestyle and ageing, along with inflammatory responses due to dysbiosis, may contribute to the pathogenesis of AD. A safe and effective way is the modulation of GM which is the core of bidirectional pathway of communication. In this review, we discuss the cross-talk between microbiota-gut-brain axis and possibility for probiotics, prebiotics, nutrition, and faecal microbiota transplantation to be used in conjunction with conventional therapies for this fatal and progressive condition.

Keywords

Gut microbiome · Alzheimer's disease · Gut-brain axis · Faecal microbiota transplantation · Probiotics

T. Soni · I. K. Jawanda · S. Kumari · V. Prabha (✉)
Department of Microbiology, Panjab University, Chandigarh, India

R. Sobti et al. (eds.), *Role of Microbes in Sustainable Development*,
https://doi.org/10.1007/978-981-99-3126-2_22

22.1 Introduction

The modern scientific doctrine delineates the human gastrointestinal microbiota as a 'virtual organ', bestowing the concept of humans as "superorganism". Among the astounding array of microbial world, the human body hosts billions of microorganisms including archaea, bacteria, viruses, and fungi. The gut microbiota (GM) is a collection of complex ecological colonies of microbes that live in the GI tract. While the human body expresses around 20,000 eukaryotic genes, the gut microbiome expresses over 3.3 million prokaryotic genes, providing insight into the extent of bacterial presence and possible impact on the host (NIH 2012). The primary bacteria in the gut include *Actinobacteria, Proteobacteria, Bacteroides,* and *Firmicutes,* followed by *Bifidobacterium, Peptococcus, Clostridium, Eubacterium, Provetella,* and others (Dahiya et al. 2017; Etxeberria et al. 2013). Furthermore, the composition of the GM varies depending on one's life stage. *Firmicutes* and *Bacteroidetes* populations are shown to be higher in elderly people than they are in younger people (Doifode et al. 2021). The GM signifies a broad and dynamic group that impacts the host's health as well as metabolic processes and defensive mechanisms. Majorly, the GM develops the immune system in the intestinal mucosa and protects the host from carcinogens by releasing short-chain fatty acids (SCFA).

However, alterations of the GM community can lead to metabolic disorders, known as dysbiosis, which are characterized by a reduction in the variety and abundance of commensal bacteria. Recent research indicates that this may also have an effect on the central nervous system (CNS), owing to the microbiota–gut–brain axis (MGBX) (Liu et al. 2020; Halverson and Alagiakrishnan 2020). Insights into the gut-brain connection have shown a comprehensive communication system that not only supports correct gastrointestinal homeostasis maintenance, but also has diverse consequences on higher cognitive functioning. The complexity of these interactions is enclosed in the denomination of gut-brain axis (GBA) which integrates the gut functions as well as links the emotional and cognitive centres of the brain with peripheral intestinal functions and also, mechanisms such as immune activation, entero-endocrine signalling, enteric reflex, and intestinal permeability. GBA communication is mediated by neuro-immuno-endocrine mediators. The brain governs sensory as well as the secretory activities of the GI tract, the autonomic nervous system (ANS), neuro-immune system, and neuro-endocrine pathways which are generated by the gut microbes that mediate the link between the gut and brain. However, the mechanism behind neurological dysfunction and intestinal dysbiosis still remains vague.

Gut dysbiosis may have a role in the development of neurocognitive disorders such as schizophrenia, depression, bipolar disorder, post-traumatic stress disorder, obsessive-compulsive disorder (OCD), anxiety, and dementia, as well as in the psychological and behavioural symptoms associated with dementia (Halverson and Alagiakrishnan 2020). Numerous cross-sectional studies have shown that disruptions in the gut microbial environment (dysbiosis) lead to Alzheimer's disease (AD) and impair memory, learning, and hippocampal plasticity (Spinelli et al. 2019). Indeed, despite the widespread belief for a long period of time that the brain was an entirely separate organ, current data indicate that the GM is the core of bidirectional pathway of communication. Hence, diet, prebiotics, probiotics, and other novel therapeutic strategies have beneficial impacts on GM modulation, causing positive influence on the host, ultimately helping the AD patients by improving the gut and general health state.

Although the mechanisms behind this relationship remain unknown, addressing the microbiome may constitute a novel diagnostic and therapeutic option for Alzheimer's disease and other neurodegenerative illnesses. Despite the fact that some publications have addressed potential microbiome-based therapeutics, a thorough overview of gut microbiota-based diagnostic and therapeutic techniques is still not available to our knowledge. We discuss the possibility for probiotics, prebiotics, nutrition, and faecal microbiota transplantation to be used in conjunction with conventional therapies for this fatal and progressive condition.

22.1.1 The Gut–Brain Axis

The gut–brain axis (GBA) is a dynamic bidirectional communication pathway that links the gastrointestinal tract (GI tract) with the central nervous system (CNS). Its primary function is to monitor and integrate intestinal functions, as well as to connect the brain's emotional and cognitive centres to peripheral intestinal mechanisms such as immune activation, intestinal permeability, enteric reflex, and enteroendocrine signalling via immune and neuroendocrine mediators. The brain influences stomach movement, sensory, and secretory functions via this communication network, and the gut signals influence brain function in turn. As a result, this link is crucial for maintaining gut homeostasis and for understanding the genesis of a variety of psychiatric dysfunctions.

Numerous methods of communication between the brain and the gut microbiota have been proposed:

- Through the outgoing and incoming ganglia of the vagus nerve, or via hormones generated by 5HT (endocrine cells) (Margolis et al. 2021). The blood-brain barrier serves as a protective barrier against blood-borne infections entering the central nervous system in either direction.
- Via the production of bioactive peptides (short-chain fatty acids) and a variety of other metabolites, as well as through microorganism-mediated regulation of transmitters (Margolis et al. 2021; Heijtz et al. 2011).
- Through the HPA's release of cortisol in response to stress, which impacts intestinal motility, and mucus secretion, which affects the CNS through stress hormone regulation (Heijtz et al. 2011).
- By cytokines and chemokines that promote inflammation (Rea et al. 2016).
- TLRs and peptidoglycans (PGNs) function as sensors and implicate the immune activity to microorganisms. Thus, this immune activation in local area may result in an immunological activation in organs such as the brain, through a variety of mechanisms (Dichter et al. 2012; Banks and Erickson 2010).

Table 22.1 Major mechanisms of the bidirectional brain-gut microbiota axis

Bidirectional brain-gut microbiota axis	
Gut microbiota to brain	Brain to gut microbiota
Production, expression, and turnover of neurotransmitters and neurotrophic factor	Changes in mucus production and biofilm formation
Protection of intestinal barrier and tight junction integrity	Modification of motility
Modulation of enteric sensory afferents	Changes in the permeability of the intestine
Bacterial metabolites	Immune dysfunction
Mucosal immune regulation	

Depression and neuro-degenerative illnesses like Alzheimer's disease have been linked to this low-grade immune activation. Table 22.1 represents the main principal mechanisms involved in the bidirectional brain-gut microbiota axis (Carabotti et al. 2015). Given this intricate interaction, it's unsurprising that the gut–brain axis, and consequently the gut microbiota as the key element of this cross-talk, influences neuropsychiatric diseases directly or indirectly.

22.1.2 Alzheimer's Disease

Alzheimer's disease (AD), the most prevalent form of dementia and a frequent fatal neurodegenerative condition, is defined by gradual cognitive impairment that finally results in memory loss, thinking, reasoning, and personality abnormalities. In the earlier stage of AD, individuals may have memory lapses such as they can forget familiar terms and places, whereas the intermediate stage may last for many years, causing the patient to become irritated, angry, disordered, and develop erratic actions. However, at a later stage, individuals lose their capacity to communicate, react to their surroundings, and eventually lose control of their movement. Individuals need intensive care as memory and cognitive problems continue to deteriorate owing to major personality changes. The incidence of AD continues to rise globally, posing a significant challenge to the twenty-first century's healthcare system. Alzheimer's disease is believed to account for between 60% and 80% of dementia cases (Alzheimer's Association (Alzheimer's Disease Report) 2020). There are already 50 million AD sufferers worldwide, and the incidence of the disease increases every 5 years after the age of 65. According to the 2016 world AD estimate, approximately 131 million people would suffer from AD by 2050, causing it to be one of the future's significant global health issues (Prince et al. 2013).

Alzheimer's disease (AD) was initially characterized in 1907 by Alois Alzheimer, a German Bavarian physician and neurologist. Since the discovery of AD 110 years ago, the amyloid and tau hypothesis have been the most widely acknowledged. The senile plaque (SP) is considered as the pathogenic aetiology of AD. It is created by gradual deposition of amyloid beta (A) and tangles of hyperphosphorylated tau protein neurofibrils in the hippocampus (Reitz et al. 2011). At least in part, synaptic dysfunction and neuronal death are caused by an overactive or non-resolving immune response and infections. This may change neuroinflammation, calcium homeostasis, and vascular degeneration, ultimately resulting in neuronal death. Neuronal loss, neuropil thread loss, and synaptic dysfunction are all strongly associated with AD.

Alzheimer's disease is caused by an abnormality in the amyloid precursor protein (APP), a vital component of the protein synthesis process. The principal causes of early onset of AD include mutations in APP, Presenilin 1 (PSEN1), and Presenilin 2 (PSEN 2) (Dorszewska et al. 2016). Additionally, lifestyle factors including ageing, food, and environment all have a role in the late emergence of AD, as does overexpression of the Apolipoprotein (Apo) E4 gene. A development of amyloid plaques is a frequent indication of Alzheimer's disease. APP enters the endosomal compartment through trans-Golgi networks and is endocytosed by clathrin. A portion of APP is recycled back to the cell surface through the endosome during this operation. The control of APP on the cell surface happens through a non-amyloidogenic process involving the action of secretase at the N-terminal end of the A domain. As a consequence, 83 amino acids are formed that are both APP and membrane-tethered and have a carboxy terminal end (CTF). Thus, CTF-83 will be further cleaved by secretase to create the APP and P3 intracellular domains. Endosomal APP will be incorporated into the amyloidogenic cascade. The secretase interacts with the extracellular domain of APP in this pathway, resulting in membrane-bound 99 amino acids consisting of CTF-(C99) and APP. Further cleavage of C99 by secretase results in the formation of a soluble fragment and an intracellular domain of APP. A peptide oligomerizes in the presence of transition metal ions such as Fe_{2+} and Cu_{2+}, resulting in the formation of H_2O_2. This increases lipid peroxidation, which finally results in the formation of 4-hydroxynonenal (4HNE) (Evrard et al. 2018; Steiner et al. 2018; Cheignon et al. 2018).

Alternatively, impairments in glucose and glutamate transport increase Ca^{2+} inflow, resulting in the formation of inositol 1,4,5-trisphosphate (IP3), which stimulates Ca^{2+} efflux from endoplasmic reticulum storage. Calpain activation activates cyclin-dependent kinase (CDK5), resulting in tau hyperphosphorylation and the development of neurofibrillary tangles (NFTs), which disrupt microtubule assembly and hamper axonal transport. This ultimately results in neuronal and synaptic dysfunction, which ultimately results in neuronal death.

The amplified influx of calcium ions and reactive oxygen species (ROS) into the mitochondria induces the production of mitochondrial cyclophilin D. This results in the release of proapoptotic molecules such as apoptosis inducing factor (AIF) and cytochrome c. Caspase cascades will eventually be initiated, resulting in neuronal cell death (Redza-Dutordoir and Averill-Bates 2016). All of these factors induce astrocytes to produce cytokines, chemokines, and acute-phase proteins, activating microglial cells. Stimulated microglia and astrocytes cause neuroinflammation in the brain, which contributes to the pathophysiology of Alzheimer's disease (Rubio-Perez and Morillas-Ruiz 2012). Recent results reveal that AD patients have a large number of peripheral immune cells and that these immune cells are obviously involved in local inflammation. Regardless of the source of peripheral inflammation, whether it is obesity or systemic inflammation, these pro-inflammatory cytokines are capable of readily breaching the blood-brain barrier (BBB) and activating further brain-specific inflammatory responses. As a consequence, the BBB becomes more porous and permeable, allowing for the passage of peripheral immune cells. As a consequence, an abnormally large number of microglia will be activated. This is originally shown in Alzheimer's disease patients as decreased hippocampal-dependent learning (Kinney et al. 2018).

22.1.3 GM Dysbiosis and AD

Intestinal dysbiosis is an imbalance of microbiota which is induced by overgrowth of bad bacteria|| in the gut, which may result in adverse effects such as improper synthesis of necessary metabolites or even the origin of toxic compounds (Weiss and Hennet 2017; Pascale et al. 2018). While the makeup of a "healthy microbiota" is yet to be defined, it is well established that a stable environment between the microbes and host is required to perform important immunological and metabolic activities. Dysbiosis has been linked to the development of a variety of disorders in recent years, including diabetes, cancer, chronic fatigue syndrome, irritable bowel syndrome, anxiety, autoimmune diseases, depression, Parkinson's disease, multiple sclerosis, and many other neurological illnesses (Cenit et al. 2017; Gotkine et al. 2020; Sun and Shen 2018; Gerhardt and Mohajeri 2018; Jiang et al. 2015; Levy et al. 2017; Schwabe and Jobin 2013; Abenavoli et al. 2019; Galicia-Garcia et al. 2020; Varesi et al. 2021; Ling et al. 2014). Numerous recent investigations have shown that changes in the gut microbiota have a direct effect on cognitive impairment, actively contributing to the development and progression of AD (Cattaneo et al. 2017; Vogt et al. 2017; Haran et al. 2019; Liu et al. 2019; Ling et al. 1891). In general, AD patients have a lower gut microbial diversity, with a large alteration toward pro-inflammatory taxa which is consistent with the findings of both mice and human ageing (Pascale et al. 2018; Cattaneo et al. 2017). For example, 97 individuals [33 AD, 32 MCI (mild cognitive impairment), and 32 controls] were used to perform faecal microbiota 16S rRNA sequencing, a noteworthy decline in the population of *Firmicutes* was observed in patients with neurodegeneration compared to healthy subjects, while *Proteobacteria, Gamma-proteobacteria,* and *Enterobacteria* abundance increased. Also, a significant difference in *Enterobacteriaceae* was also seen in AD patients and MCI patients, signifying a gradual shift in the makeup of the gut microbiota as dementia progresses (Liu et al. 2019).

Similarly, a substantial drop in *Firmicutes* and *Bifidobacteria* in faecal samples from AD patients, countered by an increase in *Bacteroidetes* species in the same group of people, was documented (Vogt et al. 2017). Zhuang et al. (2018) also demonstrated changes in the composition of the gut microbiota during neurodegeneration with the comparison of 43 AD patients to age- and sex-matched controls: rise in *Bacteroidetes* and decline in *Actinobacteria* were observed which was paralleled with the rise of *Ruminococcaceae, Lactobacillaceae*, and *Enterococcaceae*, and reduced *Lanchnospiraceae,* and *Bacteroidaceae*. In contrast to these findings, another research revealed reduced *Bacteroides, Lachnospira,* and *Ruminiclostridium* levels and increased *Prevotella*levels (Zhuang et al. 2018). Although reductive, this variation may be explained in part by the subjects' varying geographical origins, since regional identity has been shown to have a significant effect on gut microbiota composition as well as other co-morbidities (He et al. 2018). In this regard, bigger studies are undoubtedly necessary to develop consistent and repeatable insertion criteria, perhaps avoiding the perplexing influence of other co-morbidities.

The GM is associated with a variety of inflammatory and metabolic processes. Dysbiosis impairs the production of signalling proteins, which have an effect on metabolic processes involved in the course of AD. Ageing modifies the GM composition (pro-inflammatory bacteria outnumber anti-inflammatory bacteria) and generates localized systematic inflammation, impairing the GIT's permeability and blood-brain barrier function. In AD participants, the *Ruminococcaceae, Peptostreptococcaceae, Bifidobacteriaceae, Mogibacteraceae,Turicibacteraceae,* and *Clostridiaceae*families were found to be less prevalent than the *Bacteroidaceae, Gemellaceae*, and *Rikenellaceae* groups (Vogt et al. 2017). More precisely, dysbiosis promotes the growth of pro-inflammatory bacteria (such as *Shigella, Pseudomonas, Verrucomicrobia, Escherichia,*and *Proteobacteria),* while decreasing the population of anti-inflammatory bacteria (such as *Bacillus sp., Bacteroides sp.,* and *Eubacterium sp.).*

In an investigation of Mild Cognitive Impairment (MCI) and AD patients, decreased microbial diversity was detected, with increasing expansion of *Gammaproteobacteria, Enterobacteriaceae, and Enterobacteriales* seen in contrast to controls. Additionally, patients had increased biosynthesis and glycan metabolism, decreased activity of immunological pathways, and decreased levels of *Lachnospiraceae*, *Firmicutes*, *Ruminococcaceae,* and *Clostridiaceae* (Liu et al. 2019). However, considerable changes in the phylum (e.g. *Bacteroidetes* and *Firmicutes*) were observed (Lee et al. 2018), as well as rise in the *Clostridium leptum* group was observed (Brandscheid et al. 2017). Furthermore, dysbiosis may contribute to the development of AD symptoms such as insulin resistance and oxidative stress (Liu et al. 2020).

Theoretically, it has been suggested that the GM may control the CNS's oxidative status through the metabolites generated. Reduced butyrate levels may exacerbate mitochondrial dysfunction, which results in the generation of reactive oxygen species (Mottawea et al. 2016). Further, Cerovic et al. (2019) claimed that gut dysbiosis results in both central and peripheral pathogenic processes that may contribute to the risk of Alzheimer's disease. In 5xFAD animals, dysbiosis was related with the advancement of the CCAAT, asparagine endopeptidase pathway, and enhancer binding protein, which were implicated in AD pathogenesis by cleavage of both A precursor and Tau proteins (Chen et al. 2020). Li et al. (2020) recently examined the involvement of dysbiosis in AD using RNA sequencing, Y maze, transcriptome sequencing, Gene Expression Omnibus, and quantitative reverse transcriptase PCR methods on APPswe/PS1E9 transgenic and wild-type mice. They discovered a dramatically altered GM composition, impaired cognitive performance, and increased amyloid formation. Thus, GM dysfunction is related with cognitive impairment and may contribute to amyloid deposition in the brain through activation of mitogen-activated protein kinase signalling pathways (which regulate a variety of cellular processes, including differentiation, programmed cell death, proliferation, and stress responses) (Guo et al. 2020).

An increasing body of data shows that dysbiotic gut bacteria contribute to the early phases of AD pathogenesis by causing ageing of immune system, neuroinflammation, oxidative stress, i.e. pro-oxidant–antioxidant imbalance, and cytokine production (Brandscheid et al. 2017). It was documented by Cattaneo *et al.* that people suffering from AD have an increased number of pro-inflammatory *Escherichia/Shigella* endobacteria and a decreased number of anti-inflammatory *E. rectale* endobacteria, and that this alteration is directly linked with amyloidosis (Cattaneo et al. 2017). Furthermore, when 108 nursing home resident's stool samples were analysed using metagenomic sequencing, it was revealed that a drop in butyrate-synthetizing bacteria was accompanied by an increase in pro-inflammatory bacteria, possibly impairing systemic inflammation in AD patients (Haran et al. 2019). These findings indicate a strong link between microbiota dysregulation and intestinal inflammation as it has been associated with low levels of P-glycoprotein expression, a critical component for intestinal homeostasis (Haran et al. 2019). The aforementioned findings provide more credence to the notion that changes in the makeup of the gut microbiota correlate with changes in intestinal function. Indeed, variations in the population of the gut microbiota may alter the body's tryptophan and serotonin levels, as well as the production of many important brain chemicals, including brain-derived neurotrophic factor (BDNF), dopamine, and norepinephrine (Morris et al. 2017; Lee et al. 2019; Doifode et al. 2020). As previously stated, the gut microbiota also has a positive function in the formation of SCFAs such as acetate, propionate, and butyrate, which are required in energy generation, gut epithelial homeostasis, and immunological modulation (Parada Venegas et al. 2019). The synthesis is disrupted due to dysbiosis, the deposition of Aß plaques, metabolic dysfunction, and impaired regulation of microglia which are encouraged, hence driving cognitive impairment (Colombo et al. 2021; Morrison and Preston 2016; Wenzel et al. 2020). Furthermore, a reduction in butyrate-producing bacteria has been associated with T cell dysregulation, epithelial barrier leakage ("leaky gut"), and an increase in bacterial translocation in AD patients (Morris et al. 2017; Köhler et al. 2016; Fuke et al. 2019).

When LPS generated by Gram-negative endobacteria circulates, TLR4 is activated and the breakdown of the BBB is accelerated, resulting in neuroinflammation (Browne et al. 2013; Bonfili et al. 2021). Toxic compounds, such as amyloid and trimethylamine N-oxide, may also be produced in the intestines due to dysbiosis in the gut. TMAO is a microbial metabolite which has been associated with rise in beta amyloid formation, platelet hyperactivity, peripheral immunological activity activation, an increase in oxidative stress, and intestinal mucosal barrier failure (Gamba et al. 2012; Brandscheid et al. 2017; Gao et al. 2019; Colciaghi et al. 2004; Evin 2012). In addition, another possibility is that the ability of certain endobacteria to produce gaseo-transmitter molecules, for example, hydrogen (H2), methane (CH4), nitric oxide (NO), ammonia (NH3), and hydrogen sulphide (H2S), seems to be critical for normal neuronal function, and its change contributes to the pathogenesis of AD (Szabo 2010; Oleskin and Shenderov 2016).

In general, these findings show that the interaction between gut microbiota-brain connections is much more complex than previously thought, and that only a comprehensive knowledge of it may give insight into novel diagnostic and therapeutic strategies. It is noteworthy that the microbiota is more medically accessible and modifiable than the human genome. This fact provides a promising opportunity for preventing or treating neuropsychiatric conditions.

22.2 Potential Therapeutic Strategies

22.2.1 Probiotics

In 1965, Lilly and Stillwell described "probiotics" as "living microorganisms with minimal or no pathogenicity that have beneficial effects on the host's health" (Lilly and Stillwell 1965). By boosting cell viability, improving immune response, and decreasing pro-inflammatory cytokines synthesis, probiotics may help to maintain intestinal ecosystem homeostasis and control intestinal epithelial functioning in animal and human models (Martinez et al. 2013). Probiotic bacteria have been demonstrated to be capable of regulating immunological responses through Th2, Th17, Th1, Treg cells, and activation of B and NK cells when the gut microbial environment is modulated.

Additionally, probiotics have been shown to have an effect on the gut–brain axis. Proteins and neurotransmitters like glutamate, serotonin, brain-derived neurotrophic factor, and gamma-aminobutyric acid play critical roles in our central nervous system and their functions can be modulated by a novel family of probiotics called psychobiotics which has the potential to be used in the treatment of mental health issues. Probiotics exert their beneficial effects in a number of ways, but the particular mechanism by which they do so is uncertain. This occurs as a consequence of bacteriocin, short chain fatty acid production, nutritional competition, modulation of the immune system, and activation of the gut-brain axis.

SCFAs are saturated fatty acids that are produced in the stomach in response to the fibre content of the food. During fermentation *Lactobacillus, Bifidobacterium, Bacteriodes, Eubacterium, and Clostridium* species create acetate, butyrate, and propionate (Verbeke et al. 2015). SCFAs may influence brain function in three unique ways: via immune modulation, endocrine regulation, or neuronal factors. SCFAs function as endocrine signalling molecules via altering intestinal hormone output. Propionate and acetate greatly increase the production of Peptide YY (PYY) and Glucagon-like Peptide-1 (GLP-1) through the G-protein-coupled receptor in murine colonic cells (Psichas et al. 2014). GLP-1 is generated and released by neurons and intestinal enteroendocrine L-cells inside the solitary tract nucleus of the brainstem. GLP-1 functions as a neuroprotective factor in the brain, preventing apoptosis of neurons. PYY acts as a depressant in the intestines. Neuropeptide Y dramatically increases in the cerebral cortex of mice with Alzheimer's disease and imparts protective role by activating the PI3K-XBP1-induced Gip78/BiP pathway and inhibiting caspase-3 and caspase-4 activity. It also reduces oxidative stress by inhibiting A-induced lipid peroxidation as well as modulates BDNF levels.

Additionally, SCFAs may operate as a modulator of neurotransmitters and neurotrophic factors. The gut microbiota has been demonstrated to create substrates for neurotransmitters or to stimulate the production and secretion of numerous neurotransmitters through food metabolism, or even both (Lee et al. 2020). Through secretory enterochromaffin (EC) cells, neurotransmitter antecedents encourage the synthesis of neurotransmitters like 5-HT and GABA. Yano and colleagues (2015) discovered that propionate and butyrate influence the manufacture of 5-HT insera and colonic ECs of the host (Yano et al. 2015). Additionally, EC cells synthesize neuromodulators metabolites including histamine, tryptophan, and PYY. Certain neuroactive metabolites and neurotransmitter precursors may cross the blood-brain barrier and play a role in the production of brain's neurotransmitter and transmission in the central nervous system. Certain gut microbes have a direct effect on vagal nerve transmission, stimulating the vagus dorsal motor nucleus (DMV).

By boosting *Bacteroides and Actinobacteria* in the gut microflora of an AD animal model, a probiotic combination enhanced long-term neural plasticity and memory (Distrutti et al. 2014). Mitochondrial malfunction, elevated oxygen radicals' generation, and increased apoptosis have all been linked in the aetiology of Alzheimer's disease. Numerous studies have shown the significance of hydroxyl radical, nitric oxide, superoxide anion, and hydrogen peroxide in neurotoxicity caused by oxidative stress in Alzheimer's disease (Xie et al. 2002; van Dyke 1997). Recently, research conducted on mutant AD mice shown that administering a probiotic formulation (SLAB51) dramatically decreases cell damage through SIRT-1-dependent processes (Bonfili et al. 2018). Additionally, the incorporation of a multi-species *Bifidobacterium* and *Lactobacillus* probiotic strain has been shown to be capable of changing certain cerebral metabolites (O'Hagan et al. 2017). Brain inflammation was similarly reduced after the integration of probiotic with the short *Bifidobacterium* A1 strain (Kobayashi et al. 2017). Additionally, combining *Bacillus longum, Lactobacillus fermentum, Lactobacillus acidophilus*, and *Bacillus lactis* decreased learning impairment and oxidative stress in rats given intra-hippocampal injections of A 1-42 (Athari Nik Azm et al. 2018). While these evidences suggest that probiotic bacteria might play a critical role in two-way interaction in between the brain and gut and promote the possibility of probiotics boosting cognitive function, human investigations in people with AD or MCI have shown mixed findings.

Even though multiple studies have shown the importance of gut microbiota in psychiatric and neurological illnesses, the pathway by which probiotics work and their effects remain unknown, as well there are significant loopholes and contradictions. Therefore, the scope of human study should be broadened to include assessment of the gut by identifying probiotic bacteria which may help us to understand the microbiome composition of particular patient populations and determine the safety of strains that have the potential to have a significant impact on the gut–brain axis.

22.2.2 Prebiotics

Fibres like oligofructose and inulin, as well as polyphenols like FA found in food, may promote a healthy intestinal flora by encouraging the development of microbiome and preventing pathogenic bacteria from proliferating (Lupien-Meilleur et al. 2016; Constante et al. 2017). These dietary components, termed prebiotics, have been found to offer a variety of beneficial effects that help to slow the onset of neurological disorder.

Fructooligosaccharide (FOS) generated from the enzymatic breakdown of inulin present in many fruits and vegetables is a well-studied prebiotic. As a nutritional supplement, FOS enhances the growth of *Bifidobacterium* and *Lactobacillus* as a substrate for bacterial reproduction. Numerous organizations have explored the impact of FOS on cognitive problems, and their findings in pre-clinical mouse experiments support its use as a possible nutraceutical to treat neurodegeneration. Sun et al. found that feeding FOS to genetically modified AD mice increased GLP-1, a molecule that quickly passes the BBB and increases pancreatic insulin production, and satiety (Sun et al. 2018). This rise in GLP-1 prevents resistance in CNS insulin, hence slowing neurodegeneration caused by AD patients' poor glucose metabolism.

Xylooligosaccharides (XOS) are comparable oligosaccharides that have been widely explored and shown similarly promising outcomes. This prebiotic is derived from b-1,4 connected xylose units, which combine to produce xylan oligomers. XOS is the most prevalent natural polymer in the kingdom plantae and can be synthesized organically from biomass of sugar cane, bamboo sprouts, vegetables, and honey (Ávila et al. 2020). Its accessibility, along with known anti-inflammatory qualities, prompted multiple investigations investigating its possible benefits on mentally incapacitated patients. The group of researchers explored this hypothesis by providing XOS supplements to APP/PS1 mice with hepatectomy-induced postoperative cognitive impairment (POCD), a frequent comorbidity of AD (Han et al. 2020). Typical symptoms of POCD include loss of memory, inability to maintain balance, and impairment of executive functioning as a result of neuroinflammation and a breakdown of the BBB. Post-operative microbiome research revealed substantial increase in three species (*Ruminococcaceae, Rodentibacter*, and *Bacteroides*) and decrease in two (*Muribaculaceae* and *Faecalibaculum*). Increased Ruminococcaceae levels have been linked to cognitive failure in AD-induced rats' faeces; however, this association is not totally consistent across investigations. On the other hand, it has been shown that increasing Faecalibaculum significantly decreases faecal SCFA levels in AD patients, restricting systemic and thus neuroinflammation (Biagi et al. 2010; Claesson et al. 2011). Prebiotics supplementation to operated mice dramatically reduced post-operative microbiota variations, most notably in the *Muribaculaceae, Ruminococcaceae, Faecalibaculum, Bacteroidetes,* and *Lactobacillus* taxa. Additionally, increasing the diversity of the gut microbiota with XOS medication reduced gut inflammation by reducing levels of immune cytokines, which had previously been raised by hepatectomy.

Lactulose, the first commercially available prebiotic that is widely consumed as dietary ingredient, has shown the learning retrieval and to impede short-term memory in Alzheimer's disease rats. Lactulose's pharmacological approach in AD was recently discovered. Lee et al. discovered that lactulose pre-treatment may change the insulin sensitivity and gut microbiota, as well as boost autophagy pathway expression and reduce neuroinflammation (Lee et al. 2021). This implies that lactulose's neuroprotective impact is mediated partially by autophagy activity.

Sodium oligomannate (GV-971), another prebiotic, was approved for the first time in China in November 2019 for the management of mild to severe Alzheimer's disease (AD) with the goal of improving cognitive performance. GV-971 is a combination of acidic linear oligosaccharides generated from marine brown algae that is administered orally. GV-971 may block peripheral infiltration of immune cells into the brain caused by metabolites, restore dysbiosis in the gut microbiota, correct cognitive impairment, and suppress neuroinflammation (Wang et al. 2019). Additionally, GV-971 may readily traverse the BBB, binding directly to Ab and inhibiting the production of Ab fibrils. While prebiotics have been shown to be beneficial in AD, their combination with probiotics may provide higher advantages in terms of reducing AD symptoms.

22.2.3 Prebiotics Formulated with the Probiotics

In 1995, Gibson and Roberfroid initially proposed synbiotics and amassed a collection of studies demonstrating the advantages of combined pre- and probiotic supplement. A transgenic humanized *Drosophila melanogaster* model with a BACE1-APP-induced AD phenotype was used in one of these research studies (Westfall et al. 2019). The polyphenol-rich plant prebiotic Triphala (TFLA), a symbiotic (combining strain and substrate), was given to the insects instead of the three metabolically active probiotics (*Lactobacillus plantarum, Bifidobacterium longum*, and *Lactobacillus fermentum*). The probiotic strains used in the combination were chosen for their potential to create a secondary supply of antioxidants, reduce neuroinflammation, and prevent Ab aggregation (Nimgampalle and Yellamma 2017; Li et al. 2019).

Lactobacillus plantarum increases the therapeutic capabilities of *Lactobacillus fermentum* by metabolizing FA to vanillic acid and caffeic, chemicals that further boost the mixture's anti-inflammatory activities and provide neuroprotection towards fibril development. Last but not the least, gallic acid, another organic acid found in high quantity in the TFLA, has been demonstrated to decrease neurodegeneration and support a healthy gut microbiome (Vidhya Rekha et al. 2019). The results showed that a combinatorial strategy was more beneficial than the collection of its components. All biomarkers of Alzheimer's disease (AD) were improved by the synbiotic mixture over the course of the trial, which indicates that the combination might be a possible treatment. Indeed, when Drosophila melanogaster was treated with TFLA or probiotics, survival rose significantly, but the impact was highest when both were combined as a synbiotic. Individual and combination treatments both improved motility and Ab accumulation to a comparable extent. Intriguingly, as compared to either the probiotic combination or prebiotic alone, the synbiotic impact was not significantly superior on any specific test, but maintained a consistently high benefit when compared to its competitors. As assumption implies, AD lacks a single targetable symptom or biomarker, implying that several causes may impact its start. While both probiotics and prebiotics have showed efficacy on their own, the consistency of a combinatorial therapy may prove to be a more beneficial option for such a complex condition. Overall, synbiotic compositions enhance the bioavailability of microbially generated antioxidant and anti-inflammatory metabolites providing a pharmacological tool to augment the advantages of microbiome regulation on host physiology.

22.2.4 Diet

Diet has the greatest influence on gut microbiota composition throughout the course of one's lifetime. Diet, gut flora, and the host all have a significant role in health. Gut microbiota composition is influenced by a person's diet throughout their lives. Microbial composition and diversity may be influenced by a person's dietary choices, which have been widely studied (Thelen and Brown-Borg 2020; Berding et al. 2021). Numerous researches have shown a relationship between dietary habits and the prevalence of AD, establishing nutrition as a modifiable risk factor (Hu et al. 2020; Berding et al. 2021). Poor dietary habits and ageing, in combination with dysbiosis-induced inflammatory responses, may all influence the aetiology of AD.

Anti-inflammatory diets like the Mediterranean diet (MedDiet) have been demonstrated to lessen the prevalence of many chronic illnesses. According to a study by Valls-Pedret et al., the MedDiet is linked to improved cognitive performance in the aged people (Valls-Pedret et al. 2015). Epidemiological evidence shows that a decreased incidence of Alzheimer's disease is connected with a Mediterranean diet heavy in fruits, vegetables, legumes, and grains and light in sweets, high-fat dairy, and meat (Yusufov et al. 2017). In the EPIC-Spain Dementia Cohort, obedience to the MedDiet was linked to a 20% decreased overall risk of Alzheimer's disease in Mediterranean research encompassing over 16,000 middle people followed up for over 20 years (Andreu-Reinón et al. 2021). Firmicutes/Bacteroidetes ratio and SCFA levels rise after 1 year of MedDiet adherence, which is related with lower inflammatory markers in the gut microbiota. Research done by Ghosh et al. demonstrated MedDiet to be helpful in modifying the microbiota of the digestive tract as well as boosting cognitive performance (Ghosh et al. 2020). This microbiome alteration is predominantly driven by an increase in fibre consumption, vitamins, and minerals in the MedDiet treatment group, whereas alterations in controls were connected with an elevation of dietary fat (monounsaturated fat and saturated fatty acids). One of the top five preventive variables against Alzheimer's disease and cognitive decline is MedDiet.

The ketogenic diet (KD) is a nutritional programme that is high in fat and low in protein and carbohydrates (ideally, 4% carbohydrates, 6% protein, and 90% fat). It was introduced in 1994 as a therapy for epilepsy, and multiple studies have repeatedly demonstrated its efficacy (Ułamek-Kozioł et al. 2019). KD has recently been explored in vitro and in vivo as a possible therapy for additional neurological illnesses, such as AD and PD (Włodarek 2019; Rusek et al. 2019). Due to the sugar deficiency, the body breaks down and oxidizes lipids, producing ketone bodies, which are utilized as another source of energy to glucose by a variety of organs, along with the brain (Masino and Rho 2012). Ketone bodies have been shown to influence channel modulation, decrease neuroinflammation and oxidative stress, increase BDNF (brain-derived neurotrophic factor), improve mitochondrial function, neurotransmission, decrease in the accumulation of amyloid, and enhance learning and memory abilities in mice (Yin et al. 2015). In humans, Randomized Clinical trials (RCT) findings indicated that KD may be advantageous for persons with moderate cognitive impairment or Alzheimer's disease (Lilamand et al. 2021; Grammatikopoulou et al. 2020). Relative to the ketogenic diet in terms of mechanism of action, the modified Atkins diet and the medium-chain triglyceride (MCT) diet/supplementation are effective in reversing cognitive decline in AD and PD symptoms such as mood swings in depression, fatigue, epileptic seizures, and daytime sleepiness (Włodarek 2021; Taylor et al. 2018). Additionally, since the customized Atkins diet does not limit protein consumption like the KD plan does, it provides far greater nutritional versatility than the conventional KD. As a whole, ketone body-producing diets may be a promising treatment for Alzheimer's, but more research is needed to uncover the protective mechanisms in humans and the disadvantages including a lack of flexibility and unpredictability in the dietary regimen and a paucity of vitamin- and antioxidant-rich plant-based foods (Bostock et al. 2017).

Modifications in the microbiome composition as a consequence of KD were first shown in experimental mice fed with KD diet and exhibited an elevation in *Lactobacillus* and *Akkermansia*, as well as an enhancement in vascular brain function (Ma et al. 2018). An interventional analysis in which mild cognitive impaired (MCI) patients were given a customized Medi-KD illustrated that the diet has a beneficial impact on the composition of the intestinal microbiota, increasing the abundance of *Christensenellaceae, Enterobacteriaceae,* and *Akkermansia*, as well as SCFA production, resulting in amelioration of symptoms related to cognition (Nagpal et al. 2019). While recent research in an AD mice indicated that KD may promote gut dysbiosis, a carbohydrate-rich diet seemed to ameliorate the microbiota profile by decreasing *Proteobacteria* and increasing *Bacteroidetes* (Park et al. 2020; Gupta et al. 2015). Dietary patterns that allow for the ingestion of unrefined carbs while still producing ketone bodies, such as intermittent fasting, may be a viable preventive dietary approach for dementia (Nagpal et al. 2019).

In conclusion, dietary treatments are typically safer and more beneficial than drug-based therapy due to their low cost and ease of administration, which alleviates the load on careers of patients with AD.

22.2.5 Faecal Microbiota Transplantation (FMT)

Using a faecal microbiota transplantation (FMT) treatment, a donor's faeces are delivered into a recipient's digestive system through colonoscopy, nasogastric tube, or oral tablets in an effort to alter the microbiota composition of the recipient's intestines. A new therapeutic for a wide range of neurological illnesses might soon be a revolutionary treatment for FMT-mediated eubiosis reprogramming of the gut microbiota (Surawicz et al. 2013).

Reductions in neurogenesis, reduced BDNF expression, and higher memory impairment were seen when AD-model donor faeces were transplanted into healthy mice (Kim et al. 2021; Wang et al. 2021). A plaques deposition was also found. Senescence-accelerated or senescence-accelerated-resistant mice were used to create FMTs in germ-free (GF) mice, and the GF mice who received the senescence-accelerated-resistant donors' microbiota had a better profile than the senescence-accelerated recipients (Zhan et al. 2018). Another study found an increase in the production of plaques in GF mice given faeces from APPPS1 transgenic mice with cerebral A-deposition. An AD patient's FMT into GF mice resulted in an increase in cognitive decline and a reduction in microbiota-derived metabolites critical for the nervous system function (Fujii et al. 2019).

So yet, just two case studies with encouraging results in humans have been conducted. Hazan et al. found that after receiving FMT from an 85-year-old woman (recipient's wife), an 82-year-old man's AD symptoms (cognitive function, memory, and mood) improved (Hazan 2020). In a second case study, comparable improvements in cognitive function, microbial diversity, and SCFA production were seen in a 90-year-old woman with AD and severe *C. difficile* infection who received FMT from a 27-year-old healthy guy (Park et al. 2021).

Despite the obvious use of FMT in the treatment of AD, significant barriers remain. Standardization of therapy regimens, timing and duration of administration, long and short dangers, and criteria for inclusion should all be reviewed and resolved (Shanahan and Quigley 2014; Tan et al. 2020). In summary, while the encouraging conclusions derived in mice demonstrate that the gut microbiota plays a role in the development and progression of neurological illnesses, more human research is necessary before recommending FMT as an adjunctive treatment for AD.

22.3 Conclusions and Future Perspective

Without a dispute, gut dysbiosis is critical for altering the microbiota–gut–brain axis and actively contributes to the etiopathogenesis of AD. The advent of multi-omics approaches has resulted in the identification of a growing number of important functional microorganisms related with AD pathology and altered brain processes. These microorganisms may be exploited as targets for the non-invasive diagnosis and subsequent therapy of Alzheimer's disease. The gut microbiota contributes to the pathophysiology of AD via a variety of mechanisms, including oxidative stress, Ab abnormalities, neuroinflammation, and neurotransmitter dysregulation.

However, the precise functions and processes of the gut microbiota in individuals with Alzheimer's disease remain unknown. The majority of microbiome research can only demonstrate relationships between gut microbiota and AD, but cannot demonstrate a causal link between particular bacteria and AD pathology or brain dysfunctions. Thus, it is critical to regress microbiome research from relationships to cultures in order to establish causation. There is no doubt that culturomics has given a fresh method for identifying new bacterial species from microbiome research. In addition, to establish the true roles and pathways of bacteria in Alzheimer's disease, it is necessary to employ germ-free animals, particular gene knockout mice, or AD humanized animal models. For AD therapy, new techniques that target the gut bacteria may be created when a thorough understanding of their involvement in the disease's pathogenesis has been established.

References

Abenavoli L, Scarpellini E, Colica C, Boccuto L, Salehi B, Sharifi-Rad J, Aiello V, Romano B, de Lorenzo A, Izzo AA et al (2019) Gut microbiota and obesity: a role for probiotics. Nutrients 11: 2690

Alzheimer's Association (Alzheimer's Disease Report) (2020) Alzheimer's disease facts and figures. https://www.alz.org/alzheimer_s_dementia. Accessed 20 Dec 2020

Andreu-Reinón ME, Chirlaque MD, Gavrila D, Amiano P, Mar J, Tainta M et al (2021) Mediterranean diet and risk of dementia and Alzheimer's disease in the EPIC-Spain dementia cohort study. Nutrients 13:700. https://doi.org/10.3390/nu13020700

Athari Nik Azm S, Djazayeri A, Safa M, Azami K, Ahmadvand B, Sabbaghziarani F, Sharifzadeh M, Vafa M (2018) Lactobacilli and Bifidobacteria ameliorate memory and learning deficits and oxidative stress in -amyloid (1–42) injected rats. Appl Physiol Nutr Metab 43:718–726

Ávila PF, Franco Cairo JPL, Damasio A, Forte MBS, Goldbeck R (2020) Xylooligosaccharides production from a sugarcane biomass mixture: effects of commercial enzyme combinations on bagasse/straw hydrolysis pretreated using different strategies. Food Res Int 128:108702. https://doi.org/10.1016/j.foodres.2019.108702

Banks WA, Erickson MA (2010) The blood–brain barrier and immune function and dysfunction. Neurobiol Dis 37:26–32

Berding K, Vlckova K, Marx W, Schellekens H, Stanton C, Clarke G et al (2021) Diet and the microbiota-gut-brain axis: sowing the seeds of good mental health. Adv Nutr 12:1239. https://doi.org/10.1093/advances/nmaa181. [Epub ahead of print]

Biagi E, Nylund L, Candela M, Ostan R, Bucci L, Pini E et al (2010) Through ageing, and beyond: gut microbiota and inflammatory status in seniors and centenarians. PLoS One 5:e10667. https://doi.org/10.1371/journal.pone.0010667

Bonfili L, Cecarini V, Cuccioloni M, Angeletti M, Berardi S, Scarpona S, Rossi G, Eleuteri AM (2018) SLAB51 probiotic formulation activates SIRT1 pathway promoting antioxidant and neuroprotective effects in an AD mouse model. Mol Neurobiol 55:7987–8000

Bonfili L, Cecarini V, Gogoi O, Gong C, Cuccioloni M, Angeletti M, Rossi G, Eleuteri AM (2021) Microbiota modulation as preventative and therapeutic approach in Alzheimer's disease. FEBS J 288:2836–2855

Bostock ECS, Kirkby KC, Taylor BVM (2017) The current status of the ketogenic diet in psychiatry. Front Psych 8:43

Brandscheid C, Schuck F, Reinhardt S, Schäfer K-H, Pietrzik CU, Grimm M, Hartmann T, Schwiertz A, Endres K (2017) Altered gut microbiome composition and tryptic activity of the 5xFAD Alzheimer's mouse model. J Alzheimers Dis 56:775–788

Browne TC, McQuillan K, McManus RM, O'Reilly J-A, Mills KHG, Lynch MA (2013) IFN-γ production by amyloid β–specific Th1 cells promotes microglial activation and increases plaque burden in a mouse model of Alzheimer's disease. J Immunol 190:2241–2251

Carabotti M, Scirocco A, Maselli MA, Severi C (2015) The gut-brain axis: interactions between enteric microbiota, central and enteric nervous systems. Ann Gastroenterol 28(2):203–209

Cattaneo A, Cattane N, Galluzzi S, Provasi S, Lopizzo N, Festari C, Ferrari C, Guerra UP, Paghera B, Muscio C et al (2017) Association of brain amyloidosis with pro-inflammatory gut bacterial taxa and peripheral inflammation markers in cognitively impaired elderly. Neurobiol Aging 49:60–68

Cenit MC, Sanz Y, Codoñer-Franch P (2017) Influence of gut microbiota on neuropsychiatric disorders. World J Gastroenterol 23:5486–5498

Cerovic M, Forloni G, Balducci C (2019) Neuroinflammation and the gut microbiota: possible alternative therapeutic targets to counteract Alzheimer's disease? Front Aging Neurosci 11:284

Cheignon C, Tomas M, Bonnefont-Rousselot D, Faller P, Hureau C, Collin F (2018) Oxidative stress and the amyloid beta peptide in Alzheimer's disease. Redox Biol 14:450–464

Chen C, Ahn EH, Kang SS, Liu X, Alam A, Ye K (2020) Gut dysbiosis contributes to amyloid pathology, associated with C/EBPβ/AEP signaling activation in Alzheimer's disease mouse model. Sci Adv 6:6

Claesson MJ, Cusack S, O'Sullivan O, Greene-Diniz R, de Weerd H, Flannery E et al (2011) Composition, variability, and temporal stability of the intestinal microbiota of the elderly. Proc Natl Acad Sci U S A 108(Suppl. 1):4586–4591. https://doi.org/10.1073/pnas.1000097107

Colciaghi F, Marcello E, Borroni B, Zimmermann M, Caltagirone C, Cattabeni F, Padovani A, di Luca M (2004) Platelet APP, ADAM 10 and BACE alterations in the early stages of Alzheimer disease. Neurology 62:498–501

Colombo AV, Sadler RK, Llovera G, Singh V, Roth S, Heindl S, Sebastian Monasor L, Verhoeven A, Peters F, Parhizkar S et al (2021) Microbiota-derived short chain fatty acids modulate microglia and promote Aβ plaque deposition. elife 10:e59826

Constante M, Fragoso G, Lupien-Meilleur J, Calv A, Santos M (2017) Iron supplements modulate colon microbiota composition and potentiate the protective effects of probiotics in dextran sodium sulfate-induced colitis. Inflamm Bowel Dis 23:753–766. https://doi.org/10.1097/MIB.0000000000001089

Dahiya DK, Puniya M, Shandilya UK, Dhewa T, Kumar N, Kumar S, Puniya AK, Shukla P (2017) Gut microbiota modulation and its relationship with obesity using prebiotic fibers and probiotics: a review. Front Microbiol 8:563

Dichter GS, Damiano CA, Allen JA (2012) Reward circuitry dysfunction in psychiatric and neurodevelopmental disorders and genetic syndromes: animal models and clinical findings. J Neurodev Disord 4:19

Distrutti E, O'Reilly J-A, McDonald C, Cipriani S, Renga B, Lynch MA, Fiorucci S (2014) Modulation of intestinal microbiota by the probiotic VSL#3 resets brain gene expression and ameliorates the age-8. related deficit in LTP. PLoS One 9:e106503

Doifode T, Giridharan VV, Generoso JS, Bhatti G, Collodel A, Schulz PE, Forlenza OV, Barichello T (2020) The impact of the microbiota-gut-brain axis on Alzheimer's disease pathophysiology. Pharmacol Res 164:105314

Doifode T, Giridharan VV, Generoso JS, Bhatti G, Collodel A, Schulz PE, Forlenza OV, Barichello T (2021) The impact of the microbiota-gut-brain axis on Alzheimer's disease pathophysiology. Pharmacol Res 105314:105314

Dorszewska J, Prendecki M, Oczkowska A, Dezor M, Kozubski W (2016) Molecular basis of familial and sporadic Alzheimer's disease. Curr Alzheimer Res 13(9):952–963

Etxeberria U, Fernández-Quintela A, Milagro FI, Aguirre L, Martínez JA, Portillo MP (2013) Impact of polyphenols and polyphenol-rich dietary sources on gut microbiota composition. J Agric Food Chem 61:9517–9533
Evin G (2012) Platelets and Alzheimer's disease: potential of APP as a biomarker. World J Psychiatr 2:102–113
Evrard C, Kienlen-Campard P, Coevoet M, Opsomer R, Tasiaux B, Melnyk P, Octave J-N, Buée L, Sergeant N, Vingtdeux V (2018) Contribution of the endosomal-lysosomal and proteasomal systems in Amyloid-β precursor protein derived fragments processing. Front Cell Neurosci 12: 435
Fujii Y, Nguyen TTT, Fujimura Y, Kameya N, Nakamura S, Arakawa K, Morita H (2019) Fecal metabolite of a Gnotobiotic mouse transplanted with gut microbiota from a patient with Alzheimer's disease. Biosci Biotechnol Biochem 83:2144–2152
Fuke N, Nagata N, Suganuma H, Ota T (2019) Regulation of gut microbiota and metabolic endotoxemia with dietary factors. Nutrients 11:2277
Galicia-Garcia U, Benito-Vicente A, Jebari S, Larrea-Sebal A, Siddiqi H, Uribe KB, Ostolaza H, Martín C (2020) Pathophysiology of type 2 diabetes mellitus. Int J Mol Sci 21:6275
Gamba P, Testa G, Sottero B, Gargiulo S, Poli G, Leonarduzzi G (2012) The link between altered cholesterol metabolism and Alzheimer's disease. Ann N Y Acad Sci 1259:54–64
Gao Q, Wang Y, Wang X, Fu S, Zhang X, Wang R-T, Zhang X (2019) Decreased levels of circulating trimethylamine N-oxide alleviate cognitive and pathological deterioration in transgenic mice: a potential therapeutic approach for Alzheimer's disease. Aging 11:8642–8663
Gerhardt S, Mohajeri M (2018) Changes of colonic bacterial composition in Parkinson's disease and other neurodegenerative diseases. Nutrients 10:708
Ghosh TS, Rampelli S, Jeffery IB, Santoro A, Neto M, Capri M et al (2020) Mediterranean diet intervention alters the gut microbiome in older people reducing frailty and improving health status: the NU-AGE 1-year dietary intervention across five European countries. Gut 69:1218–1228. https://doi.org/10.1136/gutjnl-2019-319654
Gotkine M, Kviatcovsky D, Elinav E (2020) Amyotrophic lateral sclerosis and intestinal microbiota—toward establishing cause and effect. Gut Microbes 11:1833–1841
Grammatikopoulou MG, Goulis DG, Gkiouras K, Theodoridis X, Gkouskou KK, Evangeliou A, Dardiotis E, Bogdanos DP (2020) To keto or not to keto? A systematic review of randomized controlled trials assessing the effects of ketogenic therapy on Alzheimer disease. Adv Nutr Int Rev J 11:1583–1602
Guo YJ, Pan WW, Liu SB, Shen ZF, Xu Y, Hu LL (2020) ERK/MAPK signalling pathway and tumorigenesis. Exp Ther Med 19:1997–2007
Gupta S, Allen-Vercoe E, Petrof EO (2015) Fecal microbiota transplantation: in perspective. Ther Adv Gastroenterol 9:229–239
Halverson T, Alagiakrishnan K (2020) Gut microbes in neurocognitive and mental health disorders. Ann Med 52:423–443
Han D, Li Z, Liu T, Yang N, Li Y, He J et al (2020) Prebiotics regulation of intestinal microbiota attenuates cognitive dysfunction induced by surgery stimulation in APP/PS1 mice. Aging Dis 11:1029–1045. https://doi.org/10.14336/ad.2020.0106
Haran JP, Bhattarai SK, Foley SE, Dutta P, Ward DV, Bucci V, McCormick BA (2019) Alzheimer's disease microbiome is associated with dysregulation of the anti-inflammatory P-glycoprotein pathway. MBio 10:e00632–e00619
Hazan S (2020) Rapid improvement in Alzheimer's disease symptoms following fecal microbiota transplantation: a case report. J Int Med Res 48:0300060520925930
He Y, Wu W, Zheng H-M, Li P, McDonald D, Sheng H-F, Chen M-X, Chen Z-H, Ji G-Y, Zheng Z-D-X et al (2018) Regional variation limits applications of healthy gut microbiome reference ranges and disease models. Nat Med 24:1532–1535
Heijtz RD, Wang S, Anuar F, Qian Y, Bjorkholm B, Samuelsson A, Hibberd ML, Forssberg H, Pettersson S (2011) Normal gut microbiota modulates brain development and behavior. Proc Natl Acad Sci U S A 108:3047–3052

Hu EA, Wu A, Dearborn JL, Gottesman RF, Sharrett AR, Steffen LM et al (2020) Adherence to dietary patterns and risk of incident dementia: findings from the atherosclerosis risk in communities study. J Alzheimers Dis 78:827–835. https://doi.org/10.3233/jad-200392

Jiang H, Ling Z, Zhang Y, Mao H, Ma Z, Yin Y, Wang W, Tang W, Tan Z, Shi J et al (2015) Altered fecal microbiota composition in patients with major depressive disorder. Brain Behav Immun 48:186–194

Kim N, Jeon SH, Ju IG, Gee MS, Do J, Oh MS, Lee JK (2021) Transplantation of gut microbiota derived from Alzheimer's disease mouse model impairs memory function and neurogenesis in C57BL/6 mice. Brain Behav Immun 98:357–365

Kinney JW, Bemiller SM, Murtishaw AS, Leisgang AM, Salazar AM, Lamb BT (2018) Inflammation as a central mechanism in Alzheimer's disease. Alzheimers Dement (N Y) 4:590

Kobayashi Y, Sugahara H, Shimada K, Mitsuyama E, Kuhara T, Yasuoka A, Kondo T, Abe K, Xiao J (2017) Therapeutic potential of Bifidobacterium breve strain A1 for preventing cognitive impairment in Alzheimer's disease. Sci Rep 7:1–10

Köhler C, Maes M, Slyepchenko A, Berk M, Solmi M, Lanctôt K (2016) Carvalho, the gut-brain axis, including the microbiome, leaky gut and bacterial translocation: mechanisms and pathophysiological role in Alzheimer's disease. Curr Pharm Des 22:6152–6166

Lee L-H, Ser H-L, Khan TM, Long M, Chan K-G, Goh B-H, Ab Mutalib N-S (2018) IDDF2018-ABS-0239 dissecting the gut and brain: potential links between gut microbiota in development of Alzheimer's disease? Gut 67:A18

Lee H-J, Lee K-E, Kim J-K, Kim D-H (2019) Suppression of gut dysbiosis by Bifidobacterium longum alleviates cognitive decline in 5XFAD transgenic and aged mice. Sci Rep 9:1–12

Lee J, d'Aigle J, Atadja L, Quaicoe V, Honarpisheh P, Ganesh BP, Hassan A, Graf J, Petrosino JF, Putluri N et al (2020) Gut microbiota-derived short-chain fatty acids promote post-stroke recovery in aged mice. Circ Res 127:453–465

Lee YS, Lai DM, Huang HJ, Lee-Chen GJ, Chang CH, Hsieh-Li HM et al (2021) Prebiotic lactulose ameliorates the cognitive deficit in Alzheimer's disease mouse model through macro autophagy and chaperone mediated autophagy pathways. J Agric Food Chem 69:2422–2437. https://doi.org/10.1021/acs.jafc.0c07327

Levy M, Kolodziejczyk AA, Thaiss CA, Elinav E (2017) Dysbiosis and the immune system. Nat Rev Immunol 17:219–232

Li F, Ding Z, Ke W, Xu D, Zhang P, Bai J et al (2019) Ferulic acid esteraseproducing lactic acid bacteria and cellulase pre-treatments of corn stalk silage at two different temperatures: ensiling characteristics, carbohydrates composition and enzymatic saccharification. Bioresour Technol 282:211–221. https://doi.org/10.1016/j.biortech.2019.03.022

Li Z, Zhu H, Guo Y, Du X, Qin C (2020) Gut microbiota regulate cognitive deficits and amyloid deposition in a model of Alzheimer's disease. J Neurochem 155:448–461

Lilamand M, Mouton-Liger F, Paquet C (2021) Ketogenic diet therapy in Alzheimer's disease: an updated review. Curr Opin Clin Nutr Metab Care 24:372–378

Lilly DM, Stillwell RH (1965) Probiotics: growth-promoting factors produced by microorganisms. Science 147:747–748

Ling Z, Zhu M, Yan X, Cheng Y, Shao L, Liu X, Jiang R, Wu S (1891) Structural and functional dysbiosis of fecal microbiota in Chinese patients with Alzheimer's disease. Front Cell Dev Biol 2021:8

Ling Z, Liu X, Jia X, Cheng Y, Luo Y, Yuan L, Wang Y, Zhao C, Guo S, Li L et al (2014) Impacts of infection with different toxigenic *Clostridium difficile* strains on Faecal microbiota in children. Sci Rep 4:7485

Liu P, Wu L, Peng G, Han Y, Tang R, Ge J, Zhang L, Jia L, Yue S, Zhou K et al (2019) Altered microbiomes distinguish Alzheimer's disease from amnestic mild cognitive impairment and health in a Chinese cohort. Brain Behav Immun 80:633–643

Liu S, Gao J, Zhu M, Liu K, Zhang H-L (2020) Gut microbiota and dysbiosis in Alzheimer's disease: implications for pathogenesis and treatment. Mol Neurobiol 57:5026–5043

Lupien-Meilleur J, Roy D, Lagace L (2016) Viability of probiotic bacteria in a maple sap beverage during refrigerated storage. LWT Food Sci Technol 74:160. https://doi.org/10.1016/j.lwt.2016.07.045

Ma D, Wang AC, Parikh I, Green SJ, Hoffman JD, Chlipala G, Murphy MP, Sokola BS, Bauer B, Hartz AMS et al (2018) Ketogenic diet enhances neurovascular function with altered gut microbiome in young healthy mice. Sci Rep 8:1–10

Margolis KG, Cryan JF, Mayer EA (2021) The microbiota-gut-brain axis: from motility to mood. Gastroenterology 160:1486–1501

Martinez FAC, Balciunas EM, Converti A, Cotter PD, de Souza Oliveira RP (2013) Bacteriocin production by Bifidobacterium Spp. a review. Biotechnol Adv 31:482–488

Masino SA, Rho JM (2012) Mechanisms of ketogenic diet action. Epilepsia 51:85

Morris G, Berk M, Carvalho A, Caso JR, Sanz Y, Walder K, Maes M (2017) The role of the microbial metabolites including tryptophan catabolites and short chain fatty acids in the pathophysiology of immune-inflammatory and Neuroimmune disease. Mol Neurobiol 54: 4432–4451

Morrison DJ, Preston T (2016) Formation of short chain fatty acids by the gut microbiota and their impact on human metabolism. Gut Microbes 7:189–200

Mottawea W, Chiang C-K, Mühlbauer M, Starr AE, Butcher J, Abujamel T, Deeke SA, Brandel A, Zhou H, Shokralla S et al (2016) Altered intestinal microbiota–host mitochondria crosstalk in new onset Crohn's disease. Nat Commun 7:13419

Nagpal R, Neth BJ, Wang S, Craft S, Yadav H (2019) Modified Mediterranean-ketogenic diet modulates gut microbiome and short-chain fatty acids in association with Alzheimer's disease markers in subjects with mild cognitive impairment. EBioMedicine 47:529–542

NIH (2012) Human microbiome project defines normal bacterial makeup of the body. USDHHS

Nimgampalle M, Yellamma K (2017) Anti-Alzheimer properties of probiotic, lactobacillus plantarum MTCC 1325 in Alzheimer's disease induced albino rats. J Clin Diagn Res 11: KC01–KC05. https://doi.org/10.7860/JCDR/2017/26106.10428

O'Hagan C, Li JV, Marchesi JR, Plummer S, Garaiova I, Good MA (2017) Long-term multi-species *lactobacillus* and *Bifidobacterium* dietary supplement enhances memory and changes regional brain metabolites in middle-aged rats. Neurobiol Learn Mem 144:36–47

Oleskin AV, Shenderov BA (2016) Neuromodulatory effects and targets of the SCFAs and gasotransmitters produced by the human symbiotic microbiota. Microb Ecol Health Dis 27:634

Parada Venegas D, de la Fuente MK, Landskron G, González MJ, Quera R, Dijkstra G, Harmsen HJM, Faber KN, Hermoso MA (2019) Short chain fatty acids (SCFAs)-mediated gut epithelial and immune regulation and its relevance for inflammatory bowel diseases. Front Immunol 10: 277

Park S, Zhang T, Wu X, Yi Qiu J (2020) Ketone production by ketogenic diet and by intermittent fasting has different effects on the gut microbiota and disease progression in an Alzheimer's disease rat model. J Clin Biochem Nutr 67:188–198

Park S-H, Lee JH, Shin J, Kim J-S, Cha B, Lee S, Kwon KS, Shin YW, Choi SH (2021) Cognitive function improvement after fecal microbiota transplantation in Alzheimer's dementia patient: a case report. Curr Med Res Opin 37:1739–1744

Pascale A, Marchesi N, Marelli C, Coppola A, Luzi L, Govoni S, Giustina A, Gazzaruso C (2018) Microbiota and metabolic diseases. Endocrine 61:357–371

Prince M, Bryce R, Albanese E, Wimo A, Ribeiro W, Ferri CP (2013) The global prevalence of dementia: a systematic review and metaanalysis. Alzheimers Dementia 9:63–75

Psichas A, Sleeth ML, Murphy KG, Brooks L, Bewick GA, Hanyaloglu AC, Ghatei MA, Bloom SR, Frost G (2014) The short chain fatty acid propionate stimulates GLP-1 and PYY secretion via free fatty acid receptor 2 in rodents. Int J Obes 39:424–429

Rea K, Dinan TG, Cryan JF (2016) The microbiome: a key regulator of stress and neuroinflammation. Neurobiol Stress 4:23–33

Redza-Dutordoir M, Averill-Bates DA (2016) Activation of apoptosis signalling pathways by reactive oxygen species. Biochim Biophys Acta, Mol Cell Res 1863:2977–2992

Reitz C, Brayne C, Mayeux R (2011) Epidemiology of Alzheimer disease. Nat Rev Neurol 7:137–152
Rubio-Perez JM, Morillas-Ruiz JM (2012) A review: inflammatory process in Alzheimer's disease, role of cytokines. Sci World J 2012:756357, 1
Rusek M, Pluta R, Ułamek-Kozioł M, Czuczwar SJ (2019) Ketogenic diet in Alzheimer's Disease. Int J Mol Sci 20:3892
Schwabe RF, Jobin C (2013) The microbiome and cancer. Nat Cancer 13:800–812
Shanahan F, Quigley EMM (2014) Manipulation of the microbiota for treatment of IBS and IBD—challenges and controversies. Gastroenterology 146:1554–1563
Spinelli M, Fusco S, Grassi C (2019) Brain insulin resistance and hippocampal plasticity: mechanisms and biomarkers of cognitive decline. Front Neurosci 13:788
Steiner H, Fukumori A, Tagami S, Okochi M (2018) Making the final cut: pathogenic amyloid-β peptide generation by γ-secretase. Cell Stress 2:292–310
Sun M-F, Shen Y-Q (2018) Dysbiosis of gut microbiota and microbial metabolites in Parkinson's disease. Ageing Res Rev 45:53–61
Sun M, Wu W, Chen L, Yang W, Huang X, Ma C et al (2018) Microbiotaderived short-chain fatty acids promote Th1 cell IL-10 production to maintain intestinal homeostasis. Nat Commun 9: 3555. https://doi.org/10.1038/s41467-018-05901-2
Surawicz CM, Brandt LJ, Binion DG, Ananthakrishnan AN, Curry SR, Gilligan PH, McFarland LV, Mellow M, Zuckerbraun BS (2013) Guidelines for diagnosis, treatment, and prevention of clostridium difficile infections. Am J Gastroenterol 108:478–498
Szabo C (2010) Gaseotransmitters: new frontiers for translational science. Sci Transl Med 2:59ps54
Tan P, Li X, Shen J, Feng Q (2020) Fecal microbiota transplantation for the treatment of inflammatory bowel disease: an update. Front Pharmacol 11:1409
Taylor MK, Sullivan DK, Mahnken JD, Burns JM, Swerdlow RH (2018) Feasibility and efficacy data from a ketogenic diet intervention in Alzheimer's disease. Alzheimers Dementia Transl Res Clin Interv 4:28–36
Thelen M, Brown-Borg HM (2020) Does diet have a role in the treatment of Alzheimer's disease? Front Aging Neurosci 12:617071. https://doi.org/10.3389/fnagi.2020.617071
Ułamek-Kozioł M, Czuczwar SJ, Januszewski S, Pluta R (2019) Ketogenic diet and epilepsy. Nutrients 11:2510
Valls-Pedret C, Sala-Vila A, Serra-Mir M, Corella D, De La Torre R, Martínez- González M et al (2015) Mediterranean diet and age-related cognitive decline: a randomized clinical trial. JAMA Intern Med 175:1094–1103. https://doi.org/10.1001/jamainternmed.2015.1668
van Dyke K (1997) The possible role of peroxynitrite in Alzheimer's disease: a simple hypothesis that could be tested more thoroughly. Med Hypotheses 48:375–380
Varesi A, Deumer U-S, Ananth S, Ricevuti G (2021) The emerging role of gut microbiota in Myalgic encephalomyelitis/chronic fatigue syndrome (ME/CFS): current evidence and potential therapeutic applications. J Clin Med 10:5077
Verbeke KA, Boobis AR, Chiodini A, Edwards CA, Franck A, Kleerebezem M, Nauta A, Raes J, Tol EAF, van Tuohy KM (2015) Towards microbial fermentation metabolites as markers for health benefits of prebiotics. Nutr Res Rev 28:66
Vidhya Rekha U, Anita M, Bhuvaneswarri J, Jayamathi G, Sadhana K, Ramya V et al (2019) Free radical scavenging potentials of Triphala: a medicinal herb used in Indian ayurvedic system of medicine. Drug Invent Today 12:407–411
Vogt NM, Kerby RL, Dill-McFarland KA, Harding SJ, Merluzzi AP, Johnson SC, Carlsson CM, Asthana S, Zetterberg H, Blennow K et al (2017) Gut microbiome alterations in Alzheimer's disease. Sci Rep 7:1–11
Wang X, Sun G, Feng T, Zhang J, Huang X, Wang T et al (2019) Sodium oligomannate therapeutically remodels gut microbiota and suppresses gut bacterial amino acids-shaped neuroinflammation to inhibit Alzheimer's disease progression. Cell Res 29:787–803. https://doi.org/10.1038/s41422-019-0216-x

Wang M, Cao J, Gong C, Amakye WK, Yao M, Ren J (2021) Exploring the microbiota-Alzheimer's disease linkage using short-term antibiotic treatment followed by fecal microbiota transplantation. Brain Behav Immun 96:227–238

Weiss GA, Hennet T (2017) Mechanisms and consequences of intestinal dysbiosis. Cell Mol Life Sci 74:2959–2977

Wenzel TJ, Gates EJ, Ranger AL, Klegeris A (2020) Short-chain fatty acids (SCFAs) alone or in combination regulate select immune functions of microglia- like cells. Mol Cell Neurosci 105: 103493

Westfall S, Lomis N, Prakash S (2019) A novel synbiotic delays Alzheimer's disease onset via combinatorial gut-brain-axis signaling in Drosophila melanogaster. PLoS One 14:e0214985. https://doi.org/10.1371/journal.pone.0214985

Włodarek D (2019) Role of ketogenic diets in neurodegenerative diseases (Alzheimer's disease and Parkinson's disease). Nutrients 11:169

Włodarek D (2021) The possibility of use of the ketogenic diet and medium chain triglycerides supplementation in the support therapy of Alzheimer disease. Curr Opin Clin Nutr Metab Care 24:385–391

Xie Z, Wei M, Morgan TE, Fabrizio P, Han D, Finch CE, Longo VD (2002) Peroxynitrite mediates neurotoxicity of amyloid-peptide1–42-and lipopolysaccharide-activated microglia. J Neurosci 22:3484–3492

Yano JM, Yu K, Donaldson GP, Shastri GG, Ann P, Ma L, Nagler CR, Ismagilov RF, Mazmanian SK, Hsiao EY (2015) Indigenous bacteria from the gut microbiota regulate host serotonin biosynthesis. Cell 161:264–276

Yin JX, Maalouf M, Han P, Zhao M, Gao M, Dharshaun T, Ryan C, Whitelegge J, Wu J, Eisenberg D et al (2015) Ketones block amyloid entry and improve cognition in an Alzheimer's model. Neurobiol Aging 39:25–37

Yusufov M, Weyandt LL, Piryatinsky I (2017) Alzheimer's disease and diet: a systematic review. Int J Neurosci 127:161–175. https://doi.org/10.3109/00207454.2016.1155572

Zhan G, Yang N, Li S, Huang N, Fang X, Zhang J, Zhu B, Yang L, Yang C, Luo A (2018) Abnormal gut microbiota composition contributes to cognitive dysfunction in SAMP8 mice. Aging 10:1257–1267

Zhuang Z-Q, Shen L-L, Li W-W, Fu X, Zeng F, Gui L, Lü Y, Cai M, Zhu C, Tan Y-L et al (2018) Gut microbiota is altered in patients with Alzheimer's disease. J Alzheimers Dis 63:1337–1346

Part IV

Association of Phages and Fungi with Gut Microbiome

Fungi as a Treasure Trove of Bioactive Compounds for Human Health

23

Divjot Kour, Sofia Shareif Khan, Tanvir Kaur, Rubee Devi, Raheshwari Negi, Ajar Nath Yadav, and Amrik Singh Ahluwalia

Abstract

Fungi, especially the endophytic fungi which live in association with the plants, have been known to mimic their hosts in production of structurally diverse biologically active compounds. Bioactive compounds are of great interest as new lead structures for medicine and plant protection. Furthermore, they play an important role in community balancing and mediating interactions between microorganisms and their hosts. Bioactive compounds are of much importance in human health as they act as antimicrobial agents, anti-cancer compounds, antioxidants, anti-parasitic, and anti-diabetic agents. Thus, the research on the chemistry of the natural products has increased tremendously in the recent years due to the increased demands of the compounds exhibiting potential pharmaceutical properties or due to their economic value as cosmetics, drugs, and chemicals. The present chapter describes about the bioactive compounds from fungi and their major roles in human health.

D. Kour
Department of Microbiology, Akal College of Basic Sciences, Eternal University, Baru Sahib, Sirmour, India

S. S. Khan
Department of Biotechnology, Shri Mata Vaishno Devi University, Katra, Jammu and Kashmir, India

T. Kaur · R. Devi · R. Negi · A. N. Yadav
Department of Biotechnology, Dr. Khem Singh Gill Akal College of Agriculture, Eternal University, Baru Sahib, Sirmour, India

A. S. Ahluwalia (✉)
Department of Botany, Akal College of Basic Sciences, Eternal University, Baru Sahib, Sirmour, India

R. Sobti et al. (eds.), *Role of Microbes in Sustainable Development*, https://doi.org/10.1007/978-981-99-3126-2_23

Keywords

Anti-cancer · Antimicrobials · Bioactive compounds · Fungi · Human health

23.1 Introduction

There has been a dramatic shift in recent years toward a more sustainable, eco-friendly, and natural way of life. Many researchers believe that there is an alarming increase in drug resistance worldwide, and the problem is rapidly escalating, rendering current antimicrobial agents futile (Ayukekbong et al. 2017; Dadgostar 2019; Vasan et al. 2019). Antibiotic-resistant pathogens infect approximately two million people worldwide each year, resulting in at least 23,000 deaths (Prestinaci et al. 2015). Antimicrobial resistance (AMR), according to the World Health Organization (WHO), has emerged as one of the most serious public health concerns of the twenty-first century (Manganyi and Ateba 2020). Furthermore, given the numerous disadvantages and side effects of current antimicrobial agents, it is not surprising that a large proportion of people, particularly those in developing countries, are turning to naturally available bioactive alternatives for primary healthcare (Manganyi and Ateba 2020).

Microorganisms have been extremely beneficial to mankind, both environmentally and economically. They have distinct characteristics that are supported by their ability to grow quickly. The bacterial and the fungal communities have fascinated the scientific community and researchers as producers of novel and precious bioactive compounds and further they can be preserved indefinitely, assuring perennial availability of the biologically active compounds (Lam 2006; Uzma et al. 2019). Their easy genetic manipulations to obtain improved yields of desired biologically active compounds further give additional benefits (Kharwar et al. 2011).

Bioactive compounds exhibit broad pharmaceutical properties including cardiovascular, anti-thrombotic, anti-cancer, antidiabetic, anti-glycaemic, and antihypertensive (Basit et al. 2021). They are also utilized as favoured medicines, made synthetically to cure different diseases with minimum side effects (Chang et al. 2013). Because of their benefits to plants and human health, these are currently in high demand in naturopathy and pharmaceuticals.

Fungal communities have been widely investigated as a storehouse of bioactive compounds. The first bioactive compound, penicillin from *Penicillium notatum,* was discovered by Alexander Fleming in 1928. Soon thereafter, there was an explosion of new antimicrobial compounds from fungi (Bhardwaj and Agrawal 2014). Griseofulvin from the *Penicillium griseofulvum* is an effective antifungal agent (Oxford et al. 1939). Cyclosporine, an effectual immunosuppressive drug, has been reported from fungi *Tolypocladium niveum* and *Cylindrocarpon lucidum* (Webber 1981). In addition, taxol, an effective anti-cancer agent, has been obtained from *Taxomyces andereanae* (Stierle et al. 1993). Thus, fungi are continuously being used for the search of potential pharmaceutical products.

23.2 Fungi and Bioactive Compounds

Fungi, particularly endophytes, have recently piqued the interest of the microbial chemistry community due to their high potential for contributing to the discovery of new bioactive compounds (Verma et al. 2022). It has been suggested that the close biological association between endophytic microbes and their plant host results in the production of a diversity of biologically active molecules when compared to epiphytic or soil microbes (Strobel 2003). Furthermore, because this relationship is symbiotic, endophytic bioactive compounds are less toxic to the cell because they do not kill the eukaryotic host system. This is chiefly important to the medical community as potential drugs may not adversely affect human cells (Alvin et al. 2014). A diverse range of the endophytic fungi belonging to Ascomycota, Basidiomycota, Mucoromycota, and Oomycota have been reported from diverse crops. The species of *Aspergillus, Fusarium, Penicillium,* and *Piriformospora* have been reported as endophytes from different plants (Nisa et al. 2015). Some niche-specific endophytic fungal strains have also been reported, such as *Penicillium brevicompactum* and *Penicillium glabrum* from *Hordeum vulgare*; *Chaetomium, Cryptococcus, Berkleasmium*, and *Gibberella zeae* from *Zea mays*; and *Diaporthe phaseolorum, Gibberella moniliformis, Diaporthe helianthi, Leptospora rubella, Didymella bryoniae,* and *Guignardia vaccinii* from *Triticum aestivum* (Kousar et al. 2022; Rana et al. 2019a, b).

Endophytic fungi are known to be the significant producers of bioactive compounds. These compounds have gained much attention as therapeutic agents for the treatment of many types of the cancers. These compounds include anhydrofusarubin, asperpyrone A, camptothecin, chaetomugilides, cladodosporol H, diosgenin, fusarubin, hypericin, lijiquinone, myrotheciumone, paclitaxel, podophyllotoxin, swainsonine, and vinblastine. Chatterjee et al. (2019) reported the bactericidal activity and antioxidant potential of *Alternaria alternata* isolated from *Azadirachta indica*. The study concluded that the reported strain could be a good source of bioactive compounds of medicinal use. The study of Gauchan et al. (2020) reported the antimicrobial and cytotoxic potential of crude extract of *Alternaria alternate, Alternaria brassicae, and Cladosporium cladosporioides* and concluded that further identification of the bioactive compounds can be a fascinating source for novel pharmaceutical agents. Chatterjee et al. (2022) reported the potential of *Alternaria tenuissima* for production of bioactive compounds. The study of Kumar and Prasher (2022) reported the antibacterial activity of crude extract of *Fomitopsis meliae*. The study further reported the presence of 40 compounds with predominance of dodecane, ethyl 2-thiopheneacetate, tetradecane, hexadecane, octadecane, benzaldehyde, 4-(1-methylethyl)-, and griseofulvin. Thus, bioactive compounds are important targets for drug discovery because they play an important role in the development of novel chemical compounds (Kousar et al. 2022).

23.3 Biological Properties of Bioactive Compounds from Fungi in Human Health

23.3.1 Antibacterial Agents

Bacterial resistance has emerged as a driving force behind the global search for unknown antibacterial agents. Over the past 50 years, antibacterial agents have helped establish modern medicine and saved many lives. A public health problem that further highlights the need for antibacterial drugs is the challenge in treating bacterial infections. Antibacterial agents are those compounds that prevent the growth of bacteria, with many disrupting essential biological processes thereby causing rapid death of bacterial cell (Langeveld et al. 2014). Over the past 40 years, antibiotics have been heavily used by humans, not only through direct consumption, but also through their widespread use in animal feed (Guzman 2014). Bacteria are a promising option in drug discovery due to the unique characteristics and vast number of metabolites produced with bioactive potential. Many types of bacterial infection have been caused such as bloodstream infections, gastrointestinal diseases, urinary tract infections, nosocomial, meningitis, and bacteremia (Janny et al. 2013). A large number are antibacterial drugs manufactured by pharmaceutical industries due to reduction of bacterial infection. Up to the end of the 1980s, pharmaceutical research and subsequently industry ensured that the market had steady supplies of new antibacterial agent. These drugs frequently employed a novel mode of action that allowed for the avoidance of drug resistance. On the other hand, from the 1990s, a small number of new classes of drugs were presented, including the oxazolidinones and lipopeptides, which were marketed for systemic administration to treat infections brought by multi-resistant Gram-positive/Gram-negative bacteria. These bacteria have shown several antibiotics including amikacin, amoxicillin-clavulanic acid, aztreonam, cefazolin, ceftazidime, cefotaxime, cephalosporin, cefuroxime, ertapenem, imipenem, fosfomycin, fluoroquinolones, gentamicin, meropenem, and piperacillin-tazobactam (Sibero et al. 2017). Various fungal species have been reported for having antibacterial activity against different human pathogens.

In a study, antibacterial activity against multidrug-resistant and highly human pathogenic bacteria, including *Staphylococcus aureus, Salmonella typhi, Staphylococcus epidermidis*, and *Escherichia coli,* has been reported (Ingle et al. 2008). In another study, fungi *Trichoderma harzianum* (CBMAI 43), *Guignardia* sp. (CBMAI 69), and *Phomopsis* sp. (CBMAI 164) inhibited the growth of human pathogenic bacteria *Escherichia coli*, *Pseudomonas aeruginosa*, *Salmonella choleraesuis,* and *Staphylococcus aureus* (Sette et al. 2006). In a study, *Alternaria alternata* and *Novosphingobium oryzae* inhibited growth of pathogenic bacteria causing fecal-oral contamination in humans (Gond et al. 2012). Fungi *Penicillium citrinum, Penicillium nigricans,* and *Aspergillus niger* isolated from stem and barks of (*Alstonia boonei*-Ahun, *Enantia chlorantha*-Awopa and *Kigelia africana*-Pandoro) were reported to have antibacterial activity against the human pathogens *Pseudomonas aeruginosa*, *Escherichia coli*, *Enterococcus faecalis*, and *Candida albicans* (Adeyemi 2015). In a report, *Talaromyce spurpureogenus* was found to have antibacterial activity against the pathogenic bacteria including *Staphylococcus aureus, Bacillus cereus*, *Salmonella enterica*, *Pseudomonas aeruginosa*, and *Escherichia coli* (Hu et al. 2019). The compound (Violaceol I and Violaceol II) produced by endophytic fungus, *Trichoderma polyalthiae,* was having antibacterial activity against the human pathogenic Gram-positive bacteria including *Staphylococcus aureus*, *Bacillus cereus*, *Bacillus subtilis*, and *Staphylococcus saprophyticus*, and Gram-negative bacteria *Shigella sonnei*, and *Salmonella typhimurium* (Nuankeaw et al. 2020). Fungi *Fusarium solani, Trichoderma viride, Penicillium citrinum,* and *Cunninghamella elegans* were having antibacterial activity against human pathogenic bacteria including *Staphylococcus aureus*, *Salmonella typhi*, *Serratia marcescens*, *Klebsiella pneumoniae*, *Shigella* sp., and *E. coli* (Zainee et al. 2021). The compound Dodecane, Ethyl 2-thiopheneacetate, Tetradecane, Hexadecane, Octadecane, Benzaldehyde, 4-(1-methylethyl), and Griseofulvin associated with the endophytic fungi *Fomitopsis meliae* isolated from *Dilleniaindica* L. was found to have antibacterial activity against human pathogenic bacteria *Escherichia coli, Staphylococcus aureus, Pseudomonas aeruginosa,* and *Bacillus subtilis* (Kumar and Prasher 2022).

23.3.2 Antifungal Agents

Nowadays, fungal infection of skin is one of the most common dermatological problems worldwide. It has increased intensely during the previous 30 years and it was determined that 40 million people suffer from fungal infections. The killing or inhibition of fungal growth is referred to as an antifungal action. Superficial and subcutaneous fungal infections affect the skin, keratinous tissues, and mucous membranes. The primary afflicted disorders include dermatophytic infections, superficial candidiasis of the mouth, skin, or genital tract, and infections caused by Malassezia, such as *Pityriasis versicolor* and *Malassezia folliculitis*. The pathogens, *Candida albicans* and *Aspergillus* sp. are the most frequently found and playing role in causing systemic fungal infections. Amphotericin B, known to cause considerable nephrotoxicity, was the only drug available for the treatment of severe fungal infections for almost 30 years. Major improvements in our ability to safely and effectively treat local and systemic fungal infections were made possible by the approval of imidazoles and triazoles in the late 1980s and early 1990s. The triazoles, especially fluconazole, have an excellent safety profile, which has contributed to their widespread use. There are other numerous antifungal agents used clinically to treat fungal infections, i.e. azoles, allylamines, echinocandins, griseofulvin, and flucytosine. The alternative approaches are needed to inhibit the viral infection to treat virus-infected disease. Several studies have been reported in which fungal species are having antiviral activity against the virus. The compound resveratrol, Amphotericin B, showed antifungal activity against the human pathogenic fungi *Candida albicans*, *Saccharomyces cerevisiae*, and *Trichosporon beigelii* (Jung et al. 2005). In a study, *Alternaria* sp. was having antifungal activity against the human pathogenetic fungi *Paracoccidioides brasiliensis* (Johann et al. 2012). In another study, the endophytic fungi *Alternaria alternate, Fusarium oxysporum, Trichoderma atroviride, Aurobasidium* sp., *Penicillium viridicatum*, and *Cladosporium porophorum* were having antifungal activity against the human pathogenic fungi *Aspergillus niger* and *Aspergillus fumigatus* (Erfandoust et al. 2020). The antifungal components of Ou-gon were having antifungal activity against human pathogenic fungi *Trichophyton rubrum*, *Trichophyton mentagrophytes*, *Aspergillus fumigatus*, and *Candida albicans* (Da et al. 2019). The endophytic fungi *Cryptococcus neoformans* and *Trichophyton rubrum* were having the antifungal activity against the *Candida* sp. (Santos et al. 2012). The fungi *Pleurotus ostreatu* showed antifungal activity against human pathogenic fungi *Penicillium digitatum* (Liu et al. 2020).

23.3.3 Antiviral Agents

Viral infections have killed millions of people worldwide during the course of human civilization, making the development of antiviral medications increasingly necessary. Viruses are microscopic organisms that reproduce only in living cells, and they are the world's largest leading cause of morbidity and mortality in humans (Andersen et al. 2020). Previous attention was given to the "one drug, one virus" concept, which focuses on eliminating virus-specific aspects. The fighting against human life-threatening diseases is aided by modern antiviral medications and vaccines. Additionally, the development of viral medication resistance diminishes the therapeutic efficacy of the existing antiviral treatments, raising serious concerns for global public health (Linnakoski et al. 2018). Ideal antiviral medications would essentially have high potency against the targeted virus strains, but low negative effects on the host cells. This demonstrates that antiviral medications' modes of action are typically aimed at preventing or suppressing infection by targeting viral proteins or the host cellular components that viruses utilize to replicate and take over cellular functions (Mason et al. 2018). It is necessary to look for, find, and develop novel, affordable, and more effective antiviral medications as well as vaccines to address this issue. Isoindolones compounds, namely emeriphenolicins A and D, emerimidines A and B, as well as other compounds, including austin, austinol, aspernidines A and B, acetoxydehydroaustin, and dehydroaustin were discovered in the endophytic fungus *Emericella* sp. (HK-ZJ) from the mangrove plant *Aegiceras corniculatum*.

23.3.3.1 Biologically Active Compounds of Basidial Fungi and Their Efficacies Against Some Viruses Pathogenic for Humans

23.3.3.1.1 Influenza Virus

The influenza virus is the most well-known and widespread of the more than 100 viruses that cause infectious illnesses of the upper respiratory tract. Every year, influenza outbreaks result in 3.5 million cases of serious disease and

300–500 thousand deaths worldwide (Teplyakova and Kosogova 2015). New influenza strains that are causing epidemics point mutations in the surface glycoproteins HA (hemagglutinin) and NA (neuraminidase) cause a virus to develop every 1–2 years. Long-lasting immunity to the virus cannot be developed because the influenza virus can get beyond the human immune system's defences. Entire continents can be affected by influenza pandemics that kill a large portion of the population (Teplyakova and Kosogova 2015). One of the top tasks in the field of public health is searching for new, powerful anti-influenza preventative and therapeutic medications. It is recognized that fungus can produce substances that suppress influenza viruses. An aqueous extract of the fungus *Rozites caperatus* contained a protein-like component that stopped the reproduction of influenza virus type A (Piraino 2005). Triterpenoids, specifically ganodermadiol, lucidodiol, and aplanoxin acid G, were found to be the extract of *Ganoderma pfeifferi* that had the most antiviral efficacy against influenza virus type A and HSV-1 (Niedermeyer et al. 2005). *Inonotus hispidus* fungus (Bull.) P. Karst. ethanol extract contains the isoprenoid compounds, hispidin and hispolon, which have antiviral action against influenza virus types A and B. Both the extracts of fruiting body and fungus mycelium showed antiviral activity (NA et al. 2003). Melanin from the chaga mushroom was tested for its anti-influenza activity, and the results revealed that at the same dilution (1:3000), the neutralization index for the human influenza virus strain A/Aichi/2/68 (H3N2) was higher than the avian influenza virus A/chicken/Kurgan/05/2005 (H5N1) (6.3 lg and 3.0 lg, respectively). It is undoubtedly related to the biological traits of different influenza virus strains (Teplyakova et al. 2013). The cytopathic effect (CPE) test in a bioassay demonstrated the effectiveness of the fungal extracts against influenza A viral (H1N1) (Zhang et al. 2011).

23.3.3.1.2 Human Immunodeficiency Virus

The combination ARV (antiretroviral) therapy is the only treatment that can stop HIV infection from developing and the disease from progressing. ARV therapy is an etiotropic treatment for HIV infection that targets to suppress the human immunodeficiency virus's replication. By using ARV medications, we can stop HIV replication and lower the blood viral content (Gashnikova et al. 2012). This leads to the full or partial restoration of the CD4 lymphocyte subpopulation, which enhances the quality of life and extends the life expectancy of HIV-infected patients by preventing the onset, facilitating the course of, or contributing to the elimination of opportunistic infections. Despite the abundance of anti-HIV treatments now in development, there is a problem with antiviral therapy effectiveness. The capacity of HIV-1 to acquire resistance to antiviral medications, drug toxicity, and drug expense is the key barrier to this problem's resolution (Baryshev et al. 2014). Because of this, the need for the development of efficient and affordable antiviral medications that don't harm humans remains urgent. Human immunodeficiency virus can be inhibited by some fungal compounds. HIV-1 can be inhibited in vitro by polysaccharides from PSK krestin and PSP from *Trametes versicolor*. They demonstrate an immunostimulatory effect, krestin supporting immune system killer cells, and the polysaccharide protein complex inhibiting HIV-1 gp120 attachment to CD4 surface receptor and activation of HIV reverse transcriptase (KAC and POHLEVEN 2005). Velutin and flammulin proteins isolated from the winter mushroom *Flammulina velutipes* are cytotoxic and render ribosomes inactive. Velutin blocks reverse transcriptase of HIV-1 (Wang and Ng 2001). In MT-4 cell culture, several triterpenes from the fungus *Ganoderma lucidum*, such as ganoderic acid B, are effective against human immunodeficiency virus type 1 (Yang et al. 2007).

23.3.3.1.3 Hepatitis Virus

All hepatitis viruses cause hepatitis in humans despite belonging to various taxons and having varied biochemical and molecular properties. Virus-related hepatitis B and C are among the top ten killers of people, along with chronic liver illnesses. 350 million people worldwide have hepatitis B, compared to 170 million who have hepatitis C. Hepatitis B virus infection affects around 2 billion people globally. Hepadnaviridae, a family of DNA viruses, cause liver disorders in both humans and animals, including HBV (hepatitis B virus). Oriental medicine has historically employed the basidiomycetes *Cordyceps sinensis*, *Grifola frondoza*, and *Lentinus edodes* to treat liver illness. However, there are still very few research studies on the antiviral action of basidiomycete metabolites against the hepatitis virus. The study we've done allows us to identify two potential pathways for basidiomycetes in hepatitis therapy. Additionally, it was discovered that using pure *P. ostreatus* lectins as an adjuvant (1 mg/mL) improved the immunogenicity of the DNA vaccine against hepatitis B (Gao et al. 2013). In another study, I. *obliquus* aqueous extract fractions were found to have virucidal effects on the hepatitis C virus, reducing infectious characteristics by 100 times in just 10 min. Both preventative usage (24 h prior to infection) and therapeutic use (at the time of infecting pig embryo renal canals) of fungal extracts revealed antiviral qualities (Shibnev et al. 2011).

23.3.3.1.4 Herpesvirus

There is a global spread of herpesvirus infections. The most common site for genital herpes infection is HSV-2 (herpes simplex virus type 2) (Weiss 2004). The prevalence of genital herpes ranges from 80 to 200 cases per 100,000 people depending on the country (Johns and Cunningham 2004). HSV-2 produces widespread internal organ lesions and generalized herpetic infection in immunocompromised people, often with catastrophic results. Acyclovir (Zovirax, Virolex), a synthetic derivative of deoxy-guanidine, is the most often used antiviral for treating herpesvirus infections. After phosphorylation, it can inhibit viral DNA polymerase and viral DNA synthesis (Ho et al. 2021). Drug-resistant herpesvirus strains have appeared as a result of its extensive use. In a study, the production of herpes simplex virus type 1 virions from Vero cells with an infectious virus titer of 2.0104 PFU/mL was entirely suppressed by Lentinus edodes (shiitake mushroom) extract at a dosage of 0.3 mg/mL (Sarkar et al. 1993). In another study, after reaching the stationary phase of growth, a nucleoside was discovered in the fungus Macrocystidia cucumis (Pers.) Joss' culture media. The isolated purine nucleoside was effective in combating HSV-1 (Saboulard et al. 1998). In a different study, researchers from Belarus give information on carotenoids found in the sulfur-yellow Laetiporus sulphureus that have efficacy against HSV-1 (Kapich et al. 2004).

23.3.3.1.5 Poliovirus

The virus that causes polio in humans, known as poliovirus (*Poliovirus hominis*), is a member of the enterovirus family Picornaviridae, which also includes the Coxsackie and ECHO viruses. It comes in three distinct types (I, II, and III), with type I being the most prevalent. In 2007, a study examined antiviral efficacy of polysaccharides and aqueous and ethanolic extracts from *Agaricus brasiliensis* fruiting bodies against poliovirus type 1. It is believed that the active components operate at the first stage of poliovirus replication because the evaluated compounds demonstrated antiviral action (Faccin et al. 2007). In tests examining the antiviral properties of extracts from fresh fruiting bodies of 121 species of basidiomycetes against the poliovirus, four species demonstrated action such as *Mycena pura* (Pers.) P. Kumm. (1.25–1.75 mg/mL), *Lactarius torminosus* (Schaeff.) Gray (0.5–2.5 mg/mL), *Lepista inversa* (Scop.) Pat. (= *Lepista flaccida* (Sowerby) Pat.) (1.0–4.5 mg/mL), and *Clitocybe nebularis* (Batsch) P. Kumm. (1.0–5.0 mg/mL). These fungus exhibited vesicular stomatitis virus activity (Amoros et al. 1997).

23.3.4 Anti-Cancer Agents

The biggest challenge in the twenty-first century is without a doubt cancer as it is a worldwide health issue that affects everyone, irrespective of age, gender, race, wealth, or socioeconomic level, and that is endangering the research community and the medical system. Approximately ten million fatalities per year are attributed to cancer, making it the second biggest cause of death in the world (Prajapati et al. 2021). The main form of treatment for cancer is a combination of clinical techniques, such as chemotherapy or radiation and surgery (Subramaniam et al. 2019). The chemotherapeutic drugs used today to treat cancer give patients short-term relief and lengthen their lives, but they also have drawbacks like a lack of selectivity, side effects, and high prices that not only lower patients' life quality, but are also out of reach for billions of patients in developing countries (Reis-Mendes et al. 2018). The development of multidrug resistance and the recurrence of cancer are also by far the most serious issues in the management of this complex disorder with a complicated genesis and course (Cree and Charlton 2017). There is undoubtedly a need for innovative, promising, secure, and less hazardous treatments made from natural substances to combat this condition, which is getting worse.

Numerous phytochemicals are being used to treat this well-known disease as a result of research findings that they have anti-cancer properties (Bhadresha et al. 2022; Patel et al. 2018). Regardless of the fact that phytochemicals are the most substantial source of possible drugs, interest in using them in drug research has recently decreased due to some remarkable difficulties with the plant as a drug molecule origin, like its unusual ecosystems, slow growth rate, limited yield, and non-reproducibility of the preferred phytochemicals, as well as an unfaltering danger from civilization (Garcia-Oliveira et al. 2021). On the other hand, due to their prevalence, high biodiversity, and distinctive structure of their generated bioactive components, fungi have garnered a great deal of interest as a source of medicinal medicines. They regularly produce the desired chemicals and are frequently simple to develop in a fermenter under regulated conditions (Abdel-Razek et al. 2020).

One of the advantages of fungi is that their metabolites have been employed in medical uses; however, not all fungi can provide these advantages, and some fungal species can infect people and animals with diseases. The fungal secondary metabolites (SMs) with antibiotic and anti-cancer properties are regulated by biosynthetic gene clusters (BGCs) (Tran et al. 2019). In addition, numerous endophytic fungus secondary metabolites showed action against the tumour cell lines (Sharma et al. 2016). For instance, Paclitaxel, a diterpenoid derivative, was effective in treating ovarian and breast cancer. *Taxomyces andreanae*, a fungal endophyte, was the source of this anti-cancer substance (El-Sayed et al. 2020). Therefore, fungi play a crucial role in human health. Fungi can also create organic acids, enzymes, and other substances that may be useful in addition to fungal metabolites (Hyde et al. 2018).

Aspergillus, *Penicillium*, and *Talaromyces* species are filamentous fungus families that serve as factories for producing antibacterial, antifungal, and anticancer medications (Al-Fakih and Almaqtri 2019). Because of this, fungi are a valuable source for the discovery of natural compounds (Hyde et al. 2019). Contextually, irofulven, a semi-synthetic derivative of illudin S extracted from Omphalotus illudens, demonstrated anti-cancer activity in clinical trials against a number of cancers, including those of the brain, breast, colon, lungs, ovaries, pancreas, and sarcoma (Hyde et al. 2019). The anti-cancer effects of pure vinblastine from the mycelia of *Nigrospora sphaerica* on breast cancer cell lines has been studied (Ayob et al. 2017). *Penicillium* sp. natural extract produced two novel compounds, penicillatides A and B, as well as cyclo(d-Pro-l-Phe) cyclo(R-Pro-S-Phe) and cyclo(d-Pro-S-Phe), making members of the genus Penicillium one of the most studied fungi by herbal product chemists (R-Pro-R-Phe). The compounds' cytotoxic and antiproliferative effects on three human cancer cell lines, as well as their antimicrobial efficacy against a number of pathogens, were assessed. According to Diaa and Abdulrahman, compounds 2–4 exhibited varying cytotoxic and antibacterial properties (Youssef and Alahdal 2018).

L-asparaginase was isolated from *Aspergillus niger* using an enzyme that had impacts on various cancer cell lines (Vala et al. 2018). Children with acute clinical types of leukaemia are treated with this enzyme (Munir et al. 2019). Additionally, it was previously discovered that *Penicillium rubens,* which was removed from garden soil in the Madurai area of Tamil Nadu, produces a very promising anti-cancer metabolite. The proportion of HepG2, HeLa, and MCF-7 cancer cells that were treated with the bioactive fraction (P5) of *P. rubens* after 96 h varied between 40 and 50% (Venkatachalam and Nadumane 2021). When applied to endophytic fungi, the indole derivatives varioloid A and varioloid B had lethal effects on human cancer cell lines. *Paecilomyces variotii*, an endophytic fungus of marine algae, provided the compounds. In the same circumstance, marine algae-endophytic *Aspergillus ochraceus* species produces insulicolide A that has anti-cancer properties (Deshmukh et al. 2018). Medicinal mushrooms have important health benefits and exhibit a wide range of pharmacological activities, including antibacterial, antiallergic, antifungal, antioxidative, anti-inflammatory, cytotoxic, antiviral, anti-depressant, antihyperlipidemic hepatoprotective, immunomodulating, antidiabetic, and digestive, activities that are osteoprotective, nephroprotective, hypotensive, and neuroprotective (Venturella et al. 2021). Some anti-cancer compounds obtained from fungi are: Irofulven, Aphidicolin, Anti-Cancer Polyketide Derivative Products, Nitrogen-Containing Products, L-Asparaginase, and Derivatives of Anti-cancer Terpenoid Products (Alhasan and AL-abedi 2021).

23.3.4.1 Irofulven

Irofulven's chemical makeup is a derivative of illudin S, a semi-synthetic product that comes from the Omphalotus illudens toxin. Irofulven is a member of the acylfulvene family of anti-cancer agents. Most cancer cells undergo apoptosis, or "programmed cell death," as part of Irofulven's mechanism of action. Irofulven is undergoing phase 2 trials for advanced epithelial ovarian cancer that is hormone-refractory, recurring or refractory, recurrent malignant glioma, and incurable liver cancer. Aggregate chemotherapy is being evaluated in ongoing phase 1 research. Low white blood cell and platelet counts, vomiting, nausea, exhaustion, and visual impairment are some of the side effects of irofulven (Topka et al. 2018).

23.3.4.2 Aphidicolin

Aphidicolin was derived from *Cephalosporium aphidicola*; the fungal species is now known as *Akanthomyces muscarius*; nonetheless, *Nigrospora sphaerica* also provides this anti-cancer compound (Ayob et al. 2017). The substance is a type of tetracyclic diterpene that has the ability to compete with enzymes like DNA polymerase for binding sites. Clinical trials with aphidicolin produced no noteworthy findings (Ayob et al. 2017).

23.3.4.3 Anti-Cancer Polyketide Derivative Products

Among the most useful natural products are polyketides, which contain a variety of bioactive substances like antibiotics, anti-cancer agents, antifungal agents, and immunosuppressants (Musiol-Kroll and Wohlleben 2018). Numerous enzymes take part in the production of non-reduced or partially reduced polyketides (Klejnstrup et al. 2012). An illustration of the substance is lovastatin, whose IC50 values were reported to range from 2 to 39 μM and which showed apoptotic induction in certain ovarian cancer cell lines (Martirosyan et al. 2010). Simvastatin was found to suppress the apoptotic induction of lung, breast, and melanoma cancers by researchers, which led to optimistic results that led to the clinical use of this medication for the treatment of cancer (Relja et al. 2010).

23.3.4.4 Nitrogen-Containing Products

These compounds are nitrogen-containing natural products that fungi can generate. In general, the compounds are formulations that contain the amino acid basic

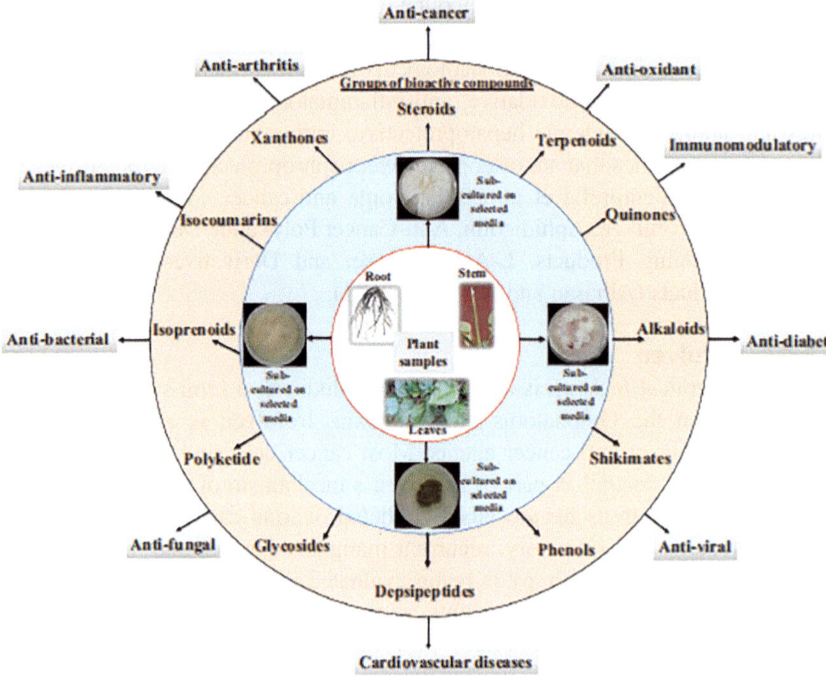

Fig. 23.1 Bioactive compounds derived from fungal endophytes with pharmacological relevance. Fungal endophytes harbour inside almost every plant tissue. Fungal endophytes can be isolated from stem, root, and leaves by culturing in the selected growth media. Fungal endophytes are potent source of number of bioactive compounds which can be utilized in the treatment of number of human health ailments (Rai et al. 2021)

components that are frequently combined into substances with sophisticated heteroaromatic structures, such as diketopiperazines, benzodiazepines, and quinazolines (Frisvad et al. 2004). The most frequently used endophytes for the isolation of numerous unique types of bioactive compounds are fungi. It is possible for fungi like *Taxomyces andreanae*, which was obtained from the yew plant, to create the anti-cancer compound paclitaxel. Other specialized fungal endophytic species produce a diverse range of anti-cancer substances, including torreyanic acid, vincristine, podophyllotoxin, camptothecin, and vinblastine (Ejaz et al. 2020) (Fig. 23.1).

23.3.4.5 Derivatives of Anti-Cancer Terpenoid Products

The biggest class of secondary metabolites, terpenes, are made up mostly of five-carbon isoprene devices that can be put together in countless ways to create various compounds. Terpenoids are a modified class of terpenes with separate functional groups and oxidized methyl groups changed or deleted at various places. Terpenes are simple hydrocarbons. Depending on the number of carbon atoms present, terpenoids are classified as monoterpenes, sesquiterpenes, diterpenes, sesterpenes, and triterpenes. The majority of terpenoids with different structural variations are physiologically active and are used all over the world to treat a variety of disorders. Several terpenoids inhibited certain human most cancer cells as anti-cancer medications, such as Taxol and its variants (Perveen 2021).

23.3.4.6 L-Asparaginase

L-asparaginase (ASNase), also known as L-asparagine amidohydrolase, was initially classified as a hydrolase enzyme. Through catalytic and hydrolyzing activities, this enzyme can transform asparagine into L-aspartic acid and ammonia (Paul and Tiwary 2020). L-asparaginase is therefore employed in cancer treatment clinical trials, where the therapeutic action of this enzyme depends on various degrees of metabolic pathways in both normal and malignant cells. *Aspergillus flavus, Aspergillus fumigatus, Alternaria tenuissima, Aspergillus oryzae, Fusarium oxysporum, Trichophyton rubrum, Rhizoctonia solani, Fusarium fujikuroi, Penicillium expansum, Pyrenophora triticirepentis*, and *Fusarium graminearum* are a few species of fungi that produce L-asparaginase (Paul and Tiwary 2020).

The potential of fungal metabolites that could move to clinical trials is Cotylenin A: It is helpful in retinoid-resistant leukaemia differentiation therapy. Myriocin: The S1P signalling pathway, which is involved in actin cytoskeleton rearrangements and increased cell proliferation (cell growth kinetics), is inhibited by myriocin and numerous of its analogues (cell migration kinetics). Palmarumycin: At least in preclinical studies, palmarumycin and its analogues show great promise in the treatment of several cancer types. Apicidin: HDAC inhibitor. Chaetocin: It exhibits anti-cancer action in a number of ways. Destruxin: It exhibits intriguing anti-cancer characteristics, such as regulation of the Wnt/beta-catenin pathway (Kornienko et al. 2015).

Even though there are a lot of anti-cancer hits in collections of metabolites isolated from fungi, there are many technological, biotechnological, and physiological problems that make it hard to study and sell fungal metabolites. Due to these limiting constraints, the "true" metabolic capacity of fungi, which is often much greater than that of bacteria and plants, is currently only partially realized. Since plants can be easily cultivated and harvested as a crop, and since bacteria can be produced more easily than fungi in bioreactors via liquid shaking fermentation, it is often easier to procure and supply materials with a botanic or bacterial origin than with a fungal one. Therefore, the quantity and expense of a metabolite produced by a fungus may actually act as a barrier to the development of clinical research. In fact, promising metabolites are frequently produced in lab settings by fungi in milligram quantities, usually only enough for the preliminary anti-cancer bioassays, or are available commercially as pricey biochemical reagents.

It is astonishing that relatively fewer of these compounds have reached clinical trials given the remarkable diversity of fungal species and the structures (metabolites) they create. This is especially true given the pressing need for novel cancer chemotherapeutics. Furthermore, the technical difficulties involved in the cultivation and biochemically guided isolation of metabolites, as well as estimates that only a small portion of fungi have been studied, suggest that this area of natural product research still has a lot of potential in the domain of anti-cancer research. Additionally, these creatures might come to be thought of as Nature's own "bio-combinatorial reactors" as a result of the discovery and use of novel methods to encourage the fungi to increase the structural diversity of the metabolites they generate. The biggest benefit of this strategy would be that each metabolite produced in this way would likewise be "privileged" because of its biological origin. In conclusion, despite the fact that there are currently no clinical anti-cancer agents based on fungal metabolites used in cancer patients' chemotherapy, it appears that it will be a long time before compounds from this category of natural products are incorporated into the pharmaceutical arsenal against cancer.

23.3.5 Antioxidant Agents

There has been a significant change in recent decades toward a more natural, eco-friendly, and sustainable way of life. According to a majority of scientists, antibiotic resistance is alarmingly rising and becoming a major global issue, rendering the present antimicrobial drugs useless (Ayukekbong et al. 2017; Vasan et al. 2019). Antibiotic-resistant bacteria are present in two million people worldwide, causing at least 23,000 fatalities per year (Prestinaci et al. 2015). The World Health Organization (WHO) claims that antimicrobial resistance (AMR) has become one of the most important public health issues of the twenty-first century (Manganyi and Ateba 2020). Furthermore, it is not surprising that a sizable fraction of people, particularly those who reside in developing countries, are using naturally accessible bioactive alternatives for their basic healthcare given the numerous drawbacks and side effects connected with current antimicrobial drugs.

More than 80% of the population in underdeveloped nations, particularly those in Africa, Asia, and Latin America, use medicinal plants to satisfy their basic healthcare needs and wellness (Manganyi and Ateba 2020). There are over 400,000 different plant species on the earth, and the bulk of them are useful for treating a wide range of diseases. This has spurred interest in research aimed at finding "perfect" bioactive agents that might be helpful to humans since they have a wide range of biological properties (Li and Weng 2017; Othman et al. 2019). However, inappropriate use and over-propagation could put the plants in danger of going extinct. Extensive research has demonstrated that endophytic fungi can colonize plant tissues, offer protection, and are a rich source of naturally occurring bioactive chemicals (Rodriguez and Redman 2008). Researchers can now easily discover, identify, and recognize primary and secondary metabolites with high antioxidant capacity, antibacterial, anti-cancer, and anti-inflammatory characteristics from affordable and readily accessible non-traditional sources such as fungi, thanks to recent advances in biotechnology, microbiology, and genetic research.

Free radicals may be created through the chemical process of oxidation, which involves the loss of electrons from an atom. Free radicals are unstable molecules that are created naturally during chemical processes like digestion. These free radicals may take part in a cascade of events that could possibly harm human cells (Elochukwu 2015). This is mostly caused by the imbalance that results in cell damage when an atom loses an electron. Therefore, cells exposed to oxidative stress may experience a variety of illnesses, including chronic difficulties in humans (Rahal et al. 2014). Numerous studies have shown that oxidative stress exposure to cells causes cellular degeneration, as well as cancer, atherosclerosis, coronary heart disease, diabetes, Alzheimer's disease, hepatic and kidney damage, as well as other neurological illnesses (Nasri and Rafieian-Kopaei 2013).

Reactive oxygen species (ROS) have been associated with a variety of diseases, and antioxidant compounds are used to prevent, treat, and combat these diseases. Antioxidant agents have also shown very good efficiency against ROS-related damage. It has been suggested that ROS strengthens the immune system by facilitating cell signalling. Antioxidant substances are used for good in many industries, including the food, pharmaceutical, and agricultural sectors. Despite the health risks linked to oxidative stress, there is a significant need for safer, more effective, and affordable natural antioxidants. According to reports, novel natural bioactive molecules protect cells from oxidative damage by avoiding or lowering reactive oxygen species and free radicals. Numerous investigations have shown that compounds with antioxidant activity include phenolic acids, phenylpropanoids, flavonoids, lignin, melanin, and tannins (Patipong et al. 2019; Smith et al. 2015).

Fungi have been kown to produce a wide range of beneficial antioxidant secondary metabolites, such as steroids, polyketides, terpenes, and phenolic compounds (Cui et al. 2005; Schuemann and Hertweck 2009). It is now widely recognized that biological characteristics like anticarcinogenicity and antiaging effects result from this capacity to inhibit cellular oxidation (Cook and Samman 1996). Due to the phenolic, terpenoid, and polysaccharide content of *Pleurotus* and other higher Basidiomycetes, these organisms have been recognized as providers of antioxidants (Valentão et al. 2005). Antioxidants are essential for the upkeep of cellular functions as well as for overall health and wellbeing. This kind of biological function is an essential aspect of life. Antioxidant supplementation is advised because the body's production of antioxidants is not always sufficient to prevent cellular harm. Fungi arc ideal ambassadors for powerful health-promoting organisms since they are a natural source of antioxidants and have a high level of biological activity. *Grifola frondosa, Aspergillus fumigates, Monascus purpureus, P. flavigenum* (CML2965), *Lentinula edodes, Pleurotus* sp., and *Trametes versicolor* are some of the fungi having reported antioxidant potential (Arora and Chandra 2011; Smith et al. 2015; Tavares et al. 2018).

Numerous mushrooms are said to contain antioxidant characteristics that allow them to combat free radicals. The antioxidant components of mushrooms, which include polysaccharides, tocopherols, phenolics, carotenoids, ergosterol, and ascorbic acid among others, are present in the fruit bodies, mycelium, and culture of the mushroom. An important benefit of extracting antioxidant compounds from mushrooms is the ability to alter fruit bodies or mycelium to create active chemicals in a short amount of time. To prevent oxidation related harm to thc human body, antioxidant substances can be isolated and utilized as functional additives, or mushrooms can be included in our diet as an alternative source of food. *Agaricus arvensis, Agaricus romagnesii, Agrocybe cylindracea, Agaricus silvicola, Boletus badius, Calocybe gambosa, Cantharellus clavatus, Clitocybe alexandri, Fistulina hepatica, Ganoderma applanatum, Hericium erinaceus, Hypholoma capnoides, Lactarius deliciosus, Lactarius volemus, Pleurotus eryngii, and Suillus luteus* are some of the mushrooms that produce antioxidant compounds (Sánchez 2017).

In conclusion, numerous compounds with promising antioxidant activity have been discovered. These molecules can be used in a variety of industries, including food, cosmeceuticals, nutraceuticals, and pharmaceuticals. The abundance of active metabolites already described can be further enhanced by additional research in the field of fungal antioxidants. Additionally, research aimed at creating new genetic and metabolic tools for their overproduction can significantly increase the market's supply of novel fungal antioxidants.

23.3.6 Antiparasitic Agents

Every year, parasites are known to cause deaths of millions of humans. The diseases caused by the parasites are the major cause of mortality in developing countries. According to a report of World Health Organization (WHO), malaria, leishmaniases, and Chagas disease are the major diseases of concern. Malaria causing protozoa named *Plasmodium falciparum* has reported for causing 1–3 million deaths and have 400 million cases annually (Mor 2009). On the other hand, leishmaniases is caused to 12 million people and untreated patients 100% killed by the pathogen known as *Leishmania* sp. Another fatal parasitic disease which harms the health of humans is trypanosomiasis which is caused by *Trypanosoma cruzi*. Trypanosomiasis has annual cases more than 18 million (Sartorelli et al. 2009). The treatment for the diseases caused by the parasites is urgent need as some have no appropriate treatment and some antiparasitic drugs have stronger side effects. Apart from the drug, vaccinations are also not available to control the diseases (Lenzi et al. 2018).

Fungi, one of the important living microbes, could play an important role in the controlling the pathogens. Various fungal species have been reported for having antiparasitic activity against different human parasites. In a report, the compound (14-norpseurotin A, pseurotin A, FD-838, pseurotin D, and fumoquinone B) isolated from *Aspergillus* sp. was reported for having antileishmanial activities (Martínez-Luis et al. 2012). A compound namely, C3 epimer of ganoderic acid T from *Ganoderma boninense,* was reported for having antiparasitic activity against *Plasmodium falciparum* (Isaka et al. 2013). In another report, *Penicillium* sp. was found to have antiparasitic activity against pathogens including *Plasmodium falciparum* and *Trypanosoma cruzi* (Higginbotham et al. 2014). Antiparasitic activity against leishmaniases causing agent, i.e. *Leishmania* donovani, was reported by fungal (*Trichosporum* sp.) compound known as (6-S)-3-(1,3-dihydroxypropyl)-6-(2-methylpropyl)piperazine-2,5-dione (Metwaly et al. 2015).

Similarly, the growth of another causal species (*Leishmania amazonensis*) causing leishmaniases was inhibited by the *Trametes versicolor* compound Trametenolic acid B (Leliebre-Lara et al. 2016). *Astraeus asiaticus* compound Trametenolic acid B was reported having antiparasitic activity against malaria causing parasite *Plasmodium falciparum* (Isaka et al. 2017). Fungi (*Mortierella parvispora, Pseudogymnoascus destructans*, and *Penicillium chrysogenum)* isolated from the soil of Antarctic were reported having antiparasitic activity (Gomes et al. 2018). The compound, (9Z, 11E)-13-oxooctadeca-9,11-dienoic acid of *Penicillium herquei* associated with the Ghanaian mangrove, was reported having antiparasitic activity against trichomoniasis causal agent *Trichomonas mobilensis* (Hayibor et al. 2019). The fungal (*Aspergillus fumigatus, A. phoenicis A. terreus, A. ochraceus, and Cunninghamella echinulata,* and *C. elegans*) obtained kaurane, labdane, and clerodane-type diterpenes were reported for having antiparasitic activity against schistosomiasis causing pathogen *Schistosoma mansoni* (Oliveira et al. 2020). The compound harzialactone A associated with marine-derived fungal species *Paecilomyces* sp. was found to have antileishmanial activity (Braun et al. 2021). French Guiana termites mutualistic fungus *Pseudallescheria boydii* was reported for having antiprotozoal activities against *Trypanosoma brucei* and *Plasmodium falciparum* (Sorres et al. 2022).

23.3.7 Acetylcholinesterase Inhibitors

Alzheimer's disease is a degenerative and progressive disease characterized by poor judgment, memory loss, deterioration of language, and impaired visuospatial skills. This disease is caused by the deficiency of neurotransmitter acetylcholine (ACh) in brain. Normally, the ACh has a very short half-life because of large amounts of acetylcholinesterase enzyme (AChE), the enzyme which is released from the nerve ending. This enzyme AChE hydrolyzes the ester bond in the molecule and results in the stimulatory activity loss. The reversible inhibition of this enzyme leads to an increase of neurotransmitter concentration in synaptic cleft which may positively affect Alzheimer's disease patients. The AChE inhibition is the most attractive approach to treat the Alzheimer disease cognitive symptoms (Singh et al. 2012). Many drugs, namely, rivastigmine, eserine, donepezil, tacrine, and galanthamine, are known to inhibit the AChE, but it has several clinical limitations such as short half-lives, low bioavailability, and unfavourable side effects like hepatotoxicity (Melzer 1998). Alternative approach is needed urgently to inhibit the enzyme acetylcholinesterase to treat Alzheimer disease. The fungi have known to produce acetylcholinesterase inhibitors which are an alternative approach to treat Alzheimer disease.

In a report, marine sponge-associated fungi belonging to genera *Acremonium* sp., *Aspergillus candidus*, and *Fusarium* sp. were reported for producing inhibitors of acetylcholinesterase (Thirunavukkarasu et al. 2012). In another investigation, a compound from fungus *Paecilomyces lilacinus* known as paecilomide was reported as a acetylcholinesterase inhibitor (Teles and Takahashi 2013). The compounds of endophytic fungus *Phomopsis* sp., i.e. cytochalasins J, 5′-epialtenuene, the mycotoxins alternariol monomethyl ether, alternariol and cytosporonc C, were found to have AChE inhibition activity (Chapla et al. 2014). Similarly, compounds such as AS-186c, talaromycesone A and isopentenyl xanthenone, talaroxanthenone isolated from fungus marine fungus *Talaromyces* sp. were reported for having AChE inhibitory activity (Wu et al. 2015). The endophytic fungi, namely, *Colletotrichum, Phyllosticta, Hypoxylon, Xylaria,* and *Nigrospora,* were reported for having AChE inhibitory activity (Wang et al. 2016). The AChE inhibitory activity was also reported from mangrove endophytic fungus *Aspergillus* sp. compounds isoaustinol, dehydroaustin, and dehydroaustinol (Long et al. 2017). In another investigation, endophytic fungal species *Colletotrichum* sp. was reported having AChE inhibitory active metabolites, i.e. 1-*O*-methylemodin, 5-methoxy-2-methyl-3-tricosyl-1,4-benzoquinone and (3β,5α,6α, 22*E*)-3-hydroxy-5,6-epoxy-7-one-8(14),22-dien-ergosta (Li et al. 2019). Cyclohexanoids from fungal endophyte *Saccharicola* sp. were reported for having AChE inhibition activity (Chapla et al. 2020). 19-hydroxy-19,20-dihydrophomacin C from *Westerdykella nigra*, the endophytic fungus was reported for having AChE inhibition activity (Sallam et al. 2021). In another investigation, *Cyathus africanus* was reported for producing thirteen different cyathane diterpenoids which were having AChE inhibitory activity, and among them, (12*S*)-11α,14α-epoxy-13α,14β,15-trihydroxycyath-3-ene have highest inhibitory activity (Yu et al. 2022).

23.4 Conclusions

Endophytic fungi are an untapped reservoir of unique metabolites with considerable pharmaceutical and therapeutical potential. Bioactive compounds from fungi have the potential to be an important source of drug formulation or novel drug discovery. These biologically active compounds play an important role in human health care, such as cancer, diabetes, microbe-related disease, oxidative stress, and inflammation. As new ecological groups of fungi and novel chemistries are discovered, fungal bioactive compounds are alive and well for development into new drugs.

Acknowledgement Authors thankfully acknowledge the infrastructural facilities of Eternal University, Baru Sahib during preparation of this chapter.

References

Abdel-Razek AS, El-Naggar ME, Allam A, Morsy OM, Othman SI (2020) Microbial natural products in drug discovery. Process 8:470

Adeyemi AI (2015) Isolation and screening of endophytic fungi from three plants used in traditional medicine in Nigeria for antimicrobial activity. Int J Green Pharm 9:58–62

Al-Fakih AA, Almaqtri WQA (2019) Overview on antibacterial metabolites from terrestrial *Aspergillus* spp. Mycology 10:191–209

Alhasan DA, AL-Abedi HF (2021) Anticancer compounds from fungi. Iraqi J Cancer Med Genet 14:26

Alvin A, Miller KI, Neilan BA (2014) Exploring the potential of endophytes from medicinal plants as sources of antimycobacterial compounds. Microbiol Res 169:483–495

Amoros M, Boustie J, Py M, Hervé V, Robin V (1997) Antiviral activity of Homobasidiomycetes: evaluation of 121 basidiomycetes extracts on four viruses. Int J Pharmacogn 35:255–260
Andersen PI, Ianevski A, Lysvand H, Vitkauskiene A, Oksenych V, Bjørås M, Telling K, Lutsar I, Dumpis U, Irie Y (2020) Discovery and development of safe-in-man broad-spectrum antiviral agents. Int J Infect Dis 93:268–276
Arora DS, Chandra P (2011) Antioxidant activity of *Aspergillus fumigatus*. Int Sch Res Notices 2011:1
Ayob FW, Simarani K, Zainal Abidin N, Mohamad J (2017) First report on a novel *Nigrospora sphaerica* isolated from *Catharanthus roseus* plant with anticarcinogenic properties. Microb Biotechnol 10:926–932
Ayukekbong JA, Ntemgwa M, Atabe AN (2017) The threat of antimicrobial resistance in developing countries: causes and control strategies. Antimicrob Resist Infect Control 6:1–8
Baryshev PB, Bogachev VV, Gashnikova NM (2014) HIV-1 genetic diversity in Russia: CRF63_02A1, a new HIV type 1 genetic variant spreading in Siberia. AIDS Res Hum Retrovir 30:592–597
Basit A, Shah ST, Ullah I, Ullah I, Mohamed HI (2021) Microbial bioactive compounds produced by endophytes (bacteria and fungi) and their uses in plant health. In: Plant growth-promoting microbes for sustainable biotic and abiotic stress management. Springer, Singapore, pp 285–318
Bhadresha KP, Jain NK, Rawal RM (2022) Assessing the protective effect of *Moringa oleifera* extract against bone metastasis: an in vitro simulated digestion approach. Nutr Cancer 74:1023–1036
Bhardwaj A, Agrawal P (2014) A review fungal endophytes: as a store house of bioactive compound. World J Pharm Pharm Sci 3:9
Braun GH, Ramos HP, Candido ACBB, Pedroso RCN, Siqueira KA, Soares MA, Dias GM, Magalhães LG, Ambrósio SR, Januário AH, Pietro RCLR (2021) Evaluation of antileishmanial activity of harzialactone a isolated from the marine-derived fungus *Paecilomyces* sp. Nat Prod Res 35:1644–1647. https://doi.org/10.1080/14786419.2019.1619725
Chang CL, Lin Y, Bartolome AP, Chen Y-C, Chiu S-C, Yang W-C (2013) Herbal therapies for type 2 diabetes mellitus: chemistry, biology, and potential application of selected plants and compounds. Evid Based Complement Alternat Med 2013:1
Chapla VM, Zeraik ML, Ximenes VF, Zanardi LM, Lopes MN, Cavalheiro AJ, Silva DHS, Young MCM, Fonseca LM, Bolzani VS, Araújo AR (2014) Bioactive secondary metabolites from *Phomopsis* sp., an endophytic fungus from Senna spectabilis. Molecules 19:6597–6608
Chapla VM, Honório AE, Gubiani JR, Vilela AFL, Young MCM, Cardoso CL, Pavan FR, Cicarelli RM, Michel Pinheiro Ferreira P, Bolzani VS, Araujo AR (2020) Acetylcholinesterase inhibition and antifungal activity of cyclohexanoids from the endophytic fungus *Saccharicola* sp. Phytochem Lett 39:116–123. https://doi.org/10.1016/j.phytol.2020.07.016
Chatterjee S, Ghosh R, Mandal NC (2019) Production of bioactive compounds with bactericidal and antioxidant potential by endophytic fungus Alternaria alternata AE1 isolated from *Azadirachta indica* a Juss. PLoS One 14:e0214744
Chatterjee S, Ghosh S, Mandal NC (2022) Potential of an endophytic fungus *Alternaria tenuissima* PE2 isolated from *Psidium guajava* L. for the production of bioactive compounds. S Afr J Bot 150:658–670. https://doi.org/10.1016/j.sajb.2022.08.016
Cook NC, Samman S (1996) Flavonoids—chemistry, metabolism, cardioprotective effects, and dietary sources. J Nutr Biochem 7:66–76
Cree IA, Charlton P (2017) Molecular chess? Hallmarks of anti-cancer drug resistance. BMC Cancer 17:1–8
Cui Y, Kim D-S, Park K-C (2005) Antioxidant effect of *Inonotus obliquus*. J Ethnopharmacol 96:79–85
Da X, Nishiyama Y, Tie D, Hein KZ, Yamamoto O, Morita E (2019) Antifungal activity and mechanism of action of Ou-gon (Scutellaria root extract) components against pathogenic fungi. Sci Rep 9:1–12
Dadgostar P (2019) Antimicrobial resistance: implications and costs. Infect Drug Resist 12:3903

Deshmukh SK, Prakash V, Ranjan N (2018) Marine fungi: a source of potential anticancer compounds. Front Microbiol 8:2536

Ejaz M, Javed S, Hamza M, Tabassum S, Abubakar M, Ullah I (2020) Fungal endophytes are effective alternatives and novel sources of anticancer drugs. Punjab Univ J Zool 35:13–24

Elochukwu C (2015) Generation and reaction of free radicals in the human body: a major cause of aging and chronic degenerative diseases. EC Nutr 1:132–136

El-Sayed AS, El-Sayed MT, Rady AM, Zein N, Enan G, Shindia A, El-Hefnawy S, Sitohy M, Sitohy B (2020) Exploiting the biosynthetic potency of taxol from fungal endophytes of conifers plants; genome mining and metabolic manipulation. Molecules 25:3000

Erfandoust R, Habibipour R, Soltani J (2020) Antifungal activity of endophytic fungi from Cupressaceae against human pathogenic *Aspergillus fumigatus* and *Aspergillus niger*. J Mycol Med 30:100987

Faccin L, Benati F, Rincão V, Mantovani M, Soares S, Gonzaga M, Nozawa C, Carvalho Linhares R (2007) Antiviral activity of aqueous and ethanol extracts and of an isolated polysaccharide from *Agaricus brasiliensis* against poliovirus type 1. Lett Appl Microbiol 45:24–28

Frisvad JC, Smedsgaard J, Larsen TO, Samson RA (2004) Mycotoxins, drugs and other extrolites produced by species in *Penicillium* subgenus *Penicillium*. Stud Mycol 49:201–241

Gao W, Sun Y, Chen S, Zhang J, Kang J, Wang Y, Wang H, Xia G, Liu Q, Kang Y (2013) Mushroom lectin enhanced immunogenicity of HBV DNA vaccine in C57BL/6 and HBsAg-transgenic mice. Vaccine 31:2273–2280

Garcia-Oliveira P, Otero P, Pereira AG, Chamorro F, Carpena M, Echave J, Fraga-Corral M, Simal-Gandara J, Prieto MA (2021) Status and challenges of plant-anticancer compounds in cancer treatment. Pharmaceuticals 14:157

Gashnikova N, Bogachev V, Baryshev P, IuV M, Savochkina E, Nla C (2012) Prevalence of mutations responsible for resistance to antiretroviral preparations among HIV-1 variants circulating in Novosibirsk region. Zh Mikrobiol Epidemiol Immunobiol 6:56–60

Gauchan DP, Kandel P, Tuladhar A, Acharya A, Kadel U, Baral A, Shahi AB, García-Gil MR (2020) Evaluation of antimicrobial, antioxidant and cytotoxic properties of bioactive compounds produced from endophytic fungi of Himalayan yew (*Taxus wallichiana*) in Nepal. F1000Res 9:1

Gomes ECQ, Godinho VM, Silva DAS, de Paula MTR, Vitoreli GA, Zani CL, Alves TMA, Junior PAS, Murta SMF, Barbosa EC, Oliveira JG, Oliveira FS, Carvalho CR, Ferreira MC, Rosa CA, Rosa LH (2018) Cultivable fungi present in Antarctic soils: taxonomy, phylogeny, diversity, and bioprospecting of antiparasitic and herbicidal metabolites. Extremophiles 22:381–393. https://doi.org/10.1007/s00792-018-1003-1

Gond SK, Mishra A, Sharma VK, Verma SK, Kumar J, Kharwar RN, Kumar A (2012) Diversity and antimicrobial activity of endophytic fungi isolated from Nyctanthes arbor-tristis, a well-known medicinal plant of India. Mycoscience 53:113–121

Guzman JD (2014) Natural cinnamic acids, synthetic derivatives and hybrids with antimicrobial activity. Molecules 19:19292–19349

Hayibor K, Kwain S, Osei E, Nartey AP, Tetevi GM, Owusu KB-A, Camas M, Camas AS, Kyeremeh K (2019) Ghanaian mangrove wetland endophytic fungus, *Penicillium herquei* strain BRS2A-AR produces (9Z, 11E)-13-oxooctadeca-9, 11-dienoic acid with activity against trichomonas mobilensis. Int J Biol Chem Sci 13:1918–1937

Higginbotham S, Wong WR, Linington RG, Spadafora C, Iturrado L, Arnold AE (2014) Sloth hair as a novel source of fungi with potent anti-parasitic, anti-cancer and anti-bacterial bioactivity. PLoS One 9:e84549

Ho SA, Slavin M, Roberts JA, Yong M (2021) Optimization of ganciclovir use in allogeneic hematopoietic cell transplant recipients–the role of therapeutic drug monitoring. Expert Rev Anti-Infect Ther 19:707–718

Hu X, Saravanakumar K, Jin T, Wang M-H (2019) Mycosynthesis, characterization, anticancer and antibacterial activity of silver nanoparticles from endophytic fungus *Talaromyces purpureogenus*. Int J Nanomedicine 14:3427

Hyde K, Chaiwan N, Norphanphoun C, Boonmee S, Camporesi E, Chethana K, Dayarathne M, De Silva N, Dissanayake A, Ekanayaka A (2018) Mycosphere notes 169–224. Mycosphere 9:271–430

Hyde KD, Xu J, Rapior S, Jeewon R, Lumyong S, Niego AGT, Abeywickrama PD, Aluthmuhandiram JV, Brahamanage RS, Brooks S (2019) The amazing potential of fungi: 50 ways we can exploit fungi industrially. Fungal Divers 97:1–136

Ingle A, Gade A, Pierrat S, Sonnichsen C, Rai M (2008) Mycosynthesis of silver nanoparticles using the fungus *Fusarium acuminatum* and its activity against some human pathogenic bacteria. Curr Nanosci 4:141–144

Isaka M, Chinthanom P, Kongthong S, Srichomthong K, Choeyklin R (2013) Lanostane triterpenes from cultures of the basidiomycete *Ganoderma orbiforme* BCC 22324. Phytochemistry 87:133–139. https://doi.org/10.1016/j.phytochem.2012.11.022

Isaka M, Palasarn S, Sommai S, Veeranondha S, Srichomthong K, Kongsaeree P, Prabpai S (2017) Lanostane triterpenoids from the edible mushroom *Astraeus asiaticus*. Tetrahedron 73:1561–1567. https://doi.org/10.1016/j.tet.2017.01.070

Janny S, Bert F, Dondero F, Nicolas Chanoine MH, Belghiti J, Mantz J, Paugam-Burtz C (2013) Fatal *Escherichia coli* skin and soft tissue infections in liver transplant recipients: report of three cases. Transpl Infect Dis 15:E49–E53

Johann S, Rosa LH, Rosa CA, Perez P, Cisalpino PS, Zani CL, Cota BB (2012) Antifungal activity of altenusin isolated from the endophytic fungus *Alternaria* sp. against the pathogenic fungus *Paracoccidioides brasiliensis*. Rev Iberoam Micol 29:205–209

Johns S, Cunningham A (2004) Vaccine prophylaxis of genital and neonatal herpes caused by herpes simplex virus. Sex Transm Infect 1:46–49

Jung HJ, Hwang IA, Sung WS, Kang H, Kang BS, Seu YB, Lee DG (2005) Fungicidal effect of resveratrol on human infectious fungi. Arch Pharm Res 28:557–560

Kac J, Pohleven F (2005) Screening of selected wood-damaging fungi for the HIV-1 reverse transcriptase inhibitors. Acta Pharma 55:69–79

Kapich A, Gvozdkova T, Kvacheva Z, Nikolaeva S, Shishkina L, Galkin S, Hatakka A, Konoplya E, Vereshako G, Hodosovskaya A (2004) Antioxidant, radioprotective and antiviral properties of mycelial extract of the fungus *Laetiporus sulphureus*. Adv Med Mycol 3:146

Kharwar RN, Mishra A, Gond SK, Stierle A, Stierle D (2011) Anticancer compounds derived from fungal endophytes: their importance and future challenges. Nat Prod Rep 28:1208–1228

Klejnstrup ML, Frandsen RJ, Holm DK, Nielsen MT, Mortensen UH, Larsen TO, Nielsen JB (2012) Genetics of polyketide metabolism in *Aspergillus nidulans*. Meta 2:100–133

Kornienko A, Evidente A, Vurro M, Mathieu V, Cimmino A, Evidente M, van Otterlo WA, Dasari R, Lefranc F, Kiss R (2015) Toward a cancer drug of fungal origin. Med Res Rev 35: 937–967

Kousar R, Naeem M, Jamaludin MI, Arshad A, Shamsuri AN, Ansari N, Akhtar S, Hazafa A, Uddin J, Khan A (2022) Exploring the anticancer activities of novel bioactive compounds derived from endophytic fungi: mechanisms of action, current challenges and future perspectives. Am J Cancer Res 12:2897

Kumar V, Prasher IB (2022) Antimicrobial potential of endophytic fungi isolated from Dillenia indica L. and identification of bioactive molecules produced by *Fomitopsis meliae* (Undrew.) Murril. Nat Prod Res 36:1–5. https://doi.org/10.1080/14786419.2022.2043855

Lam KS (2006) Discovery of novel metabolites from marine actinomycetes. Curr Opin Microbiol 9: 245–251

Langeveld WT, Veldhuizen EJ, Burt SA (2014) Synergy between essential oil components and antibiotics: a review. Crit Rev Microbiol 40:76–94

Leliebre-Lara V, Monzote Fidalgo L, Pferschy-Wenzig E-M, Kunert O, Nogueiras Lima C, Bauer R (2016) In vitro Antileishmanial activity of sterols from *Trametes versicolor* (Bres. Rivarden). Molecules 21:1045

Lenzi J, Costa TM, Alberton MD, Goulart JAG, Tavares LBB (2018) Medicinal fungi: a source of antiparasitic secondary metabolites. Appl Microbiol Biotechnol 102:5791–5810. https://doi.org/10.1007/s00253-018-9048-8
Li F-S, Weng J-K (2017) Demystifying traditional herbal medicine with modern approach. Nat Plants 3:1–7
Li Z, Ma N, Zhao P-J (2019) Acetylcholinesterase inhibitory active metabolites from the endophytic fungus *Colletotrichum* sp. YMF432. Nat Prod Res 33:1794–1797. https://doi.org/10.1080/14786419.2018.1434648
Linnakoski R, Reshamwala D, Veteli P, Cortina-Escribano M, Vanhanen H, Marjomäki V (2018) Antiviral agents from fungi: diversity, mechanisms and potential applications. Front Microbiol 9:2325
Liu Q, Kong W, Hu S, Kang Y, Zhang Y, Ng TB (2020) Effects of Oudemansiella radicata polysaccharide on postharvest quality of oyster mushroom (*Pleurotus ostreatus*) and its antifungal activity against *Penicillium digitatum*. Postharvest Biol Technol 166:111207
Long Y, Cui H, Liu X, Ze X, Wen S, She Z, Huang X (2017) Acetylcholinesterase inhibitory Meroterpenoid from a mangrove endophytic fungus *Aspergillus* sp. 16-5c. Molecules 22:727
Manganyi MC, Ateba CN (2020) Untapped potentials of endophytic fungi: a review of novel bioactive compounds with biological applications. Microorganisms 8:1934
Martínez-Luis S, Cherigo L, Arnold E, Spadafora C, Gerwick WH, Cubilla-Rios L (2012) Antiparasitic and anticancer constituents of the endophytic fungus *Aspergillus* sp. strain F1544. Nat Prod Commun 7:1934578X1200700207
Martirosyan A, Clendening JW, Goard CA, Penn LZ (2010) Lovastatin induces apoptosis of ovarian cancer cells and synergizes with doxorubicin: potential therapeutic relevance. BMC Cancer 10:1–13
Mason S, Devincenzo JP, Toovey S, Wu JZ, Whitley RJ (2018) Comparison of antiviral resistance across acute and chronic viral infections. Antivir Res 158:103–112
Melzer D (1998) New drug treatment for Alzheimer's disease: lessons for healthcare policy. BMJ 316:762–764
Metwaly AM, Ghoneim MM, Musa A (2015) Two new antileishmanial diketopiperazine alkaloids from the endophytic fungus *Trichosporum* sp. Derpharmachemica 7:322
Mor A (2009) Multifunctional host defense peptides: antiparasitic activities. FEBS J 276:6474–6482. https://doi.org/10.1111/j.1742-4658.2009.07358.x
Munir N, Zia MA, Sharif S, Tahir IM, Jahangeer M, Javed I, Riaz M, Sarwar MU, Akram M, Shah SMA (2019) L-Asparaginase potential in acrylamide mitigation from foodstuff: a mini-review. Prog Nutr 21:498–506
Musiol-Kroll EM, Wohlleben W (2018) Acyltransferases as tools for polyketide synthase engineering. Antibiotics 7:62
NA AA, Mothana R, Lesnau A, Pilgrim H, Lindequist U (2003) Antiviral activity of *Inonotus hispidus*. Fitoterapia 74:483–485
Nasri H, Rafieian-Kopaei M (2013) Oxidative stress and aging prevention. Int J Prev Med 1:1101–1102
Niedermeyer TH, Lindequist U, Mentel R, Gördes D, Schmidt E, Thurow K, Lalk M (2005) Antiviral terpenoid constituents of *Ganoderma pfeifferi*. J Nat Prod 68:1728–1731
Nisa H, Kamili AN, Nawchoo IA, Shafi S, Shameem N, Bandh SA (2015) Fungal endophytes as prolific source of phytochemicals and other bioactive natural products: a review. Microb Pathog 82:50–59
Nuankeaw K, Chaiyosang B, Suebrasri T, Kanokmedhakul S, Lumyong S, Boonlue S (2020) First report of secondary metabolites, Violaceol I and Violaceol II produced by endophytic fungus, *Trichoderma polyalthiae* and their antimicrobial activity. Mycoscience 61:16–21
Oliveira LC, Porto TS, Junior AHC, Santos MFC, Ramos HP, Braun GH, de Lima Paula LA, Bastos JK, Furtado NAJC, Parreira RLT, Veneziani RCS, Magalhães LG, Ambrósio SR (2020) Schistosomicidal activity of kaurane, labdane and clerodane-type diterpenes obtained by fungal transformation. Process Biochem 98:34–40. https://doi.org/10.1016/j.procbio.2020.07.020

Othman L, Sleiman A, Abdel-Massih RM (2019) Antimicrobial activity of polyphenols and alkaloids in middle eastern plants. Front Microbiol 10:911
Oxford AE, Raistrick H, Simonart P (1939) Studies in the biochemistry of micro-organisms: griseofulvin, C17H17O6Cl, a metabolic product of *Penicillium griseofulvum* Dierckx. Biochem J 33:240
Patel S, Waghela B, Shah K, Vaidya F, Mirza S, Patel S, Pathak C, Rawal R (2018) Silibinin, a natural blend in polytherapy formulation for targeting Cd44v6 expressing colon cancer stem cells. Sci Rep 8:1–13
Patipong T, Hibino T, Waditee-Sirisattha R, Kageyama H (2019) Induction of antioxidative activity and antioxidant molecules in the halotolerant cyanobacterium *Halothece* sp. PCC7418 by temperature shift. Nat Prod Commun 14:1934578X19865680
Paul V, Tiwary BN (2020) An investigation on the acrylamide mitigation potential of l-asparaginase from *Aspergillus terreus* BV-C strain. Biocatal Agric Biotechnol 27:101677
Perveen S (2021) Introductory chapter: terpenes and terpenoids. In: Terpenes and terpenoids-recent advances. IntechOpen, London
Piraino FF (2005) The development of the antiviral drug RC 28 from *Rozites caperata* (Pers.: Fr.) P. Karst Int J Med Mushrooms 7:356
Prajapati J, Goswami D, Rawal RM (2021) Endophytic fungi: a treasure trove of novel anticancer compounds. Curr Res Pharmacol Drug Discov 2:100050
Prestinaci F, Pezzotti P, Pantosti A (2015) Antimicrobial resistance: a global multifaceted phenomenon. Pathog Global Health 109:309–318
Rahal A, Kumar A, Singh V, Yadav B, Tiwari R, Chakraborty S, Dhama K (2014) Oxidative stress, prooxidants, and antioxidants: the interplay. Biomed Res Int 2014:1
Rai N, Kumari Keshri P, Verma A, Kamble SC, Mishra P, Barik S, Kumar Singh S, Gautam V (2021) Plant associated fungal endophytes as a source of natural bioactive compounds. Mycology 12:139–159. https://doi.org/10.1080/21501203.2020.1870579
Rana KL, Kour D, Sheikh I, Dhiman A, Yadav N, Yadav AN, Rastegari AA, Singh K, Saxena AK (2019a) Endophytic fungi: biodiversity, ecological significance, and potential industrial applications. In: Yadav AN, Mishra S, Singh S, Gupta A (eds) Recent advancement in white biotechnology through fungi: volume 1: diversity and enzymes perspectives. Springer International Publishing, Cham, pp 1–62. https://doi.org/10.1007/978-3-030-10480-1_1
Rana KL, Kour D, Sheikh I, Yadav N, Yadav AN, Kumar V, Singh BP, Dhaliwal HS, Saxena AK (2019b) Biodiversity of endophytic fungi from diverse niches and their biotechnological applications. Adv Endophyt Fung Res 1:105–144
Reis-Mendes A, Alves M, Carvalho F, Remião F, Bastos ML, Costa VM (2018) Pixantrone, a new anticancer drug with the same old cardiac problems? An in vitro study with differentiated and non-differentiated H9c2 cells. Interdiscip Toxicol 11:13
Relja B, Meder F, Wilhelm K, Henrich D, Marzi I, Lehnert M (2010) Simvastatin inhibits cell growth and induces apoptosis and G0/G1 cell cycle arrest in hepatic cancer cells. Int J Mol Med 26:735–741
Rodriguez R, Redman R (2008) More than 400 million years of evolution and some plants still can't make it on their own: plant stress tolerance via fungal symbiosis. J Exp Bot 59:1109–1114
Saboulard D, Gaspar A, Roussel B, Villard J (1998) New antiherpetic nucleoside from a basidiomycete. Comptes Rendus de l'Académie des Sciences-Series III-Sciences de la Vie 321:585–591
Sallam A, Sabry MA, Galala AA (2021) Westalsan: a new acetylcholine esterase inhibitor from the endophytic fungus *Westerdykella nigra*. Chem Biodivers 18:e2000957. https://doi.org/10.1002/cbdv.202000957
Sánchez C (2017) Reactive oxygen species and antioxidant properties from mushrooms. Synth Syst Biotechnol 2:13–22
Santos SN, Ferraris FK, de Souza AO, Henriques MG, Melo IS (2012) Endophytic fungi from *Combretum leprosum* with potential anticancer and antifungal activity. Symbiosis 58:109–117

Sarkar S, Koga J, Whitley R, Chatterjee S (1993) Antiviral effect of the extract of culture medium of *Lentinus edodes* mycelia on the replication of herpes simplex virus type 1. Antivir Res 20:293–303

Sartorelli P, Carvalho CS, Reimão JQ, Ferreira MJP, Tempone AG (2009) Antiparasitic activity of biochanin a, an isolated isoflavone from fruits of *Cassia fistula* (Leguminosae). Parasitol Res 104:311–314. https://doi.org/10.1007/s00436-008-1193-z

Schuemann J, Hertweck C (2009) Biosynthesis of fungal polyketides. Physiol Genet 1:331–351

Sette LD, Passarini MRZ, Delarmelina C, Salati F, Duarte M (2006) Molecular characterization and antimicrobial activity of endophytic fungi from coffee plants. World J Microbiol Biotechnol 22: 1185–1195

Sharma D, Pramanik A, Agrawal PK (2016) Evaluation of bioactive secondary metabolites from endophytic fungus *Pestalotiopsis neglecta* BAB-5510 isolated from leaves of Cupressus torulosa D. Don. 3 Biotech 6:1–14

Shibnev V, Mishin D, Garaev T, Finigenova N, Botikov A, Deryabin P (2011) Antiviral activity of extracts of the fungus *Inonotus obliquus* against infection by the hepatitis C virus in cell cultures. Byulleten'Eksperimental'noy biologii i meditsiny 151:549–551

Sibero MT, Sabdaningsih A, Cristianawati O, Nuryadi H, Radjasa OK, Sabdono A, Trianto A (2017) Isolation, identification and screening antibacterial activity from marine sponge-associated fungi against multidrug-resistant (MDR) *Escherichia coli*. In: IOP conference series: earth and environmental science, vol 1. IOP Publishing, Bristol, p 012028

Singh B, Thakur A, Kaur S, Chadha BS, Kaur A (2012) Acetylcholinesterase inhibitory potential and insecticidal activity of an endophytic *Alternaria* sp. from *Ricinus communis*. Appl Biochem Biotechnol 168:991–1002. https://doi.org/10.1007/s12010-012-9835-0

Smith H, Doyle S, Murphy R (2015) Filamentous fungi as a source of natural antioxidants. Food Chem 185:389–397

Sorres J, Hebra T, Elie N, Leman-Loubière C, Grayfer T, Grellier P, Touboul D, Stien D, Eparvier V (2022) Antiparasitic Ovalicin derivatives from *Pseudallescheria boydii*, a mutualistic fungus of French Guiana termites. Molecules 27:1182

Stierle A, Strobel G, Stierle D (1993) Taxol and taxane production by *Taxomyces andreanae*, an endophytic fungus of Pacific yew. Science 260:214–216

Strobel GA (2003) Endophytes as sources of bioactive products. Microb Infect 5:535–544

Subramaniam S, Selvaduray KR, Radhakrishnan AK (2019) Bioactive compounds: natural defense against cancer? Biomol Ther 9:758

Tavares DG, Barbosa BVL, Ferreira RL, Duarte WF, Cardoso PG (2018) Antioxidant activity and phenolic compounds of the extract from pigment-producing fungi isolated from Brazilian caves. Biocatal Agric Biotechnol 16:148–154

Teles APC, Takahashi JA (2013) Paecilomide, a new acetylcholinesterase inhibitor from *Paecilomyces lilacinus*. Microbiol Res 168:204–210. https://doi.org/10.1016/j.micres.2012.11.007

Teplyakova T, Kosogova T (2015) Fungal bioactive compounds with antiviral effect. J Pharm Pharmacol 3:357

Teplyakova T, Puchkova L, Kosogova T, Bulychev L, Shishkina L, Mazurkova N, Gashnikova N, Balakhnin S, Kabanov A, Kazachinskaya E (2013) A melanin-based antiviral drug. RU Patent 1:2480227

Thirunavukkarasu N, Suryanarayanan TS, Girivasan KP, Venkatachalam A, Geetha V, Ravishankar JP, Doble M (2012) Fungal symbionts of marine sponges from Rameswaram, Southern India: species composition and bioactive metabolites. Fungal Divers 55:37–46. https://doi.org/10.1007/s13225-011-0137-6

Topka S, Khalil S, Stanchina ED, Vijai J, Offit K (2018) Preclinical evaluation of enhanced irofulven antitumor activity in an ERCC3 mutant background by in vitro and in vivo tumor models. Cancer Res 78:3258–3258

Tran PN, Yen M-R, Chiang C-Y, Lin H-C, Chen P-Y (2019) Detecting and prioritizing biosynthetic gene clusters for bioactive compounds in bacteria and fungi. Appl Microbiol Biotechnol 103: 3277–3287

Uzma F, Mohan CD, Siddaiah CN, Chowdappa S (2019) Endophytic fungi: promising source of novel bioactive compounds. In: Singh BP (ed) Advances in endophytic fungal research: present status and future challenges. Springer International Publishing, Cham, pp 243–265. https://doi.org/10.1007/978-3-030-03589-1_12

Vala AK, Sachaniya B, Dudhagara D, Panseriya HZ, Gosai H, Rawal R, Dave BP (2018) Characterization of L-asparaginase from marine-derived aspergillus Niger AKV-MKBU, its antiproliferative activity and bench scale production using industrial waste. Int J Biol Macromol 108:41–46

Valentão P, Andrade PB, Rangel J, Ribeiro B, Silva BM, Baptista P, Seabra RM (2005) Effect of the conservation procedure on the contents of phenolic compounds and organic acids in chanterelle (*Cantharellus cibarius*) mushroom. J Agric Food Chem 53:4925–4931

Vasan N, Baselga J, Hyman DM (2019) A view on drug resistance in cancer. Nature 575:299–309

Venkatachalam P, Nadumane VK (2021) Modulation of Bax and Bcl-2 genes by secondary metabolites produced by *Penicillium rubens* JGIPR9 causes the apoptosis of cancer cell lines. Mycology 12:69–81

Venturella G, Ferraro V, Cirlincione F, Gargano ML (2021) Medicinal mushrooms: bioactive compounds, use, and clinical trials. Int J Mol Sci 22:634

Verma A, Gupta P, Rai N, Tiwari RK, Kumar A, Salvi P, Kamble SC, Singh SK, Gautam V (2022) Assessment of biological activities of fungal endophytes derived bioactive compounds isolated from *Amoora rohituka*. J Fungi 8:285

Wang H, Ng TB (2001) Isolation and characterization of velutin, a novel low-molecular-weight ribosome-inactivating protein from winter mushroom (*Flammulina velutipes*) fruiting bodies. Life Sci 68:2151–2158

Wang Y, Lai Z, Li X-X, Yan R-M, Zhang Z-B, Yang H-L, Zhu D (2016) Isolation, diversity and acetylcholinesterase inhibitory activity of the culturable endophytic fungi harboured in *Huperzia serrata* from Jinggang Mountain. China. World J Microbiol Biotechnol 32:20. https://doi.org/10.1007/s11274-015-1966-3

Webber J (1981) A natural biological control of Dutch elm disease. Nature 292:449–451

Weiss H (2004) Epidemiology of herpes simplex virus type 2 infection in the developing world. Herpes J IHMF 11:24A–35A

Wu B, Ohlendorf B, Oesker V, Wiese J, Malien S, Schmaljohann R, Imhoff JF (2015) Acetylcholinesterase inhibitors from a marine fungus *Talaromyces* sp. strain LF458. Mar Biotechnol 17:110–119. https://doi.org/10.1007/s10126-014-9599-3

Yang M, Wang X, Guan S, Xia J, Sun J, Guo H, Guo D-a (2007) Analysis of triterpenoids in Ganoderma lucidum using liquid chromatography coupled with electrospray ionization mass spectrometry. J Am Soc Mass Spectrom 18:927–939

Youssef DT, Alahdal AM (2018) Cytotoxic and antimicrobial compounds from the marine-derived fungus, *Penicillium* species. Molecules 23:394

Yu M, Kang X, Li Q, Liang Y, Zhang M, Gong Y, Chen C, Zhu H, Zhang Y (2022) Thirteen cyathane diterpenoids with acetylcholinesterase inhibitory effects from the fungus *Cyathus africanus*. Phytochemistry 193:112982. https://doi.org/10.1016/j.phytochem.2021.112982

Zainee N, Ibrahim N, Hidayah N, Rozaimi M (2021) Variation in antibacterial properties of endophytic fungi isolated from Phaeophytes and Rhodophytes of Johor, Malaysia. J Environ Biol 42:840–848

Zhang G, Sun S, Zhu T, Lin Z, Gu J, Li D, Gu Q (2011) Antiviral isoindolone derivatives from an endophytic fungus *Emericella* sp. associated with *Aegiceras corniculatum*. Phytochemistry 72:1436–1442

Reminiscing Phages in the Era of Superbugs 24

Parakriti Gupta, Lipika Singhal, and Varsha Gupta

Abstract

Bacteriophages are among the chief alternative therapies for use against multidrug-resistant (MDR) organisms. Bacteriophages were discovered in 1915 and noted to have clinical role in 1919; however, the simultaneous discovery of antibiotics led to the beginning of golden era of antibiotics, halting the phage approach. Subsequently, indiscriminate over-the-counter use of antibiotics led to emergence of MDR, drying up of antibiotic pipeline, and lack of armamentarium against superbugs, reviving the phage therapy. Phages being highly specific, with better penetration, less propensity to alter the gut microbiome, and lacking adverse effects of hypersensitivity and resistance, have become the hotspot of research for clinical and biomedical use. Phages can be either be administered as a cocktail of polyphages or personalized monophage or sequential therapy. Recently, phage-derived enzymes are being administered, with an advantage of mass production using recombinant technology and direct target action, eliminating the opportunity to acquire resistance. Phage-antibiotic synergy approach is being adopted more recently, and the same has led to the escalated use of compassionate phage therapy in patients with pan-drug-resistant infections. Despite plethora of advantages, phages also suffer from few disadvantages in being specific that they cannot be used under 'one fits all' approach. Moreover, resistance has also been reported for phages; however, this acquisition of resistance by phages is known to have inverse effect on the fitness of bacterial host. Phage therapy needs to be regulated stringently and information of phages ready for clinical use should be made available online, to abridge the laboratory-bedside gap.

P. Gupta · L. Singhal · V. Gupta (✉)
Department of Microbiology, Government Medical College Hospital, Chandigarh, India

R. Sobti et al. (eds.), *Role of Microbes in Sustainable Development*,
https://doi.org/10.1007/978-981-99-3126-2_24

Keywords

Bacteriophages · Phage therapy · Antimicrobial resistance · PageXchange

24.1 Introduction and History of Bacteriophages

Bacteriophages are most abundant ubiquitous organisms that can infect and replicate within the bacterial host. It has been estimated that ~10^{31}–10^{32} bacteriophages are present on Earth at any given point of time and bacteriophages in oceans outnumber stars in the universe (Chanishvili 2012). Bacteriophages have always been the fundamental players involved in growth and transmission of bacteria, killing ~20–40% of marine bacteria every 24 h (Suttle 2007). The annotation of phages dates back to ancient biblical times, though, the first formal evidence was reported in 1915 and it was Fredrick Twort who described the lytic zone associated with infection. However, source of infection was identified by Felix d'Herelle, who further coined the term 'bacteriophage' that means 'bacterium eater' (Wittebole et al. 2014). Subsequently, he conceived the novel idea of therapeutic usage of phages in clinical practice, and in 1919, the first clinical use of bacteriophages was documented for pediatric patients suffering from bacillary dysentery in Paris (Chanishvili 2012; Sulakvelidze et al. 2001). However, the initial experiments were poorly controlled and confronted many fiascoes. The reports of earlier failure could be attributed to lesser level of understanding about biological nature of phages, crude purification methods, and poor delivery systems. Phage therapy was welcomed by few and criticized by many, but despite the mixed responses, pioneers of phage therapy continued the research and irrevocably, commercialization of phages started in Europe and US in late 90 s (Sulakvelidze et al. 2001). One of the most remarkable landmarks in the history of phage therapy was observed in 1923, when Eliava Institute was established in Tbilisi by Giorgi Eliava that marked the phase of extensive research on phages in the Soviet Union during that era. The Institute is still considered one of the pioneer institutes for phage research. The successful usage of phages was reported in 1931 trial where cohort of 118 control subjects and 73 experimental subjects were given phage treatment, and 90% reduction in mortality was noted (Chanishvili 2012). Another good example of early phage treatment success was in 1946 where fifty-six patients with typhoid fever were treated with "type-specific phage" that is, phage's grown on the patients' own isolates. All received phage intravenously once. Thereafter, all but 3 had negative blood cultures within 24 h, became afebrile, and normalized clinically (Knouf et al. 1946).

Nevertheless, simultaneous research and discovery of antibiotics was underway and it was with the rise of antibiotic era in the 1940s and discovery of penicillin during the Second World War that phage therapy met a halt. However, former Soviet Union and Eastern Europe continued the research and first phage genome was sequenced in 1977, following which resurgence of phage research and animal studies was reinitiated (Summers 2012).

Today when less than a century has elapsed since the golden era of antibiotics began, footsteps of these pioneer institutes are being trailed by other countries to combat the impending danger of multidrug resistance (MDR) and superbugs (Thiel 2004; Miêdzybrodzki et al. 2012). The constraint of newer antibiotics in the dry pipeline further augmented the phage research and the first commercial phage-based biocontrol product was approved for use against *L.monocytogenes* in 2006, which was soon followed by controlled clinical trials. The clinical trials have been completed for MDR-*Pseudomonas aeruginosa* chronic otitis, *A.baumannii* septicemia, and wound infections (Wright et al. 2009; Schooley et al. 2017; Jault et al. 2019). Presently, varied studies targeting dysentery, cholera, sepsis, meningitis, typhoid, diabetic foot ulcer, ileo-cecitis, and chronic otitis have been conducted (Lin et al. 2017) and many more are underway.

24.2 Structure of Bacteriophage

Bacteriophages are diverse creatures that contain either DNA or RNA genome, encapsulated by a proteinaceous capsid. Capsid imparts either icosahedral, filamentous, or head-tail shape to the phage. Phages are considered nonliving owing to their inability to reproduce independently without bacterial host. Bacteriophage consists of a polyhedral head with collar and tail. Head of bacteriophage is composed of ~2000 capsomeres enclosing the genetic material, which is followed below by a short collar and helical tail. The tail consists of inner hollow tube surrounded by a contractile sheath with 24 annular rings. There is a basal plate with tail fibers at every corner, at the distal end of tail, which help in attachment to the bacterial cell. However, all phages don't bind to all the bacteria and are rather highly specific and infect only those host cells that have complementary receptors (Gordillo Altamirano and Barr 2019).

24.2.1 Development Cycle

Bacteriophages are classified into two types on the basis of their development cycle as virulent or temperate. The virulent phages are those that have lytic developmental cycle and temperate ones follow the lysogenic phase. Lytic phase involves destruction of the cell by bacteriophage, whereas lysogenic phase involves insertion and integration of phage into bacterial genetic material, thereby inhibiting future infection of that bacterial cell to identical phages. These lytic phages adsorb to specific receptors on bacterial host, located on either cell wall or capsule or appendages. Subsequent to adherence, bacterial host receptors and phages undergo a lock-and-key conformation, and as a consequence, phage ejects out its genetic material into the host. Thereafter, phage takes possession of the bacterial replication repertoire and exploits it for formation of phage progeny. The same is continued till phage-encoded proteins are synthesized to effectively lyse the cell. On the contrary, lysogenic or temperate phages are not used in phage therapy owing to their risk of conversion by which bacterial host can acquire newer pathogenic traits. Moreover, lysogenic phages are prone to development of rapid homo-immunity (Gordillo Altamirano and Barr 2019).

24.3 Resurgence of Phage Therapy

The research on phages met with a roadblock after the golden era of antibiotics began in the 1940s. This golden era continued for ~four decades before the pipeline for antibiotics started drying out. Meanwhile, antimicrobial resistance was gradually increasing, and finally, by the dawn of antibiotic era, resistance had outperformed and was accepted as one of the major threats of public health concerns. As per one of the recent reviews, it has been estimated that antimicrobial resistance might toll ~ten million lives by year 2050, costing ~100 trillion US$ (World Health Organization 2015). Thereby, in order to combat the rising antimicrobial resistance, newer antimicrobials and other novel approaches are the need of the hour. Bacteriophage therapy is one such novel approach which is presently undergoing the revival owing to better understanding of biology, immunology, and genetics of phages.

24.4 Phages as an Alternative to Antibiotics

Phages exhibit numerous unique characteristics that can be exploited for their use in combating the emerging antimicrobial resistance. Phages are nonliving organisms in contrast to the chemical nature of antibiotics. Phages are known to be specific and interact only with those bacterial hosts that have complementary receptors, thereby narrowing their spectrum of action. This targeted approach spares the commensal microbiome from collateral damage and selection of antimicrobial variants. Recently, it has been identified that phages can 'jump' from one host to another, which is facilitated by gut microbiota (De Sordi et al. 2017). Unlike antibiotics that can be bactericidal or static, phages impart their action via 'kill-the-winner' immunophage synergy and target to reduce the host abundance prior to the attainment of equilibrium (Maslov and Sneppen 2017). Another substantial advantage of phages over antibiotics is the absence of hypersensitivity reactions to phage therapy. The dose of phages doesn't need to be increased in severe infections as these have the potential to reproduce exponentially and can penetrate the biofilms well. Furthermore, by virtue of genetic diversity and ubiquitous nature of phages, there can never be a dearth of phages and their delivery can be tailored as per the need via modification of delivery systems. Phages were endorsed to be qualified as medicinal product for use in humans in June, 2015 at EMA workshop (Debarbieux et al. 2015). Phages have an added advantage over antibiotics in terms of better penetration and attainment of therapeutic dosage. Certain antibiotics are unable to reach therapeutic levels at varied organ systems like mycobacterial infection in lungs, daptomycin in respiratory tract infections, vancomycin in osteomyelitis, and cefixime in meningitis. The same can be overcome by phages as they exhibit their action by a different mechanism. Moreover, phages can circumvent the therapeutic failure due to mutation or antibiotic resistance (Strydom et al. 2019; Romero-Calle et al. 2019). Lytic phages are the main target of interest for phage therapy; however, a few studies have used lysogenic phages wherein lytic phages to bacteria like *Clostridium difficile* haven't been isolated and in conditions when there is time constraint in emergency situations (Hargreaves and Clokie 2014; Philipson et al. 2018).

24.5 Approaches to Phage Therapy

With the revival of phage therapy, two pronounced approaches have been delineated (Pirnay et al. 2011). The conventional commonly used approach is the "prêt-à-porter" that is based on the principle of 'one-size-fits-all' or polyphage therapy. In this approach, broad-spectrum cocktail of phages is administered to the patient for coverage of both gram-negative and gram-positive bacteria. This approach was initially difficult to prepare, standardize, and purify for broad range as compared to narrower range. Cocktail phage therapy is now available commercially in Georgia and Russia (Kutter et al. 2010). On the contrary, the second approach is "sur-mesure" which is a personalized one or monophage therapy (Brives and Pourraz 2020). This includes selection of specific phages from the phage bank that infect the target bacteria most efficiently. The personalized phage therapy is more sustainable as no selection pressure is there. A recent approach combines usage of both monophage and polyphage therapy in a sequential manner. The sequential therapy is not only feasible, but also helps combat the emergence of resistance to phages. Though there are very few Phase II clinical trials unraveling the efficacy of phage therapy, but none of the studies conducted till date have reported any major side effects due to phage therapy. On the contrary, the results of PhagoBurn trial turned out to be less promising than contemplated, which was later attributed to reduction in titer of phages with time, due to improper manufacturing, administration, and storage, which was further ascribed to poor understanding pertaining the same (Jault et al. 2019; Alsaadi et al. 2021). Many phage therapies have been approved by food and drug administration for use in food industry against *Listeria monocytogenes*, *Pseudomonas syringae,* MRSA, *Salmonella* spp., and *E. coli* O157: H7 (Monk et al. 2010; Voelker 2019).

A newer approach using phage-derived enzymes, rather than using the whole phage, has also been defined. These enzymes have functional homology to lysozyme and are responsible for hydrolyzing cell wall during release of phage progeny. Two major enzymes being targeted for the same include holins and lysins. Of these, holins act as the molecular clock during lytic cycle and lysins are involved in bacterial cell lysis and are, thereby, preferred more. Combination of lysins and antibiotics has demonstrated significant reduction in in vitro and ex vivo *C.difficle* colon model (Lin et al. 2017). These enzymes can be produced in large numbers using recombinant techniques and are easy to be commercialized. Holins, lysins, and artelysins have shown good results in elimination or decolonization of *P.aeruginosa, A.baumannii,* and *S.aureus* (Lin et al. 2017). Another fascinating characteristic of these phage-derived enzymes is their direct action on cell wall, thereby eliminating the chances of development of resistance.

Phage-antibiotic synergy is the more acceptable modality of treatment, wherein both antibiotic and phages are administered to the patient with the intention to have better results as 'two is better than one' with different mechanisms of action (Gordillo Altamirano and Barr 2019). Despite lack of completed clinical trials, phages have been used for treatment of MDR cases, under the 'Helinski Declaration of Ethical Principles for Medical Research Involving Human Subjects' and this therapy is referred as 'compassionate phage therapy', in which drugs outside the clinical trial are used for patients with no other available therapeutic options. Approximately 26 reports of compassionate therapy with phages have been reported in patients with failed antibiotic therapy (McCallin et al. 2019; Schmidt 2019; Kuipers et al. 2019; BBC n.d.; CIDRAP 2022; Elpais 2022a, b). Poly-phage or cocktail phages have always remained the preferred choice for compassionate therapy, until few days back when mono-phage therapy targeting *Mycobacterium chelonae* was successfully used in a 56-yer old male with chronic kidney disease to relieve the present symptoms of skin and soft tissue infection, in Boston (Elpais 2022b). Lately, Food and Drug Administration (FDA) has approved the compassionate use of phage therapy in patients with secondary infections and in patients with coronavirus disease (FDA n.d.). The success of this compassionate therapy has further escalated the role of personalized usage of phages. The most imperative factor for compassionate and personalized use of phage therapy is presence of stringent regulatory framework in place for every country.

24.6 Disadvantages of Phage Therapy

The specific nature of phages in itself is one of prime shortcomings, despite being chief advantage as well. Phages, being highly specific, cannot be used in clinical practice till the causative pathogen is identified, whereas phages are known to be of more use if prescribed within 6 h of infection (Gordillo Altamirano and Barr 2019). Monophage therapy will have a limited role in polymicrobial infections like burns. Since phages cannot survive on their own, once the host is killed, phages also cease to unveil their action. Few animal studies have also shown that phages have the propensity to translocate across the intestinal epithelium, and subsequently, enter the circulation (Lin et al. 2017). Another speculated shortcoming of phage therapy is the 'leaky gut', as has been shown in animal models that led to a dysfunctional state of intestinal barrier. Furthermore, despite the widespread use of lytic phages, they are also known to harbor ~50% of genes that code for cryptic function or have a role in less understood physiology of bacterial host. Moreover, in case of abortive infection, the bacterial host continues to harbor the genes of unknown function (Philipson et al. 2018). Besides these shortcomings, the major concern with phage therapy is resistance. Phages are prone to resistance like the antibiotics (Tetz et al. 2016).

24.7 Resistance to Phages

Phages and bacteria coevolve by mechanism termed 'antagonistic coevolution', in which both bacteria and the phage are involved in host-parasite relationship. Bacteria impose selection of more effective phages, which in turn selects out the resistant bacterial hosts. However, this resistance can be easily overcome by phages (Buckling et al. 2002). The most common mechanisms known to impart resistance to phages include receptor modification used for adsorption by phages, superinfection exclusion systems that prevent viral entry, and CRISP-Cas systems that identify the foreign encountered DNA and degrade them. Recently, restriction-modification systems have also been noted in phages that target only phage DNA for action of restriction enzymes, sparing the host DNA (Labrie et al. 2010).

In contrast to the fear of resistance to phages, a recent study has deciphered an appealing observation that acquisition of phage resistance is inversely proportional to fitness of the bacterial host. It has been noted that there was ~36% reduction in bacterial growth when the same was cocultivated with lytic phages (Gomez et al. 2011). Similar results were obtained in another study with *S.aureus* and *Salmonella enterica* and impaired capsular production was noted in these bacteria (Capparelli et al. 2010). The clinical translation of the same was demonstrated with *A.baumannii* that reverted the sensitivity to minocycline, along with decrease in the bacterial growth (Schooley et al. 2017). Another example was witnessed with a clinical strain of *P.aeruginosa* obtained from a graft that was resistant to ceftazidime and in vitro results of combination therapy of ceftazidime with phage OMKO1 (that directs evolution of *P.aeruginosa*) showed significant reduction in bacterial count in graft (Chan et al. 2018). Phage resistance has been noted to be species-specific, non-phylogenetically conserved evolutionary trait (Gordillo Altamirano and Barr 2019).

24.8 Regulatory Framework

Another major concern pertaining phage therapy is regulation of the therapy. We need to have stringent regulatory authorities in place to have safe, efficient, sustainable, standardized, and ethically approved supply of phages. PageXchange is one such global regulatory authority that has launched phage governance platform for the same (Pirnay et al. 2020). 'Magistral phages' is another such regulatory framework in Belgium to incorporate personalized phage therapy (Pirnay et al. 2018). FDA has also recommended following good manufacturing practice guidelines and infrastructure. Phage banks and phage directories are also increasing and their easy online availability facilitates the researchers, companies, regulators, and clinicians to effectively communicate and expedite the journey of phages from laboratories to bedside (Alsaadi et al. 2021; Würstle et al. 2022; Gelman et al. 2021).

24.9 Future Prospects

Since we are living in an era of bioinformatics, bioengineering, and –omics, phage therapy can be modulated to deliver in a better manner. Bioengineering of phages can broaden the host range and moderate delivery systems. Phages can also be exploited for vaccine development. Phage display is one of the emerging strategies for antigen expression and delivery of phages (Gordillo Altamirano and Barr 2019). However, the main concern is the long road from phage research to bedside. Though few of the patients have undergone phage treatment, it still remains less accessible to one and all, especially in developing countries. In order to lessen this wide gap, phage-libraries (@phagedirectory) are being set up, via which patient can directly contact the phage researchers. Similarly, phagesforglobalhealth.org aims to develop programs for better comprehension and teaching phage biology to researchers across the globe (Gordillo Altamirano and Barr 2019). Phage therapy, if utilized appropriately, can prove to be one of the superlative answers to emerging global antimicrobial resistance.

References

Alsaadi A, Beamud B, Easwaran M, Abdelrahman F, El-Shibiny A, Alghoribi MF et al (2021) Learning from mistakes: the role of phages in pandemics. Front Microbiol 12:653107

BBC (n.d.) My husband squeezed my hand to say he wanted to live, then I found a way to save him—BBC News. https://www.bbc.com/news/stories-50221375

Brives C, Pourraz J (2020) Phage therapy as a potential solution in the fight against AMR: obstacles and possible futures. Palgrave Commun 6:100

Buckling A, Rainey PB (2002) Antagonistic coevolution between a bacterium and a bacteriophage. Proc Biol Sci 269:931–936

Capparelli R, Nocerino N, Lanzetta R, Silipo A, Amoresano A, Giangrande C, Becker K, Blaiotta G, Evidente A, Cimmino A, Iannaccone M, Parlato M, Medaglia C, Roperto S, Roperto F, Ramunno L, Iannelli D (2010) Bacteriophage-resistant Staphylococcus aureus mutant confers broad immunity against staphylococcal infection in mice. PLoS One 5:e11720

Chan BK, Turner PE, Kim S, Mojibian HR, Elefteriades JA, Narayan D (2018) Phage treatment of an aortic graft infected with Pseudomonas aeruginosa. Evol Med Public Health 2018:60–66

Chanishvili N (2012) Phage therapy—history from Twort and d'Herelle through Soviet experience to current approaches. Adv Virus Res 83:3–40

CIDRAP (2022) AMR Action Fund announces its first investments|CIDRAP. https://www.cidrap.umn.edu/news-perspective/2022/04/amr-action-fund-announces-its-first-investments

De Sordi L, Khanna V, Debarbieux L (2017) The gut microbiota facilitates drifts in the genetic diversity and infectivity of bacterial viruses. Cell Host Microbe 22:801–808

Debarbieux L et al (2015) A bacteriophage journey at the European medicines agency. FEMS Microbiol Lett 363:fnv225

Elpais (2022a) Bacteriophage therapy: How a virus helped a woman survive a dangerous antibiotic-resistant infection|USA|EL PAÍS English. https://english.elpais.com/usa/2022-01-27/how-a-virus-helped-a-woman-survive-a-dangerous-antibiotic-resistant-infection.html

Elpais (2022b) Mycobacterium chelonae: virus saves man from antibiotic-resistant bacteria|Society|EL PAÍS English Edition. https://english.elpais.com/society/2022-05-04/virus-saves-man-from-antibiotic-resistant-bacteria.html

FDA (n.d.) FDA allows phage therapy for COVID19 patients. Bacteriophage.news. https://www.bacteriophage.news/fda-phage-therapy-for-covid19-patients

Gelman D, Yerushalmy O, Alkalay-oren S, Rakov C, Ben-porat S, Khalifa L et al (2021) Personal view clinical phage microbiology: a suggested framework and recommendations for the in-vitro matching steps of phage therapy. Lancet Microbe 2(10):e555–e563

Gomez P, Buckling A (2011) Bacteria-phage antagonistic coevolution in soil. Science 332:106–109

Gordillo Altamirano FL, Barr JJ (2019) Phage therapy in the Postantibiotic era. Clin Microbiol Rev 32(2):e00066–e00018

Hargreaves KR, Clokie MRJ (2014) Clostridium difficile phages: still difficult? Front Microbiol 5: 184

Jault P, Leclerc T, Jennes S, Pirnay JP, Que YA, Resch G et al (2019) Efficacy and tolerability of a cocktail of bacteriophages to treat burn wounds infected by Pseudomonas aeruginosa (PhagoBurn): a randomised, controlled, double-blind phase 1/2 trial. Lancet Infect Dis 19:35–45

Knouf E, Ward W et al (1946) Treatment of typhoid fever with type specific bacteriophage. J Am Med Assoc 132:134–138

Kuipers S, Ruth MM, Mientjes M, de Sévaux RGL, van Ingen J (2019) A dutch case report of successful treatment of chronic relapsing urinary tract infection with bacteriophages in a renal transplant patient. Antimicrob Agents Chemother 64(1):e01281–e01219. https://doi.org/10.1128/AAC.01281-19. Published 2019 Dec 20

Kutter E, De Vos D, Gvasalia G, Alavidze Z, Gogokhia L, Kuhl S, Abedon ST (2010) Phage therapy in clinical practice: treatment of human infections. Curr Pharm Biotechnol 11:69–86

Labrie SJ, Samson JE, Moineau S (2010) Bacteriophage resistance mechanisms. Nat Rev Microbiol 8:317–327

Lin DM, Koskella B, Lin HC (2017) Phage therapy: an alternative to antibiotics in the age of multi-drug resistance. World J Gastrointest Pharmacol Ther 8(3):162–173. https://doi.org/10.4292/wjgpt.v8.i3.162

Maslov S, Sneppen K (2017) Population cycles and species diversity in dynamic kill-the-winner model of microbial ecosystems. Sci Rep 7:39642

McCallin S, Sacher JC, Zheng J, Chan BK (2019) Current state of compassionate phage therapy. Viruses 11(4):343. Published 2019 Apr 12

Miêdzybrodzki R, Borysowski J, Weber-Dabrowska B, Fortuna W, Letkiewicz S, Szufnarowski K et al (2012) Clinical aspects of phage therapy. Adv Virus Res 83:73–121

Monk AB, Rees CD, Barrow P, Hagens S, Harper DR (2010) Bacteriophage applications: where are we now? Lett Appl Microbiol 51:363–369

Philipson CW, Voegtly LJ, Lueder MR, Long KA, Rice GK, Frey KG, Biswas B, Cer RZ, Hamilton T, Bishop-Lilly KA (2018) Characterizing phage genomes for therapeutic applications. Viruses 10:188

Pirnay JP (2020) Phage therapy in the year 2035. Front Microbiol 11:1171. Published 2020 Jun 3

Pirnay JP, De Vos D, Verbeken G, Merabishvili M, Chanishvili N, Vaneechoutte M et al (2011) The phage therapy paradigm: prêt-à- porter or Sur-mesure? Pharm Res 28:934–937

Pirnay JP, Verbeken G, Ceyssens PJ, Huys I, De Vos D, Ameloot C et al (2018) The magistral phage. Viruses 10:E 64

Romero-Calle D, Guimarães Benevides R, Góes-Neto A, Billington C (2019) Bacteriophages as alternatives to antibiotics in clinical care. Antibiotics (Basel) 8(3):138. Published 2019 Sep 4

Schmidt C (2019) Phage therapy's latest makeover. Nat Biotechnol 37(6):581–586

Schooley RT, Biswas B, Gill JJ, Hernandez-Morales A, Lancaster J, Lessor L et al (2017) Development and use of personalized bacteriophage-based therapeutic cocktails to treat a patient with a disseminated resistant Acinetobacter baumannii infection. Antimicrob Agents Chemother 61:e0095417

Strydom N, Gupta SV, Fox WS, Via LE, Bang H, Lee M, Eum S, Shim T, Barry CE 3rd, Zimmerman M, Dartois V, Savic RM (2019) Tuberculosis drugs' distribution and emergence of resistance in patient's lung lesions: a mechanistic model and tool for regimen and dose optimization. PLoS Med 16(4):e1002773

Sulakvelidze A, Alavidze Z, Morris JG Jr (2001) Bacteriophage therapy. Antimicrob Agents Chemother 45:649–659

Summers WC (2012) The strange history of phage therapy. Bacteriophage 2(2):130–133

Suttle CA (2007) Marine viruses—major players in the global ecosystem. Nat Rev Microbiol 5: 801–812

Tetz G, Tetz V (2016) Bacteriophage infections of microbiota can lead to leaky gut in an experimental rodent model. Gut Pathog 8:33

Thiel K (2004) Old dogma, new tricks—21st century phage therapy. Nat Biotechnol 2:31–36

Voelker R (2019) FDA approves bacteriophage trial. JAMA 321(7):638

Wittebole X, De Roock S, Opal SM (2014) A historical overview of bacteriophage therapy as an alternative to antibiotics for the treatment of bacterial pathogens. Virulence 5:226–235

World Health Organization (2015) Global action plan on antimicrobial resistance. WHO Press, Geneva

Wright A, Hawkins CH, Änggård EE, Harper DR (2009) A controlled clinical trial of a therapeutic bacteriophage preparation in chronic otitis due to antibiotic-resistant Pseudomonas aeruginosa; a preliminary report of efficacy. Clin Otolaryngol 34(4):349–357

Würstle S, Stender J, Andr J, Vogele K, Rothe K, Willy C et al (2022) Practical assessment of an interdisciplinary bacteriophage delivery pipeline for personalized therapy of gram-negative bacterial infections. Pharmaceuticals (Basel) 15(2):186

The Potential of Bacteriophages in Treating Covid-19-Associated Secondary Infections 25

Anshika Sharma, Isra Ahmad Farouk, Mohammad Khusni Bin Ahmat Amin, Kaveesha Senasinghe, Vincent T. K. Chow, and Sunil Kumar Lal

Abstract

Bacteriophages are viruses that have the ability to infect a variety of bacterial species. Attention towards bacteriophage therapy has recently resurfaced due to the rise of antibiotic and multidrug-resistant bacteria which are a significant cause of morbidity and mortality in many diseases. Coronavirus disease 2019 (COVID-19) is a disease caused by the severe acute respiratory syndrome coronavirus 2 (SARS-CoV-2) and has rapidly emerged and spread leading to a pandemic in the year 2020. This respiratory virus infection begins in the lungs, however, can develop severe secondary bacterial infections that result in death of a patient. Much of the current research and development on COVID-19 largely focus on the primary cause of the disease which is the SARS-CoV-2 virus and hence neglecting research on secondary infections. In this review we discuss COVID-19-related secondary infections and the issues surrounding antibiotic resistance. Additionally, we discuss the bacterial profiles of these secondary infections and the potential use of novel bacteriophages for therapy to enhance the recovery of COVID-19 patients with severe secondary bacterial infections such as pneumonia. Lastly, the safety concerns surrounding bacteriophage therapy are addressed.

Anshika Sharma, Isra Ahmad Farouk and Mohammad Khusni Bin Ahmat Amin contributed equally with all other contributors.

A. Sharma · I. A. Farouk · M. K. B. A. Amin · K. Senasinghe · S. K. Lal (✉)
School of Science, Monash University, Bandar Sunway, Selangor DE, Malaysia
e-mail: sunil.lal@monash.edu

V. T. K. Chow
Department of Microbiology and Immunology, Yong Loo Lin School of Medicine, National University of Singapore, Singapore, Singapore

R. Sobti et al. (eds.), *Role of Microbes in Sustainable Development*,
https://doi.org/10.1007/978-981-99-3126-2_25

Keywords

Bacteriophage therapy · Novel bacteriophages · Multidrug resistance · SARS-CoV-2 · COVID-19 · Secondary infections · Pneumonia

Highlights

- Bacteriophage therapy has been considered a therapeutic strategy to combat multidrug-resistant bacteria. The potential of using bacteriophages to combat bacterial pneumonia that results from SARS-CoV-2 infection is described in this review.
- Causative bacteria for COVID-19-associated pneumonia primarily include Staphylococcus aureus, *Streptococcus pneumoniae,* and *Klebsiella pneumonia.*
- Repurposing some already available bacteriophage therapies could be a beneficial strategy to combat COVID-19-associated secondary bacterial infections.
- This review covers phage therapy safety parameters and describes in depth phage modifications that can be made to reduce host immune response to phage therapy. The current stance on phage therapy from both the industrial and regulatory framework perspective is also discussed.

25.1 Introduction

Bacteriophages or phages are viruses present abundantly in the natural ecosystem that infect bacteria and utilize the bacteria as a host for their reproduction (Clokie et al. 2011). Phages can be ubiquitously found in the biosphere, such as in the ocean, soil, and sewage, and have been discovered to outnumber the bacterial population (Abedon et al. 2011). Presently, bacteriophage therapy, which was long forgotten, receives attention again as a therapeutic strategy to combat challenging multidrug-resistant infections as well as serve as an alternative to antibiotic usage (Sulakvelidze et al. 2001).

Antibiotic resistance is a natural phenomenon that has been remarkably accelerated due to the inappropriate utilization of antibiotics by humans. Antibiotics that were once effective against life-threatening bacterial infection have shown a decrease in efficacy due to the rise of antimicrobial resistance, leading to prolonged hospital stays and expensive healthcare (Tagliaferri et al. 2019; World Health Organization, Antibiotic Resistance 2020). The World Health Organization states that this is a post-antibiotic era, where minor infections and wounds can potentially be lethal (World Health Organization, Antibiotic Resistance 2020). In 2019, the UN Interagency Coordination Group on Antimicrobial Resistance warned that by 2050, over ten million individuals could succumb to antibiotic-resistant diseases per year and that the economic impact caused by antibiotic-resistant diseases is projected to be catastrophic, with approximately 24 million individuals anticipated to be pushed into extreme poverty (Ginsburg and Klugman 2020). Majority of multidrug-resistant bacteria belong to the ESKAPE group of pathogens, comprising *Enterococcus faecium, Staphylococcus aureus, Klebsiella pneumoniae, Acinetobacter baumannii, Pseudomonas aeruginosa,* and *Enterobacter species*. ESKAPE pathogens are also the leading cause of nosocomial infections around the world and are known to pose one of the greatest challenges in clinical practice (Santajit and Indrawattana 2016; Kortright et al. 2019).

The recent outbreak of the COVID-19, a respiratory infection caused by the SARS-CoV-2, has been associated with an increase in pneumonia cases and other secondary bacterial infections. The increase in antibiotic use during the current COVID-19 pandemic poses a serious threat towards the exacerbation of global antimicrobial resistance (Ginsburg and Klugman 2020). To this end, this review discusses in detail the association between COVID-19 and secondary infections and the potential of phage therapy in treating COVID-19-related secondary infections, and lastly, addresses the safety concerns surrounding phage therapy.

25.2 Severe Acute Respiratory Syndrome Coronavirus 2 (SARS-CoV-2)

25.2.1 Epidemiology

The unprecedented outbreak of COVID-19 was first reported in December 2019 in Wuhan, the largest city in Central China (Cui et al. 2019; Coronaviridae Study Group of the International Committee on Taxonomy of Viruses 2020; Lai et al. 2020). Subsequently in March 2020, the World Health Organization (WHO) acknowledged and declared the COVID-19 a pandemic (Eurosurveillance Editorial Team 2020). As of January 16, 2021, 93,844,190 confirmed cases of COVID-19 have been reported, affecting 219 countries and territories, with death toll estimated to be approximately 2.0 million globally (Hopkins 2021; World Health Organization 2020a, b; Worldometer 2021). Clinical evidence has proven that the mortality rates of COVID-19 are directly proportional to the increase in age, with the elderly and those with underlying health comorbidities shown to be more vulnerable to the disease (Hong et al. 2020; Lee et al. 2020; Xia et al. 2020; Geiss 2020).

25.2.2 Clinical Feature of COVID-19

Patients infected with SARS-CoV-2 may develop different clinical characteristics and severity of the disease, ranging from mild symptoms to chronic respiratory impairment. In humans, COVID-19 complications primarily cause regular symptomatical manifestations such as fever, fatigue (tiredness), myalgia, dyspnea, and dry cough, similar to common coronaviruses infections (Chen et al. 2020; Guan et al. 2020; Huang et al. 2020; Wang et al. 2020a). Meanwhile, the same studies also reported that the irregular symptoms included atypical sputum production, headache, hemoptysis, diarrhea, anorexia, sore throat, abdominal pain, chills, and nausea (Chen et al. 2020; Guan et al. 2020; Huang et al. 2020; Wang et al. 2020a). In most common coronavirus infections, the signs of diseases appear after 14 days of viral incubation especially in the upper respiratory tract and gastrointestinal system (Dahiya et al. 2020; Samanta et al. 2020; Yang and Tu 2020; Zhong et al. 2020). Interestingly, the most recent clinical diagnosis of COVID-19 patients is a new onset of taste disorder (gustatory dysfunctions) including ageusia (complete loss), hypogeusia (partial loss), or dysgeusia (altered taste sensation) that compromises around 71% to 88.8% (Lechien et al. 2020; Lozada-Nur et al. 2020; Yan et al. 2020). Some patients develop a condition known as long COVID or post-acute COVID-19 wherein COVID-19 symptoms continue for longer than usual after initially contracting the virus, with effects observed on different organs (Taquet et al. 2021).

25.2.3 COVID-19 Associated with Co/Secondary Bacterial Infections

The co/secondary bacterial infection in COVID-19-infected patients may enhance the morbidity and mortality of viral infection, but yet, still poorly understood. Although it is unknown whether secondary infection contributes towards persistent and prolonged sequelae of post-acute COVID-19, it is evident that secondary bacterial infection is associated with worse outcomes in COVID-19 than influenza patients (Shafran et al. 2021; Nalbandian et al. 2021). In addition, steroids are often the preferred choice of treatment for COVID-19 patients with severe inflammatory response. Improper usage of steroids may also increase patient susceptibility to secondary bacterial or fungal infections (Perappadan 2021).

Several recent reports have described the increase in antimicrobial-resistant of COVID-19 patients (Chowdhary et al. 2020; Hughes et al. 2020; Tiri et al. 2020; Li et al. 2020) which might be due to the overuse of the drugs during the pandemics (Rossato et al. 2020). Among the findings were two retrospective studies that revealed the incidence of patients colonized by Carbapenem-Resistant Enterobacteriaceae (CRE) escalated from 6.7% in 2019 to 50% in March—April 2020, Carbapenem-Resistant A. baumanni and K. pneumoniae were 91.2% and 75.5%, respectively, and Methicillin resistance of Staphylococcus aureus and Coagulase-negative staphylococci were 100% (Li et al. 2020; Tiri et al. 2020).

Since the first outbreak in China, it was reported that around 15% of hospitalized COVID-19 patients in Wuhan had been clinically diagnosed with secondary bacterial infections. A higher number of incidents were reported among the non-survivors compared to the survivors (50% vs. 1%) (Ruan et al. 2020). Bacterial co/secondary infections have been reported to be common among respiratory-related diseases. For instance, it was confirmed that up to 75% of patients diagnosed with Influenza virus infection also suffered from the bacterial co/secondary infection (Zambon 2001; Gill et al. 2010; Weinberger et al. 2012). Through screening of COVID-19 patient samples, several species of bacteria were identified, including *Enterobacter cloacae, Acinetobacter baumanni, Serratia marscecens, Streptococcus pneumoniae,* etc., which have the potential to exuberate the clinical manifestation of symptoms and further lead to complications such as the development of bronchitis (Table 25.1) (Hughes et al. 2020; Tiri et al. 2020; Li et al. 2020; Rossato et al. 2020; Ruan et al.

Table 25.1 Reported bacterial or fungal profiles of Covid-19 secondary infections

No. of patients	Country of study origin	Sample types	Bacteria identified	Fungal identified	Identification method	References
836	United Kingdom	Blood, sputum, bronchoalveolar lavage, urine	*S. aureus, Pseudomonas* spp. *Enterobacter* spp. *Klebsiella* spp. *Serratia* spp. *Candida* spp./*Yeast aspergillus* spp.	*Candida* spp.: *Candida albicans, Aspergillus fumigatus*	Standard microbiology procedures for respiratory cultures	Hughes et al. (2020)
260	Egypt	Sputum or endotracheal aspirates	*Klebsiella pneumoniae, Acinetobacter baumannii, and Staphylococcus aureus* (methicillin-resistant)	*Candida* spp.: *Candida albicans, Candida glabrata*	Standard microbiology procedures for bacterial identification. VITEK2* system for antimicrobial sensitivity and PCR* for antibiotic-resistant gene identification	Ramadan et al. (2020)
20	China	Lung tissue from deceased Covid-19 patients	*Acinetobacter, Chryseobacterium, Burkholderia, Brevundimonas, Sphingobium, Enterobacteriaceae, mycobacterium, Prevotella*	*Cutaneotrichosporon, Issatchenkia, Wallemia, Cladosporium, Alternaria, Dipodascus, Mortierella, aspergillus, Naganishia, Diutina, and Candida*	Pair-ended sequencing using Illumina MiSeq platform	Fan et al. (2020)
989	Spain	Blood, pleural sterile fluids, sputum, bronchoalveolar lavage, urine	*Streptococcus pneumoniae, Staphylococcus aureus, Haemophilus influenzae, Moraxella catarrhalis, Pseudomonas aeruginosa, Escherichia coli, Klebsiella pneumoniae, Enterococcus faecium, Proteus mirabilis, Citrobacter koseri,*	*Aspergillus fumigatus, Candida albicans*	Standard microbiology procedures, rapid standard F S. pneumoniae ag fluorescent immunoassay (urine sample)	Garcia-Vidal et al. (2020)

(continued)

Table 25.1 (continued)

No. of patients	Country of study origin	Sample types	Bacteria identified	Fungal identified	Identification method	References
			Enterococcus faecalis, Streptococcus anginosus			
731	Italy	Bronchoalveolar lavage, blood	*Staphylococcus aureus, Staphylococcus capitis, Staphylococcus epidermidis, Staphylococcus haemolyticus, Staphylococcus hominis, Enterococcus faecalis, Enterococcus faecium, Clostridium* spp., *Acinetobacter baumannii, Enterobacter cloacae, Escherichia coli, Klebsiella oxytoca, Klebsiella pneumoniae, Pseudomonas aeruginosa, Pseudomonas putida, Stenotrophomonas, maltophilia, Pantoea agglomerans, Fusobacterium nucleatum*	*Candida albicans, Candida parapsilosis, Aspergillus spp.*	Standard microbiology procedures for blood and respiratory cultures, galactomannan assay (for invasive aspergillosis)	Ripa et al. (2020)

2020; Zambon 2001; Gill et al. 2010; Weinberger et al. 2012; Du et al. 2020; Garcia-Vidal et al. 2020; Lansbury et al. 2020; Ramadan et al. 2020; Ripa et al. 2020). Despite the positive co/secondary infection findings, the threats and degree of coinfection in COVID-19 patients are heterogeneous across studies due to the demographic factor, the strategy of antimicrobial treatment, patients comorbidities, the efficiency of intervention's management, differences between the healthcare system, and sensitivity of the bacterial culture methods (Chen et al. 2020; Guan et al. 2020; Huang et al. 2020; Wang et al. 2020a, b; Dahiya et al. 2020; Samanta et al. 2020; Yang and Tu 2020; Zhong et al. 2020; Lechien et al. 2020; Lozada-Nur et al. 2020; Yan et al. 2020; Taquet et al. 2021; Shafran et al. 2021; Nalbandian et al. 2021; Perappadan 2021; Chowdhary et al. 2020; Hughes et al. 2020; Tiri et al. 2020; Li et al. 2020; Rossato et al. 2020; Ruan et al. 2020; Zambon 2001; Gill et al. 2010; Weinberger et al. 2012; Du et al. 2020; Garcia-Vidal et al. 2020; Lansbury et al. 2020; Ramadan et al. 2020; Ripa et al. 2020; Arentz et al. 2020).

25.2.4 COVID-19 Pneumonia: Severe Respiratory Illness

COVID-19 is a pulmonary-related disease that conspicuously targets the respiratory tract and eventually causes inflammation in the lungs of an infected person. Gradually this phenomenon leads to fluid and pus deposition in the alveoli and represses the oxygen supply throughout the body. Later, this process develops into a potentially fatal complication known as pneumonia (Manohar et al. 2020). A preceding virus of SARS-CoV-2 infection in patients will eventually impair the innate and adaptive antibacterial host defenses in order to exploit and compromise the immunological barrier to cause severe viral pneumonia (Ginsburg and Klugman 2020; Santajit and Indrawattana 2016; Kortright et al. 2019; Cui et al. 2019; Coronaviridae Study Group of the International Committee on Taxonomy of Viruses 2020; Lai et al. 2020; Eurosurveillance Editorial Team 2020; Hopkins 2021; World Health Organization 2020a, b; Worldometer 2021; Hong et al. 2020; Lee et al. 2020; Xia et al. 2020; Geiss 2020; Chen et al. 2020; Guan et al. 2020; Huang et al. 2020; Wang et al. 2020a; Dahiya et al. 2020; Samanta et al. 2020; Yang and Tu 2020; Zhong et al. 2020; Lechien et al. 2020; Lozada-Nur et al. 2020; Yan et al. 2020; Taquet et al. 2021; Shafran et al. 2021; Nalbandian et al. 2021; Perappadan 2021; Chowdhary et al. 2020; Hughes et al. 2020; Tiri et al. 2020; Li et al. 2020; Rossato et al. 2020; Ruan et al. 2020; Zambon 2001; Gill et al. 2010; Weinberger et al. 2012; Du et al. 2020; Garcia-Vidal et al. 2020; Lansbury et al. 2020; Ramadan et al. 2020; Ripa et al. 2020; Arentz et al. 2020; Wang et al. 2020b; Manohar et al. 2020; Richardson et al. 2020; Yang et al. 2020). Indeed, there are growing numbers of evidence indicating that hyper-inflammation plays a major role in COVID-19 disease progression and increases the severity and mortality of the disease (Afrin et al. 2020; Gustine and Jones 2020; Manson et al. 2020; Reddy et al. 2020). Notably, researchers have classified the medium of COVID-19—pneumonia infection into 3 main sources that are community-acquired (CA), hospital acquired (HA), and ventilator-associated pneumonia (VAP) (Garcia-Vidal et al. 2020; Lansbury et al.

2020; Ramadan et al. 2020; Ripa et al. 2020; Arentz et al. 2020; Wang et al. 2020b; Manohar et al. 2020; Richardson et al. 2020; Yang et al. 2020; Afrin et al. 2020; Gustine and Jones 2020; Manson et al. 2020; Reddy et al. 2020; Cilloniz et al. 2016; Dudoignon et al. 2020). According to a study conducted on 74 patients hospitalized with COVID-19, around 40.5% were reported with CA pneumonia, 25% with VA pneumonia, and 9% with HA pneumonia (Garcia-Vidal et al. 2020). Interestingly, several studies showed that the hospitalized COVID-19 patients had lower bacterial pneumonia infection compared to other respiratory diseases such as influenza H1N1 or H3N2 (Garcia-Vidal et al. 2011; Martin-Loeches et al. 2017; Burrell et al. 2018; Schauwvlieghe et al. 2018).

The most common causative agents of secondary infections in patients with COVID-19 pneumonia are *Staphylococcus aureus, Streptococcus pneumoniae,* and *Klebsiella pneumonia* (Garcia-Vidal et al. 2020; Lansbury et al. 2020; Ramadan et al. 2020; Ripa et al. 2020; Arentz et al. 2020; Wang et al. 2020b; Manohar et al. 2020). The pathophysiology of severe viral pneumonia is more likely to develop acute respiratory distress syndrome (ARDS) (Ranieri et al. 2012). Several studies indicated that critical patients with confirmed SARS-CoV-2 infection were vulnerable to develop ARDS and received mechanical ventilation (Huang et al. 2020; Wang et al. 2020a, b; Dahiya et al. 2020; Samanta et al. 2020; Yang and Tu 2020; Zhong et al. 2020; Lechien et al. 2020; Lozada-Nur et al. 2020; Yan et al. 2020; Taquet et al. 2021; Shafran et al. 2021; Nalbandian et al. 2021; Perappadan 2021; Chowdhary et al. 2020; Hughes et al. 2020; Tiri et al. 2020; Li et al. 2020; Rossato et al. 2020; Ruan et al. 2020; Zambon 2001; Gill et al. 2010; Weinberger et al. 2012; Du et al. 2020; Garcia-Vidal et al. 2011, 2020; Lansbury et al. 2020; Ramadan et al. 2020; Ripa et al. 2020; Arentz et al. 2020; Manohar et al. 2020; Richardson et al. 2020; Yang et al. 2020; Afrin et al. 2020; Gustine and Jones 2020; Manson et al. 2020; Reddy et al. 2020; Cilloniz et al. 2016; Dudoignon et al. 2020; Martin-Loeches et al. 2017; Burrell et al. 2018; Schauwvlieghe et al. 2018; Ranieri et al. 2012; Gibson et al. 2020). Regardless of the different features of a typical ARDS with COVID-19 pneumonia, both conditions contributed to a poor prognosis of infected individuals. In the early incident of COVID-19's outbreak in China, Huang et al. confirmed that all 41 patients had pneumonia infection and 29% (12 of 41) of patients manifested ARDS, followed by other clinical complications (Huang et al. 2020). Nevertheless, at the end of the study, 68% of patients were discharged, while 15% of patients died. Another research was conducted in Washington state and the findings were in concordance with the previous study. Common symptoms were identified in all 21 patients; however, 15 patients had developed ADRS and required oxygen therapy. After all, only 9.5% of patients were discharged from the ICU, 24% remained critically ill, and 67% of patients died (Arentz et al. 2020). Because of the poor clinical outcomes of COVID-19 pneumonia, there is a growing number of studies and persistent efforts by the life sciences industry and professionals to reduce the burden of this life-threatening disease across the globe. The present and potential treatments especially the implementation of phage therapy to combat COVID-19 pneumonia will be further appraised in the next section of this review.

25.3 Phage Therapy in Treating Secondary Infections

The consideration and testing of the potential of phages to achieve therapeutic success in treating bacterial infections has been ongoing for many years and is still being researched today. Their application has been tested against diverse human bacterial infections, including secondary infections which arose from various diseases including cystic fibrosis (CF), viral infections, cancer, gastrointestinal disorders, surgical wound infections, and more (Furfaro et al. 2018). In this section, we discuss some of the *in vivo* and clinical successes that have been reported for such studies and the potential for using bacteriophage therapy for COVID-19 secondary infections. Phage therapy in the past has been applied to secondary infections caused by osteoarthritis (arthritis in protective joints), wound infections, gastrointestinal infections, and more. Table 25.2 includes some examples of secondary infections that have been treated with phage therapy recently.

25.3.1 Phage Therapy and Treatment of Pulmonary Infections

The diversity of bacteria and infection sites may require different routes of administration of bacteriophage therapy to be effective. Topical and intravenous phage therapy has shown potential and success for treatment of burn-mediated wounds and diabetic ulcers (Duplessis and Biswas 2020). Other clinical studies have chosen to administer phage therapy intravenously as it has proven to be the only route that efficiently targets the bacteria of interest. The therapeutic potential of bacteriophages in mouse models with acute pulmonary infection (*Burkholderia cenocepacia*) found that systemic phage administration was more effective in treating pulmonary infection in mice as compared to phage inhalational, which may suggest that phages in circulation (intravenous) have better access to bacteria in the lungs as compared to topical phages (inhaled) (Carmody et al. 2010). As phage therapy is regaining recognition, more research has been conducted in relation to deciphering the dynamics of the therapy as little information or models are currently available for clinicians to follow when selecting phage dosages or administration methods, for patients. A recent study described the pharmacokinetics and pharmacodynamics of an antipseudomonal phage (øPEV20) in rats (Lin et al. 2020a). This study provided a technical model that is robust and can be recommended for usage in future preclinical and clinical investigations (Lin et al. 2020a). Prior to clinical trials, *in vitro* and *in vivo* studies provide indicative results of the efficiency of bacteriophages against specific bacterial targets. Table 25.3 identifies some of the recently reported *in vitro* and *in vivo* studies that have specifically focused on bacterial species that cause lung infections and severe pneumonia.

Recently in May 2019, a case study published reported the successful outcome of the adjunctive usage of bacteriophage therapy (using phage: AB-PA01, combination lytic phages produced by AmpliPhi Biosciences Cornoporation) to treat a severely ill cystic fibrosis patient. The patient, a 26-year-old female, was suffering from chronic multidrug-resistant *Pseudomonas aeruginosa* lung infection. Although she was

Table 25.2 Case-based hospital application of bacteriophages on patients with severe persistent infections

Bacteria	Diseases/condition causing pneumonia studied	Number of subjects observed	Mode of delivery of bacteriophage	Reported adverse effects	Outcome of study	References
Pseudomonas aeruginosa and *Burkholderia dolosa*	Multidrug resistance infection of lung transplant recipients	3	Intravenous (AB-PO01/ navy phage cocktail/ single lytic phage)	Unclear as underlying disease and infection do not allow for direct identification	Patients overall tolerated phage treatment when used alongside antibiotics with two out of three patients recovered	Aslam et al. (2019a)
Pseudomonas aeruginosa	Multidrug resistant pneumonia in cystic fibrosis patient	1	Intravenous (AB-PO01)	Lack of adverse effects and patient successfully recovered	Patient tolerated phage therapy when used with antibiotics and showed recovery. Patient did not develop recurring infection 100 days after administration	Law et al. (2019)
Staphylococcus aureus	*S.aureus* bacteraemia patients	13	Intravenous Myoviridae bacteriophages (AB-SA01)	No adverse effects related to phage therapy were observed	62% of patients showed clinical improvement, 38% died during surgery or bacteraemic shock. Study summarized that AB-SA01 administered this way was safe	Petrovic Fabijan et al. (2020)
Methicillin-resistant *Staphylococcus aureus* (MRSA)	Patient with severe chronic MRSA infected prosthetic joint	1	Intra-articular and intravenous (SaGR51)	No adverse effects from bacteriophage therapy were observed. Recovery	Salvaging of the patients' prosthesis was not possible due to severe bone erosion from the infection.	Doub et al. (2020)

				resumed after first dosage of therapy	After 3 days of phage and antibiotic combination, infection was completely healed	
Klebsiella pneumoniae	Patient with prosthetic knee infection with progressive clinical worsening and allergies to antibiotics	1	Intravenous (KpJH46Φ2)	Patient was given 40 intravenous doses and did not experience adverse effects	Patient was able to tolerate phage treatment along with some antibiotics and local symptoms were resolved	Cano et al. (2020)
Pseudomonas aeruginosa	Patients with burn wound infections	13	Topical application of 12 natural lytic bacteriophage cocktail (PP1131)	No adverse effects from bacteriophage treatment were observed	The dosage of PP1131 given was too low to decrease bacterial burden in the wounds in the same pace as standard of care methods. Increased phage concentrations in future studies have been suggested	Jault et al. (2019)
Escherichia coli, Staphylococcus aureus, and *Pseudomonas aeruginosa*	Patients with chronic nonhealing wounds	20	Topical application of customized bacteriophage cocktails (designed based on patient wound biopsies)	No adverse effects after bacteriophage treatment was given	Significant improvement of wound healing was achieved. No signs of infection were seen after 2 to 5 doses of treatment. Complete healing was achieved by 21 days	Gupta et al. (2019)
Staphylococcus aureus	Nonischemic cardiomyopathy patient with prolonged	1	Intravenous administration of	No clinical or adverse effects observed post-bacteriophage therapy	Patient showed clinical improvement and was reported fit for heart	Aslam et al. (2019b)

(continued)

Table 25.2 (continued)

Bacteria	Diseases/condition causing pneumonia studied	Number of subjects observed	Mode of delivery of bacteriophage	Reported adverse effects	Outcome of study	References
	infection of left ventricular assist device		AB-SA01 bacteriophage		transplant 1 week after bacteriophage therapy with antibiotics. Patient was reported to be healthy with no infection recurrence even 7 months posttransplant	
Staphylococcus aureus, Pseudomonas aeruginosa and enterococcus faecalis	Patients with severe musculoskeletal infections	4	Applications through draining system. BFC-1 bacteriophage cocktail against *S. aureus* and *P. aeruginosa*. Pyo bacteriophage was used for *E. faecalis* as no specific phage exists for these bacteria	No adverse effects related to bacteriophage therapy were observed	The treatment was reported to be effective when couples with antibiotics after a single dose. The draining system protocol was generally well-tolerated by patients. Nursing staff would require special training to apply this method	Onsea et al. (2019)
Staphylococcus aureus, Enterococcus faecium, Pseudomonas aeruginosa, Klebsiella	Patients with immunosuppression after organ transplantation had infections of vascular grafts, implanted	8	Customized bacteriophage preparations, administered either locally, orally or via inhalation (case-dependent)	No severe adverse effects were observed	All phage treatments were coupled with antibiotic treatment. Eradication of target bacteria was achieved in 7 of the 8 patients. The study concludes	Rubalskii et al. (2020)

pneumoniae, and *Escherichia coli*	medical devices, or surgical wounds				that bacteriophage therapy effectively treats bacterial infections that result from cardiothoracic surgery	
Staphylococcus aureus	Patient with prosthetic heart endocarditis	1	Intravenous AB-SA01 (AmpliPhi biosciences)	Overall well-tolerated and no adverse effects observed	Bacteriophage treatment was coupled with antibiotics. Blood cultures of patient tested negative at onset of bacteriophage therapy, and all other indicators of infection were reduced within 24-h	Gilbey et al. (2019)

Table 25.3 *In vitro* and *in vivo* reports on bacteriophages for pneumonia application

Bacteria targeted	Disease studied	Type of study	Bacteriophage evaluated	Study outcome	References
Carbapenem-resistant Acinetobacter baumannii (CRAB)	Acute pneumonia	Mouse model (*in vivo*)	A novel *A. baumannii* lytic phage (YMC 13/03/R2096 ABA BP (phage Bφ-R2096)	Bφ-R2096 phage inhibited bacterial growth in a dose-dependent manner. Bacterial clearance in the lungs of the mice was observed after 3 days postinfection. No mortality or serious adverse effects were reported after phage infection	Jeon et al. (2019)
Extensively drug-resistant *Pseudomonas aeruginosa* (XDR-PA)	Acute pneumonia	*In vitro*, in silico, and *in vivo*	Two novel *P. aeruginosa* bacteriophages (phages) (Bφ-R656 and Bφ-R1836)	Both novel phages studies exhibited bacteriolytic activity against XDR-PA strains from acute pneumonia patients. In addition, survival of *Galleria mellonella* larvae and mice was greatly improved according to the study	Jeon and Yong (2019)
Staphylococcus aureus	Rabbit necrotizing pneumonia	*In vivo*	A novel phage VB-SavM-JYL01	This pneumonia affects rabbits at a high rate. Intranasal administration efficiently improved rabbit survival rate with significant lung condition improvement as compared to the control group	Ji et al. (2019)
Streptococcus pneumoniae	*S. pneumoniae* infections	*In vitro*, in silico, and *in vivo*	Endolysin Cpl-1 derived from Cp-1 phage packaged in chitosan nanoparticles	The nanoparticles carrying endolysin Cp-1 exhibited mucoadhesive nature and revealed to have biocompatible nature. The results of the study confirmed significant potential of increased bioavailability of endolysin against *S. pneumoniae*	Gondil et al. (2020)

Klebsiella pneumoniae	*Klebsiella pneumoniae* infection in mice model	*In vivo*	A novel lytic phage (VTCCBPA43)	VTCCBPA43 was found to have high tolerance for temperature (80 ° C) and had a highly narrow host range. A single intranasal application reduced lung bacterial load significantly	Anand et al. (2020)
Multidrug-resistant (MDR) *Klebsiella pneumoniae*	Lobar pneumonia induced by *K. pneumoniae* in mice	*In vivo*	Bacteriophage (vB_KpnM-Teh.1)	A single dose of vB_KpnM-Teh.1 was administered to the mice simultaneously at the time of infection or 24 h postinfection. The results of the study concluded that this isolated phage has the potential to be used both as a preventative and as a curative treatment for pulmonary infections caused by MDR *K. pneumoniae*	Soleimani Sasani and Eftekhar (2020)
Methicillin-resistant (LAC) *Staphylococcus aureus*	Lethal murine pneumonia	*In vivo*	SAL200 endolysin novel candidate drug	Upon single intranasal administration, bacterial loads of the treated mice group were significantly lower (10 folds) than that of the control group. The study suggests that SAL200 could be used as a potential adjunct treatment against severe pneumonia caused by *S. aureus*	Bae et al. (2019)
Pseudomonas aeruginosa	Acute lung infection	*In vitro* and *in vivo*	Phage PEV20-ciprofloxacin combination	Inhalable powder of PEV20 phage coated with ciprofloxacin was evaluated for activity against *P. aeruginosa* in mice model of acute lung infection. Bacterial load was tested and reported to be highly	Lin et al. (2020b)

(continued)

Table 25.3 (continued)

Bacteria targeted	Disease studied	Type of study	Bacteriophage evaluated	Study outcome	References
				reduced 24 h posttreatment. Lung inflammation was also significantly reduced	
Pseudomonas aeruginosa	Lung infection	*In vitro*	Phage vB_PaeP_4024 (φ24), vB_PaeP_4054 (φ54) and vB_PaeS_4069 (φ69) with ciprofloxacin	Phage/ciprofloxacin combination did not induce inflammatory response in tested cell line (Calu-3 cell line and isogenic CFTR knock down cell line (cftr-)). The study concludes that this combination successfully protects the cells from *P. aeruginosa* infection and may be considered in the future for clinical trials for cystic fibrosis lung infection patients	Luscher et al. (2020)

administered multiple antibiotics over weeks of treatment, little to no improvement was initially seen. Later during her treatment, the hospital was given approval to test bacteriophage therapy along with antibiotics. The patient was reported to have a lack of adverse events and there was no recurrence of symptoms or infection within 100 days following treatment, ultimately achieving full recovery (Law et al. 2019). The Food and Drug Administration (FDA) approves bacteriophage therapy on a case-to-case basis, such as the abovementioned case, therefore allowing for more opportunity for patients in severe conditions such as multidrug-resistant pneumonia to potentially recover (Voelker 2019).

25.3.2 The Application of Bacteriophages for Their Potential Therapeutic Usage to Combat COVID-19 Secondary Infections

Recently, many researchers have begun discussions to seriously consider the usage of bacteriophages for their application to combat COVID-19 using a variety of approaches. It was highlighted that the decrease of bacterial infection burden in patients suffering from secondary infections is ideal for the improvement of the immune system of COVID-19 patients, hence allowing for more potential for recovery (Wojewodzic 2020). In severe cases of COVID-19, secondary bacterial infections resulting in pneumonia are usually the most difficult consequences to overcome, which highly contributes to the mortality rates of patients (Manohar et al. 2020). This is mainly due to the high amount of bacterial growth in the pulmonary region. To potentially combat this, some have suggested a combination of natural bacteriophages which have specificity to some of the bacterial species in the patient's lungs to be coupled with the usage of antibiotics or other therapies for a more efficient recovery (Wojewodzic 2020; U.S National Library of Medicine 2020a). This treatment approach of phage therapy is currently being tested in a clinical trial (clinical trial identifier number: NCT04636554) as a form of personalized treatment for COVID-19 patients with bacterial coinfections of *Acinetobacter baumannii, Pseudomonas aeruginosa,* or *Staphylococcus aureus* (U.S National Library of Medicine 2020a). The feasibility of development, production, and administration of phage therapy for COVID-19 is not yet thoroughly established. This process requires the attention of researchers in the field for additional trials and experiments to be conducted to achieve better detailed workflows. In Fig. 25.1, we have given a generalized outline for the steps involved in using bacteriophages for COVID-19-associated bacterial pneumonia. Although many factors are to be considered such as their safety and efficacy, it is indeed worth considering, as their potential has been made evident with past successful *in vivo* and clinical studies in other infections. It is especially important to highlight that in severe disease conditions, testing bacteriophage alongside other treatments may give patients a better chance at fighting the disease. We, the authors, would like to highlight that most of the bacteria causing secondary infections in COVID-19 patients reported (Table 25.1) have existing bacteriophages (such as SaGR51, KpJH46Φ2, PP1131, BFC-1) that may be considered for phage treatment of these patients. In addition, current clinical trials

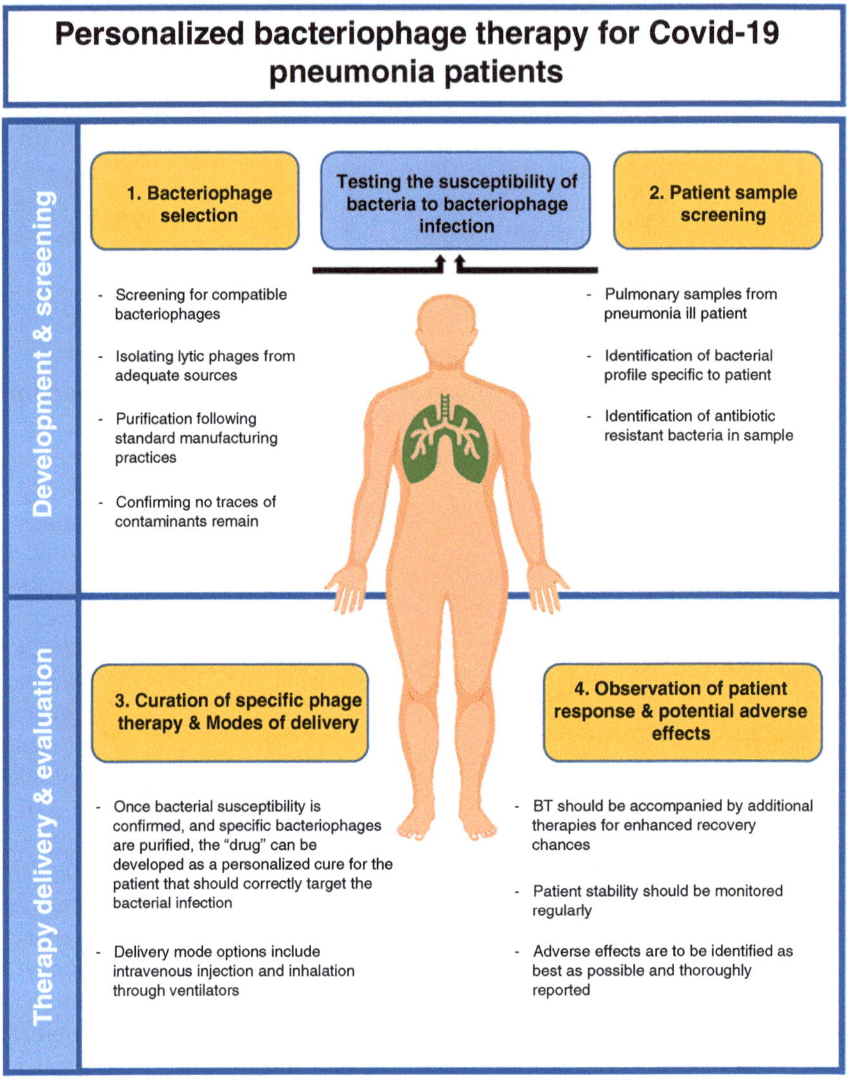

Fig. 25.1 Illustration of the suggested generalized workflow for the development, screening, therapy, and evaluation of personalized bacteriophage therapy for Covid-19 pneumonia patients

involving the usage of bacteriophages to target pulmonary infections have been listed in Table 25.4. It is evident that the potential of bacteriophage therapy to target pulmonary infections has been seriously considered, and therefore, we recommend combining bacteriophage therapy alongside antibiotics and additional necessary medical aids or treatments (such as ventilators, antiviral drugs) to be applied to COVID-19 patients with severe pulmonary bacterial infections. Combining other therapies with bacteriophages may provide better improvement results as

Table 25.4 Clinical trials testing bacteriophage efficiency in pulmonary infections

Bacteria targeted	Institution	Clinical trial identification/ phase	Patient's profile	No. of participants enrolled	Mode of delivery	Study's Approach	References
A. baumannii, P. aeruginosa or S. aureus	Adaptive phage therapeutics, Inc.	NCT04636554 (phase I)	Covid-19 patients with bacterial coinfections	Not provided	Intravenous phage	Bacterial susceptibility to phage will be custom assessed for each patient	U.S National Library of Medicine (2020a)
A variety of gram-positive and gram-negative aerobes and anaerobes	Northern state medical university	NCT04325685 (phase I)	Patients 18 years and older undergoing invasive mechanical ventilation beyond 48 h	Not provided	Oropharyngeal decontamination with bacteriophage during mechanical ventilation	Prevention of ventilator-associated pneumonia in affected patients via oropharyngeal decontamination	U.S National Library of Medicine (2020b)
Pseudomonas aeruginosa	University hospital, Montpellier	NCT01818206 (not applicable)	Sputum samples from cystic fibrosis (CF) patients	59	Cocktail of 10 bacteriophages applied to the sputum samples	The study aimed to evaluate the efficiency of the phage cocktail against the bacteria found in the sputum samples	U.S National Library of Medicine (2013)
Pseudomonas aeruginosa	Armata pharmaceuticals, Inc.	NCT04596319 (phase 1b/2a)	Patients with cystic fibrosis and chronic pulmonary *Pseudomonas aeruginosa* infection	48	Bacteriophage AP-PA02 administered by inhalation	This trial aims to evaluate the safety, tolerability, and phage recovery profile of the phage candidate	U.S National Library of Medicine (2020c)

bacteriophage therapy alone may not be enough to provide significant improvement in severely ill patients.

25.4 Phage Therapy Safety Parameters

Phage therapy is generally regarded as safe, as it does not directly affect eukaryotic cells, poses low inherent toxicity, has a narrower potential for inducing resistance when compared to antibiotic therapy, and results in minimal disruption of normal flora owing to their host specificity (Loc-Carrillo and Abedon 2011; Hyman 2019). Phages are able to evade bacterial defenses in a few ways as to ensure phage propagation (Samson et al. 2013). Some of these mechanisms include modifying their receptor-binding protein to absorb evolving bacterial populations and possessing degrading enzymes that can hydrolyze capsular polysaccharides of bacterial receptors (Samson et al. 2013). Immune response to phage therapy, however, is a major concern. It is important to highlight that phage particles contain a protein envelope which encapsulates its nucleic acids, which may be immunogenic and may lead to hypersensitivity reactions, such as anaphylaxis (Hargreaves and Clokie 2014). Clinical studies are, however, continuously evaluating the changes of inflammatory cytokines in response to phage therapy. A study by Dufour et al. 2019, examined the impact of specific phage therapy on murine acute pneumonia and found that the treatment was not associated with over inflammation, rather resulted in lower inflammation, provided faster correction of blood cell count and showed a more desired effect as compared to antibiotics (Dufour et al. 2019). In addition, phage-triggered production of antiphage antibodies by B lymphocytes has been reported to result in phage inactivation, thus hindering clinical therapeutic success (Van Belleghem et al. 2018). This issue can be addressed through careful selection of phage type, formulation, and dosage as well as through collation and analysis of results obtained from elaborate phage therapy clinical trials. Conjugation of phage particles with polyethylene glycol (PEG) has also successfully shown to reduce phage immunogenicity, thus protecting phages from neutralization by the host immune defense mechanisms, consequently increasing its half-life in circulation (Kim et al. 2008; Singla et al. 2016; Malik et al. 2017). Furthermore, it is highly recommended that phage preparations undergo a series of purification steps prior to its therapeutic application, for the removal of host bacterial components such as endotoxins produced by Gram negative bacteria, which may lead to critical side effects in the patient (Szermer-Olearnik and Boratyński 2015). More recently, spray drying phage formulation has turned out to be an attractive technique adopted to enhance phage stability, prolong shelf-life, and provide ease of handling (Zhang et al. 2020). In addition, according to a study by Puapermpoonsiri et al., encapsulation of dried phage formulations by biodegradable polymers and subsequent processing into microspheres, which could be nebulized further, allows for controlled release of phages, thereby improving therapeutic efficacy and minimizing host immunogenic responses (Puapermpoonsiri et al. 2009).

Induction of immune response by phages, however, has also been reported to be a key factor in determining the effectiveness of phage therapy (Hodyra-Stefaniak et al. 2015; Roach et al. 2017). Recognition of phage nucleic acids as pathogen-associated molecular patterns (PAMPs) by pattern recognition receptors (PRR) triggers the recruitment of phagocytes to the site of infection (Carroll-Portillo and Lin 2019). Adsorption of phage onto bacterial surface subsequently allows for ease of infection-site recognition by the immune system, thus eliciting a phenomenon known as immunophage synergy wherein phagocytes aid the phage activity in bacterial clearance through phagocytosis (Roach et al. 2017; Carroll-Portillo and Lin 2019; Kurzępa et al. 2009). According to a study by Roach et al., the synergy between the host immune system and the bacteriophage is essential for successful phage therapy against an acute respiratory pathogen (Roach et al. 2017). A side effect of phagocytosis is the production of reactive oxygen species (ROS). However, phages have been documented to indirectly inhibit ROS production via unknown phage-bacteria interactions, which attributes to the reduction in the oxidative stress and tissue damage, that is otherwise caused by ROS (Jończyk-Matysiak et al. 2017).

Multiple studies have reported that phages exert strong anti-inflammatory effects as a result of their interaction with the immune system (Van Belleghem et al. 2018; Kim et al. 2008; Singla et al. 2016; Malik et al. 2017; Szermer-Olearnik and Boratyński 2015; Zhang et al. 2020; Puapermpoonsiri et al. 2009; Hodyra-Stefaniak et al. 2015; Roach et al. 2017; Carroll-Portillo and Lin 2019; Kurzępa et al. 2009; Jończyk-Matysiak et al. 2017; Górski et al. 2017, 2019a; Sausset et al. 2020). Neutrophil degranulation is an immune response exerted in response to certain viral infections. Borysowski et al. reported that purified staphylococcal phage did not trigger neutrophil degranulation *in vitro* (Borysowski et al. 2017). Consequently, no toxic inflammatory mediators were released from neutrophils. Furthermore, Pabary et al. demonstrated that when phages were injected to mice with *Pseudomonas aeruginosa* lung infection, a reduction in inflammatory cytokine, neutrophil levels, and bacterial count in bronchoalveolar fluid was observed, indicating that phages do not aggravate inflammation (Pabary et al. 2016). Tothova et al. further added that phage therapy in a Cronobacter urinary tract infection model diminished the production of pro-inflammatory cytokines (Tothova et al. 2011). Multiple studies have shown evidence suggesting that the severity of COVID-19 is associated with increased levels of pro-inflammatory cytokines (Khadke et al. 2020; Mustafa et al. 2020; Ragab et al. 2020; Ye et al. 2020). Immune factors known to be activated in response to bacterial infections G-CSF and IFN- γ were both reported to be elevated in Covid-19 patient samples (Huang et al. 2020; Wang et al. 2020a, b; Dahiya et al. 2020; Samanta et al. 2020; Yang and Tu 2020; Zhong et al. 2020; Lechien et al. 2020; Lozada-Nur et al. 2020; Yan et al. 2020; Taquet et al. 2021; Shafran et al. 2021; Nalbandian et al. 2021; Perappadan 2021; Chowdhary et al. 2020; Hughes et al. 2020; Tiri et al. 2020; Li et al. 2020; Rossato et al. 2020; Ruan et al. 2020; Zambon 2001; Gill et al. 2010; Weinberger et al. 2012; Du et al. 2020; Garcia-Vidal et al. 2011, 2020; Lansbury et al. 2020; Ramadan et al. 2020; Ripa et al. 2020; Arentz et al. 2020; Manohar et al. 2020; Richardson et al. 2020; Yang et al. 2020; Afrin et al. 2020; Gustine and Jones 2020; Manson et al. 2020; Reddy et al. 2020; Cilloniz et al.

2016; Dudoignon et al. 2020; Martin-Loeches et al. 2017; Burrell et al. 2018; Schauwvlieghe et al. 2018; Ranieri et al. 2012; Gibson et al. 2020; Furfaro et al. 2018; Duplessis and Biswas 2020; Carmody et al. 2010; Lin et al. 2020a; Law et al. 2019; Voelker 2019; Wojewodzic 2020; U.S National Library of Medicine 2020a; Loc-Carrillo and Abedon 2011; Hyman 2019; Samson et al. 2013; Hargreaves and Clokie 2014; Dufour et al. 2019; Van Belleghem et al. 2018; Kim et al. 2008; Singla et al. 2016; Malik et al. 2017; Szermer-Olearnik and Boratyński 2015; Zhang et al. 2020; Puapermpoonsiri et al. 2009; Hodyra-Stefaniak et al. 2015; Roach et al. 2017; Carroll-Portillo and Lin 2019; Kurzępa et al. 2009; Jończyk-Matysiak et al. 2017; Górski et al. 2017, 2019a; Sausset et al. 2020; Borysowski et al. 2017; Pabary et al. 2016; Tothova et al. 2011; Khadke et al. 2020; Mustafa et al. 2020; Ragab et al. 2020; Ye et al. 2020; Costela-Ruiz et al. 2020). The suppression of pro-inflammatory cytokines through the use of phage therapy for the treatment of COVID-19-associated secondary infections may potentially aid in improving patient recovery time. The interferon landscape across the respiratory tracts of COVID-19 patients is complex, with IFN-I and IFN-III exhibiting varying expression levels during early and late infection stages. In the lower respiratory tract, it seems that IFN-III gets overexpressed, while in the upper respiratory tract IFN-1 gets overexpressed in more severe cases (Sposito et al. 2021). The interferon signaling, evasion, and application in COVID-19 are adequately described by Park and Iwasaki (Park and Iwasaki 2020). An *in vitro* study of mice fed with oral bacteriophage T7 observed very minimal changes in serum IFN- γ and other inflammatory cytokines (Park et al. 2014). This study, however, did not specifically test for IFN-I and IFN-III. Further studies examining potential variations in IFN-I and IFN-III could provide greater insight into the potential application of bacteriophages for severe COVID-19 therapy.

25.5 Current Stance on Phage Therapy

The adaptability of phage therapy is reflected in the various strategies that have been employed in various case studies and clinical trials, as illustrated in Fig. 25.2 (Pires et al. 2016; Reuter and Kruger 2020; Wei et al. 2020). According to a study by Gu Liu et al., phage-antibiotic synergy can resuscitate an ineffective antibiotic for previously resistant bacteria as well as lower the minimum inhibitory concentration of antibiotic required to inhibit the growth of a given bacterial strain (Gu Liu et al. 2020). Phages can also be genetically engineered to improve therapeutic efficacy. For instance, the modification of tail fibers and base plates of phages allows for the formulation of a more uniform group of phages that have a shifted or broadened host range. Mahichi et al. transferred long tail fiber genes of *E. coli* phage IP008 with a broader host range to phage T2 of high lytic activity but narrower host range, thus resulting in a chimeric phage with a broader host range and higher virulence (Mahichi et al. 2009). Furthermore, Ando et al. demonstrated that modifying collar genes of phages facilitates its adsorption to host bacteria and alters its host specificity (Ando et al. 2015). Lastly, the discovery of phage-derived antimicrobials, such as

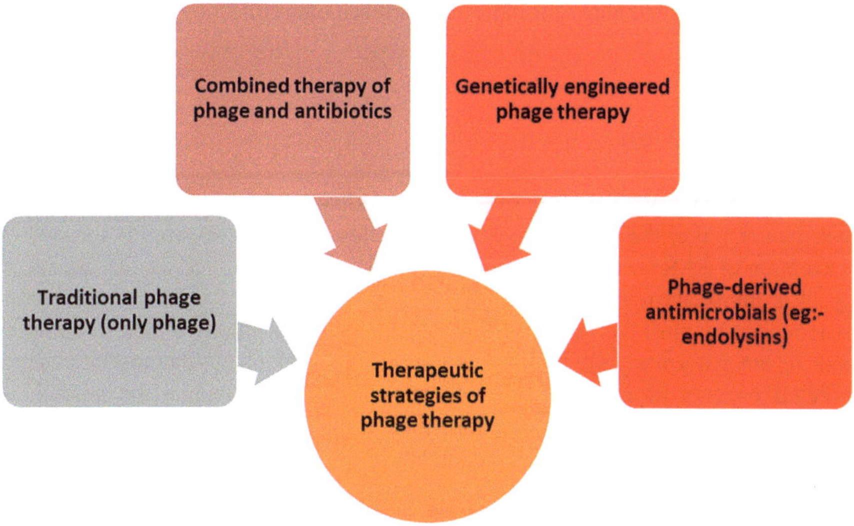

Fig. 25.2 A simplified schematic representation of the various therapeutic strategies of phages used in case studies and clinical trials so far

endolysins, could also potentially replace or augment antibiotics in treating bacterial infections (Cooper et al. 2018).

Although phage therapy has been used for over a decade and has substantially gained momentum over the years, there are no strict regulatory guidelines established governing the use of phages for the treatment of bacterial infections in humans. At present, the regulatory framework surrounding the use of phage therapy in humans varies across different countries, with most countries suggesting compassionate use (Brives and Pourraz 2020). For instance, the Food and Drug Administration in the United States allows the application of phage therapy to critically ill patients, following the emergency investigational new drug pathway (Górski et al. 2019b). On the other hand, the regulatory framework for phage therapy in Belgium is more flexible, which allows the preparation of personalized phage formulations at a pharmacy as per physician's prescription (Pirnay et al. 2018). The current regulatory framework for industrial drug development and medical products is not well suited for phage-based therapeutics, owing to its live nature (Fauconnier 2017; Sharma et al. 2017; Bretaudeau et al. 2020). The legal framework requires extensive modifications and further investigations as to whether phage therapy medical products require a marketing authorization. Lack of accepted regulations and a proper framework for the development and marketing of phage-based therapeutics impedes pharmaceutical companies from investing into the business (Cooper et al. 2016). Therefore, it is strongly recommended that strict international regulatory guidelines/recommendations for the development, marketing, and usage of phage-based therapeutics be established in order to increase user acceptance, improve public awareness, and allow for more standardized data collection on the efficacy of phage therapy.

25.6 Conclusion

Severity of the COVID-19 disease is partially due to secondary bacterial infections that manifest after the primary pathogen infection. As most research is focused on prevention of SARS-CoV-2 infection or drug development to reduce disease severity, it is important to highlight that reducing the severity of secondary infections experienced by most COVID-19 patients may potentially improve their recovery as seen in other hospital case studies who applied bacteriophage therapy to severely affected patients of other diseases. Bacteriophages may reduce the amount of immunogenic bacteria in an infection, thus reducing the burden of the disease experienced by the patient. Novel and already available bacteriophages are relatively easy to isolate and develop for therapeutic purposes. They can be highly specific to their bacterial targets which could significantly reduce the potential of adverse effects experienced by patients. In a time where therapy is in desperate demand, bacteriophage therapy deserves a wider chance of consideration, especially for severely ill patients with bacterial coinfections who may have no other option for therapy.

Acknowledgments None.

Conflict of Interest The authors declare no conflict of interest.

Author Contributions AS, IAF, MKBAA, and KS contributed to the preparation and drafting of the manuscript. Critical revision and editing of the article were done by AS and SKL.

Funding This manuscript was generated through internal funding from the School of Science (SoS), Monash University, Malaysia, and in part by the SoS Strategic Grant 2020.

References

Abedon ST, Kuhl SJ, Blasdel BG, Kutter EM (2011) Phage treatment of human infections. Bacteriophage 1:66–85. https://doi.org/10.4161/bact.1.2.15845

Afrin LB, Weinstock LB, Molderings GJ (2020) Covid-19 hyperinflammation and post-Covid-19 illness may be rooted in mast cell activation syndrome. Int J Infect Dis 100:327–332. https://doi.org/10.1016/j.ijid.2020.09.016

Anand T, Virmani N, Kumar S, Mohanty AK, Pavulraj S, Bera BC, Vaid RK, Ahlawat U, Tripathi BN (2020) Phage therapy for treatment of virulent Klebsiella pneumoniae infection in a mouse model. J Glob Antimicrob Resist 21:34–41. https://doi.org/10.1016/j.jgar.2019.09.018

Ando H, Lemire S, Pires DP, Lu TK (2015) Engineering modular viral scaffolds for targeted bacterial population editing. Cell Syst 1:187–196. https://doi.org/10.1016/j.cels.2015.08.013

Arentz M, Yim E, Klaff L, Lokhandwala S, Riedo FX, Chong M, Lee M (2020) Characteristics and outcomes of 21 critically ill patients with COVID-19 in Washington state. JAMA 323:1612. https://doi.org/10.1001/jama.2020.4326

Aslam S, Courtwright AM, Koval C, Lehman SM, Morales S, Furr CLL, Rosas F, Brownstein MJ, Fackler JR, Sisson BM, Biswas B, Henry M, Luu T, Bivens BN, Hamilton T, Duplessis C, Logan C, Law N, Yung G, Turowski J, Anesi J, Strathdee SA, Schooley RT (2019a) Early clinical experience of bacteriophage therapy in 3 lung transplant recipients. Am J Transplant 19: 2631–2639. https://doi.org/10.1111/ajt.15503

Aslam S, Pretorius V, Lehman SM, Morales S, Schooley RT (2019b) Novel bacteriophage therapy for treatment of left ventricular assist device infection. J Heart Lung Transplant 38:475–476. https://doi.org/10.1016/j.healun.2019.01.001

Bae JY, Il Jun K, Kang CK, Song K-H, Choe PG, Bang J-H, Kim ES, Park SW, Bin Kim H, Kim N-J, Park WB, Oh M (2019) Efficacy of intranasal administration of the recombinant endolysin SAL200 in a lethal murine Staphylococcus aureus pneumonia model. Antimicrob Agents Chemother 63:1–8. https://doi.org/10.1128/AAC.02009-18

Borysowski J, Międzybrodzki R, Wierzbicki P, Kłosowska D, Korczak-Kowalska G, Weber-Dąbrowska B, Górski A (2017) A3R phage and Staphylococcus aureus lysate do not induce neutrophil degranulation. Viruses 9:36. https://doi.org/10.3390/v9020036

Bretaudeau L, Tremblais K, Aubrit F, Meichenin M, Arnaud I (2020) Good manufacturing practice (GMP) compliance for phage therapy medicinal products. Front Microbiol 11:1161. https://doi.org/10.3389/fmicb.2020.01161

Brives C, Pourraz J (2020) Phage therapy as a potential solution in the fight against AMR: obstacles and possible futures. Palgrave Commun 6:100. https://doi.org/10.1057/s41599-020-0478-4

Burrell A, Huckson S, Pilcher DV (2018) ICU admissions for sepsis or pneumonia in Australia and New Zealand in 2017. N Engl J Med 378:2138–2139. https://doi.org/10.1056/NEJMc1717178

Cano EJ, Caflisch KM, Bollyky PL, Van Belleghem JD, Patel R, Fackler J, Brownstein MJ, Horne B, Biswas B, Henry M, Malagon F, Lewallen DG, Suh GA (2020) Phage therapy for limb-threatening prosthetic knee Klebsiella pneumoniae infection: case report and in vitro characterization of anti-biofilm activity. Clin Infect Dis 73:1–8. https://doi.org/10.1093/cid/ciaa705

Carmody LA, Gill JJ, Summer EJ, Sajjan US, Gonzalez CF, Young RF, LiPuma JJ (2010) Efficacy of bacteriophage therapy in a model of Burkholderia cenocepacia pulmonary infection. J Infect Dis 201:264–271. https://doi.org/10.1086/649227

Carroll-Portillo A, Lin HC (2019) Bacteriophage and the innate immune system: access and signaling. Microorganisms 7:625. https://doi.org/10.3390/microorganisms7120625

Chen N, Zhou M, Dong X, Qu J, Gong F, Han Y, Qiu Y, Wang J, Liu Y, Wei Y, Xia J, Yu T, Zhang X, Zhang L (2020) Epidemiological and clinical characteristics of 99 cases of 2019 novel coronavirus pneumonia in Wuhan, China: a descriptive study. Lancet (London, England) 395: 507–513. https://doi.org/10.1016/S0140-6736(20)30211-7

Chowdhary A, Tarai B, Singh A, Sharma A (2020) Multidrug-resistant Candida auris infections in critically ill coronavirus disease patients, India, April-July 2020. Emerg Infect Dis 26:2694–2696. https://doi.org/10.3201/eid2611.203504

Cilloniz C, Martin-Loeches I, Garcia-Vidal C, San Jose A, Torres A (2016) Microbial etiology of pneumonia: epidemiology, diagnosis and resistance patterns. Int J Mol Sci 17:2120. https://doi.org/10.3390/ijms17122120

Clokie MRJ, Millard AD, Letarov AV, Heaphy S (2011) Phages in nature. Bacteriophage 1:31–45. https://doi.org/10.4161/bact.1.1.14942

Cooper CJ, Khan Mirzaei M, Nilsson AS (2016) Adapting drug approval pathways for bacteriophage-based therapeutics. Front Microbiol 7:1–15. https://doi.org/10.3389/fmicb.2016.01209

Cooper C, Koonjan S, Nilsson A (2018) Enhancing whole phage therapy and their derived antimicrobial enzymes through complex formulation. Pharmaceuticals 11:34. https://doi.org/10.3390/ph11020034

Coronaviridae Study Group of the International Committee on Taxonomy of Viruses (2020) The species severe acute respiratory syndrome-related coronavirus: classifying 2019-nCoV and naming it SARS CoV 2. Nat Microbiol 5:536 544. https://doi.org/10.1038/s41564-020-0695-z

Costela-Ruiz VJ, Illescas-Montes R, Puerta-Puerta JM, Ruiz C, Melguizo-Rodríguez L (2020) SARS-CoV-2 infection: the role of cytokines in COVID-19 disease. Cytokine Growth Factor Rev 54:62–75. https://doi.org/10.1016/j.cytogfr.2020.06.001

Cui J, Li F, Shi ZL (2019) Origin and evolution of pathogenic coronaviruses. Nat Rev Microbiol 17: 181–192. https://doi.org/10.1038/s41579-018-0118-9

Dahiya DS, Kichloo A, Albosta M, Pagad S, Wani F (2020) Gastrointestinal implications in COVID-19. J Investig Med 68:1397–1401. https://doi.org/10.1136/jim-2020-001559

Doub JB, Ng VY, Johnson AJ, Slomka M, Fackler J, Horne B, Brownstein MJ, Henry M, Malagon F, Biswas B (2020) Salvage bacteriophage therapy for a chronic MRSA prosthetic joint infection. Antibiotics 9:241. https://doi.org/10.3390/antibiotics9050241

Du R-H, Liang L-R, Yang C-Q, Wang W, Cao T-Z, Li M, Guo G-Y, Du J, Zheng C-L, Zhu Q, Hu M, Li X-Y, Peng P, Shi H-Z (2020) Predictors of mortality for patients with COVID-19 pneumonia caused by SARS-CoV-2: a prospective cohort study. Eur Respir J 55:2000524. https://doi.org/10.1183/13993003.00524-2020

Dudoignon E, Caméléna F, Deniau B, Habay A, Coutrot M, Ressaire Q, Plaud B, Berçot B, Dépret F (2020) Bacterial pneumonia in COVID-19 critically ill patients: a case series. Clin Infect Dis 72:1–2. https://doi.org/10.1093/cid/ciaa762

Dufour N, Delattre R, Chevallereau A, Ricard JD, Debarbieux L (2019) Phage therapy of pneumonia is not associated with an overstimulation of the inflammatory response compared to antibiotic treatment in mice. Antimicrob Agents Chemother 63:1–12. https://doi.org/10.1128/AAC.00379-19

Duplessis CA, Biswas B (2020) A review of topical phage therapy for chronically infected wounds and preparations for a randomized adaptive clinical trial evaluating topical phage therapy in chronically infected diabetic foot ulcers. Antibiotics 9:377. https://doi.org/10.3390/antibiotics9070377

Eurosurveillance Editorial Team (2020) Note from the editors: World Health Organization declares novel coronavirus (2019-nCoV) sixth public health emergency of international concern. Euro Surveill 25:200131e. https://doi.org/10.2807/1560-7917.ES.2020.25.5.200131e

Fan J, Li X, Gao Y, Zhou J, Wang S, Huang B, Wu J, Cao Q, Chen Y, Wang Z, Luo D, Zhou T, Li R, Shang Y, Nie X (2020) The lung tissue microbiota features of 20 deceased patients with COVID-19. J Infect 81:e64–e67. https://doi.org/10.1016/j.jinf.2020.06.047

Fauconnier A (2017) Regulating phage therapy. EMBO Rep 18:198–200. https://doi.org/10.15252/embr.201643250

Furfaro LL, Payne MS, Chang BJ (2018) Bacteriophage therapy: clinical trials and regulatory hurdles. Front Cell Infect Microbiol 8:376. https://doi.org/10.3389/fcimb.2018.00376

Garcia-Vidal C, Barba P, Arnan M, Moreno A, Ruiz-Camps I, Gudiol C, Ayats J, Orti G, Carratala J (2011) Invasive aspergillosis complicating pandemic influenza a (H1N1) infection in severely immunocompromised patients. Clin Infect Dis 53:e16–e19. https://doi.org/10.1093/cid/cir485

Garcia-Vidal C, Sanjuan G, Moreno-García E, Puerta-Alcalde P, Garcia-Pouton N, Chumbita M, Fernandez-Pittol M, Pitart C, Inciarte A, Bodro M, Morata L, Ambrosioni J, Grafia I, Meira F, Macaya I, Cardozo C, Casals C, Tellez A, Castro P, Marco F, García F, Mensa J, Martínez JA, Soriano A, Rico V, Hernández-Meneses M, Agüero D, Torres B, González A, de la Mora L, Rojas J, Linares L, Fidalgo B, Rodriguez N, Nicolas D, Albiach L, Muñoz J, Almuedo A, Camprubí D, Angeles Marcos M, Camprubí D, Cilloniz C, Fernández S, Nicolas JM, Torres A (2020) Incidence of co-infections and superinfections in hospitalized patients with COVID-19: a retrospective cohort study. Clin Microbiol Infect 27:83. https://doi.org/10.1016/j.cmi.2020.07.041

Geiss B (2020) Older people are at more risk from COVID-19 because of how the immune system ages. The Conversation Media Group Ltd. https://theconversation.com/older-people-are-at-more-risk-from-covid-19-because-of-how-the-immune-system-ages-133899. Accessed 4 Dec 2020

Gibson PG, Qin L, Puah SH (2020) COVID-19 acute respiratory distress syndrome (ARDS): clinical features and differences from typical pre-COVID-19 ARDS. Med J Aust 213:54–56. e1. https://doi.org/10.5694/mja2.50674

Gilbey T, Ho J, Cooley LA, Petrovic Fabijan A, Iredell JR (2019) Adjunctive bacteriophage therapy for prosthetic valve endocarditis due to Staphylococcus aureus. Med J Aust 211:142. https://doi.org/10.5694/mja2.50274

Gill JR, Sheng Z-M, Ely SF, Guinee DG, Beasley MB, Suh J, Deshpande C, Mollura DJ, Morens DM, Bray M, Travis WD, Taubenberger JK (2010) Pulmonary pathologic findings of fatal 2009 pandemic influenza a/H1N1 viral infections. Arch Pathol Lab Med 134:235–243. https://doi.org/10.1043/1543-2165-134.2.235

Ginsburg AS, Klugman KP (2020) COVID-19 pneumonia and the appropriate use of antibiotics. Lancet Glob Health 8:e1453–e1454. https://doi.org/10.1016/S2214-109X(20)30444-7

Gondil VS, Dube T, Panda JJ, Yennamalli RM, Harjai K, Chhibber S (2020) Comprehensive evaluation of chitosan nanoparticle based phage lysin delivery system; a novel approach to counter S. pneumoniae infections. Int J Pharm 573:118850. https://doi.org/10.1016/j.ijpharm.2019.118850

Górski A, Dąbrowska K, Międzybrodzki R, Weber-Dąbrowska B, Łusiak-Szelachowska M, Jończyk-Matysiak E, Borysowski J (2017) Phages and immunomodulation. Future Microbiol 12:905–914. https://doi.org/10.2217/fmb-2017-0049

Górski A, Międzybrodzki R, Jończyk-Matysiak E, Żaczek M, Borysowski J (2019a) Phage-specific diverse effects of bacterial viruses on the immune system. Future Microbiol 14:1171–1174. https://doi.org/10.2217/fmb-2019-0222

Górski A, Międzybrodzki R, Węgrzyn G, Jończyk-Matysiak E, Borysowski J, Weber-Dąbrowska B (2019b) Phage therapy: current status and perspectives. Med Res Rev 40:459–463. https://doi.org/10.1002/med.21593

Gu Liu C, Green SI, Min L, Clark JR, Salazar KC, Terwilliger AL, Kaplan HB, Trautner BW, Ramig RF, Maresso AW (2020) Phage-antibiotic synergy is driven by a unique combination of antibacterial mechanism of action and stoichiometry. mBio 11:1–19. https://doi.org/10.1128/mBio.01462-20

Guan W, Ni Z, Hu Y, Liang W, Ou C, He J, Liu L, Shan H, Lei C, Hui DSC, Du B, Li L, Zeng G, Yuen K-Y, Chen R, Tang C, Wang T, Chen P, Xiang J, Li S, Wang J, Liang Z, Peng Y, Wei L, Liu Y, Hu Y, Peng P, Wang J, Liu J, Chen Z, Li G, Zheng Z, Qiu S, Luo J, Ye C, Zhu S, Zhong N (2020) Clinical characteristics of coronavirus disease 2019 in China. N Engl J Med 382:1708–1720. https://doi.org/10.1056/NEJMoa2002032

Gupta P, Singh HS, Shukla VK, Nath G, Bhartiya SK (2019) Bacteriophage therapy of chronic nonhealing wound: clinical study. Int J Low Extrem Wounds 18:171–175. https://doi.org/10.1177/1534734619835115

Gustine JN, Jones D (2020) Immunopathology of hyperinflammation in COVID-19. Am J Pathol 191:4. https://doi.org/10.1016/j.ajpath.2020.08.009

Hargreaves KR, Clokie MRJ (2014) Clostridium difficile phages: still difficult? Front Microbiol 5: 1–14. https://doi.org/10.3389/fmicb.2014.00184

Hodyra-Stefaniak K, Miernikiewicz P, Drapała J, Drab M, Jończyk-Matysiak E, Lecion D, Kaźmierczak Z, Beta W, Majewska J, Harhala M, Bubak B, Kłopot A, Górski A, Dąbrowska K (2015) Mammalian host-versus-phage immune response determines phage fate in vivo. Sci Rep 5:14802. https://doi.org/10.1038/srep14802

Hong H, Wang Y, Chung H-T, Chen C-J (2020) Clinical characteristics of novel coronavirus disease 2019 (COVID-19) in newborns, infants and children. Pediatr Neonatol 61:131–132. https://doi.org/10.1016/j.pedneo.2020.03.001

Hopkins J (2021) University & Medicine, COVID-19 Dashboard by the Center for Systems Science and Engineering (CSSE) at John Hopkins University (JHU). https://coronavirus.jhu.edu/map.html. Accessed 6 Jan 2021

Huang C, Wang Y, Li X, Ren L, Zhao J, Hu Y, Zhang L, Fan G, Xu J, Gu X, Cheng Z, Yu T, Xia J, Wei Y, Wu W, Xie X, Yin W, Li H, Liu M, Xiao Y, Gao H, Guo L, Xie J, Wang G, Jiang R,

Gao Z, Jin Q, Wang J, Cao B (2020) Clinical features of patients infected with 2019 novel coronavirus in Wuhan, China. Lancet (London, England) 395:497–506. https://doi.org/10.1016/S0140-6736(20)30183-5

Hughes S, Troise O, Donaldson H, Mughal N, Moore LSP (2020) Bacterial and fungal coinfection among hospitalized patients with COVID-19: a retrospective cohort study in a UK secondary-care setting. Clin Microbiol Infect 26:1395–1399. https://doi.org/10.1016/j.cmi.2020.06.025

Hyman P (2019) Phages for phage therapy: isolation, characterization, and host range breadth. Pharmaceuticals 12:35. https://doi.org/10.3390/ph12010035

Jault P, Leclerc T, Jennes S, Pirnay JP, Que YA, Resch G, Rousseau AF, Ravat F, Carsin H, Le Floch R, Schaal JV, Soler C, Fevre C, Arnaud I, Bretaudeau L, Gabard J (2019) Efficacy and tolerability of a cocktail of bacteriophages to treat burn wounds infected by Pseudomonas aeruginosa (PhagoBurn): a randomised, controlled, double-blind phase 1/2 trial. Lancet Infect Dis 19:35–45. https://doi.org/10.1016/S1473-3099(18)30482-1

Jeon J, Yong D (2019) Two novel bacteriophages improve survival in galleria mellonella infection and mouse acute pneumonia models infected with extensively drug-resistant Pseudomonas aeruginosa. Appl Environ Microbiol 85:1–16. https://doi.org/10.1128/AEM.02900-18

Jeon J, Park J-H, Yong D (2019) Efficacy of bacteriophage treatment against carbapenem-resistant Acinetobacter baumannii in galleria mellonella larvae and a mouse model of acute pneumonia. BMC Microbiol 19:70. https://doi.org/10.1186/s12866-019-1443-5

Ji Y, Cheng M, Zhai S, Xi H, Cai R, Wang Z, Zhang H, Wang X, Xue Y, Li X, Sun C, Feng X, Lei L et al (2019) Preventive effect of the phage VB-SavM-JYL01 on rabbit necrotizing pneumonia caused by Staphylococcus aureus. Vet Microbiol 229:72–80. https://doi.org/10.1016/j.vetmic.2018.12.021

Jończyk-Matysiak E, Weber-Dąbrowska B, Owczarek B, Międzybrodzki R, Łusiak-Szelachowska M, Łodej N, Górski A (2017) Phage-phagocyte interactions and their implications for phage application as therapeutics. Viruses 9:150. https://doi.org/10.3390/v9060150

Khadke S, Ahmed N, Ahmed N, Ratts R, Raju S, Gallogly M, de Lima M, Sohail MR (2020) Harnessing the immune system to overcome cytokine storm and reduce viral load in COVID-19: a review of the phases of illness and therapeutic agents. Virol J 17:154. https://doi.org/10.1186/s12985-020-01415-w

Kim K-P, Cha J-D, Jang E-H, Klumpp J, Hagens S, Hardt W-D, Lee K-Y, Loessner MJ (2008) PEGylation of bacteriophages increases blood circulation time and reduces T-helper type 1 immune response. Microb Biotechnol 1:247–257. https://doi.org/10.1111/j.1751-7915.2008.00028.x

Kortright KE, Chan BK, Koff JL, Turner PE (2019) Phage therapy: a renewed approach to combat antibiotic-resistant bacteria. Cell Host Microbe 25:219–232. https://doi.org/10.1016/j.chom.2019.01.014

Kurzępa A, Dąbrowska K, Skaradziński G, Górski A (2009) Bacteriophage interactions with phagocytes and their potential significance in experimental therapy. Clin Exp Med 9:93. https://doi.org/10.1007/s10238-008-0027-8

Lai C, Shih T, Ko W, Tang H, Hsueh P (2020) Severe acute respiratory syndrome coronavirus 2 (SARS-CoV-2) and coronavirus disease-2019 (COVID-19): the epidemic and the challenges. Int J Antimicrob Agents 55:105924. https://doi.org/10.1016/j.ijantimicag.2020.105924

Lansbury L, Lim B, Baskaran V, Lim WS (2020) Co-infections in people with COVID-19: a systematic review and meta-analysis. SSRN Electron J 81(2):266. https://doi.org/10.2139/ssrn.3594598

Law N, Logan C, Yung G, Furr C-LL, Lehman SM, Morales S, Rosas F, Gaidamaka A, Bilinsky I, Grint P, Schooley RT, Aslam S (2019) Successful adjunctive use of bacteriophage therapy for treatment of multidrug-resistant Pseudomonas aeruginosa infection in a cystic fibrosis patient. Infection 47:665–668. https://doi.org/10.1007/s15010-019-01319-0

Lechien JR, Chiesa-Estomba CM, De Siati DR, Horoi M, Le Bon SD, Rodriguez A, Dequanter D, Blecic S, El Afia F, Distinguin L, Chekkoury-Idrissi Y, Hans S, Delgado IL, Calvo-Henriquez C,

Lavigne P, Falanga C, Barillari MR, Cammaroto G, Khalife M, Leich P, Souchay C, Rossi C, Journe F, Hsieh J, Edjlali M, Carlier R, Ris L, Lovato A, De Filippis C, Coppee F, Fakhry N, Ayad T, Saussez S (2020) Olfactory and gustatory dysfunctions as a clinical presentation of mild-to-moderate forms of the coronavirus disease (COVID-19): a multicenter European study. Eur Arch Otorhinolaryngol 277(8):2251–2261. https://doi.org/10.1007/s00405-020-05965-1

Lee P-I, Hu Y-L, Chen P-Y, Huang Y-C, Hsueh P-R (2020) Are children less susceptible to COVID-19? J Microbiol Immunol Infect 53:371–372. https://doi.org/10.1016/j.jmii.2020.02.011

Li J, Wang J, Yang Y, Cai P, Cao J, Cai X, Zhang Y (2020) Etiology and antimicrobial resistance of secondary bacterial infections in patients hospitalized with COVID-19 in Wuhan, China: a retrospective analysis. Antimicrob Resist Infect Control 9:153. https://doi.org/10.1186/s13756-020-00819-1

Lin YW, Chang RY, Rao GG, Jermain B, Han M-L, Zhao JX, Chen K, Wang JP, Barr JJ, Schooley RT, Kutter E, Chan H-K, Li J (2020a) Pharmacokinetics/pharmacodynamics of antipseudomonal bacteriophage therapy in rats: a proof-of-concept study. Clin Microbiol Infect 26:1229–1235. https://doi.org/10.1016/j.cmi.2020.04.039

Lin Y, Quan D, Chang RYK, Chow MYT, Wang Y, Li M, Morales S, Britton WJ, Kutter E, Li J, Chan H-K (2020b) Synergistic activity of phage PEV20-ciprofloxacin combination powder formulation—a proof-of-principle study in a P. aeruginosa lung infection model. Eur J Pharm Biopharm 158:166–171. https://doi.org/10.1016/j.ejpb.2020.11.019

Loc-Carrillo C, Abedon ST (2011) Pros and cons of phage therapy. Bacteriophage 1:111–114. https://doi.org/10.4161/bact.1.2.14590

Lozada-Nur F, Chainani-Wu N, Fortuna G, Sroussi H (2020) Dysgeusia in COVID-19: possible mechanisms and implications. Oral Surg Oral Med Oral Pathol Oral Radiol 130:344–346. https://doi.org/10.1016/j.oooo.2020.06.016

Luscher A, Simonin J, Falconnet L, Valot B, Hocquet D, Chanson M, Resch G, Köhler T, van Delden C (2020) Combined bacteriophage and antibiotic treatment prevents Pseudomonas aeruginosa infection of wild type and cftr- epithelial cells. Front Microbiol 11:1. https://doi.org/10.3389/fmicb.2020.01947

Mahichi F, Synnott AJ, Yamamichi K, Osada T, Tanji Y (2009) Site-specific recombination of T2 phage using IP008 long tail fiber genes provides a targeted method for expanding host range while retaining lytic activity. FEMS Microbiol Lett 295:211–217. https://doi.org/10.1111/j.1574-6968.2009.01588.x

Malik DJ, Sokolov IJ, Vinner GK, Mancuso F, Cinquerrui S, Vladisavljevic GT, Clokie MRJ, Garton NJ, Stapley AGF, Kirpichnikova A (2017) Formulation, stabilisation and encapsulation of bacteriophage for phage therapy. Adv Colloid Interf Sci 249:100–133. https://doi.org/10.1016/j.cis.2017.05.014

Manohar P, Loh B, Nachimuthu R, Hua X, Welburn SC, Leptihn S (2020) Secondary bacterial infections in patients with viral pneumonia. Front Med 7:2000524. https://doi.org/10.3389/fmed.2020.00420

Manson JJ, Crooks C, Naja M, Ledlie A, Goulden B, Liddle T, Khan E, Mehta P, Martin-Gutierrez L, Waddington KE, Robinson GA, Ribeiro Santos L, McLoughlin E, Snell A, Adeney C, Schim van der Loeff I, Baker KF, Duncan CJA, Hanrath AT, Lendrem BC, De Soyza A, Peng J, J'Bari H, Greenwood M, Hawkins E, Peckham H, Marks M, Rampling T, Luintel A, Williams B, Brown M, Singer M, West J, Jury EC, Collin M, Tattersall RS (2020) COVID-19-associated hyperinflammation and escalation of patient care: a retrospective longitudinal cohort study. Lancet Rheumatol 2:e594–e602. https://doi.org/10.1016/S2665-9913(20)30275-7

Martin-Loeches I, Schultz MJ, Vincent J-L, Alvarez-Lerma F, Bos LD, Solé-Violán J, Torres A, Rodriguez A (2017) Increased incidence of co-infection in critically ill patients with influenza. Intensive Care Med 43:48–58. https://doi.org/10.1007/s00134-016-4578-y

Mustafa MI, Abdelmoneim AH, Mahmoud EM, Makhawi AM (2020) Cytokine storm in COVID-19 patients, its impact on organs and potential treatment by QTY code-designed detergent-free chemokine receptors. Mediat Inflamm 2020:1–7. https://doi.org/10.1155/2020/8198963

Nalbandian A, Sehgal K, Gupta A, Madhavan MV, McGroder C, Stevens JS, Cook JR, Nordvig AS, Shalev D, Sehrawat TS, Ahluwalia N, Bikdeli B, Dietz D, Der-Nigoghossian C, Liyanage-Don N, Rosner GF, Bernstein EJ, Mohan S, Beckley AA, Seres DS, Choueiri TK, Uriel N, Ausiello JC, Accili D, Freedberg DE, Baldwin M, Schwartz A, Brodie D, Garcia CK, Elkind MSV, Connors JM, Bilezikian JP, Landry DW, Wan EY (2021) Post-acute COVID-19 syndrome. Nat Med 27:601–615. https://doi.org/10.1038/s41591-021-01283-z

Onsea J, Soentjens P, Djebara S, Merabishvili M, Depypere M, Spriet I, De Munter P, Debaveye Y, Nijs S, Vanderschot P, Wagemans J (2019) Bacteriophage application for difficult-to-treat musculoskeletal infections: development of a standardized multidisciplinary treatment protocol. Viruses 11:891. https://doi.org/10.3390/v11100891

Pabary R, Singh C, Morales S, Bush A, Alshafi K, Bilton D, Alton EWFW, Smithyman A, Davies JC (2016) Antipseudomonal bacteriophage reduces infective burden and inflammatory response in murine lung. Antimicrob Agents Chemother 60:744–751. https://doi.org/10.1128/AAC.01426-15

Park A, Iwasaki A (2020) Type I and type III interferons—induction, signaling, evasion, and application to combat COVID-19. Cell Host Microbe 27:870–878. https://doi.org/10.1016/j.chom.2020.05.008

Park K, Cha KE, Myung H (2014) Observation of inflammatory responses in mice orally fed with bacteriophage T7. J Appl Microbiol 117:627–633. https://doi.org/10.1111/jam.12565

Perappadan BS (2021) Doctors warn of severe post-COVID secondary infections

Petrovic Fabijan A, Lin RCY, Ho J, Maddocks S, Ben Zakour NL, Iredell JR, Khalid A, Venturini C, Chard R, Morales S, Sandaradura I, Gilbey T (2020) Safety of bacteriophage therapy in severe Staphylococcus aureus infection. Nat Microbiol 5:465–472. https://doi.org/10.1038/s41564-019-0634-z

Pires DP, Cleto S, Sillankorva S, Azeredo J, Lu TK (2016) Genetically engineered phages: a review of advances over the last decade. Microbiol Mol Biol Rev 80:523–543. https://doi.org/10.1128/MMBR.00069-15

Pirnay J-P, Verbeken G, Ceyssens P-J, Huys I, De Vos D, Ameloot C, Fauconnier A (2018) The magistral phage. Viruses 10:64. https://doi.org/10.3390/v10020064

Puapermpoonsiri U, Spencer J, van der Walle CF (2009) A freeze-dried formulation of bacteriophage encapsulated in biodegradable microspheres. Eur J Pharm Biopharm 72:26–33. https://doi.org/10.1016/j.ejpb.2008.12.001

Ragab D, Salah Eldin H, Taeimah M, Khattab R, Salem R (2020) The COVID-19 cytokine storm; what we know so far. Front Immunol 11:1–4. https://doi.org/10.3389/fimmu.2020.01446

Ramadan HK-A, Mahmoud MA, Aburahma MZ, Elkhawaga AA, El-Mokhtar MA, Sayed IM, Hosni A, Hassany SM, Medhat MA (2020) Predictors of severity and co-infection resistance profile in COVID-19 patients: first report from upper Egypt. Infect Drug Resist 13:3409–3422. https://doi.org/10.2147/IDR.S272605

Ranieri VM, Rubenfeld GD, Thompson BT, Ferguson ND, Caldwell E, Fan E, Camporota L, Slutsky AS (2012) Acute respiratory distress syndrome. JAMA 307:2526–2533. https://doi.org/10.1001/jama.2012.5669

Reddy K, Rogers AJ, McAuley DF (2020) Delving beneath the surface of hyperinflammation in COVID-19. Lancet Rheumatol 2:e578–e579. https://doi.org/10.1016/S2665-9913(20)30304-0

Reuter M, Kruger DH (2020) Approaches to optimize therapeutic bacteriophage and bacteriophage-derived products to combat bacterial infections. Virus Genes 56:136–149. https://doi.org/10.1007/s11262-020-01735-7

Richardson S, Hirsch JS, Narasimhan M, Crawford JM, McGinn T, Davidson KW, Barnaby DP, Becker LB, Chelico JD, Cohen SL, Cookingham J, Coppa K, Diefenbach MA, Dominello AJ, Duer-Hefele J, Falzon L, Gitlin J, Hajizadeh N, Harvin TG, Hirschwerk DA, Kim EJ, Kozel ZM, Marrast LM, Mogavero JN, Osorio GA, Qiu M, Zanos TP (2020) Presenting

Characteristics, comorbidities, and outcomes among 5700 patients hospitalized with COVID-19 in the New York City area. JAMA 323:2052. https://doi.org/10.1001/jama.2020.6775
Ripa M, Galli L, Poli A, Oltolini C, Spagnuolo V, Mastrangelo A, Muccini C, Monti G, De Luca G, Landoni G, Dagna L, Clementi M, Rovere Querini P, Ciceri F, Tresoldi M, Lazzarin A, Zangrillo A, Scarpellini P, Castagna A, Andolina A, Redaelli MB, Bigai G, Bigoloni A, Borio G, Bossolasco S, Bruzzesi E, Calabrò MG, Calvisi S, Campochiaro C, Canetti D, Canti V, Castellani J, Castiglioni B, Cavalli G, Cavallo L, Cernuschi M, Chiurlo M, Cilla M, Cinel E, Cinque P, Conte C, Da Prat V, Danise A, De Lorenzo R, Dell'Acqua A, Dell'Acqua R, Della Torre E, Della Torre L, Di Terlizzi G, Dumea I, Farolfi F, Ferrante M, Frangi C, Fumagalli L, Gallina G, Germinario B, Gianotti N, Guffanti M, Hasson H, Lalla F, Lanzillotta M, Li Voti R, Messina E, Molinari C, Moizo E, Montagna M, Morsica G, Nozza S, Pascali M, Patrizi A, Pieri M, Poloniato A, Prestifilippo D, Ramirez G, Ranzenigo M, Sapienza J, Seghi F, Tambussi G, Tassan Din C, Turi S, Uberti-Foppa C, Vinci C (2020) Secondary infections in patients hospitalized with COVID-19: incidence and predictive factors. Clin Microbiol Infect 2:2–8. https://doi.org/10.1016/j.cmi.2020.10.021
Roach DR, Leung CY, Henry M, Morello E, Singh D, Di Santo JP, Weitz JS, Debarbieux L (2017) Synergy between the host immune system and bacteriophage is essential for successful phage therapy against an acute respiratory pathogen. Cell Host Microbe 22:38–47.e4. https://doi.org/10.1016/j.chom.2017.06.018
Rossato L, Negrão FJ, Simionatto S (2020) Could the COVID-19 pandemic aggravate antimicrobial resistance? Am J Infect Control 48:1129–1130. https://doi.org/10.1016/j.ajic.2020.06.192
Ruan Q, Yang K, Wang W, Jiang L, Song J (2020) Clinical predictors of mortality due to COVID-19 based on an analysis of data of 150 patients from Wuhan, China. Intensive Care Med 46:846–848. https://doi.org/10.1007/s00134-020-05991-x
Rubalskii E, Ruemke S, Salmoukas C, Boyle EC, Warnecke G, Tudorache I, Shrestha M, Schmitto JD, Martens A, Rojas SV, Ziesing S, Bochkareva S, Kuehn C, Haverich A (2020) Bacteriophage therapy for critical infections related to cardiothoracic surgery. Antibiotics 9:232. https://doi.org/10.3390/antibiotics9050232
Samanta J, Dhar J, Khaliq A, Kochhar R (2020) 2019 novel coronavirus infection: gastrointestinal manifestations. J Dig Endosc 11:13–18. https://doi.org/10.1055/s-0040-1712077
Samson JE, Magadán AH, Sabri M, Moineau S (2013) Revenge of the phages: defeating bacterial defences. Nat Rev Microbiol 11:675–687. https://doi.org/10.1038/nrmicro3096
Santajit S, Indrawattana N (2016) Mechanisms of antimicrobial resistance in ESKAPE pathogens. Biomed Res Int 2016:1–8. https://doi.org/10.1155/2016/2475067
Sausset R, Petit MA, Gaboriau-Routhiau V, De Paepe M (2020) New insights into intestinal phages. Mucosal Immunol 13:205–215. https://doi.org/10.1038/s41385-019-0250-5
Schauwvlieghe AFAD, Rijnders BJA, Philips N, Verwijs R, Vanderbeke L, Van Tienen C, Lagrou K, Verweij PE, Van de Veerdonk FL, Gommers D, Spronk P, Bergmans DCJJ, Hoedemaekers A, Andrinopoulou E-R, van den Berg CHSB, Juffermans NP, Hodiamont CJ, Vonk AG, Depuydt P, Boelens J, Wauters J (2018) Invasive aspergillosis in patients admitted to the intensive care unit with severe influenza: a retrospective cohort study. Lancet Respir Med 6: 782–792. https://doi.org/10.1016/S2213-2600(18)30274-1
Shafran N, Shafran I, Ben-Zvi H, Sofer S, Sheena L, Krause I, Shlomai A, Goldberg E, Sklan EH (2021) Secondary bacterial infection in COVID-19 patients is a stronger predictor for death compared to influenza patients. Sci Rep 11:1–8. https://doi.org/10.1038/s41598-021-92220-0
Sharma S, Chatterjee S, Datta S, Prasad R, Dubey D, Prasad RK, Vairale MG (2017) Bacteriophages and its applications: an overview. Folia Microbiol 62:17–55. https://doi.org/10.1007/s12223-016-0471-x
Singla S, Harjai K, Katare OP, Chhibber S (2016) Encapsulation of bacteriophage in liposome accentuates its entry in to macrophage and shields it from neutralizing antibodies. PLoS One 11: e0153777. https://doi.org/10.1371/journal.pone.0153777

Soleimani Sasani M, Eftekhar F (2020) Potential of a bacteriophage isolated from wastewater in treatment of lobar pneumonia infection induced by Klebsiella pneumoniae in mice. Curr Microbiol 77:2650–2655. https://doi.org/10.1007/s00284-020-02041-z

Sposito B, Broggi A, Pandolfi L, Crotta S, Ferrarese R, Sisti S, Clementi N, Ambrosi A, Liu E, Frangipane V, Saracino L, Marongiu L, Facchini FA, Bottazzi A, Fossali T, Colombo R, Clementi M, Tagliabue E, Pontiroli AE, Meloni F, Wack A, Mancini N, Zanoni I (2021) Severity of SARS-CoV-2 infection as a function of the interferon landscape across the respiratory tract of COVID-19 patients. bioRxiv. https://doi.org/10.1101/2021.03.30.437173

Sulakvelidze A, Alavidze Z, Morris JG (2001) Bacteriophage therapy. Antimicrob Agents Chemother 45:649–659. https://doi.org/10.1128/AAC.45.3.649-659.2001

Szermer-Olearnik B, Boratyński J (2015) Removal of endotoxins from bacteriophage preparations by extraction with organic solvents. PLoS One 10:e0122672. https://doi.org/10.1371/journal.pone.0122672

Tagliaferri TL, Jansen M, Horz H-P (2019) Fighting pathogenic bacteria on two fronts: phages and antibiotics as combined strategy. Front Cell Infect Microbiol 9:22. https://doi.org/10.3389/fcimb.2019.00022

Taquet M, Dercon Q, Luciano S, Geddes JR, Husain M, Harrison PJ (2021) Incidence, co-occurrence, and evolution of long-COVID features: a 6-month retrospective cohort study of 273,618 survivors of COVID-19. PLoS Med 18:e1003773. https://doi.org/10.1371/journal.pmed.1003773

Tiri B, Sensi E, Marsiliani V, Cantarini M, Priante G, Vernelli C, Martella LA, Costantini M, Mariottini A, Andreani P, Bruzzone P, Suadoni F, Francucci M, Cirocchi R, Cappanera S (2020) Antimicrobial stewardship program, COVID-19, and infection control: spread of carbapenem-resistant Klebsiella Pneumoniae colonization in ICU COVID-19 patients. What did not work? J Clin Med 9:1–9. https://doi.org/10.3390/jcm9092744

Tothova L, Celec P, Babickova J, Gajdosova J, Al-Alami H, Kamodyova N, Drahovska H, Liptakova A, Turna J, Hodosy J (2011) Phage therapy of Cronobacter-induced urinary tract infection in mice. Med Sci Monit 17:BR173–BR178. https://doi.org/10.12659/MSM.881844

U.S National Library of Medicine (2013) Bacteriophage effects on Pseudomonas aeruginosa (MUCOPHAGES). https://clinicaltrials.gov/ct2/show/NCT01818206?term=phage&cond=lung&draw=2. Accessed 23 Dec 2020

U.S National Library of Medicine (2020a) Personalized phage treatment in Covid-19 patients with bacterial co-infections microbials for pneumonia or bacteremia/septicemia. https://clinicaltrials.gov/ct2/show/NCT04636554?term=phage&cond=pneumonia. Accessed 23 Dec 2020

U.S National Library of Medicine (2020b) The effect of supraglottic and oropharyngeal decontamination on the incidence of ventilator-associated pneumonia (SGDC-VAP). https://clinicaltrials.gov/ct2/show/NCT04325685?term=phage&cond=pneumonia. Accessed 25 Dec 2020

U.S National Library of Medicine (2020c) Ph 1/2 study evaluating safety and tolerability of inhaled AP-PA02 in subjects with chronic Pseudomonas aeruginosa lung infections and cystic fibrosis (SWARM-Pa). https://clinicaltrials.gov/ct2/show/NCT04596319?term=phage&cond=lung&draw=2. Accessed 23 Dec 2020

Van Belleghem J, Dąbrowska K, Vaneechoutte M, Barr J, Bollyky P (2018) Interactions between bacteriophage, bacteria, and the mammalian immune system. Viruses 11:10. https://doi.org/10.3390/v11010010

Voelker R (2019) FDA approves bacteriophage trial. JAMA 321:638. https://doi.org/10.1001/jama.2019.0510

Wang D, Hu B, Hu C, Zhu F, Liu X, Zhang J, Wang B, Xiang H, Cheng Z, Xiong Y, Zhao Y, Li Y, Wang X, Peng Z (2020a) Clinical characteristics of 138 hospitalized patients with 2019 novel coronavirus-infected pneumonia in Wuhan, China. JAMA 323:1061–1069. https://doi.org/10.1001/jama.2020.1585

Wang Z, Yang B, Li Q, Wen L, Zhang R (2020b) Clinical features of 69 cases with coronavirus disease 2019 in Wuhan, China. Clin Infect Dis 71:769–777. https://doi.org/10.1093/cid/ciaa272

Wei J, Peng N, Liang Y, Li K, Li Y (2020) Phage therapy: consider the past, embrace the future. Appl Sci 10:7654. https://doi.org/10.3390/app10217654

Weinberger DM, Simonsen L, Jordan R, Steiner C, Miller M, Viboud C (2012) Impact of the 2009 influenza pandemic on pneumococcal pneumonia hospitalizations in the United States. J Infect Dis 205:458–465. https://doi.org/10.1093/infdis/jir749

Wojewodzic MW (2020) Bacteriophages could be a potential game changer in the trajectory of coronavirus disease (COVID-19). PHAGE 1:60–65. https://doi.org/10.1089/phage.2020.0014

World Health Organization (2020a) Novel coronavirus (2019-nCoV) situation report, p 11

World Health Organization (2020b) WHO Coronavirus Disease (COVID-19) Dashboard. https://covid19.who.int/. Accessed 3 Dec 2020

World Health Organization, Antibiotic Resistance (2020). https://www.who.int/news-room/fact-sheets/detail/antibiotic-resistance. Accessed 4 Dec 2020

Worldometer (2021) COVID-19 coronavirus pandemic. https://www.worldometers.info/coronavirus/. Accessed 16 Jan 2021

Xia W, Shao J, Guo Y, Peng X, Li Z, Hu D (2020) Clinical and CT features in pediatric patients with COVID-19 infection: different points from adults. Pediatr Pulmonol 55:1169–1174. https://doi.org/10.1002/ppul.24718

Yan CH, Faraji F, Prajapati DP, Boone CE, DeConde AS (2020) Association of chemosensory dysfunction and COVID-19 in patients presenting with influenza-like symptoms. Int Forum Allergy Rhinol 10:806–813. https://doi.org/10.1002/alr.22579

Yang L, Tu L (2020) Implications of gastrointestinal manifestations of COVID-19. Lancet Gastroenterol Hepatol 5:629–630. https://doi.org/10.1016/S2468-1253(20)30132-1

Yang X, Yu Y, Xu J, Shu H, Xia J, Liu H, Wu Y, Zhang L, Yu Z, Fang M, Yu T, Wang Y, Pan S, Zou X, Yuan S, Shang Y (2020) Clinical course and outcomes of critically ill patients with SARS-CoV-2 pneumonia in Wuhan, China: a single-centered, retrospective, observational study. Lancet Respir Med 8:475–481. https://doi.org/10.1016/S2213-2600(20)30079-5

Ye Q, Wang B, Mao J (2020) The pathogenesis and treatment of the `cytokine storm' in COVID-19. J Infect 80:607–613. https://doi.org/10.1016/j.jinf.2020.03.037

Zambon MC (2001) The pathogenesis of influenza in humans. Rev Med Virol 11:227–241. https://doi.org/10.1002/rmv.319

Zhang Y, Zhang H, Ghosh D (2020) The stabilizing excipients in dry state therapeutic phage formulations. AAPS PharmSciTech 21:133. https://doi.org/10.1208/s12249-020-01673-5

Zhong P, Xu J, Yang D, Shen Y, Wang L, Feng Y, Du C, Song Y, Wu C, Hu X, Sun Y (2020) COVID-19-associated gastrointestinal and liver injury: clinical features and potential mechanisms. Signal Transduct Target Ther 5:256. https://doi.org/10.1038/s41392-020-00373-7

Part V

Diverse Roles of Microbiome

Role of Microbes in Production of Vaccines 26

Varsha Gupta, Ivneet Kour, and Lipika Singhal

Abstract

The human race has been seeking ways to protect themselves in order to prevent the emergence of deadly diseases from centuries. As it is known, the practice of inoculation for the prevention of disease is of considerable antiquity. A vaccination is a substance that is given to a person to boost the immune system. It can be administered through various routes to prevent and to protect the community from contracting an infectious disease. Microorganisms are closely related to vaccination. To get employed in vaccinations, the disease-causing bacteria are either attenuated or heated/chemically treated. The genetic makeup of the microorganisms used in DNA vaccine production is also crucial. Additionally, in order to boost immune response and memory against the disease-causing pathogen, microorganisms are also utilized as vaccine vectors. As part of this chapter, microbes are discussed as they are used for the development of the first generation of vaccines to the current generation of DNA and mRNA vaccines. Also discussed are the subunit and conjugate vaccines made from microorganisms.

Keywords

Microbes · Prevention · Emerging infectious diseases and vaccines

V. Gupta · I. Kour · L. Singhal (✉)
Department of Microbiology, Government Medical College Hospital, Chandigarh, India

R. Sobti et al. (eds.), *Role of Microbes in Sustainable Development*,
https://doi.org/10.1007/978-981-99-3126-2_26

26.1 Introduction

26.1.1 Background

The human race has been seeking ways to protect themselves in order to prevent the emergence of deadly diseases from centuries. As it is known, the practice of inoculation for the prevention of disease is of considerable antiquity. Immunization has a long history, beginning with experiments and from taking risks to the global roll-out of a vaccine in the midst of a pandemic which was unprecedented (Riedel 2005). In spite of the fact that vaccine production can take a long time, it isn't difficult to produce them. However, the process of vaccine production requires extensive testing so as to ensure that there are no unintended consequences from newer ones. Therefore, technical, ethical, and regulatory hurdles must be overcome before vaccines can be produced and rolled out to the general population at global level (Robinson 2016).

26.1.2 History of Vaccine

A history of efforts to induce immunity to smallpox dates back to the fifteenth century, when smallpox crusts were inhaled or inserted into small cuts by both Chinese and Turks; what is called as "Variolation" (Miller 2018). After these crude attempts, Lady Mary Wortley Montagu was the first to encourage further immunization experiments since 1718, and Edward Jenner's experiments with cowpox to stimulate immunity against smallpox were far more notable than these earlier experiments (Clem 2011).

The new understanding of disease and immunity offered by Pasteur went beyond attenuating or weakening microbes to enhancing the ability of the immune system to prevent the disease itself without causing it.

This balance must be taken into account in determining the efficacy of a live vaccine. The first laboratory-developed vaccination against the bacterial disease chicken cholera was created by Louis Pasteur in 1879 (Pasteur 1881). Based on Robert Koch's (1843–1910) pivotal 1876 proof of the causal agent for anthrax, he created a vaccination for veterinary anthrax in 1881. In 1885, Pasteur created the first therapeutic vaccination to stop postexposure disease in addition to the rabies vaccine. Scientists concentrated their efforts on isolating diseases and creating vaccinations against them after Pasteur's achievements (Witebsky 1967).

26.2 Definition

When exposed to a pathogen later, a vaccination can be used to safely elicit an immune response that gives protection against primary infection or disease. A vaccination increases immunity by preventing the onset of an infectious disease in a person. Immune responses are sparked by vaccinations by weak or dormant components of a particular organism known as antigen (Siegrist and Lambert 2016). Clinical trials that link immune responses to the vaccine's antigen to clinical end points determine the degree of protection that a vaccination confers (such as the prevention of infections, a reduction in disease severity, or a reduction in hospitalization). Finding an immune response that is correlated with protection helps speed up the development and availability of novel vaccines.

26.2.1 The Concept of Modern Vaccines

Modern vaccines contain either the blueprint for making antigen or the whole antigen itself. This weakened version will not cause the disease in a person receiving the vaccine, but rather will cause their immune system to respond to the pathogen even if exposed to it, for the first time (Gomez et al. 2013). There are various ways of using microbes in vaccines either to make vaccines, sometimes as carrier of important genes or antigens (World Health Organization 2022a). Genetic material from microorganisms is also used for the generation of DNA vaccines. In addition, microorganisms are also used as vectors for vaccination in order to increase the immune response and memory against disease-causing pathogens (Centers for Disease Control and Prevention 2012).

26.3 Role of Microbes in Manufacture of Vaccines

The Global Alliance for Vaccines and Vaccination (GAVI), whose goal is to increase access to immunization in developing nations, has made significant strides in improving access to and uptake of vaccines in the developing world over the past 10 years. Through GAVI, 72 nations are collaborating to provide, manage, and effectively cover everyone under various global vaccination schemes, increase the funding of health systems, and achieve full immunization coverage with conventional vaccines (WHO Global Health Workforce Alliance 2020).

26.3.1 Immune Response and Microbes

To induce an immune reaction, an antigen is required. This step includes the generation of the pathogen itself for subsequent inactivation or isolation of a subunit or generation of a recombinant protein derived from the pathogen. On the other hand, viruses are grown on cells, primary cells like chicken fibroblasts, or on continuous cell lines such as MRC-5. The bacterial pathogens are grown in bioreactors to maximize the antigen yield while maintaining the integrity of the antigen. The recombinant proteins can also be manufactured in bacteria, yeast, or cell culture. It's also necessary to protect the seed cultures, cell lines, and bacterial cultures used for viral production (Casto and Brunell 1991).

26.3.2 Steps in the Manufacturing of Vaccines

The first step in manufacturing of vaccine is that a "master cell bank" (MCB) must be established, which is a collection of viable cells that will form the basis of all future production. In addition to its high performance, it is free of adventitious agents. A working cell bank is prepared from this bank and then used as the routine culture for future production lots. An alteration in the vaccine's seed can be as challenging as developing a new product. The next step is to release the antigen from the substrate and finally purification of the antigen is done through column chromatography and ultrafiltration in case of recombinant vaccines; however, in case of inactivated vaccines, the virus can simply be inactivated and no further purification is required. Lastly, adjuvants are added to enhance the immune response in order to preserve and stabilize so as to prolong the vaccine shelf-life before delivering it (European Medicines Agency 2022).

26.4 Role of Microbes in Vaccine Development

The vaccines not only protect the individual, but also provide herd immunity (Giubilini 2020). As part of its 2017 Annual Review of diseases prioritized under the Research and Development Blueprint, the World Health Organization has identified the top seven priority pathogens. In addition to Crimean-Congo hemorrhagic fever viruses (CCHFV), Middle East respiratory syndrome coronaviruses (MERS-CoV), severe acute respiratory syndrome coronaviruses (SARS-CoV), Ebola and Marburg viruses (EBOV), Rift Valley fever viruses (RVFV), and Nipah viruses (NiV), these diseases can trigger outbreaks and even pandemics. So, to avoid an uncontrollable pandemic in the near future, scientists have been developing vaccines against Chikungunya virus and Zika virus which are also among these priority pathogens (Mehand et al. 2018).

The importance of microbes in developing vaccines has not decreased over centuries despite changes in the process of vaccine development. There are many ways in which microbes can be used to develop effective vaccines (Khan et al. 2022). Vaccines can be classified into several broad categories, which are briefly described below, depending on the principle of vaccine development.

26.5 Classification of Vaccines

26.5.1 Live Attenuated Vaccines

The concept of vaccination likely came from a practice known as variolation, in which people were given little doses of poison or toxins to render them immune to the harmful consequences of disease (Boylston 2012). Edward Jenner tested a similar idea by injecting animal pox virus into human beings to prevent small pox in people. He then proposed the idea that a substance that is harmful to animals but not to humans could nonetheless cause important immunological reactions (Plotkin 2014). Following the eradication of small pox in 1980, the vaccinia vaccine was among the most effective vaccinations. According to the World Health Organization, another grave human illness of the time, poliomyelitis, has also been all but eradicated worldwide, thanks to Albert Sabin's live attenuated vaccine (WHO) (Minor 2012). This live attenuated tuberculosis vaccine is made from an attenuated strain of tuberculosis called (Bacillus Calmette-Guerin) BCG, which is an avirulent strain that triggers an immune response sufficient to protect against tuberculosis in contrast to the contagious strain of *Mycobacterium tuberculosis* (Tangy and Naim 2005). Vaccines with attenuated properties have shown significant advantages over vaccines with killed components (Luca and Mihaescu 2013). Although attenuated vaccines have been shown to be quite effective, there are many problems with their efficacy. Since the attenuated version of the virus can change into a pathogenic form, risking the lives of those who received the vaccination, a live attenuated vaccine necessitates considerable knowledge of the pathophysiology of the virus (Pennington et al. 2016). The appearance of paralytic polio in kids who had received oral polio vaccinations is a well-known illustration of virulence reversal (OPV). A Stalk inactivated polio vaccine (IPV) was taken into consideration for the treatment of polio when virulence reappeared with changes in the attenuated strains (Galen et al. 2016). Additionally, after injection, certain live vaccinations show either brief immunity durability or partial immunity.

26.5.2 Killed Vaccines

Another popular method of vaccination is the use of killed vaccines that require microbes which are metabolically active. In this, the microbes are rendered inactivated by genetic engineering, but they can still elicit immune responses in humans despite being incapable of growth and pathogenesis. The advantage of killed vaccines is that there is no possibility of reversing the virulence as these microbes are modified to introduce an absolute block on the DNA replication, which eliminates the possibility of growth and pathogenesis. Thus, killed vaccines provide an advantage over the live vaccines (Baicus 2012). An example of killed vaccine is the cholera vaccine with presence or absence of the B subunit of the cholera toxin, which can be administered as an oral vaccine. Formalin-inactivated whole-cell pertussis vaccine is another example of a killed vaccine (Dubensky et al. 2012).

26.5.3 Purified Proteins and Polysaccharides Vaccines

Without introducing the actual organism, these vaccines produce enough immunogenicity. The host immune system's primary site of recognition for most of these vaccinations is the polysaccharide capsule that surrounds them. It was shown that the polysaccharide capsule component of the first polysaccharide vaccine for meningococcus induced a significant degree of immunological response without triggering pathogenesis (Holmgren et al. 1992). Several other polysaccharide vaccines, including those for typhoid, pneumococcus, and influenza, were introduced based on the same principles (Reckseidler-Zenteno et al. 2010). Without exposing the body to the actual pathogen, these vaccinations induce enough immunogenicity. The majority of these vaccinations are included in polysaccharide capsules, which act as the main immune system of the host recognition site. A B cell-mediated immune response is the basis of how the polysaccharide vaccine functions. Schneerson and associates created a novel immunization strategy for *Haemophilus influenzae* in 1980. They enhanced the immunogenicity and efficacy of the vaccine component by conjugating the polysaccharide capsule to a protein subunit. Later, scientists exploited this concept to create more potent meningococcal and pneumococcal vaccinations for all age groups.

Additionally, vaccines made of proteins can produce toxic substances. It is a toxin that has been rendered inactive and boosts immune responses, but has no toxicity (Schneerson et al. 1980). For instance, poisons are rendered inactive, while yet maintaining their capacity to stimulate the formation of antibodies against them in order to create toxoid vaccines. The most popular and effective immunization up to this point has been the tetanus toxoid shot. In recent years, protein-based vaccines against a range of diseases have grown in popularity (Kumar et al. 2022). As an illustration, an acellular pertussis vaccine uses microbe-specific proteins that might trigger an immune response. After the influenza virus was artificially produced and digested to remove the essential protein components, a similar method was employed to separate the main immunogens from it (Sato and Sato 1999).

26.5.4 Genetically Modified Vaccines

These vaccines entail genetically altering pathogens as well as altering other species' pathogen recognition antigens (PRA) in order to administer the vaccination to the host body without affecting the pathogen. The DNA sequence for the pathogen's surface antigen was inserted into a yeast cell to produce many copies of the antigen, which elicit an appropriate immune response in the host body. This procedure was used to create the first effectively genetically designed vaccination against hepatitis B (Cate et al. 1977). Other examples of genetically engineered vaccinations are meningococcal protein vaccines, cholera vaccines, human papillomavirus (HPV) quadrivalent or bivalent for females in the reproductive age group, and vaccines against Lyme disease (Valenzuela et al. 1982).

26.5.5 Viral Vectored Vaccines

Another type of vaccination known as vectorized vaccines has been created with the recent introduction of genetic engineering in the manufacture of vaccines. These vaccines result in the production of a carrier virus that carries a particular gene from the target pathogen, such as an adenovirus or poxvirus. Although these carriers are not pathogenic, they can express a significant number of interesting genes. Once these vectors have begun to express the inserted genes, a T cell response and protective immunity are produced in response to these antigens (Germanier and Füer 1975). Adenovirus-based vaccines are created by manipulating the E1A and E1B area of the virus' genetic code. However, because adenoviruses are innate, the host cells must be coaxed into expressing adenoviral receptors on their surfaces in order for the immune system to identify them (Ramezanpour et al. 2016). Due to their capacity to maintain physical and genetic integrity while avoiding integration into the host genome, nonreplicating adenovirus vectors (AdV) were used to create vaccines in the wake of the Covid pandemic. A recombinant, nonreplicating viral vector vaccine called ChAdOx1 nCoV-19 was created utilizing a chimpanzee adenovirus-encapsulated plasmid vector and expressed the S protein. Because the gene that causes the virion to assemble has been removed, making it safer for the host cell, these modified adenoviruses cannot reproduce in humans. The Covid vaccines are produced by AstraZeneca and Janssen/Johnson & Johnson using viral vector systems. Following that, a number of studies are currently concentrating on the creation of viral vector vaccines against various diseases like Ebola, Zika, and HIV. Viral vectors have been investigated for gene therapy, cancer treatment, and molecular biology studies in addition to their usage in vaccinations (Capasso et al. 2014).

26.6 Development of the mRNA Vaccines

As a general rule, vaccine development takes approximately 15 years for clinical trials to be approved through sequential procedures. Despite this, the fastest vaccine developed so far was for mumps in the 1960s, which took approximately 4 years to develop. As a result of reverse vaccinology, vaccine development for recent pandemic has sped up. By saving millions of lives in Covid-19, this technique has proven beneficial for mankind (World Health Organization 2022b). The mRNA class of vaccines is capable to imitate the antigen structure and therefore does not pose a risk of infection or insertional mutagenesis. The mRNA can also elude the anti-vector immunity after repeated vaccinations (Cohen 2020). MRNA vaccine delivery methods are crucial as these molecules must penetrate lipid membranes, localize in the cytoplasm, and trigger translation-to-transcription processes. The manufacturer of the BNT162b2 vaccine, Pfizer, uses lipid nanoparticles (LNP) to encapsulate mRNA in the vaccine, which has been shown to effectively deliver mRNA in vivo and protect it from degradation by nucleases (Padda and Parmar 2022).

Almost 2 years after the discovery of SARS-CoV-2, rapidly advancing technology has made it possible to produce a COVID-19 vaccine. Vaccines for COVID-19 are being developed at a rate unprecedented in history, some using platforms never used in humans before this pandemic. As a matter of concern, a number of new variants have arisen despite the rapid production and distribution of vaccines to all corners of the world, with threats that vaccines might fail to provide the population with effective immunity against these variants (Rijkers et al. 2021).

Outbreaks like influenza A, Zika virus, severe acute respiratory syndrome (SARS), dengue fever, etc. are a major threat to humankind. Among the parasitic diseases, malaria is also a global threat that requires immediate effective vaccines (Karanja and Kiboi 2016). With the advent of new technologies, the time will soon come when vaccines will be used against vector-borne diseases also (Weaver et al. 2018).

26.7 Recent Advances

Worldwide, extensive research is being done to improve the current and develop innovative vaccine production techniques. The use of modified vectorized vaccines, which are currently the subject of intensive research, is one of the potential strategies. Since there are many different viral vectors available today, scientists have also been able to learn a lot about how to manipulate them and how they work as immunogens. The benefit of viral vectors is the confirmation and accurate expression of the target antigen within the host, the capacity to imitate the infection-related conditions that naturally occurred, and the induction of both cellular and humoral immunity against this target antigen (Rauch et al. 2018).

The creation of DNA or RNA that encodes for an antigen is another novel vaccine development process. These nucleic acid sequences are taken up by the host cell, which then expresses these antigens on the cell surface, making them simple prey for cellular and humoral immune responses. However, when given using these traditional ways, DNA vaccines are less immunogenic. The desired antigen may only then be translated from RNA by means of DNA transcription. The RNA vaccine, on the other hand, can function effectively after crossing the plasma membrane because the endoplasmic reticulum may utilize it to create the necessary protein. It is clear that using various microorganisms in the manufacturing of vaccines has a direct impact on how those vaccines are produced (Leitner et al. 1999).

26.8 Conclusion

A known pathogen can unexpectedly harm all of humanity due to a very small mutation, making changes and improvements in this field of medicine of utmost importance in light of the emerging and impending pandemic threats from known or unknown pathogens. The development of vaccines is without a doubt one of the most significant advances in medical science. Therefore, it is crucial that vaccines be used as effectively as possible during outbreaks. Further research on the bacteria is necessary to rule out any potential negative long-term impacts on the host organism. It must be made sure that these bacteria are never permitted to initiate pathogenesis on their own or to affect the person's innate immunity. It is crucial to realize that using microbes in the production of vaccines will help people fight future outbreaks and pandemics more effectively, along with ongoing research and future developments in this area.

References

Baicus A (2012) History of polio vaccination. World J Virol 1:108–114

Boylston A (2012) The origins of inoculation. J R Soc Med 105:309–313

Capasso C, Garofalo M, Hirvinen M, Cerullo V (2014) The evolution of adenoviral vectors through genetic and chemical surface modifications. Viruses 6:832–855

Casto DT, Brunell PA (1991) Safe handling of vaccines. Pediatrics 87:108–112

Cate TR, Couch RB, Kasel JA, Six HR (1977) Clinical trials of monovalent influenza a/New Jersey/76 virus vaccines in adults: reactogenicity, antibody response, and antibody persistence. J Infect Dis 136(Suppl):S450–S455

Centers for Disease Control and Prevention (2012) National vaccine program office vaccine fact sheets; [cited 2012 Mar 7]. http://www.cdc.gov/vaccines/pubs/vis/vis-facts.htm#basic

Clem AS (2011) Fundamentals of vaccine immunology. J Glob Infect Dis 3:73–78

Cohen S (2020) The fastest vaccine in history. UCLA Health; [cited 2022 Nov 22]. https://connect.uclahealth.org/2020/12/10/the-fastest-vaccine-in-history/

Dubensky TW, Skoble J, Lauer P, Brockstedt DG (2012) Killed but metabolically active vaccines. Curr Opin Biotechnol 23:917–923

European Medicines Agency (2022) Ich guideline Q5A(R2) on viral safety evaluation of biotechnology products derived from cell lines human or animal origin—scientific guideline European Medicines Agency; [cited 2022 Nov 7]. https://www.ema.europa.eu/en/ich-guideline-q5ar2-viral-safety-evaluation-biotechnology-products-derived-cell-lines-human-animal

Galen J, Buskirk A, Tennant S, Pasetti M (2016) Live attenuated human salmonella vaccine candidates: tracking the pathogen in natural infection and stimulation of host immunity. EcoSal Plus 7:7

Germanier R, Füer E (1975) Isolation and characterization of gal E mutant ty 21a of salmonella typhi: a candidate strain for a live, oral typhoid vaccine. J Infect Dis 131:553–558

Giubilini A (2020) The ethics of vaccination. Saint Philip Street Press, p 138

Gomez PL, Robinson JM, Rogalewicz JA (2013) Vaccine manufacturing. Vaccines, pp 44–57

Holmgren J, Svennerholm AM, Jertborn M, Clemens J, Sack DA, Salenstedt R et al (1992) An oral B subunit: whole cell vaccine against cholera. Vaccine 10:911–914

Karanja JK, Kiboi NG (2016) Current milestones towards development of a fully deployable anti-malaria vaccine-future hope for malaria-free world: a review. J Vaccine 7:7

Khan YH, Saifullah A, Mallhi TH (2022) Bacterial vaccines. In: Rezaei N (ed) Encyclopedia of infection and immunity. Oxford, Elsevier, Amsterdam, pp 530–544

Kumar S, Basu M, Ghosh P, Ansari A, Ghosh MK (2022) COVID-19: clinical status of vaccine development to date. Br J Clin Pharmacol 89:114. https://doi.org/10.1111/bcp.15552

Leitner WW, Ying H, Restifo NP (1999) DNA and RNA-based vaccines: principles, progress and prospects. Vaccine 18:765–777

Luca S, Mihaescu T (2013) History of BCG vaccine. Maedica (Bucur) 8:53–58

Mehand MS, Al-Shorbaji F, Millett P, Murgue B (2018) The WHO R&D blueprint: 2018 review of emerging infectious diseases requiring urgent research and development efforts. Antivir Res 159:63–67

Miller G (2018) The adoption of inoculation for smallpox in England and France. The adoption of inoculation for smallpox in England and France. University of Pennsylvania Press, Philadelphia, PA

Minor PD (2012) The polio-eradication programme and issues of the end game. J Gen Virol 93:457–474

Padda IS, Parmar M (2022) COVID (SARS-COV-2) vaccine. In: StatPearls. StatPearls Publishing, Treasure Island (FL); [cited 2022 Nov 22]. http://www.ncbi.nlm.nih.gov/books/NBK567793/

Pasteur L (1881) De l'atténuation des virus et de leur retour à la virulence. Gauthier-Villars, p 7

Pennington SH, Thompson AL, Wright AKA, Ferreira DM, Jambo KC, Wright AD et al (2016) Oral typhoid vaccination with live-attenuated salmonella Typhi strain Ty21a generates Ty21a-responsive and heterologous influenza virus-responsive CD4+ and CD8+ T cells at the human intestinal mucosa. J Infect Dis 213:1809–1819

Plotkin S (2014) History of vaccination. Proc Natl Acad Sci U S A 111:12283–12287

Ramezanpour B, Haan I, Osterhaus A, Claassen E (2016) Vector-based genetically modified vaccines: exploiting Jenner's legacy. Vaccine 34:6436–6448

Rauch S, Jasny E, Schmidt KE, Petsch B (2018) New vaccine technologies to combat outbreak situations. Front Immunol 9:1963

Reckseidler-Zenteno SL, Viteri DF, Moore R, Wong E, Tuanyok A, Woods DE (2010) Characterization of the type III capsular polysaccharide produced by Burkholderia pseudomallei. J Med Microbiol 59:1403–1414

Riedel S (2005) Edward Jenner and the history of smallpox and vaccination. Proc (Bayl Univ Med Cent) 18:21–25

Rijkers GT, Weterings N, Obregon-Henao A, Lepolder M, Dutt TS, van Overveld FJ et al (2021) Antigen presentation of mRNA-based and virus-vectored SARS-CoV-2 vaccines. Vaccines (Basel) 9:848

Robinson JM (2016) Vaccine production: main steps and considerations. In: The vaccine book, pp 77–96

Sato Y, Sato H (1999) Development of acellular pertussis vaccines. Biologicals 27:61–69

Schneerson R, Barrera O, Sutton A, Robbins JB (1980) Preparation, characterization, and immunogenicity of Haemophilus influenzae type b polysaccharide-protein conjugates. J Exp Med 152:361–376

Siegrist CA, Lambert PH (2016) How vaccines work. In: Bloom BR, Lambert PH (eds) The vaccine book, 2nd edn. Academic Press, London, pp 33–42

Tangy F, Naim HY (2005) Live attenuated measles vaccine as a potential multivalent pediatric vaccination vector. Viral Immunol 18:317–326

Valenzuela P, Medina A, Rutter WJ, Ammerer G, Hall BD (1982) Synthesis and assembly of hepatitis B virus surface antigen particles in yeast. Nature 298:347–350

Weaver SC, Charlier C, Vasilakis N, Lecuit M (2018) Zika, chikungunya, and other emerging vector-borne viral diseases. Annu Rev Med 69:395–408

WHO Global Health Workforce Alliance (2020) GAVI—The global alliance for vaccines and immunizations; [cited 2020 Dec 23]. www.who.int/workforcealliance/members_partners/member_list/gavi/en/

Witebsky E (1967) Immunology: views of the past and visions for the future. Can Med Assoc J 97: 1371–1377

World Health Organization (2022a) How are vaccines developed?; [cited 2022 Nov 22]. https://www.who.int/news-room/feature-stories/detail/how-are-vaccines-developed

World Health Organization (2022b) AstraZeneca ChAdOx1-S/nCoV-19 [recombinant], COVID-19 vaccine; [cited 2022 Nov 22]. https://www.who.int/publications/m/item/chadox1-s-recombinant-covid-19-vaccine

Microbial-Induced Calcite Precipitation Approach Towards Sustainable Development

27

Inderpal Devgon, Khushboo, Rohan Samir Kumar Sachan, Nisha, Abhishek Rana, and Arun Karnwal

Abstract

The creation of sustainable environment has revolutionized the world in the past two decades. The area of biomineralization sector has also grown to provide promising results as a potential environment eco-friendly process to produce. Microbial-Induced Calcite Precipitation played a major role in promotion of biomineralization, and many patents have been generated in the past few years. Even now studies related to optimization, economic, and eco-friendly perspective are carried out and compared with other sustainable techniques to tackle environmental pollutants. In this chapter, MICP-based bio-cement has been focused as an alternative of cement produced using non-renewable sources. The pros of using bio-cement against cement have been discussed along with impact on conventional cement market. Lastly, different stages of bio-cement production using MICP process have been discussed to provide a step forward in sustainability.

Keywords

Bio-cement · Sustainability · Cement · Microorganism

I. Devgon · Khushboo · R. S. K. Sachan · Nisha · A. Karnwal (✉)
School of Bioengineering and Biosciences, Lovely Professional University, Phagwara, Punjab, India
e-mail: arun.20599@lpu.co.in

A. Rana
Jindal Global Law School, OP Jindal Global University, Sonepat, Haryana, India

R. Sobti et al. (eds.), *Role of Microbes in Sustainable Development*,
https://doi.org/10.1007/978-981-99-3126-2_27

27.1 Introduction

Cement is the primary material used in construction, providing mechanical properties such as compressive strength to buildings (Mukherjee et al. 2019). But along with the benefits, there are some disadvantages. Cement production requires high-temperature conditions maintained by burning fossil fuels and other materials leading to the release of gasses and chemicals in the environment that cause pollution. Nitrogen oxides, sulphur dioxide, carbon monoxide, ammonia, mercury, HCL, hydrogen fluoride, etc. are the primary pollutants of cement production (Mishra et al. 2022). Cement manufacturing consumes between 15% and 8% of the energy utilized in the cement industry and accounts for 5% to 8% of anthropogenic CO_2 emissions (Martuscelli et al. 2020). Higher demand for cement has generated new ideas, such as calcite precipitation using microbes. Microorganisms like bacteria, algae, and fungi are exploited in microbially induced calcite precipitation.

MICP is also used for increasing soil adhering, which can be done using chemical treatments, e.g. chemical grouting, which has harmful effects on the environment, so MICP can be an alternative and sustainable approach towards soil adhering using microorganisms (Sharma et al. 2021), e.g. a new method called MICP, which plugs concrete's pores with bacteria, algae, and fungi, extends the life of concrete structures by enhancing their mechanical qualities (Kaur et al. 2021). Different ways, like using photosynthetic microbes, can produce bio-cement. The second method uses sulphate-reducing microbes, and the third method uses urease-positive (Seifan et al. 2016), which utilizes CO_2, while calcite precipitation gives a sustainable approach.

Since it doesn't produce CO_2 during production, bio-cement offers a more environmentally friendly alternative to regular cement. The CO_2 is used up in the mineralization of calcium carbonate ($CaCO_3$) when the metabolic conversion of calcium salts creates it. Additionally, these minerals can increase the strength and longevity of cementitious materials. Research on the microbiological precipitation of $CaCO_3$ by urea hydrolysis has been conducted with applications in soils to develop novel building materials and fill concrete cracks. Due to its environmental benefit, it has proven superior to many older methods (Martuscelli et al. 2020). Agricultural waste can also produce bio-cement by providing bacteria with a substrate for growth (Joshi et al. 2020), which makes it an eco-friendly approach as it will also help to solve the problem of using rice husk or other agricultural waste.

27.2 Types of Sustainable Materials

A sustainable substance is typically created from natural or recycled resources with minimal energy use. It has little impact on the environment and uses few non-renewable resources. Therefore, a material can be sustainable if produced with negligible adverse effects on the environment and human health. This allows the resources from which it was created to remain accessible to future generations. The building industry, in particular, has given "green" materials a lot of attention in recent years. Research organizations routinely develop unique, eco-friendly materials with fascinating thermal and acoustic properties. Ceramic materials, some metals and their alloys, some types of bioactive glass, polymeric materials and their composites, and materials made from animal origin are among the several sustainable biomaterials (Asdrubali et al. 2012; Biswal et al. 2020; Desarnaulds et al. 2005; Kumar Gupta et al. 2016).

However, a material regarded as sustainable must follow specific pointers mentioned below:

1. Single-use materials that can't be composted or recycled ought to be avoided.
2. Products made from renewable resources are always preferred above those based on fossil fuels and vice versa.
3. Sustainability needs to be preserved throughout the lifespan of manufactured goods and future uses of product components, resource recovery from waste, and feedstock expansion.
4. Sustainability must incorporate social, health, economic, and environmental challenges.
5. The items are made and created to be naturally recyclable, reusable, or compostable.
6. The federal and state governments should support sustainable agricultural infrastructure for farmers and other communities.
7. Genetically engineered species should be outlawed and not used to produce feedstock.
8. It is desirable to utilize chemicals that adhere to the 12 principles of green chemistry in agricultural fields.
9. It is necessary to restrict the use of chemicals and nanomaterial products that, throughout their life cycles, pose a risk to the environment and the general public's health.
10. The various production-process steps must be decentralized.

27.2.1 Biopolymeric Materials

The scientific community has shifted its focus to using ecologically friendly materials to minimize the environmental impact of conventional synthetic plastics due to advancements in polymer research and engineering. Biopolymers exhibit great potential for use in various industries, e.g. food, electronics, etc. They are also environmentally friendly, chemically adaptable, sustainable, biocompatible, biodegradable, naturally functional, and eco-friendly (Arif et al. 2022).

27.2.2 Bio-Cement/Biomaterials

Around the world, the building sector significantly impacts social and economic growth. However, it has a significant carbon footprint because it uses a lot of energy during every stage of manufacturing, from gathering raw materials to building structures. It quickens the process of global warming, ice cap melting, and the subsequent rise in sea levels. The world's population is suffering due to this environmental issue. The building materials that release the most carbon dioxide (CO_2) during production are burnt clay brick, concrete, and cement mortar. The MICP applications can enhance comprehension of the function of bio-cementation in the construction of environmentally friendly structures. The scientific data can provide a clear path for further investigation into the mechanism of microorganism bio-cementation in creating tangible bio-based building materials equivalent to conventional resources (Choi et al. 2017; Iqbal et al. 2021).

27.2.3 Biofuels

More than any other sector, transportation has seen an increase in greenhouse gas (GHG) emissions. The sector heavily depends on fossil fuels, which accounted for 96.3% of all transportation fuels in 2018. Transport is responsible for 23% of all energy-related CO_2 emissions and 15% of global GHG emissions. Biofuels are frequently seen as practical substitutes for petroleum-based fuels that can help reduce dependence on them and fight climate change. Bioenergy is one of the various resources that can be exploited to supply the energy demand. Bioenergy obtained from organic materials is referred to as biomass. There has been a lot of algae used in the creation of bioenergy. The energy received is often used in biofuels and electricity. On a global scale, bioenergy contributes around 10% of the energy supplies as thermal energy for cooking and heating. Biomass sources for bioenergy include plants, animals, agricultural waste, algae, and industrial waste (Jeswani et al. 2020; Singh et al. 2022).

27.3 Microbial-Induced Calcite Precipitation as a Coping Approach

Microbial-induced calcite precipitation uses several organisms like bacteria, algae, and fungi. Microorganisms which have the property to hydrolyse urea into ammonium and can precipitate calcium into calcium carbonate. The different tests help select microorganism for MICP, urease assay, and calcite precipitation test, and furthermore, scientific techniques like XRD, SEM, FTIR, etc. help select microorganism capable of MICP (Abo-El-Enein et al. 2012). So, urease-positive microorganisms are exploited for this technique (Kumar et al. 2020). These ureolytic bacteria, in the presence of urea and calcium, can produce calcite. This technique can also be used for binding loose soil and can increase soil water retention properties.

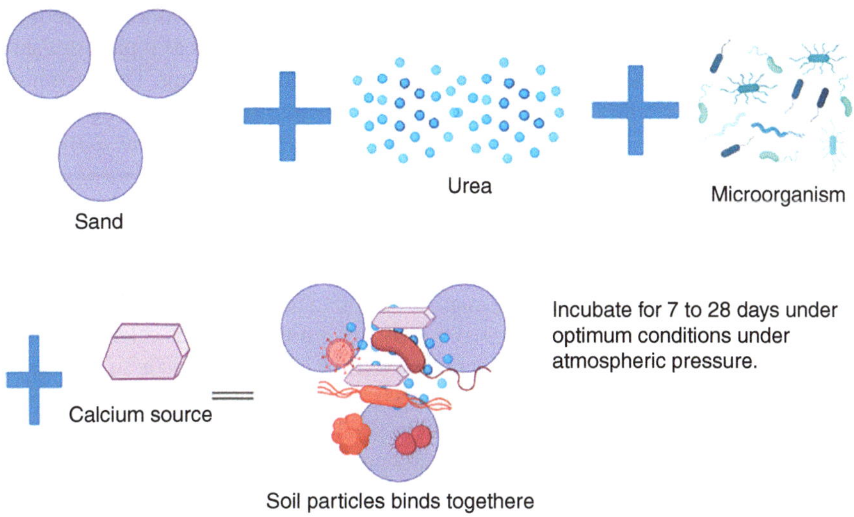

Fig. 27.1 Representation of process for bio-cement

These microorganisms can also remove cations from wastewater (Cheng et al. 2020). MICP can be obtained using plant-based urease enzyme, but it is easy to use non-pathogenic microorganisms, which take less time and can utilize agricultural waste as a substrate. Also, municipal sludge can be used whose burial or spreading on land utilizes a large amount of energy and currency (Parmar and Marjadi 2017). Few bacteria present have ureolytic activity and can precipitate calcite, e.g. B. subtilis and B. sphaericus, which shows promising results in MICP. These non-pathogenic microorganisms convert urea into ammonium and convert calcium into calcium carbonate ($CaCO_3$). Bacteria can grow within 24 h of incubation and are easy to cultivate, making them a good candidate for MICP (Sharma et al. 2021). Organisms induce calcite in MICP by the hydrolysis of urea by the enzyme urease, which results in the production of carbonate ions in the presence of ammonium.

The pH rise caused by the ammonium (NH4+) released from the urea hydrolysis results in calcium carbonate precipitation (calcite). When calcium ions (Ca^{2+}) from the supplied calcium chloride react with carbonate ions (CO_3^-) from the urea hydrolysis, calcite is precipitated.

Parameters including cell number, pH, and ammonia are monitored over a consistent period of time to ascertain the relationship between the creation of calcium carbonate and the parametric changes during the MICP process. Calcite precipitation is indicated by a change in pH or the generation of ammonia in a sample. The calcite ($CaCO_3$) formed is responsible for improving the engineering properties of soil (Fig. 27.1).

MICP has one more property that is self-healing which solves the problem of early cracking. In cement produced using technique MICP, cracks get repaired automatically when spore of microorganisms gets in touch with water entering through those cracks or is provided by nutrition from outside to get it repaired.

Using microorganisms for crack repair can save a lot of labour, time, and money; because of this advantage, this technique is getting considerations as it provides safe, natural, and pollution-free solution to a big issue, i.e. early cracking. This technique provides great microcrack filling capacity, strong bond, high compatibility, and is favourable to thermal changes, and is a sustainable approach (Luo et al. 2018).

27.4 Microbial-Induced Calcite Precipitation Process

In the recent years, MICP is gaining much attention to researchers for the bio-cement formation worldwide. One of the most widely known methods for calcite (calcium carbonate) production by MICP is the urease activity existing in bacteria. Biological control and biological induction are the two main mechanisms involved in the calcium carbonate precipitation. In biologically controlled mechanisms, the organisms primarily affect the nucleation and growth of mineral particles. However, the unique form of minerals has been synthesized from the organisms which are not related to the environment conditions. On the other hand, bio-induced mineralization mechanisms involved the calcium carbonate production by bacteria. The minerals produced by this mechanism are mostly affected by the environmental conditions and no other mechanisms involved in it. There are some key factors that affect the calcium carbonate precipitation process such as the pH value, calcium ion concentration, bacterial species, bacterial concentration, temperature, and many other factors (Zhang et al. 2021; Castro-Alonso et al. 2019).

The presence of microorganisms has been found everywhere on the earth where they undergo numerous metabolic processes. These processes increase the total carbonate content and pH levels and result in the carbonate biomineral precipitation. The following vital metabolic strategies/ processes involved in MICP are as follows: urea hydrolysis, denitrification, photosynthesis, dissimilatory sulfate reduction, and ammonification of amino acids in which strains having urea hydrolytic action show highest calcite precipitation (~20–80%) (Jain et al. 2021; Achal et al. 2011).

27.4.1 Urea Hydrolysis

In this process, the microbial enzyme called urease hydrolysed the urea and leads to ammonia and carbamic acid formation which was further hydrolysed to produce biocarbonate ions. The alkalinity of the environment increases and results in hydroxyl and ammonium ions formation due to reaction between ammonia and water (Van Tittelboom et al. 2010). By measuring the free Cu ion concentration in the soil solution, a recent study showed the inhibitory effect of Cu in the soil. Due to the presence of a vital component, calcium, caused by microbially generated calcium carbonate precipitation, urea hydrolysis is inhibited (MICP). The free Cu ion concentration in the soil solution was found to be 0.39 mg/L, which is half of the maximal inhibitory concentration (Chung et al. 2020). One more study demonstrated the role of S. pasteurii urease activity on the bio-treated coarse sand specimens. The bacterial urease activity was decreased up to the desired volume for the MICP procedure from generation to generation and this results in the mean particle size of very coarse sand at 1820 μm. This study spotlights the importance of urease activity for the quality and quantity of the cements made with calcium carbonate.

$$\begin{aligned}
&CO(NH_2)_2 + H_2 \rightarrow NH_2COOH + NH_3 \\
&NH_2COOH + H_2O \rightarrow NH_3 + H_2CO_3 \\
&H_2CO_3 \leftrightarrow HCO_3^- + H+ \\
&NH_3 + 2H_2O \rightarrow 2NH_4^+ + 2OH \\
&HCO_3^- + H^+ + 2NH_4^+ + 2OH \rightarrow CO_{32} - + 2NH_4^+ + 2H_2O \\
&Ca^{2+} + Cell \rightarrow Cell - Ca^{2+} \\
&Cell - Ca^{2+} + CO_{32}^- \rightarrow Cell - CaCO_3
\end{aligned}$$

27.4.2 Denitrification

In the presence of organic matter, nitrate is denitrified to produce alkalinity, carbon dioxide, and nitrogen gas. Moreover, bicarbonate ion formation takes place when carbon dioxide equilibrates with water (Martin et al. 2013). The denitrification-based MICP process has been observed for the cementation of soils. The sand samples were treated with different levels of denitrification, and after conducting several triaxial tests, result in the improvement of the dilatancy and peak drained strength of the tested samples. Moreover, the resulting calcium carbonate precipitation amount and rate can be as high as 4.61% by weight and 0.129% per day comparison to other literature study (Gao et al. 2022).

$$(CH_3COOH)_2Ca + NO_3^- \rightarrow CaCO_3 + 4/5N_2 + 3CO_2 + 3H_2O + OH.$$

27.4.3 Photosynthesis

This process includes photosynthetic microorganisms such as cyanobacteria and microalgae cause calcium carbonate precipitation due to exchange of bicarbonate and carbonate ions. Numerous settings, including freshwater, marine, hot springs, and terrestrial locations, have been found to contain various types of carbonate minerals, the majority of which are generated by microbial photosynthesis (Zhu and Dittrich 2016; Rodríguez-Martínez et al. 2012; Abdel-Basset et al. 2021).

$$Ca^{2+} + 2HCO_3^- \rightarrow CaCO_3 + CO_2 + H_2O$$
$$Ca^{2+} + HCO_2^- + OH^- \rightarrow CaCO_3 + 2H_2O$$
$$2HCO_2^- \leftrightarrow CO_2 + CO_{32}^- + H_2O$$

27.4.4 Dissimilatory Sulfate Reduction

In this process, the sulfate-reducing bacteria produce hydrogen sulfide and bicarbonate ions that are responsible for reduction of sulfate under both aerobic and anaerobic circumstances (Castro-Alonso et al. 2019).

$$6CaSO_4 + 4H_2O + 6CO_2 \rightarrow CaCO_3 + 4H_2S + 2S + 11O_2.$$

27.4.5 Ammonification of Amino Acids

The production of ammonia and CO_2 takes place by the microbial activities during metabolism of amino acids. Ammonium hydrolysis results in the production of ammonium and hydroxide ions all over the cell which increases the system pH and bicarbonate ions formed by CO_2 (Zhu and Dittrich 2016) (Fig. 27.2).

$$\text{Amino acids} + O_2 \rightarrow NH_3 + CO_2 + H_2O$$
$$NH_3 + H_2O \rightarrow NH_4 + OH^-$$
$$CO_2 + OH \rightarrow HCO_3^-$$
$$Ca^{2+} + HCO_3^- \rightarrow CaCO_3 + H^+$$

27.5 Market and Sustainable Potential of MICP

27.5.1 Market Potential

The market for bio-cement is expanding rapidly due to urbanization, industrialization, and the global construction sector in not just the developed economies but even in the developing and least developed countries. Bio-cement market's fastest-growing region is Asia-Pacific, which is directly related to the region's expanding construction sector. While the market may be broken down into many categories (intrinsic, capsule-based, and vascular healing), there are other divisions, such as civil and residential. China is the centre of this rapidly growing global construction sector, accounting for 20% of all construction investment globally. China plans to invest more than US$13 trillion in construction projects by 2030.

Fig. 27.2 Schematic representation of Microbially induced calcium carbonate precipitation (MICP) Strategies

The Australian government announced $50 million in May 2020 for their programme for smart suburbs and cities. They aim to rely upon cutting-edge technological solutions for growing urban difficulties in cities, suburbs, and towns, while enhancing their productivity, sustainability, and liveability. These initiatives anticipate strengthening the region's building sector and raising demand for bio-cement.

Favourable incentives by the governments are being extended in many nations to support the construction of green buildings, which directly propels the bio-cement market. As an illustration, the state governments of numerous Indian states, such as Punjab, Rajasthan, West Bengal, Uttar Pradesh, Maharashtra, Andhra Pradesh, and Haryana, among others, offer additional Floor Area Ratios (FAR) for green building projects that receive the Indian Green Building Council's Gold or higher rating (Global Bio-cement Market Demand, Analysis, Trends and Forecast - ResearchCMFE n.d.).The world market for self-healing concrete was estimated at USD 24.60 billion in 2019 and is anticipated to increase at a CAGR of 37.0% from 2020 to 2027. Over the projected period, rising demand for dependable and long-lasting projects, including infrastructure, commercial, industrial, and residential, is anticipated to fuel the market for self-healing concrete. An increase in the desire for less structural upkeep of buildings, together with the expanding global construction sector, is an additional factor predicted to promote market expansion (Self-Healing Concrete Market Size, and Share Report, 2020–2027 n.d.).

Some of the top players in the bio-cement market include Biomason, Bio-Cement Technologies Basilisk, Halliburton Technology India Private Limited, Acciona Infrastructures S.A., Avecom N.V., Comercializadora Espanola De Innovaciones Y Materiales, COWI A/S, Devan-Micropolis, and Fusion. The bio-cement market is expanding due to the constantly expanding construction sector and the increasing demand for green buildings around the globe. The construction sector can use bio-cement for various purposes, such as reinforcing and stabilizing soil, coating techniques for buildings and walls, and stabilizing sand in earthquake-prone areas. The leading bio-cement businesses also strongly emphasize product innovation and research, broadening the market.

The leading market participants have identified the key market drivers and are concentrating on creating low-cost nutrient sources to lower the manufacturing costs of bio-cement. They are working hard to strengthen the market's motivators and remove obstacles to support market expansion. Additionally, major market players are engaging in mergers and acquisitions to enhance their corporate performance and gain an advantage over rival companies. For example, a multinational fashion company' H&M Group', and Biomason, a pioneer in environmentally responsible building, announced their strategic alliance in June 2021. Through this relationship, Biomason hopes to enhance the development of ecologically friendly retail environments and help the H&M Group achieve its sustainability objectives for circular construction.

Henceforth, the construction sector can use bio-cement for various purposes, such as reinforcing and stabilizing soil, covering walls and buildings, and stabilizing sand in earthquake-prone areas. These advantages expect to increase the demand for bio-cement in the building sector. The key players in the bio-cement sector also concentrate on product innovation and research, broadening the market (Self-Healing Concrete Market Size, Share and Industry Forecast 2025 n.d.).

27.5.2 Sustainable Potential

Simply put, sustainable development is a cleaner and more innovative way of economic development for countries worldwide. It is one of the most fundamental principles of international environmental law and has become a watchword for all ecological policymakers. The Stockholm Declaration's Principle 11, adopted in 1972, was the first document to propose sustainable development on a global scale formally. *"The environmental policies of all States should enhance and not adversely affect the present or future development potential of developing countries, nor should they hamper the attainment of better living conditions for all"*.[1]

[1]Right to development | Environment Rights. https://environment-rights.org/rights/right-to-development/

The most common definition of 'sustainable development' is that it is the "*development that meets the needs of the present, without compromising the ability of the future generations to meet their own needs*". This definition came from the World Commission on Environment and Development's (Brundtland Commission) report known as 'Our Common Future'. Since 2015, the United Nations and its member states have agreed on 17 SDG 9: "Build resilient infrastructure, promote inclusive and sustainable industrialization, and foster innovation." SDG 11: "Make cities and human settlements inclusive, safe, resilient, and sustainable." SDG 12: "Ensure sustainable consumption and production patterns." SDG 13: "Take urgent action to combat climate change and its impacts."

Environmental governance could be arbitrary, discretionary, subjective, and unpredictable without an environmental law and the enforcement of legal rights and obligations. Largely, ecological problems, including climate change, biodiversity loss, water scarcity, air pollution, and soil degradation, are intimately related to poverty and growing social inequality. Conflicts over natural resources and environmental crimes exacerbate these problems.

At least 40% of internal disputes during the past 60 years had some connection to natural resources. When using natural resources causes economic hardship, damage to the environment, or an uneven distribution of benefits, the likelihood of violent conflict rises. Managing natural resources in a transparent, ethical, and consistent manner with the rule of law could provide sustainable development and a foundation for peace and justice (Environmental Rule of Law I UNEP - UN Environment Programme n.d.).

MICP, as a sustainable approach, has reduced its environmental carbon footprint by switching to commercial-grade calcium chloride for MICP. All MICP routes experienced a decline of between 18% and 49.62%. The available calcium chloride is a crucial element for all processes. To further improve the sustainability of calcium carbonate (bio-cement) produced by MICP, future studies should look into alternatives to sources of laboratory-grade calcium. The carbon footprint created by the given organic carbon source (urea or nitrate) during MICP via urea hydrolysis or denitrification is more significant than calcium chloride. All approaches, except denitrification, saw a considerable reduction in embodied energy (between 43 and 95%). This decline is mainly attributable to the use of commercial-grade calcium chloride in place of laboratory-grade calcium chloride. The fuel utilized in the MICP process is 91% less than the conventional carbonation method.

The calcium source was a significant factor in the carbon footprint, embodied energy, and cost for all MICP metabolic routes (calcium chloride). The carbon footprint was reduced by between 26.7 and 82% when commercial calcium chloride was used in place of laboratory-grade calcium chloride. To further improve the sustainability of carbonates produced using the MICP process, it is evident that alternatives to laboratory-grade calcium sources should be looked into and added in subsequent trials (Porter et al. 2021).

27.6 Conclusion

Microorganisms have always come to rescue humans almost in every aspect in their life to provide a sustainable and meaningful life ahead. Researchers have formulated many sustainable molecules derived by microbes to be applied to our daily life, e.g. biofuel, biogas, bioplastics, bio-cement, etc. However, the biggest challenge or gap faced by them is the applicability of bio-products in real life.

There is need to widen research in pilot scale for the bio-products to be successfully applied in day-to-day life. There is also a need to produce bio-products using economical substrate of renewable sources to provide an alternative of non-renewable sources. Finally, the biggest challenge still faced is the awareness among humans to utilize bio-sustainable products in daily life. To make a sustainable future in coming years, these challenges need to be addressed and worked upon.

References

Abdel-Basset R, Hasssan EA, Grossart HP (2021) Manifestations and environmental implications of microbially-induced calcium carbonate precipitation (MICP) by the cyanobacterium Dolichospermum flosaquae. Biogeosci Discuss:1–22

Abo-El-Enein SA, Ali AH, Talkhan FN, Abdel-Gawwad HA (2012) Utilization of microbial induced calcite precipitation for sand consolidation and mortar crack remediation. HBRC J 8(3):185–192

Achal V, Pan X, Özyurt N (2011) Improved strength and durability of fly ash-amended concrete by microbial calcite precipitation. Ecol Eng 37(4):554–559

Arif ZU, Khalid MY, Sheikh MF, Zolfagharian A, Bodaghi M (2022) Biopolymeric sustainable materials and their emerging applications. J Environ Chem Eng 10(4):108159. https://doi.org/10.1016/J.JECE.2022.108159

Asdrubali F, Schiavoni S, Horoshenkov KV (2012) A review of sustainable materials for acoustic applications. Build Acoust 19(4):283–312. https://doi.org/10.1260/1351-010X.19.4.283/ASSET/1351-010X.19.4.283.FP.PNG_V03

Biswal T, BadJena SK, Pradhan D (2020) Sustainable biomaterials and their applications: a short review. Mater Today Proc 30:274–282. https://doi.org/10.1016/J.MATPR.2020.01.437

Castro-Alonso MJ, Montañez-Hernandez LE, Sanchez-Muñoz MA, Macias Franco MR, Narayanasamy R, Balagurusamy N (2019) Microbially induced calcium carbonate precipitation (MICP) and its potential in bioconcrete: microbiological and molecular concepts. Front Mater 6:126

Cheng L, Kobayashi T, Shahin MA (2020) Microbially induced calcite precipitation for production of "bio-bricks" treated at partial saturation condition. Construct Build Mater 231:117095

Choi SG, Chu J, Brown RC, Wang K, Wen Z (2017) Sustainable biocement production via Microbially induced calcium carbonate precipitation: use of limestone and acetic acid derived from pyrolysis of lignocellulosic biomass. ACS Sustain Chem Eng 5(6):5183–5190. https://doi.org/10.1021/ACSSUSCHEMENG.7B00521/ASSET/IMAGES/MEDIUM/SC-2017-005218_0009.GIF

Chung H, Kim SH, Nam K (2020) Inhibition of urea hydrolysis by free cu concentration of soil solution in microbially induced calcium carbonate precipitation. Sci Total Environ 740:140194. (Chung et al., 2020)

Desarnaulds V, Costanzo E, Carvalho A, Arlaud B (2005) Sustainability of acoustic materials and acoustic characterization of sustainable materials. In: Proceedings of the 12th international congress on sound and vibration 2005, vol 1. ICSV, Lisbon

Environmental Rule of Law I UNEP - UN Environment Programme (n.d.). https://www.unep.org/explore-topics/environmental-rights-and-governance/what-we-do/promoting-environmental-rule-law-0. Accessed 30 Oct 2022

Gao Y, Wang L, He J, Ren J, Gao Y (2022) Denitrification-based MICP for cementation of soil: treatment process and mechanical performance. Acta Geotech 17:1–17

Global Bio-cement Market Demand, Analysis, Trends and Forecast - ResearchCMFE (n.d.). https://www.researchcmfe.com/blog/global-biocement-market-demand-analysis-trends-and-forecast. Accessed 29 Oct 2022

Iqbal DM, Wong LS, Kong SY (2021) Bio-cementation in construction materials: a review. Materials 14(9):2175. https://doi.org/10.3390/MA14092175

Jain S, Fang C, Achal V (2021) A critical review on microbial carbonate precipitation via denitrification process in building materials. Bioengineered 12(1):7529–7551

Jeswani HK, Chilvers A, Azapagic A (2020) Environmental sustainability of biofuels: a review. Proc R Soc A 476(2243):20200351. https://doi.org/10.1098/RSPA.2020.0351

Joshi K, Meher MK, Poluri KM (2020) Fabrication and characterization of bioblocks from agricultural waste using fungal mycelium for renewable and sustainable applications. ACS Appl Bio Mater 3(4):1884–1892

Kaur P, Singh V, Arora A (2021) Microbial concrete—a sustainable solution for concrete construction. Appl Biochem Biotechnol 194:1401–1416

Kumar Gupta G, De S, Franco A, Balu AM, Luque R (2016) Sustainable biomaterials: current trends, challenges and applications. Molecules 21(1):E48. https://doi.org/10.3390/MOLECULES21010048

Kumar A, Dikshit R, Gupta N, Jain A, Dey A, Nandi A, Rajendra A et al (2020) Bacterial growth induced biocementation technology, 'Space-Brick'—a proposal for experiment at microgravity and planetary environments. In: bioRxiv

Luo J, Chen X, Crump J, Zhou H, Davies DG, Zhou G, Jin C et al (2018) Interactions of fungi with concrete: significant importance for bio-based self-healing concrete. Construct Build Mater 164: 275–285

Martin D, Dodds K, Butler IB, Ngwenya BT (2013) Carbonate precipitation under pressure for bioengineering in the anaerobic subsurface via denitrification. Environ Sci Technol 47(15): 8692–8699

Martuscelli C, Soares C, Camões A, Lima N (2020) Potential of fungi for concrete repair. Proc Manuf 46:180–185

Mishra UC, Sarsaiya S, Gupta A (2022) A systematic review on the impact of cement industries on the natural environment. Environ Sci Pollut Res 29:1–12

Mukherjee S, Sahu RB, Mukherjee J, Sadhu S (2019) Application of microbial-induced carbonate precipitation for soil improvement via ureolysis. In: Ground improvement techniques and geosynthetics. Springer, Singapore, pp 85–94

Parmar S, Marjadi D (2017) Bio cementation: a novel technique and approach towards sustainable material. World J Res Rev 4(3):262839

Porter H, Mukherjee A, Tuladhar R, Dhami NK (2021) Life cycle assessment of biocement: an emerging sustainable solution? Sustainability (Switzerland) 13(24). https://doi.org/10.3390/SU132413878/S1

Rodríguez-Martínez M, Sánchez F, Walliser EO, Reitner J (2012) An upper Turonian fine-grained shallow marine stromatolite bed from the munecas formation, northern Iberian ranges, Spain. Sediment Geol 263:96–108

Seifan M, Samani AK, Berenjian A (2016) Bioconcrete: next generation of self-healing concrete. Appl Microbiol Biotechnol 100(6):2591–2602

Self-Healing Concrete Market Size & Share Report, 2020–2027 (n.d.). https://www.grandviewresearch.com/industry-analysis/self-healing-concrete-market. Accessed 30 Oct 2022

Self-Healing Concrete Market Size, Share and Industry Forecast 2025 (n.d.). https://www.alliedmarketresearch.com/self-healing-concrete-market. Accessed 30 Oct 2022

Sharma M, Satyam N, Reddy KR (2021) Investigation of various gram-positive bacteria for MICP in Narmada sand, India. Int J Geotech Eng 15(2):220–234

Singh AR, Singh SK, Jain S (2022) A review on bioenergy and biofuel production. Mater Today Proc 49:510–516. https://doi.org/10.1016/J.MATPR.2021.03.212

Van Tittelboom K, De Belie N, De Muynck W, Verstraete W (2010) Use of bacteria to repair cracks in concrete. Cem Concr Res 40(1):157–166. (Van Tittelboom et al., 2010)

Zhang J, Shi X, Chen X, Huo X, Yu Z (2021) Microbial-induced carbonate precipitation: a review on influencing factors and applications. Adv Civil Eng 2021:9974027

Zhu T, Dittrich M (2016) Carbonate precipitation through microbial activities in natural environment, and their potential in biotechnology: a review. Front Bioeng Biotechnol 4:4

Microbial Functional Foods and Nutraceuticals

28

Ananya Rana, Vikram Kumar, Neetu Kumra Taneja, and Tejpal Dhewa

Abstract

Bacteria, yeast, and microalgae serve as catalysts in producing several food components, such as enzymes and nutraceuticals. The current fad for natural ingredients has significantly increased the demand for microbially-derived flavours, colours, and enzymes and their large-scale bioprocessing. It is, therefore, crucial that microbes are exploited as bio-factories for the continuous production of their metabolites like organic acids, enzymes, proteins, vitamins, antibiotics, and hydrocolloids. It has been shown that lactic acid bacteria, in particular *Lactococcus lactis*, make the best cell factories to manufacture these vital nutraceuticals. In recent years, researchers have discovered that bacteria is also a significant source of terpenes, a crucial component of many medications and food additives. Furthermore, diseases including protein energy malnutrition (PEM), anaemia, diarrhoea, cancer, obesity, ulcerative colitis, Crohn's disease, irritable bowel syndrome, and gluten treatment-resistant celiac can all benefit from the use of microorganisms as adjuvant therapy.

Keywords

Bacteria · Nutraceuticals · Organic acids · Anemia · Cancer

A. Rana · V. Kumar · N. K. Taneja
Department of Basic and Applied Sciences, National Institute of Food Technology and Entrepreneurship Management, Sonepat, Haryana, India

T. Dhewa (✉)
Department of Nutrition Biology, Central University of Haryana, Jant-Pali, Mahendergarh, Haryana, India

R. Sobti et al. (eds.), *Role of Microbes in Sustainable Development*,
https://doi.org/10.1007/978-981-99-3126-2_28

28.1 Introduction

Functional foods are referred to as conventional foods or dietary ingredients that supply vital nutritional constituents, such as carbohydrates, proteins, fats, vitamins, and minerals (Galanakis 2021). They might provide other health advantages that can be used to prevent or reverse a diseased state. Bioactive substances, naturally occurring molecules produced from plants, animals, or microorganisms and advantageous for human health, may be present in functional foods. Since ancient times, cyanobacteria have been used as potential nutritional supplements (Pai et al. 2022). Spirulina maxima was first used for food purposes in 1521 when it was taken from Lake Texcoco and sold dry in the markets of Mexico City. The first Spirulina products were produced commercially here in the 1970s as well. Since 1940, members of the Kanembu tribe in Africa who reside near Lake Chad have utilised Spirulina as a protein supplement (Saha et al. 2015). The use of bacteria, yeast, and microalgae can accelerate the production of food components, enzymes, and nutraceuticals. Recent interest in microbial flavour, colours, and bioprocessing employing enzymes has increased due to the current vogue for natural ingredients (McNeil et al. 2013). The importance of microbial synthesis of chemicals, including organic acids, enzymes, proteins, vitamins, antibiotics, and hydrocolloids, has not changed. It has been shown that lactic acid bacteria (LAB), especially *Lactococcus lactis*, are the optimal cell factories for the manufacture of these vital nutraceuticals (Angelin and Kavitha 2020b). Metabolic engineering and mutations may be used for some food-grade microorganisms for overexpression, improvement of nutraceuticals, or introducing their production in microbes for the first time as newly engineered strains. Microbes are responsible for the generation of food flavours, flavonoids, carotenoids, terpenoids, enzymes, organic acids that are utilized in food, food preservatives such as bacteriocins, amino acids, live probiotic cells, nutraceuticals, and other derivatives used in foods and pharmaceuticals (Lyu et al. 2022). They are also used to produce microbial polysaccharides in food, produce xylitol, and other polyols such as prebiotics and oligosaccharides. Scientists recently discovered that microorganisms could potentially be a significant source of terpenes, which are a key component of many pharmaceuticals and food additives (Gültekin 2019). The previous studies identified 262 gene sequences in genomic databases of different bacteria. Terpenes synthases, enzymes that catalyse the synthesis of terpenes, are to be encoded. The researcher subsequently isolated 13 previously unidentified bacterial terpenes using a number of those enzymes (Boutanaev et al. 2015). The terpene compounds are produced using a bio-refinery made of genetically modified *Streptomyces* bacteria.

Furthermore, for conditions like protein energy malnutrition (PEM), anaemia, diarrhoea, cancer, irritable bowel syndrome, ulcerative colitis, and gluten-treatment-resistant celiac, microorganisms can be used as an adjuvant therapy (Masyita et al. 2022). Numerous diseases, including physiologic and lifestyle disorders like diabetes, high blood pressure, cardiovascular, neurological, etc., are largely caused by free radicals. Free radicals catalyse harmful oxidative processes that result in the generation of toxic lipid peroxides. Additionally, they prevent the electron transport chain

(ETC) from functioning, damage DNA and proteins, and render the cell fetal (Pizzino et al. 2017). Therefore, looking for newer, less harmful alternatives to allopathic medications is critical. With the aid of their enzyme "racemases", the bacteria synthesise amino acids and contribute to lowering D-amino acid toxicity. However, the fact that D-amino acids are dangerous to life on Earth is perpetually in flux produced through bacterial growth and geological racemization (Panjiar et al. 2017).

28.2 Functional Foods and Nutraceuticals

Functional foods are believed to offer advantages above and beyond basic nutrition and may help to lower or minimise the risk of developing certain illnesses and other health issues. In addition to providing essential nutrients (such as vitamins and minerals), functional foods are believed to have the additional health benefits (Hasler 2002). When consumed regularly as a part of a varied diet, functional foods may help lower or prevent the risk of developing certain diseases and other health conditions (Table 28.1). The idea first emerged in Japan in the 1980s, when the Ministry of Health and Welfare established a regulatory framework to certify specific meals with proven health advantages for enhancing the country's ageing population (Shimizu 2003). Now known as items for specified health use, these foods are qualified to display a distinctive seal. Functional foods are any modified food or food ingredients that may give health benefits in addition to the traditional nutrients it provides, according to the Food and Nutrition Board of the National Academy of Sciences, which issued this definition in 1994 (Hasler 2002).

In 1989, the Foundation of Innovation Medicine (FIM), Crowford, New Jersey, combined the words "nutrition" and "pharmaceutical"to get the phrase "nutraceuticals." Health Canada has altered its meaning, defining nutraceuticals as a substance that have been separated or purified from foods, typically offered in therapeutic forms unrelated to food, and proven to provide physiological benefits or offer protection from chronic disease, for example, β-carotene and lycopene (Zhao et al. 2021). Functional foods and nutraceuticals both contain active ingredients with physiological actions that promote a better and happier way of life. While the administration differs between nutraceuticals and functional foods, both share the same biological and chemical processes (Mahajan and Rishi 2016). The body requires precise amounts of vitamins, lipids, proteins, carbs, and other nutrients from functional food to ensure a healthy existence. Functional food is referred to be nutraceutical if it helps with the prevention or disorders other than anaemia. Citrus fruits and fortified dairy products (like milk) are examples of nutraceuticals (e.g. orange juice) (Cencic and Chingwaru 2010).

Over the millennia, our forefathers have advocated the idea that certain diets might fend off disease. In the Old Testament, proverb 24:13, the wise Solomon emphasised the primary significance of honey in nourishment and as a treatment for health issues, saying, "Eat honey, my son, since it is good" (Jalgaonkar et al. 2019). Like cinnamon, valued higher than gold in ancient Egypt, ginseng has been used as a

Table 28.1 Functional food components, natural sources, and its beneficial effects on human health (Jalgaonkar et al. 2019)

Natural food sources	Functional food components	Health benefits
Spinach, tomatoes, mango, pumpkin, carrots	B-carotene	Reduced the risk of developing cancer or heart related disease
Papaya, watermelon, red cabbage, mango, tomato and its products	Lycopene	Reduced chronic eye disease (associated with age related macular degeneration)
Citrus fruit, corn, spinach, papaya, broccoli, cabbage, collards, turnip greens	Zeaxanthin	Reduced the risk of breast and prostate cancer arises due to misbalancing cellular signaling pathways
Millet, sorghum, triticale, quinoa, cereal grains	Whole grains	Reduced the chances of the coronary heart disease (CHD), contribute to maintain well and healthy weight
Bran, nuts, fruit skins, vegetables (potatoes green beans	Insoluble fibres	Contributing to maintain healthy digestive system, reduced the risk of breast and colonic cancer
Cheese, beef, some mushroom species (*Agaricus bisporus*), meat products	Conjugated linoleic acid	Contributing body building aid
Fish oil, algal oil, fatty fish (salmon)	Long-chain ɯ-30fatty acids	Reduced the risk of autoimmune disease like rheumatoid arthritis, CHD, enhance infants cognitive
Kidney beans, beetroot, fruits, tomatoes	Anthocyanins (cyaniding, malvidin)	Help in improving cholesterol level as well as blood sugar
Citrus fruits	Flavanones	Protection against free radicals, reduced the risk of many chronic disease as well as cancer, prevention against cardiovascular disease
Low fat dairy products, fish, foods and drinks, leafy vegetables	Calcium	Provide bone strength, help to maintain body weight
Onions, plums, pears, coffee, apples, whole grain	Caffeic acid	Prevent against free radicals, improves heart health
Cereal grain, meat, pork	Thiamin	Improve mental health, metabolism regulation
Fish, dairy products, meat, nuts	Riboflavin	Increase cellular growth, precursors for coenzymes FMN and FAD
Broccoli, beans, green leafy vegetables, carrots	Folate or folic acids	Reduced the risk of neural tube defects, helps to reduce anemia, skin disorder
Eggs, almonds, nuts, dairy products, oysters	Biotin	Help to treat skin disorder, metabolic regulation, enhance hair growth
Meat, beef liver, fish nuts, bananas	Pyridoxine	Regulate immune system, enhance energy level

traditional medicine for thousands of years in China. Due to the focus on economic development, the standard of living in terms of earnings, expenses, and lifestyle has significantly increased today (Metwaly et al. 2021). The nutraceuticals market has grown due to the notion that they are essential to wellness and a viable alternative to contemporary medications. Numerous reported data have proposed the application of dietary supplements as healing agents in recent years due to their beneficial effects, keeping the end in mind of validating their success in extending life expectancy, improving the lifestyle quality for the elderly, and, last but not least, lowering the overall cost of health care (Alali et al. 2021).

28.3 Microbes as a Potential Source for the Production of Nutraceutical Ingredients

Nutraceuticals, as opposed to functional foods and drugs, are now more focused on both structural and functional aspects of various bioactive substances which have in the long run therapeutic or physiological benefits in addition to merely nutritional or direct pharmacological effects (Hoti et al. 2022). They can originate from a multitude of sources, including plants (such as phytochemicals and vitamins), animals (such as polysaccharides), microbes (such as poly amino-acids) (Table 28.2), and the marine sources (such as glucosamine and chitosan) (Wang et al. 2016). Numerous people use nutraceuticals to promote health or prevent disease, particularly in the prevention of ageing-related illnesses such as cancer, diabetes, inflammation, oxidative stress, depression, osteoporosis, gastrointestinal disorders, and cardiovascular diseases (Das et al. 2012). The need for an interest in nutritional supplements to maintain human wellness has significantly fueled the market's expansion. Although the nutraceuticals market is growing, be satisfied with the output of traditional nutraceuticals industries (Das et al. 2012). The effectiveness of inventory control, the low amount and purity of the nutraceuticals, the price and access to raw materials, and direct extraction procedures are all the factors that restrict their use. Although a different method, chemical synthesis can only produce simple biochemicals and is impractical for synthesising complicated biochemicals, especially the ones that are chemically unfavourable (Atanasov et al. 2015). Microbes and microbial-based metabolic engineering provide a sustainable solution to clarify these problems.

28.3.1 Microbes as Producers of Amino Acids

According to studies, bacteria are capable of producing some important amino acids. The productivity of strains that produce amino acids is constantly increasing via metabolic engineering, particularly *Corynebacterium glutamicum* and *Escherichia coli* strains. Combining intracellular metabolite sensing with traditional mutagenesis and screening has sped up both processes (Wendisch 2014). In order to actualise a variable feedstock concept, synthetic biology technologies have made it possible to

Table 28.2 Microbial production of nutraceuticals in different metabolically engineered strains

Name of the nutraceuticals	Name of the strain	Production concentration (mg/L)	References
Apigenin (flavones)	*Escherichia coli*	13	Miyahisa et al. (2006)
Curcumin (curcuminoid)	*Escherichia coli*	0.6	Wang et al. (2015b)
Caffeic acid	*Escherichia coli*	766.68	Huang et al. (2013)
Catechin (flavanonol)	*Escherichia coli*	910.9	Zhao et al. (2015)
Genistein glucosides	*Escherichia coli*	37.29	Pandey et al. (2014)
Genistein	*Saccharomyces cerevisiae*	Not defined	Li et al. (2014)
Daidzin	*Saccharomyces cerevisiae*	Not defined	Li et al. (2014)
Galactooligosaccharides	*Lactococcus lactis*	197,000	Yu and O'Sullivan (2014)
Heparosan	*Escherichia coli*	1880	Zhang et al. (2012)
Hyaluronan	*Streptomyces albulus*	6200	Weissmann and Meyer (1954)
Scleroglucan	*Sclerotium rolfsii*	22,320	Survase et al. (2007)
Cyanophycin	*Escherichia coli*	120,000	Miyahisa et al. (2006)
Zeaxanthin	*Escherichia coli*	43.46	Li et al. (2015)
Poly-γ-glutamic acid	*Bacillus subtilis* BL53	17,000	Da Silva et al. (2014)
Poly-ε-L-lysine	*Streptomyces* sp. M-Z 18	35,140	Chen et al. (2012)

access additional carbon sources. New amino acid synthesis mechanisms and fermentative strategies for producing novel substances are made from amino acids or their metabolic substrates (Carruthers and Lee 2022). These include amino acids, dipeptides, acetylated amino acids, diamines, diacids, and keto acids. Numerous bacteria have been shown to synthesise various types of amino acids. Amino acids that resemble mycosporines and mycosporines Shinorine (mycosporines, glycine, and serine), as well as Porphyra, are mycosporine-like amino acids (MAAs) (Wendisch 2014). A chemical that absorbs UV light is 334 (mycosporines, glycine, and threonine) produced by marine microbes, fungi, cyanobacteria, and microalgae (Geraldes and Pinto 2021). These species can be protected from UV radiation from the environment using MAAs. Another recent study discovered that a cytosolic tyrosine participates in a mechanism resembling that of microbes that help plants produce the amino acid phenylalanine: phosphoryl pyruvate aminotransferases (Geraldes and Pinto 2021). All living things need phenylalanine as a component of their proteins, and plants use it as a starting point for thousands of other metabolites. Phenylalanine cannot be synthesised by animals. Thus, they must either directly or indirectly acquire it from plants (Wendisch 2014).

28.3.2 Microbes as the Source of Vitamins

A high vitamin intake is related to better health as humans and other organisms require various vitamins to sustain metabolic processes. Their primary function is to act as co-enzymes and enter biological reactions (Tardy et al. 2020). The B-group or B-complex vitamins, including thiamin (B1), riboflavin (B2), niacin (B3), pyridoxine (B6), and pantothenic acid (B5), cobalamin (B7 or H), biotin (B7 or H), and folate (B11-B9 or M) (B12) are all synthesised by microorganisms (Rodríguez et al. 1989). These compounds are soluble and significantly impact the metabolic processes, specifically the protein metabolism within cells, plus fats (riboflavin and pyridoxine). B-group vitamins are typically found in a variety of foods, but because cooking and food processing may easily remove or destroy them, it is usual for people to be deficient in them (Bonku and Yu 2020). Due to this, a number of nations have passed legislation requiring the fortification of particular vitamins and minerals in various foods. Microbes possess the simplicity and adaptability required for efficient vitamin synthesis. Enzymes involved in the manufacturing pathway of different vitamins can be over-expressed in a variety of microorganisms by genetic engineering (Bonku and Yu 2020). In addition to having a higher yield than chemical synthesis and the capacity to grow on inexpensive feedstock, microbial production of vitamins also claims a significantly reduced energy input need. Lactic acid bacteria (LAB) used in foods to produce, release, or increase particular nutraceutical substances is made possible by some of its key characteristics (Kennedy 2016), including their biosynthetic ability and metabolic plasticity.

Innovative approaches are being used for food production by fermentation that enhances nutritional and/or health-enhancing properties by careful selection and utilization of LAB that produces nutraceuticals. LAB-promoted biosynthesis can produce fermented milk or bread with higher concentrations of B-group vitamins (such as folate and riboflavin) (Thakur et al. 2016). In order to produce innovative fermented foods by LAB that produces nutraceuticals, vitamin synthesis by lactobacilli is increasingly being focused on (Capozzi and Russo 2012; Leblanc et al. 2011). Numerous lactobacilli include the genes necessary for riboflavin (vitamin B2) production; however, the genetic ability to do so varies depending on the species and/or strain. It has been demonstrated that some bacteria, such as *L. plantarum* WCFS1, have interruptions in the rib operon, which prevent them from producing riboflavin.

On the other hand, due to their ability to synthesise riboflavin, *Lactobacillus plantarum* NCDO 1752, *Lactobacillus plantarum* JDMI, and other strains recently isolated from products produced from cereals can survive without this vitamin (Thakur et al. 2016). The generation of vitamins by microorganisms still faces several significant problems. Vitamin or synthesis pathway intermediates can cause allosteric inhibition, for example, can lower yield in engineered microorganisms expressing every enzyme required to produce the needed vitamin (Vijendra et al. 2022).

28.3.3 Microbial Production of Polysaccharide

Numerous bacteria, yeast, and fungus may synthesise polysaccharides, which are sugar polymers with a variety of structures. They can also be taken from plant and animal tissues. Due to their favourable health effects, polysaccharides produced from microbes have been recognised as a source of nutraceuticals (Angelin and Kavitha 2020a). Xanthan, gellan, dextran, and alginate are a few examples of bacterial polysaccharides that can be produced and purified by microbes for commercial use. Exo-polysaccharides (EPS) are widely used in dairy products and are preferentially synthesised through metabolic engineering of *Lactococcus* and *Streptococcus* (Gupta and Diwan 2017). The wide immunostimulating, anticancer, antibacterial, antioxidant, hypercholesterolemic, and hypoglycemic effects of bioactive fungal polysaccharides, in particular, suggest tremendous promise in both nutraceuticals and pharmaceutical applications. Scleroglucan is a potential anticancer and anti-viral polysaccharide secreted by *Sclerotium rolfsii* mycelia, and its yields may be increased by L-lysine with uridine monophosphate (UMP) addition. A number of animal polysaccharides, such as hyaluronic acid (HA) (Lemieszek and Rzeski 2012), chondroitin, and heparosan, synthesised by microbial hosts rather than being extracted from animal tissues. Bacteria have produced hyaluronic acid in a variety of hosts, including *Escherichia coli, Lactococcus lactis*, and *Streptomyces albulus.* Engineered *Escherichia coli* may manufacture medicinally significant polysaccharides like heparosan and chondroitin, reaching comparatively high titers of 1.88 g/L and 2.4 g/L, respectively (Wang et al. 2016).

28.3.4 Microbes as producers of Alkaloids

Alkaloids are nitrogenous chemicals produced from amino acids that have significant medicinal benefits, such as anticancer and anti-malarial properties. Due to their complicated structures and protracted biosynthesis pathways, plants have been the only source of alkaloids for a very long period (Bufo and Karaman 2019). The top three recognised alkaloids are monoterpene indole alkaloids (MIAs) obtained from glucosinolates and tryptophan, and benzylisoquinoline alkaloids (BIAs) synthesised from tyrosine. In recent years, the explicating of BIA pathways in both *Escherichia coli* and *Saccharomyces cerevisiae* was made possible by the reconstruction of numerous BIA biosynthetic pathways (Ehrenworth et al. 2015). Bacterial consortia comprising *Escherichia coli* producing (S)-reticulate and 7.2 mg/L of magnoflorine and 8.3 mg/L of scoulerine were produced by *Saccharomyces cerevisiae* expressing either CYP80G2 monooxygenase or the berberine bridge enzyme (BBE). *Saccharomyces* cerevisiae was also made to create (R, S)-reticuline, and it was further developed to manufacture salutaridine from (S)-reticuline and scoulerine, tetrahydrocolumbamine, and tetrahydroberberine from (R)-reticuline (Bird and Facchini 2001). However, in microorganisms, there is little metabolic engineering of MIA alkaloids. By inserting 21 new genes and removing three others from the yeast genome, de novo manufacture of the MIA alkaloid strictosidine has just been accomplished in yeast. Glucosinolates are natural substances generated from amino acids that are sulfur-rich. Tryptophan is used to make the compound indolylglucosinolate (IG), has recently been synthesised in *Saccharomyces cerevisiae* (Wang et al. 2015a) by introducing eight plant genes into the genome. This is a rare instance of glucosinolate production in microorganisms. Although plant platforms are better suited for producing scalable amounts of alkaloids, metabolic engineering in microorganisms has demonstrated possibilities for producing these plant-derived compounds at a low cost (Mora-Pale et al. 2013).

28.4 Probiotic-Based Functional Foods

There are already more than 500 probiotic food products available into the market on a global scale in the last several years, and this number is steadily rising. The probiotic foods made from the fermentation of fruits and vegetables, grains, and animal products are attracting interest from both consumers and scientists (Rezac et al. 2018). Probiotics are now commercially available in items including ice cream, milk, sour milk, and fruit juices. The most recent probiotic-containing foods are mayonnaise edible spreads, cheese and cheese-based dips, meat-based goods, and cheese. The probiotic bacteria are artificially infused into the food during the manufacturing of probiotic foods (Terpou et al. 2019). Most culture preparations are produced for various uses as either highly concentrated freeze-dried cultures or freeze-dried powders, both of which are commercially available in extremely concentrated forms. Due to the production of many metabolic components, including the production of byproducts such as acetic acid during fermentation by *Bifidobacterium* spp., the scent and flavour of food products may change when probiotic microorganisms are added (Vijayendra and Gupta 2014). The amount of active cells or total viable cells/ml/g of the item right before consumption by the individual determines how successful probiotic food products are as medicines. To retain consumer confidence in probiotic food products, we must make sure that a high probiotic survival rate is maintained throughout the during production and over the storage of the products. (Mohammadi et al. 2011). The survival of the LAB in various food products throughout production, processing and storage is influenced by a number of factors. Chemicals like bacteriocins, hydrogen peroxide, colouring agents, and artificial flavours are among the factors that have been identified. Titratable acidity, water activity, molecular oxygen, and the amount of sugar and salt are a few examples of food parameters (Misra et al. 2022). Processing parameters include incubation temperature, the rate at which food products cool after high-heat treatment, storage procedures, and packing materials and methods. Microbiological parameters include the percentage (%) of inoculums and the symbiotic count.

Infants that get probiotic supplements from the *Lactobacillus casei* bacterial strain had high levels of circulating immunoglobulin A, which is correlated with a shorter duration of rotavirus-induced diarrhoea (Aspri et al. 2020). Additionally, the circulating blood granulocytes' nonspecific immunological phagocytic activity is noticeably enhanced by the addition of probiotic foods containing *Bifidobacterium bifidum* and *Lactobacillus acidophilus*. According to a study, probiotic yoghurt consumption can increase blood mononuclear cells' ability to produce cytokines (Gill and Guarner 2004). Only when the food contains the necessary minimum viable microbe count at the moment of intake, probiotics' claimed health benefits can be realised. The food sector frequently uses the lowest advised level, which is 10^6 CFU/mL at the time of ingestion (Cristofori et al. 2021). The United States Food and Drug Administration (US FDA) recommends probiotic food items include a minimum of 10^6 CFU/mL. A daily dose of 10^8 to 10^9 probiotic bacteria is necessary to accomplish probiotic action in humans, depending on the amount consumed and considering storage conditions (Zommiti et al. 2020).

28.4.1 Health Promoting Properties of Probiotic Functional Foods

Probiotics should be consumed first by healthy individuals because they improve general health and protect against various ailments. Improving health is the most sensible and preventive move we can take to safeguard ourselves against various illnesses (Amara and Shibl 2015). However, how do probiotics enhance health? The topic of how probiotics directly or indirectly benefit our health will be discussed in the following paragraphs. The various microorganisms found in human bodies collaborate to carry out various tasks. Our digestive system's functions are its most important ones (Chandarakesan et al. 2018). These microbes improve the consumption and digestion of particular foods and significantly lighten the load on our digestive system. These bacteria simplify the processes that the human body must go through in order to change complicated dietary structures. The following facts and points can be used to summarise how probiotics relate to improving human health:

I. Probiotic microbes are functional and user-friendly.
II. Probiotic microorganisms aid in the fermentation of our food, turning it from a complex form into simpler byproducts, and they can improve our health in a variety of ways.
III. They have the ability to competitively inhibit harmful microorganisms and seize control of our digestive system.
IV. Numerous variables, including poor diet, ageing, alcohol consumption, and a number of others, may result in a decline in the viable count of probiotic microorganisms. Probiotics should therefore be a part of our everyday diets as a supplement.
V. They are expected to be significantly impacted in certain circumstances, such as after taking antibiotics. Thus, they should be taken orally in large quantities with food.
VI. They deliver essential beneficial consequences to the organism.
VII. They eliminate the harmful microorganisms' negative effects.

VIII. Because they make it possible for any amount of food to be properly digested and metabolised, they reduce the amount of food that the human body needs.
IX. They lessen the additional stress placed on the digestive tract of humans.
X. They reduced the impact of the initial attack of hazardous substances on our cells by forming a biofilm.
XI. By using lactose intolerance, that are genetically determined traits, probiotics can often compensate for the deficiencies in our genetic makeup.

28.5 The Mechanisms Through Which Functional Foods and Nutraceuticals Exert Health Advantages

Numerous studies have demonstrated the neuroprotective effects of nutraceuticals and functional foods, which work at several biochemical and metabolic levels. In particular, they can guard against cell death, oxidative stress, toxic levels of tau and amyloid-b, mitochondrial damage, and oxidative stress (Atlante et al. 2020). They have been shown to have a significant impact on the composition of the gut microbiota, which has helped researchers understand how the diversity of bacteria affects the production and accumulation of harmful proteins in the brain. It is more than just a diet to follow; it is a way of life that plays a significant part in the prevention and treatment of numerous diseases as well and reduced the risk of developing diabetes, heart disease, cancer, and other of our most dreaded illnesses (Zhang et al. 2015). The same is true in terms of preventing Alzheimer's disease (AD). Owing to its high metabolic activity and amount of oxidisable components, the brain is especially vulnerable to oxidative damage, including the lipid milieu of the myelin membrane found in neuronal cells (Gella and Durany 2009). Therefore, diets high in antioxidants may enhance cognitive performance. According to the studies, a number of berries have significant antioxidant capacities because they contain tannins, anthocyanins, and phenols, which can boost the plasticity of the hippocampus and aid in learning and memory function. Alpha lipoic acid is additionally essential for preserving mitochondrial energy balance and enhancing cognitive abilities, which may be reported in vegetables like spinach and broccoli (Kelly et al. 2018).

Additionally, the antioxidant-rich beverages such as green and both types of black tea include epigallocatechin gallate, both types of black tea contain epigallocatechin gallate, which has been demonstrated to indirectly lessen the development of amyloid plaques, one of the AD's defining characteristics (Gella and Durany 2009). When taken in and absorbed by the gut, nutraceuticals function similarly to pharmaceuticals, providing preventative and/or adjuvant activity in the process of healing in the healing phase and increasing our resistance to dangerous environmental chemicals. Plants produce compounds having high antioxidant activity as a defence mechanism against insects, and they're mostly found in fruits and vegetables (Atlante et al. 2020). As a result, they are of considerable importance to humans via the positive effects that may be utilised even at low dosages. The fact that the growing number of innovative studies suggest that generations may

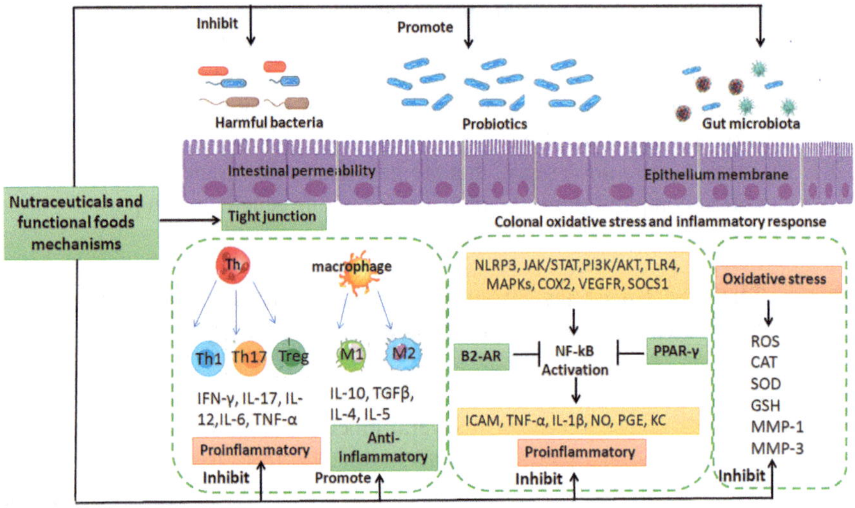

Fig. 28.1 Some of the proposed mechanisms by which functional foods and nutraceuticals exert their beneficial effects

experience the effects of diet on mental health, which is a fascinating idea is an intriguing aspect of the situation, despite the fact that the molecular processes underpinning the impact of nutrition on epigenetic are still not fully understood (Stevens et al. 2018). The ability of a dietary element to influence non-genetic events—even those that nonetheless result in potentially heritable phenotypic changes—opens a therapeutic pathway that may be useful in the management, if not only the control, of the illness development (Franzago et al. 2020). The antioxidant, anti-estrogenic, anti-inflammatory, immunomodulatory, and anticarcinogenic effects of these nutrients make them significant for the human body (Fig. 28.1) even though they are not necessary for survival, unlike vitamins, for example. Additionally, they are less potent as bioactive substances than pharmaceuticals, but because they are routinely ingested in large quantities as part of a meal, they can nevertheless have important long-term physiological impacts (Cory et al. 2018). Thus, phytonutrients or phytochemicals may have an anti-chronic inflammatory effect and may aid in the prevention of age- and chronic-related disorders, including atherosclerosis and arthritis. By modifying the makeup of the microbiota, the aforementioned bioactive substances have profound effects on the gut environment (Yin et al. 2019). The biological functions of functional dietary components can also differ since the gut microbiota itself can impact their metabolites and relative effects, which can even vary from person to person. Although numerous animal and in vitro researches have been conducted to examine the bioactivity of various functional foods (nutraceuticals), clinical trials involving humans are currently rare and inconclusive (Valdes et al. 2018). It must be remembered that any chemical changes to the initial bioactive component that take place during digestion or storage could significantly alter both the compound's bioactivity and bioavailability.

To further explain this idea, it should be remembered that while a compound's antioxidant activity, as measured in a test tube, can be used to gauge a product's quality, this does not necessarily mean that it will perform well in vivo or within a living being. Pomegranate ellagitannins are the subjects in this instance (ETs) (Sadeer et al. 2020). Pomegranates are one of the most potent in vitro antioxidants because of the high antioxidant potential of ETs. In contrast, these substances are not absorbed in living organisms and are heavily degraded by the intestinal microbiota into urolithins, which are no longer endowed with the ETs' initial antioxidant capacity (Asgary et al. 2014). Onion, garlic, grapes, rosemary, broccoli, spinach, turmeric, parsley, and other foods have significant antioxidant properties and guard against a number of neurodegenerative illnesses, including AD. For instance, antioxidant compounds found in odourless old garlic extracts (AGE) help to reduce oxidative damage (Atlante et al. 2020). AGE has been demonstrated to prevent damage to cells, cellular damage, brought on by the buildup of -amyloid peptide (Aβ) in vitro and to mitigate the toxicity of Aβ.

Additionally, glycogen synthase kinase-3 activity is regulated to control tau protein hyperphosphorylation, usually in relation to the aetiology of AD (GSK) (Diering and Freeman 2018). Garlic's ability to penetrate through the way of blood-brain barrier (BBB), a characteristic that further supports its promise as a medicinal substance, has been used to investigate the neuroprotective effects of AGE on dementia caused by AD. A bioactive component's distinct ability to cross or not the BBB can differ more that this property is now deserving attention in the AD therapy strategy (Mathew and Biju 2008).

28.6 Applications of Microbial Functional Foods and Nutraceuticals

A set of compounds with recognised nutritional functions, including vitamins, minerals, amino acids, and fatty acids, are referred to as "nutrients," naturally occurring in our foods and have a wide range of possible health advantages (Chen et al. 2018). Most vegetables, entire grains, dairy products, fruits, and meat and poultry from animals all include vitamins that are beneficial for managing diabetes, cancer, cataracts, heart disease, stroke, and other conditions (Asif 2014). The minerals found in plants, animals, and dairy products are beneficial for treating anaemia and osteoporosis, strengthening bones, teeth, and muscles, and enhancing nerve signals and heart rhythm. The term "nutraceuticals" describes a large group of substances (Atlante et al. 2020). Some important nutraceuticals and functional foods and their applications are described as follows:

Dietary Fibres Dietary fibres are chemicals of the vegetable origin that are found in meals that increase the amount of intestinal contents without being digested in the large intestine. In chemistry, the term "dietary fibre" refers to carbohydrate polymers having at least a percentage of 3's polymerization that is not absorbed or processed in the small intestine (Lunn and Buttriss 2007). Fruit, barley, oats, lignin, cellulose, and pectin are among examples, and other items. An appropriate intake of these fibres through food is linked to a decreased risk of gastrointestinal problems, colon cancer, hypertension, diabetes, obesity, and cardiovascular disease (Nirmala Prasadi and Joye 2020).

Probiotics Probiotics are living bacteria found in food that have a variety of positive effects on the human body. The enhancement of the intestinal microbiota has been attributed to the benefits of probiotic-rich diets that are prepared either naturally or through commercial fermentation techniques (Markowiak and Ślizewska 2017).

Polyunsaturated Fatty Acids Polyunsaturated fatty acids found in oily fish, flax seeds, soybeans, and other foods include omega 3 fatty acids such as -linolenic acid, eicosapentaenoic acid, and docosahexaenoic acid. Omega 6 fatty acids, such as α-linoleic acid and arachidonic acid, are also present in foods like corn, soybean, sunflower, and wheat (Balić et al. 2020).

Prebiotics Prebiotics are non-living substances, in contrast to probiotics. These short-chain polysaccharides with distinctive chemical compositions particularly fructose-based oligosaccharides that are present in food either naturally or artificially yet nonetheless have a variety of beneficial effects on our bodies while not being digested by people. The microbiota's composition and function fluctuate in the colon, affecting the development and activity of particular bacteria that benefit the host's health (Megur et al. 2022). Prebiotics act as the nutrition of probiotics and promote their growth in the gastrointestinal tract. The roots of chicory, bananas, tomatoes, alliums, beans, etc., are a few examples of these edibles (Markowiak and Śliżewska 2017).

Polyphenols In order to defend themselves against reactive oxygen species and photosynthetic stress, plants synthesise phytochemicals called polyphenols. Others include the anti-inflammatory and antioxidant characteristics of flavonoids, anthocyanins, and phenolic acids (Panche et al. 2016), which can be found in a range of foods.

Additionally, some nutraceuticals or functional foods are found in many food products, such as quercetin, which is found in red grapes, broccoli, onions, and other foods. This makes mentioning the foods that are higher in a substance or family of substances while discussing nutraceuticals worthwhile. The exact quantity of nutraceuticals or functional foods in food is affected by several factors, including the soil on which plants grow and how animals are fed. As a result, eating a gorgeous red tomato that has ripened in the sun or merely a rosé tomato is not the same (Das et al. 2012). In actuality, adverse reactions and toxicity are frequently recorded due to the consumption of nutraceuticals and functional foods. Perhaps, this contamination comes from pesticides, other hazardous plants, metals, fertilisers, etc. (Colombo et al. 2020). The possible health advantages of functional foods are mentioned below:

(a) Enhancing the nutritional value of food.
(b) Aid in extending the life.
(c) Assist in preventing specific medical conditions.
(d) May appear more "natural" than conventional treatment and be associated with less negative side effects.
(e) Foods that are nutrient rich for the elderly are one example of how to improve the nutritional content of food for those with special needs.
(f) Be readily available and reasonably priced.

28.7 Future Perspectives

The realm of "neuro-nutraceuticals" currently offers strong research ideas, as was extensively noted in this study, and in the near future, it may open interesting windows on effective and secure treatments, even in individuals with dementia who have already been diagnosed. We hope that several food-borne chemicals will eventually help AD therapy because of the preclinical evidence that they may be beneficial in preventing and treating cognitive decline. To validate their positive effects on humans, additional research is required. Currently, adopting an active lifestyle along with a diet that fits the criteria of the Mediterranean diet (perhaps enhanced with walnuts and olive oil) is the strongest suggestion in terms of behaviour and nutrition—when a healthy cognitive reserve still exists. Recent research by Ravi et al. highlights the beneficial, mild, and preventive anti-AD properties of the Mediterranean diets, which are mostly constituted of fruits, vegetables, and omega-3 fatty acids. We can battle this debilitating neurological condition by making the necessary lifestyle changes. Maintaining a healthy diet is vital for all age groups. However, it is crucial to encourage specific nutritional practices among the elderly owing to the possibility of malnutrition, decreased absorption, appetite loss, and chewing difficulty, all of which can impair a person's nutritional status. In reality, nutritional intake and assimilation can be reduced in ageing individuals, which may help to explain the contradictory outcomes of clinical research involving dietary treatments. These techniques aim to fundamentally restore the variety of gut flora in the elderly in an effort to lessen certain negative effects of ageing.

28.8 Conclusion

Due to the high demand for value-added nutraceuticals to protect people from diseases, the technologies based on metabolic engineering of microorganisms have made amazing strides in manufacturing nutraceuticals over the past few decades. The preparation and manufacture of complex natural chemicals like carotenoids on an industrial scale using only basic carbon sources were made possible, thanks to metabolic engineering in microorganisms. Novel or more complex value-added substances could be created in microbes naturally, particularly with the clarification of biosynthetic pathways and the development of synthetic biology technologies. The effectiveness of enzymes produced by plants or animals, which restricts the production of the target molecules, is one of the main drawbacks of microbial platforms. While various non-microbial manufacturing methods, such as those using algae and plant-based cell cultures, have also been developed to produce naturally occurring or synthetic value-added nutraceuticals and functional foods. Therefore, even though there is no question about the role that nutraceuticals and functional foods play in preventing, limiting, and treating numerous diseases, it is important to remember that everyone responds differently to nutraceuticals. The assimilation of an individual's nutraceuticals and susceptibility to a specific disease will depend on their lifestyle, environmental exposures, and genetic makeup.

References

Alali M, Alqubaisy M, Aljaafari MN, Alali AO, Baqais L, Molouki A, Abushelaibi A, Lai KS, Lim SHE (2021) Nutraceuticals: transformation of conventional foods into health promoters/disease preventers and safety considerations. Molecules 26(9):2540. https://doi.org/10.3390/molecules26092540

Amara AA, Shibl A (2015) Role of probiotics in health improvement, infection control and disease treatment and management. Saudi Pharmaceut J 23(2):107–114. https://doi.org/10.1016/j.jsps.2013.07.001

Angelin J, Kavitha M (2020a) Exopolysaccharides from probiotic bacteria and their health potential. Int J Biol Macromol 162:853–865. https://doi.org/10.1016/j.ijbiomac.2020.06.190

Angelin J, Kavitha M (2020b) Since January 2020 Elsevier has created a COVID-19 resource centre with free information in English and Mandarin on the novel coronavirus COVID- 19. The COVID-19 resource centre is hosted on Elsevier Connect, the company's public news and information, January

Asgary S, Javanmard S, Zarfeshany A (2014) Potent health effects of pomegranate. Adv Biomed Res 3(1):100. https://doi.org/10.4103/2277-9175.129371

Asif M (2014) The prevention and control the type-2 diabetes by changing lifestyle and dietary pattern. J Educ Health Promot 3(1):1. https://doi.org/10.4103/2277-9531.127541

Aspri M, Papademas P, Tsaltas D (2020) Review on non-dairy probiotics and their use in non-dairy based products. Fermentation 6(1):1–20. https://doi.org/10.3390/fermentation6010030

Atanasov AG, Waltenberger B, Pferschy-Wenzig EM, Linder T, Wawrosch C, Uhrin P, Temml V, Wang L, Schwaiger S, Heiss EH, Rollinger JM, Schuster D, Breuss JM, Bochkov V, Mihovilovic MD, Kopp B, Bauer R, Dirsch VM, Stuppner H (2015) Discovery and resupply of pharmacologically active plant-derived natural products: a review. Biotechnol Adv 33(8):1582–1614. https://doi.org/10.1016/j.biotechadv.2015.08.001

Atlante A, Amadoro G, Bobba A, Latina V (2020) Functional foods: an approach to modulate molecular mechanisms of Alzheimer's disease. Cell 9(11):1–62. https://doi.org/10.3390/cells9112347

Balić A, Vlašić D, Žužul K, Marinović B, Mokos ZB (2020) Omega-3 versus Omega-6 polyunsaturated fatty acids in the prevention and treatment of inflammatory skin diseases. Int J Mol Sci 21(3):741. https://doi.org/10.3390/ijms21030741

Bird DA, Facchini PJ (2001) Berberine bridge enzyme, a key branch-point enzyme in benzylisoquinoline alkaloid biosynthesis, contains a vacuolar sorting determinant. Planta 213(6):888–897. https://doi.org/10.1007/s004250100582

Bonku R, Yu J (2020) Health aspects of peanuts as an outcome of its chemical composition. Food Sci Human Wellness 9(1):21–30. https://doi.org/10.1016/j.fshw.2019.12.005

Boutanaev AM, Moses T, Zi J, Nelson DR, Mugford ST, Peters RJ, Osbourn A (2015) Investigation of terpene diversification across multiple sequenced plant genomes. Proc Natl Acad Sci U S A 112(1):E81–E88. https://doi.org/10.1073/pnas.1419547112

Bufo SA, Karaman R (2019) Herbivores, cancerous cells and pathogens. Toxins 11(656):1–28

Capozzi V, Russo P (2012) Lactic acid bacteria producing B-group vitamins: a great potential for functional cereals products. Appl Microbiol Biotechnol 96:1383. https://doi.org/10.1007/s00253-012-4440-2

Carruthers DN, Lee TS (2022) Translating advances in microbial bioproduction to sustainable biotechnology. Front Bioeng Biotechnol 10(August):1–22. https://doi.org/10.3389/fbioe.2022.968437

Cencic A, Chingwaru W (2010) The role of functional foods, nutraceuticals, and food supplements in intestinal health. Nutrients 2(6):611–625. https://doi.org/10.3390/nu2060611

Chandarakesan A, Muruhan S, Sayanam RRA (2018) Morin inhibiting photocarcinogenesis by targeting ultraviolet-B-induced oxidative stress and inflammatory cytokines expression in Swiss albino mice. Int J Nutr Pharmacol Neurol Dis 8:41–46. https://doi.org/10.4103/ijnpnd.ijnpnd

Chen XS, Ren XD, Dong N, Li S, Li F, Zhao FL, Tang L, Zhang JH, Mao ZG (2012) Culture medium containing glucose and glycerol as a mixed carbon source improves ε-poly-l-lysine production by Streptomyces sp. M-Z18. Bioprocess Biosyst Eng 35(3):469–475. https://doi.org/10.1007/s00449-011-0586-z

Chen Y, Michalak M, Agellon LB (2018) Importance of nutrients and nutrient metabolism on human health. Yale J Biol Med 91(2):95–103

Colombo F, Di Lorenzo C, Biella S, Vecchio S, Frigerio G, Restani P (2020) Adverse effects to food supplements containing botanical ingredients. J Funct Foods 72(February):103990. https://doi.org/10.1016/j.jff.2020.103990

Cory H, Passarelli S, Szeto J, Tamez M, Mattei J (2018) The role of polyphenols in human health and food systems: a mini-review. Front Nutr 5(September):1–9. https://doi.org/10.3389/fnut.2018.00087

Cristofori F, Dargenio VN, Dargenio C, Miniello VL, Barone M, Francavilla R (2021) Anti-inflammatory and immunomodulatory effects of probiotics in gut inflammation: a door to the body. Front Immunol 12(February):1–21. https://doi.org/10.3389/fimmu.2021.578386

Da Silva SB, Cantarelli VV, Ayub MAZ (2014) Production and soptimisation of poly-γ-glutamic acid by Bacillus subtilis BL53 isolated from the Amazonian environment. Bioprocess Biosyst Eng 37(3):469–479. https://doi.org/10.1007/s00449-013-1016-1

Das L, Bhaumik E, Raychaudhuri U, Chakraborty R (2012) Role of nutraceuticals in human health. J Food Sci Technol 49(2):173–183. https://doi.org/10.1007/s13197-011-0269-4

Diering M, Freeman M (2018) 乳鼠心肌提取 HHS public access. Physiol Behav 176(1):139–148. https://doi.org/10.3233/JAD-160612.Amyloid

Ehrenworth AM, Sarria S, Peralta-Yahya P (2015) Pterin-dependent mono-oxidation for the microbial synthesis of a modified monoterpene indole alkaloid. ACS Synth Biol 4(12):1295–1307. https://doi.org/10.1021/acssynbio.5b00025

Franzago M, Santurbano D, Vitacolonna E, Stuppia L (2020) Genes and diet in the prevention of chronic diseases in future generations. Int J Mol Sci 21(7):2633. https://doi.org/10.3390/ijms21072633

Galanakis CM (2021) Functionality of food components and emerging technologies. Foods 10(1):1–26. https://doi.org/10.3390/foods10010128

Gella A, Durany N (2009) Oxidative stress in Alzheimer disease. Cell Adh Migr 3(1):88–93. https://www.ncbi.nlm.nih.gov/pmc/articles/PMC2675154/pdf/cam0301_0088.pdf

Geraldes V, Pinto E (2021) Mycosporine-like amino acids (Maas): biology, chemistry and identification features. Pharmaceuticals 14(1):1–17. https://doi.org/10.3390/ph14010063

Gill HS, Guarner F (2004) Probiotics and human health: a clinical perspective. Postgrad Med J 80(947):516–526. https://doi.org/10.1136/pgmj.2003.008664

Gültekin F (2019) Food additives and microbiota. North Clin Istanb 7(2):192–200. https://doi.org/10.14744/nci.2019.92499

Gupta P, Diwan B (2017) Bacterial exopolysaccharide mediated heavy metal removal: a review on biosynthesis, mechanism and remediation strategies. Biotechnol Rep 13:58–71. https://doi.org/10.1016/j.btre.2016.12.006

Hasler CM (2002) Functional foods: benefits, concerns and challenges—a position paper from the American Council on Science and Health. J Nutr 132(12):3772–3781. https://doi.org/10.1093/jn/132.12.3772

Hoti G, Matencio A, Pedrazzo AR, Cecone C, Appleton SL, Monfared YK, Caldera F, Trotta F (2022) Nutraceutical concepts and dextrin-based delivery systems. Int J Mol Sci 23(8):4102. https://doi.org/10.3390/ijms23084102

Huang Q, Lin Y, Yan Y (2013) Caffeic acid production enhancement by engineering a phenylalanine over-producing Escherichia coli strain. Biotechnol Bioeng 110(12):3188–3196. https://doi.org/10.1002/bit.24988

Jalgaonkar K, Mahawar MK, Bibwe B, Nath P, Girjal S (2019) Chapter 17: Nutraceuticals and functional foods. In: Trends & prospects in processing of horticultural crops. Today & Tomorrow's Printers and Publishers, New Delhi, pp 231–250

Kelly E, Vyas P, Weber JT (2018) Biochemical properties and neuroprotective effects of compounds in various species of berries. Molecules 23(1):26. https://doi.org/10.3390/molecules23010026

Kennedy DO (2016) B vitamins and the brain: mechanisms, dose and efficacy—a review. Nutrients 8(2):68. https://doi.org/10.3390/nu8020068

Leblanc JG, Laiño JE, del Valle MJ, Vannini V, van Sinderen D, Taranto MP, de Valdez GF, de Giori GS, Sesma F (2011) B-group vitamin production by lactic acid bacteria - current knowledge and potential applications. J Appl Microbiol 111(6):1297–1309. https://doi.org/10.1111/j.1365-2672.2011.05157.x

Lemieszek M, Rzeski W (2012) Anticancer properties of polysaccharides isolated from fungi of the basidiomycetes class. Wspolczesna Onkologia 16(4):285–289. https://doi.org/10.5114/wo.2012.30055

Li J, Li Z, Li C, Gou J, Zhang Y (2014) Molecular cloning and scharacterisation of an isoflavone 7-O-glucosyltransferase from Pueraria lobata. Plant Cell Rep 33(7):1173–1185. https://doi.org/10.1007/s00299-014-1606-7

Li XR, Tian GQ, Shen HJ, Liu JZ (2015) Metabolic engineering of Escherichia coli to produce zeaxanthin. J Ind Microbiol Biotechnol 42(4):627–636. https://doi.org/10.1007/s10295-014-1565-6

Lunn J, Buttriss JL (2007) Carbohydrates and dietary fibre. Nutr Bull 32(1):21–64. https://doi.org/10.1111/j.1467-3010.2007.00616.x

Lyu X, Lyu Y, Yu H, Chen WN, Ye L, Yang R (2022) Biotechnological advances for improving natural pigment production: a state-of-the-art review. Bioresour Bioprocess 9(1). https://doi.org/10.1186/s40643-022-00497-4

Mahajan M, Rishi M (2016) Research and Reviews in Biotechnology and Biosciences, vol Functional foods and nutraceuticals: an overview. Review article, pp 1–11. http://biotechjournal.in/images/paper_pdffiles/Fun-5d3196efd9d79.pdf

Markowiak P, Ślizewska K (2017) Effects of probiotics, prebiotics, and synbiotics on human health. Nutrients 9(9):1021. https://doi.org/10.3390/nu9091021

Masyita A, Mustika Sari R, Dwi Astuti A, Yasir B, Rahma Rumata N, Emran TB, Nainu F, Simal-Gandara J (2022) Terpenes and terpenoids as main bioactive compounds of essential oils, their roles in human health and potential application as natural food preservatives. Food Chem X 13 (January):100217. https://doi.org/10.1016/j.fochx.2022.100217

Mathew B, Biju R (2008) Neuroprotective effects of garlic: a review. Libyan J Med 3(1):23–33. https://doi.org/10.4176/071110

McNeil B, Archer D, Giavasis I, Harvey L (2013) Microbial production of food ingredients, enzymes and nutraceuticals. In: Microbial production of food ingredients, enzymes and nutraceuticals, March 2013, pp 1–610. https://doi.org/10.1533/9780857093547

Megur A, Daliri EBM, Baltriukienė D, Burokas A (2022) Prebiotics as a tool for the prevention and treatment of obesity and diabetes: classification and ability to modulate the gut microbiota. Int J Mol Sci 23(11):6097. https://doi.org/10.3390/ijms23116097

Metwaly AM, Ghoneim MM, Eissa IH, Elsehemy IA, Mostafa AE, Hegazy MM, Afifi WM, Dou D (2021) Traditional ancient Egyptian medicine: a review. Saudi J Biol Sci 28(10):5823–5832. https://doi.org/10.1016/j.sjbs.2021.06.044

Misra S, Pandey P, Dalbhagat CG, Mishra HN (2022) Emerging technologies and coating materials for improved probiotication in food products: a review. Food Bioproc Tech 15(5):998–1039. https://doi.org/10.1007/s11947-021-02753-5

Miyahisa I, Funa N, Ohnishi Y, Martens S, Moriguchi T, Horinouchi S (2006) Combinatorial biosynthesis of flavones and flavonols in Escherichia coli. Appl Microbiol Biotechnol 71(1): 53–58. https://doi.org/10.1007/s00253-005-0116-5

Mohammadi R, Mortazavian AM, Khosrokhavar R, Da Cruz AG (2011) Probiotic ice cream: viability of probiotic bacteria and sensory properties. Ann Microbiol 61(3):411–424. https://doi.org/10.1007/s13213-010-0188-z

Mora-Pale M, Sanchez-Rodriguez SP, Linhatrdt RJ, Dordick JS, Koffas MAG (2013) Metabolic engineering and in vitro biosynthesis of phytochemicals and non-natural analogues. Plant Sci 210:10–24. https://doi.org/10.1016/j.plantsci.2013.05.005

Nirmala Prasadi VP, Joye IJ (2020) Dietary fibre from whole grains and their benefits on metabolic health. Nutrients 12(10):1–20. https://doi.org/10.3390/nu12103045

Pai S, Hebbar A, Selvaraj S (2022) A critical look at challenges and future scopes of bioactive compounds and their incorporations in the food, energy, and pharmaceutical sector. Environ Sci Pollut Res 29(24):35518–35541. https://doi.org/10.1007/s11356-022-19423-4

Panche AN, Diwan AD, Chandra SR (2016) Flavonoids: an overview. J Nutr Sci 5:e47. https://doi.org/10.1017/jns.2016.41

Pandey RP, Parajuli P, Koirala N, Lee JH, Park YI, Sohng JK (2014) Glucosylation of isoflavonoids in engineered Escherichia coli. Mol Cells 37(2):172–177. https://doi.org/10.14348/molcells.2014.2348

Panjiar N, Mishra S, Yadav AN, Verma P (2017) Functional foods from cyanobacteria: an emerging source for functional food products of pharmaceutical importance. In: Microbial functional foods and nutraceuticals. Wiley, Hoboken. NJ, pp 21–37. https://www.researchgate.net/profile/Ajar_Nath_Yadav/publication/320335917_Functional_Foods_from_Cyanobacteria_An_Emerging_Source_for_Functional_Food_Products_of_Pharmaceutical_Importance/links/59fedd580f7e9b9968c4fe77/Functional-Foods-from-Cyanobacteria-

Pizzino G, Irrera N, Cucinotta M, Pallio G, Mannino F, Arcoraci V, Squadrito F, Altavilla D, Bitto A (2017) Oxidative stress: harms and benefits for human health. Oxid Med Cell Longev 2017: 8416763. https://doi.org/10.1155/2017/8416763

Rezac S, Kok CR, Heermann M, Hutkins R (2018) Fermented foods as a dietary source of live organisms. Front Microbiol 9(AUG):1785. https://doi.org/10.3389/fmicb.2018.01785

Rodríguez E, Arqués JL, Rodríguez R, Nuñez M, Medina M, Talarico TL, Casas IA, Chung TC, Dobrogosz WJ, Axelsson L, Lindgren SE, Dobrogosz WJ, Kerkeni L, Ruano P, Delgado LL, Picco S, Villegas L, Tonelli F, Merlo M, Masuelli M et al (1989) We are IntechOpen, the world's leading publisher of Open Access books Built by scientists, for scientists TOP 1%. INTECH 32(tourism):137–144. https://www.intechopen.com/books/advanced-biometric-technologies/liveness-detection-in-biometrics

Sadeer NB, Montesano D, Albrizio S, Zengin G, Mahomoodally MF (2020) The versatility of antioxidant assays in food science and safety—chemistry, applications, strengths, and limitations. Antioxidants 9(8):1 39. https://doi.org/10.3390/antiox9080709

Saha SK, McHugh E, Murray P, Walsh DJ (2015) Microalgae as a source of nutraceuticals. In: Phycotoxins: chemistry and biochemistry, 2nd edn, pp 255–291. https://doi.org/10.1002/9781118500354.ch12

Shimizu T (2003) Health claims on functional foods: the Japanese regulations and an international comparison. Nutr Res Rev 16(2):241–252. https://doi.org/10.1079/nrr200363

Stevens AJ, Rucklidge JJ, Kennedy MA (2018) Epigenetics, nutrition and mental health. Is there a relationship? Nutr Neurosci 21(9):602–613. https://doi.org/10.1080/1028415X.2017.1331524

Survase SA, Saudagar PS, Singhal RS (2007) Enhanced production of scleroglucan by Sclerotium rolfsii MTCC 2156 by use of metabolic precursors. Bioresour Technol 98(2):410–415. https://doi.org/10.1016/j.biortech.2005.12.013

Tardy AL, Pouteau E, Marquez D, Yilmaz C, Scholey A (2020) Vitamins and minerals for energy, fatigue and cognition: a narrative review of the biochemical and clinical evidence. Nutrients 12(1):228. https://doi.org/10.3390/nu12010228

Terpou A, Papadaki A, Lappa IK, Kachrimanidou V, Bosnea LA, Kopsahelis N (2019) Probiotics in food systems: significance and emerging strategies towards improved viability and delivery of enhanced beneficial value. Nutrients 11(7):1591. https://doi.org/10.3390/nu11071591

Thakur K, Tomar SK, De S (2016) Lactic acid bacteria as a cell factory for riboflavin production. J Microbial Biotechnol 9(4):441–451. https://doi.org/10.1111/1751-7915.12335

Valdes AM, Walter J, Segal E, Spector TD (2018) Role of the gut microbiota in nutrition and health. BMJ (Online) 361:36–44. https://doi.org/10.1136/bmj.k2179

Vijayendra SVN, Gupta RC (2014) Performance evaluation of bulk freeze dried starter cultures of dahi and yoghurt along with probiotic strains in sstandardised milk of cow and buffalo. J Food Sci Technol 51(12):4114–4119. https://doi.org/10.1007/s13197-013-0944-8

Vijendra V, Ananya M, Jayesh R (2022) Riboswitch—mediated regulation of riboflavin biosynthesis genes in prokaryotes. 3 Biotech 12:278. https://doi.org/10.1007/s13205-022-03348-3

Wang J, Guleria S, Koffas MAG, Yan Y (2015a) Microbial production of value-added nutraceuticals. Curr Opin Biotechnol 37:97

Wang S, Zhang S, Xiao A, Rasmussen M, Skidmore C, Zhan J (2015b) Metabolic engineering of Escherichia coli for the biosynthesis of various phenylpropanoid derivatives. Metab Eng 29:153–159. https://doi.org/10.1016/j.ymben.2015.03.011

Wang J, Guleria S, Koffas MAG, Yan Y (2016) Microbial production of value-added nutraceuticals. Curr Opin Biotechnol 37:97–104. https://doi.org/10.1016/j.copbio.2015.11.003

Weissmann B, Meyer K (1954) The structure of Hyalobiuronic acid and of hyaluronic acid from umbilical cord. J Am Chem Soc 76(7):1753–1757. https://doi.org/10.1021/ja01636a010

Wendisch VF (2014) Microbial production of amino acids and derived chemicals: synthetic biology approaches to strain development. Curr Opin Biotechnol 30:51–58. https://doi.org/10.1016/j.copbio.2014.05.004

Yin R, Kuo HC, Hudlikar R, Sargsyan D, Li S, Wang L, Wu R, Kong AN (2019) Gut microbiota, dietary phytochemicals, and benefits to human health. Curr Pharmacol Rep 5(5):332–344. https://doi.org/10.1007/s40495-019-00196-3

Yu L, O'Sullivan DJ (2014) Production of galactooligosaccharides using a hyperthermophilic β-galactosidase in spermeabilised whole cells of Lactococcus lactis. J Dairy Sci 97(2):694–703. https://doi.org/10.3168/jds.2013-7492

Zhang C, Liu L, Teng L, Chen J, Liu J, Li J, Du G, Chen J (2012) Metabolic engineering of Escherichia coli BL21 for biosynthesis of heparosan, a bioengineered heparin precursor. Metab Eng 14(5):521–527. https://doi.org/10.1016/j.ymben.2012.06.005

Zhang YJ, Li S, Gan RY, Zhou T, Xu DP, Li HB (2015) Impacts of gut bacteria on human health and diseases. Int J Mol Sci 16(4):7493–7519. https://doi.org/10.3390/ijms16047493

Zhao S, Jones JA, Lachance DM, Bhan N, Khalidi O, Venkataraman S, Wang Z, Koffas MAG (2015) Improvement of catechin production in Escherichia coli through combinatorial metabolic engineering. Metab Eng 28:43–53. https://doi.org/10.1016/j.ymben.2014.12.002

Zhao M, Chen S, Ji X, Shen X, You J, Liang X, Yin H, Zhao L (2021) Current innovations in nutraceuticals and functional foods for intervention of non-alcoholic fatty liver disease. Pharmacol Res 166(January):105517. https://doi.org/10.1016/j.phrs.2021.105517

Zommiti M, Feuilloley MGJ, Connil N (2020) Update of probiotics in human world: a nonstop source of benefactions till the end of time. Microorganisms 8(12):1–33. https://doi.org/10.3390/microorganisms8121907

Synthesis of Nanoparticles by Microbes 29

Simran Rani, Pradeep Kumar, Priyanka Dahiya, Priya, Amita Suneja Dang, and Pooja Suneja

Abstract

Modern life is not conceivable without the involvement of nanoparticles. Major aspects of science are dependent on nanotechnology for varied applications. The nanoscale size of nanoparticles, larger surface/volume ratio, characteristic structures, and similar dimensions to biomolecules led to their application in biomedical sciences. In the due course of development, nanoparticle synthesis has come a long way from conventional physical and chemical methods to latest method involving environment-friendly biological synthesis. Microorganisms like bacteria, fungi, blue-green algae, and yeast have been screened and adopted for biological synthesis of variety of metallic nanoparticles. This chapter discusses the detailed mechanism of microbe-mediated biosynthesis of nanoparticles, their applications in healthcare mainly as antimicrobials, and attempts at optimizing the factors influencing the quality of nanoparticles along with the newer dimensions and limitations of this arena.

Keywords

Nanoparticles · Green synthesis · Biogenic synthesis · Optimization

S. Rani · P. Kumar · P. Dahiya · Priya · P. Suneja (✉)
Department of Microbiology, Maharshi Dayanand University, Rohtak, Haryana, India

A. S. Dang
Centre for Medical Biotechnology, Maharshi Dayanand University, Rohtak, Haryana, India

R. Sobti et al. (eds.), *Role of Microbes in Sustainable Development*,
https://doi.org/10.1007/978-981-99-3126-2_29

29.1 Introduction

Nanotechnology is the branch of science and engineering that involves the design, production, and application of structure, and systems by manipulating atoms and molecules at the nanoscale, i.e. having at least one dimension in the order of 1 nm to 100 nm. It is concerned with the creation of nanoparticles (NPs) that exhibits unique chemical properties, controlled monodispersity, and more reactivity than parent metal atom owing to their high surface-to-volume ratio (Sahoo et al. 2021). NP consists of three layers: a surface layer equipped with a wide range of molecules, surfactants, metal ions, and polymers; a shell layer, above the core, which is the central portion of NP; and the NP itself (Shin et al. 2016). These are classified into different types based on their morphology, size, and shape. Organic NPs include ferritin, micelles, dendrimers, and liposome. Being biodegradable and non-toxic, they are widely used in drug delivery. Inorganic NPs are biocompatible, hydrophilic, and non-toxic besides being more stable than organic ones, and include both metallic as well as non-metallic oxide NPs. Ceramic NPs are non-metallic solids having wide applications in the photodegradation of dyes, photocatalysis, catalysis, and imaging. Bio-nanoparticles are an aggregate of atoms/ions/molecules, which are prepared in the living system naturally (Ijaz et al. 2020).

NPs have been reported to improve crop yield, seed germination, and total biomass of crops (Ali et al. 2021). Metal-based NPs, in particular, have gained significance in the agriculture sector as nanobiofertilizers and nanobiopesticides. Different types of metallic NPs have been created as they exhibit unique physical, chemical, and biological properties and have broad-spectrum applications in areas such as drug delivery, gene transfer, management of insect pests and agriculture and as antibacterial agents (Ghosh et al. 2008; Rai et al. 2011).

There have been innumerable attempts on researching and analyzing the most economical and efficient ways to synthesize the NPs. These are synthesized by physical methods such as the ball milling method, pulsed wire discharge method, and pulse laser ablation. Chemical production of NPs is an expensive, harmful, time-consuming, and low-yielding process; whereas, physical methods require sophisticated and high power consuming equipment (Khodashenas and Ghorbani 2014). Recently, the focus has shifted towards reducing the production cost and to enhance the properties of NPs mediated by various biological systems, which are eco-friendly and safer to use. It involves the production of NPs using a living source, bacteria, fungi, plant extract, etc., and is termed a biological or 'Green' synthesis (Mukunthan and Balaji 2012). This chapter describes the biogenic synthesis of NPs via microorganisms, related aspects such as the mechanism taken up by them to reduce metal ions into its nano-form, and factors influencing the same. In addition, it cites recent developments happening in this particular field along with future prospects.

29.2 'Green' Synthesis of NPs

Conventional physical and chemical methods of synthesizing the NPs are associated with their respective limitations. Physically synthesized NPs are formed by a top-down approach in which bulk material is broken down into nano-size particles. Although it is a fast manufacturing process, a high requirement of energy makes it unsuitable for large-scale production. It also leads to imperfection in surface structure, which affects other properties of the material (Kandasamy and Prema 2015). Chemical methods viz., chemical reduction, microemulsion, and solvothermal decomposition, employ a bottom-up approach using organic or inorganic chemicals acting as reducing (sodium citrate, sodium borohydride, and sodium hydroxide) and stabilizing/capping (PVA, PVP, polyethylene glycol, and polymethacrylic acid) agents (Satyanarayana and Reddy 2018). In the chemical and physical method of NPs synthesis, highly toxic, expensive, and hazardous chemicals are used which are harmful to the environment. The biological method is preferred, as it does not involve toxic chemicals for the synthesis of NPs. It is a non-toxic and energy-efficient method. Herein, NPs are formed by the bottom-up/constructive method by reducing and stabilizing metabolites present in the microbial supernatant/microbial cell or plant extract; thereby, there is no requirement of adding these agents separately (Salem and Fouda 2021). In addition to this, the biogenic method is environmentally safe, and is being seen as a more sustainable way of NPs production as compared to the other conventional methods in use.

29.3 Microbe-Mediated Biogenic Synthesis of NPs and Their Applications

Production of NPs through microorganism is considered one of the most efficient ways of synthesizing NPs, as they are less toxic and covers a wider area of applications. Different types of microbes utilize different kinds of mechanisms to synthesize NPs (Rani et al. 2022). Reducing enzymes, mainly reductase, and stabilizing or capping agents present in microbes help in the conversion of metal into its nanoscaled particles (Rao et al. 2017). Stabilizing and capping agents prevent aggregation, therefore, lead to the formation of stable NPs. On the basis of the location of NPs formation, it can be differentiated as intracellular and extracellular synthesis. The reduction of metal ions into their nanoform within the cell is referred to as the intracellular mode of synthesis. It starts by binding oppositely charged metal ions to the carboxylate groups of the metabolites like polypeptides, and enzymes, present on the microbial cell wall (Zhang et al. 2011). This binding is necessary to facilitate the trapping of metal ions inside the cell, which are reduced to their atomic form by the reducing agents such as NADH-dependent reductase, present in the microbial cell. Then, these atomic nuclei undergo growth leading to the formation of aggregates and ultimately NPs after being stabilized by amino acids such as tryptophan, cysteine, tyrosine, and proteins/peptides (Iravani 2014; Shedbalkar et al. 2014). NPs are found to accumulate in the cytoplasm as well as

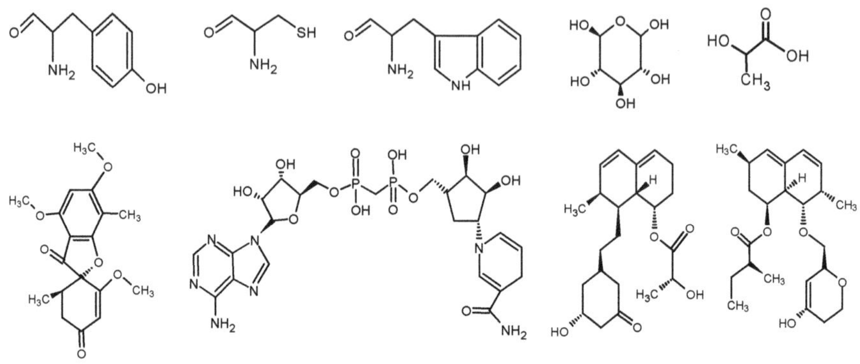

Fig. 29.1 Structures of the chemical compounds involved in the synthesis of nanoparticles from microbes (tyrosine, cysteine, tryptophan, glucose, lactic acid, griseofulvin, NADPH, mevastatin, and lovastatin)

cell wall (Mukherjee et al. 2001; Yusof et al. 2019). The extracellular mode of synthesis is facilitated by the extracellular enzymes that are either located on the cell membrane or released into the growth medium and mediate the reduction of metals from their ionic to atomic form followed by nucleation to form NPs. A few extracellular proteins also acts as capping agents and stabilize the NPs thus formed (Yusof et al. 2019). A number of microbial metabolites have been found to involve in the extracellular production of NPs (Ovais et al. 2018a, b; Ghosh et al. 2022). Barabadi et al. (2017) and Qamar and Ahmad (2021) discussed a few compounds that are involved in the synthesis of NPs from bacteria and fungi. Chemical structures of some of them have been depicted in Fig. 29.1.

There have been a number of attempts at synthesizing the metallic NPs from the microorganisms in recent times, some of which have been discussed below. Copper NPs are well known for their anti-microbial, anti-oxidant, and catalytic activity (Reddy Mallem 2022). Shantkriti and Rani (2014) reported the synthesis of CuNPs by incubating cell-free culture supernatant of non-pathogenic bacteria *P. fluorescens* with $CuSO_4$ (Copper sulfate) solution for 90 min. Lv et al. (2018) used *Shewanella loihica* PV-4 to produce CuNPs having antibacterial activity against *E. coli.* Noman et al. (2020) studied the production of CuNPs from the bacteria *Escherichia* sp. within a size range of 22.3 to 39 nm that acts as a photo catalyst during dye degradation. ZnONPs, widely used in antibacterial creams, lotions, ointments, mouthwashes, and paints, also act as antimicrobial agents and biofilm growth inhibitors (Mandal 2016). Spherical Zinc NPs within a size range of 14.39–37.85 nm were reported to be synthesized from the supernatant of a microalgae, *Cladophora glomerata* (Abdulwahid et al. 2019). A yeast, *Pichia kudriavzevii*, was also employed to prepare hexagonal zinc NPs by Moghaddam et al. (2017). Nadeem et al. (2018) reported bacterial strains, *Bacillus subtilis,* and *Lactobacillus* sp., as a source for the bio-production of TiO_2 NPs. Khan et al. (2021) reported that iron oxide magnetic nanoparticles synthesized by *Pseudomonas aeruginosa* are used in many applications, such as magnetic resonance imaging, diagnostics, and therapeutics. Magnetic NPs with narrow-size distribution help in maintaining the temperature needed as per the exact calculations for cancer treatment. Iron NPs synthesized using *Proteus vulgaris* ATCC-29905 proved to be excellent anticancer and antimicrobial agents (Nadeem et al. 2021). Spherical AgCl (silver chloride) NPs with a centered cubic crystal structure and a mean particle size of around 10–50 nm synthesized by cell-free culture supernatants of *Pantoea agglomerans* and *Raoultella planticola* exhibited antimicrobial activity against *Staphylococcus aureus, Streptococcus pyogenes, Salmonella,* and *Bacillus amyloliquefaciens* (Ghiuta et al. 2021). Alsamhary (2020) studied the antibacterial efficacy of *Bacillus subtilis*-mediated silver NPs to treat MDR (multidrug-resistant) microorganisms. Jebur and Abd (2021) synthesized MgO (magnesium oxide) NPs having different average diameters by two *Streptococcus* sp., *S. salivarius* and *S. mutans*. These NPs showed antibacterial efficacy against MDR, Gram-positive bacterium, *Acinetobacter baumannii* with a MIC (minimum inhibitory concentration) of 250 μg/mL. In a similar study, Fouda et al. (2021) investigated the antimicrobial potential of MgO-NPs (Source-*Penicillium chrysogenum*) which exhibited a zone of inhibitions at 200 μg/mL against *Staphylococcus aureus, Bacillus subtilis, Pseudomonas aeruginosa, Escherichia coli,* and *Candida albicans*.

29.4 Factors Affecting the Microbial Synthesis of NPs and Optimization Studies

Microbial species, the concentration of metallic precursor solution, incubation and reaction time, varying concentration of supernatant, pH, and temperature, are the factors that influence the quantity and quality of NPs significantly (Patra and Baek 2014). Not just the metabolites composition of each microbial species is different, but the pathways taken up to synthesize the NPs are also distinct. HR-TEM analysis revealed variations in the size and shape of NPs when the same microbe took up extracellular and intracellular modes of synthesis (Mohd Yusof et al. 2020). Different precursors of similar metals are used along with its varying concentration that influences the characteristics of NPs viz., shape and size. Therefore, a number of studies have been taken up to optimize these parameters in order to obtain NPs of desired quality and yield. One factor at a time (OFAT) is one of the most common approaches to carry this out that involves varying one factor at a time while keeping other factors as standard. Ebadi et al. (2019) optimized the synthesis of ZnONPs from the cell extract of cyanobacterium *Nostoc* sp. EA03. The maximum quantity of ZnONPs was obtained at a slightly alkaline pH (9) and 1000 μL cell extract concentration. Similarly, Bukhari et al. (2021) optimized the parameters that came out to be the concentration of Cu metal precursor solution (5 mM), reaction time (60 min.), filtrate to substrate ratio (1:1), and pH (7), from marine *Streptomyces* sp. Table 29.1 depicts various metallic NPs that have been optimized for different parameters affecting their biological synthesis from the microbes.

Table 29.1 Optimized parameters for the microbial synthesis of metallic NPs

NP	Source	Optimized conditions	Size (nm)	Applications	Reference
Ag	*Bacillus subtilis*	T (37 °C), pH (7.5)	2–26	Antibacterial properties against *Escherichia coli*, and *Staphylococcus aureus*	Yu et al. (2021)
	Streptomyces sp. SSUT88A	T (37 °C), pH (7), incubation time (5 days)	74.12	Antimicrobial activity against *Pseudomonas aeruginosa*	Rosyidah et al. (2021)
	Chlorella vulgaris	pH (12), incubation time (24 h), concentration of $AgNO_3$ (3 mM), extract ratio (8:2)	55	Photocatalytic degradation of methylene blue	Rajkumar et al. (2021)
	Aspergillus sydowii	T (50 °C), pH (8.5), substrate concentration (1.5 mM)	1–24	Antifungal activity against clinical pathogens	Wang et al. (2021)
	Cedecea sp.	T (50 °C), incubation time (8 h), silver salt concentration (1.5 mM)	10–40	Antimicrobial activity against *Staphylococcus epidermidis* and *Staphylococcus aureus*	Singh et al. (2021)
	Kocuria sp.	T (55 °C), pH (9.8), incubation time (15 h), Silver nitrate concentration (1 mM)	46.73	Antibacterial potential against *Escherichia coli*, *Salmonella typhimurium*	Kumar et al. (2022)
Au	*Agaricus bisporus*	T (28 °C), pH (7), reaction time (72 h), concentration of gold chloride (1:9)	10–50	Antibacterial potential against human pathogens	Krishnamoorthi et al. (2021)
Se	*Lactobacillus paracasei* HM1	T (35 °C), pH (6), reaction time (32 h), agitation speed (160 rpm)	56.91	Antifungal activity against animal pathogenic fungi	El-Saadony et al. (2021)
MgO	*Aspergillus terreus*	T (35 °C), pH (8), concentration	8–38	Inhibiting the growth of pathogenic	Saied et al. (2021)

(continued)

Table 29.1 (continued)

NP	Source	Optimized conditions	Size (nm)	Applications	Reference
		of Mg $(NO_3)_2.6H_2O$ (3 mM)		microbes, tanning effluent treatment, and chromium ion removal	
TiO	*Synechocystis* NCCU-370	pH (7), temperature (30 °C), reaction time (12 h)	73.39	Antimicrobial activity against *Candida albicans*	Siddiqui et al. (2022)
FeO	*Alcaligenes* sp.	T (70 °C), pH (13), concentration of $FeCl_3$ (2 mM), incubation time (2 h)	50–7	Degradation of malachite green	Sharma et al. (2022)
	Purpureocillium lilaceinum MW831030.1	pH (9), concentration of ferrous sulphate solution (3 mM), incubation time (3 days)	57.9	Dye removal (Safranine) and antibacterial activity against *Pseudomonas aeruginosa*	Hammed et al. (2022)

29.5 Bimetallic NPs: A New Dimension of Microbe-Based Nanotechnology

Bimetallic NPs, composed of two metals, are prepared by reducing the ions of two different metals simultaneously. These have been reported to improve the catalysts property of the original single metal that creates a new property (Ramsurn and Gupta 2013). The synergy existing between these metals is the major cause of their multi-functionality. They act as effective catalyzers in a variety of reactions owing to their large surface area. Bimetallic nanostructures comprise segregated and mixed nanostructures. Two different metals possess ordered and random arrangements of atoms, in the case of mixed structures like intermetallic and alloy nanoparticles, respectively. Segregated nanostructures, on the other hand, usually involve the formation of an initial structure of one metal type followed by the addition of the second metal. Thus, these are further classified as subcluster structures (separate distribution of two metals with a shared interface), core–shell structures (metal core is surrounded by a shell of second metal), multishell core–shell structures (shells possess alternative arrangement forming a shape like onion rings), and multiple core materials coated by a single shell (Ferrando et al. 2008; Srinoi et al. 2018).

Exhibiting higher activity, selectivity, and stability, they have wide applications in imaging and biomedical devices including nano-medicines (Erkey 2011; Loza et al. 2020). Bimetallic nanocomposites act as sensors and help in the early diagnosis of the disease. They have the ability to detect even 2–3 cancerous cells present in the body and are emerging as a bright option in the diagnostics field (Sharma et al. 2019). They also demonstrate greater antioxidant activity and dye removal efficacy used for wastewater treatment as compared to monometallic NPs (Riaz et al. 2022). Merugu et al. (2021) used the toddy palm extract to synthesize copper-zinc bimetallic NPs having anti-bacterial activity against *Alcaligenes faecalis, S. aureus, K. pneumoniae*, and *Clostridium perfringens*. Similarly, Riaz et al. (2022) took up the synthesis of monometallic as well as bimetallic/alloy NPs of zinc and copper from the aqueous extract of *Mirabilis Jalapa*. An equal volume of the plant extract was added to the aqueous solution of zinc chloride and copper sulfate (precursor salts of both metals) at varying concentrations. The continuous change in color from royal blue to sea green indicated the formation of nanoparticles. Fawzy et al. (2021) suspended cell pellet as well as cell-free supernatant of *Pseudomonas fluorescens* (PS1and PS2) into $ZnSO_4$ and $CuSO_4$ solution wherein color change indicated the formation of bimetallic (Cu-Zn) NPs. Iron-manganese NPs synthesized using bacteria-fermented supernatant containing auxin complex compounds (indole-3-acetic acid) as reducing agents were evaluated as plant nanofertilizer. Spherical NPs showed a positive impact on seed germination, root development, and fresh weight of *Zea mays* (de França Bettencourt et al. 2020). They also act as promising antimicrobial agents besides displaying their tendencies in several industries affected by microbes such as water, food, textiles, and oil and gas (Arora et al. 2020).

29.6 Conclusion

Microbial synthesis of NPs is expected to take a leap forward in the near future owing to its environment friendliness, lesser cost of production, and biocompatibility provided it address a few limitations (Ovais et al. 2018a, b). A slower reduction process, reproducibility issues, and requirements of entirely aseptic infrastructure, besides unelucidated mechanisms underlying the same, can halt the synthesis as well as the application process (Rani et al. 2022). Therefore, besides focusing on exploring various microbial taxa for the synthesis of metallic NPs, researchers should also focus on the details of the reduction and stabilization process. In addition, the procedures involved in the extraction and purification of NPs from microbes needs to be studied in order to facilitate the full-fledged use of this application.

References

Abdulwahid KE, Dwaish AS, Dakhil OA (2019) Green synthesis and characterization of zinc oxide nanoparticles from *Cladophora glomerata* and its antifungal activity against some fungal isolates. Plant Arch 19:3527–3532. http://plantarchives.org/19-2/3527-3532(5932).pdf. Accessed 26 Nov 2021

Ali A, Shah T, Ullah R, Zhou P, Guo M, Ovais M, Rui Y et al (2021) Review on recent progress in magnetic nanoparticles: synthesis, characterization, and diverse applications. Front Chem 9: 629054

Alsamhary KI (2020) Eco-friendly synthesis of silver nanoparticles by Bacillus subtilis and their antibacterial activity. Saudi J Biol Sci 27(8):2185–2191

Arora N, Thangavelu K, Karanikolos GN (2020) Bimetallic nanoparticles for antimicrobial applications. Front Chem 8:412

Barabadi H, Ovais M, Shinwari ZK, Saravanan M (2017) Anti-cancer green bionanomaterials: present status and future prospects. Green Chem Lett Rev 10(4):285–314

Bukhari SI, Hamed MM, Al-Agamy MH, Gazwi HS, Radwan HH, Youssif AM (2021) Biosynthesis of copper oxide nanoparticles using Streptomyces MHM38 and its biological applications. J Nanomater 2021:1

de França Bettencourt GM, Degenhardt J, Torres LAZ, de Andrade Tanobe VO, Soccol CR (2020) Green biosynthesis of single and bimetallic nanoparticles of iron and manganese using bacterial auxin complex to act as plant bio-fertilizer. Biocatal Agric Biotechnol 30:101822

Ebadi M, Zolfaghari MR, Aghaei SS, Zargar M, Shafiei M, Zahiri HS et al (2019) A bio-inspired strategy for the synthesis of zinc oxide nanoparticles (ZnO NPs) using the cell extract of cyanobacterium: Nostoc sp. EA03: from biological function to toxicity evaluation. RSC Adv 9:23508–23525. https://doi.org/10.1039/c9ra03962g

El-Saadony MT, Saad AM, Taha TF, Najjar AA, Zabermawi NM, Nader MM, AbuQamar SF, El-Tarabily KA, Salama A (2021) Selenium nanoparticles from lactobacillus paracasei HM1 capable of antagonizing animal pathogenic fungi as a new source from human breast milk. Saudi J Biol Sci 28(12):6782–6794

Erkey C (2011) Synthesis of nanostructured composites of metals. In: Supercritical fluid science and technology, vol 1. Elsevier, Amsterdam, pp 79–120

Fawzy H, Mohamed Mahmoud S, Ahmed Hanafy M, Hassan Bakr M, Mohamed Mahmoud AE, Abdel-Alim Ali M, Sayed Barakat O (2021) Production of zinc and copper as nanoparticles by green synthesis using *Pseudomonas fluorescens*. Pak J Biol Sci 24(4):445–453

Ferrando R, Jellinek J, Johnston RL (2008) Nanoalloys: from theory to applications of alloy clusters and nanoparticles. Chem Rev 108(3):845–910

Fouda A, Hassan SED, Abdel-Rahman MA, Farag MM, Shehal-deen A, Mohamed AA, Azab MS et al (2021) Catalytic degradation of wastewater from the textile and tannery industries by green synthesized hematite (α-Fe2O3) and magnesium oxide (MgO) nanoparticles. Curr Res Biotechnol 3:29–41

Ghiuta I, Croitoru C, Kost J, Wenkert R, Munteanu D (2021) Bacteria-mediated synthesis of silver and silver chloride nanoparticles and their antimicrobial activity. Appl Sci 11(7):3134

Ghosh P, Han G, De M, Kim CK, Rotello VM (2008) Gold nanoparticles in delivery applications. Adv Drug Deliv Rev 60(11):1307–1315

Ghosh S, Sarkar B, Kaushik A, Mostafavi E (2022) Nanobiotechnological prospects of probiotic microflora: synthesis, mechanism, and applications. Sci Total Environ 838:156212

Hammed A, Polunin Y, Voronov A, Pryor S (2022) Tubular electrosynthesis of silica-coated magnetite and evaluation of magnetic nanobiocatalyst efficacy during biomass hydrolysis. Bioprocess Biosyst Eng 45(8):1311–1318

Ijaz I, Gilani E, Nazir A, Bukhari A (2020) Detail review on chemical, physical and green synthesis, classification, characterizations and applications of nanoparticles. Green Chem Lett Rev 13(3): 223–245

Iravani S (2014) Bacteria in nanoparticle synthesis: current status and future prospects. Int Sch Res Not 2014:1–18. https://doi.org/10.1155/2014/359316

Jebur YM, Abd FG (2021) Biosynthesis of MgO nanoparticles by using streptococcus species and its antibacterial activity. Biochem Cell Arch 21:2557

Kandasamy S, Prema RS (2015) Methods of synthesis of nano particles and its applications. J Chem Pharm Res 7(3):278–285

Khan AA, Khan S, Khan S, Rentschler S, Laufer S, Deigner HP (2021) Biosynthesis of iron oxide magnetic nanoparticles using clinically isolated *Pseudomonas aeruginosa*. Sci Rep 11(1):1–10

Khodashenas B, Ghorbani HR (2014) Synthesis of copper nanoparticles: an overview of the various methods. Korean J Chem Eng 31(7):1105–1109

Krishnamoorthi R, Bharathakumar S, Malaikozhundan B, Mahalingam PU (2021) Mycofabrication of gold nanoparticles: optimization, characterization, stabilization and evaluation of its antimicrobial potential on selected human pathogens. Biocatal Agric Biotechnol 35:102107

Kumar M, Upadhyay LS, Kerketta A, Vasanth D (2022) Extracellular synthesis of silver nanoparticles using a novel bacterial strain Kocuria rhizophila BR-1: process optimization and evaluation of antibacterial activity. BioNanoScience 12(2):423–438

Loza K, Heggen M, Epple M (2020) Synthesis, structure, properties, and applications of bimetallic nanoparticles of noble metals. Adv Funct Mater 30(21):1909260

Lv Q, Zhang B, Xing X, Zhao Y, Cai R, Wang W, Gu Q (2018) Biosynthesis of copper nanoparticles using Shewanellaloihica PV-4 with antibacterial activity: novel approach and mechanisms investigation. J Hazard Mater 347:141–149

Mandal BK (2016) Scopes of green synthesized metal and metal oxide nanomaterials in antimicrobial therapy. In: Nanobiomaterials in antimicrobial therapy. William Andrew Publishing, Norwich, NY, pp 313–341

Merugu R, Gothalwal R, Deshpande PK, De Mandal S, Padala G, Chitturi KL (2021) Synthesis of Ag/Cu and Cu/Zn bimetallic nanoparticles using toddy palm: investigations of their antitumor, antioxidant and antibacterial activities. Mater Today Proc 44:99–105

Moghaddam AB, Moniri M, Azizi S, Rahim RA, Ariff AB, Saad WZ et al (2017) Biosynthesis of ZnO nanoparticles by a new *Pichia kudriavzevii* yeast strain and evaluation of their antimicrobial and antioxidant activities. Molecules 22:872. https://doi.org/10.3390/molecules22060872

Mohd Yusof H, Abdul Rahman NA, Mohamad R, Zaidan UH, Samsudin AA (2020) Biosynthesis of zinc oxide nanoparticles by cell-biomass and supernatant of *lactobacillus plantarum* TA4 and its antibacterial and biocompatibility properties. Sci Rep 10:1–13. https://doi.org/10.1038/s41598-020-76402-w

Mukherjee P, Ahmad A, Mandal D, Senapati S, Sainkar SR, Khan MI et al (2001) Bioreduction of AuCl4- ions by the fungus, *Verticillium* sp. and surface trapping of the gold nanoparticles formed. Angew Chem Int Ed 40:3585–3588. https://doi.org/10.1002/1521-3773(20011001)40:19<3585::AID-ANIE3585>3.0.CO;2-K

Mukunthan KS, Balaji S (2012) Cashew apple juice (Anacardium occidentale L.) speeds up the synthesis of silver nanoparticles. Int J Green Nanotechnol Biomed 4:71–79. https://doi.org/10.1080/19430892.2012.676900

Nadeem M, Tungmunnithum D, Hano C, Abbasi BH, Hashmi SS, Ahmad W, Zahir A (2018) The current trends in the green syntheses of titanium oxide nanoparticles and their applications. Green Chem Lett Rev 11(4):492–502

Nadeem M, Khan R, Shah N, Bangash IR, Abbasi BH, Hano C, Celli J (2021) A review of microbial mediated iron nanoparticles (IONPS) and its biomedical applications. Nanomaterials 12(1):130

Noman M, Shahid M, Ahmed T, Niazi MBK, Hussain S, Song F, Manzoor I (2020) Use of biogenic copper nanoparticles synthesized from a native Escherichia sp. as photocatalysts for azo dye degradation and treatment of textile effluents. Environ Pollut 257:113514

Ovais M, Khalil AT, Ayaz M, Ahmad I, Nethi SK, Mukherjee S (2018a) Biosynthesis of metal nanoparticles via microbial enzymes: a mechanistic approach. Int J Mol Sci 19(12):4100

Ovais M, Khalil AT, Islam NU, Ahmad I, Ayaz M, Saravanan M, Shinwari ZK, Mukherjee S (2018b) Role of plant phytochemicals and microbial enzymes in biosynthesis of metallic nanoparticles. Appl Microbiol Biotechnol 102(16):6799–6814

Patra JK, Baek KH (2014) Green Nanobiotechnology: factors affecting synthesis and characterization techniques. J Nanomater 2014:219. https://doi.org/10.1155/2014/417305

Qamar SUR, Ahmad JN (2021) Nanoparticles: mechanism of biosynthesis using plant extracts, bacteria, fungi, and their applications. J Mol Liq 334:116040

Rai M, Gade A, Yadav A (2011) Biogenic nanoparticles: an introduction to what they are, how they are synthesized and their applications. In: Metal nanoparticles in microbiology, pp 1–14. https://doi.org/10.1007/978-3-642-18312-6_1

Rajkumar R, Ezhumalai G, Gnanadesigan M (2021) A green approach for the synthesis of silver nanoparticles by Chlorella vulgaris and its application in photocatalytic dye degradation activity. Environ Technol Innov 21:101282

Ramsurn H, Gupta RB (2013) New and future developments in catalysis: Chapter 15. In: Hydrogenation by nanoparticle catalysts. Elsevier, Amsterdam

Rani S, Kumar P, Dahiya P, Dang AS, Suneja P (2022) Biogenic synthesis of zinc nanoparticles, their applications, and toxicity prospects. Front Microbiol 13:824427

Rao BG, Mukherjee D, Reddy BM (2017) Novel approaches for preparation of nanoparticles. In: Nanostructures for novel therapy. Elsevier, Amsterdam, pp 1–36

Reddy Mallem SP (2022) Nano/micro and bio-inspired materials on wide-bandgap-semiconductor-based optoelectronic/power devices. Crystals 12(1):67

Riaz T, Assey N, Javed M, Shahzadi T, Zaib M, Shahid S, Fatima U (2022) Biogenic plant mediated synthesis of monometallic zinc and bimetallic copper/zinc nanoparticles and their dye adsorption and antioxidant studies. Inorg Chem Commun 140:109449

Rosyidah A, Nantapong N, Chudapongse N, Weeranantanapan O, Limphirat W (2021) Optimization of silver nanoparticles synthesis by the green method using Streptomyces sp. SSUT88A and their antimicrobial activity against Pseudomonas aeruginosa. In: IOP conference series: earth and environmental science, vol 948. IOP Publishing, Bristol, p 012085

Sahoo SK, Panigrahi GK, Sahoo A, Pradhan AK, Dalbehera A (2021) Bio-hydrothermal synthesis of ZnO–ZnFe2O4 nanoparticles using Psidium guajava leaf extract: role in waste water remediation and plant immunity. J Clean Prod 318:128522

Saied E, Eid AM, Hassan SED, Salem SS, Radwan AA, Halawa M, Saleh FM, Saad HA, Saied EM, Fouda A (2021) The catalytic activity of biosynthesized magnesium oxide nanoparticles (Mgo-nps) for inhibiting the growth of pathogenic microbes, tanning effluent treatment, and chromium ion removal. Catalysts 11(7):821

Salem SS, Fouda A (2021) Green synthesis of metallic nanoparticles and their prospective biotechnological applications: an overview. Biol Trace Elem Res 199:344–370. https://doi.org/10.1007/s12011-020-02138-3

Satyanarayana T, Reddy SS (2018) A review on chemical and physical synthesis methods of nanomaterials. Int J Res Appl Sci Eng Technol 6(1):2885–2889

Shantkriti S, Rani P (2014) Biological synthesis of copper nanoparticles using Pseudomonas fluorescens. Int J Curr Microbiol App Sci 3(9):374–383

Sharma G, Kumar A, Sharma S, Naushad M, Dwivedi RP, ALOthman ZA, Mola GT (2019) Novel development of nanoparticles to bimetallic nanoparticles and their composites: a review. J King Saud Univ Sci 31(2):257–269

Sharma B, Kumari N, Mathur S, Sharma V (2022) Kinetics and optimization of azo dye Decolorisation using green synthesized iron-oxide nanoparticles: a pilot scale study. Int J Environ Res 16(4):1–18

Shedbalkar U, Singh R, Wadhwani S, Gaidhani S, Chopade BA (2014) Microbial synthesis of gold nanoparticles: current status and future prospects. Adv Colloid Interface Sci 209:40–48. https://doi.org/10.1016/j.cis.2013.12.011

Shin WK, Cho J, Kannan AG, Lee YS, Kim DW (2016) Cross-linked composite gel polymer electrolyte using mesoporous methacrylate-functionalized SiO2 nanoparticles for lithium-ion polymer batteries. Sci Rep 6(1):1–10

Siddiqui T, Khan NJ, Asif N, Ahamad I, Yasin D, Fatma T (2022) Screening, characterisation and bioactivities of green fabricated TiO2 NP via cyanobacterial extract. Environ Sci Pollut Res 29(26):39052–39066

Singh P, Pandit S, Jers C, Joshi AS, Garnæs J, Mijakovic I (2021) Silver nanoparticles produced from Cedecea sp. exhibit antibiofilm activity and remarkable stability. Sci Rep 11(1):1–13

Srinoi P, Chen YT, Vittur V, Marquez MD, Lee TR (2018) Bimetallic nanoparticles: enhanced magnetic and optical properties for emerging biological applications. Appl Sci 8(7):1106

Wang D, Xue B, Wang L, Zhang Y, Liu L, Zhou Y (2021) Fungus-mediated green synthesis of nano-silver using aspergillus sydowii and its antifungal/antiproliferative activities. Sci Rep 11(1):1–9

Yu X, Li J, Mu D, Zhang H, Liu Q, Chen G (2021) Green synthesis and characterizations of silver nanoparticles with enhanced antibacterial properties by secondary metabolites of Bacillus subtilis (SDUM301120). Green Chem Lett Rev 14(2):190–203

Yusof NAA, Zain NM, Pauzi N (2019) Synthesis of ZnO nanoparticles with chitosan as stabilizing agent and their antibacterial properties against gram-positive and gram-negative bacteria. Int J Biol Macromol 124:1132–1136. https://doi.org/10.1016/j.ijbiomac.2018.11.228

Zhang X, Yan S, Tyagi RD, Surampalli RY (2011) Synthesis of nanoparticles by microorganisms and their application in enhancing microbiological reaction rates. Chemosphere 82:489–494. https://doi.org/10.1016/j.chemosphere.2010.10.023

Microbial Biopharmaceuticals in Urolithiasis Management and Treatment 30

Gupta Shruti and Kanwar Shamsher Singh

Abstract

Urolithiasis is a potentially brutal condition characterized by the presence of insoluble deposits or stones in the kidneys and urinary tract. Among the various urological diseases, urolithiasis is ranked third in its prevalence. Around one in 11 people over the globe is affected by a highly pervasive urolithiasis which accelerates to excessive pain and chronic kidney diseases, thereby affecting people's health and quality of life. Oxalate that is endogenously produced by the liver as well as consumed in the diet is the major component of nearly 80% of kidney stones. Although, till date several drugs and therapies have been recommended and assessed both in vitro and in vivo for targeting kidney stones, but microbial drugs for stone dissolution have rarely been explored. Therefore, specific drugs for stone prevention and urolithiasis are still limited, and hence, it becomes essential to discuss microbial biopharmaceuticals such as oxalate degrading enzymes, probiotics and postbiotics that provide new strategies for prevention and treatment of CaOx kidney stones.

Keywords

Urolithiasis · Calcium oxalate stones · Oxalate decarboxylase · Probiotics · Postbiotics · *Oxalobacter*

G. Shruti · K. S. Singh (✉)
Department of Biotechnology, Himachal Pradesh University, Summer Hill, Shimla, Himachal Pradesh, India

R. Sobti et al. (eds.), *Role of Microbes in Sustainable Development*,
https://doi.org/10.1007/978-981-99-3126-2_30

30.1 Introduction

Oxalate is a small dicarboxylic acid that is likely to contribute to chronic kidney diseases (CKD), end-stage renal disease (ESRD)- associated cardiovascular diseases, polycystic kidney diseases and/ or poor renal allograft survival (Alshaikh and Hassan 2021). However, the elevated oxalate level in urine known as hyperoxaluria is one of the major risk factors for recurrent urolithiasis and progressive nephrocalcinosis (Shah et al. 2021). Hyperoxaluria elicits supersaturation of calcium oxalate salts and crystal formation in the urine, thereby resulting in urolithiasis and deposition of calcium oxalate crystals in kidney parenchyma, a condition termed as oxalate nephropathy (Buysschaert et al. 2020). Besides calcium oxalate (which accounts for more than 75% of renal stones), kidney stones are also composed of calcium hydroxyl phosphate (brushite or calcium hydroxyapatite) magnesium ammonium phosphate (struvite or triple phosphate), urate and cystine (Gupta and Kanwar 2018). Risk factors such as diet, environmental factors, socio-economic status, drugs, family or personal history of kidney stones, host genetics are responsible for occurrence of different types of kidney stones (Gupta and Kanwar 2020a). Being a multifaceted disease, renal stone formation is enhanced by high dietary oxalate intake, low fluid intake, high salt intake and consumption of high animal protein intake. Besides these, obesity, diabetes, hypertension and hyperlipidemia are other factors known to contribute to calcium oxalate stone formation (Thakore and Liang 2021). Kidney stones also have a considerably high recurrence rate with almost half of the patients displaying a recurrent episode in 10 years and about 90% within 30 years (Alelign et al. 2021).

Apart from the formation of stones in the kidney, oxalate crystals can destruct epithelium in the oral cavity and gastrointestinal tract, causing inflammation, diarrhoea and gastric haemorrhage (Gupta and Kanwar 2020b). Chronic kidney stone disease may ultimately advance to kidney failure and is often associated with other co-morbidities such as asthma, cardiovascular diseases, metabolic syndromes as well as end stage renal diseases (ESRD) like renal cancers (Gupta and Kanwar 2020a). Oxalate stones prompt renal inflammation and through overexpression of inflammatory bodies like interleukins (IL) and activating NOD like receptors proteins (NLRP3)/inflammatory bodies that cause necrosis of renal tubular epithelial cells thereby leading to oxalate nephropathy (Gupta and Kanwar 2022).

With its tendency to relapse, severity, increase in morbidity and reduced age of diagnosis, the management of urolithiasis has become an un-met real challenge in the field of urology. Conventional methods being employed for the treatment of kidney stones including drug therapy and surgical methods such as extracorporeal shock wave lithotripsy, ureteroscopy lithotripsy and percutaneous nephrolithotomy give unsatisfactory results and have several shortcomings (Lin et al. 2017; Paul et al. 2018). The exposure to the shock waves leads to severe side effects including acute renal injury, decreased renal function and increased rate of stone recurrence as well as being highly expensive. Dietary management of oxalate stones through restricting oxalate intake in diet might not be a reliable approach to prevent recurrent stones in many patients, as this may lead to nutritional deficiencies (Tavasoli et al. 2020). For

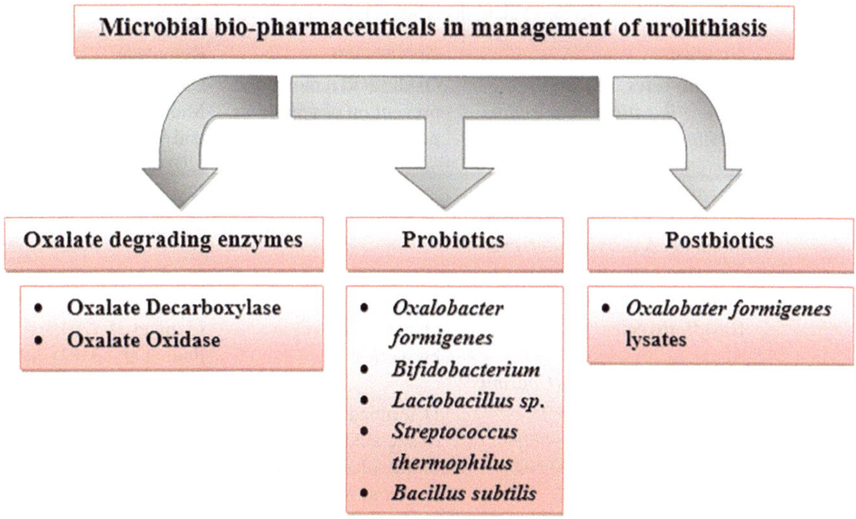

Fig. 30.1 Microbial biopharmaceuticals in management of urolithiasis

absolute treatment and prevention of kidney stones, the ideal strategy is to target the initial processes of stone formation like oxalate supersaturation or crystallization events. Oxalate degrading microbial enzymes such as oxalate decarboxylase (OxDC) and the oxalate oxidase (OxOx) act directly on oxalate and therefore represent biochemical factors useful in dissolving or reducing the size of CaOx stones already formed in the urinary tract. Using such enzymatic therapies to treat kidney stones would be highly relevant in patients, especially those who are poor candidates for surgery, have rapidly recurrent formation of stones and/or have received shockwave lithotripsy. However, oxalate degrading probiotic bacteria colonization in the gut or their products (postbiotics) can reduce the chances for renal stone formation and also manage hyperoxaluria (Fig. 30.1). Therefore, in the present study, we discuss the potential of oxalate degrading microbial products or probiotic bacteria as potential futuristic remedies for treatment and prevention of urolithiasis.

30.2 Oxalate Decarboxylase: A Novel Therapeutic in Management of Calcium Oxalate Urolithiasis

Oxalate decarboxylase catalyses the disproportionation reaction of oxalate monoanions into formate and CO_2 (Zhu et al. 2018). It is a member of the cupin superfamily of proteins and exists as a homogenous polymerase with manganese ions (Lin et al. 2017). O_2 and Mn^{2+} ions act as the cofactors for this enzyme, and it displays a strong pH dependency between pH 5 and pH 6. It was originally discovered in mycelia extracts of the white-rot fungi *Trametes hirsuta* (*Coriolus*

Table 30.1 Microbial diversity of oxalate decarboxylase and their applications

Class	Organism	Applications	Reference
Fungi	**Ascomycetes** *Aspergillus niger, Aspergillus clavatus, Aspergillus fumigates, Aspergillus phoenicis, Aspergillus oryzae, Aspergillus flavus, Aspergillus nidulans, Aspergillus terreus, Myrothecium verrucaria, Penicillium chrysogenum*	Oxalic acid removal for prevention of scaling in pulp and paper industry	Cassland et al. (2010), Makela et al. (2010) and Winestrand et al. (2014)
	Basidiomycetes *Agaricus bisporus, Coprinopsis cinerea (Coprinus cinereus), Ceriporiopsis subvermispora, Dichomitus squalens, Flammulina* spp., *Flammulina velutipes, Gloeophyllum trabeum, Heterobasidion annosum, Laccaria bicolor, Phanerochaete chrysosporium, Phanerochaete sanguinea, Pleurotus ostreatus, Postia placenta, Schizophyllum commune Serpula lacrymans, Trametes ochracea, Trametes versicolor, Trametes hirsuta*	Oxalate removal in breweries	Liang et al. (2015), Paul et al. (2017) and Conter et al. (2019)
		Determination of oxalic acid level in blood, plasma, urine, wort and beer	Gupta and Kanwar (2020a) and Vos et al. (2021)
		OxDC expressing plants: Chickling vetch (*Lathyrus sativus*) Lettuce (*Lactuca sativa*) Tomato (*Solanum lycopersicum*) Tobacco (*Nicotiana tabacum*)	Lambein et al. (2019) and Kumar et al. (2019)
Bacteria	*Agrobacterium tumefaciens, Bacillus subtilis, Pandorea* spp., *Synechocystis* spp., *Thermotoga maritima, Photorhabdus luminescens*	Treatment of hyperoxaluria, urolithiasis and nephrocalcinosis	Paul et al. (2019) and Chellappan et al. (2020)
		Depletion of dietary oxalate with enzyme formulation	Gupta and Kanwar (2020a)
		OxDC-expressing probiotic lactic acid bacteria	Mathivanan (2021)

hirsutus) and *Flammulina velutipes* (*Collybia velutipes*) (Shimazono 1955). Other than these, OxDC has been reported in several other microbial species (Table 30.1). *Bacillus subtilis* oxalate decarboxylase is the best oxalate metabolizing enzyme characterized till date. The enzyme is coded by the *Yvrk* gene and is effectuated by the acidic pH (5.0) of the growth media (Qi et al. 2017). Also, the *B. subtilis yoaN* (renamed as *oxdD*) and *yxaG* genes might also encode for enzymes exhibiting the OxDC activity (Lee et al. 2014). Another structural genomics study of proteins encoded by putative open reading frames in *Thermatoga maritima* and *Photorhabdus luminescens* has identified an oxalate-binding monocupin as a putative bacterial OxDC (Chellappan et al. 2020). Our recent investigation described a potent 49 kDa OxDC from a Gram-negative bacteria *Pseudomonas* sp. OXDC12 may serve as a prospective therapeutic agent for hyperoxaluria or kidney stones as it significantly inhibited the formation of calcium oxalate crystals under in vitro conditions (Gupta and Kanwar 2021). Knowledge-based modelling of the enzyme displayed a β barrel core in each of the two domains organized in the hexameric state. A cluster of three histidine, suitably juxtaposed to coordinate a divalent metal ion exists in both the domains. It was hypothesized that the histidine clusters were involved in the catalytic mechanism of the enzyme, possibly through coordination of a metal cofactor. The histidine triad forms an optimal topology for binding of Mn^{2+} ions due to the orientation of the imidazole rings in the two domains of the OxDC. The histidine-coordinated divalent Mn ion in the two domains provide scope for binding of ligands (oxalate and formate). The Mn bound oxalate is degraded by OxDC through C-C bond cleavage involving the PCET reaction resulting in the formation of formate (Gupta and Kanwar 2022).

The enzyme functions in a distinctive manner by firstly catabolizing free oxalate to minimize oxalate's bioavailability which lowers its levels below supersaturation and permits the dissolution of precipitated crystals naturally and/or secondly, within the stone matrix, it dissolves the calcium oxalate mineral precipitates slowly thereby directly acting on calcium oxalate crystals in kidney stones (Peck et al. 2016). OxDC displays good prospects in the deduction of excess dietary oxalate as well as in the dissolution of CaOx stones in the kidneys as the enzyme can degrade oxalate into formate and carbon dioxide. Studies in the hyperoxaluria model of mice have revealed that oral therapy using a cross-linked formulation of OxDC could reduce the amount of oxalate in urine, suggesting a potential approach for treatment of hyperoxaluria (Langman et al. 2016). Previously, the heterologous expression of this enzyme in *Lactobacillus plantarum* (*L. plantarum*) was developed and utilized as a potential probiotic for depletion of intestinal dietary oxalate (Albert et al. 2020). In a recent double blind, placebo controlled, randomized, crossover study, administration of OxDC orally in healthy volunteers on a high oxalate diet significantly reduced urinary oxalate levels without affecting urine creatinine or other solutes coupled to supersaturation of CaOx, (Quintero et al. 2020). Fungal OxDC expressed in transgenic plants might significantly reduce nutritional stress for herbivores because of diminished oxalate content in the feed (Chakraborty et al. 2013).

30.3 Oxalate Oxidase

Oxalate oxidase (OXO) also known as the oxalic acid oxidase pertains to the oxido reductase enzyme family with an enzyme commission number of 1.2.3.4. The enzyme requires FAD and Mn^{2+} as cofactors and plays role in glyoxylate and decarboxylate mechanism (Arumugam et al. 2020). The enzyme catalyses oxidation of oxalate to CO_2 with the reduction of molecular oxygen to hydrogen peroxide. The OXO activity was initially observed in plants due to a functionally diverse superfamily of proteins known as cupins which play a role in cell wall synthesis, floral induction, fungal defence and salt tolerance. However, studies have reported extracellular OXO activity in various microbial strains such as *Ceriporiopsis subvermispora* (a white rot basidiomycete), the bacterium *Pseudomonas* sp. OX-53 and *Ochrobactrum intermedium* CL6 (an endophytic bacterium; Kizhakedathil et al. 2020). Lately, Graz and colleagues detected the oxalate oxidase activity in nine basidiomycete fungi which can be used in the diagnosis and prevention of kidney diseases caused by the precipitation of calcium oxalates (Grąz et al. 2023). Various commercial kits have been prepared using OXO to monitoring oxalate levels in blood and urine that helps control hyperoxaluria and prevents urolithiasis. In a recent investigation, Kizakedathil and his co-workers for the first time reported the activity of OXO in decomposing calcium oxalate crystals. The thermophilic OXO was produced from the fungal strain *Fusarium oxysporum* RBP3 isolated from rotting taro corn (Kizhakedathil et al. 2020). Concurrently, Arumugam and his team extracted an oxalate oxidase from the bacterium *Staphylococcus pasture* which when analysed for its efficiency against the formation of kidney stones exhibited significant degradation of calcium oxalate kidney stones under in vitro conditions (Arumugam et al. 2020).

30.4 Probiotic Management of Kidney Stones

Using probiotic therapy for treatment and prevention of kidney stones and hyperoxaluria is a new frontline area for researchers. It has been observed that aerotolerant *Lactobacillus* and obligatory anaerobe *Bifidobacterium* residing in the intestine show oxalate-degrading activity, which was considered useful for the prevention of CaOx stone formation (Abratt and Reid 2010; Akulenko et al. 2020). Studies confirmed that through treatment with *Bifidobacterium lactis* DSM 10140, *Bifidobacterium longum* MB 282 and *Bifidobacterium adolescentis* MB 238 strains, the degradation of oxalate could be achieved up to 61%, 35.2% and 57%, respectively (Gupta and Kanwar 2020a). Both *Lactobacillus* and *Bifidobacterium* spp. are known as "generalized oxalobacters" as they do not use oxalate as a sole source of carbon and degrade it only in the presence of glucose and lactose (Sadaf et al. 2017). It has been experimentally demonstrated that *Lactobacillus acidophilus* NCFM contains genes that code for the oxalyl CoA decarboxylase (oxc) and formyl CoA transferase (frc) enzymes and constitute the functional oxalate catabolising operon. Various native sources such as milk, yogurt, pickles, cucumber and dieffenbachia plant were figured out to contain natural population of *Lactobacillus* and *Oxalobacter* probiotics used in the prevention of kidney stones (Gomathi et al. 2014). According to a report, "Oxadrop" the brand name of a mixed culture of *Bifidobacterium infantis, L. acidophilus, Streptococcus thermophilus* and *Lactobacillus brevis* did not produce any effect on the inhibition of kidney stone formation when given along with a low oxalate diet; however, with a normal diet, it reduced oxalate excretion (Miller and Dearing 2013).

Oxalobacter formigenes is an anaerobic oxalate degrading bacterium in the mammalian gut that relies exclusively on oxalate for growth (Fig. 30.2). It has the greatest reservoir of oxalate metabolizing genes (Nazzal et al. 2021). Results from

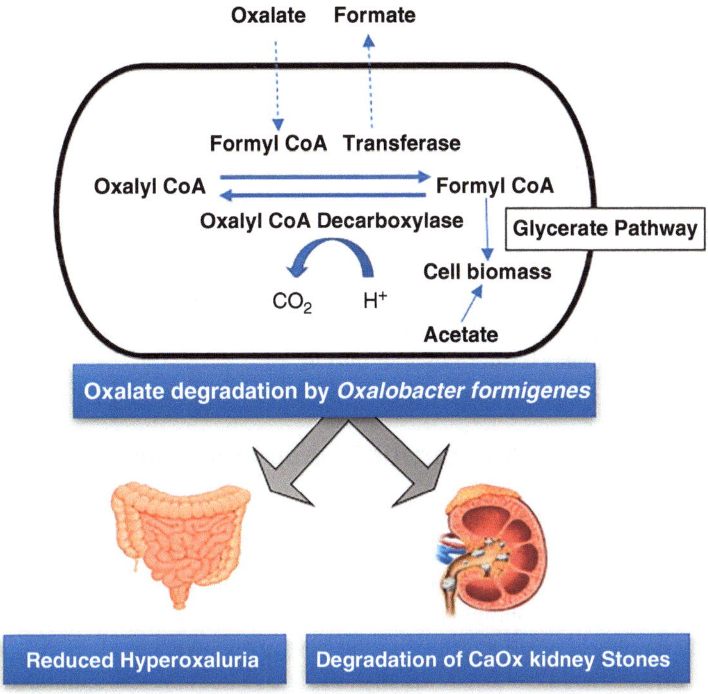

Fig. 30.2 Oxalate metabolism by *Oxalobacter formigenes*

numerous ongoing clinical trials studying the effect of *Oxalobacter formigenes* in patients with urolithiasis have made apparent that *O. formigenes* intake decreases oxalate levels in patients with kidney stones. Investigations demonstrated that urinary oxalate levels were lower in patients positive for *O. formigenes*. The dearth of this bacterium in the gut increases the likelihood of hyperoxaluria and recurrent urolithiasis (Wigner et al. 2022). In another clinical study, Nazal and co-workers illustrated that *O. formigenes* colonization was remarkably suppressed in subjects exposed to antibiotics; however, it remained stable in controls (Nazzal et al. 2021). Further, a cohort study with a large number of patients showed a 70% decreased risk of recurrent kidney stone formation on colonization with *O. formigenes* (Daniel et al. 2021). The study conducted by Verhulst and his team on animal models displayed beneficial effects of *O. formigenes* treatment on nephrocalcinosis and ethylene glycol-induced hyperoxalemia and thus supports the use of *O. formigenes* as potential therapeutics in primary hyperoxaluria (Verhulst et al. 2022).

Recently, it was manifested that after administration of *Lactobacillus,* the oxalate content and stone size were significantly reduced in patients with hyperoxaluria. Expression of OxDC by *Lactobacillus* was considered responsible for oxalate degradation. Besides *Lactobacillus* regulates the expression of tight junction protein and the body's immune response thereby strengthening the intestinal barrier, playing an important role in maintaining the balance of intestinal flora. Given the advantages of *Lactobacillus* and OxDC mentioned above, it has turned out into a promising and potential strategy for the treatment of CaOx stones patients (Wu et al. 2022). Al and co-workers demonstrated that colonization of the intestinal tract of *D. melanogaster* with *Bacillus subtilis* 168 (BS168) exerted promising outcomes by reducing stone burden in dissected Malpighian tubules and faecal excreta. Similarly, in vitro experimentation involving pre-treatment of MDCK cell lines with BS168 prevented CaOx crystal adhesion and aggregation thereby suggesting that BS168 provides a novel curative which could significantly reduce the incidences of recurrent CaOx nephrolithiasis in high-risk patients (Al et al. 2020). However, due to bioavailability issues or viability of the probiotic formulations, these attempts of human colonization with oxalate degrading probiotic bacteria should not be considered successful so far and hence demand further research into it.

30.5 Postbiotics: A Futuristic Therapy for Urolithiasis

Dysbiosis in the intestinal microbiota and urinary tract microbiome can result in sever kidney stone disease. The use of prebiotics, probiotics and postbiotics to alter the gut microbiome has attracted recent interest. Postbiotics are substances produced by or released by metabolic activity of the microorganisms, which directly or indirectly employ a positive effect on the host (Żolkiewicz et al. 2020). Although studies demonstrated reduced urinary oxalate excretion on administration of *O. formigenes* in the gut, however, it was also exhibited that colonization of the bacterium inside the gut depended on dietary calcium intake. The colonization of the bacterium was persistent only when dietary calcium was low, i.e. there was less calcium available to bind oxalate. This reduces the beneficial effects of *O. formigenes* probiotic therapy on oxalate degradation as this would require a low calcium diet. These benefits of probiotic *O. formigenes* could be reinstated by the postbiotic *O. formigenes* lysates (Favero et al. 2022). Investigations revealed that enteric coated encapsulated *O. formigenes* freeze diet lysates twice daily for 5 days reduced urinary oxalate excretion by 50% as well as in rats with renal insufficiency and hyperoxaluria it supported colonic oxalate secretion. In this study, freeze-dried lysates of *O. formigenes* strain, oxalyl CoA and thiamine pyrophosphate in the ratio of 8:1:1 was formulated and used in the form of gelatin capsules. Also enhanced uptake of oxalate was observed in Cacoo 2BEE cells when incubated in O. formigenes conditioned medium as compared to control cells grown in non-*O. formigenes* conditioned medium (Arvans et al. 2017). Although additional studies are necessary to verify the efficacy of *O. formigenes* postbiotic however, the results obtained in this work braced the concept that the postbiotic could maintain a balance renal and enteric oxalate (Hatch et al. 2006).

30.6 Conclusion

Over the years, hyperoxaluria or kidney stones have evolved from an acute disease to a chronic condition which leads to recurrent stone formation and severe renal disorders for which satisfactory therapeutic options are still awaited. Renal calculi or kidney stones with a worldwide prevalence of 1–20% are predominantly made up of calcium oxalate crystals and are found to a greater extent in males than in females. Drugs and surgical intervention of kidney stones are available; however, the side effects posed by these therapies as well as recurrence rate of stones have stipulated for better and safer options. Microbial products such as oxalate degrading enzymes, i.e. oxalate decarboxylase and oxalate oxidase have been explored in the recent past as potential therapeutics for kidney stones and hyperoxaluria. Moreover, colonization of the gut with whole *O. formigenes* (probiotic) or treatment with *O. formigenes* lysates (postbiotics) have been demonstrated to reduced urinary oxalate excretion in animal models. Therefore, further advancements in this filed can guarantee superior and definitive therapeutics for urolithiasis.

Acknowledgements The authors are grateful to the Department of Biotechnology, Himachal Pradesh University, Shimla, for the providing various resources. We are also thankful to the Department of Biotechnology, Government of India, for providing the necessary support to the authors.

Conflict of Interest Further, the authors have no conflict of interest among themselves at their place of work or with the institution.

References

Abratt VR, Reid SJ (2010) Oxalate degradation bacteria of the human gut as probiotic in the management of kidney stones diseases. Adv Appl Microbiol 72:63–87

Akulenko I, Skovorodka M, Serhiichuk T, Tolstanova G (2020) The oxalate-degrading activity of *lactobacillus* spp. isolated from different sources as the potential probiotic modulators for oxalate homeostasis. J Microbiol Exp 8:118–123

Al KF, Daisley BA, Chanyi RM, Bjazevic J, Razvi H, Reid G, Burton JP (2020) Oxalate-degrading *Bacillus subtilis* mitigates urolithiasis in a *Drosophila melanogaster* model. Msphere 5(5): e00498–e00420

Albert A, Paul E, Rajakumar S, Saso L (2020) Oxidative stress and endoplasmic stress in calcium oxalate stone disease: the chicken or the egg? Free Radic Res 54:244–253

Alelign T, Tessema TS, Debella A, Petros B (2021) Evaluations of the curative efficacy of G. fruticosus solvent extracts in experimentally induced nephrolithiatic Wistar male rats. BMC Complement Med Ther 21:1–14

Alshaikh AE, Hassan HA (2021) Gut–kidney axis in oxalate homeostasis. Curr Opin Nephrol Hypertens 30:264–274

Arumugam N, Veerakumar SK, Saravanan S, Vasudevan Y, Raj PJS (2020) Oxalate oxidase: an antilithiatic enzyme from *Staphylococcus pasteuri* and its degradation studies. Int J Pharmaceut Res 12(3)

Arvans D, Jung YC, Antonopoulos D, Koval J, Granja I, Bashir M, Karrar E, Roy-Chowdhury J, Musch M, Asplin J, Chang E, Hassan H (2017) *Oxalobacter formigenes*-derived bioactive factors stimulate oxalate transport by intestinal epithelial cells. J Am Soc Nephrol 28:876–887. https://doi.org/10.1681/ASN.2016020132

Buysschaert B, Aydin S, Morelle J, Gillion V, Jadoul M, Demoulin N (2020) Etiologies, clinical features, and outcome of oxalate nephropathy. Kidney Int Rep 5:1503–1509. https://doi.org/10.1016/j.ekir.2020.06.021

Cassland P, Sjöde A, Winestrand S, Jönsson LJ, Nilvebrant NO (2010) Evaluation of oxalate decarboxylase and oxalate oxidase for industrial applications. Appl Biochem Biotechnol 161:255–263

Chakraborty N, Ghosh R, Ghosh S, Narula K, Tayal R, Datta A, Chakraborty S (2013) Reduction of oxalate levels in tomato fruit and consequent metabolic remodeling following overexpression of a fungal oxalate decarboxylase. Plant Physiol 162:364–378

Chellappan S, Mathivanan S, Thippeswamy R, Nagesh M, Savithri HS, Murthy MRN (2020) Crystal structure of oxalate decarboxylase from *Photorhabdus luminescens*, a symbiotic bacterium associated with entomopathogenic nematodes. Curr Sci 119:1349–1356

Conter C, Oppici E, Dindo M, Rossi L, Magnani M, Cellini B (2019) Biochemical properties and oxalate-degrading activity of oxalate decarboxylase from *Bacillus subtilis* at neutral pH. Int Union Biochem Mol Biol Life 71:917–927

Daniel SL, Moradi L, Paiste H, Wood KD, Assimos DG, Holmes RP, Knight J (2021) Forty years of *Oxalobacterformigenes*, a gutsy oxalate-degrading specialist. Appl Environ Microbiol 87: e00544-21

Favero C, Giordano L, Mihaila SM, Masereeuw R, Ortiz A, Sanchez-Niño MD (2022) Postbiotics and kidney disease. Toxins 14:623

Gomathi S, Sasikumar P, Anbazhagan K, Sasikumar S, Kavitha M, Selvi MS, Selvam GS (2014) Screening of indigenous oxalate degrading lactic acid bacteria from human faeces and south Indian fermented foods: assessment of probiotic potential. Sci World J 2014:648059

Grąz M, Ruminowicz-Stefaniuk M, Jarosz-Wilkołazka A (2023) Oxalic acid degradation in wood-rotting fungi. Searching for a new source of oxalate oxidase. World J Microbiol Biotechnol 39: 1–9

Gupta S, Kanwar SS (2018) Kidney stones: mechanism of formation, pathogenesis and possible treatments. J Biomol Biochem 2:1–5

Gupta S, Kanwar SS (2020a) The influence of dysbiosis on kidney stones that risk up renal cell carcinoma (RCC). Semin Cancer Biol 70:134–138

Gupta S, Kanwar SS (2020b) Therapeutic applications of microbial enzymes in the management of kidney stone diseases. In: Aroera NK et al (eds) Microbial enzymes: roles and applications in industries, microorganisms for sustainability, 11th edn. Springer Nature, Singapore. https://doi.org/10.1007/978-981-15-1710-5_13

Gupta S, Kanwar SS (2021) Optimization of growth conditions for oxalate decarboxylase production from *pseudomonas* sp. OXDC12 and in vitro inhibition of calcium oxalate crystallization by oxalate decarboxylase. Curr Biotechnol 10:133–143

Gupta S, Kanwar SS (2022) Intestinal Dysbiosis induced chronic inflammation that risks up oxalate nephropathy. J Cell Sci Therapy 13:342

Hatch M, Cornelius J, Allison M, Sidhu H, Peck A, Freel RW (2006) Oxalobacter Sp. reduces urinary oxalate excretion by promoting enteric oxalate secretion. Kidney Int 69:691–698

Kizhakedathil MPJ, Bose R, Belur PD (2020) Calcium oxalate degrading thermophilic oxalate oxidase from newly isolated fusarium oxysporum RBP3. Biocatal Agric Biotechnol 25:101583

Kumar V, Irfan M, Datta A (2019) Manipulation of oxalate metabolism in plants for improving food quality and productivity. Phytochemistry 158:103–109

Lambein F, Travella S, Kuo YH, Van Montagu M, Heijde M (2019) Grass pea (*Lathyrus sativus* L.): orphan crop, nutraceutical or just plain food? Planta 250:821–838

Langman CB, Grujic D, Pease RM, Easter L, Nezzer J, Margolin A, Brettman L (2016) A double-blind, placebo controlled, randomized phase 1 cross-over study with alln-177, an orally administered oxalate degrading enzyme. Am J Nephrol 44:150–158

Lee E, Jeong BC, Park YH, Kim HH (2014) Expression of the gene encoding oxalate decarboxylase from *Bacillus subtilis* and characterization of the recombinant enzyme. BMC Res Notes 7:598

Liang X, Moomaw EW, Rollins JA (2015) Fungal oxalate decarboxylase activity contributes to S clerotinia sclerotiorum early infection by affecting both compound appressoria development and function. Mol Plant Pathol 16(8):825–836

Lin R, He J, Wu J, Cai X, Long H, Chen S, Liu H (2017) Chemical modification of oxalate decarboxylase to improve adsorption capacity. Int J Biol Macromol 98:495–501

Makela MR, Hildén K, Lundell TK (2010) Oxalate decarboxylase: biotechnological update and prevalence of the enzyme in filamentous fungi. Appl Microbiol Biotechnol 87:801–814

Mathivanan K, Chandirika JU, Vinothkanna A, Govindarajan RK, Meng D, Yin H (2021) Characterization and biotechnological functional activities of exopolysaccharides produced by Lysinibacillus fusiformis KMNTT-10. J Polym Environ 29:1742–1751

Miller AW, Dearing D (2013) The metabolic and ecological interactions of oxalate-degrading bacteria in the mammalian gut. Pathogens 2:636–652

Nazzal L, Francois F, Henderson N, Liu M, Li H, Koh H, Wang C, Gao Z, Perez GP, Asplin JR et al (2021) Effect of antibiotic treatment on *Oxalobacter formigenes* colonization of the gut microbiome and urinary oxalate excretion. Sci Rep 11:16428

Paul E, Sasikumar P, Gomathi S, Abhishek A, Govindan SS (2017) Recombinant lactic acid bacteria secreting OxdC as a novel therapeutic tool for the prevention of kidney stone disease. In: Grumezescu AM (ed) Multifunctional systems for combined delivery, biosensing and diagnostics. Elsevier, Amsterdam, pp 327–345

Paul E, Albert A, Ponnusamy S, Mishra SR, Vignesh AG, Sivakumar SM, Sivasamy G, Sadasivam SG (2018) Designer probiotic *Lactobacillusplantarum* expressing oxalate decarboxylase developed using group II intron degrades intestinal oxalate in hyperoxaluric rats. Microbiol Res 215: 65–75

Paul E, Albert A, Ponnusamy S, Venkatesan S, Sadasivam SG (2019) Chromosomal integration of heterologous oxalate decarboxylase in *lactobacillus plantarum* WCFS1 using mobile genetic element Ll. LtrB. Arch Microbiol 201:467–476

Peck AB, Canales BK, Nguyen CQ (2016) Oxalate-degrading microorganisms or oxalate-degrading enzymes: which is the future therapy for enzymatic dissolution of calcium-oxalate uroliths in recurrent stone disease? Urolithiasis 44:45–50

Qi Z, Yu J, Shen L, Yu Z, Yu M, Du Y, Zhang R, Song T, Yin X, Zhou Y, Li H, Wei Q, Liu Y (2017) Enhanced resistance to rice blast and sheath blight in rice (*Oryza sativa* L.) by expressing the oxalate decarboxylase protein bacisubin from *Bacillus subtilis*. Plant Sci 265:51–60

Quintero E, Bird VY, Liu H, Stevens G, Ryan AS, Buzzerd S, Klimberg IW (2020) A prospective, double-blind, randomized, placebo-controlled, crossover study using an orally administered oxalate decarboxylase (OxDC). Kidney360 1:1284–1290

Sadaf H, Raza SI, Hassan SW (2017) Role of gut microbiota against calcium oxalate. Microb Pathog 109:287–291

Shah A, Leslie SW, Ramakrishnan S (2021) Hyperoxaluria. In: StatPearls. StatPearls, Treasure Island, FL

Shimazono H (1955) Oxalic acid decarboxylase, a new enzyme from the mycelium of wood destroying fungi. J Biochem 42:321–340

Tavasoli S, Alebouyeh M, Naji M, Shakiba Majd G, Shabani Nashtaei M, Broumandnia N, Basiri A (2020) Association of intestinal oxalate-degrading bacteria with recurrent calcium kidney stone formation and hyperoxaluria: a case-control study. BJU Int 125:133–143. https://doi.org/10.1111/bju.14840

Thakore P, Liang TH (2021) Urolithiasis. In: StatPearls [internet]. StatPearls Publishing

Verhulst A, Dehmel B, Lindner E et al (2022) *Oxalobacter formigenes* treatment confers protective effects in a rat model of primary hyperoxaluria by preventing renal calcium oxalatedeposition. Urolithiasis 50:119–130. https://doi.org/10.1007/s00240-022-01310-9

Vos AM, Bleichrodt RJ, Herman KC, Ohm RA, Scholtmeijer K, Schmitt H, Lugones LG, Wösten HA (2021) Cycling in degradation of organic polymers and uptake of nutrients by a litter-degrading fungus. Environ Microbiol 23:224–238

Wigner P, Bijak M, Saluk-Bijak J (2022) Probiotics in the prevention of the calcium oxalate urolithiasis. Cell 11:284. PMID: 35053400; PMCID: PMC8773937. https://doi.org/10.3390/cells11020284

Winestrand S, Sjöde A, Cassland P, Hildén L, Nilvebrant NO, Jönsson LJ (2014) Evaluation of oxalate decarboxylases in industrial bleaching filtrates and in pulp-mill experiments. Ind Biotechnol 10:126–129

Wu F, Cheng Y, Zhou J, Ye P, Liu X, Zhang L, Lin R, Xiang S, Liu Z, Wang C (2022) Manipulating oxalate decarboxylase provides the basis of antilithic therapy by acting on the gut microbiota. bioRxiv

Zhu W, Reinhardt LA, Richards NGJ (2018) Second-shell hydrogen bond impacts transition-state structure in *Bacillus subtilis* oxalate decarboxylase. Biochemistry 57:3425–3432

Żolkiewicz J, Marzec A, Ruszczyński M, Feleszko W (2020) Postbiotics—a step beyond pre-and probiotics. Nutrients 12:2189

Use of Yeast in the Welfare of Human and Their Applications

31

Nishu Lohan and Sukesh Chander Sharma

Abstract

The usage of yeasts in human nutrition together with food technology, as an alternative source of protein to substitute the requirements in a world of low agriculture production and rapidly multiplying population, makes yeast incredibly important. The use of yeasts, for SCP production, is more accessible, as they can be effortlessly reproduced utilizing inexpensive raw materials and readily picked due to their bigger cell sizes and flocculation capabilities. One of the benefits of using yeast is that they hold a lesser quantity of nucleic acids than bacteria, so it can be used by human. Yeast biomass along with biomass-enriched yeast through trace minerals proved to be a non-traditional alternative source of protein. And it is applicable in enhancing animal fitness and growth performance and characterizes a premium supplement to the human diet owing to the high content of minerals, proteins, and vitamins, too. Yeast can also be used as a trace element supplement in food, feed, and medicine, known as trace element enriched yeast; it is specified for its capability to accumulate metal ions and to form organically bounded microelements. This review will explore the potential applications of yeast in human welfare.

31.1 Introduction

Yeast is a unicellular, chemoorganotrophic eukaryotic microorganism categorized as a kingdom fungus member. The earliest yeast evolved hundreds of millions of years ago. Yeasts are very frequent in the environment, grow excellent in a slightly acidic

N. Lohan · S. C. Sharma (✉)
Department of Biochemistry, Panjab University, Chandigarh, India
e-mail: sukeshcs@pu.ac.in

R. Sobti et al. (eds.), *Role of Microbes in Sustainable Development*,
https://doi.org/10.1007/978-981-99-3126-2_31

pH to neutral pH, and are generally isolated from sugar-rich materials (like exudates from plants, the skin of berries, and fruits). Yeasts are also found in the gut flora of mammals and some insects and indeed deep-sea environments host an array of yeasts. Yeast may have sexual and asexual (budding and fission) reproductive cycles. The prevalent mode of vegetative growth in yeast is asexual reproduction by budding (Balasubramanian et al. 2004); few yeasts also reproduce by fission. It was reported that yeasts (terrestrial as well as marine yeast) can produce many bioactive substances, such as amino acids, enzymes, phytase, glutathione, glucans, killer toxins, and vitamins with probable applications in the various domains of food, pharmaceutical, cosmetic, and chemical industries (Zaky et al. 2014). Therefore, over the centennial, yeasts turned out to be extensively used in different industries, such as baking, brewing, wine, bioethanol, bioremediation, foods, and nutritional supplements, probiotics, scientific research, genetically engineered biofactories, and pharmaceutical protein production. Certain strains of some yeasts species produce proteins called yeast killer toxins that permit them to exclude hustling strains. They have the potential as natural antimicrobials in food and for biological control of plant pathogens and therapeutic agents against animal and human infections. The yeast-killer toxins harm the bacteria that usually produce antibiotics to kill the parasite; as a result, the yeast-killer toxin can be used as bio preservative and it has been anticipated as a possible alternative for chemical food preservatives (their potential toxicity comprises a direct risk to public health, has supported research into alternative and secure approaches of food preservation, i.e., biopreservation). They may also have medical uses in treating yeast infections. Among the potential microbial antagonists, yeasts have been highly studied since they own several features that make them a suitable biocontrol agent (Santos et al. 2004). Many yeast species have simple nutritional needs, and they can colonize dry surfaces for long times and can grow quickly on affordable substrates in bioreactors (Chanchaichaovivat et al. 2007). Thus, accompanying its nutritional value, yeast can be used as a dietetic supplement. Dietary supplements are becoming progressively applicable and popular widespread, further booming this multi-billion-dollar industry, while consumers struggle for self-optimization, longevity, health, and standard of living.

31.2 The Nutrition Aspects of Yeast for Human Health

Yeast biomass contributes to the bioavailability of bread which is now widely used around the world. *Saccharomyces cerevisiae* contains higher vitamins, amino acids, and mineral contents. Yeasts contain minerals such as chromium, iron, magnesium, and zinc, B complex vitamins such as riboflavin, folic acid, niacin, and biotin, many essential amino acids such as valine, lysine, and leucine, nucleic acids, and various phenolics such as protocatechuic acid, cinnamic acid, and gallic acid (Demirgul et al. 2022). In addition to essential amino acids, yeast extracts are also rich in flavor-enhancer amino acids such as alanine, alanine, aspartic acid, and glutamic acid. Today, yeast SCP is considered a potential protein source for humans as well as animals. Yeast also holds some primary and secondary metabolites, they may function as neurohormones and antioxidants, opening a new scenario of applications from nutritional supplements to functional foods, that are significant for human nutrition and health. The acceptability of a particular microorganism as food or feed depends on its nutritional value and safety (including the presence of toxins, nucleic acid content, and residual undesirable compounds such as heavy metals). Presently, the only species completely acceptable as food for humans is *S. cerevisiae* also known as baker's and brewer's yeasts; therefore, *S. cerevisiae* has become an essential component of human civilization.

31.3 Applications of Yeast for Human Welfare

31.3.1 Food-Grade Yeast

Food-grade yeasts can be accustomed as sources of high nutritional value proteins (SCP), vitamins, and enzymes, through usages in the health food industry as nutritive supplements, food subsidiaries, flavoring agents, and conditioners, for the production of microbiology media, together with livestock feeds. Yeasts are encompassed in starter cultures, for the production of precise sorts of fermented foods like fermented meat, vegetable products, sourdoughs, bread, cheese, vinegar, etc. Varied microorganisms are used for human consumption widespread as SCP or as ingredients of conventional food starters, including fungi (*Penicillium*, *Aspergillus*, etc.), bacteria (*Alcaligenes*, *Lactobacillus*, *Cellulomonas*, etc.), algae (*Spirulina*, *Laminaria*, *Chlorella*, *Rhodymenia*, etc.), and yeasts (*Saccharomyces*, *Candida*, *Torulopsis*, *Pichia*, and *Kluyveromyces*) (Bekatorou et al. 2006). Among the yeast species, *Saccharomyces cerevisiae* and *Candida utilis* have been completely acceptable for human consumption. The highly frequent food-grade yeast is *Saccharomyces cerevisiae*, also known as baker's yeast, which is used widely for the production of bread and baking products.

31.3.2 Probiotic Yeast

Probiotics are vital and beneficial organisms that offer health benefits to the host organisms. *Saccharomyces boulardii*, the only patented strain belonging to yeast genera for human use, has been broadly evaluated for their probiotic effect. Probiotic yeasts have promising antimicrobial, antioxidant, and anticancer properties, cholesterol assimilation, and immunomodulatory effects and can also be utilized as biotherapeutics. Yeasts are broadly used in biotechnological and industrial applications for the production of fermented food products, including enzymes, acids, and vitamins on large scale. To act constructively, probiotics should be in the living condition to create symbiotic stability in the host alimentary tract. Being resistant to antibiotics proves to be a beneficial characteristic of a probiotic organism. Hence, it seems to be an important attribute for their use as a probiotic. The major threat associated with antibiotic-resistant *Lactobacillus* strains is their capability to transfer the resistant gene to pathogenic bacteria. The transfer of genes between yeast and bacteria is rare; therefore, their application as probiotics is secure and advisable (Shruthi et al. 2022). *Saccharomyces cerevisiae var boulardii* is the ultimate common human yeast probiotic that is available in the market presently. Yeasts promote both human and animal health, also intensify the bioavailability of minerals through hydrolysis of phytate, folate biofortification, detoxification of fungal toxins, and xenobiotics. Yeast probiotics produce different VOCs (volatile organic compounds, which are metabolites with low molecular weight and high vapor pressure, and low polarity), mycotoxins (killer toxins), and antimicrobials which show the antagonistic effect against pathogenic fungi and bacteria. Yeast's genera that produce killer toxins include *Saccharomyces, Candida, Torulopsis, Cryptococcus, Kluyveromyces, Debaryomyces, Pichia, Cryptococcus, and Zygosaccharomyces*. Phagocytic cells produce a high level of microbicidal ROS (reactive oxygen species) like superoxide anion and hydrogen peroxide, to attack invading foreign pathogens during phagocytosis. However, excessive relative oxygen species are generated, which can cause the oxidation of membrane phospholipids, mutations in DNA, and damage to proteins. In this process, superoxide dismutase and catalase are the crucial enzymes that convert ROS into less reactive oxygen species (Angulo et al. 2019). The well-known mechanisms of the killer toxins are they disturb the cell cycle and stop DNA synthesis and block the synthesis of the cell wall by inhibiting β 1,3-glucan component (present in a cell wall) (Liu et al. 2015). These killer toxins and volatile organic compounds have antagonistic effects against the opens. Thus, probiotic yeast metabolites show virtuous antioxidant properties.

31.3.3 Yeast Extract

Yeast extract containing water-soluble mixtures of the yeast cell that remains behind the disruption and dumping of the cell wall (Khan et al. 2020) and is accustomed in various foods like noodles, sauces, gravies, soups, bouillons, and meatballs to enhance their flavor and aroma. Since it improves the umami taste of these foods, yeast extract is used as an alternative to monosodium glutamate (MSG, E621), which owns enticed consumers' response in current periods (Wang et al. 2019). It can likewise demote the use of NaCl in foods due to its salty flavor (Desmond 2006). Besides its flavor-enhancing property, yeast extract contributes to the nutritional value of the foods it uses because of its rich nutritional composition (Filipović et al. 2020). In addition to its rich nutritive content, yeast extracts retain the potential to be used in operating foods due to their DPPH radical scavenging activities, immune regulation properties, antioxidant activities, and anti-inflammatory properties. Yeast extracts are positively valuable ingredients for vegetarian as well as traditional diets.

31.3.4 Pulse-Yeast-Based Formulations for Celiac People

Safe, functional, and gluten-free snacks are made from pulse-yeast-based formulations. The term "nutritional yeast" refers to the deactivated form of *Saccharomyces cerevisiae*, which is rich in proteins, vitamins (mainly vitamins of B-complex), minerals (Mg, Zn, Fe, and Cr), and low in sodium and fat. The novel-developed extruded gluten-free snacks are advisable for celiac people. As nutritional yeasts are gluten-free, they are a suitable food source for the AC patients (Pérez-Torrado et al. 2015). Therefore, the consumption of nutritional yeast-enriched products, like ready-to-eat snacks developed from pulses flours, is a healthy substitute for celiac people, athletes, vegans, as well as young people during the puberty period. It is the first time that nutritional yeast is used in the production of novel "gluten-free" flavorful expanded snacks, which may contribute to an increase in the consumption of pulses, mainly in teens and children, and it is a novel alternative for celiac people. The improved expanded snack-type products are developed from high-protein nutritional yeast flours, and novel formulations of lentils, under suitable environments of temperature and moisture. The developed novel snack-type products contain increased content of beneficial α-galactosides (having prebiotic properties) and significantly low content of undesirable lectins, inositol phosphates, and trypsin inhibitors. Therefore, the final product is considered a safe, functional gluten-free food that could be suitable for celiac as well as the general population.

31.3.5 Role of Minerals-Enriched Yeast in Maintaining the Health of Living Organisms

31.3.5.1 The Function of Trace Elements

Micronutrients play a crucial role in maintaining the health of living things. The WHO categories the essential trace elements into three classes. The first class contains the following essential trace elements: chromium (Cr), selenium (Se), zinc (Zn), iodine (I), iron (Fe), copper (Cu), molybdenum (Mo), and cobalt (Co). The second class contains boron (B), manganese (Mn), vanadium (V), silicon (Si), and nickel (Ni). The third class includes those trace elements which are probably toxic but may have necessary roles for the human body at low dosages, including fluorine (F), lithium (Li), cadmium (Cd), lead (Pb), mercury (Hg), arsenic (As), aluminum (Al), and tin (Sn). Yeast is not only used as single-cell proteins and fermentation products, but also used as a trace element supplement that is broadly used in feed, food, and medicine. Trace-element-enriched yeast, namely, chromium-, selenium-, zinc-, and iron-enriched yeast, is an impressive microelement supplement since it is safer than its inorganic and organic equivalents and is more effective and more environmentally friendly. Yeast is known for its capacity to accumulate metal ions and to form an organically defined microenvironment.

31.3.5.1.1 Chromium

Chromium functions in maintaining the regular metabolism of glucose, lipid, and protein in the body. It can maintain the dynamic equilibrium of blood glucose and improves levels of lipid and blood glucose to some limit (McCarty 2000; Via et al. 2008). Chromium is a significant active element of glucose tolerance factor (GTF), which enhances the sensitivity of tissue receptors regarding insulin (Anderson 2000). GTF can control the metabolism of biological macromolecules, deposition of muscle tissue, and use of cholesterol by accelerating the tendency between the insulin and tissue receptors (Moeini et al. 2011). At the same time, chromium is also known to be a promising anti-heat stress element that can demote the use of antibiotics because of its strong antioxidant effect, which prohibits ROS from demolishing lipid film structures, leading to lipid peroxidation and cell damage (Bin-Jumah et al. 2020). In the case of metallic chromium, tetravalent chromium is found as the most toxic form. In contrast with the tetravalent form, the toxicity of trivalent chromium is too low, and the quick reduction of tetravalent chromium by intracellular reducing agents (such as glutathione, ascorbic acid, and cysteine) to trivalent chromium generates ROS and chromium in the pentavalent and hexavalent forms. Prolonged exposure to chromium in its hexavalent state is affiliated with an elevated threat of lung cancer.

31.3.5.1.1.1 Chromium Enriched Yeast

Chromium (III), as an essential trace element, is a significant element of human glucose tolerance factor and can improve hyperglycemia by lowering fasting blood glucose and enhancing glucose tolerance (Sundaram et al. 2013). Chromium (III) shortage augments the threat of lipid and glucose metabolism disorders (Li et al. 2019). Dietary chromium (III) addition can accelerate glucose tolerance and insulin sensitivity, hence decreasing the risk of atherosclerotic complexities and hyperglycemia. However, inorganic chromium (III) is challenging to be immersed in the gastrointestinal tract and perhaps adverse to human fitness because it can be oxidized into toxic hexavalent chromium (Guo et al. 2020). In comparison with inorganic chromium (III), organic chromium (III) attracted increasing attention lately owing to its fairly low toxicity, high bioavailability, and broad range of health advantages (Kralovec et al. 2009). After estimating the pharmacological properties of certain synthesized organic chromium supplements, such as chromium histidinate, chromium propionate, polysaccharides-chromium (Hayat et al. 2020), and YCr is chromium (III) rich yeast preparation, only YCr reported with no hostile physiological effects in a sequence of animal toxicological investigations or human clinical efficacy trials (Hou et al. 2019). A 16S rDNA amplicon sequencing displayed that dietary YCr intervention possesses advantageous effects on enhancing fasting glucose tolerance, decreasing hepatic TG, serum TC, hepatic glycogen levels, and blood pressure in patients with type 2 diabetes (Hosseinzadeh et al. 2013). Therefore, it can be served as a functional component to prevent hyperglycemia and hyperlipidemia, thus suggesting that chromium (III)- enriched yeast have preventative effects on high-fat and high-fructose diet (HFHFD). Chromium-fortified yeast elevates liver glycogen synthesis and glucose tolerance. Liver metabolomics dissection exposed that dietary YCr intermediate substantially regulated the levels of certain biomarkers involved in glycerophospholipid metabolism, purine and pyrimidine metabolism, citrate cycle, phenylalanine, tryptophan, and tyrosine biosynthesis, and so on. Furthermore, dietary YCr intervention controlled the mRNA levels of main genes connected with glucose, bile acids, fatty acids, and cholesterol, metabolism in the liver. Liver metabolomics was further used to examine the preventive mechanism of YCr on hyperlipidemia and hyperglycemia. Outcomes displayed that dietary YCr intervention significantly elevated the levels of acetylcarnitine, arachidonate, retinol, and ergothioneine, but decrease the levels of taurochenodeoxycholate, kynurenine, and 4-pyridoxic acid, in the liver. Retinol (essential fat-soluble micronutrient) was found to be critically interrelated with plasma cholesterol level, hampering lipid over-accretion in mammals, and therefore is advantageous for reducing the occurrence of atherosclerosis. Additionally, retinol can be metabolized into retinoic acid in the liver, which further provokes liver lipolysis and decreases triglyceride levels (Wolf 2010). The presence of arachidonic acid can elevate the biosynthesis of long-chain unsaturated fatty acids, and it diminished the concentration of free fatty acids in the liver (Hatziantoniou et al. 2004). Ergothioneine is known as a powerful scavenger of discrete reactive oxygen species (ROS), that may evade the destruction of cellular constituents and manage the function of a mitochondrion (Kushairi et al. 2020). In addition, acetylcarnitine defends the liver against oxidative stress by enhancing the activities of glutathione peroxidase. The 4-pyridoxic acids (synthesized from pyridoxal via enzyme aldehyde oxidase) are a potent positive indicator of vascular risk score (Obeid et al. 2019). Kynurenine is a direct precursor of kynurenic acid, reported being increased in the mice suffering from a tumor. Taurochenodeoxycholate is a potential marker of liver injury and is synthesized in the gut through microbial fermentation and is closely connected with glycolysis and fatty acid biosynthesis (Wang et al. 2018).

31.3.5.1.2 Selenium

Selenium is a constituent of selenocysteine and a necessary element of enzymes selenases and selenoproteins such as phospholipid hydrogen glutathione peroxidase (PHGPx), thioredoxin reductase (TrxR), and glutathione peroxidase (GPx) (Yadav et al. 2015). Selenium in the form of selenocysteine is the active component of several antioxidant enzymes in the body. Selenium also has been connected to human antioxidant activity, and anti-inflammatory and anti-virus properties (Zoidis et al. 2018). Selenium insufficiency can induce tumor diseases, skeletal muscle, myocardial necrosis, and a decrease in immune function (Zoidis et al. 2018).

31.3.5.1.2.1 Selenium-Enriched Yeast

In S. cerevisiae, the enzymes glutathione synthase (GSH2) and glutamylcysteine synthetase (GSH1) catalyze the regular synthesis of glutathione (Li et al. 2004). Glutathione is composed of three amino acids, namely, glycine, cysteine, and glutamate. The thiol group of cysteine in glutathione binds to heavy metals to conquer the toxicity induced by metals so that they cannot disintegrate cellular metabolic activities (Valko et al. 2016). Glutathione not only binds to selenium through cysteine residues but also is closely relevant to the production of Nano-selenium. In S. cerevisiae, tetravalent selenium ions readily react with reduced glutathione to form selenide glutathione (GS-Se-SG) and glutathione disulfide (GSSG) (Herrero and Wellinger 2015). Selenide glutathione can be further converted back into glutathione. The enzyme superoxide dismutase catalyzes the conversion of glutathione into elemental selenium (Se0). It was reported that this is the key pathway used by S. cerevisiae to reduce selenite to form organic elemental nano-selenium (Kessi and Hanselmann 2004). Selenium integrated as methyl selenic acid might participate in apoptosis, cell cycle arrest, angiogenesis, and alternative mechanisms for cancer prevention. The organic form of selenium acts as a chemo protectant against many types of cancer, it involves the following mechanism of action- regulation of cell cycle and apoptosis, specific inhibition of tumor cell growth, effect on DNA repair, and protection of selenoproteins from antioxidants. Selenium may have a necessary role in immune function, male reproduction, mammalian development, and decelerating the aging process.

31.3.5.1.3 Iron

Iron is an important component of hemoglobin and its deficiency can cause anemia. Iron is involved in transporting oxygen through red blood cells, so iron deficiency may provoke chronic inflammation (Cappellini et al. 2017). Iron participates in many essential biochemical processes, such as the synthesis of amino acids, lipids deoxyribonucleotides, and sterols. Iron is an essential element of cytochrome enzymes; therefore, it participates in the electron transport chain and is crucial to oxidative phosphorylation and redox reactions involved in the respiratory chain (Puig et al. 2017). A normal grain-based diet is generally unable to fulfill the daily requirement of iron; thus, it can lead to iron deficiency.

31.3.5.1.3.1 Iron-Enriched Yeast

Iron deficiency can be treated through oral supplementation. But, the use of dietary supplements is also associated with the risk of side effects, the most common of which are gastrointestinal symptoms such as nausea, vomiting, abdominal pain, constipation, flatulence, and diarrhea, occurring in up to 40% of patients (Jankowska et al. 2016). Additionally, these preparations are characterized by having a metallic taste and low absorption of iron in the intestines (Shubham et al. 2020). Yeasts produce metal-protein complexes called metalloproteins, which are highly absorbed by the human body. The use of iron-enriched yeast in PEF conditions for the preparation of dough increased the iron content in the flatbreads to almost 386 mg/100 g dry mass (Nowosad and Sujka 2021). The flatbread with yeast P has the highest iron bioavailability, which is associated with the highest iron element content in the product. The studies proved that rather than adding iron compounds directly to the product, we can use yeast enriched with iron ions to prepare flatbread. The prepared flatbread contains about 386 mg/100 g dry mass of iron with good potential bioavailability of the element. Thus, it is acceptable to consumers (without metallic aftertaste) and has a moderate glycemic index.

31.3.5.1.4 Zinc

The element zinc contributes to **cell growth** and **apoptosis**. Zinc is abundant in the brains of mammals, and insufficient intake of zinc can affect intellectual development and immunity among children (Brown et al. 2002). Zinc is also important for the central nervous system. About 10% of the human proteome consists of zinc bound proteins involved in gene regulation and their functions. Zinc deficiency activates inflammatory responses which trigger ROS-induced oxidative stress (Hosseinzadeh et al. 2013). Prolonged and severe zinc deprivation critically impairs the immune system and increases incidences of intestinal infection (Hosseinzadeh et al. 2013), being responsible for about 10% of diarrheal diseases worldwide (Maares et al. 2022). Dietary Zn availability and absorption are reduced in the presence of phytates, and high dietary intakes of calcium, phosphorus, and copper (Šillerová et al. 2012).

31.3.5.1.4.1 Zinc-Enriched Yeast

Multiple inorganic and organic zinc sources are available on the global market as its supplementation compensates for insufficient zinc status and plays a positive role in the treatment of diseases such as diarrhea and the common cold (Haase et al. 2008).

The study demonstrated that *S. pastorianus* Rh was well suited to generate zinc-enriched biomass as a readily bioavailable organic source of zinc. This was highlighted by its characterization regarding zinc enrichment and distribution as well as the intestinal zinc bioaccessibility and bioavailability, which was comparable to commonly used organic and inorganic zinc supplements. Most importantly, zinc released from ZnYeast after digestion was available for biological processes upon its uptake by human enterocytes.

According to our zinc bioaccessibility and bioavailability screening, zinc-enriched *S. pastorianus* Rh released about 40% of its zinc content into the intestinal liquid after gastrointestinal digestion, having a bioavailability similar to other commonly used inorganic and organic zinc supplements.

Considering its zinc content (5.9 $\pm$ 1.0 mg zinc/g yeast) and intestinal accessibility of 40%, already 1.1–1.5 g of Zn Yeast would be sufficient to restore daily endogenous zinc losses (Haase et al. 2020).

31.4 Industrial Applications of Yeast

S. cerevisiae becomes an essential component of human civilization since it has been extensively used in food and beverage fermentation of sugar and is the oldest and largest industrial application of yeast. The yeast species that has been most closely connected with humankind is Saccharomyces cerevisiae which has long been used for brewing, distilling (for both industrial and drinkable alcohol), winemaking, and baking bread. A wide diversity of pure yeast cultures, mainly Saccharomyces (*S. cerevisiae, S. bayanus, S. uvarum, S. oviformis, S. carlsbergensis, S. chevalieri, S. diastaticus, S. fructuum, S. pasteurianus, S. sake, S. vini,* etc.) (Bekatorou et al. 2006) are prompted industrially for the use in induced wine fermentations, coinciding to the industrial requirements for fermentation efficacy and productiveness. Their operational purposes are to constantly metabolize wort (an herbaceous plant) ingredients into carbon dioxide, ethanol, and other fermentable products to produce spirits and beer with good quality, reliability, and portability. The industrial uses of yeast can be categorized as yeast extracts, potable ethanol, industrial ethanol, Baker's yeast (biomass, flavoring, and carbon dioxide), heterologous peptides and proteins, and a plethora of medicinal applications. Baker's and brewer's yeast are distinct strains of *Saccharomyces cerevisiae*; they are rather different from each other. Baker's yeast functions as the leavening of dough in baking which causes the bread to rise by converting fermenting sugar (present in dough) into carbon dioxide and ethanol. Sourdough appetizers can likewise be accustomed to the conditioning of dough, development time of bread and improvement preservation time, and baking products with unique organoleptic properties. Brewer's yeast is used in brewing several types of beers via fermentation. Nutritional brewer's yeast is a type of inactive yeast (quiescent cells without any leavening power), the residues that are left after the brewing process. It is a great source of protein, and it is used as a nutrient supplement rich in B vitamins, Fe, Mn, K, Cu, Cr, P, Ca, Mg, and Zn (Bekatorou et al. 2006). It has been extensively for its medicinal properties such it is frequently used for the treatment of diabetes (regulation of insulin levels), loss of appetite, chronic acne, diarrhea, etc. It is also confined as a dietary supplement for healthy hair and nails. Another nutritious food that can be used as a dietary supplement is *Torula* yeast, which is highly nutritious and digestible, containing minerals and vitamins (mainly pantothenic acid niacin and B vitamins) more than 50% of protein (rich in valine, glutamic acid, lysine, and threonine) (Olvera-Novoa et al. 2002), *Torula* yeast can be used as a food additive or flesh substitute in several refined foods, in sauces, soups, seasonings, spices, dips, etc. It is also used in diet food and vegetarian, in doughs, baby food, meat products, etc. Similarly, probiotic yeasts, like S. cerevisiae, are live microbial feed supplements that benefit the host animal by improving its intestinal microbial balance. Probiotic properties of yeasts have been studied and revealed their ability to survive through the gastrointestinal (GI) tract and interact antagonistically with GI pathogens such as *Shigella*, *Escherichia coli*, and *Salmonella*. Specifically, *S. boulardii* a thermophilic, non-pathogenic yeast is used as an animal feed probiotic supplement and therapeutic agent for the treatment of various gut disorders like diarrhea (Lourens-Hattingh and Viljoen 2001). This yeast is safe, it is resistant to antibiotics, attains high cell numbers in the intestine in little time, does not eternally inhabit the intestine, and is immediately cleared following the halt of administration. Its probiotic effects are also intensified by its ability to produce polyamines, which are mixtures that vigorously influence cell growth and differentiation. *Saccharomyces* makes positive contributions to the fermentation of beverages and foods more so than many other yeast species; thus, *S. cerevisiae* has a high commercial significance.

References

Anderson RA (2000) Chromium in the prevention and control of diabetes. Diabete Metab 26(1): 22–27

Angulo M, Reyes-Becerril M, Cepeda-Palacios R, Tovar-Ramírez D, Esteban MÁ, Angulo C (2019) Probiotic effects of marine Debaryomyces hansenii CBS 8339 on innate immune and antioxidant parameters in newborn goats. Appl Microbiol Biotechnol 103(5):2339–2352. https://doi.org/10.1007/s00253-019-09621-5

Balasubramanian MK, Bi E, Glotzer M (2004) Comparative analysis of cytokinesis in budding yeast, fission yeast and animal cells. Curr Biol 14(18):806–818. https://doi.org/10.1016/j.cub.2004.09.022

Bekatorou A, Psarianos C, Koutinas AA (2006) Production of food grade yeasts. Food Technol Biotechnol 44(3):407–415

Bin-Jumah M, Abd El-Hack ME, Abdelnour SA, Hendy YA, Ghanem HA, Alsafy SA, Khafaga AF, Noreldin AE, Shaheen H, Samak D, Momenah MA, Allam AA, AlKahtane AA, Alkahtani S, Abdel-Daim MM, Aleya L (2020) Potential use of chromium to combat thermal stress in animals: A review. Sci Total Environ 707:135996. https://doi.org/10.1016/j.scitotenv.2019.135996

Brown KH, Peerson JM, Rivera J, Allen LH (2002) Effect of supplemental zinc on the growth and serum zinc concentrations of prepubertal children: A meta-analysis of randomized controlled trials1-3. Am J Clin Nutr 75(6):1062–1071. https://doi.org/10.1093/ajcn/75.6.1062

Cappellini MD, Comin-Colet J, de Francisco A, Dignass A, Doehner W, Lam CS, Macdougall IC, Rogler G, Camaschella C, Kadir R, Kassebaum NJ, Spahn DR, Taher AT, Musallam KM (2017) Iron deficiency across chronic inflammatory conditions: international expert opinion on definition, diagnosis, and management. Am J Hematol 92(10):1068–1078. https://doi.org/10.1002/ajh.24820

Chanchaichaovivat A, Ruenwongsa P, Panijpan B (2007) Screening and identification of yeast strains from fruits and vegetables: potential for biological control of postharvest chilli anthracnose (*Colletotrichum capsici*). Biol Control 42(3):326–335. https://doi.org/10.1016/j.biocontrol.2007.05.016

Demirgul F, Simsek O, Sagdic O (2022) Amino acid, mineral, vitamin B contents and bioactivities of extracts of yeasts isolated from sourdough. Food Biosci 50(PA):102040. https://doi.org/10.1016/j.fbio.2022.102040

Desmond E (2006) Reducing salt: a challenge for the meat industry. Meat Sci 74(1):188–196. https://doi.org/10.1016/j.meatsci.2006.04.014

Filipović V, Filipović J, Vucurović V, Radovanović V, Košutić M, Novković N, Vukelić N (2020) The effect of yeast extract addition on bread quality parameters. J Serb Chem Soc 85(6): 737–750. https://doi.org/10.2298/JSC191024137F

Guo WL, Chen M, Pan WL, Zhang Q, Xu JX, Lin YC, Li L, Liu B, Bai WD, Zhang YY, Ni L, Rao PF, Lv XC (2020) Hypoglycemic and hypolipidemic mechanism of organic chromium derived from chelation of Grifthe Ola frondosa polysaccharide-chromium (III) and its modulation of intestinal microflora in high fat-diet and STZ-induced diabetic mice. Int J Biol Macromol 145: 1208–1218. https://doi.org/10.1016/j.ijbiomac.2019.09.206

Haase H, Overbeck S, Rink L (2008) Zinc supplementation for the treatment or prevention of disease: current status and future perspectives. Exp Gerontol 43(5):394–408. https://doi.org/10.1016/j.exger.2007.12.002

Haase H, Ellinger S, Linseisen J, Neuhäuser-Berthold M, Richter M (2020) Revised D-A-CH-reference values for the intake of zinc. J Trace Elem Med Biol 61(October):126536. https://doi.org/10.1016/j.jtemb.2020.126536

Hatziantoniou S, Giamarellos-Bourboulis EJ, Skiathitis S, Demetzos C, Donta I, Papaioannou GT, Dionyssiou-Asteriou A, Karayannacos PE, Giamarellou H (2004) Rapid alterations of serum fatty acids with the intravenous administration of an arachidonate solution. Prostaglandins Leukot Essent Fatty Acids 70(5):465–468. https://doi.org/10.1016/j.plefa.2003.09.002

Hayat K, Bodinga BM, Han D, Yang X, Sun Q, Aleya L, Abdel-Daim MM, Yang X (2020) Effects of dietary inclusion of chromium propionate on growth performance, intestinal health, immune response and nutrient transporter gene expression in broilers. Sci Total Environ 705:135869. https://doi.org/10.1016/j.scitotenv.2019.135869

Herrero E, Wellinger RE (2015) Yeast as a model system to study metabolic impact of selenium compounds. Microbial Cell 2(5):139–149. https://doi.org/10.15698/mic2015.05.200

Hosseinzadeh P, Djazayery A, Mostafavi SA, Javanbakht MH, Derakhshanian H, Rahimiforoushani A, Djalali M (2013) Brewer's yeast improves blood pressure in type 2 diabetes mellitus. Iran J Public Health 42(6):602–609

Hou Y, Hou Y, Yao L, Chen S, Fan J, Qian L (2019) Effects of chromium yeast, tributyrin and bile acid on growth performance, digestion and metabolism of Channa argus. Aquacult Res 50(3): 836–846. https://doi.org/10.1111/are.13954

Jankowska EA, Drozd M, Ponikowski P (2016) Current approach to heart failure. Springer, Cham, pp 535–547. https://doi.org/10.1007/978-3-319-45237-1

Kessi J, Hanselmann KW (2004) Similarities between the abiotic reduction of selenite with glutathione and the dissimilatory reaction mediated by Rhodospirillum rubrum and Escherichia coli. J Biol Chem 279(49):50662–50669. https://doi.org/10.1074/jbc.M405887200

Khan MA, Javed MM, Ahmed A, Zahoor S, Iqbal K (2020) Process optimization for the production of yeast extract using fresh Baker's yeast. Pak J Biochem Biotechnol 1(2):1–10. https://doi.org/10.52700/pjbb.v1i2.15

Kralovec JA, Potvin MA, Wright JHD, Watson LV, Ewart HS, Curtis JM, Barrow CJ (2009) Chromium(III)-docosahexaenoic acid complex: synthesis and characterization. J Funct Foods 1(3):291–297. https://doi.org/10.1016/j.jff.2009.06.001

Kushairi N, Phan CW, Sabaratnam V, Naidu M, David P (2020) Dietary amino acid ergothioneine protects HT22 hippocampal neurons against H2O2-induced neurotoxicity via antioxidative mechanism. PharmaNutrition 13(April):100214. https://doi.org/10.1016/j.phanu.2020.100214

Li Y, Wei G, Chen J (2004) Glutathione: a review on biotechnological production. Appl Microbiol Biotechnol 66(3):233–242. https://doi.org/10.1007/s00253-004-1751-y

Li L, Xu JX, Cao YJ, Lin YC, Guo WL, Liu JY, Bai WD, Zhang YY, Ni L, Liu B, Rao PF, Lv XC (2019) Preparation of Ganoderma lucidum polysaccharide-chromium (III) complex and its hypoglycemic and hypolipidemic activities in high-fat and high-fructose diet-induced pre-diabetic mice. Int J Biol Macromol 140:782–793. https://doi.org/10.1016/j.ijbiomac.2019.08.072

Liu GL, Chi Z, Wang GY, Wang ZP, Li Y, Chi ZM (2015) Yeast killer toxins, molecular mechanisms of their action and their applications. Crit Rev Biotechnol 35(2):222–234. https://doi.org/10.3109/07388551.2013.833582

Lourens-Hattingh A, Viljoen BC (2001) Growth and survival of a probiotic yeast in dairy products. Food Res Int 34(9):791–796. https://doi.org/10.1016/S0963-9969(01)00085-0

Maares M, Keil C, Pallasdies L, Schmacht M, Senz M, Nissen J, Kieserling H, Drusch S, Haase H (2022) Zinc availability from zinc-enriched yeast studied with an in vitro digestion/Caco-2 cell culture model. J Trace Elem Med Biol 71(January):126934. https://doi.org/10.1016/j.jtemb.2022.126934

McCarty MF (2000) Towards practical prevention of type 2 diabetes. Med Hypotheses 54(5): 786–793. https://doi.org/10.1054/mehy.1999.0952

Moeini MM, Bahrami A, Ghazi S, Targhibi MR (2011) The effect of different levels of organic and inorganic chromium supplementation on production performance, carcass traits and some blood parameters of broiler chicken under heat stress condition. Biol Trace Elem Res 144(1–3): 715–724. https://doi.org/10.1007/s12011-011-9116-8

Nowosad K, Sujka M (2021) The use of iron-enriched yeast for the production of flatbread. Molecules 26(17):5204. https://doi.org/10.3390/molecules26175204

Obeid R, Geisel J, Nix WA (2019) 4-Pyridoxic acid/pyridoxine ratio in patients with type 2 diabetes is related to global cardiovascular risk scores. Diagnostics 9(1):1–12. https://doi.org/10.3390/diagnostics9010028

Olvera-Novoa MA, Martínez-Palacios CA, Olivera-Castillo L (2002) Utilization of torula yeast (Candida utilis) as a protein source in diets for tilapia (Oreochromis mossambicus Peters) fry. Aquacult Nutr 8(4):257–264. https://doi.org/10.1046/j.1365-2095.2002.00215.x

Pérez-Torrado R, Gamero E, Gómez-Pastor R, Garre E, Aranda A, Matallana E (2015) Yeast biomass, is an optimised product with myriad applications in the food industry. Trends Food Sci Technol 46(2):167–175. https://doi.org/10.1016/j.tifs.2015.10.008

Puig S, Ramos-Alonso L, Romero AM, Martínez-Pastor MT (2017) The elemental role of iron in DNA synthesis and repair. Metallomics 9(11):1483–1500. https://doi.org/10.1039/c7mt00116a

Santos A, Sánchez A, Marquina D (2004) Yeasts as biological agents to control Botrytis cinerea. Microbiol Res 159(4):331–338. https://doi.org/10.1016/j.micres.2004.07.001

Shruthi B, Deepa N, Somashekaraiah R, Adithi G, Divyashree S, Sreenivasa MY (2022) Exploring biotechnological and functional characteristics of probiotic yeasts: a review. Biotechnol Rep 34 (February):e00716. https://doi.org/10.1016/j.btre.2022.e00716

Shubham K, Anukiruthika T, Dutta S, Kashyap AV, Moses JA, Anandharamakrishnan C (2020) Iron deficiency anemia: a comprehensive review on iron absorption, bioavailability and emerging food fortification approaches. Trends Food Sci Technol 99(January):58–75. https://doi.org/10.1016/j.tifs.2020.02.021

Šillerová S, Lavová B, Urminská D, Poláková A, Vollmannová A, Harangozo Ľ (2012) Copper enriched yeast saccharomyces cerevisiae as a potential supplement in nutrition. J Microbiol Biotechnol Food Sci 1:696–702

Sundaram B, Aggarwal A, Sandhir R (2013) Chromium picolinate attenuates hyperglycemia-induced oxidative stress in streptozotocin-induced diabetic rats. J Trace Elem Med Biol 27(2): 117–121. https://doi.org/10.1016/j.jtemb.2012.09.002

Valko M, Jomova K, Rhodes CJ, Kuča K, Musílek K (2016) Redox- and non-redox-metal-induced formation of free radicals and their role in human disease. Arch Toxicol 90(1):1–37. https://doi.org/10.1007/s00204-015-1579-5

Via M, Scurlock C, Raikhelkar J, Di Luozzo G, Mechanick JI (2008) Chromium infusion reverses extreme insulin resistance in a cardiothoracic ICU patient. Nutr Clin Pract 23(3):325–328. https://doi.org/10.1177/0884533608318676

Wang W, Zhao J, Gui W, Sun D, Dai H, Xiao L, Chu H, Du F, Zhu Q, Schnabl B, Huang K, Yang L, Hou X (2018) Tauroursodeoxycholic acid inhibits intestinal inflammation and barrier disruption in mice with non-alcoholic fatty liver disease. Br J Pharmacol 175(3):469–484. https://doi.org/10.1111/bph.14095

Wang S, Tonnis BD, Wang ML, Zhang S, Adhikari K (2019) Investigation of monosodium glutamate alternatives for content of umami substances and their enhancement effects in chicken soup compared to monosodium glutamate. J Food Sci 84(11):3275–3283. https://doi.org/10.1111/1750-3841.14834

Wolf G (2010) Retinoic acid activation of peroxisome proliferation-activated receptor d represses obesity and insulin resistance. Nutr Rev 68(1):67–70. https://doi.org/10.1111/j.1753-4887.2009.00261.x

Yadav A, Kon K, Kratosova G, Duran N, Ingle AP, Rai M (2015) Fungi as an efficient mycosystem for the synthesis of metal nanoparticles: progress and key aspects of research. Biotechnol Lett 37(11):2099–2120. https://doi.org/10.1007/s10529-015-1901-6

Zaky AS, Tucker GA, Daw ZY, Du C (2014) Marine yeast isolation and industrial application. FEMS Yeast Res 14(6):813–825. https://doi.org/10.1111/1567-1364.12158

Zoidis E, Seremelis I, Kontopoulos N, Danezis GP (2018) Selenium-dependent antioxidant enzymes: actions and properties of selenoproteins. Antioxidants 7(5):1–26. https://doi.org/10.3390/antiox7050066

Photoautotrophic Microbes with Potential for a Super Health Food on This Planet

32

Amrik Singh Ahluwalia, Kawalpreet Kaur Bhatia, Divjot Kour, Kanwaljit Kaur Ahluwalia, Ajar Nath Yadav, and M. C. Sidhu

Abstract

Photoautotrophic microbes are distributed globally and have been here since long (3.5 billion years). These are among the early organisms which included cyanobacteria and various types of algae. These organisms have been found to be rich in important components of food and nutraceuticals which are being probed further and used globally, in addition, for many new applications for human beings. These vary from prokaryotic to eukaryotic in their cell structure and may or may not have chloroplasts and mitochondria enclosed in a membrane in all of these organisms. These are rich in protein, vitamins, polyunsaturated fatty acids, iron, beta-carotene, and antioxidants due to which the World Health Organisation labelled them as the super health food on this planet. These have been found to be hepatoprotectant and anticancerous too. Apart from this, their role from quenching the carbon dioxide respired out to providing oxygen for our survival will also be discussed briefly.

A. S. Ahluwalia (✉)
Department of Botany, Eternal University, Baru Sahib, Himachal Pradesh, India
e-mail: pvc@eternaluniversity.edu.in

K. K. Bhatia
Department of Botany, SGGSC, Sector 26, Chandigarh, India

D. Kour
Department of Microbiology, Eternal University, Baru Sahib, Himachal Pradesh, India

K. K. Ahluwalia
Department of Zoology, Eternal University, Baru Sahib, Himachal Pradesh, India

A. N. Yadav
Department of Biotechnology, Eternal University, Baru Sahib, Himachal Pradesh, India

M. C. Sidhu
Department of Botany, Panjab University, Chandigarh, India

R. Sobti et al. (eds.), *Role of Microbes in Sustainable Development*,
https://doi.org/10.1007/978-981-99-3126-2_32

Keywords

Algae · Cyanobacteria · Food · Microbes · Nutraceutical · Photoautotrophs

32.1 Introduction

The blue-green algae (Cyanobacteria) are pioneer in the evolution of eukaryotes during the Cretaceous period. Though these evolved to different groups of organisms, still these maintained their own presence to date too. Algae and plants are broadly similar in producing similar storage products but differ in the level of their differentiation (Kaur et al. 2021). Algae lack typical root, stem, leaves, flowers, and fruits but instead some brown algae have holdfast, stipe and blade-like structures as the advanced ones as these show some advancement over more primitive parts.

To combat food scarcity and to ensure a nutritious food supply for a sustainable living of an increasing population, microalgae provide adequate alternative (Majit and Dua 1992; Mandotra et al. 2020; Ścieszka and Klewicka 2019). They can lay aside the vast quantities of food reserves on which the animal world is entirely and irrevocably dependent. Algae are expected to initiate all animal food chains in the sea and freshwater. Further, these organisms act as the important nutraceuticals, pharmaceuticals, and cosmeceuticals and also provide protection from harmful wavelengths of radiations.

Algae alone produce 50% of the oxygen for the sustenance of different life forms including the mankind. Attempt has been made to highlight the role of algal species to achieve some of the sustainable development goals like SDG-2 (Zero Hunger), SDG-6 (clean water and sanitation), SDG-7 (affordable and clean energy), SDG-12 (responsible consumption and production), SDG-14 (life below water), and SDG-15 (life on land) (Mandotra et al. 2020; Pal and Bose 2022).

32.2 As Source of Food, Nutraceuticals, Pharmaceuticals, and Cosmeceuticals

These microbial organisms are diversified organisms which are active photosynthetically. These play an important role in the food industry as food supplements in addition to functional food. During last few years, the range of such species as food supplements has increased significantly and their applications are becoming more diverse. Their judicious use can contribute at large to the sustainable development goals which state that by 2030, hunger must be taken care of sustainable development goals (SDG-2) and ensuring its access to such people, particularly the poor and people in vulnerable situations, including infants, for safe, nutritious and sufficient food throughout the year (Amita 2010). The most commonly consumed macro-algae include the red algae *Porphyra yezoensis, Gracilaria* sp., *Chondrus crispus,* and *Palmaria palmata* (dulse), the brown algae *Laminaria, Undaria,* and *Macrocystis* sp., and the green algae *Caulerpa racemosa,* species of *Codium* and *Ulva*

(Borowitzka 1998). However, microalgae are richer source of proteins, lipids, fatty acids, vitamins, and minerals (Bhatia et al. 2014, 2016). In addition, many PUFA (polyunsaturated fatty acid) like DHA (docosahexaenoic acid), EPA (eicosapentaenoic acid), linolenic acid, and linoleic acid are highly valuable in human diet (Swanson et al. 2012; Torres-Tiji et al. 2020). Algae like *Spirulina maxima, Spirulina platensis, Chlorella vulgaris, Scenedesmus* sp., and *Porphyra yezoensis* are rich source of protein (Araújo and Peteiro 2021; Bhatia et al. 2014; Michalak et al. 2020). In addition, these provide biocolors and their food colouring products are extensively used in food industry. For example, β-carotene from *Dunaliella* sp. is used to enhance colour of flesh of salmon fish (Borowitzka 2013; Pulz and Gross 2004) reported the use of astaxanthin from *Haematococcus* sp. to enhance the colour of fish muscle and brighten shell colour as well. Recently, Reliance company has got a patent for faster growth of such useful species. Spirulina market size was worth US$410.2m in 2021 and it is pegged to be US$ 989.6m by 2028. North America has been the largest consumer of *Spirulina* products. It is projected as an immune booster and antiviral. Hence, during Covid outbreak, there was an increased demand of *Spirulina* as a protein source with anti-oxidising and antiviral properties. Even Spirulina beverage market size has been at US$15.5m in 2019 and increased during Corona to US$17.25m in 2020.

Microalgae are in the vanguard of emerging nutritional supplements which have myriad of therapeutic benefits due to its rich macro- and micronutrient diversity. These are continuously been acknowledged as quality dietary supplement and nutraceuticals. *Spirulina (Arthrospira platensis* and *Arthrospira maxima*) and *Chlorella* powder are considered safe for human consumption. Dried *Spirulina* biomass contains balanced proteins with essential amino acids in balanced proportions and also rich in antioxidant pigments phycocyanin, phycoerythrin and allophycocyanin. According to the available literature, anti-inflammatory, antioxidant and membrane-stabilizing properties of *Spirulina* are believed to be effective to treat diabetes, hypertension, inflammation, liver diseases, allergic rhinitis (Bhatia et al. 2016) an also act as antiviral and immunostimulatory (Tiwari et al. 2022). *Chlorella* improves total cholesterol, diastolic and systolic blood pressure and blood glucose (Fallah et al. 2018). In addition, *Chlorella* supplement can decrease wound healing time and increase immunity (Merchant and Andre 2001) and act as antioxidant and hepatoprotective (Mobin and Alam 2017). Thus, microalgae are an alternative food and nutraceutical source to brace SDG-3 and its sustainable cultivation practices assist SDG-12, SDG-14 and SDG-15 (Fig. 32.1).

32.3 Cost-Effective Cultivation

The cost of large-scale production is one of the barriers to popularise these organisms among masses. Increased interest in last few decades to use microalgae as food has resulted in major research for increased cultivation of microalgae. However, the cost of cultivation is directly proportional to the final product being produced (Hoffman et al. 2017). Food, feed and supplements from microalgae

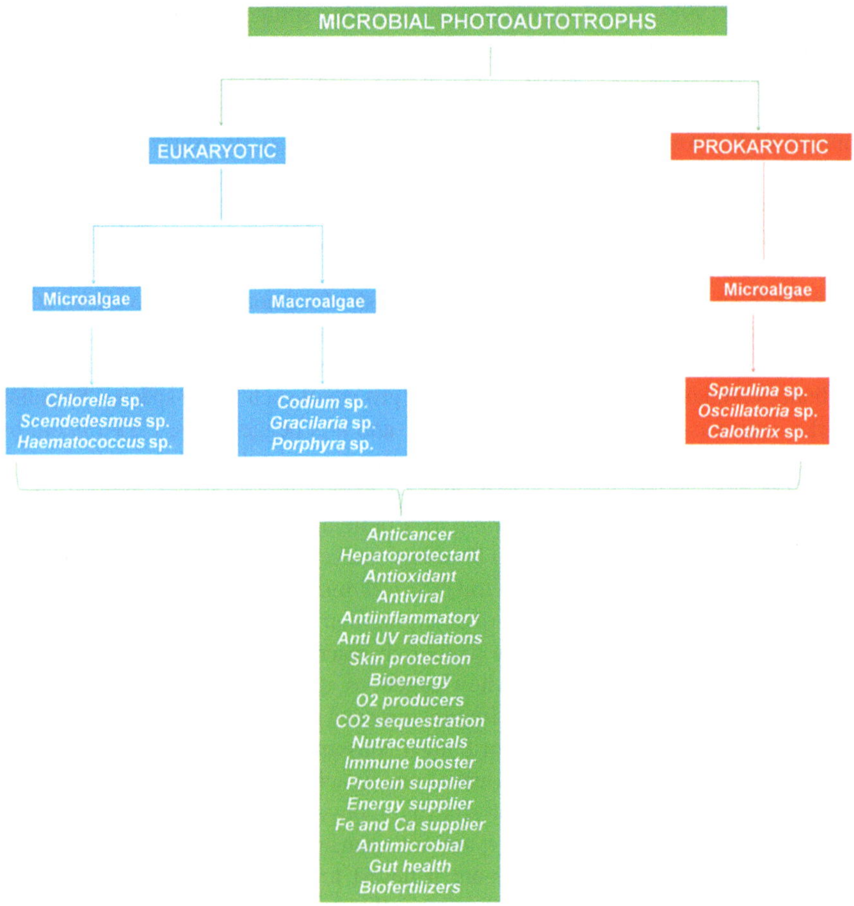

Fig. 32.1 The cosmeceutical, neutraceutical, and pharmaceutical properties of microbial photoautotrophs

require low cost methods for producing biomass (Ray et al. 2019) while its pharmaceutical use requires photobioreactors for culturing microalgae for quality and purity. Several steps have been taken to reduce cost per yield by maximizing nutrient load, carbon dioxide augmentation, retention time, mixing, and culture recycling. Efforts regarding production, optimization and developing novel strains will improve cost effectiveness.

32.4 Harvesting

Harvesting cost is a major challenge faced by microalgal growers. Many factors contribute in high harvesting costs such as low cell concentrations, small cell size, culture density, and buoyancy (Montemezzani et al. 2017). Various methods of harvesting are divided into two parts depending on its cost (1) low cost: gravity harvesting, dissolved air floatation, and chemical flocculation, etc. and (2) high cost: membrane filtration and centrifugation. New technologies and more research in effective advances would greatly impact the microalgal production at the commercial level.

32.5 Processing

The cost of downstream processing technique depends upon the specificity of microalgae product, e.g., nutraceutical requires extraction and purification while biogas production depends on simple anaerobic digestion of biomass for processing. The availability of many microalgae products in the market proves that downstream processing can be cost effective. However, industrial level extraction and purification is still crucial in sustainable bio-economy of microalgae (Khanra et al. 2022).

Continuous improvement and research is required to manage cost effectiveness for establishing successful microalgae industry. Effective steps are needed to be taken in order to make it affordable and accessible to all the nations. Research capital has to be invested in making this rapid development in microalgal technology searching new effective strains of useful organisms and involving a variety of innovative measures.

32.6 Cultivation of Photoautotrophs (Algae and Cyanobacteria) Involves Indoor and Outdoor Culture Facilities

In outdoor cultivation, energy requirement is lower as solar light provides enough energy for photosynthesis and production of useful biomolecules for human consumption. This also adds value for our climate stability as these organisms take up carbon dioxide (fix carbon) biologically to produce renewable feedstock and reducing the foot print of carbon dioxide during the process (Singh and Ahluwalia 2013). In addition, these provide oxygen in plenty without which human beings cannot survive. *Spirulina* species have been reported to have 20 times more protein/hectare than any other crop, require lesser water than many crops, recycle about 6.4 tons of carbon dioxide/hectare, and produce nearly 16.8 tons of oxygen. Some algae like *Nannochloris* sp. showed a high photosynthetic rate with increasing height and temperatures (up to 45 °C) and achieved an outdoor biomass productivity of 43 g/m^2/d in summer (Paul et al. 2022). The application of red light/far red light cycle in green algae, an enhanced productivity especially in *Haematococcus* sp., has been achieved. These algae have a great potential for commercial cultivation to produce food ingredients, nutraceuticals, and vegetable protein.

Cyanobacteria differentiate specialized cells called heterocysts which are the major sites of nitrogen fixation to provide bio-fertilizer to crops therapy reducing the use of chemical fertilizers (Ahluwalia 2004). This in turn provides healthy vegetables and grains for human consumption, in addition to proteins, useful fatty acids, and antioxidants.

32.7 Protection from UV Radiations

Some Cyanobacteria (*Scytonema* sp., *Microcoleus vaginatus,* etc.) has the ability to produce compounds like Scytonemin and mycosporine-like amino acids (MAA) that absorbed strong sunlight in ultraviolet range (370–384 nm) (Fuentes-Tristan et al. 2019). Hence, these are utilized to protect human skin from strong sunlight and ultraviolet radiations. Among the three spectral forms of UV like UV-A, UV-B and UV-C, the most harmful response is from UV-B radiations. UV-C is absorbed by Ozone, and hence, its negative effect is mitigated. Therefore, the effect of UV-B radiation is taken care of by algal sunscreens through MAA absorbing between 310 and 360 nm (La Barre et al. 2014). Thus, MAAs can protect the human skin from ill effects of UV radiations. Members of Cyanobacteria are used in the preparation of many products meant for the protection of humans from such radiations (Fuentes-Tristan et al. 2019).

32.8 Cyanotoxins

These toxins are produced by some Cyanobacteria and broadly are of two types like neurotoxins and hepatotoxins (Svrcek and Smith 2004). Neurotoxins comprise alkaloids of low molecular weight which can block transmission of signal from neuron to neuron and neuron to muscle in man and animals. It can result in staggering, muscle twitching, gasping, etc. and prove fatal due to respiratory arrest at high concentrations. Cyanotoxins can be anatoxins and hepatotoxins. Cyanobacteria produce anatoxins in *Anabaena* sp., *Oscillatoria* sp., and other species. Hepatotoxins are terpenoids in nature and usually cause bleeding in the liver by inhibiting protein phosphatase 1 and 2A (Cross 1997). These can cause vomiting, diarrhoea, and weakness. These further can be either microcystins and nodularins named after the species which produce them. Bloom formation of toxin producing algae and Cyanobacteria can be harmful to humans. Poisoning may result rarely with cyanotoxins. Of course presence of such organisms in water bodies produce a musty/muddy odour. These compounds can be significant in food, beverage, and perfume industries.

32.9 Source of Healthy Food and Fodder

Keeping in the view the population increase and the fear of climate change, scientists have been on the look out to find alternative sources of nutritious food. Hence, the use of algae as promising nutritious organisms appeared on the scene. A vast spectrum of distribution of algae and cyanobacteria exist globally and researchers are optimistic that some of these organisms will provide enough biomass to feed the growing population. San Francisco-based start-up is making bacon from the seaweeds. Algae may help fill the nutritional gaps in future food markets. Unique feature of these organisms is that they can grow using sun light in marine water as also in freshwater or terrestrial habitats and hence do not compete with traditional crops for land. Therefore, market scenario for use of seaweeds as food products is rosy and positive. It is estimated that in 2020, the macro-algal seaweed market was worth 40 billion dollars which is likely to be more than double in next 5–7 years. In addition, there are microalgae too which form the major part of primary producers to carry on the food chain for human food. Most of the contribution for food comes from members of brown algae, red algae, and green algae; important food supplements, however, are contributed by Cyanobacterial species.

32.10 Nutraceuticals

Spirulina is one of the Cyanobacteria which has been well recognized globally as an emerging source of nutritional supplement with a diversity of macro-nutrients and micro-nutrients. This organism has a rich therapeutic chemical profile. It is important even in metabolic disorders. The role as hepato-protectant has been known. Hence, *Spirulina* is useful in liver disorders and hepatic toxicity. Even hypolipidemic and hypocholesterolemic nature of this organism plays a significant role in the mitigation of non-alcoholic fatty liver disease. It acts by lowering the raised activity of liver enzymes. It can be useful as anti-inflammatory and antioxidant and has membrane stabilizing properties. *Spirulina* intake has been safer in different forms, i.e., capsules, powder, tablets, pastries, *Spirulina*-filled chocolate blocks, etc. It has been without any adverse effect; though in very few cases, temporary giddiness has been indicated. Experiments for its role in cancer mitigation and radiation protection have indicated effective positive results. In China, *Spirulina* industry is supported by the State Science and Technology Commission as a natural strategic program.

It is available as also many other algal products at health food stores in the developed countries. *Nostoc* sp. is known as 'hair vegetable' due to its hair-like appearance (Su et al. 2005). Chinese add spiritual value to this food. It has been in use there as food for the last 2000 years. In Japan also, Cyanobacterial dishes have been known since ancient times which mainly comprised of *Aphanothece sacrum, Nostoc verrucosum,* and *N. commune*. In Mexico City, such food has been reported since 1521. They were collecting it from lakes and selling in the form of small cakes, which curdles and make bread with a flavour like cheese. However, another cyanobacterium *Aphanizomenon flos-aquae* has been harvested from lakes in California, USA, for the last about 40 years and sold as food and as health supplement (Carmichael et al. 2000).

Algal blooms refer to excessive growth of these organisms which affect the quality of drinking water, fish stocks, killing due to species of *Chaetoceros, Chrysochromulina, Spirogyra,* etc. clogging the gills of fish and oxygen deprivation towards end of the bloom season. Cyanobacterial blooms also result in the production of phyco-toxins which kill most of their predators. Filter feeding fish get more of toxins accumulated and may harm humans and animals on their consumption. Neurotoxins can be anatoxin and saxitoxin which can block the communication signal among neurons. Hepatotoxins are microcystin and nodularin which also inhibit phosphatase 1 and phosphatase 2A. Fish poisoning has been known from toxins produced by various algal groups, i.e., dinoflagellates, Bacillarioplyceae, Raphidophytes, Prymnesiophyceae, etc. They produce a variety of toxins like dinophysistoxin, ciguatoxin, brevetoxins, saxitoxins, pteriatoxin, neurotoxin, yessotoxin, etc.

32.11 Red Algae

These algae produce polysaccharides like agar and carrageenan (Haslin et al. 2001). Commercially species of *Gelidium, Pterocladia, Gracilaria,* etc. are employed for agar extraction through a defined process. This is used in food preparation, technology, and pharmaceutical industry. It is important in canning of fish and meat, manufacture of processed cheese, puddings, creams, and jellies. Some species are used as laxative, stabilizer for emulsions, and cosmetic preparations for skin ointments. Carrageenans inhibit replication and reverse transcriptase in human immunodeficiency virus (HIV) (Bourgougnon et al. 1996). Carrageenan-based vaginal microbicide called 'Carraguard' has been reported to block HIV and other sexually transmitted diseases. 'Carraguard' has been undergoing clinical trials in South Africa (Burges Watson and Stratford 2008).

Acknowledgements The authors thank the University/College Authorities for the infrastructural support during the preparation of this chapter.

References

Ahluwalia A (2004) Cyanobacterial differentiation. Pb Univ Res J (Sci) 54:7–11

Amita M (2010) Effect of processing on functional properties of spirulina protein preparations. Afr J Microbiol Res 4:055–060

Araújo R, Peteiro C (2021) Algae as food and food supplements in Europe. Publications Office of the European Union, Luxembourg

Bhatia KK, Puri S, Ahluwalia A (2014) Efficacy and safety of spirulina in biomedical field: evidence based critical appraisal. Vegeto Int J Plant Res 27(3):104–116

Bhatia KK, Puri S, Kaur R, Ahluwalia KK, Ahluwalia AS (2016) Efficacy of spirulina as hepatoprotectant: a review. Vegetos 29:129–136

Borowitzka MA (1998) Algae as food. In: Microbiology of fermented foods. Springer, Berlin, pp 585–602

Borowitzka MA (2013) High-value products from microalgae—their development and commercialisation. J Appl Phycol 25:743–756. https://doi.org/10.1007/s10811-013-9983-9

Bourgougnon N, Lahaye M, Quemener B, Chermann J, Rimbert M, Cormaci M, Furnari G, Kornprobst J (1996) Annual variation in composition andin vitro anti-HIV-1 activity of the sulfated glucuronogalactan from *Schizymenia dubyi* (Rhodophyta, Gigartinales). J Appl Phycol 8:155–161

Burges Watson D, Stratford E (2008) Feminizing risk at a distance: critical observations on the constitution of a preventive technology for HIV/AIDS. Soc Cult Geogr 9:353–371

Carmichael WW, Drapeau C, Anderson DM (2000) Harvesting of Aphanizomenon flos-aquae Ralfs ex born. & Flah. Var. flos-aquae (cyanobacteria) from Klamath Lake for human dietary use. J Appl Phycol 12:585–595

Cross DM (1997) Analytical methods for cyanobacterial toxins. University of Bath, Bath

Fallah AA, Sarmast E, Dehkordi SH, Engardeh J, Mahmoodnia L, Khaledifar A, Jafari T (2018) Effect of *Chlorella* supplementation on cardiovascular risk factors: a meta-analysis of randomized controlled trials. Clin Nutr 37:1892–1901

Fuentes-Tristan S, Parra-Saldivar R, Iqbal HM, Carrillo-Nieves D (2019) Bioinspired biomolecules: Mycosporine-like amino acids and scytonemin from Lyngbya sp. with UV-protection potentialities. J Photochem Photobiol B Biol 201:111684

Haslin C, Lahaye M, Pellegrini M, Chermann J-C (2001) In vitro anti-HIV activity of sulfated cell-wall polysaccharides from gametic, carposporic and tetrasporic stages of the Mediterranean red alga *Asparagopsis armata*. Planta Med 67:301–305

Hoffman J, Pate RC, Drennen T, Quinn JC (2017) Techno-economic assessment of open microalgae production systems. Algal Res 23:51–57

Kaur S, Srivastava A, Ahluwalia AS, Mishra Y (2021) Cyanobacterial blooms and cyanotoxins: occurrence and detection. In: Mandotra SK, Upadhyay AK, Ahluwalia AS (eds) Algae: multifarious applications for a sustainable world. Springer Singapore, Singapore, pp 339–352. https://doi.org/10.1007/978-981-15-7518-1_15

Khanra A, Vasistha S, Rai MP, Cheah WY, Khoo KS, Chew KW, Chuah LF, Show PL (2022) Green bioprocessing and applications of microalgae-derived biopolymers as a renewable feedstock: circular bioeconomy approach. Environ Technol Innov 28:102872

La Barre S, Roullier C, Boustie J (2014) Mycosporine-like amino acids (MAAs) in biological photosystems. In: Outstanding marine molecules, pp 333–360

Majit K, Dua S (1992) Ahluwalia a biochemical studies on spirulina proteins. In: *Spirulina* ETTA Nat symp MCRC, pp 78–84

Mandotra SK, Upadhyay AK, Ahluwalia AS (2020) Algae: multifarious applications for a sustainable world. Springer Nature, Berlin

Merchant RE, Andre CA (2001) A review of recent clinical trials of the nutritional supplement *Chlorella pyrenoidosa* in the treatment of fibromyalgia, hypertension, and ulcerative colitis. Altern Ther Health Med 7:79–92

Michalak I, Dmytryk A, Chojnacka K (2020) Algae cosmetics. In: Encyclopedia of marine biotechnology, vol 1. Wiley, Hoboken, NJ, pp 65–85

Mobin S, Alam F (2017) Some promising microalgal species for commercial applications: a review. Energy Procedia 110:510–517

Montemezzani V, Duggan IC, Hogg ID, Craggs RJ (2017) Screening of potential zooplankton control technologies for wastewater treatment high rate algal ponds. Algal Res 22:1–13

Pal I, Bose C (2022) *Spirulina*—a marine miracle for sustainable food system. In: Marine biology research, pp 1–17

Paul K, Gaikwad M, Choudhary P, Mohan N, Pai P, Patil SD, Pawar Y, Chawande A, Banerjee A, Nagle V (2022) Year round sustainable biomass production potential of Nannochloris sp. in

outdoor raceway pond enabled through strategic photobiological screening. Photosynth Res 154:303–328
Pulz O, Gross W (2004) Valuable products from biotechnology of microalgae. Appl Microbiol Biotechnol 65:635–648
Ray M, Kumar N, Kumar V, Negi S, Banerjee C (2019) Microalgae: a way forward approach towards wastewater treatment and bio-fuel production. In: Applied microbiology and bioengineering. Elsevier, Amsterdam, pp 229–243
Ścieszka S, Klewicka E (2019) Algae in food: a general review. Crit Rev Food Sci Nutr 59:3538–3547
Singh UB, Ahluwalia AS (2013) Microalgae: a promising tool for carbon sequestration. Mitig Adapt Strat Glob Chang 18:73–95. https://doi.org/10.1007/s11027-012-9393-3
Su J, Jia S, Qiao C, Kim J-G, Hong W-H, Cho K-A, Choi D (2005) Cultivation of Nostoc flagelliforme on solid medium. Kor J Environ Biol 23:135–140
Svrcek C, Smith DW (2004) Cyanobacteria toxins and the current state of knowledge on water treatment options: a review. J Environ Eng Sci 3:155–185
Swanson D, Block R, Mousa SA (2012) Omega-3 fatty acids EPA and DHA: health benefits throughout life. Adv Nutr 3:1–7
Tiwari O, Ghosh D, Debnath S, Sahu M, Vanitha K (2022) Multifaceted utilization of microalgal biomass towards industrial applications. In: Biotechnology for waste biomass utilization. Apple Academic Press, Palm Bay, FL, pp 79–127
Torres-Tiji Y, Fields FJ, Mayfield SP (2020) Microalgae as a future food source. Biotechnol Adv 41:107536

Autopsy and COVID-19

33

Masatoshi Watanabe, Eri Usugi, Miki Usui, Akinobu Hayashi, Yoshifumi Hirokawa, and R. C. Sobti

Abstract

The current coronavirus disease 2019 (COVID-19) outbreak caused by severe acute respiratory syndrome (SARS)-coronavirus-2 (CoV-2) has caused a global economic and healthcare crisis. The routes of transmission, signs and symptoms, incubation period, pathogenesis, and pathophysiology of the disease have been studied extensively. In such situations, autopsies remain important as they provide insight into the possible mechanisms of SARS-CoV-2 activity and the extent of organ involvement during infection. In this chapter, we describe the

M. Watanabe (✉)
Department of Oncologic Pathology, Mie University Graduate School of Medicine, Tsu, Mie, Japan

Bio-Bank Center, Mie University Hospital, Tsu, Japan

Department of Diagnostic Pathology, Mie University Hospital, Tsu, Japan
e-mail: mawata@med.mie-u.ac.jp

E. Usugi
Bio-Bank Center, Mie University Hospital, Tsu, Japan

M. Usui
Department of Diagnostic Pathology, Mie University Hospital, Tsu, Japan

A. Hayashi
Department of Oncologic Pathology, Mie University Graduate School of Medicine, Tsu, Mie, Japan

Department of Diagnostic Pathology, Mie University Hospital, Tsu, Japan

Y. Hirokawa
Department of Oncologic Pathology, Mie University Graduate School of Medicine, Tsu, Mie, Japan

R. C. Sobti
Panjab University, Chandigarh, Chandigarh, India
e-mail: rcsobti@pu.ac.in

R. Sobti et al. (eds.), *Role of Microbes in Sustainable Development*,
https://doi.org/10.1007/978-981-99-3126-2_33

relationship between autopsy and COVID-19, autopsy findings in patients with COVID-19, the relationship between vaccinations and deaths, and biosafety.

Keywords

COVID-19 · SARS-Co-2 · Autopsy · Vaccine

33.1 Introduction

The current coronavirus disease 2019 (COVID-19) outbreak is caused by severe acute respiratory syndrome (SARS)-coronavirus-2 (CoV-2) and has caused a global health and economic crisis. SARS-CoV-2 is a new member of the betacoronavirus genus of the family Coronaviridae that emerged in Wuhan, China, at the end of 2019 and caused the current SARS-CoV-2 outbreak (Hu et al. 2021). This outbreak has lasted for longer than 2 years despite personal preventive behavior, public health measures, and vaccines, and this virus will likely establish a niche in humans and coexist with us.

The routes of transmission, signs and symptoms, incubation period, pathogenesis, and pathophysiology of the disease have been studied extensively (Hu et al. 2021). However, performing autopsies to provide insight into the possible mechanisms of SARS-CoV-2 activity and the extent of organ involvement targeted by this virus remains crucial (Sperhake 2020; Khatam-Lashgari et al. 2021; Bryce et al. 2021).

In the current review, we describe the relationship between autopsy and COVID-19, as well as that between COVID-19 vaccination and death.

33.2 Autopsy

There are two main types of autopsies, namely clinical and forensic. In this chapter, we focus on clinical autopsies, especially hospital-based autopsies performed by pathologists. Autopsy (derived from Greek autopsia) means "to see for oneself" and is synonymous with the terms postmortem, postmortem examination, and necropsy (Burton 2001; Hanzlick and Hutchins 1999). Its long history began with mummification and human dissection in 3000 BC and persisted in ancient Greece, where Hirophilus performed a live dissection of a human being, discovering the duodenum. This led to Rokitansky's work on gross pathology and Virchow's systematic method of autopsy based on cellular pathology, leading to autopsy of modern pathology based on immunology and molecular biology (Van den Tweel and Taylor 2013). Autopsies are important and necessary to define the exact cause of death, offering useful clinical and epidemiologic information, as well as pathophysiological findings to further provide diagnostic and therapeutic tools. Pathology departments with high-level pathological activities are also expected to feature not only a center of provision of samples such as biopsies, surgical parts, or autopsies, but also tissue processing centers and histochemical, immunohistochemical, and molecular techniques (da Silva et al. 2016). Autopsies yield advantages besides direct clinical or administrative advantages, such as educational and epidemiological value. Autopsy is the gold standard for quality control of clinical management; thus, the number of significant diagnostic discrepancies can be reduced, and patient care can be optimized (Marshall and Milikowski 2017; Kurz et al. 2021).

However, the global clinical autopsy rate has declined for several decades. The statistics are unclear and not based on national statistics in any country. Autopsy was performed in 40–60% of all hospital deaths in the United States before 1970 (Shojania and Burton 2008; Hoyert 2011). Currently, this rate has dropped to <5% (Shojania and Burton 2008; Hoyert 2011). In Japan, the rate was 54.1% in 1958, below 10% in 2004 and 2.8% in 2018 (Irie et al. 2021). Another group reported that the mean autopsy rate of all university hospitals in Japan was maintained at 45% until 1983, then began declining steadily and reached 6.8% in 2021 (Yang et al. 2016). A 30% decline in the overall number of autopsies performed between 2005 and 2014 was observed in Germany. The reduction in autopsy numbers is not only a recent phenomenon observed in 1983. This decline appears to be attributable to several factors such as costs, cultural reasons, and diagnostic progress (Waidhauser et al. 2021). However, a significant increase in autopsy rates can be achieved through intensive and structured cooperation between clinicians and pathologists and increasing physicians' motivation to obtain consent to perform an autopsy (Waidhauser et al. 2021). A meta-analysis of 5863 hospital autopsies revealed a 5.5–100% prevalence of misdiagnoses, especially for diseases such as myocardial infarction, pulmonary embolism, and pneumonia (Winters et al. 2012). A recent comparative retrospective analysis of 1112 cases revealed that 73.9% of patients had no discrepancies between autopsy and clinical diagnosis (Kurz et al. 2021). Age, cardiovascular diseases, and duration of hospital stay also significantly affected discrepancies in ante- and post-mortem diagnoses. Thus, autopsies play an important role in assessing discrepant diagnoses.

33.3 COVID-19, Biosafety and Autopsy

Coronaviruses are considered inconsequential pathogens that cause the common cold in humans. However, in the twenty-first century, two highly infectious coronaviruses appeared: SARS-CoV, which was first reported in November 2002 in Guangdong, China, and resulted in 8098 laboratory-confirmed cases with a global case fatality rate of 9.6%. SARS-CoV is highly contagious and primarily transmitted via respiratory droplets, leading to the highest transmission rates of SARS occurring in healthcare facilities (Mann et al. 2020). Infected patients present with myalgia, malaise, fever, chills, cough, dyspnea, and respiratory distress. The routes of transmission include respiratory droplets, fomities, and fecal-oral routes. The other is Middle East respiratory coronavirus (MERS-CoV), which was first reported in Saudi Arabia in 2012 with 2521 laboratory-confirmed cases and a fatality rate of 36%. Infected patients presented with fever, cough, chills, sore throat, myalgia, arthralgia, dyspnea, pneumonia, and acute renal failure. The route of transmission is through respiratory droplets and fomites (Mann et al. 2020). SARS-CoV-2 is the third most recently discovered coronavirus disease (COVID-19). In 2019, SARS-CoV-2 was detected in sewage samples as early as March and November in Europe and South America, respectively. However, the index case is generally attributed to China. A mysterious pneumonia outbreak occurred that was characterized by fever, dry cough, fatigue, and occasional gastrointestinal symptoms. On March 11, 2020, the World Health Organization (WHO) declared the novel coronavirus disease (COVID-19) outbreak a global pandemic (Khatam-Lashgari et al. 2021; Mann et al. 2020; Kumar et al. 2021). The number of COVID-19 cases continues to increase worldwide, with more than 70 million cases and 1.6 million deaths worldwide as of December 15, 2020. A systemic review reported that fever, cough, loss of appetite, shortness of breath, loss of taste, and sputum production are common symptoms. Gastrointestinal symptoms, such as diarrhea, nausea/vomiting, and abdominal pain, have also been reported in patients. ARDS, acute respiratory failure, acute cardiac injury, and acute renal failure are common complications in patients. SARS-CoV-2 is thought to spread through respiratory droplets, unprotected direct contact with patients, and contact with contaminated objects. Medical professionals are at a particularly high risk of exposure as they directly handle contagious materials from patients with COVID-19. This risk is elevated due to the varying disease severity in patients with COVID-19 (Khatam-Lashgari et al. 2021; Kumar et al. 2021).

The risk of infectious disease transmission has been recognized by pathologists, clinicians, and other individuals in close proximity to autopsies. Other hazards include toxic chemicals such as formalin and radiation from radionucleotides used for therapy and diagnosis. Pathogens may be transmitted during autopsy after direct cutaneous inoculation, contact with droplets, or aerosol exposure (Nolte et al. 2002). For example, pathologists have died of streptococcal sepsis after sustaining minor cutaneous injuries during autopsies in individuals with the same disease. Pathologists are also recognized as a high-risk group for occupationally acquired hepatitis B virus infection because of their exposure to blood. Mycobacterium tuberculosis is transmitted through autopsy-generated aerosols. Autopsy-transmitted tuberculosis has broken out in medical facilities such as the Syracuse Medical Examiner's Office, Los Angeles Coroner's Office, and University of Arkansas School of Medicine.

Biological safety levels have been established in both biomedical and microbiological laboratories (NIH Guidelines n.d.). Biological agents are classified as RG1-RG4 (Loibner et al. 2021). As risk group classifications vary between countries, Table 33.1 shows the risk groups defined by the NIH guidelines and WHO (NIH Guidelines n.d.; WHO 2004). Biosafety levels (BSLs) prescribe procedures and containment levels for particular microorganisms or material (including research involving recombinant or synthetic nucleic acid molecules), are classified as 1–4, and are established by the CDC/NIH. Agents with increased relative risk require higher degrees of containment when they are worked with. Laboratories and other workplaces are also characterized by BSLs at each level, affording a greater degree of containment. BSLs do not always correspond to RGs. A

Table 33.1 Basic classification of risk groups

RG	NIH Guideline (April 2019)	WHO Laboratory Biosafety Manual 3rd
RG1	Agents not associated with disease in healthy adult humans	(No or low individual and community risks). A microorganism unlikely to cause human or animal disease
RG2	Agents associated with disease that is rarely serious in humans and for which preventative or therapeutic interventions are often available	(Moderate individual risk; low community risk). A pathogen that can cause human or animal disease but is unlikely to be a serious hazard to laboratory workers, the community, livestock, or the environment. Laboratory exposures may cause serious infection, but effective treatment and preventive measures are available and the risk of spread of infection is limited
RG3	Agents associated with seriousness or lethal human disease for which preventive or therapeutic interventions may be available (high individual risk, but low community risk)	(High individual risk; low community risk). A pathogen that usually causes serious human or animal disease but does not ordinarily spread from one infected individual risk to another). Effective treatment and preventive measures are available
RG4	Agents are likely to cause serious or lethal human disease for which preventive or therapeutic interventions are not usually available (high individual and community risk)	(High individual and community risk). A pathogen that usually causes serious human or animal disease and can be readily transmitted from one individual to another, directly or indirectly. Effective treatment and preventive measures are not usually available

RG risk group

risk assessment determines the degree of correlation between the agent's risk group classification and BSL (NIH Guidelines n.d.).

Strong suggestions have been made regarding the biosafety of autopsies under such circumstances, e.g., that BSL 3 should be considered the standard for every autopsy facility (Nolte et al. 2002). The guidelines for safe autopsies are based on either national or international systems that group infectious diseases depending on factors such as severity, transmission rate, and treatment options. The WHO, CDC, European Center for Disease Prevention and Control (ECDC), and Royal College of Pathologists have provided autopsy recommendations, all of which emphasize the importance of a high BSL (Khatam-Lashgari et al. 2021).

SARS-CoV-2 has been classified as an RG-3 pathogen by the European Commission and several other organizations. However, the mandatory requirements for BSL-3 do not include further specified PPE requirements because BSL-3 requirements are defined and implemented at different levels in each country. Differing requirements include guidelines published by the WHO or European Commission (Loibner et al. 2021). The CDC has provided guidelines that specifically address biosafety requirements for the collection and handling of postmortem specimens (Loibner et al. 2021; Kritselis and Remick 2020). According to their recommendations, all autopsies would be safest if performed at BSL-3 facilities, and if these are unavailable, autopsies can be performed using the barrier precautions of BSL-2 accompanied by negative airflow and the respiratory precautions of BSL-3 (Kritselis and Remick 2020). During autopsy, the examiners should wear a fluid-resistant jumpsuit or shirts and pants that cover the body from the neck to the feet and arms, waterproof aprons and sleeves, closed-toed shoes covered with fluid-proof shoe covers or booties, and a surgical cap or hood bonnet covering the entire head. Double gloving is recommended with cut-resistant gloves of fine-woven steel covered by rubber gloves. At the very least, examiners should also wear a disposable N-95 respirator, which was approved by the National Institute for Occupational Safety and Health. Those without N-95 equipment are required to wear a purified air-powdered respirator. When removing the brain, aerosol generation should be minimized by avoiding the use of oscillating saws, moistening the bone before cutting, and using vacuum attachments if an oscillating saw is used (Kritselis and Remick 2020). Various pieces of information about autopsy management have been presented and accumulated to date to implement strict infection prevention and control measures to curb the spread of COVID-19 (Basso et al. 2020; Keten et al. 2020; Rani 2020; Skok et al. 2021; Pomara et al. 2021a).

Looking back on the first phase of COVID-19, many governments imposed severe restrictions on the corpse management of subjects with COVID-19 related deaths (Pomara et al. 2021a). These restrictions were imposed without any individualized risk assessment and performed prudently to reduce the risk of infection. Recommendations by the Autopsy Work Group of the Spanish Society of Anatomical Pathology and the Royal College of Pathologists were more restrictive than the WHO guidelines, which led to difficulties in performing autopsies (Pomara et al. 2021a). In practice, few standard BSL-3 or BSL-4 autopsy facilities exist as they are expensive to build, operate, and maintain worldwide. The uncertainty in the potential exposure risk during autopsy and constrained PPE supply have also resulted in fewer autopsies being performed. Moreover, most hospitals likely did not perform autopsies on patients who died from COVID-19 since the cause of death is often related to respiratory failure (Bhatt et al. 2022). In such situations, the few reported early autopsies of patients with COVID-19 revealed that the respiratory organs were the major affected organs in most cases; however, various lesions were observed in other organs, as well as comorbidity (Pomara et al. 2021a; Salerno et al. 2020), suggesting that autopsy practice is an investigative tool to define an effective treatment to reduce mortality.

Interestingly, Davis et al. reported the risk of COVID-19 transmission during autopsy. Given that one person involved in COVID-19 autopsies acquired COVID-19 infection after approximately 675 exposures, performing autopsies with the recommended PPE appears to present an exceedingly low risk of COVID-19 transmission to autopsy personnel (Davis and Williamson 2020).

Several studies have suggested a so-called cytokine storm and immune dysregulation as a rationale for treating severe forms of COVID-19 with various immune system modulators and anticoagulants; however, studies on the underlying mechanisms of COVID-19 pathogenesis remain surprisingly scarce (Layne et al.

2022). Layne et al. posed outstanding questions for pathology studies of COVID-19 (Layne et al. 2022), i.e., can biomarkers be identified and used to predict progression to severe COVID-19? and other 4 questions. These findings indicate that more autopsy studies are required to understand the pathogenesis of severe COVID-19.

33.4 Pathological Findings from Autopsies

The index case was reported on December 8, 2019, and the WHO China Country Office was informed of several cases of pneumonia of unknown etiology detected in Wuhan. Eventually, the novel COVID-19 outbreak, which turned into a pandemic, was declared by the WHO. As mentioned above, severe restrictions on the corpse management of subjects who died of COVID-19 related causes were imposed. In China, upon the outbreak of COVID-19, over 60 pathologists and technicians assembled a COVID-19 Pathology Team in Wuhan and Chongqing, China. As of April 22, 2020, the COVID-19 Pathology Team had performed systematic autopsies on 37 deceased patients with COVID-19 (Bian 2020). The PubMed database was searched for articles written in English from December 1, 2019, to December 31, 2020. In May 2020, some English case reports began to be published in China. The reports found during the search primarily focused on pulmonary pathology and originated from Spain, Switzerland, Italy, USA (Conde et al. 2020; COVID-19 Autopsy 2020; Suess and Hausmann 2020; Aguiar et al. 2020; Barton et al. 2020; Carsana et al. 2020), showing histologically diffuse alveolar damage and vascular thrombi.

Several autopsy-based reports, studies, and systematic reviews of COVID-19 have been conducted to date, in which numerous pieces of information regarding the organ and tissue tropism of SARS-CoV-2 and morphological features of COVID-19 were accumulated (Maiese et al. 2021; Caramaschi et al. 2021; Chawla et al. 2022; Menezes et al. 2022; Jonigk et al. 2022).

Pathological findings of the respiratory system were accumulated since the beginning of the pandemic. Pulmonary manifestations are the primary cause of death in patients with COVID-19 since SARS-CoV-2 infects the nasopharynx and lungs, resulting in various symptoms of varying severity. These patients usually presented with peripheral lung ground-glass opacities on CT, consistent with autopsy findings. Macroscopic examination revealed that the lungs appeared heavy and edematous. On microscopic examination, both exudative and proliferative diffuse alveolar damage are in difficult stages, with hyaline membrane formation, hemorrhage, edema, and pneumocyte damage as clinically acute respiratory distress syndrome (ARDS). This feature resembles the pulmonary damage caused by SARS and MERS (Maiese et al. 2021; Jonigk et al. 2022). SARS-CoV-2 infection leads to angiocentric inflammation in cases of COVID-19-induced respiratory failure, as well as several angiotensin-converting enzyme 2 (ACE2)-positive endothelial cells. ACE2 regulates the renin-angiotensin-aldosterone system, which is central in angiotensin II (AngII) signaling (Aimrane et al. 2022; Lamers and Haagmans 2022). The binding of SARS-CoV-2 to ACE2 and TMPRSS2 is followed by the fusion of the

Table 33.2 Summary of the common pathological findings in the post-mortem examinations

Organ/System	Pathological findings
Lung	Diffuse alveolar damage at different stages (acute, proliferative, and fibrosis), hyaline membrane formation, interstitial/alveolar edema, interstitial lymphocytic infiltration, pneumocyte hyperplasia, pulmonary microthrombi, alveolar hemorrhage, fibrosis
Heart	Hypertrophy, fibrosis, cardiomegaly, cardiac inflammation, dilatation
Kidney	Arteriosclerosis, acute tubular injury, endothelial injury, microthrombi, glomerulopathy, focal segmental glomerulosclerosis, shock kidney
Liver	Steatosis, congestion, fibrosis, shock liver
Spleen	Congestion/hemorrhage
Gastrointestinal	Shock changes
Genitourinary	Interstitial edema, thrombosis
Central nervous system	Hypoxic-ischemic injury, olfactory bulbus edema, neuronal degeneration, necrosis, inflammatory infiltration
Skin	Perivascular mononuclear/lymphocytic infiltrate, endothelial change, fibrin microthrombi

virus with the plasma membrane and endocytosis. Several other organs express this membrane-integrated protein, and infection can spread from the pulmonary system to other organs (Jonigk et al. 2022; Aimrane et al. 2022; Lamers and Haagmans 2022). Clinically, endothelial dysfunction is a denominator of SARS-CoV-2 infection, which is histologically characterized by acute vascular inflammation and perivascular T-cell recruitment, leading to swelling and disruption of the endothelial cell barriers and an anomalous microvascular architecture. Consequently, vascular injury causes thrombosis, vasoconstriction, and intussusceptive angiogenesis. Pulmonary thrombosis in large vessels was only observed in some cases, whereas obstructed microvasculature was observed in most. An exacerbated immune response, including severe inflammation and cytokine activation, can also lead to respiratory and multiple organ failure, resulting in death in some cases. Respiratory, olfactory, and paranasal sinus epithelia with tropism to ciliated mucosa have also been observed as the nasopharyngeal and oropharyngeal tissues are the primary entry and early replication points of SARS-CoV-2.

In addition to severe respiratory damage, various systems, such as the nervous, motor, urinary, reproductive, and digestive systems are affected (Table 33.2) (Maiese et al. 2021; Caramaschi et al. 2021; Chawla et al. 2022; Menezes et al. 2022; Jonigk et al. 2022; Aimrane et al. 2022; Lamers and Haagmans 2022). These pathological findings could be attributed to patients' underlying diseases, as most patients suffered from hypertension, coronary heart disease, or diabetes mellitus.

33.5 COVID-19 Vaccines, Adverse Effects, and Autopsy

Successes in global smallpox eradication programs have given great hope for vaccines and the control of infectious diseases, leading to the development of comprehensive vaccination programs, which have become the cornerstone of public health intervention (Hardt et al. 2016).

The COVID-19 pandemic has caused a major burden on the healthcare system due to its high morbidity and mortality rates. Despite the implementation of strict public health measures, the global economy was impacted significantly. The ideal goal of the COVID-19 vaccination is to promote global herd immunity, and as a result, the main aim of COVID-19 vaccination is to minimize deaths, severe disease, and overall disease burden, leading to socioeconomic restoration. Various countries have approved COVID-19 vaccines for human use in response to the COVID-19 pandemic, and drugs are expected to be licensed. These platforms can be classified either as traditional approaches that have previously resulted in licensed vaccines, such as inactivated recombinant proteins and vectored vaccines, or as novel approaches that have not been used as licensed vaccines to date, such as RNA and DNA vaccines (Rahman et al. 2022). On March 16, 2022, the WHO validated nine COVID-19 vaccines in the Emergency Use Listing: the Pfizer/BioNTech Comirnaty vaccine (BNT162b2) on December 31, 2020, the SII/CIVISHIELD and AstraZeneca/AZD1222 vaccines (Oxford and AstraZeneca, ChAdOx1 nCoV-19) on February 16, 2021, and the Janssen/Ad26 vaccine. The COV2.S vaccine was developed by Johnson & Johnson on March 12, 2021, the Moderna COVID-19 vaccine (mRNA-1273) on 30 April, 2021, the Sinopharm COVID-19 vaccine on May 7, 2021, the Sinovac-CoronaVac vaccine on June 1, 2021, the Bharat Biotech BBV152 COVAXIN vaccine on 3 November, 2021, the Covovax (NVX-CoV2373) vaccine on December 17, 2021, and the Nuvaxovid (NVX-Co V2373) vaccine on December 21, 2021. Table 33.3 shows the vaccine platform and formulation techniques (Rahman et al. 2022). However, countries have autonomy to issue emergency use authorizations for any health product based on their national regulations and legislation. Thus, domestic emergency use authorizations were issued at each country's discretion. For example, in the USA, FDA-approved COVID-19 vaccines authorized for emergency use include the Pfizer-BioNTech, Moderna, and Janssen COVID-19 vaccines.

COVID-19 vaccines have undergone randomized clinical trials to test their quality, safety, and efficiency, and have been approved by the WHO for emergency use listing. After approval, they continued to be monitored for their ongoing safety and effectiveness. A vaccine's efficiency is measured in a controlled clinical trial and is based on how many vaccinated people developed the disease compared with the number of people who received the placebo that developed the disease. Vaccine effectiveness (VE) is a measure of how well vaccines work in the real world. Real-world effectiveness can differ from the efficiency measured in a trial because we cannot predict exactly how effective vaccination will be for a much larger and more variable population, i.e., getting vaccinated under real life conditions (https://www.who.int/news-room/feature-stories/detail/vaccine-efficacy-effectiveness-and-

Table 33.3 COVID-19 vaccines (until 20 December 2021 approved by WHO)

Vaccines	Platform and formulation techniques	Target antigen
BNT162b2	RNA, nucleotide modification of SARS-CoV-2 mRNA	Spike protein
ChAdOx1 nCoV-19	Non-replicating viral vector, use recombinant	Spike protein
ChAdOx1	Adenoviral vector encoding the spike (S) protein of the SARS-CoV-2	
Ad26.COV2.S	Non-replicating viral vector, use and replication incompetent	Spike protein
	Adenovirus type 26 (Ad26) vector which encodes the SARS-CoV	-2 spike (S) protein
mRNA-1273	RNA, encapsulation of SARS-CoV-2	Spike protein
	mRNA in lipid nanoparticle (LNP)	
Sinopharm	Inactivated SARS-CoV-2, in vitro production in Vero cells	Whole virus
CoronaVac	Inactivated SARS-CoV-2, in vitro production in Vero cells	Whole virus
BBV152	Inactivated SARS-CoV-2, in vitro production in Vero cells	Spike protein
NVX-CoV2373	Recombinant protein	Spike protein, both humoral and cell mediate

protection). A systematic review and network meta-analysis revealed that the BNT162b and mRNA-1273 prevented symptomatic COVID-19 infections more effectively than Gam-COVID-Vac, NVX-CoV23730, CoronaVac, BN02, WIV04, and Ad26.COV2.5 (Rotshild et al. 2021). The efficacy of these vaccines is reported as follows: Pfizer, 95%; Moderna, 94.1%; AstraZeneca, 70.4%; and Janssen-66.9%, indicating that they effectively reduce the incidence and severity of SARS-CoV-2 infection among the study populations. (Francis). The efficacy of the vaccines are also reported as follows: Sinopharm-Whuan, 72.51%; BBV152, 77.8%; and CoronaVac, 51% and 83.5% in Brazil and Turkey, respectively (Rahman et al. 2022).

Global real-world VE has been reported in a series of studies. Zheng et al. reported the real-world effectiveness of COVID-19 vaccines in a meta-analysis comprising 51 records (Zheng et al. 2022), showing that the VE against SARS-CoV-2 infection, COVID-19-related hospitalization, intensive care unit admission, and death were 89.1% (95% CI 85.6–92.6%), 97.2% (95% CI 96.1–98.3), 97.4% (95% CI 96.0–98.8%), and 99.0%(95% CI 98.5–99.6%), respectively. The VE against infection in the general population aged ≥16 years, the elderly, and health workers was 86.1% (95% CI 77.8–94.4%), 83.8% (95% CI 77.1–90.6%), and 95.3% (95% CI 92.0–98.6%), respectively. For those fully vaccinated against infection, the observed effectiveness of the Pfizer-BioNTech vaccines was 91.2%, that of the Moderna vaccine was 98.1%, and that of the CoronaVac vaccine was 65.7%. They concluded that COVID-19 vaccines were highly protective against SARS-CoV-2-related diseases in real-world settings (Zheng et al. 2022). Ssentongo et al. reported the real-world effectiveness of COVID-19 vaccines, including BNT162b, mRNA-1273, and AD26.COV2.S, using a meta-analysis comprising 18 records representing nearly seven million individuals. In this study, VE against all SARS-CoV-2 infections declined from 83% in the first month after completion of the original vaccination series to 22% at five months or longer. The VE against symptomatic COVID-19 declined from 94% in the first month after vaccination to 64% by the fourth month. VE against severe COVID-19 for all ages was high overall, with a level of 90% (95% CI 87.0–92.0%) at 5 months or longer after being fully vaccinated. VE against severe COVID-19 was lower in individuals aged ≥65 years and those who received the Ad26.COV2.S vaccine (Ssentongo et al. 2022). Their data show the temporal waning of VE against SARS-CoV-2 infection and symptomatic illness with preservation of VE against severe illness in most circumstances, and the need for booster vaccine doses should be considered in the context of clearly outlined international public health goals for VE outcomes.

The evolution of the SARS-CoV-2 virus during the COVID-19 pandemic has resulted in the emergence of four variants of concern (VOC), namely alpha (B.1.1.7), beta (B.1.351), gamma (P.1), and delta (B.1.617.2) (Tregoning et al. 2021). These SARS-CoV-2 VOC are more transmissible and have the potential to increase disease severity and decrease the effectiveness of COVID-19 vaccines. In Tregoning's group, a reduction in in vitro serum neutralization activity was observed in highly sensitive assays, and evidence of infection with VOCs has been observed in vaccinated populations. However, the disease severity is nevertheless much reduced, indicating that vaccines remain highly effective (Tregoning et al. 2021). Nasreen et al. reported the effectiveness of BNT162b2, mRNA-1273 and ChAdOx1 vaccines against symptomatic SARS-CoV-1 infection and COVID-19 hospitalization or death caused by the alpha (B.1.1.7), beta (B.1.351), gamma (P.1), and delta (B.1.617.2) VOC in Ontario, Canada, using a test-negative design study. Their results suggested that even a single dose of these three COVID-19 vaccine products provides considerable protection against symptomatic infection and severe outcomes caused by these four VOC, particularly in young adults, and that two doses provide even higher protection (Nasreen et al. 2022). At the time of writing, a rapid increase in COVID-19 cases due to the omicron variant (B.1.1.5290 variant of SARS-CoV-2) in highly vaccinated populations raised concerns on the effectiveness of current vaccines. A test-negative case-control study in England indicated that two doses of the BNT162b2 or ChAdOx1 nCoV-19 vaccines failed to provide adequate levels of protection against infection with the omicron variant and mild disease, and boosting with BNT162b2 or mRNA-1273 provided a substantial increase in protection against mild disease, although this waned over time, suggesting that boosters will likely offer even greater levels of protection against severe and fatal diseases (Andrews et al. 2022).

Adverse effects due to COVID-19 vaccines, including local events such as pain, redness, and swelling, as well as systemic events such as fatigue, headache, and muscle or joint pain, have been reported in randomized clinical trials. Most adverse effects were mild to moderate; however, approximately 50–90% of participants reported experiencing some adverse effects (Beatty et al. 2021). In a real-world cohort with BNT162b2, mRNA1273, and JNK-78436735, serious COVID-19 vaccine adverse effects were rare and similar to industry and government reports (Beatty et al. 2021). However, the severity of adverse effects varied between vaccine brands. For example, compared with participants receiving the BNT162b2 vaccine, those receiving mRNA-1273 had were at a two times higher risk of reporting adverse effects (odds ratio, 2.00; 95% CI 1.86–2.15; $p < 0.01$). Cases of more serious adverse events, such as anaphylaxis, myocarditis, and thrombocytopenia, have been reported, although infrequently. Sixty-six cases of anaphylaxis have been reported among 17,524,676 mRNA vaccinations in the USA as of February 2021 (Shimabukuro et al. 2021). A meta-analysis of the risk of anaphylaxis due to authorized COVID-19 vaccines revealed that the highest incidence of anaphylaxis was observed in vaccinated women compared to men aged 18–85 years. Interestingly, in this cohort of women, administration of BNT162b2 vaccines yielded the most cases of anaphylaxis. The results of this study suggest that all authorized COVID-19 vaccines may cause anaphylactic reactions, anaphylactoid reactions, anaphylactic shock, and anaphylactoid shock (Sobczak and Pawliczak 2022). In May 2021, several cases of myocarditis and pericarditis were reported by the US Center for Disease Control and Prevention Vaccine Adverse Event Reporting System following BNT162b2 or mRNA-1273 vaccination. In a descriptive study of reports of myocarditis to the Vaccine Adverse Event Reporting System (VAERS) following mRNA-based COVID-19 vaccination between December 2020 and August 2021 in 192,405,448 individuals aged >12 years in the USA, the risk of myocarditis increased across multiple age and sex strata and was highest after the second vaccination dose in adolescent males and young men (Oster et al. 2022). The European Medicines Agency concluded that a causal link exists between AZD1222 (ChAdOx1 nCoV-19) administration, blood clotting, and low platelet counts (thrombocytopenia) in an extremely small number of vaccinated individuals, leading to 30 deaths in vaccinated individuals. As a result, vaccine agencies in EU countries and the UK have issued age-based restrictions on the use of AZD1222. Extremely rare events of thrombocytopenia were also observed in the USA, reportedly due to Ad26.COV2-S vaccination. The FDA briefly paused the use of As26.COV2-S in April 2021. Thrombocytopenia appears to be more strongly associated with adenovirus-based vaccines; however, it has also been observed after mRNA vaccination. The rate of vaccine-induced immune thrombotic thrombocytopenia (VITT) differs among countries. A systematic review of published reports of VITT based on searches of PubMed, SCOPUS, and Web of Science from December 2020 to May 2021 revealed that certain demographics were associated with an increased risk of VITT and clinical presentations thereof following ChAdOx1 nCoV-19 and Ad26. COV2.S vaccination, namely young individuals, particularly women. The clinical presentation of VITT commonly includes cerebral thrombi, pulmonary embolism, and deep venous thrombosis, but other presentations are also possible, highlighting the importance of clinical vigilance in recent vaccine recipients (Elberry et al. 2022).

Coincidental adverse events, including death and related fatalities, influence the COVID-19 vaccination campaign. Hesitancy to be vaccinated against COVID-19 is generally caused by safety concerns and adverse effects, as well as mistrust of the vaccine and government. Thus, COVID-19 vaccine hesitancy must be overcome to achieve a higher percentage of vaccinated individuals. In the first month of COVID-19 vaccine safety monitoring in the USA between December 14, 2020, and January 31, 2021, a total of 13,794,904 COVID-19 vaccine doses were administered, and 6994 cases of COVID-19-associated adverse events were reported in the Vaccine Adverse Event Reporting System (VAERS) (Gee et al. 2021). Among all reports, 6354 (90.8%) were classified as non-serious and 60 (9.2%) as serious, and 113 (1.6%) deaths were reported. All-cause mortality is high in residents of long-term care facility (LTCF). Using the VAERS over roughly the same period, Lv et al. reported a mortality rate of 53.4 per million following COVID-19 vaccination among LTCF residents during the study period. This was far lower than the 2019 monthly all-cause mortality rate of 0.3% among adults aged ≥65 years. They suggested that the benefits of COVID-19 vaccines far outweighed the potential risks in older frail populations (Lv et al. 2021). Recently, while the safety of the COVID-19 vaccine booster doses was being monitored among adults between September 22, 2021, and February 6, 2022, a total of 721,562 unique v-safe registrants aged ≥18 years received a COVID-19 vaccine booster (Lv et al. 2021). A total of 39,286 cases of adverse events were reported in the VAERS. In this report, myocarditis was most often reported among men aged 18–24 years following an mRNA-1273 vaccine booster (8.7 per one million doses administered). However, there is no mention of a relationship between COVID-19 vaccines and mortality, except for the unusual case of an ongoing investigation into the relationship between the mRNA-1273 vaccine and myocarditis.

Postmortem studies in patients following the first dose of the BNT162b2 vaccine have been reported (Edler et al. 2021; Hause et al. 2022; Hansen et al. 2021) along with a series of thrombotic thrombocytopenia cases following ChAdOx1 nCoV-19 vaccine administration (Greinacher et al. 2021; Schultz et al. 2021; Pomara et al. 2021b).

Sessa et al. analyzed 17 papers published on fatal cases with post-mortem investigations (Sessa et al. 2021). A total of 38 cases were analyzed, of which 22 were related to ChAdOx1 nCoV-19 administration, 10 to BNT162b2, 4 to mRNA-1273, and 2 to Ad26.COV2.S. They suggested that autopsy is extremely useful for defining the main characteristics of VITT after ChAdOx1 nCoV-19 vaccination and myocarditis related to the BNT162b2 vaccine. In a literature review of COVID-19 vaccine-related deaths performed in accordance with the Preferred Reporting Items for Systematic Review (PRISMA) standards, 55 cases of death after COVID-19 vaccination have been reported. A causal relationship was excluded in 17 cases, not specified in 8 cases, considered possible in 15 cases, probable in one case, and very probable/demonstrated in 14 cases. The causes of death included 32 cases of VITT, 3 cases of acute myocarditis, one case of disseminated encephalomyelitis, and one case of rhabdomyolysis. They mentioned that post-mortem investigations are required to deepen the understanding of the possible pathophysiological mechanisms of fatal adverse effects due to vaccination (Maiese et al. 2022).

33.6 Summary

Conventional autopsy is a well-known crucial method for auditing the reliability of clinical diagnosis and analyzing how diseases, including infectious diseases, affect various organs and systems. With the global number of conventional autopsies decreasing yearly, this number has decreased even further during the COVID-19 pandemic. Thus, few studies exploring pathological lesions in fatal COVID-19 cases have been conducted, which have led to fundamental insights into the pathogenesis of this disease. However, a comprehensive analysis of autopsy data must be conducted to provide more detailed insights into the fatal adverse effects and deaths associated with vaccination.

References

Aguiar D, Lobrinus JA, Schibler M, Fracasso T, Lardi C (2020) Inside the lungs of COVID-19 disease. Int J Leg Med 134(4):1271–1274

Aimrane A, Laaradia MA, Sereno D, Perrin P, Draoui A, Bougadir B, Hadach M, Zahir M, Fdil N, El Hiba O, El Hidan MA, Kahime K (2022) Insight into COVID-19's epidemiology, pathology, and treatment. Heliyon 8(1):e08799

Andrews N, Stowe J, Kirsebom F, Toffa S, Rickeard T, Gallagher E, Gower C, Kall M, Groves N, O'Connell AM, Simons D, Blomquist PB, Zaidi A, Nash S, Aziz NI, Thelwall S, Dabrera G, Myers R, Amirthalingam G, Gharbia S, Barrett JC, Elson R, Ladhani SN, Ferguson N, Zambon M, Campbell CNJ, Brown K, Hopkins S, Chand M, Ramsay M, Lopez BJ (2022) COVID-19 vaccine effectiveness against the omicron (B.1.1.529) variant. N Engl J Med 386(16):1532–1546

Barton LM, Duval EJ, Stroberg E, Ghosh S, Mukhopadhyay S (2020) COVID-19 autopsies, Oklahoma, USA. Am J Clin Pathol 153(6):725–733

Basso C, Calabrese F, Sbaraglia M, Del Vecchio C, Carretta G, Saieva A, Donato D, Flor L, Crisanti A, Dei Tos AP (2020) Feasibility of postmortem examination in the era of COVID-19 pandemic: the experience of a Northeast Italy University Hospital. Virchows Arch 477(3): 341–347

Beatty AL, Peyser ND, Butcher XE, Cocohoba JM, Lin F, Olgin JE, Pletcher MJ, Marcus GM (2021) Analysis of COVID-19 vaccine type and adverse effects following vaccination. JAMA Netw Open 4(12):e2140364

Bhatt M, MovaseghiGargari M, Chand MT (2022) The importance of autopsies despite the declining number amidst the COVID-19 pandemic. Autops Case Rep 12:e2021371

Bian XW (2020) COVID-19 pathology team. Autopsy of COVID-19 patients in China. Natl Sci Rev 7(9):1414–1418

Bryce C, Grimes Z, Pujadas E, Ahuja S, Beasley MB, Albrecht R, Hernandez T, Stock A, Zhao Z, AlRasheed MR, Chen J, Li L, Wang D, Corben A, Haines GK 3rd, Westra WH, Umphlett M, Gordon RE, Reidy J, Petersen B, Salem F, Fiel MI, El Jamal SM, Tsankova NM, Houldsworth J, Mussa Z, Veremis B, Sordillo E, Gitman MR, Nowak M, Brody R, Harpaz N, Merad M, Gnjatic S, Liu WC, Schotsaert M, Miorin L, Aydillo Gomez TA, Ramos-Lopez I, Garcia-Sastre A, Donnelly R, Seigler P, Keys C, Cameron J, Moultrie I, Washington KL, Treatman J, Sebra R, Jhang J, Firpo A, Lednicky J, Paniz-Mondolfi A, Cordon-Cardo C, Fowkes ME (2021) Pathophysiology of SARS-CoV-2: the Mount Sinai COVID-19 autopsy experience. Mod Pathol 34(8):1456–1467

Burton JL (2001) The history of the autopsy. In: Burton JL, Rutty GN (eds) The hospital autopsy. Arnold, London, pp 1–6

Caramaschi S, Kapp ME, Miller SE, Eisenberg R, Johnson J, Epperly G, Maiorana A, Silvestri G, Giannico GA (2021) Histopathological findings and clinicopathologic correlation in COVID-19: a systematic review. Mod Pathol 234(9):1614–1633

Carsana L, Sonzogni A, Nasr A, Rossi RS, Pellegrinelli A, Zerbi P, Rech R, Colombo R, Antinori S, Corbellino M, Galli M, Catena E, Tosoni A, Gianatti A, Nebuloni M (2020) Pulmonary post-mortem findings in a series of COVID-19 cases from northern Italy: a two-centre descriptive study. Lancet Infect Dis 20(10):1135–1140

Chawla K, Hong S, Shields R, Ford A, Pacella G, Lurie F (2022) Multifactorial effects of COVID-19: a review of published autopsy reports. COVID 2(5):553–568

Conde PN, Monraval PA, Medina CM, Jiménez Sánchez A, Teruel JCA, Ferrando JM, Santos VP, Aranda EM (2020) Autopsy findings from the first known death from Severe Acute Respiratory Syndrome SARS-CoV-2 in Spain. Rev Esp Patol 53(3):188–192

COVID-19 Autopsy (2020) The first COVID-19 autopsy in Spain performed during the early stages of the pandemic. Rev Esp Patol 53(3):182–187

da Silva LFF, Saldiva HP, Alves VAF (2016) History and prospects of pathology in medicine. Rev Med (São Paulo) 95(Special Issue 2):68–72

Davis GG, Williamson AK (2020) Risk of coronavirus disease 2019 transmission during autopsy. Arch Pathol Lab Med 144(12):1445

Edler C, Klein A, Schröder AS, Sperhake JP, Ondruschka B (2021) Deaths associated with newly launched SARS-CoV-2 vaccination (Comirnaty®). Leg Med (Tokyo) 51:101895

Elberry MH, Abdelgawad HAH, Hamdallah A, Abdella WS, Ahmed AS, Ghaith HS, Negida A (2022) A systematic review of vaccine-induced thrombotic thrombocytopenia in individuals who received COVID-19 adenoviral-vector-based vaccines. J Thromb Thrombolysis 53(4): 798–823

Gee J, Marquez P, Su J, Calvert GM, Liu R, Myers T, Nair N, Martin S, Clark T, Markowitz L, Lindsey N, Zhang B, Licata C, Jazwa A, Sotir M, Shimabukuro T (2021) First month of COVID-19 vaccine safety monitoring - United States, December 14, 2020-January 13, 2021. MMWR Morb Mortal Wkly Rep 70(8):283–288

Greinacher A, Thiele T, Warkentin TE, Weisser K, Kyrle PA, Eichinger S (2021) Thrombotic thrombocytopenia after ChAdOx1 nCov-19 vaccination. N Engl J Med 384(22):2092–2101

Hansen T, Titze U, Kulamadayil-Heidenreich NSA, Glombitza S, Tebbe JJ, Röcken C, Schulz B, Weise M, Wilkens L (2021) First case of postmortem study in a patient vaccinated against SARS-CoV-2. Int J Infect Dis 107:172–175

Hanzlick R, Hutchins GM (1999) History repeats itself (sometimes). Autopsy Committee of the College of American Pathologists. Arch Intern Med 159(16):1837–1838

Hardt K, Bonanni P, King S, Santos JI, El-Hodhod M, Zimet GD, Preiss S (2016) Vaccine strategies: optimising outcomes. Vaccine 34(52):6691–6699

Hause AM, Baggs J, Marquez P, Myers TR, Su JR, Blanc PG, Baumblatt JAG, Woo EJ, Gee J, Shimabukuro TT, Shay DK (2022) Safety monitoring of COVID-19 vaccine booster doses among adults—United States, September 22, 2021-February 6, 2022. MMWR Morb Mortal Wkly Rep 71(7):249–254

Hoyert DL (2011) The changing profile of autopsied deaths in the United States, 1972–2007. NCHS Data Brief 67:1–8

Hu B, Guo H, Zhou P, Shi ZL (2021) Characteristics of SARS-CoV-2 and COVID-19. Nat Rev Microbiol 19(3):141–154

Irie H, Asano A, Ueno M, Ishida T (2021) Decrease in pathological autopsy cases and recent tendency. Bull Teikyo Jr Coll 22:145–154

Jonigk D, Werlein C, Acker T, Aepfelbacher M, Amann KU, Baretton G, Barth P, Bohle RM, Büttner A, Büttner R, Dettmeyer R, Eichhorn P, Elezkurtaj S, Esposito I, Evert K, Evert M, Fend F, Gaßler N, Gattenlöhner S, Glatzel M, Göbel H, Gradhand E, Hansen T, Hartmann A, Heinemann A, Heppner FL, Hilsenbeck J, Horst D, Kamp JC, Mall G, Märkl B, Ondruschka B, Pablik J, Pfefferle S, Quaas A, Radbruch H, Röcken C, Rosenwald A, Roth W, Rudelius M, Schirmacher P, Slotta-Huspenina J, Smith K, Sommer L, Stock K, Ströbel P, Strobl S, Titze U,

Weirich G, Weis J, Werner M, Wickenhauser C, Wiech T, Wild P, Welte T, von Stillfried S, Boor P (2022) Organ manifestations of COVID-19: what have we learned so far (not only) from autopsies? Virchows Arch 1:1–21

Keten D, Okdemir E, Keten A (2020) Precautions in postmortem examinations in Covid-19-related deaths: recommendations from Germany. J Forensic Leg Med 73:102000

Khatam-Lashgari A, Henningsen MJ, Olsen KB, Jacobsen C, Hasselby JP, Colville-Ebeling B, Banner J (2021) Autopsies in pandemics—a perspective on barriers and benefits. Is it time for a revival? APMIS 129(7):324–339

Kritselis M, Remick DG (2020) Universal precautions provide appropriate protection during autopsies of patients with infectious diseases. Am J Pathol 190(11):2180–2184

Kumar A, Singh R, Kaur J, Pandey S, Sharma V, Thakur L, Sati S, Mani S, Asthana S, Sharma TK, Chaudhuri S, Bhattacharyya S, Kumar N (2021) Wuhan to world: The COVID-19 pandemic. Front Cell Infect Microbiol 11:596201

Kurz SD, Sido V, Herbst H, Ulm B, Salkic E, Ruschinski TM, Buschmann CT, Tsokos M (2021) Discrepancies between clinical diagnosis and hospital autopsy: a comparative retrospective analysis of 1,112 cases. PloS One 16(8):e0255490

Lamers MM, Haagmans BL (2022) SARS-CoV-2 pathogenesis. Nat Rev Microbiol 20:270–284

Layne SP, Walters KA, Kash JC, Taubenberger JK (2022) More autopsy studies are needed to understand the pathogenesis of severe COVID-19. Nat Med 28(3):427–428

Loibner M, Langner C, Regitnig P, Gorkiewicz G, Zatloukal K (2021) Biosafety requirements for autopsies of patients with COVID-19: Example of a BSL-3 autopsy facility designed for highly pathogenic agents. Pathobiology 88(1):37–45

Lv G, Yuan J, Xiong X, Li M (2021) Mortality rate and characteristics of deaths following COVID-19 vaccination. Front Med (Lausanne) 8:670370

Maiese A, Manetti AC, La Russa R, Di Paolo M, Turillazzi E, Frati P, Fineschi V (2021) Autopsy findings in COVID-19-related deaths: a literature review. Forensic Sci Med Pathol 17(2): 279–296

Maiese A, Baronti A, Manetti AC, Di Paolo M, Turillazzi E, Frati P, Fineschi V (2022) Death after the administration of COVID-19 vaccines approved by EMA: has a causal relationship been demonstrated? Vaccines (Basel) 10(2):308

Mann R, Perisetti A, Gajendran M, Gandhi Z, Umapathy C, Goyal H (2020) Clinical characteristics, diagnosis, and treatment of major coronavirus outbreaks. Front Med (Lausanne) 7:581521

Marshall HS, Milikowski C (2017) Comparison of clinical diagnoses and autopsy findings: six-year retrospective study. Arch Pathol Lab Med 141(9):1262–1266

Menezes RG, Rizwan T, Ali SS, Hassan W, Khetpal A, Aqil M, Madadin M, Siddiqi TJ, Usman MS (2022) Postmortem findings in COVID-19 fatalities: A systematic review of current evidence. Leg Med (Tokyo) 54:102001

Nasreen S, Chung H, He S, Brown KA, Gubbay JB, Buchan SA, Fell DB, Austin PC, Schwartz KL, Sundaram ME, Calzavara A, Chen B, Tadrous M, Wilson K, Wilson SE, Kwong JC, Canadian Immunization Research Network (CIRN) Provincial Collaborative Network (PCN) Investigators (2022) Effectiveness of COVID-19 vaccines against symptomatic SARS-CoV-2 infection and severe outcomes with variants of concern in Ontario. Nat Microbiol 7(3): 379–385

NIH Guidelines (n.d.). https://osp.od.nih.gov/biotechnology/nih-guidelines/

Nolte KB, Taylor DG, Richmond JY (2002) Biosafety considerations for autopsy. Am J Forensic Med Pathol 23(2):107–122

Oster ME, Shay DK, Su JR, Gee J, Creech CB, Broder KR, Edwards K, Soslow JH, Dendy JM, Schlaudecker E, Lang SM, Barnett ED, Ruberg FL, Smith MJ, Campbell MJ, Lopes RD, Sperling LS, Baumblatt JA, Thompson DL, Marquez PL, Strid P, Woo J, Pugsley R, Reagan-Steiner S, DeStefano F, Shimabukuro TT (2022) Myocarditis cases reported after mRNA-based COVID-19 vaccination in the US from December 2020 to August 2021. JAMA 327(4):331–340

Pomara C, Salerno M, Sessa F, Esposito M, Barchitta M, Ledda C, Grassi P, Liberto A, Mattaliano AR, Rapisarda V, Ferrante M, Agodi A (2021a) Safe management strategies in clinical forensic autopsies of confirmed COVID-19 cases. Diagnostics (Basel) 11(3):457
Pomara C, Sessa F, Ciaccio M, Dieli F, Esposito M, Garozzo SF, Giarratano A, Prati D, Rappa F, Salerno M, Tripodo C, Zamboni P, Mannucci PM (2021b) Post-mortem findings in vaccine-induced thrombotic thrombocytopenia. Haematologica 106(8):2291–2293
Rahman MM, Masum MHU, Wajed S, Talukder A (2022) A comprehensive review on COVID-19 vaccines: development, effectiveness, adverse effects, distribution and challenges. Virus Dis 33(1):1–22
Rani S (2020) A review of the management and safe handling of bodies in cases involving COVID-19. Med Sci Law 60(4):287–293
Rotshild V, Hirsh-Raccah B, Miskin I, Muszkat M, Matok I (2021) Comparing the clinical efficacy of COVID-19 vaccines: a systematic review and network meta-analysis. Sci Rep 11(1):22777
Salerno M, Sessa F, Piscopo A, Montana A, Torrisi M, Patanè F, Murabito P, Volti GL, Pomara C (2020) No autopsies on COVID-19 deaths: a missed opportunity and the lockdown of science. J Clin Med 9(5):1472
Schultz NH, Sørvoll IH, Michelsen AE, Munthe LA, Lund-Johansen F, Ahlen MT, Wiedmann M, Aamodt AH, Skattør TH, Tjønnfjord GE, Holme PA (2021) Thrombosis and thrombocytopenia after ChAdOx1 nCoV-19 vaccination. N Engl J Med 384(22):2124–2130
Sessa F, Salerno M, Esposito M, Di Nunno N, Zamboni P, Pomara C (2021) Autopsy findings and causality relationship between death and COVID-19 vaccination: a systematic review. J Clin Med 10(24):5876
Shimabukuro TT, Cole M, Su JR (2021) Reports of anaphylaxis after receipt of mRNA COVID-19 vaccines in the US-December 14, 2020-January 18, 2021. JAMA 325(11):1101–1102
Shojania KG, Burton EC (2008) The vanishing nonforensic autopsy. N Engl J Med 358(9):873–875
Skok K, Vander K, Setaffy L, Kessler HH, Aberle S, Bargfrieder U, Trauner M, Lax SF (2021) COVID-19 autopsies: Procedure, technical aspects and cause of fatal course. Experiences from a single-center. Pathol Res Pract 217:153305
Sobczak M, Pawliczak R (2022) The risk of anaphylaxis behind authorized COVID-19 vaccines: a meta-analysis. Clin Mol Allergy 20(1):1
Sperhake JP (2020) Autopsies of COVID-19 deceased? Absolutely! Leg Med (Tokyo) 47:101769
Ssentongo P, Ssentongo AE, Voleti N, Groff D, Sun A, Ba DM, Nunez J, Parent LJ, Chinchilli VM, Paules CI (2022) SARS-CoV-2 vaccine effectiveness against infection, symptomatic and severe COVID-19: a systematic review and meta-analysis. BMC Infect Dis 22(1):439
Suess C, Hausmann R (2020) Gross and histopathological pulmonary findings in a COVID-19 associated death during self-isolation. Int J Leg Med 134(4):1285–1290
Tregoning JS, Flight KE, Higham SL, Wang Z, Pierce BF (2021) Progress of the COVID-19 vaccine effort: viruses, vaccines and variants versus efficacy, effectiveness and escape. Nat Rev Immunol 21(10):626–636
Van den Tweel JG, Taylor CR (2013) The rise and fall of the autopsy. Virchows Arch 462(4): 371–380
Waidhauser J, Martin B, Trepel M, Märkl B (2021) Can low autopsy rates be increased? Yes, we can! Should postmortem examinations in oncology be performed? Yes, we should! A postmortem analysis of oncological cases. Virchows Arch 478(2):301–308
WHO (2004) Laboratory biosafety manual, 3rd edn https://www.who.int/publications/i/item/9241546506
Winters B, Custer J, Galvagno SM Jr, Colantuoni E, Kapoor SG, Lee H, Goode V, Robinson K, Nakhasi A, Pronovost P, Newman-Toker D (2012) Diagnostic errors in the intensive care unit: a systematic review of autopsy studies. BBMJ Qual Saf 21(11):894–902
Yang X, Bai J, Imai H (2016) Changes in autopsy rate in Japanese university hospitals during the past 34 years. Juntendo Med J 62(3):240–247
Zheng C, Shao W, Chen X, Zhang B, Wang G, Zhang W (2022) Real-world effectiveness of COVID-19 vaccines: a literature review and meta-analysis. Int J Infect Dis 114:252–260

34 COVID-19 and Their Impacts on Aquatic Systems: Is It a Solution for Environmental Resilience?

Kajol Goria, Gagandeep Kour, and Richa Kothari

Abstract

COVID-19 pandemic has drastically affected each and every section of social, economic and industrial sectors. Sever life-threatening episodes, psychological and mental disorders have been faced by mankind. Despite several health impacts, management of generated plastic-based personal protective equipment (PPE) and single-use plastic wastes also became a challenge. In order to restrict the transmission of infectious and contagious COVID-19, complete lockdown with confined manmade activities and complete industrial shut down has imposed across the globe for several months. Restrictions imposed during lockdown might have slowed the economic growth globally, but on contrary, natural ecosystems have got some disturbance-free environment from numerous anthropogenic activities to flourish. Restricting anthropogenic activities during lockdown could have encouraged sustainable and environmental resilience with biodiversity revival and improved ecosystem health predominantly in the aquatic ecosystem.

Keywords

COVID-19 pandemic · Environmental resilience · Lockdown · Waste management

---

K. Goria · G. Kour · R. Kothari (✉)
Department of Environmental Sciences, Central University of Jammu, Samba, Jammu and Kashmir, India

R. Sobti et al. (eds.), *Role of Microbes in Sustainable Development*,
https://doi.org/10.1007/978-981-99-3126-2_34

34.1 Introduction

Incidence of coronavirus disease (COVID-19) has ruthlessly impacted mankind globally. It has stroked more than 200 countries of the world and threatened the health security of public globally (Singh 2020). COVID-19 first emerged in Wuhan, China, in the late December during the year 2019. Within few months, the virus had spread across several nations, thus becoming a serious matter of concern. The World Health Organization (WHO n.d.) stated a pandemic of the novel coronavirus disease (COVID-19) on 13 March 2020 (Cucinotta and Vanelli 2020). With the passage of time, the pandemic has been spreading among more and more communities across the globe with increasing number of deaths gradually. In addition to health damage, the pandemic has declined economic growth along with fostering many environmental mutilations. Until now, any scientifically accepted vaccines or antiviral medications and treatments which could work effectively to counter COVID-19 have not reported (Chakraborty and Maity 2020). Rapidly spreading pandemic has posed immense pressure on well-being of people, financial development, along with several environmental, and societal challenges over the human population entirely around the globe. As per the International Monetary Fund reports, the pandemic has already affected the topmost economies of developed countries that has impacted the global economy majorly. The pandemic has resulted the whole world economy probably to undergo sharp contraction by around -3% during the year 2020 which is much worse than the financial crisis raised during the year 2008–09 (IMF 2020). In this twenty-first century, six major worldwide diseases and widespread outbreaks similar to COVID-19 have previously strutted over the world viz. Severe Acute Respiratory Syndrome (SARS) (2002–2004), H1N1 influenza (2009), Middle East respiratory syndrome (MERS) (2012–2020), the West-African Ebola virus epidemic (2013–2016), the Zika fever (2015–2016), and Avian influenza (2008–2014) (Cheval et al. 2020). However, not any of them has attained that much spatial extent and the extensively wide-ranging impacts which the current coronavirus has realized quickly. To restrict the transmission of coronavirus, various preventive and protective guidelines have been issued by governments throughout the world. They recommended complete lockdown across the globe by ensuring closure of various activities like religious, commercial, industrial, communal, scientific, cultural, sports tournaments, and congregation of political masses along with the cancellation of several events like Hajj, Olympics, etc. (Verma and Prakash 2020). In the meantime, efforts such as lockdown to check covid-19 spread have shown an outstanding effect in the terms of environmental health. As a result of non-functioning and closure of industries, stopping of vehicles on roads, a significant reduction in harmful emissions including greenhouse gases into environment has noticed. Rivers like Ganges showed noteworthy upgradation in the lockdown time (Dutta et al. 2020). Several metro cities such as Hyderabad, Kolkata, Mumbai, and Chennai including the national capital New Delhi have witnessed clean air (Singh and Chauhan 2020). Although several efforts were made before the pandemic to regain environmental health, no considerable results were obtained. However, novel coronavirus pandemic had made it possible to effect environmental resilience.

Aviation emissions as per Environmental and Energy Study Institute (EESI) contributes around 2.4% of global CO_2 emissions in 2018 have found it dropping substantially (Verma and Prakash 2020).

34.2 Origin and Characteristics of Coronavirus

Coronaviruses are considered as huge assemblage of viruses coming under the family Coronaviridae that can transmit diseases between humans and animals and for that reason they are popular as zoonotic viruses (Sreenivas et al. 2020). These are basically enveloped and single-stranded positive sense RNA viruses. These are further categorized into sub-family Coronaviridae depending upon their genomic organization and phylogenetic relationships. Coronaviridae comprises four different genera or variants of coronavirus such as alpha (αCoV), Beta (βCoV), Gamma (γCoV), and Delta (δCoV) (Cui et al. 2019). These viruses are acknowledged to result infections in humans fluctuating between mild respiratory infections like common cold/flue, fever, etc. to further deadly disorders like Severe Acute Respiratory Syndrome (SARS) and Middle East Respiratory Syndrome (MERS). Presently, the new strain of coronavirus resembles share 98% of its spike protein with that of the bat coronavirus. Evolutionary history of coronaviruses showed that αCoV and βCoV have been emerged from bats and rodents, whereas γCoV and δCoV are known to be evolved from birds (Ge et al. 2017). SARS-CoV, MERS-CoV, and SARS-CoV-2 can be majorly recognized as βCoV (Chakraborty and Maity 2020). These three βCoV are known to be evolved from bats and further spread into their respective intermediate or transitional mammalian hosts. Civet cats are acknowledged as the intermediate host for SARS-CoV while in case of MERS-CoV, camels are identified as the intermediate host. These hosts ultimately spread the viruses among humans that cause mild to severe health implications (Song et al. 2019).

34.2.1 Novel Coronavirus/COVID-19 Disease

In the late December of the year 2019, a deadly infection called COVID-19 (coronavirus disease 2019) has been emerged and within few months, it spread out across the globe, thus becoming pandemic. The infection pertains to common cold, highly contagious and identified to be resulted by a novel class of corona virus, known as severe acute respiratory syndrome coronavirus 2 (SARS-CoV-2) (Bilal and Iqbal 2020). The new strain of coronavirus is deadly because infection caused by this strain has no treatment at yet, i.e., October 2020. These viruses are generally spherical in shape and bear protruding protein structures called spikes on their surface that give the virus particle a crown-like appearance (Yaqinuddin 2020). On coming in contact with the human cell, the protein spikes present on the surface of coronavirus (SARS-CoV-2) tends to bind with cellular receptor called angiotensin converting enzyme 2 on resting on the human cell membrane that permits the virus to get entered into the cell (Guo et al. 2020). The protein spikes present in SARS-CoV-

2 are found to have about 10 to 20 times more binding power than that of earlier coronaviruses like SARS (Zheng 2020). This ability of SARS-CoV-2 makes it distinguished from other available strains and also serve as a major force behind the reckless man to man spread of COVID-19. Because of this, the novel corona virus has shown its impact throughout the world irrespective of cast, color, religion, and creed. The incubation period for COVID-19 lies between 1 and 14 days (mostly 5 days) (Li et al. 2020). It is the time taken by the virus to begin to show the symptoms of disease after caught. Early indications of COVID-19 mostly comprise high body temperature, tiredness, dry cough, loss of smell as well as taste, etc. Later symptoms may comprise body aches, painful throat tenderness, breathing problem due to nasal congestion, or sometimes looseness of the bowels (Cuevas-Barragan et al. 2020). These symptoms are usually mild and begin slowly and progressively. However, these symptoms are not the sole determiners for infection of COVID-19. There are many cases where people do not get any symptoms but reported the infection of COVID-19. Out of every six COVID-19 infected people, there occurs one person with severe illness that develops major symptoms like difficulty in breathing (Patil and Jain 2020). The severity of this disease also ranges from mild to deadly effects depending upon various factors like age, immunity, or occurrence of previous health disorders (Derosa et al. 2020). Moreover, as per WHO, there occurred some cases where most people (around 70%) after taking some precautionary measures like quarantine, social distancing, and little medical care have recovered from the disease without getting any special treatment.

34.2.2 Modes of COVID-19 Infection

34.2.2.1 Droplet Transmission

Droplets are water holding entities and respiratory in origin with a diameter greater than 5 μm. Healthy person can be caught by droplets infected from diseased person within a certain range of approximately 1 m. These are produced during sneezing, coughing, and speaking to an infected person, and the infected droplets can reach to healthy person in a closed proximity. Any person in vicinity of someone with respirational indications like coughing, sneezing, etc. is likely to be exposed to potentially infectious respiratory droplets that can be accidently inhaled (Wilson et al. 2020).

34.2.2.2 Direct Contact with Surface Deposition/Contamination

Infected respiratory droplets containing coronavirus from infected person may land and place on a surface of door bells, stairs, lift push button, fruits, vegetables, etc. where the virus could persist viably. These surfaces may frequently encounter healthy persons, and thus, the virus may reach from these surfaces to hands and subsequently to eyes, nose, mouth, etc. and finally may result in contacting the healthy person with new covid-19 patients. It is also apparent to say that an infected person's surroundings can serve as an excellent source of coronavirus transmission (Morawska et al. 2020).

34.2.2.3 Fecal-Oral Transmission

Fecal stuff is also regarded as one of the leading transmitting sources in Covid-19 infected patients. Some investigations have indicated that the virus may cause several intestinal problems and thus can be found in stools or feces. However, risk of catching virus through this route appears to be quite less (Ding and Liang 2020).

34.2.2.4 Transmission Stages of COVID-19

The novel COVID-19 appeared very infectious or contagious disease and has swiftly spread worldwide. Globally, in accordance with the WHO reports, as on 11th of October 2020, there have been about 36,996,501 people infected with COVID-19, that include around 1,069,476 demises. These digits are growing rapidly with the passage of time. Researchers have claimed that the novel coronavirus involves four transmission stages to spread the disease (Fig. 34.1). These stages are as follows:

34.2.2.4.1 Stage-1: Imported Cases

When a healthy person visits abroad and gets infected with coronavirus and return to the native country or several goods imported from the infected country and they start the transmission of virus within the country. This is the early and first stage of pandemic. This also occurs when any infected person visits to another country and spread the virus among the individuals residing there.

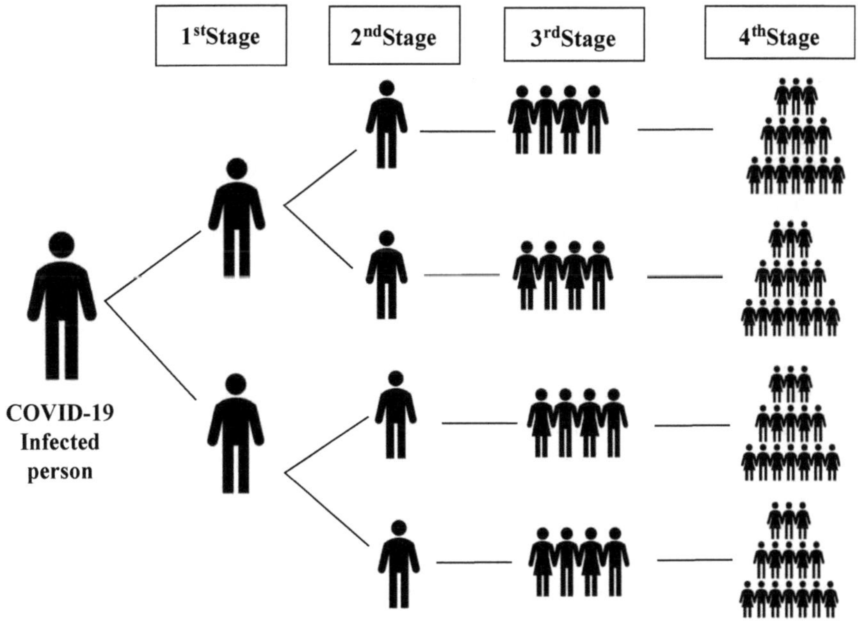

Fig. 34.1 Different stages involved in man-to-man transmission of novel coronavirus (COVID-19). (https://timesofindia.indiatimes.com/india/india-has-30-days-to-halt-onset-of-stage-iii-says-icmr-dg/articleshow/74620277.cms)

34.2.2.4.2 Stage-2: Local Transmission

This is the second stage of virus transmission which is dangerous than stage 1. It takes place locally among the masses by the infected patient. This can be illustrated as local transmission of coronavirus from corona positive persons.

34.2.2.4.3 Stage-3: Community Transmission

This is a much high-level transmission stage when a significant number of coronavirus infection cases appeared every day. In this stage, the disease spreads within the community and large areas get affected.

34.2.2.4.4 Stage-4: Transmission Out of Control

This is the final stage of coronavirus transmission when the virus acquires the shape of an epidemic and spread seamlessly out of control or with no clear endpoint. This is the stage where most of the top developed and developing nations are existing including India.

34.3 Role of Health Organizations to Deal with COVID-19

COVID-19 disease has affected the mankind overwhelmingly. Several global organizations like WHO, World Bank, World Trade Organization (WTO), and International Monetary Fund (IMF) besides many other international-, national-, central-, and state-level bodies are playing major role in supporting the whole world by providing policy advice, technical assistance, financial resources, and much more. These organizations are providing direct services to mankind to tackle the drastic situation created by novel coronavirus. These are offering help to public regardless of any profit or gain. Threat to human health is the foremost challenge of COVID-19 that the world is confronting. In this regard, WHO is the prime organization that is playing significant role in protecting human health by providing requisite guidelines to deal with the novel coronavirus and preventing human health from being exposed to virus. WHO has issued several directions in the form of reports administrating the do's and don'ts during the novel covid-19 pandemic. As per WHO, all workers working at different fields such as doctors, nurses, waste handlers, waste pickers, soldiers, and many others are required to wear appropriate personal protective equipment (PPE), that comprises protective outerwear, goggles or a face shield, a face mask, gloves, and boots (World Health Organization 2020a, b). The workers also needed to sanitize hands frequently and must avoid touching the face including eyes, nose, and mouth with unhygienic hands. WHO has also published guidelines to manage public health by implementing some social measures including social distancing, quarantine wards for patients suspected with novel coronavirus symptoms, proper hand hygiene, lockdown, etc. As vaccine is not discovered against covid-19, in order to prevent and hinder the virus infection transmission as well as to protect ourselves from virus, the usage of mask along with frequent usage of the alcohol-based hand sanitizers or hand wash are regarded as best preventive measures, following certain norms including social distancing.

There are numerous technical documents providing guidelines on specific concerns, including infection prevention and control (IPC) has been published by WHO. These documents can be accessed from https://www.who.int/emergencies/diseases/novel-coronavirus2019/technical-guidance/infection-prevention-and-control.

34.3.1 Waste Generation Due to COVID-19 Infection

Contributing towards worldwide economic recession, Covid-19 pandemic has also created a huge quantity of medical or healthcare waste. The waste generated amid Covid-19 may consists of several disposable and discarded plastic-based personal protective equipment (PPE) and one-time usable plastics that can pose enormous negative impact on health of public as well as environment (Rowan and Laffey 2020). Solid waste management has previously been a major problem for world since a long time. Presently, the rapid surge in biomedical waste because of COVID-19 pandemic is additionally intensifying the solid waste management menace. Among the various solid wastes generated due to Covid-19, biomedical wastes (BMW) are one of the most concerning wastes as these possibly can serve a potent source of infection itself and spread the contagious virus, if left untreated and improperly handled. BMW may arise from different healthcare facilities, laboratories, quarantine centers, quarantine homes, etc. (Kargar et al. 2020). These wastes may include the sharp or piercing wastes such as needles, syringes, blades, etc., pathological/anatomical wastes such as blood and other fluids, tissues, placental wastes, etc. and many other infectious wastes such as gloves, bandages, or gauges contaminated with blood and other body fluids or microbes such as bacteria, viruses, or parasites. BMW can cause generation of toxic wastes like pharmacological waste (expired and useless medications), chemical waste (inoperable and destroyed chemical constituents, unusable antiseptics, and sanitizers), and radioactive or radiation emitting wastes (including dangerous radioactive constituents, packaging wastes, glassware wastes, radionuclides enriched excretory wastes of patients undergone treatment or tested with radionuclides and sealed sources) (Anwer and Faizan 2020). However, not 100% BMW is hazardous but around 85% of the waste is found to be non-hazardous and the rest 15% is considered hazardous that includes 5% non-infectious and 10% infectious waste (Choudhary and Rai 2019). These wastes are generated in enormous quantity every day which makes it quite challenging to confront waste management problem. Unsafe disposal of various kind of wastes generated due to Covid-19 can have an instantaneous threat of resulting environmental pollution along with sprawl of numerous infectious and deadly diseases like cholera, typhoid, hepatitis, human immunodeficiency virus (HIV)/acquired immunodeficiency syndrome (AIDS), and many other respiratory complications.

34.3.2 Management and Handling of COVID-19 Waste

To combating the waste generation problem aided by Coronavirus (COVID-19) pandemic, solid waste management especially healthcare wastes handling and management has become a critical issue. Different variety of wastes has been emerged from different sources and to manage and handle such kind of waste, different management strategies are required. Generally, there occurs following steps in the management of healthcare waste as shown in Fig. 34.2. WHO has provided appropriate guidelines to manage the infectious and other wastes generated due to Covid-19 (World Health Organization 2020a, b).

34.3.2.1 Classification/Categorization of Waste

Solid wastes usually include solid materials which become unfit for use and not have any economical importance. For proper management of wastes, waste classification is the foremost step that classify solid waste based on their characteristics and source of origin. They may categorize as waste from public activities or municipalities, industries, farming activities, healthcare places, etc. In hospitals and other health centers, waste get generated consequently from curing animal or human patients as well as from other medical and testing laboratories by performing diagnostic tests like blood test, urine test, and stools tests of infected patients. Such kind of wastes are named as biomedical wastes (BMW). Huge quantity generation and transmission of BMW among the masses are majorly accompanied by healthcare centers. It is the need of the hour to manage that waste as management of it reflects an essential role in countering the spread of infections as well as supporting allied hygiene programs in healthcare places.

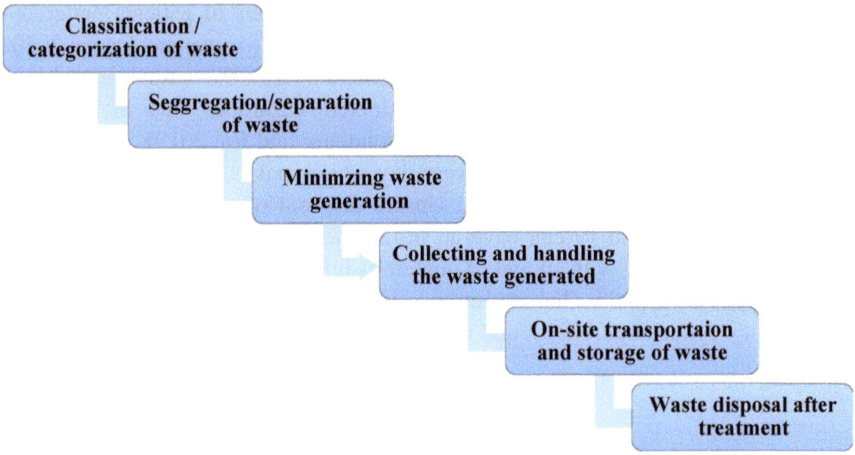

Fig. 34.2 Steps in management of healthcare waste

34.3.2.2 Waste Segregation and Waste Minimization

Measure should be adopted to minimize waste generation and wherever possible, non-infectious wastes should be segregated and reused again.

34.3.2.3 On-Site Handling of Waste

On-site healthcare waste handling and management involves treatment of generated waste within the healthcare facility premises or hospitals. Whereas in case of waste treatment where hospitals decontaminate or remediate their generated waste along with other health facilities, then the waste collected in small area is termed as cluster treatment. If any dedicated waste management unit involved in handling the wastes of various healthcare centers in a municipal or regional center, then it is referred to central treatment.

34.3.2.4 Waste Treatment and Disposal

Waste is treated finally to lessen the possible risks to be imposed due to healthcare waste and to keep the man and environment healthy and the remaining residues must be disposed of properly. Treatment of waste must be regarded in the aspect of the waste management hierarchy represented in Fig. 34.3.

34.3.2.5 Treatment of Healthcare Waste

There are five basic processes commonly used to treat harmful wastes of healthcare units which is also called as disease causing and pathological waste. These are physical treatment, chemical treatment, radiation method, biological disinfection or

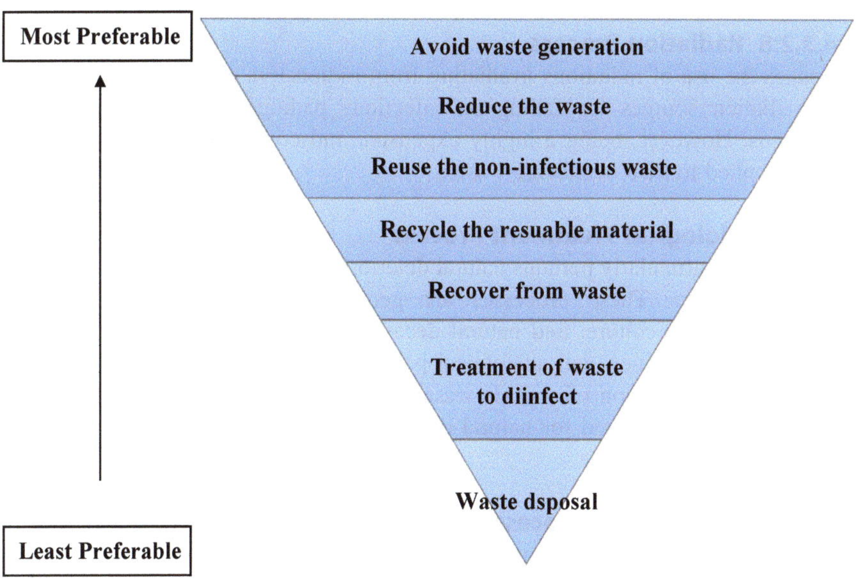

Fig. 34.3 Waste management hierarchy

degradation method, and mechanical methods. Mechanical methods are usually supplement for other treatments.

34.3.2.6 Physical Treatment Processes

The physical process of waste remediation mainly comprises heating treatment termed as thermal treatment method. The process relying on heat for destruction of infectious pathogens is commonly called as thermal treatment process. These are basically classified into two types, i.e., high-heat thermal systems that operates above 800 °C and involve combustion and/or pyrolysis treatment methods for treating healthcare waste and low-heat thermal systems that are also termed as non-burn or non-incineration treatment technologies. Such non-burn process makes use of thermal energy at temperatures capable to adequately kill pathogenic microbes, however, insufficient in carrying out waste combustion or waste pyrolysis. This is because of the fact that combustion and pyrolysis take place at the temperature lying in the range of 100 °C and 180 °C. Low-heat thermal treatment methods are often taking place in humid or dehydrated heat conditions. Humid or wet heat utilizes steam in waste sterilization which is generally carried out in devices like autoclaves or other steam-based structures. Such thermal processes are commonly termed to as wet thermal process. However, dry or dehydrated heat makes use of heat devoid of water or steam less to sterilize the waste.

34.3.2.7 Chemical Treatment Process

The waste treatment methods making the use of disinfectants like ozone gas (O_3), oxide of chlorine (ClO_2), lime solution, peracetic acid, bleach (sodium hypochlorite), or dry inorganic chemicals are referred as chemical treatment methods.

34.3.2.8 Radiation Process

It makes the use of radiations irradiating from radioactive electrons, Cobalt-60, or UV radiation sources for falling on infectious pathogenic wastes and destroy pathogens. However, being a highly expensive and costly method, it is not commonly applied to treat healthcare waste.

34.3.2.9 Biological Treatment Process

This process particularly pertains natural deterioration processes for management of biological waste. These processes are generally comprised of biodigestion, composting, vermiculture, and natural decay via burial of remains of body parts called cadavers, tissues, and anatomical parts. Some technologies add enzymes to speed up decomposition of organic waste. Composting and vermiculture methods are effectively being used for natural decomposition of wastes including placenta and hospital kitchen waste.

34.3.2.10 Mechanical Process

These involve grinding, crushing, mingling, and compaction that results in volume reduction of waste, yet incapable in destruction of harmful pathogenic microbes. This makes more surface area of waste to be exposed to treatment. These are usually supplement methods for other treatment options. These methods are required to be utilized only after the waste is fully disinfected because in case of mixing and shredding of infectious waste in an open system may increase the risk of workers exposure to pathogens.

34.3.2.11 Treatment Prior Disposal

Biomedical wastes treatment is appropriately performed through sterilization in sterilizer systems like autoclaves, hybrid steam-based units, microwave units, frictional heat unit, and dry heat units to make them free from any contamination from virus, bacteria, or other microbes. After decontamination, the waste can become suitable for environmental disposal and can be in a controlled and insulating landfill systems.

34.4 Consequences of COVID-19

COVID-19 pandemic has swallowed the entire globe and impacted the lives of people, economy, and the environment most importantly. With a view to control the transference of deadly virus, to isolate the virus affected cases, mandatory nationwide lockdown and border closures, halt on domestic and international flights were carried out in almost every nation of the world as stringent steps taken towards mitigating the pandemic. But these measures have proven to shatter the modern world economy (Ibn-Mohammed et al. 2020). Inspite of these efforts, the deadly virus became successful in establishing its roots throughout nations, thereby distressing labour-intensive system, disturbing the financial markets (Sarkis et al. 2020). Socioeconomic activities close down in many ways as millions were quarantined at home or at quarantine centre; national and international borders were locked; schools, colleges, and training institutions were closed; road and air traffic, manufacturing and travel industries paralysed; and restaurants and shopping and entertainment complexes were closed to avoid gathering. Tourist destination locations were deserted, and among all this, unemployment claims reached millions (Basilaia and Kvavadze 2020; Devakumar et al. 2020; Kraemer et al. 2020; Lokhandwala and Gautam 2020; Ibn-Mohammed et al. 2020). The COVID-19 has also negatively affected sustainable development goals as most of the targets will not be achieved by 2030 due to the pandemic and the associated impacts (Naidoo and Fisher 2020). From the very beginning when COVID-19 was declared pandemic, the scientific community has been constantly working hard to inspect the consequences of COVID-19 on different components of the environment.

34.4.1 Effects of COVID-19 on Human

COVID-19 pandemic has resulted into a state of health emergency worldwide and affected the health and medical infrastructure at the most. As many as 3.5 million people (hosts for the virus) were infected with a mortality rate of more than 3.6% as calculated for the early 3–4 months since COVID-19 was declared pandemic (Kumar et al. 2020a, b). It has brought a wave of anxiety and fear around the globe as is evident from shorter term to extended psycho-social and mental health implications. Such psychological disorders are more prominent in children and adolescents as compared to adults (Singh et al. 2020). This could be due to the number of susceptibility aspect like age, previous mental health condition, income, gender, physical activity, and current educational status (Pieh et al. 2020; Shen et al. 2020; Singh et al. 2020). Complete isolation to contain and prevent the further spread of deadly virus, no doubt brought benefits to the community at a large scale, but has created psychological problems that cannot be overlooked (Liu et al. 2020). Studies carried out on the behavioural changes of children, adolescents, and adults due to lockdown were observed and found that children in the age group of 3–6 years were more liable with noticeable symptoms like fear of family members being infected, become more clingy as compared to the children of age between 6 and 18 years. However, all the children irrespective of their age showed severe psychological conditions like increased clinginess, bad temper, inattention, disturbed sleep, poor appetite, and separation-related anxiety (Jiao et al. 2020; Singh et al. 2020; Viner et al. 2020).

Closure of schools and colleges all over the globe for a long period of time have negatively affected a vast majority (> 91%) of the world's student population with youngsters anxious about termination of examinations and academic events, lack of interaction with the peer group (Lee 2020). With the closure of educational and training institutions, children lack access to the conducive learning environment, lack of interactions with the peer group and with the teachers. Also, opportunities of development of basic skills like social conduct, communication skills is also hampered. Home confinement and lack of interactions with the fellow mates and friends, lack of physical activities may trigger outburst of temper crabbiness among teenagers resulting in conflict between parents and teenagers (Jiao et al. 2020). Excessive surfing of internet about COVID-19 also results in anxiety among adolescents. Unemployment due to loss of job created trauma and anxiety among adults.

Emerging data clearly suggests that COVID-19 infected patients associated with multiple medical conditions such as high blood pressure, disturbed blood sugar levels, cardiovascular disease, obesity, chronic pulmonary disease, asthma, chronic kidney disease, and malignancy have shown increased complications and high mortality rate (Singh and Misra 2020). The frontline healthcare workers which comes in direct contact with COVID-19 positive patients are reported to have much more chances of COVID-19 contraction as compared to general community because of inadequate supply of PPEs (Nguyen et al. 2020). Because of surge in number of COVID-19 infected patients, the health and medical infrastructure collapsed as there is acute shortage of medical facilities, ventilators, beds in the hospitals, making the situation even worse.

34.4.2 Effects of COVID-19 on Animals

There have been not much evidence that coronavirus can infect animals, and studies are still going on about the infection of animals by coronavirus. A study ranked the animals susceptible to SARS-CoV-2 with primates as most susceptible, followed by carnivores, cetaceans, and wild rodents. The laboratory rodents are at lower risk whereas most of the birds, reptiles, and amphibians were placed at the bottommost ranges in the susceptibility analysis. There is an urgency in taking preventive measures enabling limited or avoiding direct human contact with wildlife, and for monitoring of wildlife for the virus (Martínez-Hernández et al. 2020). The farm animals which are directly dependent on humans are greatly affected by COVID-19 as because of human confinement, these animals are not given proper care and suffered from shortage of food and animal diseases. Lockdown in response to COVID-19 has badly affected the food production sector especially aquaculture in many countries (Kumaran et al. 2020).

COVID-19 pandemic focuses on increased dependence on animal research as the human population is facing exceptional threat, and there is need for the development of an effective drug. Development of effective drug requires animal research. A detailed investigation on how COVID-19 and other diseases create complications and increase mortality rate, the adaptive immunity against the virus can only be authenticated in animals (Genzel et al. 2020). Animals crucially used in COVID-19-related research include rodents, ferrets, non-human primates, and pigs.

The rodents can be infected with the coronavirus responsible for the pandemic. Now, research communities are trying to develop the hamsters, mice, ferrets, and monkeys into animal models intended to develop effective COVID-19 vaccines and treatments and to know the viral pathogenesis (Cohen 2020; Genzel et al. 2020). Thus, with COVID-19, there has been increase in the use of laboratory animals in the research studies.

34.4.3 Effects of COVID-19 on Environment

Despite fall in global economy due to lockdown, many detrimental effects of COVID-19, some accidental positive implication of lockdown imposed to curb and to contain the deadly coronavirus can also be seen on the environment (Ibn-Mohammed et al. 2020). Global spread of COVID-19 in a relatively short span resulted in drastic decline in the industrial and production system and with highly reduced road traffic (Khursheed et al. 2020). As humans were locked up at their residences, the nature, environment, and wildlife started blooming. There have been reports about wild animals, peacocks, and deers roaming on the roads and along residential complex areas in cities. Reports from across the globe are indicative of resiliency and restoration in the environmental conditions as evident from improvement in air and water quality (Chen et al. 2020; Saadat et al. 2020; Lokhandwala and Gautam 2020). There is marked reduction in the release of greenhouse gases (GHGs) as the industrial units and transportation sectors were stopped and road traffic restricted. The coal dependence for energy requirement by industrial units also dropped because of COVID-19 pandemic resulting in less release of harmful gases and GHGs into the atmosphere (Ibn-Mohammed et al. 2020). The reduced vehicular traffic and industries being shut resulted in noise-free environment. $PM_{2.5}$, PM_{10}, and NO_2 concentration in the air greatly reduced during the global shutdown in response to COVID-19 (Aman et al. 2020). Studies carried in Egypt showed reduction in NO_2, CO, and GHGs emissions during the COVID-19 lockdown period. The ozone level increased by 2% over Cairo and Alexandria in Egypt (Mostafa et al. 2020). The atmospheric environment of Southeast Asia (Malaysia) as reported by a study (Kanniah et al. 2020) also showed marked reduction in $PM_{2.5}$, PM_{10}, NO_2, SO_2, and CO percentage. One can also see pollution free and clean beaches (Ibn-Mohammed et al. 2020; Mostafa et al. 2020). Despite the above-said positive impressions of COVID-19 on environment, there is a surge in the rate of municipal waste generation. The dispersal of used, discarded masks, and gloves on streets along road sides is creating environmental harm and degradation (Ragazzi et al. 2020). This problem needs to be controlled though proper management practices to prevent the future waste disaster.

34.5 Aquatic Ecosystems Resilience: Impacts Noticed

Although Covid-19 brought major setback to human life in terms of health, social, economical, and psychological aspects but has proven as a healing phase for the earth's environment. Imposing lockdown throughout the world and restrictions on numerous factories and industrial operations in an effort for Covid-19 spread prevention, natural ecosystem has realized healing effect. Generally, a number of industrial effluents and harmful discharges from different operations find their way into water bodies resulting in the deterioration of aquatic environment. Minimization of anthropogenic activities due to mandatory period of restriction has reduced the industrial discharge into water bodies and started recovering to the original, clear, and pollution-free state of aquatic systems to a greater extent, thereby encouraging the aquatic life to flourish effectively. Many evidences and observation have been found regarding enhanced water quality and rejuvenation of some river systems such as Ganga, Yamuna, Rapti, Saryu, etc. during the lockdown period (Verma and Prakash 2020). All these studies show positive impacts of COVID-19 on aquatic ecosystems as pollution-free rivers signify healthy ecosystem.

COVID-19 lockdown resulted as a blessing to the aquatic ecosystem as seen from the improved water quality index (WQI) of many rivers across the globe. Yamuna river, one of the polluted rivers of India, showed marked drop in turbidity and suspended particulate matter (SPM) due to reduction in pollution loads from anthropogenic activities (Patel et al. 2020). The period of closure also improved the surface water quality of lakes in terms of reduced SPM (Yunus et al. 2020). A case study of the Sabarmati river from Ahmadabad (India) helped to realize the consequence of anthropogenic activities on the hydrosphere on short- and long-term basis (Aman et al. 2020). The study showed average decrease in SPM concentrations by 36.48%

during lockdown when compared with pre-lockdown. However, waste water from domestic activities continued to drain the river confirming industrial sewage water and other outdoor activities as main culprits for deteriorating the quality of hydrosphere (Aman et al. 2020). Signs of restoration and significant positive enhancement in many water quality parameters are also shown by one of the polluted Indian rivers, the Ganga river. Several efforts to clean Ganga failed but due to COVID-19 lockdown with number of anthropogenic activities restricted, factories and industries being shut, the river reached its self-reclamation potential and started showing resiliency (Dutta et al. 2020). With temporary closure of industrial and commercial establishments due to coronavirus pandemic, industrial wastewater was not discharged into the river resulting in significant transformation in river water quality, giving signs of optimism from the point of restoration. Because of restriction on religious activities, the organic pollution load in terms of flowers, earthen pots, and saplings is highly reduced resulting in reduced BOD and COD values indicating improved water quality. Such transformations indicate that a periodic stoppage in the pollution load enabled the rivers to attain its self-restoration potential, indicating the anthropogenic activities as the prime and sole cause of water pollution. However, it is to be noted that these transformations are short-lived and the rivers will revert to their original, previous, and polluted state, once the restrictions are lifted.

34.6 Conclusion

COVID-19 pandemic shattered the global economy, badly affected human life in terms of psychological and mental disorders, mortality, and unemployment but has caused some accidental positive impacts on the environment. During the COVID-19 period, with humans being quarantined, stoppage of industrial and production units, the nature, environment, and wildlife started blooming. With industries and commercial units, non-functional, negligible pollutants, and industrial wastewater is discharged into rivers. With negligible pollution load and human interventions, the rivers reached their self-purification capacity. They are showing signs of resiliency and recovery. However, these improvements in the aquatic ecosystems are short-lived. Because of human negligence and improper waste management strategies, the used facemasks and PPE kits are seen floating in the water bodies. These need to be checked by imposing strict waste disposal and management rules and guidelines; otherwise, this careless attitude will further lead us to the future waste disaster. We should learn a lesson from COVID-19 that humans because of greed and careless attitude resulted in degradation of environment. With humans in confinement, locked indoor, the nature, rivers, and air started healing. This transformation is not permanent and policy makers should frame policies and guidelines so that resiliency is achieved by nature and aquatic ecosystems should not reverted.

References

Aman MA, Salman MS, Yunus AP (2020) COVID-19 and its impact on environment: improved pollution levels during the lockdown period–a case from Ahmedabad, India. Rem Sens Appl Soc Environ 20:100382. https://doi.org/10.1016/j.rsase.2020.100382

Anwer M, Faizan M (2020) Solid waste management in India under COVID19 pandemic: challenges and solutions. Int J Eng Res Technol 9:6

Basilaia G, Kvavadze D (2020) Transition to online education in schools during a SARS-CoV-2 coronavirus (COVID-19) pandemic in Georgia. Pedagogical Res 5(4):1–9. https://doi.org/10.29333/pr/7937

Bilal M, Iqbal HM (2020) Recent advances in therapeutic modalities and vaccines to counter COVID-19/SARS-CoV-2. Hum Vaccin Immunother 16:3034–3042

Chakraborty I, Maity P (2020) COVID-19 outbreak: migration, effects on society, global environment and prevention. Sci Total Environ 728:138882

Chen F, Wang M, Pu Z (2020) Effects of COVID-19 lockdown on global air quality and health. Sci Total Environ 755:142533. https://doi.org/10.1016/j.scitotenv.2020.142533

Cheval S, Mihai Adamescu C, Georgiadis T, Herrnegger M, Piticar A, Legates DR (2020) Observed and potential impacts of the COVID-19 pandemic on the environment. Int J Environ Res Public Health 17(11):4140

Choudhary S, Rai RH (2019) Management of bio-medical waste: awareness and practices in hospitals in Delhi. World J Pharmaceut Med Res 5(8):251–254

Cohen J (2020) From mice to monkeys, animals studied for coronavirus answers. Science 368:221–222. https://doi.org/10.1126/science.368.6488.221

Cucinotta D, Vanelli M (2020) WHO declares COVID-19 a pandemic. Acta BioMed 91:157–160

Cuevas-Barragan CE, Buenrostro-Nava MT, Palos-Gómez GM, Ramirez-Padilla EA, Mendoza-Macias BI, Rivas-Caceres RR (2020) Use of Nasoill® via intranasal to control the harmful effects of Covid-19. Microb Pathog 149:104504

Cui J, Li F, Shi ZL (2019) Origin and evolution of pathogenic coronaviruses. Nat Rev Microbiol 17(3):181–192. https://doi.org/10.1038/s41579-018-0118-9

Derosa L, Melenotte C, Griscelli F, Gachot B, Marabelle A, Kroemer G, Zitvogel L (2020) The immuno-oncological challenge of COVID-19. Nat Cancer 1:1–19

Devakumar D, Shannon G, Bhopal SS, Abubakar I (2020) Racism and discrimination in COVID-19 responses. Lancet 395(10231):1194. https://doi.org/10.1016/S0140-6736(20)30792-3

Ding S, Liang TJ (2020) Is SARS-CoV-2 also an enteric pathogen with potential fecal-Oral transmission: a COVID-19 virological and clinical review. Gastroenterology 159:53

Dutta V, Dubey D, Kumar S (2020) Cleaning the river ganga: impact of lockdown on water quality and future implications on river rejuvenation strategies. Sci Total Environ 743:140756

Ge XY, Yang WH, Zhou JH, Li B, Zhang W, Shi ZL, Zhang YZ (2017) Detection of alpha-and betacoronaviruses in rodents from Yunnan. China. Virol J 14(1):98

Genzel L, Adan R, Berns A, van den Beucken J, Blokland A, Boddeke EH, Bogers WM, Bontrop R, Bulthuis R, Bousema T, Clevers H (2020) How the COVID-19 pandemic highlights the necessity of animal research. Curr Biol 30:R1014. https://doi.org/10.1016/j.cub.2020.08.030

Guo YR, Cao QD, Hong ZS, Tan YY, Chen SD, Jin HJ, Yan Y et al (2020) The origin, transmission and clinical therapies on coronavirus disease 2019 (COVID-19) outbreak–an update on the status. Mil Med Res 7(1):1–10

Ibn-Mohammed T, Mustapha KB, Godsell JM, Adamu Z, Babatunde KA, Akintade DD, Acquaye A, Fujii H, Ndiaye MM, Yamoah FA, Koh SCL (2020) A critical review of the impacts of COVID-19 on the global economy and ecosystems and opportunities for circular economy strategies. Resour Conserv Recycl 164:105169. https://doi.org/10.1016/j.resconrec.2020.105169

IMF (2020) World Economic Outlook. Chapter 1: The Great Lockdown. International Monetary Fund, April 2020. https://www.imf.org/en/Publications/WEO/Issues/2020/04/14/weo-april-2020

Jiao WY, Wang LN, Liu J, Fang SF, Jiao FY, Pettoello-Mantovani M, Somekh E (2020) Behavioral and emotional disorders in children during the COVID-19 epidemic. J Pediatr 221:264. https://doi.org/10.1016/j.jpeds.2020.03.013

Kanniah KD, Zaman NAFK, Kaskaoutis DG, Latif MT (2020) COVID-19's impact on the atmospheric environment in the Southeast Asia region. Sci Total Environ 736:139658. https://doi.org/10.1016/j.scitotenv.2020.139658

Kargar S, Pourmehdi M, Paydar MM (2020) Reverse logistics network design for medical waste management in the epidemic outbreak of the novel coronavirus (COVID-19). Sci Total Environ 746:141183

Khursheed A, Alam S, Tyagi VK, Nagpure AS, Khan AA, Gaur RZ, Singh S, Bhattacharya P, Mukherjee S, Kumar M (2020) Future liasing of the lockdown during COVID-19 pandemic: the dawn is expected at hand from the darkest hour. Groundw Sustain Dev 11:100433. https://doi.org/10.1016/j.gsd.2020.100433

Kraemer MU, Yang CH, Gutierrez B, Wu CH, Klein B, Pigott DM, Du Plessis L, Faria NR, Li R, Hanage WP, Brownstein JS (2020) The effect of human mobility and control measures on the COVID-19 epidemic in China. Science 368(6490):493–497

Kumar M, Kuroda K, Dhangar K (2020a) The most eagerly awaited summer of the Anthropocene: a perspective of SARS-CoV-2 decay and seasonal change. Groundw Sustain Dev 11:100400. https://doi.org/10.1016/j.gsd.2020.100400

Kumar M, Taki K, Gahlot R, Sharma A, Dhangar K (2020b) A chronicle of SARS-CoV-2: part-I-epidemiology, diagnosis, prognosis, transmission and treatment. Sci Total Environ 734:139278. https://doi.org/10.1016/j.scitotenv.2020.139278

Kumaran M, Geetha R, Antony J, Vasagam KK, Anand PR, Ravisankar T, Angel JRJ, De D, Muralidhar M, Patil PK, Vijayan KK (2020) Prospective impact of Corona virus disease (COVID-19) related lockdown on shrimp aquaculture sector in India–a sectoral assessment. Aquaculture 531:735922. https://doi.org/10.1016/j.aquaculture.2020.735922

Lee J (2020) Mental health effects of school closures during COVID-19. Lancet Child Adolesc Health 4(6):421. https://doi.org/10.1016/S2352-4642(20)30109-7

Li G, Chang B, Li H, Wang R, Li G (2020) Precautions in dentistry against the outbreak of Corona virus disease 2019. J Infect Public Health 13:1805

Liu JJ, Bao Y, Huang X, Shi J, Lu L (2020) Mental health considerations for children quarantined because of COVID-19. Lancet Child Adolesc Health 4(5):347–349. https://doi.org/10.1016/S2352-4642(20)30096-1

Lokhandwala S, Gautam P (2020) Indirect impact of COVID-19 on environment: a brief study in Indian context. Environ Res 188:109807. https://doi.org/10.1016/j.envres.2020.109807

Martínez-Hernández F, Isaak-Delgado AB, Alfonso-Toledo JA, Muñoz-García CI, Villalobos G, Aréchiga-Ceballos N, Rendón-Franco E (2020) Assessing the SARS-CoV-2 threat to wildlife: potential risk to a broad range of mammals. Perspect Ecol Conserv 18:223. https://doi.org/10.1016/j.pecon.2020.09.008

Morawska L, Tang JW, Bahnfleth W, Bluyssen PM, Boerstra A, Buonanno G, Haworth C et al (2020) How can airborne transmission of COVID-19 indoors be minimised? Environ Int 142:105832

Mostafa MK, Gamal G, Wafiq A (2020) The impact of COVID 19 on air pollution levels and other environmental indicators—a case study of Egypt. J Environ Manage 277:111496. https://doi.org/10.1016/j.jenvman.2020.111496

Naidoo R, Fisher B (2020) Reset sustainable development goals for a pandemic world, vol 583, p 198. https://doi.org/10.1038/d41586-020-01999-x

Nguyen LH, Drew DA, Graham MS, Joshi AD, Guo CG, Ma W, Mehta RS, Warner ET, Sikavi DR, Lo CH, Kwon S (2020) Risk of COVID-19 among front-line health-care workers and the general community: a prospective cohort study. Lancet Public Health 5(9):e475–e483. https://doi.org/10.1016/S2468-2667(20)30164-X

Patel PP, Mondal S, Ghosh KG (2020) Some respite for India's dirtiest river? Examining the Yamuna's water quality at Delhi during the COVID-19 lockdown period. Sci Total Environ 744:140851. https://doi.org/10.1016/j.scitotenv.2020.140851

Patil PA, Jain RS (2020) Theoretical study and treatment of novel COVID-19. Res J Pharmacol Pharmacodyn 12(2):71–72

Pieh C, Budimir S, Probst T (2020) The effect of age, gender, income, work, and physical activity on mental health during coronavirus disease (COVID-19) lockdown in Austria. J Psychosom Res 136:110186. https://doi.org/10.1016/j.jpsychores.2020.110186

Ragazzi M, Rada EC, Schiavon M (2020) Municipal solid waste management during the SARS-COV-2 outbreak and lockdown ease: lessons from Italy. Sci Total Environ 745:141159. https://doi.org/10.1016/j.scitotenv.2020.141159

Rowan NJ, Laffey JG (2020) Unlocking the surge in demand for personal and protective equipment (PPE) and improvised face coverings arising from coronavirus disease (COVID-19) pandemic–implications for efficacy, re-use and sustainable waste management. Sci Total Environ 752: 142259

Saadat S, Rawtani D, Hussain CM (2020) Environmental perspective of COVID-19. Sci Total Environ 728:138870. https://doi.org/10.1016/j.scitotenv.2020.138870

Sarkis J, Cohen MJ, Dewick P, Schröder P (2020) A brave new world: lessons from the COVID-19 pandemic for transitioning to sustainable supply and production. Resour Conserv Recycl 159: 104894

Shen K, Yang Y, Wang T, Zhao D, Jiang Y, Jin R, Zheng Y, Xu B, Xie Z, Lin L, Shang Y (2020) Diagnosis, treatment, and prevention of 2019 novel coronavirus infection in children: experts' consensus statement. World J Pediatr 16:223–231. https://doi.org/10.1007/s12519-020-00343-7

Singh SK (2020) COVID-19: a master stroke of nature. AIMS Public Health 7(2):393

Singh RP, Chauhan A (2020) Impact of lockdown on air quality in India during COVID-19 pandemic. Air Qual Atmos Health 13(8):921–928

Singh AK, Misra A (2020) Impact of COVID-19 and comorbidities on health and economics: focus on developing countries and India. Diabetes Metab Syndr Clin Res Rev 14(6):1625–1630. https://doi.org/10.1016/j.dsx.2020.08.032

Singh S, Roy MD, Sinha CPTMK, Parveen CPTMS, Sharma CPTG, Joshi CPTG (2020) Impact of COVID-19 and lockdown on mental health of children and adolescents: a narrative review with recommendations. Psychiatry Res 293:113429. https://doi.org/10.1016/j.psychres.2020.113429

Song Z, Xu Y, Bao L, Zhang L, Yu P, Qu Y, Qin C (2019) From SARS to MERS, thrusting coronaviruses into the spotlight. Viruses 11:E59. https://doi.org/10.3390/v11010059

Sreenivas N, Chakarpani I, Kumar PA (2020) The paradigm of entry into host-concerns in developing antivirals for COVID-19. In: Coronavirus disease 2019 (COVID-19), medical virology: from pathogenesis to disease control, pp 23–31. https://doi.org/10.1007/978-981-15-4814-7_3

Verma A, Prakash S (2020) Impact of covid-19 on environment and society. J Glob Biosci 9(5): 7352–7363. https://ssrn.com/abstract=3644567

Viner RM, Russell SJ, Croker H, Packer J, Ward J, Stansfield C, Mytton O, Bonell C, Booy R (2020) School closure and management practices during coronavirus outbreaks including COVID-19: a rapid systematic review. Lancet Child Adolesc Health 4:397. https://doi.org/10.1016/S2352-4642(20)30095-X

WHO (n.d.). https://www.who.int/vietnam/news/detail/14-07-2020-q-a-how-is-covid-19-transmitted

Wilson N, Corbett S, Tovey E (2020) Airborne transmission of covid-19. BMJ 370:m3206

World Health Organization (2020a) Water, sanitation, hygiene, and waste management for the COVID-19 virus: interim guidance, 23 April 2020 (No. WHO/2019-nCoV/IPC_WASH/2020.3). World Health Organization, Geneva

World Health Organization (2020b) Water, sanitation, hygiene, and waste management for SARS-CoV-2, the virus that causes COVID-19: interim guidance, 29 July 2020 (No. WHO/COVID-19/IPC_WASH/2020.4). World Health Organization, Geneva
Yaqinuddin A (2020) Cross-immunity between respiratory coronaviruses may limit COVID-19 fatalities. Med Hypotheses 144:110049
Yunus AP, Masago Y, Hijioka Y (2020) COVID-19 and surface water quality: improved lake water quality during the lockdown. Sci Total Environ 731:139012. https://doi.org/10.1016/j.scitotenv.2020.139012
Zheng J (2020) SARS-CoV-2: an emerging coronavirus that causes a global threat. Int J Biol Sci 16(10):1678

MIX
Papier aus verantwortungsvollen Quellen
Paper from responsible sources
FSC® C105338

If you have any concerns about our products,
you can contact us on
ProductSafety@springernature.com

In case Publisher is established outside the EU,
the EU authorized representative is:
Springer Nature Customer Service Center GmbH
Europaplatz 3, 69115 Heidelberg, Germany

Printed by Libri Plureos GmbH
in Hamburg, Germany